**PEDRO BARBOSA**
*Formerly, Department of Entomology, University of Massachusetts, Amherst, Massachusetts; Presently, Department of Entomology and Economic Zoology, Rutgers University*

and

**T. MICHAEL PETERS**
*Department of Entomology, University of Massachusetts, Amherst, Massachusetts*

# READINGS IN ENTOMOLOGY

1972

W. B. SAUNDERS COMPANY • Philadelphia • London • Toronto

W. B. Saunders Company: West Washington Square
Philadelphia, Pa. 19105

12 Dyott Street
London, WC1A 1DB

1835 Yonge Street
Toronto 7, Ontario

Readings in Entomology ISBN 0-7216-1541-4



Print No: 9 8 7 6 5 4 3 2 1

# Preface

This set of readings is intended for students enrolled in introductory entomology courses. The general theme points up the number and diversity of subdivisions within the discipline. At the same time it demonstrates the interdependence of the subdivisions in problem-solving and the integrity of the discipline as a whole.

The articles provide specific research material founded on and illustrating the basic concepts of entomology. Not only can students be introduced to these basic concepts, but they can discuss the various techniques that entomologists apply and the conclusions drawn from their findings. In this way, the student's understanding of entomology is modified as he sees the subject less as a monolith of accumulated, irrefutable knowledge and more as an arena for dynamic exploration. The excitement of pioneering exploratory research is as obvious as the satisfaction that culminates a long, drawn out quest for an answer and its ultimate solution in the more advanced developmental stages. Thus the student is exposed to experimental entomology, the vital area most difficult to incorporate into traditional entomology courses. Among the more pedestrian rewards are introductions to most of the journals reporting work of entomological interest, and ready access to further information through use of the bibliographies.

The selected articles have been grouped into major accepted categories. However, most of the articles in the section on Man and Insect could be placed in one or more of the earlier sections since they essentially deal with the applied aspects of ecology, behavior, physiology, etc. Similarly, many of the articles in the "basic" sections concern "applied" problems.

We have designed the readings to complement current popular entomology texts by emphasizing common areas of weakness, but although we have stressed the areas of ecology, behavior, and economic entomology, we have attempted to present the traditionally emphasized areas as well. Therefore, the readings will benefit both classical and modern courses.

We would like to express our deep appreciation for the cooperation and enthusiasm demonstrated by the authors of the articles included. The excitement generated in us by this project was greatly increased by their helpful attitudes. We would also like to thank the societies and publishers of the journals in which the articles appeared.

Pedro Barbosa
T. Michael Peters

# Contents

## HEALTH AND WELFARE

# Part I: Introduction

# Entomology Today and Tomorrow

*By* F. S. Arant[1]

*Head, Department of Zoology-Entomology, Auburn University, Auburn, Alabama*

American entomology today stands on the cloverleaf of a vast system of highways to entomological achievement—if we, the entomologists, have the vision to plan wisely and the determination to pursue these plans vigorously.

Entomology in the United States has been largely agricultural since its beginning a little over 100 years ago (Osborn, 1952). The early insect taxonomists and naturalists, trained principally in other professions, laid a solid foundation in basics, and agricultural entomology has developed on this foundation. Crop destruction by migratory grasshoppers and damage by other pests of food, feed, and fiber crops stimulated the development of economic entomology on the state and federal levels (Howard, 1930). Thus, it is only natural that the profession of entomology has been closely associated with and fostered by the Land Grant institutions and the United States Department of Agriculture. In recent years, medical entomology has come into prominence and has made outstanding contributions to human health. Pest control in other areas has also increased in importance. However, major, perhaps overwhelming, emphasis is still on control of pests of agricultural crops.

The earliest insect control measures were crude. Controls listed by Harris (1862) included hand picking; use of soap suds, fish oil, potash, poisoned baits, lime, ashes, salt, and walnut leaves; heat treatment; burning infested parts of plants; protection of woodpeckers; and other methods. Later control recommendations included crop rotation, fall plowing, time of planting, trap crops, protection of birds and beneficial insects, and use of insecticides ranging from turpentine, kerosene, and sulphur to tobacco concoctions and arsenicals. For many years, a principal recommendation of many economic entomologists was "crop rotation and lead arsenate."

It was not until the synthetic organic insecticides came into general use that insect control reached a high level of proficiency. These synthetics have been so effective that crop yields have increased from their use to the extent of 65 to 165 per cent in many instances and quality has been improved immeasurably. Yet, the use of these insecticides has created problems. Pests are developing resistance to them so rapidly that a large part of our effort in entomology today is devoted to testing new chemicals to keep ahead of the resistance problem. In our enthusiasm we have sometimes neglected problems and hazards associated with residues; partly as a result we are in serious trouble with residues in food and feed. Many of us have also tended to forget the importance of insect bionomics and the value of beneficial species of insects. In the use of insecticides, we have sometimes created new problems requiring additional applications of chemicals in a never-ending cycle. Many of our urban and other non-farm friends are becoming increasingly critical of large-scale insecticidal control and eradication programs. Some of the critics can be ignored as fanatics. Others are intelligent taxpaying citizens who have a right to a voice in programs financed from state and federal funds.

With the foregoing comments as a background, let us examine some recent trends in populations and land uses and attempt to use these trends as guides for action programs.

### Population Changes[2]

The United States is now an urban country (Figure 1). In 1920, 48.8 per cent of the people lived in rural areas and 29.7 per cent received their incomes from farming. In 1960, 39.0 per cent of the population lived in the country and only 8.7 per cent were actually farm people. Thus, 30.3 per cent of the people were classed as rural non-farm and 61.0 per cent urban for a total of 91.3 per cent non-farm (Census of Agriculture, 1960).

Not only is the country as a whole becoming urbanized, but also the so-called rural areas are becoming urban. For example, population changes in 13 southern states are presented in Figure 2. In 1920, 74.6 per cent of the people in these states lived in rural areas and 53.5 per cent received their incomes from farming. By 1960, 42.3 per cent of the population lived in the country and only 13.4 per cent were farm people. Thus, 28.9 per cent of the people were considered rural non-farm and 57.7 per cent urban or a total of 86.6 per cent of the population living on income from sources other than farming. It is evident from these data that the "rural South" is no longer rural.

As the farm population has decreased, the number of farms has also decreased and the size of farms has increased in the southern states and throughout the United States (Figure 3). From 1920 to 1960, the number of farms decreased from 3.1 to 1.6 million in 13 southern states and from 6.4 to 3.7 million in the country as a whole. Average size of farms increased from 110 to 221 acres in the southern states and from 148 to 302 acres in the United States. These shifts represent a 51.8 per cent reduction in numbers of farms and 108.3 per cent increase in average size of farms.

### Land-Use Changes

Land uses have changed with changes in the popula-

[1] Address of the President of the Entomological Society of America at the annual meeting held in Miami, Florida, November 27-30, 1961.

[2] The writer is indebted to B. T. Lanham, Head, Department Agricultural Economics, Auburn University, for assistance in compiling data on population and land-use changes.

Reproduced with permission from *The Bulletin of the Entomological Society of America,* 8:3–8, 1962. Published by the Entomological Society of America, College Park, Maryland.

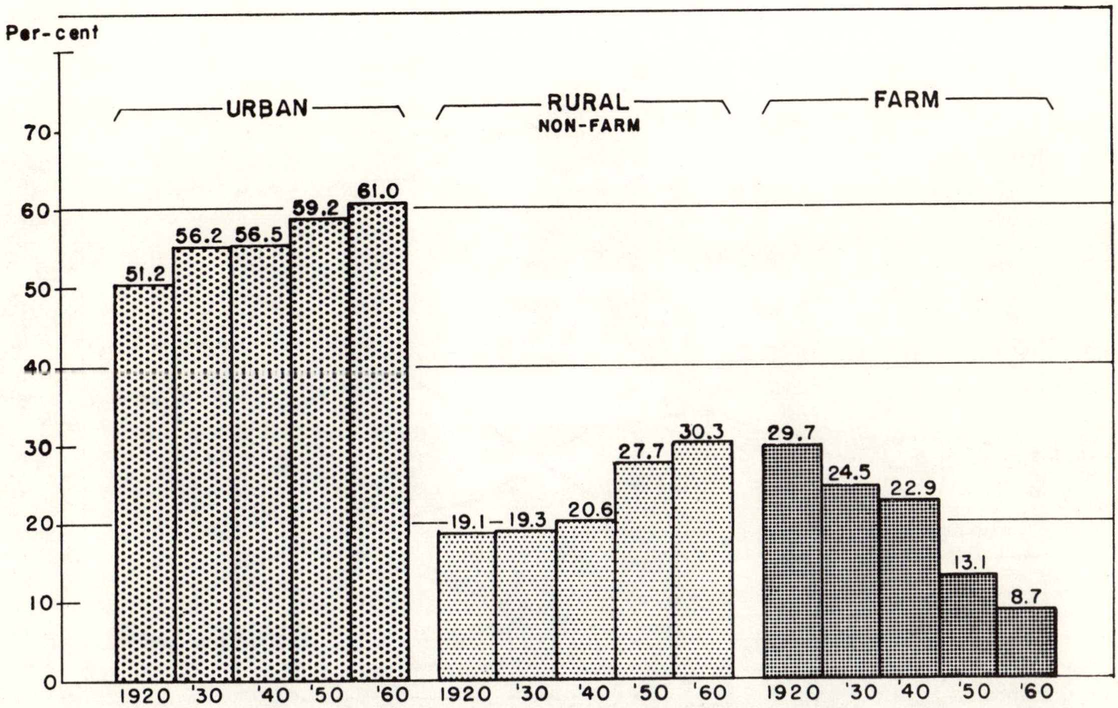


FIG. 1.—Population trends in the United States, 1920–1960.

tion and the size of farms. Acreage of crops has decreased, pasture land has increased, and recreational use of forests has become an important factor in land-use planning. Between 1920 and 1960, crops harvested decreased in acreage by 30.5 per cent in the 13 southern states and 13.4 per cent in the United States. Pasture acreage increased 58.5 and 52.4 per cent, respectively, during the same period. Adequate data on forests are

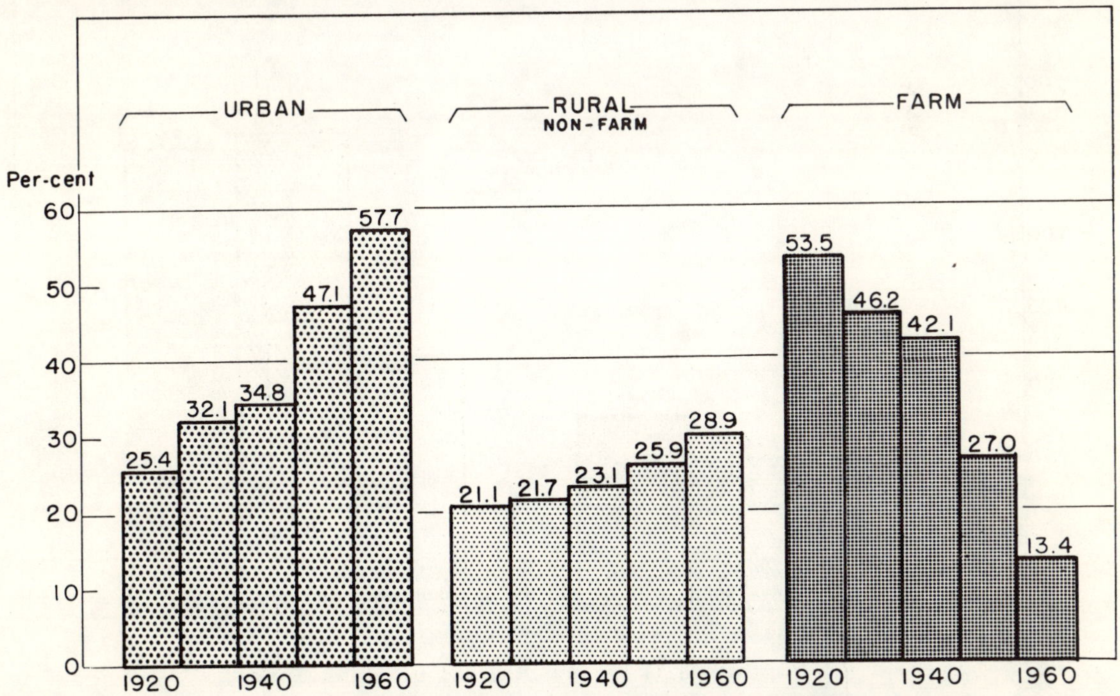


FIG. 2.—Population trends in 13 southern states, 1920–1960.

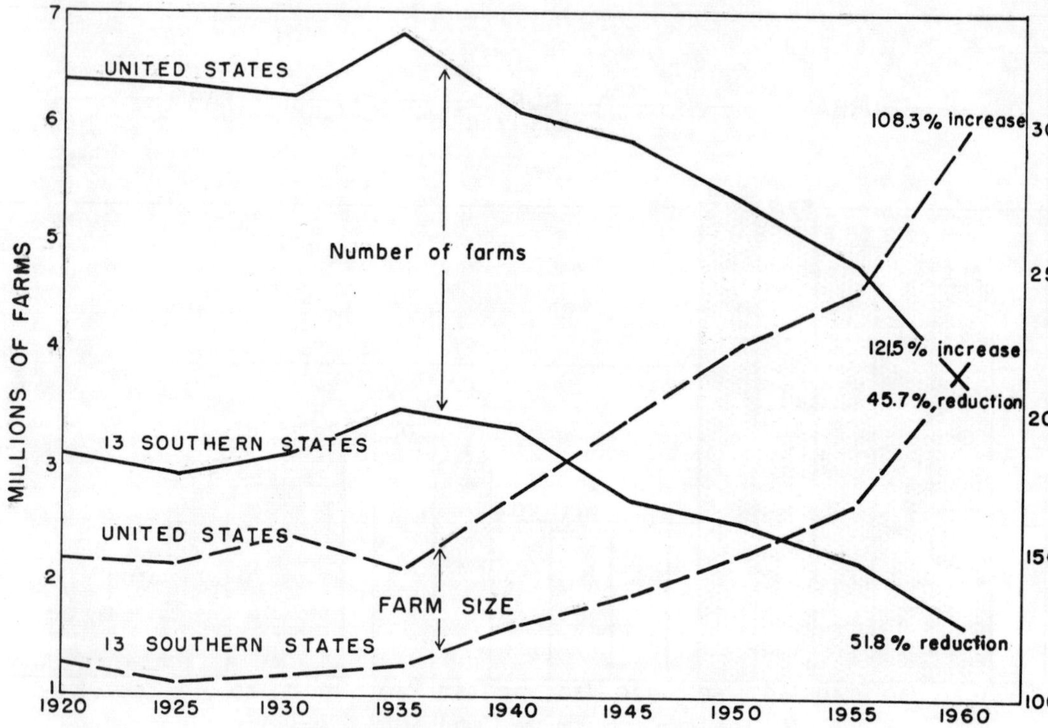


FIG. 3.—Number and size of farms in specified years.

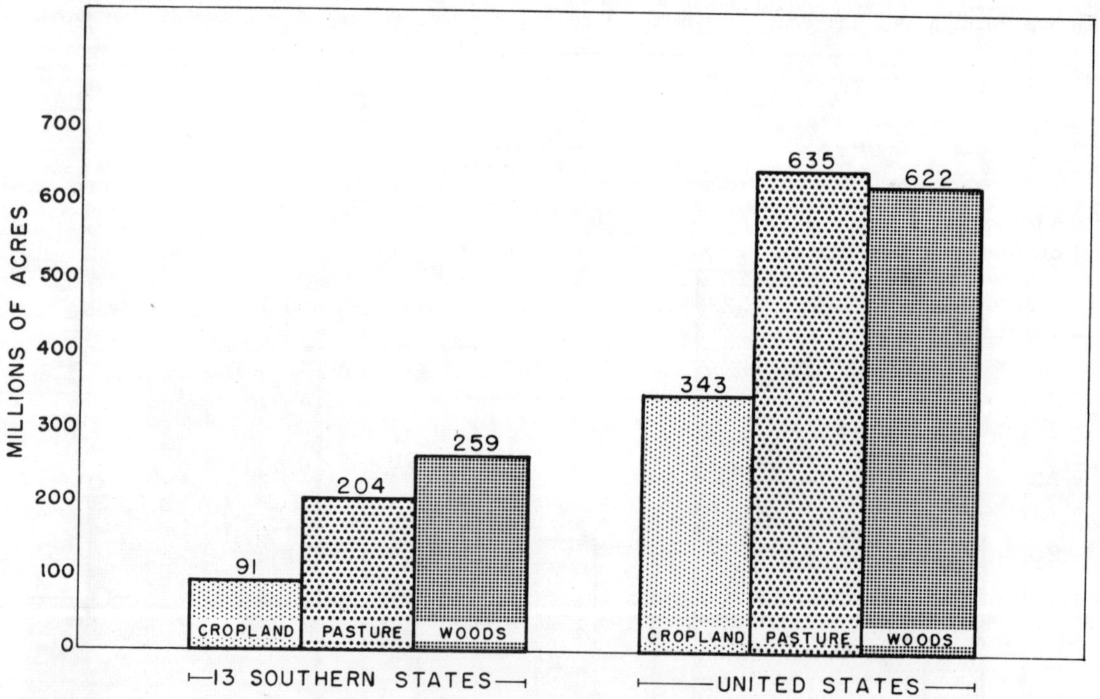


FIG. 4.—Land use in 13 southern states and the United States.

not readily available for the past 40-year period. However, average figures on the status of forest areas in relation to crops and pastures in 1945 and 1955 are presented in Figure 4. Woodlands constituted 2.8 times as much area as cropland in 13 southern states, and 1.8 times as much in the continental United States. Nearly half of the forest areas in the southern states and about two-thirds of those in the United States were owned and/or operated by persons or organizations other than farmers.

### Entomology in Relation to Agriculture

If entomology is to serve the needs of the public, can it remain predominantly agricultural? What should be our course in planning for the future?

Fewer and fewer farmers are being required to supply food and fiber for a rapidly expanding population. The consumer is demanding high standards of quality and the absence of insecticidal residues. Hence, better methods of insect control must be developed as an essential part of more efficient production of agricultural crops, and agricultural entomology must retain a high priority in program planning in entomology. However, we need to educate the public as to the service rendered by agriculturists and agricultural entomologists to all the people. We need to re-emphasize the interdependence of agricultural entomology and related industries employing many more people than are employed directly in agriculture. We need also to re-define agriculture. The public must be made aware of the fact that farming is no longer plowing and hoeing, but is a complicated, scientific enterprise affecting directly the economic status of a large segment of our population in related businesses.

### Opportunities in Non-Agricultural Entomology

Emphasis on agricultural entomology alone is not enough. Future programs must be adequate to meet the direct needs of the public both rural and urban, farm and non-farm. No other group of scientists is in so favorable a position as the entomologists to offer direct, usable service to all: farmers, home owners, housewives, gardeners, orchardists, ranchers, dairymen, poultrymen, food processors, foresters, merchants, warehousemen, transportation and shipping establishments, the armed services, health organizations, land-use recreational facilities, schools, architects, building contractors, and many others (Arant, 1961).

Increased emphasis is needed in all of these non-agricultural areas, and we are familiar with the opportunities in many. The following examples point out a few of these needs and opportunities. Home owners have problems with termites and insects of lawns and ornamental plants; housewives are troubled with such pests as cockroaches, ants, flies, silverfish, clothes moths, carpet beetles, mosquitoes and many others; amateur gardeners find insects and diseases to be a limiting factor in their operations; individuals and groups are concerned with insect pests in an estimated 14,000,000 acres of turf in golf courses, parks, airports, cemeteries, lawns, and shoulders of highways; food processors and distributors must constantly guard against insect contamination of edible products; hunters, fishermen, picnickers, and campers are subject to attack each year by flies, mosquitoes, chiggers, ticks, ants, and other pests.

We may be less familiar with equally great opportunities and needs in programs of recreational land-use facilities, public health, insect inspection and quarantine work, and entomology as a basic science.

(1) *Recreational Land-Use Facilities.*—As the population increases and more leisure time is available, the demand for recreational facilities increases. Existing parks and facilities are inadequate. The demand is increasing for use of forest areas for recreation as well as for production of timber, pulpwood, and other forest products. State and federal agencies, private companies, and individuals are responding to this growing demand by promoting recreational programs. Entomologists must work in these projects in cooperation with foresters and wildlife biologists. Specific entomological problems include control or suppression of forest insect pests; protection of campers, hikers, hunters, fishermen, and others against attacks from insects, mites, and ticks by controlling the pests or by other methods; assistance in maintaining and developing population of arthropods important as food for fish and/or game species; development of programs for protection of fish, game, and other beneficial species from deleterious effects of insecticides used in control of pests; and provide guidance for entomological nature studies in the areas.

(2) *Medical and Public Health Entomology.*—Although such insect-borne diseases as malaria, plague, and yellow fever are practically non-existent in the United States today, medical and public health entomology offers unusual opportunities for achievement in this country and in international programs. At least 46 species of vectors of diseases have now developed resistance to insecticides in 30 different countries, including the United States (Committee on Government Operations, 1959). The development of insecticide resistance by *Anopheles* mosquitoes and the frequent presence in the United States of malarial organisms in the blood of persons from endemic malarial areas of the world combine to create a potential malaria hazard in the southern United States. In the world, some 1,168,000,000 people live in areas where malaria is endemic, and malaria is still the most important disease of man. Some concern has been expressed over the rather consistent northward spread of yellow fever from South America into Central America; the principal mosquito vector, *Aedes aegypti* (L.), is common in southern United States and may migrate northward during the summer months. Some other problems in the United States include such arthropod-borne diseases as dengue, encephalitis, endemic typhus, Rocky Mountain spotted fever, Colorado tick fever, tularemia, scabies, dysenteries, and others. Since our civilian and military personnel are employed and deployed in all parts of the world, such arthropod-borne diseases as Chagas', Trachoma, Kala-azar, sleeping sickness, filariasis, and scrub typhus are of direct concern to us as well as malaria, yellow fever, plague, and other insect-borne diseases common in the United States. International programs involving control of arthropod vectors of several of these diseases are underway, and they offer challenges and opportunities to American entomologists.

(3) *Exclusion of Foreign Pests.*—Port inspection and related programs designed to protect the United States against introduction of foreign pests are extremely important. In 1960, 30,230 lots of unauthorized products were intercepted at the ports of entry, representing a 19.5 per cent increase over 1959 (Dean 1961). The rate of increase was 6½ times the rate of increase in entries of people; it represented a pest being intercepted and prevented from entering each 17 minutes during the year. In some instances this work has been considered routine, police type of employment, not particularly challenging to the scientifically trained entomologist with a high degree of technical skills. We need to readjust our thinking in this area and perhaps also our programs of training. To be sure entomologists trained in physiology and toxicology would hardly go into port inspection, or vice versa. However, an effective program of preventing the introduction of new pests should and does require a rare combination of entomological skill, public relations, intelligence, and perseverance. It must function at home

and in foreign areas. This important segment of entomology offers real opportunities to those with the vision and dedication to grasp and pursue them. The work needs to be expanded and elevated to a position of greater scientific prestige commensurate with the services rendered.

(4) *Basic Research in Entomology.*—If entomology is to serve the public, entomologists must make a greater contribution to basic scientific knowledge. No longer does the public have to be sold on the need for basic research. The public accepts research with an attitude somewhat akin to awe and expects some fundamental studies with no direct applications. It may be that the entomologists are the ones hesitant in turning to more basic studies. Perhaps we have been engaged too intensively in testing insecticides a, b, and c against pests x, y, and z to keep ahead of resistance and in putting out "brush fire" on a daily or annual basis. The prestige of the entomology profession can be raised, job opportunities increased, and services to the public enhanced by well-planned programs of basic research in insect genetics, physiology, embryology, morphology, bionomics, ecology, taxonomy, and evolution.

### Integrated Insect Control Needed

What are our most urgent immediate problems and needs? What specific action programs should we develop? In my opinion, one of our greatest needs is to develop and adopt so-called integrated programs of insect control—control utilizing insecticides[3] and integrating or coordinating their use with other methods of control.

More progress in insect control has been made since the introduction of synthetic organic insecticides than in all previous years combined. So confident have we been in the near infallibility of these synthetics that the president of the AAEE some years ago voiced a rather widely accepted viewpoint, predicting that entomology was emerging from the era of insect control and entering the era of eradication with DDT and other synthetics. However, scarcely had the prediction been made before the house fly, one of the first species tagged for eradication, developed resistance to DDT and could no longer be controlled, much less eradicated. In rapid succession numerous other insects also developed resistance until the number now exceeds 137 species (Brown 1961). Concurrently, many additional problems arose involving principally residues in food, forage, soil, and water and the destruction of beneficial insects and wildlife species. Nevertheless, insecticides are still our main defense against insect pests, and they will and should continue to be used extensively for the foreseeable future. There are unlimited opportunities for development of systemic and other insecticides with selective action on insects and other organisms. However, one of the real challenges to the economic entomologist today is to prevent any indiscriminate or excessive use of insecticides and to coordinate or integrate insecticidal control with other methods in balanced, effective, economical programs acceptable to the public.

In some instances, we pay lip service to biological, ecological, and cultural methods to supplement insecticidal control but continue to apply ever-increasing amounts of insecticides with little regard to beneficial insects or even the economy of the treatment. In some sections of the cotton belt, 30 or more applications of insecticides may be applied to cotton by some farmers in a single season, representing as much as 100 pounds of technical material per acre. In my own state, official recommendations of the Land-Grant University for 1961 list 80 insecticides for control of 141 insect pests or groups of pests; less than one page is devoted to non-chemical control as contrasted with 63 pages dealing with insecticidal measures. These examples illustrate how far some, if not most, of us are from any real programs of non-insecticidal control.

### Much Research Needed

How are we to develop coordinated control programs? Entomologists in the far west have initiated such programs and are largely responsible for the terms "integrated control." However, at least in some sections of the country, we are woefully ignorant of the techniques and mechanisms of coordinating insecticidal control with other methods. In fact, there are few workable non-insecticidal control practices available for use against many pests. All too frequently so little is known about insect parasites and predators in relation to the pest insect and to insecticidal treatments that it is impossible to utilize or even tolerate naturally occurring beneficial species in a control program.

New programs involving old and new concepts of control must be developed in conjunction with existing procedures. Precise, usable information is needed on the following subjects:

(1) *Naturally occurring parasites, predators, and diseases of insect pests.*—Utilization of beneficial insects and diseases may involve the toleration of large numbers of fairly innocuous insects as food for an adequate population of beneficial species to control a highly destructive pest at a later date. For example, thrips may cause only minor damage to cotton early in the season under many conditions, but they serve as food for beneficial insects that are effective in control of highly destructive bollworms later in the season. Other procedures could involve the use of chemicals with selective action against the pest insect, and/or proper timing to utilize the beneficial species at a critical period.

(2) *Biological agents.*—The rearing and releasing of biological agents is not a new subject. This method of control has been practiced with varying degrees of success. However, the entire field of insect pathogens is a very fertile one for investigation (Steinhaus, 1960). I doubt if we visualize the potential of new pathogens or of the possibilities in the use of new and old biological agents disseminated by insects released at critical seasons of the year or at vulnerable stages in the life cycle of the pest in a coordinated program with insecticidal control.

(3) *Hormones, anti-metabolites, sterilants, and other physiological agents.*—Disturbing or altering physiological processes affecting ecdysis, diapause, or reproduction is a new approach to insect control. Hormones, anti-metabolites, sterilants, and related substances may hold the key to preventing diapause and hibernation, to breaking of diapause at critical seasons of the year, thereby causing suicidal emergence of insects, and to sterilizing insects in the field or insects reared in the laboratory and released in the field to compete with the reproductive population.

(4) *Resistant hosts.*—The plant pathologists have far surpassed the entomologists in developing strains of host plants resistant to specific organisms, although Painter (1951) and others have demonstrated the desirability and feasibility of developing plants resistant to insect attack. Opportunities for research on plants and animals resistant to insect attack are almost unlimited.

(5) *Irradiation.*—The now classic example of the eradication of the screw-worm, *Cochliomyia hominivorax* (Coquerel), in Florida by the release of irradiated,

[3] Insecticide is used in this paper in a restricted sense to mean a toxic chemical lethal to the insect.

sterile flies is proof of the feasibility of this method. This procedure should be applicable to other species. In addition there may be many other methods of control through irradiation, such as changing the genetic characters of an insect to include a lethal factor or other genes detrimental to the species, breeding the modified strain in the laboratory, and releasing large numbers of these insects in the field. (There is also the possibility of developing super races or strains.)

(6) Insect genetics.—Although an insect, *Drosophila melanogaster* Meigen, has been used perhaps more than any other single organism in basic studies of inheritance, insect genetics has played a minor role in pest control. A major bottle-neck in research on insect resistance to insecticides is the paucity of information on heritable characters of the specific insects. Insect genetics is extremely important in studies on resistance as well as in finding lethal or other factors detrimental to the species.

(7) *Attractants and baits.*—Attractants and baits have been used for many years with a considerable degree of success. Although the extensive use of broad-spectrum highly effective synthetic insecticides has detracted somewhat from the use of baits in recent years, problems with residues have stimulated renewed interest. The use of an insecticidal bait with selective attractiveness to the Mediterranean fruit fly, *Ceratitis capitata* (Wiedemann), made possible eradication of this insect in Florida with minimum residue problems. Baits are offering real promise of alleviating the bitter controversy over control of imported fire ant, *Solenopsis saevissima richtaeri* Forel, in several southern states. Not only are attractants promising in luring the target insect to an effective insecticide, but they may be used with other chemicals, such as sterilants and anti-metabolites, that act indirectly to decimate a population of insects. Knipling (1960) has ably pointed out that indirect action may be much more effective than direct insecticidal control.

(8) *Ecological methods.*—The term ecological control is used here to designate manipulation of the environment to break the life cycle of the insect or to expose the insect to natural enemies, unfavorable climatic conditions, or other factors during vulnerable periods in the life cycle. A thorough knowledge of the bionomics of the pest insect is essential, and it is in this area that our research programs are often weak. Information is lacking on the biology and population dynamics of some of our most important pests, including such widely publicized species as boll weevil, *Anthonomus grandis* Boh. In addition, the biology of a species frequently changes as it becomes adapted through the years to the environment of new localities. Information is also often lacking on the biology and ecology of parasites, predators, and pathogens of the pests species. If ecological control is to be integrated properly with insecticidal, accelerated research is needed on the bionomics of pest species.

(9) *Insecticides of high specificity.*—Fashions in insecticides are changing rapidly. The broad-spectrum long-residual chemicals, such as DDT, at first appeared to be the ultimate achievement in efficient insecticides. Problems with residues and with destruction of beneficial species by these long residual materials have resulted in increased use of materials with shorter residual action. The acute need, however, is for higher specificity of action of insecticides. The ideal material may be systemic or nonsystemic with long residual or short residual action, but it should have a high degree of selective toxicity to the pest insect. Herbicides with highly selective action are common, and there is every reason to believe that insecticides of this type can and will be developed. Highly specific insecticides will eliminate many difficult problems involved in integrated control.

The nine subjects listed above are by no means a complete enumeration of areas where vigorous research is needed to develop more effective programs of insect control. They are indicative of types of research necessary to develop adequate, acceptable coordinated programs of insect control to serve the public—farm and non-farm, rural and urban.

### The Challenge

The challenge before us is clear: (1) We must retain and use wisely the fine synthetic organic insecticides available now and in the future; (2) we must develop through research old and also totally new non-insecticidal controls; (3) we must coordinate insecticidal control with other methods for more effective economical control of agricultural and non-agricultural pests without serious residue hazards to man, domestic animals, fish, wildlife, or beneficial insects; and (4) we must have well-planned action programs to achieve these ends. Only you and I as professional entomologists can plan and execute these programs.

## REFERENCES CITED

**Arant, F. S.** 1961. A paradox in entomology. Bull. Ent. Soc. America 7(1): 1–2.

**Brown, A. W. A.** 1961. The challenge of insect resistance. Bull. Ent. Soc. Amer. 7(1): 6–19.

**Census of Agriculture,** 1960. U. S. Department of Commerce, Bureau of the Census.

**Committee on Government Operations,** U. S. Senate, 86th Congress, 1959. Status of world health in outline text and chart. U. S. Govt. Printing Office, Washington, D. C. 81 pp.

**Dean, H. S.** 1961. Entomological contributions to human welfare through plant quarantine. Bull. Ent. Soc. Amer. 7(2): 68–71.

**Harris, T. W.** 1862. A treatise on insects injurious to vegetation (revised). Cambridge, Mass.: C. L. Flint, 640 pp.

**Howard, L. O.** 1930. A history of applied entomology, Washington, D. C.: The Smithsonian Institute Publication 3065, 564 pp.

**Knipling, E. F.** 1960. Use of insects for their own destruction. Jour. Econ. Ent. 53(3): 415–20.

**Osborn, Herbert.** 1952. A brief history of entomology. Columbus, Ohio: The Spahr and Glenn Co., 303 pp.

**Painter, R. H.** 1951. Insect resistance in crop plants, New York: The Macmillan Co., 520 pp.

**Steinhaus, Edward A.** 1960. Insect pathology: Challenge, achievement, and promise. Bull. Ent. Soc. Amer. 6(1): 9–16.

# PROBLEMS OF INSECT ECOLOGY IN DEVELOPING COUNTRIES

By B. P. UVAROV

*Anti-Locust Research Centre, London*

The present period is characterized by a rapid transition, in many developing countries, from unorganized and often primitive systems of land use to the exploitation of their natural resources on the basis of a planned approach. This involves, on one hand, the initiation of large-scale development schemes and, on the other, the introduction into peasant farming of new crops and modern practices. In most cases it results in a drastic disturbance of the environment of the local insect fauna and in consequent changes in its composition. The essential practical feature of these changes is a radical reconstruction of the injurious fauna of useful plants.

Economic entomologists in the past have had to concentrate their attention on the pests which were urgently claiming it by their depredations, and they had little time, or means, to investigate the basic reasons for certain insect species acquiring pest status. The whole history of applied entomology appears as an incessant development of its purely defensive aspects aiming only at reducing the losses, or at the best, at attempting to eliminate a pest from a country. This may well be the only approach that is justifiable in the more advanced countries where the man-made conditions have become more or less stabilized, but the situation is radically different in the vast areas of Africa, Asia, South America and Australia where a revolution in land use is in progress or even still largely a matter of the future. There it is possible, and urgently necessary, to inquire into the possibilities of preventing developments that appear likely to handicap or even to undermine the best intentions, by favouring actual or potential insect pests. The only way of clarifying the problem is to see what effects the different types of land use have had on the insect pest populations in the already developed or developing countries. This subject has already been discussed with special reference to arid lands (Uvarov 1957, 1962), and here it will be dealt with in broader geographical aspects. The problem is a vast one and no general review of it can be attempted, but certain outstanding points, with factual illustrations, will be discussed.

## FORMATION OF PEST FAUNA OF A CROP

As one of the best illustrations of the revolutionary reconstruction of a natural insect population under the impact of cultivation, a recent paper by Bey-Bienko (1961) may be taken in which the results of 2 year comparative quantitative studies of the population of virgin grass steppe and of the new wheat fields in Kazakhstan were presented (Table 1).

The significant conclusions from Table 1 are: (1) That there are more than twice as many species in the virgin steppe as there are in the wheat field. (2) That the density of population in the wheat field is almost double that of the steppe. (3) That the number of dominant and constant species in the steppe is twice that of the wheat field, but in the steppe they comprise only about half the fauna, while in wheat they amount to 94·2%. (4) That there is, therefore, a greater specific variety in the population of the virgin steppe and a better numerical balance between its species; while in wheat fields a few species

Reproduced with permission from the *Journal of Applied Ecology,* 1:159–168, 1964. Published by the British Ecological Society, Oxford, England.

become dominant and others are represented by a few individuals. (5) That most of the dominant species of the wheat field are important pests.

Such changes in the composition of the insect population occur with a startling rapidity as shown in Table 2, comparing the populations of the more important species in a virgin plot and in a field ploughed for the first time in the autumn and sown with wheat next spring.

Table 1. *A comparison of the total insect population in* 1 $m^2$ *of virgin grass steppe and of wheat field* (*Bey-Bienko* 1961)

| | Virgin steppe | Wheat field |
|---|---|---|
| No. of all species | | |
| Homoptera | 35 | 12 |
| Heteroptera | 38 | 19 |
| Coleoptera | 93 | 39 |
| Hymenoptera | 37 | 18 |
| Others | 137 | 54 |
| Total species | 330 | 142 |
| Individuals of all species per $m^2$ | 199 | 351 |
| Dominant and constant species | | |
| Number | 41 | 19 |
| Individuals per $m^2$ | 111·2 | 331·6 |
| Percentage of the population | 54·4 | 94·2 |

The last three species in Table 2 are amongst the outstanding wheat pests.

Grigor'eva (1960), in similar comparative studies, has confirmed the general conclusions of Bey-Bienko and obtained valuable quantitative evidence that parallel with an increase in the populations of injurious species there was a substantial decrease in the numbers of predacious and parasitic insects in newly cultivated fields. On the other hand, their numbers tended to increase gradually in the areas of old cultivations; in the newly cultivated areas the percentage of parasitization of noctuid caterpillars (*Hadena sordida*) was 1–7%, while in the old areas it was up to 80%. There is thus a continuous process of the evolution of the fauna of the wheat areas, the bulk of which consists of the species originally connected with wild grasses; the role of the latter as a source of the pest species, however, gradually diminishes as they become better adapted to the conditions of cultivated fields.

Table 2. *Densities of the more important species in virgin steppe and wheat sown for the first time* (*Bey-Bienko* 1961)

| | Number per $m^2$ | | Ratio of increase (×) or decrease (÷) |
|---|---|---|---|
| | Virgin steppe | Wheat field | |
| *Caeculus dubius* | 2·23 | 0·06 | ÷37·2 |
| *Sminthurus viridis* | 7·65 | 0·19 | ÷40·2 |
| *Ectobius duskei* | 2·83 | 0·03 | ÷94·3 |
| *Leptothorax nassonovi* | 16·48 | 0·03 | ÷550·0 |
| *Haplothrips tritici* | 1·07 | 300·40 | ×280·8 |
| *Phyllotreta vittula* | 0·05 | 1·03 | ×20·6 |
| *Hadena sordida* | 0·09 | 2·25 | ×25·0 |

There is vast and highly informative Russian literature on this subject in connection with the recent grandiose schemes for the development of the virgin lands (some references in Uvarov 1962), and their main general conclusions are worth considering in relation to the extension of grain crops in other countries. Such crops are, naturally, most suited to the areas of original grasslands of a type comparable to the kind of grain crop which is being introduced. While wheat and similar crops are grown mainly in the areas of original short-grass steppe, maize and sorghum belong to the tall grass savanna. It has been shown (Neiswander 1931) that out of 350 insect species associated with maize, many as serious pests, at least half are, or have been originally, grass feeders. Pests of sorghum, such as the notorious lepidopterous stem-borers, are ever present in tall savanna grasses including wild sorghum (Bowden 1954; Jepson 1954), which are their original source and not merely 'alternate hosts', as they are usually regarded by economic entomologists. The principle of a broad ecological approach to the stem-borer problem was admirably expressed by Bowden (1954), who insisted that 'the borers concerned should be considered not merely as something found in a cereal crop but as part and parcel of the fauna associated with the Gramineae of the area'.

Another important sorghum pest is the midge *Contarinia sorghicola*, causing sterility of heads in all countries where this crop is cultivated; it is a close relation of other species of the genus occurring in wild sorghum, *Andropogon*, *Panicum* and other tall savanna grasses. Some of these cause damage identical to that in cultivated sorghum (Barnes 1954).

The next tropical gramineous crop of great economic importance is sugar-cane, which is a plant of more humid habitats than sorghum and maize, but belongs essentially to the same group of large grasses. Although its native country is not known (possibly in south-eastern Asia), it is now cultivated in all sub-humid and humid tropics. As regards its pest fauna, it consists in each country of local species of insects, suggesting that they belong to the native grass fauna and have adopted sugar-cane as a more favourable food. Definite information on this point is available from tropical America, where Myers (1932, 1935) and Box (1951) found that the more important stem-borers (*Diatraea* spp.) and froghoppers (*Tomaspis* spp.) belong to the fauna of tall hydrophilous grasses (*Paspalum*, *Echinochloa*) and are widely distributed in tropical South America. An interesting exception is the stem-borer *Castnia licoides*, whose original host is not a gramineous plant but *Heliconia*, a member of Musaceae; this suggests that sometimes a wild plant not closely related to a cultivated one may be of primary importance as a pest source. Nevertheless, the few examples reviewed above, serve to extend Bowden's idea that pests of gramineous plants must be regarded as an integral part of grassland communities of the area, and the study of their relations to the crop should not be confined to the period when they cause damage. Many of them, in fact, spend part of their life-history not in the crop. For most of them the wild vegetation, where they may be less numerous than in the crop, is still the main habitat and maintains a population reserve which eludes artificial control measures. In addition, there are certain plants which grow in cultivated fields as weeds, and some of them are the original hosts of pests. Thus *Aphis mayidis*, the vector of the mosaic disease of sugar-cane, is forced to move to cane when wild grasses are removed by weeding. Some polyphagous caterpillars normally feeding on weeds in sisal and cotton plantations in the West Indies also move to the crop plant after weeding (Wolcott 1928).

Cotton, a plant extensively cultivated in many subtropical and tropical countries, has amassed a formidable number of insects associated with it (Hargreaves 1948; Pearson 1958). A perusal of their list shows that, apart from several polyphagous leaf-eating insects, notorious migrants or introduced species, the vast majority of its more specific pests are native to the particular country. Thus, the well-known American boll-weevil (*Anthonomus grandis*) spread to cultivated cotton in America from two Mexican species of wild cotton as recently as the end of the last century and became firmly established in the new conditions. In South Africa also, a boll-worm (caterpillar of *Diparopsis*) living in wild *Gossypium* and related shrubs, particularly *Hibiscus*, adopted cultivated cotton when it was introduced. Other cotton insects, e.g. the African cotton-stainers (*Dysdercus* spp.) belong to the fauna of the local Malvaceae, Bombacaceae, Sterculiaceae and Tiliaceae (all of the Malvales), and only part of their life-cycle is spent on cotton; in South America, local species of *Dysdercus* play the same part.

Still another example of a special crop is sugar-beet. Its recent introduction into Kazakhstan resulted in a rapid acquisition of a large number of outstanding pests, all of them members of the native fauna of wild Chenopodiaceae and Salsolaceae; in the areas of intensive large-scale cultivation such species continue to exist partly in the crop, partly outside it. Where the crop is irrigated many species pupating in the soil are periodically reduced, but there are always abundant reserves outside the crop (Brunner 1963). In the southern United States sugar-beet suffers severely from curly-top virus transmitted by a leafhopper (*Eutettix tenella*), the original food plants of which are *Artemisia* and salt bushes, from which it migrates for long distances to beet plantations (Lawson, Chamberlain & York 1951).

A most illuminating case is that of the sunflower, which is an American plant that has been cultivated in Russia for oil-seed for only 100 years and has acquired over fifty specialized insect pests, not counting the more polyphagous ones, all of them originally associated with wild native Compositae. A detailed analysis of the fauna of sunflower and related cultivated Compositae by Pravdin (1957) presents a remarkably detailed picture of its formation in the different natural zones of the Soviet Union in relation to the fauna of wild plants belonging to the same order.

These randomly chosen examples should be sufficient to illustrate the thesis that the pest fauna of cultivated plants is usually a product of segregation of a few native species associated with related wild plants and their adaptation to a crop which often presents very similar, but vastly improved, conditions for their existence. It is also important that there is not a complete transference of these species from the wild to cultivated plants; their wild populations still remain and serve as a reservoir from which the losses inflicted by artificial control methods within the crop are continually replaced. In the case of pests which spend only part of their life-cycle in the crop, such replacement occurs every season from the outside populations which are immune from control. The last fact will often reduce the value of studies of fluctuations in the population of a crop pest, if carried out only within the crop.

## POLYPHAGOUS AND MIGRANT INSECTS

While the preceding paragraphs are concerned with more or less specialized plant-feeding insects, a large proportion of crop pests, particularly in the warmer countries, are generalized feeders. Some of them, in addition, have seasonal migratory habits, so that their importance to crops depends on their seasonal distribution and general ecological requirements in relation to agricultural areas. The most notorious and the best studied amongst these insects are locusts, but they will not be discussed here in detail because this has been done elsewhere (Uvarov 1957), and only some essential points need to be mentioned. Firstly, there is abundant evidence of a direct ecological connection between locust or grasshopper plagues and general agricultural and pastoral development of new countries. Thus plagues of the migratory locust (*Locusta migratoria*) and the Italian locust (*Calliptamus italicus*) were particularly severe during the earlier stages of the settlement of the Ukraine during the second part of the last century, but became of secondary importance later; during recent years *C. italicus* and some grasshopper species have been very active in the virgin lands of Kazakhstan and western Siberia which are now undergoing development on the lines of the Ukraine, but on a more extensive scale and with an accelerated tempo. From the ecological point of view there are interesting long-term changes in the economic importance of locust and grasshopper species in relation to the state of general progress in land use. In North America the early settlers had to contend with vast swarms of the Rocky Mountain locust (*Melanoplus spretus*), which is now supposed to be extinct (Gurney & Brooks 1959); later on, expanding agriculture suffered continuous depredations by many grasshopper species adapted to life in arable land and feeding on a variety of crops. Finally, during more recent times, after these crop grasshoppers had become considerably reduced by a vast-scale insecticidal control, new dangers continued to arise from species native to uncultivated grasslands. These grasslands have undergone deterioration through overgrazing and have also become reduced in area, and the result is that species of grasshoppers adapted to them periodically spread into arable land, in addition to causing severe damage to pastures. Other human practices affecting locust and grasshopper populations will be mentioned below.

Outstanding examples of the importance of migrant insects are offered by some Lepidoptera, such as cotton leaf-worm (*Alabama argillacea*), adults of which annually invade the southern United States, producing there several generations and disappearing again; the species is supposed to live permanently in Central or South America, but its biology and ecology there remain almost unstudied. There are several so-called army-worms, caterpillars of several species of moths with similar habits, in most tropical countries of the world, and at least some of them are thought to originate in grasslands which have been degraded by fires, overgrazing, etc.; investigations of the biology and ecology of the African army-worms are now in progress (Brown 1962).

Apart from the obvious long-range migrants there are many polyphagous insects of various orders which undertake local movements, usually seasonal (e.g. connected with the dry period of the year in tropical countries), from natural habitats to crops; and still others which merely visit crops for feeding, while the reproductive period is spent elsewhere. Such cases suggest the importance of wild habitats adjoining crops and of their insect populations. An interesting illustration, with regard not to pests but to useful insects, has been given by Ponomareva (1959), who found that the number of wild bees pollinating lucerne, cultivated in oases of Middle Asia for seed, falls sharply when large areas are put under the crop; but it is much greater when there remain in the vicinity areas of uncultivated ground with wild vegetation.

## TYPES OF LAND DEVELOPMENT AFFECTING INSECT POPULATIONS

One of the first human activities in originally forested areas has been, and still is, the removal of trees to clear the ground for cultivation. This was done long ago in the temperate countries and it has generally resulted in profound changes of the fauna in the general direction of a spread of more xerophilous elements into the newly created open landscapes. Thus in the forest-steppe zone of Siberia the clearance of forests by fire and cutting permitted a northward penetration of steppe grasshoppers into the cleared areas used for grazing, hay and crops (Vinokurov & Rubtzov 1930; Rubtzov 1932; other references in Uvarov 1957). An important additional characteristic of such habitats is the instability of their vegetation and microclimate, which underlies the numerical fluctuations in grasshopper populations (Uvarov 1957).

In the dry Mediterranean countries, deforestation (mainly for fuel and in the interests of grazing) opened vast new areas of uncultivated land eminently suitable for the Moroccan locust (*Dociostaurus maroccanus*), which can lay eggs only in undisturbed hard soil (Dempster 1957; Merton 1959, 1961).

The effects of deforestation on insect fauna in the subtropics and tropics are little known, except in the case of tsetse flies where there is a recognized method of ecological control of some species which cannot survive outside the forest (Buxton 1955). The removal of the primary forest and clearing of patches within it for cultivation must create entirely new ecological conditions and permit a penetration of the savanna insects into newly opened areas. A gradual encroachment of semi-desert conditions upon the dry bush in West Africa results from clearing the bush for shifting cultivation; such areas become suitable for the breeding of the desert locust (*Schistocerca gregaria*) and, almost certainly, of many other pests (Uvarov 1961). In tropical southern Asia, Malaya, Indonesia and the Philippines clearing of forests for cultivation results in large areas being invaded by tall grasses, which are periodically burnt; this creates the conditions eminently suitable for the oriental migratory locust (*Locusta migratoria manilensis*) which cannot possibly exist in a forest (Uvarov 1936). In Malaya and the adjoining countries, recent clearance of forests for cropping and rubber planting and their replacement by secondary grassland led to the increasing populations of the Bombay locust (*Patanga succincta*) (Pagden 1959).

An excellent example of a development in tropical silviculture favouring a pest is quoted by Kalshoven (1952), who observed that in the original teak forests of Java the trees are seldom attacked by a scolytid beetle *Xyleborus destructor*, but when the primeval forest is replaced by plantations of teak serious infestations develop in them. Similarly, long-horned grasshoppers (*Sexava*, *Segestes*) are said to be rare in the virgin forests of Indonesia, but are serious pests of coconut palms in plantations (Leefmans 1927).

A reverse process of afforestation of the Ukrainian steppes has resulted in creating ecological conditions favourable to mesophilous insect species which previously occurred only in the relict woods in ravines and river valleys, but rapidly spread into new plantations in the steppe; in addition, some xerophilous species became adapted to the new conditions (Arnoldi 1952, 1953). Even in the Caspian sandy semi-desert where no trees originally existed, new plantations suffered heavily, particularly from melolonthid larvae, which occurred in very low populations in the open sand areas, but increased rapidly under tree cover (Zinov'eva & Rafès 1954).

Forest shelter-belts intended to ameliorate general ecoclimatic and agricultural conditions in the southern Russian steppes provided extremely instructive examples of rapid and radical effects on the insect communities connected with the original landscape. Within a short time there was seen a revolution in the fauna, with the disappearance or reduction of xerophilous species and a parallel increase in the mesophilous ones. Some xerophilous species which require shelter for hibernation (e.g. grain bugs *Eurygaster* and *Aelia*), which they used to find in the few relict woods, were now provided with most suitable conditions under the nearest shelter-belts, without the need of undertaking hazardous migrations. A critical examination of the situation created (Mel'nichenko

1949) has shown that in many cases the advantages gained by noxious insects were such as to reduce materially, and even to nullify, the increase in agricultural production due to the direct effects of shelter-belts on the crop.

Large-scale utilization of natural pastures, particularly uncontrolled grazing and over-stocking, rapidly results in great population changes. It was shown long ago (Weese 1939) that the total number of insects in a heavily overgrazed plot of a prairie in Oklahoma was four times as great as in the adjoining protected area; in the case of grasshoppers, the factor was nine times. Usually, the effect of overgrazing is a suppression of mesophilous species and their replacement by the xerophilous ones, and the grasshopper problem in particular is in many countries directly connected with overgrazing (Uvarov 1947, 1957).

High hopes are placed in many developing countries, particularly those with semi-arid and arid conditions, on the creation of new agricultural areas by irrigation. An excellent review of the impact of irrigation on insects has been provided by Smith (1959). In the isolated desert of the Salt River valley in Arizona, some seventy-three insect species became crop pests; some of them were indigenous but increased in numbers, others were introduced (McKinney 1939). Large-scale cultivation of irrigated lucerne (alfalfa) in the semi-arid southern United States provided completely different new conditions for the existence of many local noxious insects, and the same happened in South America where native grasshoppers flourished in irrigated fields (Daguerre 1940; Libermann & Schiuma 1946). In the Sudan, cotton formerly suffered from certain pests, particularly thrips (*Caliothrips*, *Hercothrips*), only when cultivated near the rivers, but when the large irrigation scheme of the Gezira came into being they became the most serious pests of cotton (Tothill 1948). A reverse effect, suppression of some insects in irrigated sugar-beet plantations in Middle Asia has already been noted above (p. 162). Rapid recent extension of irrigation in Israel which makes possible the cultivation of various crops in the desert areas throughout the year has resulted in a number of previously rare or even unknown insects becoming major pests (Rivnay 1962).

Turning from irrigation to the amelioration of swampy areas by drainage, valuable experience has been obtained in the subtropical western lowlands of the Georgian Republic (the ancient Colchis), where the original wild invertebrate populations of the main habitats (*Alnus* woods, *Molinia* marshy grassland) were studied quantitatively (Kobakhidze 1943), and then compared with the populations for 1–20 years after drainage (Kobakhidze 1959). The populations of the original swampy habitats were poor in species, a few of them specially adapted to the environment but the majority polyphagous; none occurred in sufficient quantities to make them potential pests. In the areas which have been drained but not yet cultivated there has been a reduction in numbers of insects and other invertebrates (particularly earthworms), but the cultivated areas were soon populated by invaders from the surrounding drier habitats, including a number of prominent pests.

## GENERAL CONSIDERATIONS

It should be stressed that this cursory review is obviously very incomplete, but the facts quoted should be sufficient for formulating some general principles.

First of all, it is certainly a general rule that the injurious fauna of a given crop in a given country is produced as a result of segregation of a reduced number of species already present in wild habitats, and pest status is reached by them owing to improved conditions of existence. Significant exceptions to this rule are pests of introduced plants having no near relatives in the country; an example is tea in the subtropical parts of the U.S.S.R., particularly the Caucasus, all of the important pests of which have been introduced either with the tea plant, or with other exotics, e.g. citrus, etc.; there are only a few polyphagous native insects that cause occasional damage (Kobakhidze 1954).

Secondly, those species which adapt themselves to the new conditions still maintain vast reserves of their population in the remaining untouched areas. Even in highly developed countries, e.g. the United States of America, such habitats as roadsides, boundary strips, strips along railways, etc., retain the original vegetation and insect fauna; in the British Isles, live hedges are a notorious reservoir of orchard pests.

These two points justify a view that artificial measures, whether chemical or agricultural, practised only within the crop, while essential for its protection, are by their nature only palliatives, and to be effective must be continuously repeated. Both these points are particularly relevant to the subject of this paper. In the still undeveloped areas the pest potential of the original insect fauna as a result of subsequent development is virtually unlimited and unknown. It could be predicted to some extent if the species composition, and their ecological (particularly food) requirements were known. In certain large-scale development schemes, surveys are carried out of the soils, climate, vegetation, etc., of the area (Phillips 1959), but no attempt is generally made to evaluate the insect fauna. An apparent exception to this omission is seen in the Soviet Union where biologists, and entomologists in particular, have been consistently urging the need for studies of insect populations in the areas scheduled for development. Much work on these lines has actually been carried out there, as indicated in the few examples given above, but little is known as to whether the conclusions reached and recommendations made were taken into account when the schemes were implemented.

The results of neglecting entomological aspects of development are clear in the older countries, where permanent problems of defensive pest control have been created and perennially require an expenditure on a vast and ever-increasing scale. These lessons of the past should be sufficient to be heeded in the planned development of new countries. If this is not done there are bound to be disappointments, and vast sums of money spent on producing crops, a substantial share of which will be for the benefit of insects, not man. This applies to large-scale mechanized schemes, but even more to individual farming on which the welfare of the human population in developing countries mainly depends.

## REFERENCES


**Arnoldi, K. V. (1952).** To the elucidation of zonal regularities in the formation of new groupings of insects and of the population of forest plantations by xerophilous species during the afforestation of steppes [In Russian]. *Zool. Zh., Moscow*, **31**, 329–46.

**Arnoldi, K. V. (1953).** On the forest–steppe sources and the ways of penetration of forest insects into steppe during the afforestation of steppes [In Russian]. *Zool. Zh., Moscow*, **32**, 175–94.

**Barnes, H. F. (1954).** The sorghum midge problem. *Rep. 6th Commonw. ent. Conf., London*, pp. 155–60.

**Bey-Bienko, G. Ya. (1961).** On some regularities in the changes of the invertebrate fauna during the utilization of virgin steppe [In Russian with English summary]. *Rev. Ent. U.R.S.S.* **40**, 763–75.

**Bowden, J. (1954).** The stem-borer problem in tropical cereal crops. *Rep. 6th Commonw. ent. Conf., London*, pp. 104–7.

**Box, H. E. (1951).** The history and changing status of some neotropical insect pests of sugar-cane. *Trans. IXth int. ent. Congr.*, pp. 254–59.

**Brown, E. S. (1962).** *The African army worm* Spodoptera exempta (*Walker*) (*Lepidoptera, Noctuidae*). *A Review of the Literature.* Commonwealth Institute of Entomology, London.

**Brunner, Yu. N. (1963).** To the problem of forecasting of specialized pests of sugar-beet in irrigated areas of Middle Asia and Kazakhstan [In Russian]. *5th Conf. vsesoyuz. ent. Obsch.*, pp. 75–6.

**Buxton, P. A. (1955).** The natural history of tsetse flies. *Mem. Lond. Sch. Hyg. trop. Med.* **10**, 1–816.

**Daguerre, J. B. (1940).** Observaciones biologicas sobre *Dichroplus arrogans* (Stål). *Rev. Soc. ent. Argent.* **10,** 341–6.

**Dempster, J. P. (1957).** The population dynamics of the Moroccan locust (*Dociostaurus maroccanus* Thunb.) in Cyprus. *Anti-Locust Bull.* **27,** 1–60.

**Grigor'eva, T. G. (1960).** On the method of protecting cereal cultures in the zone of development of virgin and long-fallow land cultivation [In Russian]. *Rev. Ent. U.R.S.S.* **39,** 509–20.

**Gurney, A. B. & Brooks, A. R. (1959).** Grasshoppers of the *mexicanus* group, genus *Melanoplus* (Orthoptera, Acrididae). *Proc. U.S. nat. Mus., Wash.* **110,** 1–93.

**Jepson, W. F. (1954).** *A Critical Review of the World Literature on the Lepidopterous Stalk Borers of Tropical Graminaceous Crops.* Commonwealth Institute of Entomology, London.

**Kalshoven, L. G. E. (1952).** Important outbreaks of insect pests in Indonesia. *Trans. IXth int. ent. Congr.* **2,** 229–34.

**Kobakhidze, D. N. (1943).** Analysis of terrestrial biocoenoses of the central part of the Colchis lowlands [In Russian, with summaries in Georgian and English]. *Trudy zool. Inst. Akad. Nauk Gruzin. S.S.R.* **5,** 1–189.

**Kobakhidze, D. N. (1954).** *Injurious Insects of Tea Plantations in U.S.S.R.* [In Russian]. Inst. Zool. Gruzin. Akad. Nauk, Moscow.

**Kobakhidze, D. N. (1959).** The peculiarities of the formation of entomological complexes in the new Colchis [In Russian, with English summary]. *Zool. Zh., Moscow,* **38,** 860–6.

**Lawson, F. R., Chamberlain, J. C. & York, G. T. (1951).** Dissemination of the beet leafhopper in California. *U.S. Dep. Agric. Techn. Bull.* **1030,** 1–59.

**Leefmans, S. (1927).** Gegevens over sabelsprinkhanen as cocosvijanden in Nederlandsch-Indie en hunnen parasiten. *Meded. Inst. Pl Ziekt., Weltevreden,* **72.**

**Liebermann, J. & Schiuma, R. (1946).** Las 'tucuras' más prejudiciales de nuestra agricultura y ganaderia. *Publ. Inst. Sanit. veg., Buenos Aires,* **2**(1), 1–64.

**McKinney, K. B. (1939).** Common insects attacking sugar-beet and vegetable crops in the Salt River valley in Arizona. *J. econ. Ent.* **32,** 808–10.

**Mel'nichenko, A. N. (1949).** *Forest Shelter-Belts of the Trans-Volga Steppes and their Effects on the Multiplications of Animals Useful and Harmful to Agriculture* [In Russian]. Moscow.

**Merton, L. F. H. (1959).** Studies in the ecology of the Moroccan locust (*Dociostaurus maroccanus* Thunberg) in Cyprus. *Anti-Locust Bull.* **34,** 1–123.

**Merton, L. F. H. (1961).** The Moroccan locust (*Dociostaurus maroccanus* Thunberg) in Iran. *Anti-Locust Bull.* **37,** 1–66.

**Myers, J. G. (1932).** The original habitat and hosts of three major sugar-cane pests of Tropical America (*Diatraea, Castnia* and *Tomaspis*). *Bull. ent. Res.* **23,** 257–72.

**Myers, J. G. (1935).** The ecological distribution of some South American grass and sugar-cane borers (*Diatraea* spp., Lep., Pyralidae). *Bull. ent. Res.* **26,** 335–42.

**Neiswander, C. R. (1931).** The sources of American corn insects. *Bull. Ohio agric. Exp. Stn.* **473,** 1–98.

**Pagden, H. T. (1959).** *Patanga succincta* (L.), the 'Bombay locust' in Malaya. *Bull. Dep. Agric. Malaya,* **106,** 1–31.

**Pearson, E. G. (1958).** *The Insect Pests of Cotton in Tropical Africa.* Commonwealth Institute of Entomology, London.

**Phillips, J. (1959).** *Agriculture and Ecology in Africa.* London.

**Ponomareva, A. A. (1959).** The pollination of lucerne in western Kopet-Dag [In Russian]. *Trudy Inst. Zool. Parasit. Akad. Nauk Turkmen. S.S.R.* **4,** 34–46.

**Pravdin, F. N. (1957).** Regularities in the formation of new complexes of injurious and useful insects following the introduction of Compositae in cultivation [In Russian]. *Uchen. Zap. mosk. Pedag. Inst.* **100**(5), 1–190.

**Rivnay, E. (1962).** *Field Crop Pests in the Near East.* The Hague.

**Rubtzov, I. A. (1932).** The habitats and conditions of grasshopper outbreaks in East Siberia [In Russian with English summary]. *Bull. Pl. Prot., Leningrad,* **3,** 33–130.

**Smith, R. F. (1959).** Plant diseases, insects and weeds, as affected by irrigation. Insects. *Proc. 1st Intersoc. Conf. Irrig. San Francisco,* 1957, pp. 71–6.

**Tothill, J. D.** (ed.) **(1948).** *Agriculture in the Sudan.* London.

**Uvarov, B. P. (1936).** The oriental migratory locust (*Locusta migratoria manilensis* Meyen 1835). *Bull. ent. Res.* **27,** 91–104.

**Uvarov, B. P. (1947).** The grasshopper problem in North America. *Nature, Lond.* **160,** 857–9.

**Uvarov, B. P. (1957).** The aridity factor in the ecology of locusts and grasshoppers of the Old World. *Arid Zone Res., UNESCO,* **8,** 164–98.

**Uvarov, B. P. (1961).** Insect hazards in land development. *Span, Lond.* **4,** 154–7.

**Uvarov, B. P. (1962).** Development of arid lands and its ecological effects on their insect fauna. *Arid Zone Res.* **18,** 235–48.

**Vinokurov, G. M. & Rubtzov, I. A. (1930).** Studies in the ecology of grasshoppers in the Irkutsk district [In Russian with English summary]. *Izv. Irkut. Stant. Zasch. Rast.* **2,** 3–86.

**Weese, A. O. (1939).** The effects of overgrazing on insect population. *Proc. Oklahoma Acad. Sci.* **19,** 95–9.

**Wolcott, G. N. (1928).** Increase of insect transmitted plant disease and insect damage through weed destruction in tropical agriculture. *Ecology*, **9,** 461–6.

**Zinov'eva, L. A. & Rafès, P. M. (1954).** The influence of vegetation and of soil-forming process on the cockchafer population of the Naryn sand massif [In Russian]. *Zool. Zh., Moscow*, **33,** 1081–91.

# Part II: The Structure and Identification of Insects

More insect species are known than all the species of birds, mammals, fish, amphibians, and reptiles taken collectively. In fact, of the identified species of animals, insects comprise 75% of the total. Yet entomologists continue to find and describe hundreds of new species each year.

Without some method of ordering and classifying insects, this diversity of species would be a chaotic mass of unorganized and virtually useless information. However, **taxonomy,** *the science devoted to the theory and practice of classifying organisms,* has provided us with the methods for coping with insect diversity. The primary principle is to group species into categories that reflect some type of relationship. The relationship might be based on any of a number of criteria: size, color, habitat, food habits, behavior, etc. However, **phylogenetic relationship,** based upon *descent from common ancestry,* has provided the taxonomic framework that has proven to be most useful for entomology, and indeed for all of biology. The major difficulty with phylogenetic relationship is that descent cannot be tested directly, so it must frequently be based on many characteristics and inferred relationships as represented by present similarities.

In traditional taxonomy the "experienced" taxonomist proposes a classification for an organism or group of organisms that best fits the data currently available and expresses the most probable relationships between species or groups of species. One argument against traditional taxonomy is that classifications are usually based on judgment by an experienced taxonomist. To reduce this dependence on subjective judgment, a variety of attempts have been made to introduce a more objective approach to taxonomy. As a part of this effort, a general trend toward quantification has developed. One particular school, that of numerical taxonomy, has engendered much critical appraisal either supporting or attacking its premises. Among these premises are the equal taxonomic value of all characters (distinguishing features) and the separation of empirical taxonomic procedure from phylogenetic speculations. *Numerical Taxonomy, Convergence, and Evolutionary Reduction* (W. B. Brown, Jr.) discusses some of the weaknesses in the methodology and philosophy of numerical taxonomy, thereby introducing the student to the philosophic bases of taxonomy and the frequently controversial nature of contemporary research in this traditional area of entomology.

Brown also emphasizes one of the problems that make taxonomy by traditional methods difficult. The problem is that of **convergent evolution,** *the development of similar characters by two separate lineages undergoing adaptation to similar ecological conditions.* Both convergent evolution and **mimicry** *(adaptive* *resemblance of one species to another, conferring some advantage)* lead to morphologically similar forms. Further discussion of mimicry is presented in Part VI.

The primary character used in insect taxonomy has been the structure of the insect's exoskeleton. This character has been used for two important reasons. First, it is the only character available throughout the vast collections of dried pinned specimens maintained by research institutions. Second, it has proven satisfactory in most cases, resulting in a classification that provides for a high degree of accuracy in predicting the biological characteristics of a similar insect or insect group. However, strictly morphological taxonomy may not always be sufficient.

As with a number of insect groups, the classification of North American field crickets prior to 1960 was primarily based on morphology, but allowances were not made for adequate interpretation of population differences that were detectable or suspected in the field. R. D. Alexander, in *The Taxonomy of the Field Crickets of the Eastern United States,* was able to distinguish between biologically distinct cricket populations only by including characteristics of the living insects (for example, differences in song, distribution, and reproductive compatibility) into his interpretation of field cricket taxonomy. Alexander thus amply demonstrates how the array of taxonomically useful characters other than strictly morphological ones can be exploited in living species to solve problems in insect classification.

Even though entomologists have incorporated other aspects of insect biology into determinations of taxonomy, morphological (functional structure) characteristics of the insect's exoskeleton still provide the primary clues to phylogenetic relationship. The study of morphology, of course, is concerned with much more than taxonomic criteria. It is used to learn how the insect body functions. It can be used to trace the development of insect structures by examining the variety of forms a structure takes and by comparing its presence and characteristics in both primitive and advanced species. Of particular interest are morphological studies of the more significant insect structures, as found in *The Flight Mechanics and Evolution of the Wings of Ephemeroptera, with Notes on the Archetype Insect Wing* by G. F. Edmunds and J. Traver. The authors have researched one of the most significant morphological characteristics possessed by the majority of insects, their wings and ability to fly. Wings are largely responsible for the wide-spread evolutionary success of insects because they provide a method of eluding enemies and permit movement over long distances. Great mobility not only increases the breeding range of insects but facilitates the development of specialized feeding habits such as the utilization of rich, but scattered and sporadically available, food sources such as pollen or carrion.

Using the most primitive living winged insects available, Edmunds and Traver illustrate difficulties in the physics of flight and the entomological solutions to these difficulties: fluting to strengthen the thin membranous wing, and bullae (weakened vein areas) to permit bending of the wing on the upstroke. They also demonstrate how these characteristics, so essential in the early evolution of wings, later became detrimental as flight evolved from a flapping to a sculling action.

Two aspects of insect structure and identification remain to be mentioned before reading the articles in this section. The first is **nomenclature,** *the names applied to species and higher groups.* Inasmuch as ecology, physiology, behavior, and the other areas of entomology continue to produce data that may be applied to insect taxonomy, the classification of each insect group must periodically be revised. This revision is intended to reflect more accurately the probable relationships between groups by integrating the new data with the old. Therefore, one frequently sees changes in the names applied to insect species and higher categories (Genus, family, order, etc.).

The final aspect is insect development. Insects may pass through some remarkably different stages as they develop from egg to adult. Many organ systems become drastically altered as the insect undergoes **metamorphosis,** *a major change in form within the life history of an organism,* usually during the transition from larva to adult. *Morphology of the Alimentary Canal During the Metamorphosis of Hyalophora cecropia* by K. J. Judy and L. I. Gilbert provides an example of these metamorphic changes in a primary organ system. After describing the changes from larva to pupa to adult, the authors perform some elementary surgery on the alimentary canal and provide evidence that the crop does not function as a mechanism for breaking out of the pupal skin, an idea that has been widely accepted for this type of insect. Instead, it rather passively acts to replace the volume of blood used to expand the wings of the newly emerged adult moth.

# THE TAXONOMY OF THE FIELD CRICKETS OF THE EASTERN UNITED STATES (ORTHOPTERA: GRYLLIDAE: ACHETA)[1]

RICHARD D. ALEXANDER

Department of Zoology and Entomology, The Ohio State University, Columbus 10

## INTRODUCTION

Orthopteran taxonomists are plagued with a number of genera containing species groups that defy subdivision by the time-honored criterion of morphological distinction. Although relatively few species of singing Tettigoniidae and Gryllidae have been described in the last fifty years from the eastern United States, an estimated 10 percent of the species in these groups are largely unrecognized because preserved specimens are morphologically indistinguishable. Many of these have been mentioned in the literature, and some have been recognized at one time or another as distinct species, only to be cast into synonymy or compromised as trinomials by subsequent authors. Many of them have been variously referred to as "physiological races," "ecological races," "biological races," "song forms," or by other more or less meaningless terms.

Our present knowledge of the relationships of many of these species is largely the result of the contributions of B. B. Fulton whose investigations have been characterized by field recognition of species on the basis of distributional, ecological, seasonal, and song relationships, followed by attempts to find morphological characters by which preserved specimens can be separated. Much of Fulton's work has been conducted with groups upon which supposedly exhaustive studies had already been carried out through morphological analysis of large series of preserved specimens.

However, even Fulton has hesitated to honor with formal nomenclature species which cannot be distinguished "on the pin." This stand has been supported by Hubbell (1954, 1956) and other orthopterists (personal communication), apparently on the grounds that (1) this is a relatively rare phenomenon, (2) only a small group of workers will be concerned with such situations, (3) a great deal of confusion would result if museum workers could not apply the proper specific names to specimens in their collections, and (4) in such cases the designation of the entire group by a single specific name and the use of colloquial names to distinguish the various species will suffice for most applied problems and museum collections.

The evidence indicates that the presence of morphologically indistinguishable species is not at all a rare phenomenon in the Orthoptera, and the work done by Fulton might well serve as an example for investigators puzzling over morphologically ill-defined forms in the non-singing Orthoptera with distributional, ecological, and seasonal relationships analogous to those ocurring between closely related species in the singing Orthoptera.

It is difficult to estimate the size of the group of workers which will be interested in a particular group of closely related species. There is a growing amount of published material dealing with the morphology, physiology, nutrition, and genetics of certain groups of Orthoptera in which the organisms involved were not and now perhaps cannot be identified because the investigator was not aware of the existence of several species in the group he was studying. Yet many morphologically ill-defined species are easy to separate when alive, in some cases even easier for the non-specialist than morphologically well-defined species, since their ecological, seasonal, and song relationships have been more thoroughly studied and are easier to interpret. If these closely related pairs and groups of species remain unknown to all but a handful of orthopterists, their potential usefulness to students of systematics and evolution will not be realized.

As with most biological phenomena, the problem of morphological distinctness is not always clear-cut. For example, if only one sex of a species can be identified, this still leaves specimens which cannot be determined. In the present study of the five species of *Acheta* in the eastern United States, certain pairs of species can always be distinguished "on the pin," at least in one sex. In others only certain percentages of the individuals can be distinguished. This paper is the first of a series concerning several such species groups whose relationships have been clarified through a study of sound production and associated behavior in the Orthoptera.

In addition to the type material of *A. fultoni*, representative male and female specimens and tape recordings of the calling songs of each of the species discussed in this paper will be deposited in the Museum of Zoology, Ann Arbor, Michigan. Representative male and female specimens will also be deposited in the United States National Museum and the Entomological Museum of The Ohio State University. An additional copy of the tape of the calling songs will be deposited

[1]Accepted for publication March 19, 1957.

*Author's note: Alexander and Bigelow (1960: Evol. 14:334–346) and Alexander and Walker (1962: Ann. Ent. Soc. Amer. 55:90–94), in updating the systematics and nomenclature of eastern field crickets, provide the following species list: *Gryllus assimilis* (Fabricius), the Jamaican field cricket (southern Florida only); *G. vernalis* Blatchley, the northern wood cricket; *G. fultoni* (Alexander), the southern wood cricket; *G. veletis* (Alexander and Bigelow), the spring field cricket; *G. pennsylvanicus* Burmeister, the fall field cricket; *G. rubens* Scudder, the southern field cricket; and *G. firmus* Scudder, the sand cricket.

Reproduced with permission from the *Annals of the Entomological Society of America,* 50:584–602, 1957. Published by the Entomological Society of America, College Park, Maryland.

with the Cornell Laboratory of Ornithology (Repository for the International Committee on Biological Acoustics). All the original recordings used for analysis are in the Library of Animal Sounds, Department of Zoology and Entomology, The Ohio State University.

### PRESENT STATUS OF AMERICAN *Acheta*

Native field crickets in the genus *Acheta* occur over most of North, Central, and South America, in the West Indies, the Galapagos, and on many other islands. Within this area they can be found almost everywhere except at high altitudes, in the far north, and in extremely moist situations. They are large, commonplace, easily captured insects with many prominent morphological features. Yet no satisfactory systematic analysis of even a small part of the genus has been produced. The first description of an American field cricket was published in 1775, and an almost uninterrupted flow of taxonomic papers dealing with the group has continued since. Forty-seven names had been applied to the American field crickets by 1903. Lutz in 1908, and Rehn and Hebard in 1915, after studying large series of specimens, concluded that only one highly variable species was represented. Since 1915 all the native field crickets have been lumped together under the name *Acheta assimilis* Fabricius,[2] the type of which is from Jamaica.

Practically all the early workers on this group considered morphology alone; such characters as size, coloration, wing venation, body proportions, number and relative lengths of tibial spines, and the length of the tegmina, wings, ovipositor, and hind femora were most often discussed. A wide range of variation exists in most of these characters. American field crickets range from 12.8 mm. to 28.8 mm. in body length, and from solid black to pale straw in color. In general, forms that are morphologically similar in different localities can be found in three types of habitats: (1) large, light-colored forms in sandy areas (beaches, sand dunes, and deserts), (2) small, black forms in deciduous forests, and (3) forms intermediate between these two extremes in grassy and weedy areas such as prairies, fields, pastures, roadsides, and lawns. Rehn and Hebard (1915) recognized this ecological distribution of morphologically similar forms, but because no discontinuities existed with respect to the characters they observed, described them (p. 294) as "... mere variations, the adaptations of this exceedingly plastic species to local environmental conditions." They further stated (p. 299) "... none of these are distinct either specifically or as geographic races, and really show only the various phases resultant from varied environmental conditions."

[2]The sequence of changes in the genus name used for American field crickets (*Gryllus* Linnaeus, 1758, to *Gryllulus* Uvarov, 1935, to *Acheta* Fabricius, 1775) is discussed in detail by Gurney (1950, 1951). Since completion of this manuscript the writer has been informed by Dr. Gurney (in correspondence) that Chopard (Mem. Soc. Roy. Ent. Belgique 27: 153, 1955) has suggested that the present separation of American and some other field crickets and the European house cricket in the genus *Acheta* be dropped, and that the genus name *Gryllus* again be used for all the crickets in this general group. Dr. Gurney states, "I have decided to continue using *Acheta* until going into the matter further, but I expect the changes will eventually be accepted here because the genera [*Acheta* and *Gryllus*] have been rather artificial."

Blatchley (1903) was the only investigator to supply detailed biological information along with description of new species. A number of workers, such as McNeill (1889) in Illinois, Allard (1929) in Georgia and New England, Severin (1935) in South Dakota, Ball (1942) in Arizona, and Cantrall (1943) in Michigan, noted variations in life history and song in the field crickets they observed in these different parts of the country. In 1952, B. B. Fulton published a study of the field crickets of North Carolina in which he described four populations that differed in ecology, life history, song, and distribution, but had no distinguishing morphological features. These four populations failed to interbreed in fifty attempts. In 1953, when the present study was begun, Dr. Edward S. Thomas of the Ohio State Museum told this writer that he had known for some time that at least two different kinds of field crickets occurred in Ohio, with differences in ecology and song. The present study has revealed three species in northeastern United States, one of these in addition to the four discussed by Fulton (1952), making a total of five species known in eastern United States. In view of the complexity of the situation, it is not surprising that conventional taxonomic techniques have failed in its analysis. It is hoped that the results presented here will stimulate further study of this interesting genus in the great part of its range that is yet unexplored.

### NOMENCLATURE AND SYNONYMY

There are forty-seven names available for American *Acheta* from descriptions based almost entirely on morphological characters (see Rehn and Hebard, 1915). For four of these descriptions the type locality given is simply "North America." Six descriptions were of specimens from western United States, and eight were of specimens from eastern United States. The rest were from Jamica, Bermuda, the Galapagos Islands, Guadalupe Island (Lower California), and various South American countries. Fifteen types have been examined by the writer, including all but three of the possible twelve from eastern United States. Information concerning the types of the remaining three (*G. luctuosus* Serville, *G. abbreviatus* Serville, and *G. scudderianus* Saussure) was obtained through correspondence with Dr. Lucien Chopard, Paris, and Dr. Charles Ferriere, Geneva.

In the course of this study several thousand specimens have been examined from all over the United States, and from Jamaica, Puerto Rico, Mexico, and several South American countries. The type of *A. assimilis* Fabricius (1775, Jamaica) has been lost, according to Dr. S. L. Tuxen, Copenhagen, where the Fabricius collection is on loan from the Zoological Museum at Kiel, Germany. This information has been confirmed by Dr. Tischler at Kiel. Specimens obtained from Jamaica through the kindness of Mr. Thomas Farr of the Science Institute at Kingston appear morphologically distinct from all the species discussed here, and Fabricius' description is so brief that it could be applied to almost any of the American species. From morphological comparisons alone it seems unlikely that any of our eastern species extend into Central or South America.

The types of the following species described from western United States have been seen by the writer: *G. personatus* Uhler, 1864, Kansas; *G. vocalis* Scudder, 1901, Palm Springs and Los Angeles, California; *G. integer* Scudder, 1901, West Berkeley to San Diego, California; *G. armatus* Scudder, 1902, Beaver Dam, Utah, Ehrenberg and Fort Whipple, Arizona; and *G. alogus* Rehn, 1902, Albuquerque, New Mexico. The type of the only other species described from this area, *G. lineaticeps* Stål, 1858, San Francisco, California, has been lost, according to Dr. R. Malaise of the Natural History Museum, Stockholm. *G. personatus* Uhler is quite distinct from all the eastern species, and seems to be recognizable in collections. Even Rehn and Hebard recognized the distinctiveness of this form, stating (1915, p. 294), that it ". . . constitutes the nearest approach to a geographic racial development." The taxonomic status of the other five species is not yet clear, but any that are the same as an eastern species here recognized are synonyms of a species described earlier; hence none of these western names need be considered as valid names of any of the eastern species. The synonymy which is given below is based on distribution, morphology, and life history, and discussions of the reasons for the less evident decisions will be found in the sections treating these characteristics.

### **Acheta pennsylvanica** (Burmeister)

NORTHERN FIELD CRICKET

*Gryllus pennsylvanicus* Burmeister, 1838, p. 734, North America; Zoologisches Museum, Humboldt-Universitat zu Berlin (male lectotype here designated from syntypes; only labels are type label, catalogue number 983 (same as for other specimens in series), and determination labels as follows: *pennsylvanicus* Burm.*, *abbreviatus* Serv., *pensylv.* Timerm., *Gryllus* L. Burm.).

*Gryllus luctuosus* Serville, 1839, p. 335, North America; Museum National d'Histoire Naturelle, Paris (only a single female remains).

*Gryllus abbreviatus* Serville, 1839, p. 336, North America; Museum National d'Histoire Naturelle, Paris (only a single female with broken ovipositor remains).

*Acheta nigra* Harris, 1841, p. 123, no type locality given, but apparently New England; Museum of Comparative Zoology, Harvard College, Cambridge, Massachusetts (type probably destroyed; a few remnants of two specimens, a male and a female, remain, and an entry in Harris' notebook, "Autumn of 1834," indicates these specimens were from a fall population which would preclude their being any species other than *pennsylvanicus*).

*Gryllus angustus* Scudder, 1862, p. 427, Cambridge and Cape Cod, Massachusetts; Museum of Comparative Zoology, Harvard College, Cambridge, Massachusetts.

*Gryllus neglectus* Scudder, 1862, p. 428, Massachusetts and Cape Cod, Massachusetts; Museum of Comparative Zoology, Harvard College, Cambridge, Massachusetts.

*Gryllus arenaceous* Blatchley, 1903, p. 434, sand dune region of Lake County, Indiana; Entomological Collection, Purdue University, Lafayette, Indiana.

### **Acheta firma** (Scudder)

BEACH CRICKET (Fulton, 1952)

*Gryllus firmus* Scudder, 1902, p. 294, Brookville, Indiana, Smithville and Pungo (not Dingo) Bluff, North Carolina; Georgia; Sanford and Key West, Florida; Museum of Comparative Zoology, Harvard College, Cambridge, Massachusetts. Lectotype here designated, male from Pungo Bluff, North Carolina. Scudder's Brookville, Indiana, specimen is *Acheta pennsylvanica* (Burmeister).

### **Acheta rubens** (Scudder)

TRILLER FIELD CRICKET (Fulton, 1952)

*Gryllus rubens* Scudder, 1902, p. 294, Auburn, Alabama; Museum of Comparative Zoology, Harvard College, Cambridge, Massachusetts (description based on a single female).

### **Acheta vernalis** (Blatchley)

NORTHERN WOOD CRICKET

*Gryllus assimilis vernalis* Blatchley, 1920, p. 704 (nom. nov.). Although Blatchley's holotype and allotype female, both from Crawford County, Indiana, are of this species, some of the specimens in his collection which he determined as *vernalis* are *A. fultoni* n. sp., described below.

*Gryllus americanus* Blatchley, 1903 (not Drury, 1773), p. 443, Crawford, Posey, Vigo, Putnam, Marion, Marshall, Wells, and Lake Counties, Indiana; Entomological Collection, Purdue University, Lafayette, Indiana.

### **Acheta fultoni** n. sp.

SOUTHERN WOOD CRICKET

*Holotype Male:* Collected by the author in Hocking County, Ohio, Goodhope Township, Section 18, 4 June 1955, in a patch of leaf litter just outside an oak forest bordering an abandoned hillside field dominated by *Andropogon virginicus* L. Head, pronotum, and abdomen black; tegmina and cerci a rather uniform light brown; hind femora and all tibiae and tarsi a dull reddish. Pronotum narrowest near its anterior border; head narrower than pronotum; tegmina reaching to base of cerci; underwings completely hidden by tegmina. Otherwise not distinguishable from other American *Acheta* except by body measurements given below, and as shown in Figures 1–6, 16, and 17.

*Allotype Female:* Collected with holotype; same as holotype except tegmina lack reaching to base of cerci by about four abdominal segments; ovipositor reddish brown, darker near the base.

*Measurements of Holotype and Allotype* (in millimeters, made with ocular micrometer in binocular microscope): Body length, ♂, 15.8, ♀, 16.4; pronotal length, ♂, 3.5, ♀, 3.5; pronotal width, ♂, 5.2, ♀, 5.2; head width, ♂, 4.3, ♀,

4.3; length of hind femur, ♂, 10.3, ♀ 10.6; length of tegmina, ♂, 8.3, ♀, 9.0; length of ovipositor, 12.0.

*Paratypes:* Alexander Collection (collected by the author unless otherwise indicated: OHIO: type locality, 12 May 1954, 1♂ (E. L. Sleeper); 28 June 1954, 5♂; 4 June 1955, 26♂, 7♀; Fairfield Co., Madison Twp., Section 2, 5 June 1955, 2♂; 12 May 1955, 2♂; Adams Co., Jefferson Twp., Cedar School, 8 June 1955, 1♂. ILLINOIS: Jackson Co., about one mile southeast of Carbondale, 15 June 1955, 1♂; Giant City State Park, Jackson Co., 15 June 1955, 8♂, 3♀; Wolf Lake, Union Co., 15 June 1955, 3♂, 1♀. MARYLAND: Snow Hill, Worcester Co., 21 July 1956, 1♂; Whiton, Worcester Co., 21 July 1956, 1♂. NORTH CAROLINA (B. B. Fulton): Raleigh, 27 May 1929, 1♀; 3 June 1931, 1♂, 1♀.

This species is closest to *Acheta vernalis* (Blatchley) from which it differs in being slightly larger and in having brown tegmina and reddish appendages (at the least having reddish markings on the hind femora and usually on other appendages). As shown in figures 17 and 18, the tegmina in the male and the ovipositor in the female are both slightly longer in proportion to body length than in *vernalis*. The general body form, especially the distinctively narrow, retracted appearance of the head and the anterior narrowness of the pronotum is very similar to that of *vernalis*. Specimens from North Carolina, where this species occurs in pine-deciduous forests and along their borders, average larger and darker than those from Ohio and Illinois, where the species occurs in *Andropogon*-covered hillsides and along the borders of oak forests around such fields.

The life history, ecology, and general morphological characteristics of the southern wood cricket in North Carolina were described by Fulton (1952). This species is named in honor of Dr. Fulton, whose keen investigations first aroused the interest of the writer in the singing Orthoptera, and whose prompt and excellent replies to correspondence, and expert aid in locating and collecting singing Orthoptera have been invaluable in this and other studies.

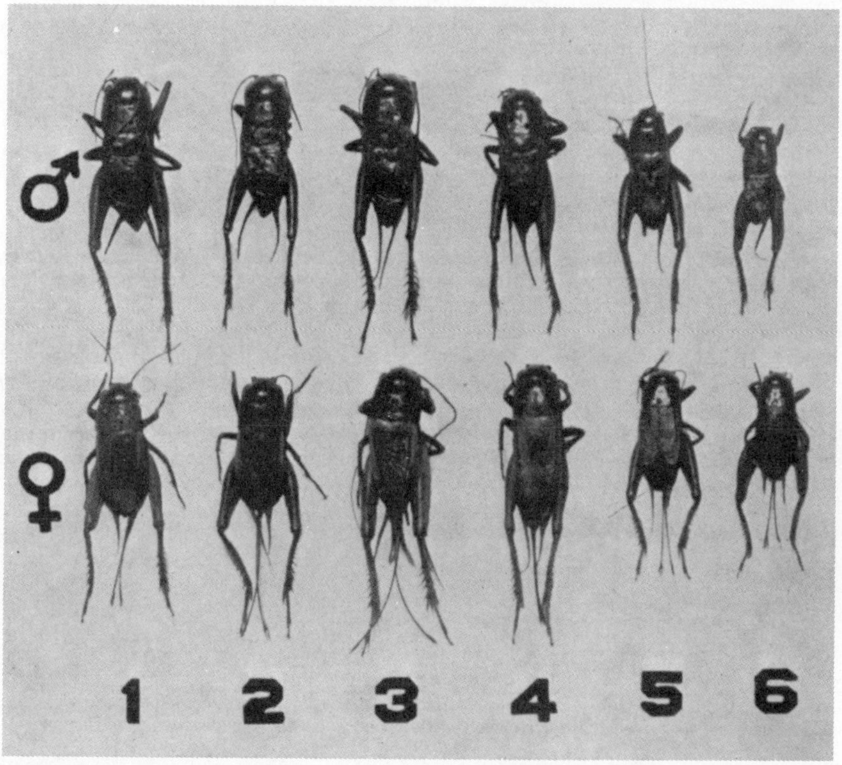


FIG. 1–6. Photographs of typical adult specimens of eastern field crickets. 1, *Acheta pennsylvanica* (Burmeister), spring brood; 2, *A. pennsylvanica* (Burmeister), fall brood; 3, *A. firma* (Scudder); 4, *A. rubens* (Scudder); 5, *A. fultoni* n. sp.; 6, *A. vernalis* (Blatchley). X 0.94.

The lack of understanding of American field crickets by early investigators is perhaps best revealed by the fact that an eastern species remains today to which none of the existing names, shuffled and reshuffled many times, seems to apply. This species was apparently first recognized by Fulton (1952) who called it the "wood cricket." The only earlier literature reference which seems to apply to this form is a remark by Rehn and Hebard (1915, p. 302) under the morphological class for which they include the names *luctuosus* Serville and *abbreviatus* Serville. They state, "The maximum of this [morphological] condition is found in material from the pine woods of southeastern United States." However, most of the specimens in the series determined by Rehn and Hebard as representative of this class are probably *rubens*. Furthermore, since field-collected specimens of this species never have the underwings exceeding the tegmina, and the single remaining female of *luctuosus* in Serville's collection is long-winged, this name could not apply. The measurements given for both *abbreviatus* and *luctuosus*, in addition to the almost solid black coloration, probably preclude their being any species but *pennsylvanica*. In correspondence, Dr. Lucien Chopard states, "I think that *abbreviatus* is the micropterous condition of *luctuosus*; both are very close to *pennsylvanicus*."

Only one name which could have been applied to a field cricket occurring in the eastern United States, *Gryllus scudderianus* Saussure, 1874, is omitted from the above synonymy. Saussure described eight species of field crickets and gave Mexico or various South American countries as type localities for all except this species, for which he gave only North America. Furthermore he named this species after Scudder, who did practically all his work on this group within the United States. In his discussion of the American species of *Gryllus*, Saussure placed *scudderianus* between *pennsylvanicus* Burmeister and *mexicanus* Saussure. In 1877 he placed *scudderianus*, along with *neglectus*, as a synonym of *abbreviatus* Serville. He described *scudderianus* from a male and a female, but according to Dr. Charles Fierriere, only a single female remains in the collection in the Natural History Museum, Geneva, Switzerland, with "Amer. Sept." and "Mr. H. deSaussure" the only information on the label. There is no indication as to how or where Saussure obtained this specimen. In general, the measurements Saussure gave for the type specimens fall somewhere near those obtained for specimens of *vernalis* and *fultoni* in the present study (body length, ♂, 14 mm., ♀, 15.0; ovipositor length, 15.0 mm.). Since both type specimens had brown tegmina and hind femora, it cannot be the same as *vernalis*. Practically all the available measurements differ from those given here for *fultoni*, and especially noticeable is the variation in proportionate length of body and ovipositor. It seems most probable that if Saussure's types came from eastern United States (and this is certainly doubtful), they represent dwarfed, atypical specimens of *pennsylvanica* (Burmeister), or *rubens* (Scudder). Very small specimens of these species have been seen, which nevertheless retain the general body proportions and other characteristics of the typical larger specimens. However, the deviation of *scudderianus* from the measurements given for species treated here, and the improbability of discovering further information about this name, in the opinion of the writer, justifies its exclusion from the present discussion.

### DISTRIBUTIONAL RELATIONSHIPS

The known distribution of the eastern field crickets, based almost entirely on the collections and "listening" records of Fulton and the writer, is shown in figure 7. The Missouri, Arkansas, Louisiana, and Mississipppi records were obtained by Thomas J. Walker on a trip through this area, June 14–17, 1956. The Fentress and Cumberland County, Tennessee records for *vernalis* are from specimens in the University of Michigan Museum of Zoology collected by T. H. Hubbell on July 19, 1924, and July 13, 1922, respectively. The Fentress County, Tennessee, and Fulton County, Georgia records for *fultoni* are from specimens in the University of Michigan Museum of Zoology collected by T. H. Hubbell on June 10, 1925, and April 28, 1938, respectively. The Quantico, Virginia record for *fultoni* is from specimens in the Ohio State Museum collected by Edward S. Thomas on May 25, 1935. The general lack of records in the Kentucky-Tennessee-Alabama area is due to a lack of collecting in that area.

The most complete portions of the records are in the northward extension of *rubens*, the southward extension of *pennsylvanica*, the northward extension of *fultoni* in Ohio and on the Delaware peninsula, and the northward extension of *firma* along the Atlantic coast. The western limits are not known for any of the species. The northward and southward extensions of *vernalis* are not known. However, all of the 740 plus specimens from Michigan in the collection at Ann Arbor are *pennsylvanica*, and no specimens of *vernalis* have been found among hundreds of specimens seen from Wisconsin, Minnesota, Canada, and the New England states.

To avoid the possibility of error, distribution records based only on specimens seen in collections have not been included for *pennsylvanica*, *rubens*, or *firma*. The morphological variability and overlap of the specimens which would be referred to these three species suggests at least the possibility that additional species not encountered in this study may occur in eastern United States. Consequently, for all specimens determined as belonging to one of these three species, either the habitat or the song is known, in addition to the collection date and locality.

### ECOLOGICAL RELATIONSHIPS

*A. pennsylvanica* occurs in all kinds of grassy situations, such as fields, pastures, weedy areas, roadsides, and lawns, and it is the most abundant and continuously distributed field cricket in northeastern United States. In southeastern United States it is replaced in similar situations by *rubens*. These are probably the only field crickets in eastern United States which occur in appropriate places and achieve sufficient abundance to become economic pests. They are also much easier to collect than any of the other species, and therefore dominate collections of

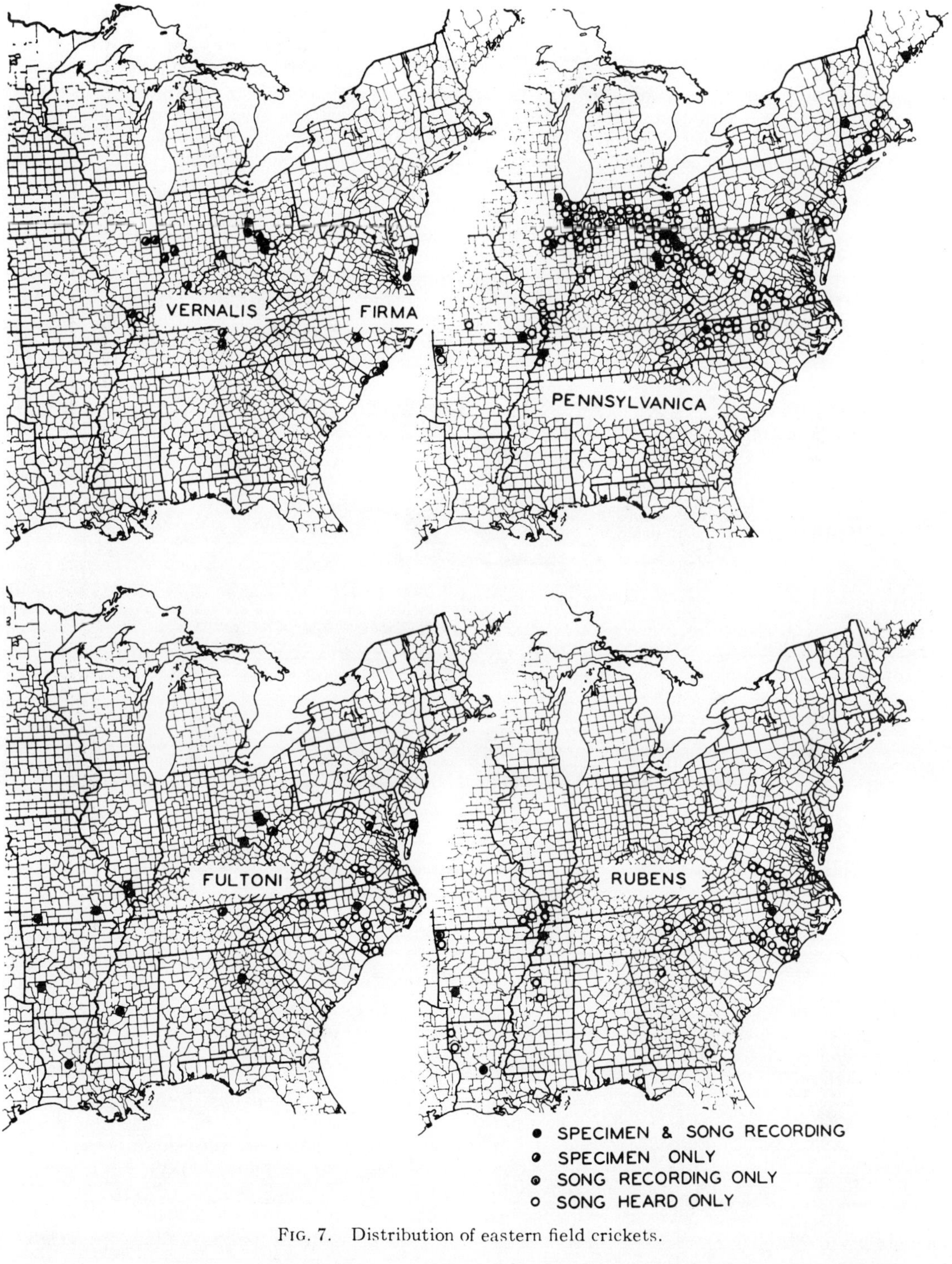


FIG. 7. Distribution of eastern field crickets.

eastern *Acheta*. This is perhaps one of the most important reasons for the confusion in early attempts to split this genus into species on the basis of morphological characters. Lutz (1908) placed great emphasis on the fact that the extremes of the characters he studied were represented in relatively few specimens, as one would expect in a single, highly variable population. He states (p. 6), "Typical specimens of *americanus* [*vernalis*] and *firmus* are rare for the same reason that very short men and very tall men are rare." The present study indicates that they are rare *in collections* because of less likelihood of encountering them and greater difficulty in capturing them. Of approximately 2700 eastern field crickets examined in six collections in eastern United States, there were about 1900 *pennsylvanica*, about 750 *rubens* and *firma* combined, 72 *vernalis*, and 24 *fultoni*. This does not include the collections of Fulton and the writer.

As stated by Fulton (1952, p. 285), *firma* inhabits "the flat sandy area bordering the ocean beaches. This may be grassy or shrubby or largely bare sand, but is not wooded." Later Fulton found chirping crickets in inland sand hills in Harnett and Hoke Counties, North Carolina, which produced fertile offspring in crosses with ocean beach crickets (personal communication). Audiospectrographic analysis has shown these island specimens to have the same song as the ocean beach crickets.

The distribution of *firma* along the Gulf coast and inland in that area is unknown. However, crickets morphologically resembling beach crickets have been seen in collections from inland locations in Mississippi, Alabama, and Georgia. Light-colored crickets occur in sandy areas around Lake Michigan and Lake Erie, along the Illinois and Mississippi Rivers, and probably in similar areas all over eastern United States. Such specimens were responsible for Blatchley's description of *arenaceous* from Lake County, Indiana. However, crossing experiments and song analysis have revealed light-colored specimens collected at Dunes State Park, Indiana, along Lake Michigan, and Cedar Point, Ohio, along Lake Erie, to be conspecific with the more commonly encountered, darker-colored specimens

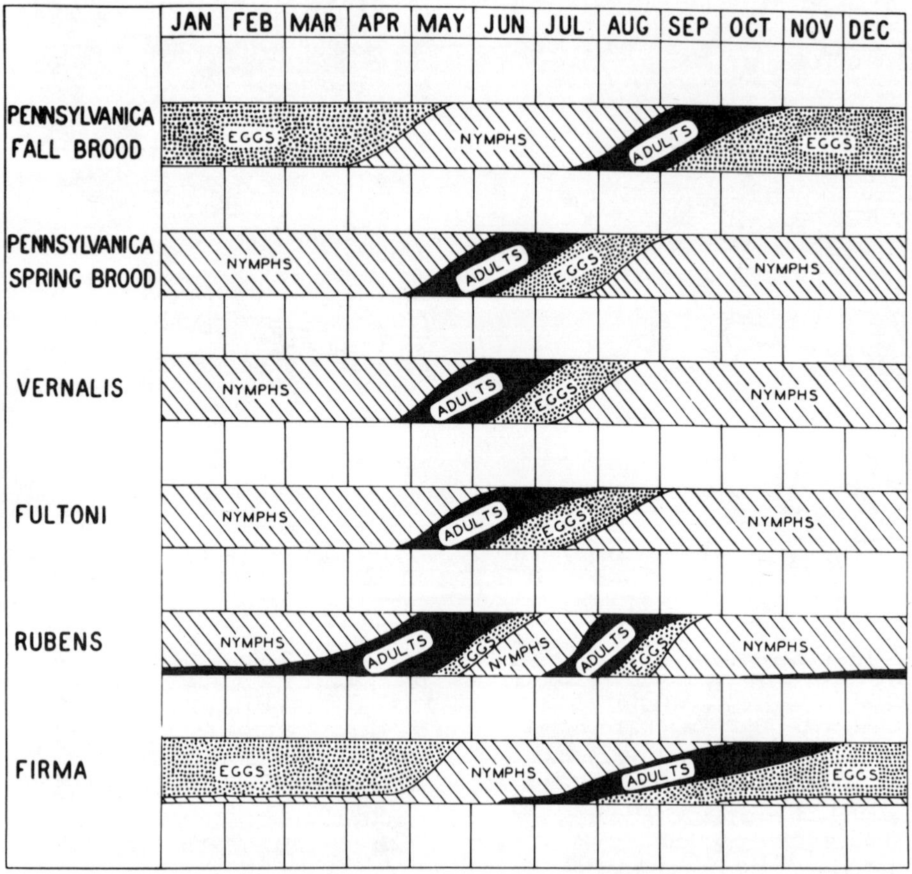


FIG. 8. Seasonal life histories of eastern field crickets.

of *pennsylvanica*. Such specimens apparently represent "adaptations . . . to local environmental conditions," such as are referred to by Rehn and Hebard (1915, p. 294). The morphological variants from inland localities in Mississippi, Alabama, and Georgia, however, remain to be checked, and may represent additional species to those discussed here.

*A. vernalis* is found in leaf litter in deciduous forests. In North Carolina and Virginia, where *vernalis* does not occur, this habitat is occupied by *fultoni*, which is abundant in deciduous and pine-deciduous forests, and along the forest borders. Both *vernalis* and *fultoni* occur in southern Ohio, Indiana, and Illinois. Here, however, *fultoni* is chiefly an inhabitant of dry, *Andropogon*-covered hillsides. In southeastern Ohio it is found almost exclusively in old fields, 10 to 30 years

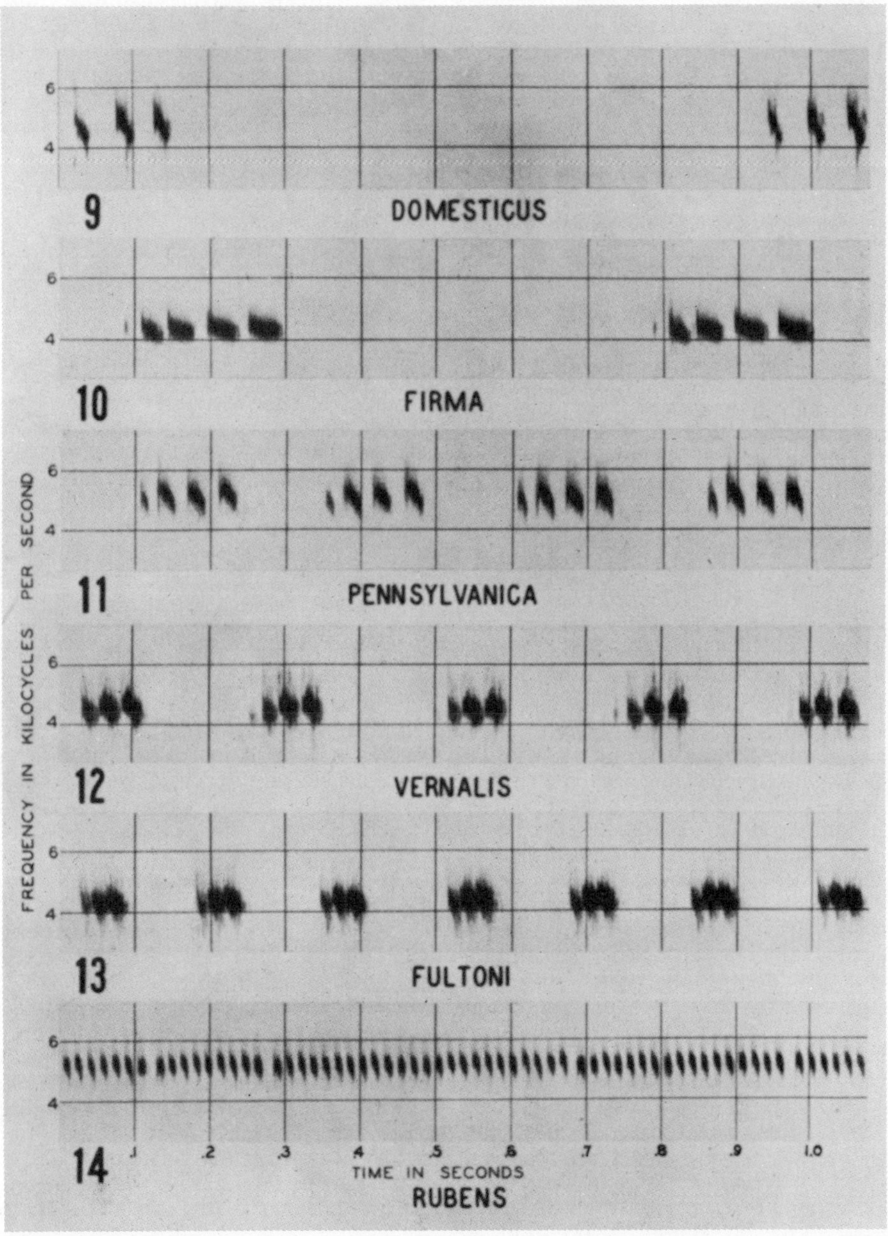


FIG. 9–14. Sample audiospectrographs of typical calling songs of *Acheta* species. 9, *Acheta domesticus* (Linnaeus); 10, *A. firma* (Scudder); 11, *A. pennsylvanica* (Burmeister), spring brood; 12, *A. vernalis* (Blatchley); 13, *A. fultoni* n. sp.; 14, *A. rubens* (Scudder).

abandoned, which are dominated by *Andropogon virginicus* L. In southwestern Ohio and west to along the Mississippi River bluffs in southern Illinois, *fultoni* occurs in the hilltop praire openings dominated by *Andropogon scoparius* Michx. In both Illinois and Ohio, *fultoni* is often more abundant along the forest borders around such fields than it is out in the fields themselves. In Maryland, Virginia, Arkansas, and Missouri, *fultoni* has been found both in *Andropogon* fields (and similar grassy areas) and in leaf litter in woods. The factors involved in this ecological variation remain to be elucidated. However, morphological comparisons, song analyses, and a single successful crossing experiment (between an Ohio female and a North Carolina male) seem to indicate that a single species is involved, with geographic variation in habitat and in certain song characteristics which will be described later.

*A. fultoni* and *pennsylvanica* overlap but little with *vernalis* where their habitats meet in Ohio. On the other hand, *fultoni* and *pennsylvanica* occur together in the same fields in southern Ohio, and appear to be randomly intermixed in such areas. *A. rubens* and *firma* have been found in mixed colonies at Kure Beach and Carolina Beach, North Carolina. In Worcester County, Maryland, *rubens, firma,* and *fultoni* were found together along the border of a wheat field, and *pennsylvanica* was collected only a few yards away along a grassy roadside.

### SEASONAL LIFE HISTORIES

The seasonal relationships of the eastern field crickets are indicated in figure 8. The diagrams for *rubens* and *firma* are redrawn from Fulton (1952). *A. fultoni* has approximately the same life history in North Carolina as in Ohio.

No way of distinguishing the spring and fall broods of *pennsylvanica* in Ohio has been found other than the differences in their life histories, and a slight difference in the relative length of the body and the ovipositor in most females (fig. 18). One brood overwinters as partly grown nymphs, one or two instars from adulthood, and produces adults which sing from early May until about mid-July. The other overwinters in the egg stage and produces adults which sing from mid-July until frost. These two broods occupy the same habitat, a fall male often moving into the same burrow or niche inhabited by a spring male earlier in the season. There is a slight overlap of adults in mid-July. Although very few adults can be seen or heard at this time, there were at least a few singing males on the Ohio State University campus every night from May to October during the three years of this study. The end of the spring brood and the beginning of the fall brood are difficult to determine, but are indicated by the gradual disappearance of singing males in familiar locations, and the appearance of numbers of singers in new locations. Adult male field crickets are sedentary, ordinarily remaining in and singing from the same location throughout their adult lives.

There are indications that offspring of the spring brood of *pennsylvanica* never mature in fall in Ohio. No adults or last instar nymphs have ever been taken in early winter. Last instar nymphs of the fall brood can be found on the Ohio State University Campus until mid-September, but later in the fall only adults of the fall brood and young nymphs of the spring brood, about three instars from adulthood, can be found. In the laboratory, the offspring of Franklin County, Ohio, spring adults mated on May 4, started maturing as early as August 12. However, a few individuals in this group had not yet reached the last nymphal instar by mid-October. Another pair of spring adults from Piatt County, Illinois, was mated on June 6, and removed from the cage on July 11. Their offspring began maturing in October, but the last did not mature until the following April. Neither *fultoni* nor *vernalis* mature in the fall in Ohio, but both produced offspring from early May matings in the laboratory which began maturing in mid-August and early September, respectively. Several factors may be involved in these discrepencies between field and laboratory observations, such as temperature, diet, photoperiod, and water availability.

Fulton's findings in North Carolina indicated that offspring of spring adults of *pennsylvanica* may sometimes mature in the fall, while Cantrall (1943) failed to obtain any fall adults from seven pairs of spring brood adults caged outside on June 9, in Michigan. On the George Reserve, Michigan, only about five percent of the total number are spring crickets, a result, Cantrall believes, of high mortality among overwintering nymphs. In central Ohio, no noticeable difference in numbers occurs between the spring and fall broods. These two broods may interbreed in mid-summer, or possibly in fall in the southern part of their range, or it may be that they have been isolated such a short time that no noticeable differences have yet appeared between them. Certainly more investigation is needed to clarify their relationship.

*A. vernalis* and *fultoni* overwinter as nymphs one or two instars from adulthood, and during the three seasons that this study has been in progress, they have matured a week or two earlier than *pennsylvanica* in southern Ohio, beginning to sing in late April or early May. The adults disappear in late July and early August. In North Carolina, Fulton (1952) says *fultoni* sings from mid-April until early August. The writer collected an adult male and heard several singing on August 10, 1955, near Raleigh, North Carolina. One of the males collected was mated with a female from Hocking County, Ohio, on August 18, and was

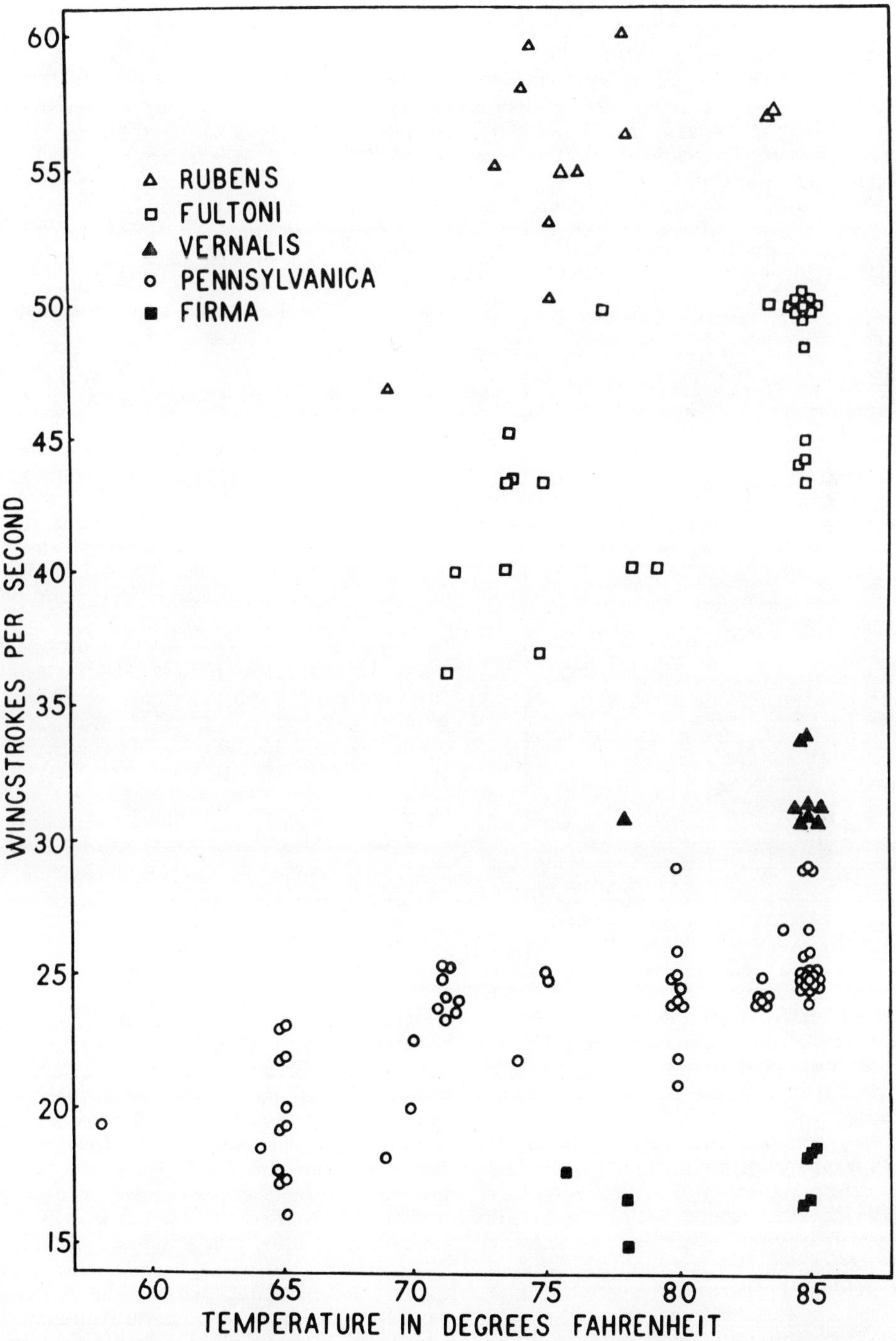


FIG. 15. The effect of temperature on rate of wing-stroke inthe calling songs of eastern field crickets.

still alive on September 30. Offspring from this mating appeared on September 30.

### SONG RELATIONSHIPS

Geographic variation in the songs of field crickets was first noted by Allard (1910). Fulton (1952) first used it as a taxonomic character in this group. In the singing Gryllidae, Tettigoniidae, and Cicadidae, song differences have been responsible for the initial recognition of many undescribed species (Davis, 1922; Fulton, 1930; Pringle, 1955; Thomas and Alexander, 1957). As remarked by Dr. Edward S. Thomas in conversation with the writer, to one who has studied these groups in the field it is almost a foregone conclusion that two insects which occur in the same area and have different songs are different species.

As with most singing Tettigoniidae and Gryllidae, male field crickets ordinarily space themselves in the field and remain in one spot all their adult lives. A large part of the time they produce a sound which is generally referred to as the calling song (also solitary or common song). This is the only sound which will be discussed here. The variety of sounds made by field crickets in other situations is discussed by Alexander (1957). The calling song attracts the adult females which move toward and locate the males. The roving females are seen much more often in the field than the sedentary males, and dominate collections for this reason.

Tape recordings and audiospectrographs have been made of the calling songs of 145 field crickets from the localities indicated in Figure 7. The recordings were made both in the laboratory and in the field, using a Magnecorder PT6A in the laboratory, and a Magnemite 610–E in the field. American Microphone Company D–33 and D–33A microphones were used in both cases, with a 24–inch parabolic reflector sometimes employed in the field. The audiospectrographs were made with a Vibralyzer (Kay Electric Company, Pinebrook, New Jersey). For further explanations of the use of this instrument and its portrayal of cricket sounds, see Borror and Reese (1952) and Alexander (1957).

Figures 9–14 show sample audiospectrographs from typical calling songs of each of the five field cricket species under consideration, and of the house cricket, *Acheta domesticus* (Linnaeus), a species introduced from Europe. The house cricket recording was made of an individual from a culture supplied to Ohio State University from the University of Missouri.

TABLE I

COMPARISON OF SONG CHARACTERISTICS IN EASTERN FIELD CRICKETS

| Species | | *firma* | *pennsylvanica* | *vernalis* | *fultoni* | *rubens* |
|---|---|---|---|---|---|---|
| Number of Individuals Recorded | | 11 | 74 | 11 | 37 | 12 |
| Pulses/Second | Range | 17–19 | 24–29 | 30–33 | 43–50 | 60 |
| (85° F.) | Mean | 18 | 25 | 31 | 49 | 60 |
| Chirps/Minute | Range | 56–168 | 120–370 | 108–246 | 176–420 | — |
| (85° F.) | Usual Number | 100–120 | 150–240 | 180–200 | 300–360 | — |
| Pulses per Chirp | Range | 4–6 | 3–7 | 2–4 | 2–4 | — |
| | Mode | 4 | 4 | 3 | 3 | — |
| Degree of Dominance of Pure Frequency | Rank from most "musical" to least "musical" | 1 | 1 | 3 | 3 | 2 |
| Intensity | Rank from loudest to softest | 2 | 1 | 4 | 4 | 3 |
| Frequency (Toothstrike Rate) | Rank from highest to lowest | 5 | 4 | 2 | 3 | 1 |

A number of analyzable characters appear in *Acheta* sounds, such as the frequency (rate of toothstrike) (individual toothstrikes are not shown in Figures 9–14), rate of pulsation (wingstroke rate), rate of chirping, regularity of chirping, number of pulses per chirp, and "musical" quality or degree of dominance of the pure frequency. Figure 15 is a scatterdiagram comparing the songs of all individuals recorded at known temperatures with respect to what is probably the most significant character, rate of wingstroke, or rate of production of sound pulses. Both laboratory and field recordings are included, and much of the infraspecific variation is probably due to error in measuring temperatures. Table I is a comparison of several characteristics of the calling songs of eastern field crickets.

All of the eastern field crickets sing both day and night, though *vernalis*, *fultoni*, and *rubens* do most of their singing at night. The songs of these three crickets are soft and somewhat less musical when compared to the loud, clear chirps of *pennsylvanica* and *firma*. *A. vernalis* and *fultoni* chirp more rapidly and more regularly than do *pennsylvanica* and *firma*, with *vernalis* being somewhat intermediate between the two extremes in this respect. Differences in qualities such as these are difficult to describe adequately, though they may actually be recognition characters for an observer distinguishing the songs of the different species. For example, *fultoni* chirps with considerably more regularity in southern Ohio than it does across the Appalachians in Virginia and North Carolina. This difference

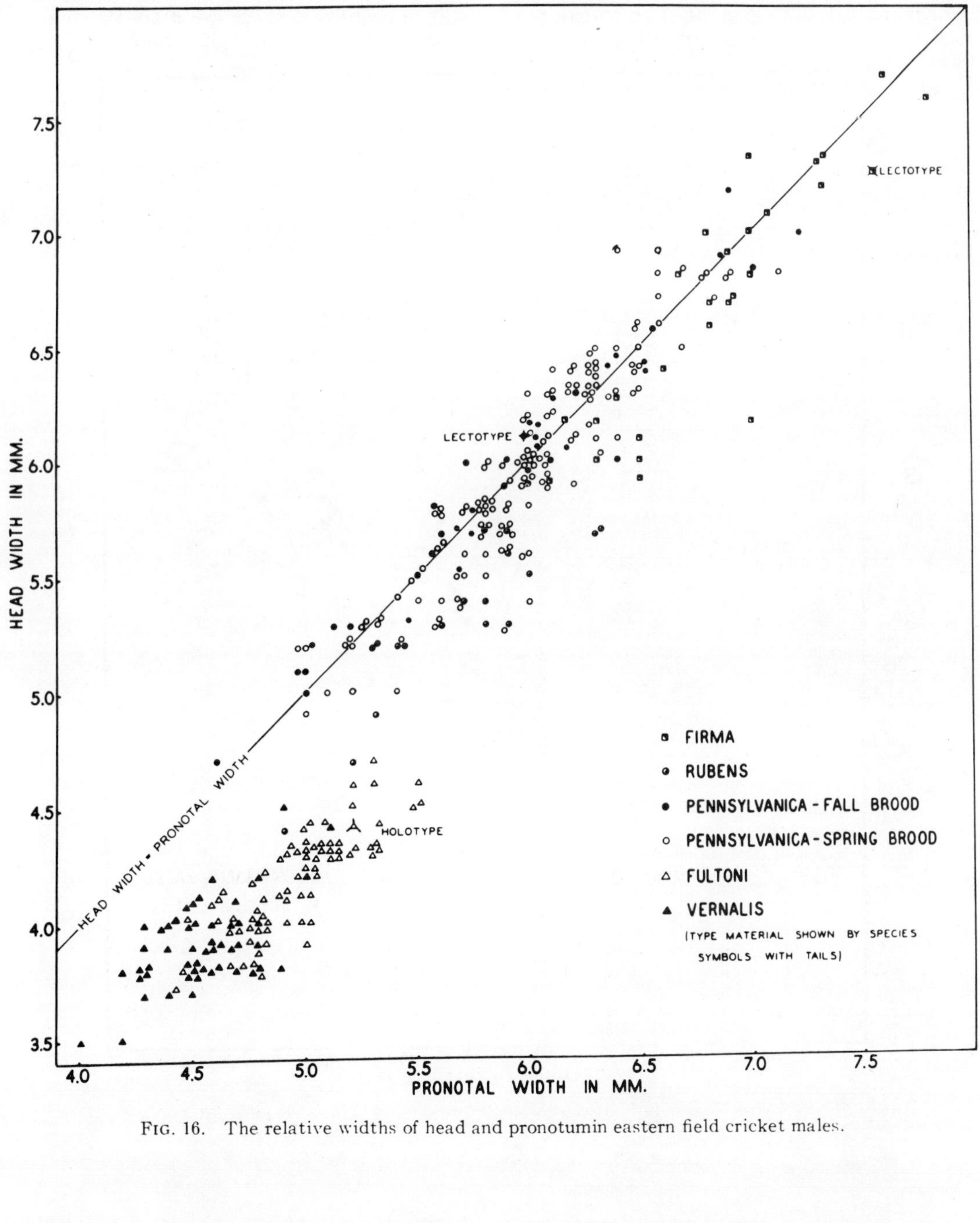


FIG. 16. The relative widths of head and pronotumin eastern field cricket males.

seems to be associated with the variation in ecology already mentioned. If uniform or rhythmical intervals in these songs are behaviorally significant, this introduction of an additional and different rhythmical element into the song of *fultoni* may be associated with its occurrence together with ***pennsylvanica***, an irregular chirper. In the south, *fultoni* ordinarily occurs together with only one other species, ***rubens***, which has a trilling song.

The chirping rate is not always the same in a single individual. One unmated male of ***pennsylvanica*** was timed singing at 120 chirps per minute shortly after molting to adulthood, and three weeks later the same male, still unmated and still caged alone, was timed singing steadily at a rate of 370 chirps per minute.

Although in certain cases other characters are more distinct and more easily used, song differences have proven better than any other single set of related characters for separating all of the *Acheta* species distinguished thus far. The recorded calling songs of all species can be separated by audiospectrographic analysis with 100 percent accuracy. Not enough courtship songs have been recorded to make comparisons, but it appears that these are more variable and less distinctive than the calling songs.

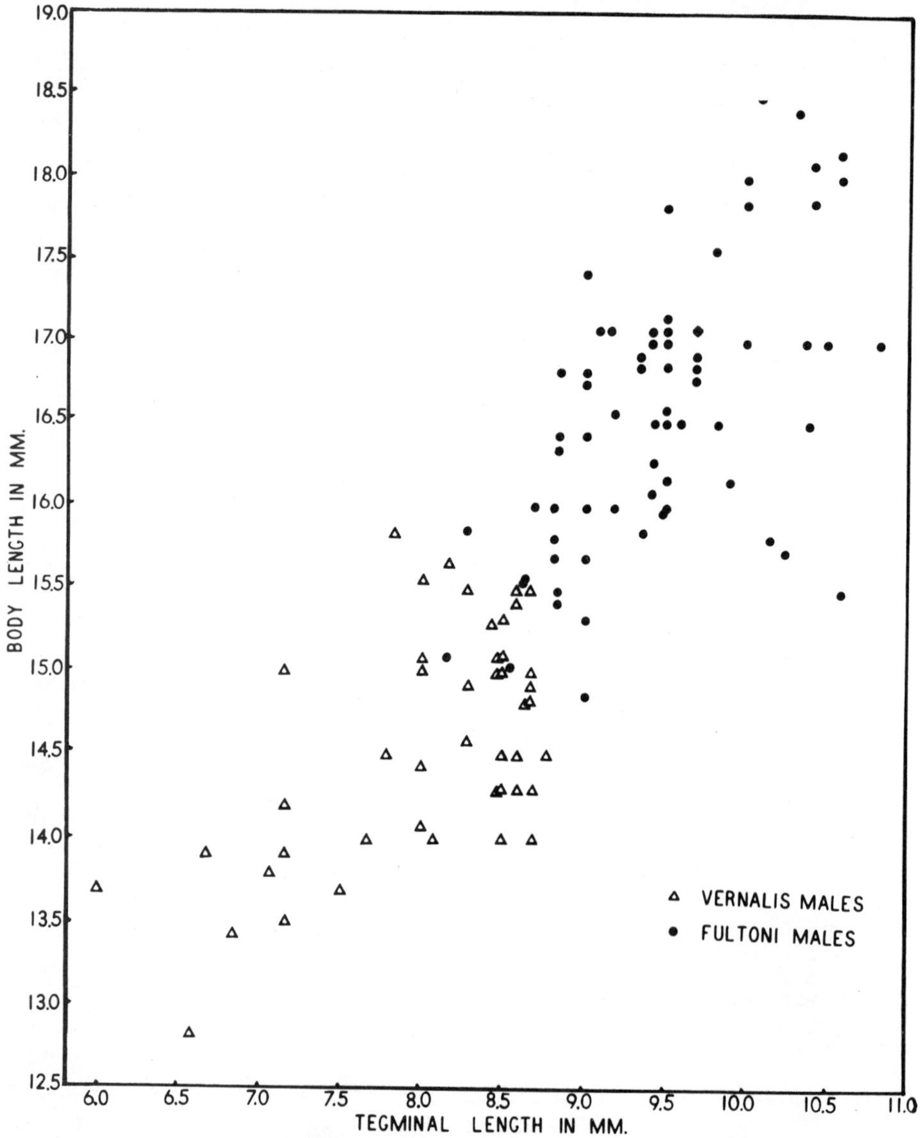


FIG. 17. The relative lengths of body and tegmina in males of *Acheta vernalis* (Blatchley) and *A. fultoni* n. sp.

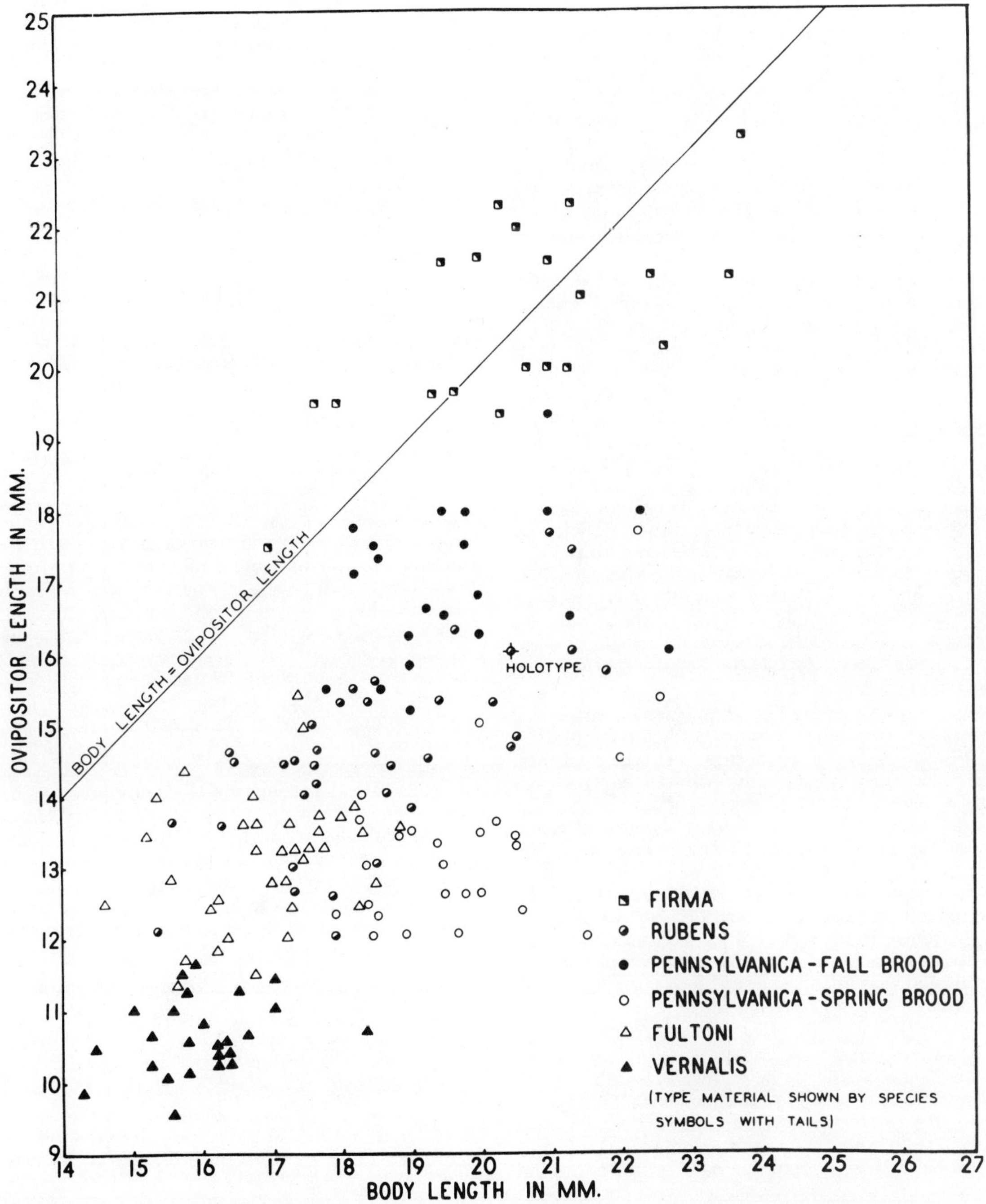


FIG. 18. The relative lengths of body and ovipositor in females of eastern field crickets.

### MORPHOLOGICAL CHARACTERISTICS

The morphological characters found to be of value in this study were body proportions, size, coloration, and the length of the tegmina, wings, and ovipositor. A large number of other characters, including wing venation, structure of the male genitalia, and the number and relative lengths of the spines on the hind tibiae were examined, and none of value was discovered. No one morphological character was found which would separate all five species. Except for *vernalis*, both sexes of which are distinguishable from all the other species, most pairs of species can be distinguished in one sex only. The scatterdiagrams comparing morphological characters (figs. 16–18) are based almost entirely on the writer's collection, with type material located as designated.

Dimorphism in the length of the hind wings is exhibited by *pennsylvanica*, *firma*, and *rubens*, while only short-winged individuals of *vernalis* and *fultoni* have been collected in the field. Some individuals of *fultoni* reared on a high-protein dog food diet and under crowded conditions developed underwings slightly longer than the tegmina, but similar conditions never produced long-winged *vernalis*. Almost 100 percent of the individuals of *pennsylvanica* and *rubens* reared under these conditions develop long wings. Some authors have considered this character to be of no value since it was discovered that long-winged parents may produce short-winged offspring and short-winged parents may produce long-winged offspring. However, the ability to develop underwings functional in flight has apparently been completely lost in *vernalis* and *fultoni*.

As shown in figure 16, all species except *pennsylvanica* and *firma* have the head invariably narrower than the pronotum. In such species, the pronotum is generally widest near its posterior margin and the head appears somewhat retracted (figs. 5, 6). In *pennsylvanica* and *firma*, especially the males, the reverse is true, with the widest part of the pronotum near its anterior border, and the head as often wider than the pronotum as not, and rounded or swollen in appearance (figs. 1–3). This character of proportionate width and general conformation of the head and pronotum readily separates the males of *fultoni* and *vernalis* from those of the other species.

With respect to coloration, *vernalis* represents an extreme, being nearly always totally black. A few specimens have small reddish areas near the base of the hind femora. *A. pennsylvanica* varies from totally black specimens to specimens with light brown tegmina, pale or reddish hind femora, and pale or reddish spots on the sides of the pronotum. The pale forms are most commonly encountered in sandy areas, such as around the Great Lakes. In the writer's collection, only one out of 280 specimens of *pennsylvanica* has reddish spots on the sides of the pronotum, while 33 out of 54 specimens of *rubens* do. At present this is the best morphological character known for separating these two species. Since the ranges of these two species overlap rather narrowly, a high proportion of specimens can be separated on the basis of distribution and the coloration of the pronotum.

*A. rubens* and *firma* are similarly marked, generally having brown tegmina and pale or reddish hind femora, in addition to generally having reddish patterning on the sides of the pronotum. *A. fultoni* also has brown tegmina and at least partially reddish hind femora. It is separable from *vernalis* in both sexes on the basis of color alone.

*A. vernalis* is the smallest of the five species, with the smallest specimens seen only 12.8 mm. in body length (male). *A. firma* is the largest, with specimens in the writer's collection up to 24 mm. in body length (female). This is probably not a good indication of the upper limit in size for *firma*, but specimens positively identified as *firma* in this study are limited to those for which data other than morphological characteristics are available. Fulton (1952) includes a small number of beach crickets in the class with a body length of 26–29 mm., and Rehn and Hebard (1915) list the range of body length for American *Acheta* as from 14.0 mm. to 28.8 mm. *A. fultoni* averages slightly larger than *vernalis*, with considerable overlap, and *pennsylvanica* and *rubens* average smaller than *firma* with considerable overlap. *A. rubens* averages slightly smaller than *pennsylvanica*, and individuals from the fall brood of *pennsylvanica* average a little smaller than those from the spring brood.

As shown in figure 18, the females of *firma* are separable from those of all the other species on the basis of relative lengths of the ovipositor and body. The fall and spring broods of *pennsylvanica* are also partially separable on this basis, and the location of the syntypes of *pennsylvanica* on this diagram indicates that these were fall specimens, which is to be expected since the fall brood is generally present in larger numbers and is better represented in collections.

### CROSSING EXPERIMENTS AND HYBRIDIZATION

The only evidence of hybridization of the northern crickets in the field, in spite of concentrated collecting along zones of contact and overlap, consists of one male collected in a field in southern Ohio in which both *fultoni* and *pennsylvanica* occur. This specimen, though having the narrow head and general body proportions of *fultoni*, had long black tegmina and fully developed underwings, and was larger than any *fultoni* collected. Its chirping had the typical regularity and speed of wing motion of *fultoni*, but the chirps were four-pulse and delivered at a regular rate of about 240 per minute (80° F.). It failed to produce offspring when mated with a *pennsylvanica* female from the spring brood.

Figure 19 is a composite diagram showing the 50 crosses and 25 controls completed by Fulton (1952) and 30 additional crosses and 34 controls completed in the present study. The three successful crosses probably do not reflect a lack of reproductive isolation in the field. Adults of *fultoni* and *pennsylvanica* occur in the same places at the same time, and yet maintain their separate identities. These crosses do indicate, however, that physiological isolation is incomplete, and that the effective isolating factors in the field are probably behavioral. This suggestion is strengthened by the fact that only one presumed hybrid has been heard or collected in the field. It seems likely that song differences are of importance in this regard, although tests have not yet been performed to determine whether or not the observed differences in song are significant with respect to the behavior of the females.

In all possible types of matings between northern species, courtship, successful copulation (transference of the spermatophore), and oviposition occurred, though not in all the individual matings. It was noted that in matings set up at the same time these events took place sooner in the controls than in the crosses. Males in cross-matings were heard singing the calling song much more than those in the control matings. This song is generally not produced in the presence of a responsive female. Adult females in cross-

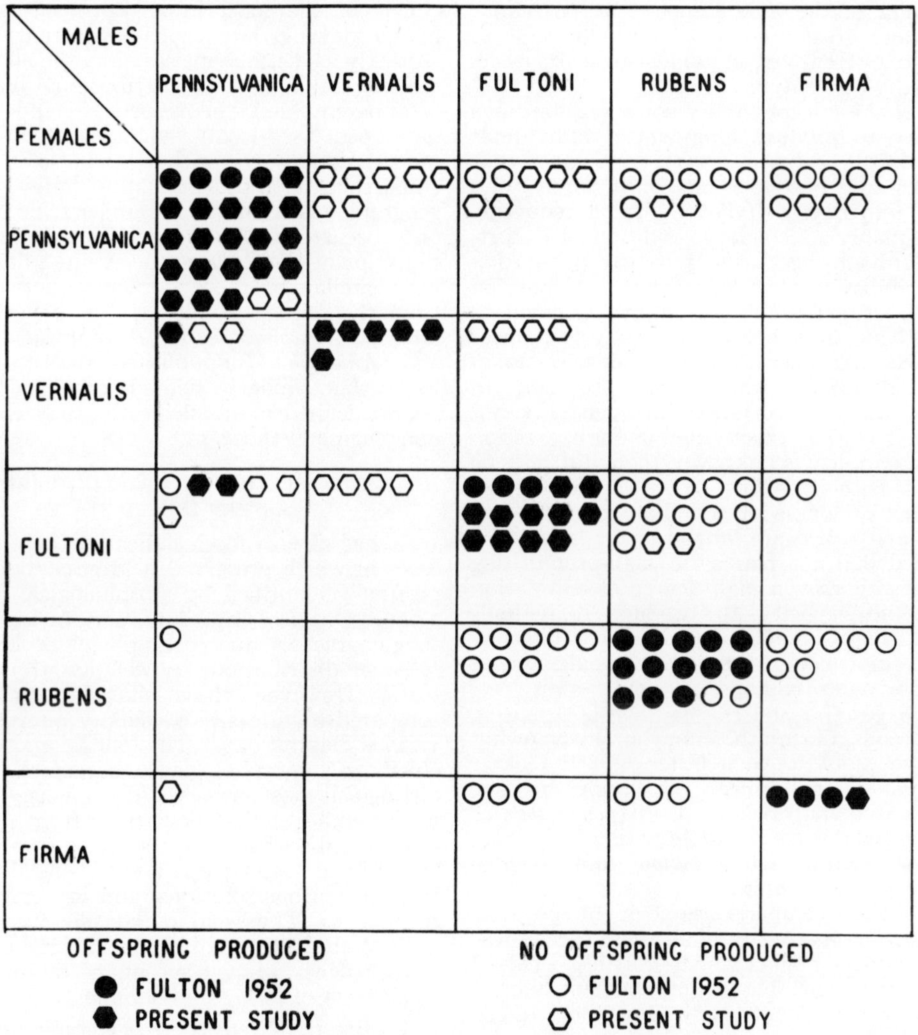


FIG. 19. Crossing experiments with eastern field crickets.

matings and virgin females caged alone often chewed their way through the cheesecloth covers of the cages and escaped, while this happened only once or twice in control cages. This increased activity could have been due to attraction to the songs of males of their own species coming from other cages in the room, or simply to a change in the physiological condition of unmated females.

#### PHYLOGENETIC IMPLICATIONS

Our knowledge of the distributional and other relationships of the *Acheta* species is too scanty to allow any great amount of speculation as to their origin. However, certain historical aspects seem worthy of mention.

Three of the crickets under consideration have the ability to produce long-winged individuals (*pennsylvanica, rubens, firma*), one does not (*vernalis*), and one does so only rarely and probably imperfectly (*fultoni*). Wing reduction here is probably a secondary condition, the short-winged forms being derived from a population with the ability to form long wings. As pointed out to the writer by Dr. Fulton, in this genus as well as others, those species occurring in pioneer associations are generally able to develop long-winged individuals, while those occurring in mature or climax associations are generally not able to do so. The exact stimulus for expression of this potentiality is unknown, though it may be connected with crowding or with diet. High proportions of long-winged individuals of *pennsylvanica* and *rubens* are formed in the laboratory in crowded cultures reared on high-protein dog food. At any rate, a high degree of correlation between the capacity to produce individuals capable of flight and the degree of instability or lack of permanence in the habitat would appear to have obvious advantages.

*A. pennsylvanica* and *rubens* have both spring and fall broods, though these may differ somewhat in their relations to each other. Of the other species, *fultoni* and *vernalis* are spring crickets, and *firma* is a fall cricket. Both *pennsylvanica* and *rubens* exhibit considerable variation in most characters studied, while *fultoni* and *vernalis*, especially, are quite uniform.

From all these comparisons one might speculate that the common ancestor of all these species was a double-brooded cricket which probably occurred in some sort of unstable or impermanent habitat and possessed the ability to produce long-winged individuals.

Settlement and clearing have undoubtedly brought about drastic changes in the relative numbers and distribution of the field crickets in eastern United States. Prior to settlement, about 95 percent of the state of Ohio was forested and five percent was prairie. Today, only about five percent of the state is forested, and the rest is chiefly involved in agriculture and perpetually in various types of grassy, weedy, or cultivated siutations. Today the range of *pennsylvanica* is fairly continuous across Ohio, Indiana, and Illinois, while *vernalis* and *fultoni* occur only in small isolated colonies. The uniformity of these last two species across a range of 300 to 400 miles from east to west suggests, however, that their distribution was once fairly continuous. It is probable that *vernalis* was once the most abundant field cricket in this area, and *fultoni* must have occurred more or less continuously across the southern half of these states.

One of the most important consequences of finally viewing the various populations of *Acheta* correctly as distinct species, is in the effect on our reconstruction of their distributional history. It is apparent that these species could not have developed as "local" adaptations to environmental conditions, with distributional relationships such as those prevailing today, and that geographic isolation on a grander scale must have once occurred. The general distributional relationships of the *Acheta* species illustrate patterns very similar to those occurring between other similarly related forms in the genera *Amblycorypha, Neoconocephalus, Pterophylla, Oecanthus,* and *Anaxipha* (unpublished data). It seems likely that similarly related forms, perhaps also poorly defined morphologically, may exist in the non-singing Orthoptera.

#### KEY TO SPECIES OF *Acheta* OCCURRING IN EASTERN UNITED STATES

As has already been indicated there are many cases in which preserved field crickets cannot be positively identified by morphological characters alone. One is tempted to write a key utilizing song characters, and geographical, ecological, and seasonal distribution, as well as structural features. However, these characteristics can be compared more easily by simply referring to the various illustrations. The following key is based chiefly on morphological characters, and will distinguish most preserved specimens. It should be remembered that deviations from the coloration described here may occur in specimens collected in molasses traps and cleaned by passing through various solutions, and in very old dried specimens. Thus, old pinned specimens of *vernalis* often develop a dull blackish red coloration, though newly pinned specimens retain the solid black color of living individuals.

1. General coloration straw or dull brownish-yellow; head and pronotum dark brown, mottled with paler areas; head with an irregular dark, transverse bar extending between the eyes near their dorsal border, narrower light bars in front of and behind this bar...... **domesticus**

1'. General coloration ranging from solid black to dark brown, tending toward reddish in paler specimens rather than toward straw-color; head and pronotum black, with or without pale markings; head without transverse bars. 2

2(1'). Pronotum of male widest near its posterior border; head of male definitely narrower than pronotum, presenting a retracted appearance, not exceeding 4.7 mm. in width; body length 12.8–19 mm.; ovipositor length 9.5–15.5 mm.; always micropterous in both sexes .......... 3

2'. Pronotum of male usually widest near or anterior to its middle; head of male as often wider than pronotum as not, presenting a swollen globose appearance, almost always over 4.7 mm. in width; body length 15–29 mm. or more; ovipositor length 12–24 mm. or more; macropterous or micropterous in either sex .............................. 4

3(2). Coloration solid black, rarely a dash of reddish on the inside or outside of base of hind femora ............................ **vernalis**

3'. Tegmina always brown; hind femora usually reddish or pale, at least on basal third; tibiae pale ............................... **fultoni**

4(2'). Found along the Atlantic Coast (and probably the Gulf Coast), and in sandy areas in the Southeast; body usually over 18 mm. in length; ovipositor 19–23 mm. or more in length and as often longer than the body as not ................................... **firma**

4'. Found in fields and grassy areas throughout eastern United States, including the areas described above; body rarely over 23 mm. in length; ovipositor rarely over 18 mm. in length, and always shorter than body .. 5

5(4'). Pronotum usually with reddish or light-colored spots on sides; southern crickets occurring northward only into the Appalachians and to the southern tips of Illinois and Missouri. **rubens**

5'. Pronotum almost always lacking light spots on sides, except in crickets collected in northern sandy areas, such as around the Great Lakes; northern crickets extending southward to just below the Appalachians ......... **pennsylvanica**

## SUMMARY

The singing Tettigoniidae and Gryllidae of eastern United States contain a number of species which have not received formal recognition by taxonomists because morphological characters are not known by which preserved specimens can be distinguished. Among these are five species of *Acheta*, four of which were shown by Fulton (1952) to be non-interbreeding populations in North Carolina. All but one pair of these species can now be separated, in at least one sex, on the basis of morphological characters whose value had previously been obscured by (1) attempts to deal with the whole genus at once, and (2) studies based on morphological relationships alone. All five species can be separated on the basis of song differences revealed by audiospectrographic analysis.

*Acheta pennsylvanica* (Burmeister) is the most abundant species in northeastern United States. It occurs in grassy and weedy areas, has a chirping song, and has two single-brooded populations, one adult in spring and one adult in fall. *A. rubens* (Scudder) is the most abundant species in southeastern United States. It also occurs in grassy and weedy areas, but it consists of a single, double-brooded population, producing both spring and fall adults, and it has a trilling song. *A. vernalis* (Blatchley) and *A. fultoni* n. sp. are chirping crickets which occur in leaf litter in woods in northeastern and southeastern United States, respectively. *A. firma* (Scudder) is a chirping fall cricket which occurs along the Atlantic coast south from Delaware, and inland in sand hills in North Carolina. North-south variations in ecology and song occur in *A. fultoni*, and certain variations in life history occur in *A. pennyslvanica*.

It is probable that at least one eastern species and several western species of *Acheta* in the United State are as yet unrecognized.

## ACKNOWLEDGMENTS

This study has been supported by the Ohio State University Development Fund, the Graduate School of The Ohio State University, and the Rockefeller Foundation. The writer is particularly indebted to Dr. B. B. Fulton of North Carolina State College and Dr. E. S. Thomas of the Ohio State Museum for information and advice. E. L. Sleeper, R. E. Woodruff, T. J. Walker, J. A. Buxton, A. N. Dowdy, D. L. Chambers, and many others assisted by collections and observations. Special thanks are due Prof. J. J. Davis of Purdue University, Dr. H. H. Ross of the Illinois Natural History Survey, Prof. J. N. Knull of The Ohio State University, Dr. J. A. G. Rehn and Mr. H. Grant, Jr. of the Academy of Natural Sciences of Philadelphia, Dr. P. J. Darlington and Dr. W. L. Brown of the Museum of Comparative Zoology, Harvard, Dr. T. E. Moore, Dr. I. J. Cantrall, and Dr. T. H. Hubbell of the University of Michigan Museum of Zoology. Appreciation is expressed to Dr. H. Schiemenz of the Zoological Museum of Humboldt University, Berlin, for the loan of type material; and to Dr. R. Malaise of the Natural History Museum, Stockholm; Dr. S. L. Tuxen of the Zoological Museum, Copenhagen; Dr. W. Tischler, Zoological Institute and Museum, University of Kiel, Germany; Dr. Lucien Chopard of the Natural History Museum, Paris; Dr. K. Princis of the Zoological Institute, Lund; and Dr. C. Ferriere of the Natural History Museum, Geneva, for supplying information concerning *Acheta* types. Dr. A. B. Gurney of the U. S. National Museum was very helpful in supplying the addresses of many of these people. Mr. T. H. Farr of the Science Museum of the Institute of Jamaica sent specimens and information on the songs and ecology of Jamaican *Acheta*. Dr. D. J. Borror of the Department of Zoology and Entomology, The Ohio State University, has supplied helpful advice and criticism throughout the study.

## REFERENCES CITED

**Alexander, R. D.** 1956. A comparative study of sound production in insects, with special reference to the singing Orthoptera and Cicadidae of the eastern United States. Pp. xviii+529; 180 fig.; bibliog. (Unpublished Ph.D. Dissertation, The Ohio State University).

1957. Sound production and associated behavior in insects. Ohio Jour. Sci. 57(2): 101–113, 13 figs., 1 text fig.

**Allard, H. A.** 1910. Some New England Orthoptera observed in late October. Ent. News 21(8): 352–7.

1929. Physiological differentiation in overwintering individual of certain musical Orthoptera. Canadian Ent. 61(9): 195–8.

**Ball, E. D.,** et al. 1942. The grasshoppers and other Orthoptera of Arizona. Univ. Ariz. Agric. Expt. Sta. Tech Bull. 93: 257–73.

**Blatchley, W. S.** 1903. The Orthoptera of Indiana. 27th Ann. Rept. Ind. Dept. Geol. and Nat. Resources: 123–471, 122 fig.

1920. Orthoptera of Northeastern America with especial reference to the faunas of Indiana and Florida. The Nature Publ. Co., Indianapolis, Ind. Pp. 1–784; 246 fig.

**Borror, D. J.,** and **C. R. Reese.** 1953. The analysis of bird songs by means of a Vibralyzer. Wilson Bull. 65(4): 271–6, 7 fig.

**Burmeister, H.** 1838. Handbuch der Entomologie. Berlin, 1839. Pp. 1–1050.

**Cantrall, I. J.** 1943. The ecology of the Orthoptera and Dermaptera of the George Reserve, Michigan. Univ. Mich. Mus. Zool. Misc. Publ. 54: 1–182, 3 fig., 2 maps, 16 tab., 10 pl.

**Davis, W. T.** 1922. An annotated list of the cicadas of Virginia with description of a new species. Jour. New York Ent. Soc. 30(1): 36–52, 4 fig.

**Drury, D.** 1773. Illustrations of Natural History. Pp. 2: 9+90; 50 illus., London.

**Fabricius, C.** 1775. Systema Entomologiae. Flensburgi et Lipsiae.

**Fulton, B. B.** 1930. A new species of *Nemobius* from North Carolina (Orthoptera: Gryllidae). Ent. News 41(2): 38–42; 1 fig.

1952. Speciation in the field cricket. Evolution 6(3): 283–95; 4 fig.

**Gurney, A. B.** 1950. The Linnaean subgeneric names of *Gryllus* (Orthoptera). Jour. Wash. Acad. Sci. 40(12): 409–13.

1951. The names of the field and house crickets. Jour. Econ. Ent. 44(4): 611.

**Harris, T. W.** 1841. A treatise on some of the insects injurious to vegetation. Orange Judd and Co., New York. Pp. xi+640, numer. fig.

**Hubbell, T. H.** 1954. The naming of geographically variant populations. Systematic Zool. 3(3): 113–21, 3 fig.

1956. Some aspects of geographic variation in insects. Ann. Rev. Ent. Annal Reviews, Inc., Stanford, Calif., pp. 71–88.

**Lutz, F. E.** 1908. The variation and correlations of certain taxonomic characters of *Gryllus*. Carnegie Inst. of Wash., Publ. No. 101: 3–63; 6 fig.; 93 tab.

**McNeill, J.** 1889. Notes upon *Gryllus* and *Oecanthus*. Ent. Amer. 5(5): 101–4.

**Pringle, J. W. S.** 1955. The songs and habits of Ceylon cicadas, with a description of two new species. Spolia Zeylonica 27(2): 229–38; 6 pl., 6 text fig.

**Rehn, J. A. G.** 1902. Notes on the Orthoptera of New Mexico and western Texas. Proc. Acad. Nat. Sci. Phila. 54: 726–7.

**Rehn, J. A. G.,** and **M. Hebard.** 1915. The genus *Gryllus* (Orthoptera) as found in America. Proc. Acad. Nat. Sci. Phila. 67: 292–322, 12 fig.

**Saussure, H. de.** 1874. Mission scientifique au Méxique et dans l'Amérique Centrale. Sixième partie. Etudes sur les insectes Orthoptères et les Myriapodes. Paris.

1877. Mélanges orthoptérologiques, 5th fascicule. Mem. Soc. de Physique et d'Hist. Nat. Genève 25(1): 1–352.

**Scudder, S. H.** 1862. Materials for a monograph of the North American Orthoptera, including a catalogue of the known New England species. Bost. Jour. Nat. Hist. 7: 409–80.

1901. The species of *Gryllus* on the west coast. Psyche 9(307): 267–70.

1902. The species of *Gryllus* found in the United States east of the Sierra Nevadas. Psyche 9(309): 291–6.

**Serville, M. A.** 1839. Histoire naturelle des insectes Orthoptères (dans les suites à Buffon). Paris, Roret. Pp. 1–776, 22 fig.

**Severin, H. C.** 1935. The common black cricket as a serious pest in South Dakota. South Dakota Expt. Sta. Bull. 295: 1–51.

**Stål, C.** 1858. [Description of *Gryllus lineaticeps*.] (*In* Virgin, Christian A., Kongliga sveska fregatten Eugenies resa omkring jorden. Del. 2. Zoologie, pp. 314–5).

**Thomas, E. S.,** and **R. D. Alexander.** 1957. *Nemobius melodius*, a new cricket from Ohio (Orthoptera: Gryllidae). Ohio Jour. Sci. (in press).

**Uhler, P. R.** 1864. Orthopterological contributions. Proc. Ent. Soc. Phila. 2: 543–55.

# Numerical Taxonomy, Convergence, and Evolutionary Reduction

WILLIAM L. BROWN, JR.

Let us be clear from the beginning—I am writing to criticize *numerical taxonomy*, not quantitative systematics. I am sure that measurement and mathematical methods will and should be much more widely employed in systematic studies. It should also be understood that *numerical taxonomy*, by which I mean specifically the doctrine and methods outlined by Sokal and Sneath (1963) in their book of this title, is to me not without interest, import, and potential usefulness for systematic biology. But I intend to show that in at least one important respect, numerical taxonomy is headed in the wrong direction.

I am not going to attempt to deal with mathematical questions, and I am going to skip over the obvious objections based on the difficulty of properly analyzing characters, so that I can get to the subject that interests me most. While considering this fault in numerical taxonomy, I hope to bring the criticism around to a more constructive position by hinting at an approach to systematics that the numerical taxonomists might find worth trying.

The aspect I have chosen to discuss deals with phyletic *convergence*, a term that I here interpret broadly to include the finer-grained phenomenon, *parallelism*. Michener and Sokal (1957) make the following statement:

> Possible parallelisms present a difficult problem. The amount of evolution, after divergence from a common stem, of two forms that have evolved in a parallel manner, would be underestimated by our method, as well as by customary systematic methods.

The authors claim that their studies of saturniid moths and megachilid bees show relatively few parallel characters.

> We believe that much of the difficulty due to parallelisms is eliminated in our study by the large number of characters used. Parallelism or convergence affecting only a few characters would not greatly influence our measure of relationship.

It hardly needs to be added that convergence or parallelism affecting a *large* proportion of the characters used would seriously upset the conclusions of the numerical taxonomist, as well as those of many a traditional phylogeny-builder. Perhaps it was uneasiness about evolutionary convergence that led Sneath and Sokal (1962) to pose this rhetorical question.

> Suppose an absurdly extreme overall convergence had occurred so that two forms were almost indistinguishable and could readily and successfully hybridize: What is the purpose of separating them on grounds of ancestry when in all other attributes they are the same?

We accept this question as "absurdly extreme" indeed (at least at the metazoan level), but we answer it anyway: We separate the two forms if we know their ancestry, because the *history* of this situation is of overriding biological importance. It should be borne in mind that systematic biology is not merely an arrangement for gathering and filing data on the diversity of living organisms; it is also, with genetics and ecology, one of our main hopes of understanding how evolution occurs. That understanding is clearly fragmentary right now, despite the advances of Darwin and the geneticists.

Let us examine the case of the convergence that may not be "absurdly" extreme but is at least extreme enough to threaten taxonomic judgment. In the group I work on (the ants), the old tribe Dacetini has contained convergent elements of three separate tribes, and genera actually repre-

Reproduced with permission from *Systematic Zoology*, 14:101–109, 1965. Published by the Society of Systematic Zoology, Washington, D.C.

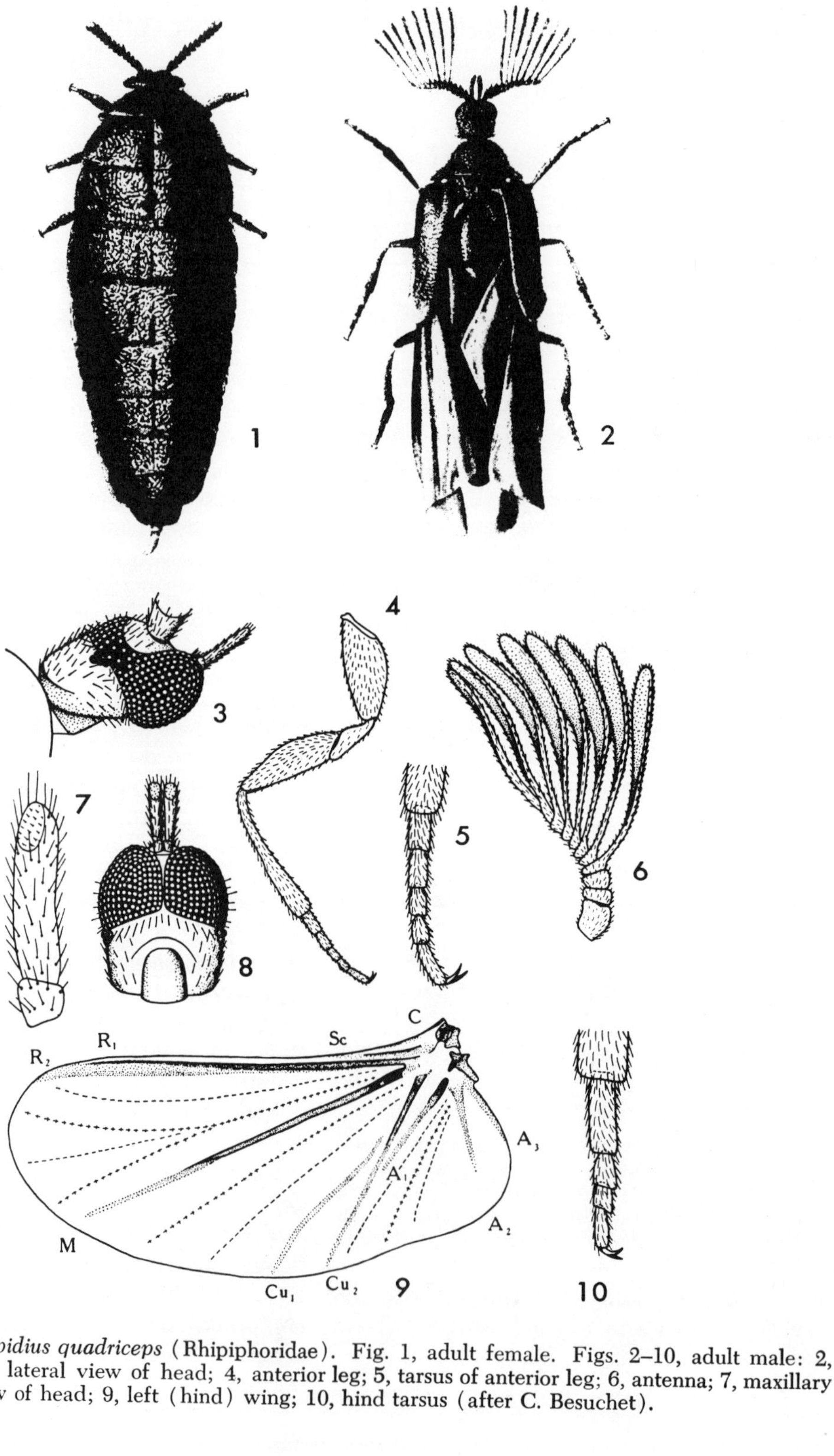


FIGS. 1–10. *Rhipidius quadriceps* (Rhipiphoridae). Fig. 1, adult female. Figs. 2–10, adult male: 2, habitus drawing; 3, lateral view of head; 4, anterior leg; 5, tarsus of anterior leg; 6, antenna; 7, maxillary palp; 8, ventral view of head; 9, left (hind) wing; 10, hind tarsus (after C. Besuchet).

senting two additional tribes might well have been placed there in recent years if the classification had been done on either the old basis or by numerical taxonomy (Brown and Wilson, 1957, 1959).

One might also point to the subfamily Dorylinae, the army ants, the different tribes of which share an impressive number of characters, both morphological and behavioral, in worker, female and male castes, in larval morphology, in the behavior, physiology and ecology of the colony taken as a whole, and in the taxa of inquilines they harbor. In spite of these detailed resemblances, all of which are no doubt adaptively correlated with the specialized army-ant way of life, I suspect that three and possibly four completely independent lineages make up the Dorylinae as the subfamily now stands in the books. This suspicion is based on major differences noted in internal characters and on other considerations too complicated to consider here; the case is currently under study. The literature of zoology, especially in recent times, is full of cases in which formerly "solid" higher taxa turn out to be polyphyletic after careful study.

The fact is that evolution into specialized adaptive zones commonly affects whole complexes of characters more inclusively than the numerical taxonomists seem to realize. I could most comfortably discuss several more interesting cases of convergence that have caused trouble in ant taxonomy, but I have chosen instead to make my point in more detail by comparing two lineages of degenerate parasitic beetles. I refer to the family Rhipiphoridae and the superfamily Stylopoidea, or Strepsiptera, both of which are regarded in the prevailing interpretations as coleopterous stocks (Crowson, 1955).

The Rhipiphoridae are an entomophagous offshoot of the primarily phytophagous (occasionally entomophagous) family Mordellidae, a major group of heteromerous beetles. The rhipiphorids reach the peak of their parasitic specialization in subfamily Rhipidiinae, which is composed of internal parasitoids of cockroaches (Blattaria). Figures 1 to 10, 13 and 16 show *Rhipidius quadriceps*. Characters especially worthy of note here are the small, narrow, hairy elytra of the male, the coarsely facetted "raspberry" eyes, the broad fan-like male hind wings with radiate venation, the uniflabellate male

TABLE 1.—A COMPARISON OF TWENTY OBVIOUS CHARACTERS OF THE ABERRANT BEETLE FAMILIES RHIPIPHORIDAE (SUBFAMILY RHIPIDIINAE) AND MENGEIDAE.

| Character | Rhipiphoridae (Rhipidiinae) | Stylopoidea (Mengeidae) |
|---|---|---|
| Spiracle on Abdom. VIII (cucujoid base char.) | 0 | 0 |
| Adult trochanters | + (all legs) | 0 in legs 1, 2? |
| ♂ tarsal segments | 5-5-4 or less | 5-5-5 |
| Wing shape fan-like, vannal area developed | + | + |
| Venation radiating | + | + |
| ♀ wingless, modified for endoparasitism | + | + |
| ♀ with legs | + | + |
| Adult raspberry eyes | + | + |
| Pronotum reduced | + | + |
| Internal parasites of: | *Blattaria* | *Thysanura* |
| Hypermetamorphosis | + | + |
| "Triungulin" primary larva (several chars.) | + | + |
| Larvae: trochanters | + | 0? |
| Elytra reduced, narrow | + | + |
| Elytra soft | + | + |
| Elytra hairy | + | + |
| Gonoforceps of ♂ genitalia | + | 0 |
| Antennal segments ♂ | 6-11 | 4-7 |
| Antennal segments ♀ | 6-11 | 4-5 |
| ♂ antennae flabellate | + | + |

antennae, the wingless, broad form of the adult female, and the peculiar series of traits marking the so-called "triungulin" first-instar larva. The adult tarsi have the heteromerous base number 5-5-4, or 5 segments on the forelegs, 5 on the middle legs, and 4 on the hind legs.

Characters of the Mengeidae, most primitive of stylopoid families, are shown in Figures 11, 12, 15, and 18. Some of the more

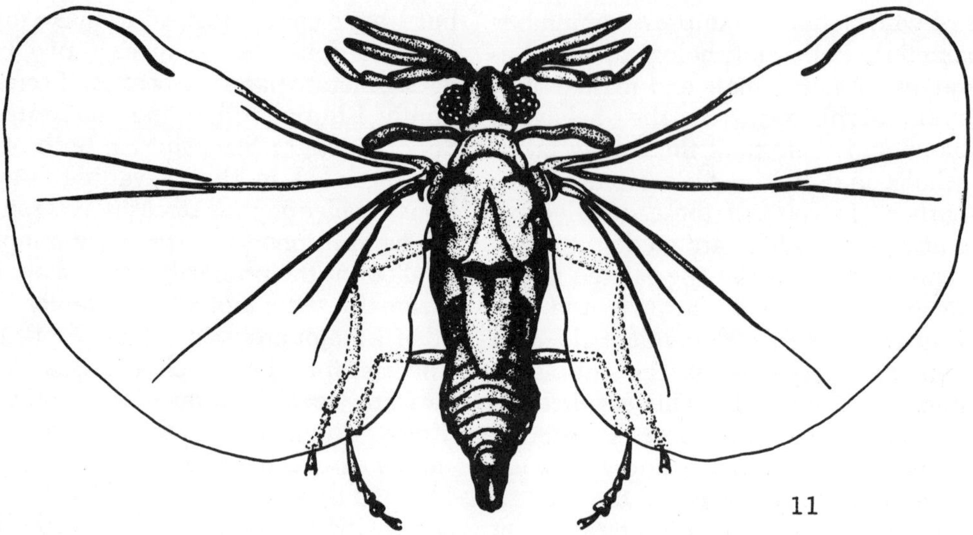

11

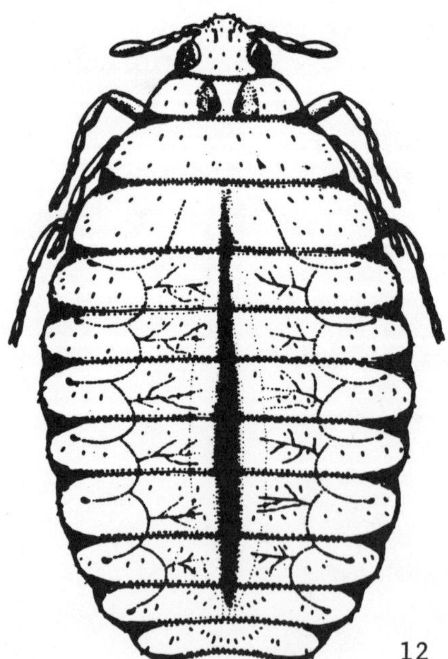

12

FIGS. 11 and 12. Representative adult Stylopoidea of the primitive family Mengeidae. Fig.11, *Triozocera mexicana*, adult male (after R. M. Bohart). Fig. 12, *Eoxenos laboulbenei*, adult female (after Parker and Smith).

striking similarities and differences between rhipidiines and mengeids are arrayed in tabular form (Table 1). This table could be extended to other similarities and differences by going into finer and finer points, and probably by studying features of internal anatomy, cytology, physiology, ecology, and behavior that are not now known. This could be done for the similar first-instar "triungulin" larvae, for example, although I have indicated in the table only the probable difference in presence vs. absence of larval trochanters. Another difference lies in the apparent absence of true antennae in the mengeid larva. So far as external character analysis goes, however, there seem to be as many points of similarity as there are of difference between rhipiphorid and mengeid triungulins.

Taking all the characters together, we can make a reasonably strong case for the relationship of Rhipidiinae and Mengeidae. In most of the few known character differences, we see that the mengeid character represents a reduction from the state of the rhipidiine character, so that one might be inclined (as some past authors have been) to derive the stylopoid line from the rhipiphorid stock. There is at least one character, however, that gives taxonomists pause: The tarsi of male mengeids are all 5-segmented. This means that the mengeid male has one more tarsal segment in each hind leg than do the rhipidiines. Furthermore, may it be noted that the Rhipiphoridae and their putative mordellid ancestors belong to a large division of the Coleoptera, the Heteromera, in which the hind tarsus never has more than 4 segments. If the numerical taxonomists will not grant us more phyletic *weight* for this little tarsal segment, at least they will have to admit that it is a trait of more than routine interest to the phylogenist.

## *Character Weighting*

The numerical taxonomists have found common cause with a scattering of recent anti-phylogenetic, and often even anti-evolutionary, writers. Some members of this uneasy alliance will accept indication of phylogenetic relationships as a *result* of systematic study, but none will condone the use of phyletic reasoning in constructing a classification. I should at this point strongly affirm my belief in phyletic data and logic both as *conclusions* of systematic investigations and as *contributory necessities* toward raising classifications in the first place. This is not to say that we should construct a complete phylogeny first, and then our classification. But judicious observation will show us, as it has shown many taxonomists in the past, that some *characters* tend to evolve in definite directions with high probability. This is why we must give these characters a greater weight in phylogenetic reasoning.

We have already seen, in the case of the tarsal segmentation of the Coleoptera (Rhipiphoridae and Stylopoidea), that phylogenetic weighting is a real component of systematic judgment. But how, exactly, does weighting work?

The answer to this question calls for an assessment of standard phylogenetic thinking—a difficult task considering that much of this thought is implicit and obscure. The taxonomist often has a feeling for the primitive or derivative aspects of the taxa he studies, but most taxonomists are hard put to set this intuition to logic, and they tend to resort to the circular reasoning about which numerical taxonomists rightly complain. It seems to me that there exist certain regularities in evolution above the species level that are the basis of all good phylogenetic reasoning, whether implicit or explicit. Perhaps the tarsal count of the beetles gives us an important clue that we can use as an opening toward a modern study of this question.

It seems that the main reason why most authors have refused to derive the stylopoids from the rhipiphorids is their feeling that the addition of a tarsal segment, to make up the five of the hind tarsus of primitive stylopoids from the four of Rhipiphoridae (and related Heteromera), is an evolutionary event of such low probability that it has received no significant consideration. But why is the simple addition of one small

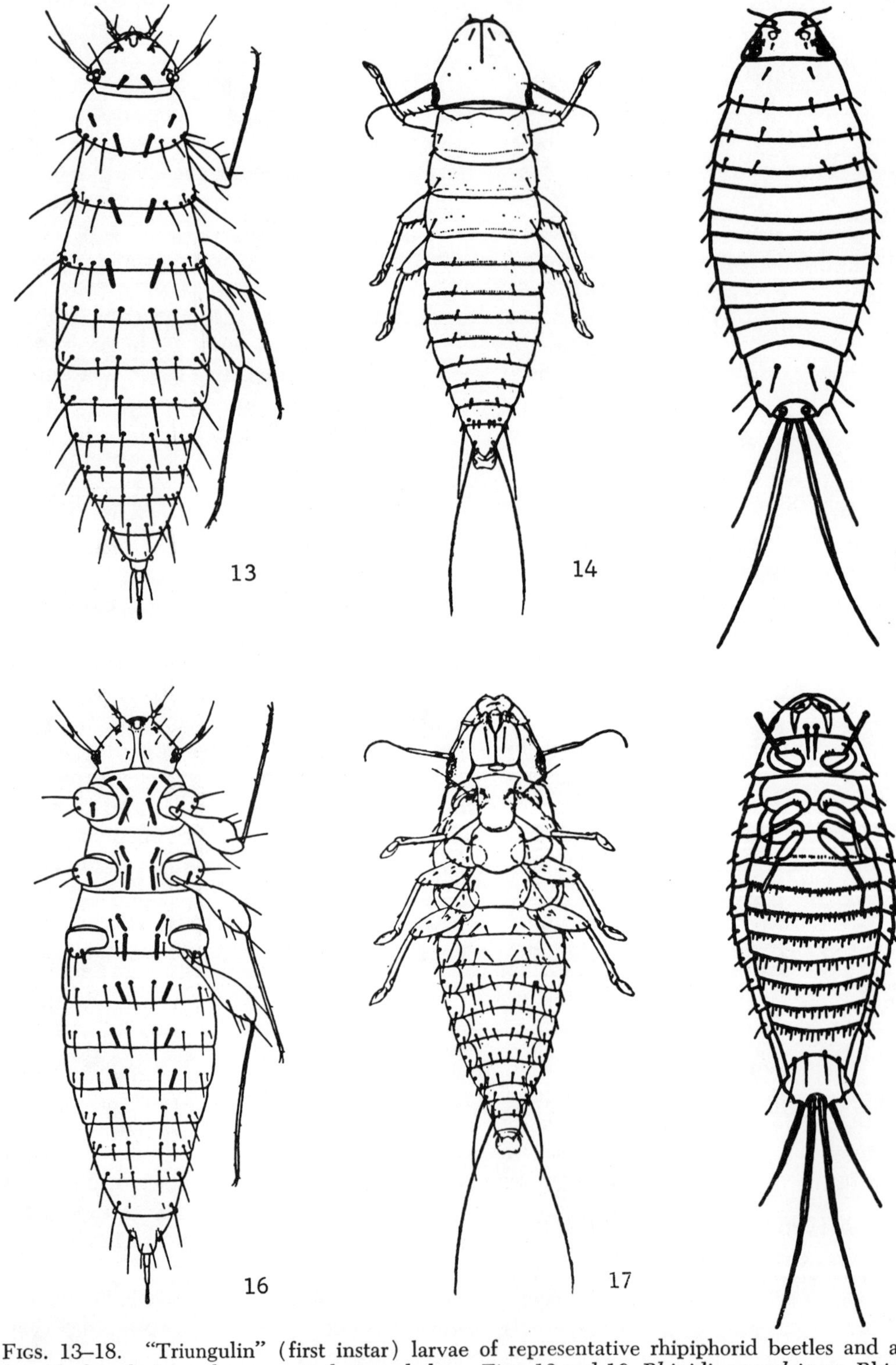


FIGS. 13–18. "Triungulin" (first instar) larvae of representative rhipiphorid beetles and a mengeid stylopoid, dorsal views above, ventral views below. Figs. 13 and 16, *Rhipidius quadriceps*, Rhipiphoridae (after Besuchet). Figs. 14 and 17, *Macrosiagon flabellatum*, Rhipiphoridae (after Grandi). Figs. 15 and 18, *Eoxenos laboulbenei*, Mengeidae (after Bohart).

segment on each hind tarsus denied by the overwhelming majority of responsible workers on beetle taxonomy? The answer is that they believe that among Coleoptera a change in the number of tarsal segments, such as appears to have taken place many times in different lines, has always resulted in a decrease. The reason for this belief is again not obvious. When asked why he believes it, the coleopterist is likely to reply something such as, "The reduced number is found most often in forms that are derivative as judged on other characters, and the larger number is found in primitive forms of the same group." In other words, he implies some sort of correlation among characters. However, when considering a restricted group, such as a genus or tribe, one often finds that it is "primitive" in one or more characters, but "derivative" in others, so that the correlation, in the strict sense of the term, may be weak or non-existent at this level.

Nevertheless, if one considers very large groups, such as most insect families, superfamilies, and orders, a correlation of a stronger sort is often so evident that we commonly are able to speak of the "lower" and "higher" genera of this or that family with some degree of confidence. But what gives us this confidence?

It seems to me that the answer lies with a rule or principle that we have all absorbed without perhaps thinking much about it in any explicit way: This we may call the *rule of evolutionary reduction.* It is exemplified by the statement, familiar, I am sure, to all college botany students, to the effect that the flowers of higher plants have tended to evolve *from a condition of large, indefinite numbers of parts toward small, definite numbers.* The rule of reduction has been explicitly recognized by plant morphologists (e.g., Eames, 1936), and is perhaps implicit in such "Laws" as those of Dollo or Meyrick, as well as in the writings of many other zoologists.

Stated in a crude and general way, the rule of reduction is simply that *a gross evolutionary change in a given character is more likely to be a reduction than an elaboration.*

To avoid the difficulties of defining "reduction" and "elaboration," we may for the moment restrict ourselves to considering only meristic characters—characters that can be counted as integral units, such as the vertebrae of a fish or the segments of an insect antenna. The rule of reduction then becomes (for metazoan animals, at least): *For macroscopic meristic characters in general, a phylogenetic change in number much more frequently results in a decrease than in an increase.* Putting it another way, we may say that for any evolutionary span of time the probability that a given meristic character will lose in number is much greater than the probability that it will gain. When we compare "lower" and "higher" genera of a family, for instance, we generally associate large counts for meristic characters in the "lower" genera, and *vice versa.* We can frequently test the *direction* of evolution in such cases by (1) the direct chronological evidence of the fossil record (e.g., the decrease in number of dermal skull bones during the fossil history of the vertebrates, or loss of some longitudinal wing veins in the record of the insects); or (2) by the interpretation of regressive "vestigial" characters.

So far as I can tell from a spot check of actual groups (Formicidae as well as Hymenoptera and Arthropoda in general, Diptera, Rhipiphoridae, and Stylopoidea as well as Coleoptera in general, Vertebrata), following generally accepted and in the main stable classifications for each group, the meristic reduction rule holds extraordinarily well. The exceptions are conspicuous because they run against the "reductive current" of the remaining characters. For instance, *Prionus* species have 12–25 antennal segments as against the "base number" of 11 for other Cerambycidae and for beetles in general; therefore, it seems clear that this is one case in which the lesser probability of a meristic increase has been realized. It should be noted that the idea of a "base number" or "primitive pattern" is very important to phylogeneticists.

Among the ants (family Formicidae), the maximum normal antennomere counts are 13 in the males, 12 in females (including workers); the maxillary and labial palpi have at most 6 and 4 segments respectively; and 5 is the invariable number of tarsomeres. It can scarcely be coincidence that wasp groups showing many other similarities to the ants also display these basic numbers in their more generalized species. In even the lowest phytophagous Hymenoptera, we find the same "basic" numbers of palpomeres (6 and 4) and tarsomeres (5), although the antennomeres are basically more numerous (and more variable in number). Against a general background of phyletic reduction throughout the order, the "second trochanter" found in many phytophagous and parasitic Hymenoptera (a piece apparently split off from the distal end of the femur, lost again in the higher families of the order) shows up as a "secondary increase" character like that of the extra antennomeres in *Prionus*. The pentamerous tarsal count, on the other hand, is the usual one among Hymenoptera, and no doubt carries back eventually to protoperlarian ancestors.

Meristic characters are not *always* so helpful, and within smaller taxa such as genera and tribes few or none of them may vary enough to give a clear picture of phyletic progression. So it is in the specialized ant genus *Strumigenys*, with over 150 known species (Brown, 1962); one cannot yet say with certainty in this genus which characters (and consequently, which species) are primitive, and which are derivative.

It is nonetheless possible to arrange the species of *Strumigenys* in branching relationship-chains, or "open-ended morphoclines," without making a hard-and-fast decision as to which free end of the chain represents the primitive condition. As evidence gathers, in the form of newly discovered species and closer investigation of the ones already known, it is often possible to "polarize" such a morphocline—that is, to show its probable evolutionary direction(s) and to point out the loose chain-end that is most likely primitive. Elements of the theory of morphoclines are set forth in an apparently neglected paper by Maslin (1952) in *Systematic Zoology*; this article is sensible reading for any systematist. But even before a happy state of phylogenetic polarity is attained, the open-ended morphocline can be a valuable basis for classifying and thinking about any taxon above the species level, as I have found in working with the aforementioned genus *Strumigenys* and other ant groups.

### *Possibilities for the Future*

It may well be that the reasoning used in constructing a morphocline, taking account of the reduction rule, is a better way to approach classification and phylogeny than the best that numerical taxonomy has to offer. One notes that numerical taxonomy has worked out rather poorly in the mosquito genus *Aedes*, where Rohlf (1963a, b) has not been able to align larval- and adult-based systems in spite of the employment of a large number of characters from both adults and larvae. Furthermore, two different methods of analysis yielded different results for Rohlf, even for the same stages of the same species-field (Barr and Chapman, 1964). One wonders how a reduction-morphocline approach might work on this collection of *Aedes* species; there has been no opportunity to try it yet.

In my opinion, numerical taxonomy's chief virtue has been its need to see the characters systematically and completely compared for all taxonomic units. The best modern form for such a comparison is a matrix in which every taxon is listed against every character used in the study. This form has been used by many taxonomists in past years, especially for the "rough work" of keymaking. It should be developed and consciously used much more widely in systematic work.

### *Summary*

1. Numerical taxonomy is inadequate to deal with systematic situations involving adaptive convergence since these

situations may commonly involve too large a proportion of taxonomic traits.

2. Examples are cited of likely convergent taxa that would probably be interpreted wrongly by numerical taxonomy, as they have been by conventional taxonomy.
3. Of these taxa, the rhipiphorid and stylopoid beetles are dealt with in detail, and their non-cognate relationship is demonstrated mainly by their difference in a minor character, segmentation of the hind tarsus.
4. Weighting of such characters is covered by a proposed *rule of evolutionary reduction*, considered to be implicit in most phylogenetic arguments. This rule is stated formally for the case of meristic characters. *For macroscopic meristic characters in general, a phylogenetic change in number much more frequently results in a decrease than in an increase.*
5. Phylogenetic studies ought to benefit from the conscious application of the reduction rule yardstick and from extension of the morphocline concept of Maslin.

### *Acknowledgments*

I have been greatly helped in writing this paper by an introduction to the literature of the Rhipiphoridae and the Stylopoidea furnished by R. B. Selander, and by ideas gained through conversations with colleagues, especially A. J. Eames, H. E. Evans, J. G. Franclemont, H. F. Howden, and E. O. Wilson. It is not implied that these individuals necessarily support all of the views expressed above.

REFERENCES

BARR, A. R., and H. C. CHAPMAN. 1964. Mosquito classification. Systematic Zool. 13:100–101.

BROWN, W. L., JR. 1962. The neotropical species of the ant genus *Strumigenys* Fr. Smith: Synopsis and keys to the species. Psyche 69:238–267.

BROWN, W. L., JR., and E. O. WILSON. 1957. *Dacetinops*, a new ant genus from New Guinea. Brev. Mus. Comp. Zool. Harv. 77:1–7.

1959. The evolution of the dacetine ants. Quart. Rev. Biol. 34:278–294.

CROWSON, R. A. 1955. The natural classification of the families of Coleoptera. Nathaniel Lloyd, London.

EAMES, A. J. 1936. Morphology of vascular plants. Lower groups. McGraw-Hill, New York.

MASLIN, T. P. 1952. Morphological criteria of phyletic relationships. Systematic Zool. 1:49–70.

MICHENER, C. D., and R. R. SOKAL. 1957. A quantitative approach to a problem in classification. Evolution 11:130–162.

ROHLF, F. J. 1963a. Congruence of larval and adult classifications in *Aedes* (Diptera: Culicidae). Systematic Zool. 12:97–117.

1963b. Classification of *Aedes* by numerical taxonomic methods (Diptera: Culicidae). Ann. Ent. Soc. Amer. 56:798–804.

SNEATH, P. H. A., and R. R. SOKAL. 1962. Numerical taxonomy. Nature 193:855–860.

SOKAL, R. R., and P. H. A. SNEATH. 1963. Numerical taxonomy. W. H. Freeman, San Francisco and London.

**WILLIAM L. BROWN, JR.** is Associate Professor of Entomology at Cornell University, Ithaca, New York. A substantially similar version of this paper was read as part of a symposium on numerical taxonomy at the XIIth International Congress of Entomology, London, July 1964, but it will be published in the Proceedings of the Congress only in abstract form.

# Morphology of the Alimentary Canal During the Metamorphosis of *Hyalophora cecropia*

## (Lepidoptera: Saturniidae)[1]

KENNETH J. JUDY[2] AND LAWRENCE I. GILBERT

Department of Biological Sciences, Northwestern University, Evanston, Illinois 60201

ABSTRACT

Changes in the anatomy of the alimentary canal are described during the life of the cecropia moth, *Hyalophora cecropia* (L.), with particular attention given to the transformations from larva to pupa and pupa to adult. During the development of the pupa, the larval foregut and hindgut empty and collapse to form narrow tubes, but the most dramatic change is the deterioration of the mass of cryptonephric tubules and the subsequent development of rectal pads in the adult rectum. In general, the changes from larva to pupa are degenerative. During development of the adult, a crop is formed from the posterior foregut. It is postulated that the crop is a compensatory mechanism for maintaining hemocoel volume when hemolymph is pumped into the wings of the adult shortly after ecdysis.

Despite the widespread use of saturniid moths such as the cecropia moth, *Hyalophora cecropia* (L.), as objects for experimentation by insect physiologists and biochemists, remarkably little is known about the morphological and histological changes that accompany metamorphosis in this insect. This paucity of knowledge is particularly true of the alimentary canal which undergoes extensive alterations during the life cycle of Lepidoptera. For the most part, descriptions of the lepidopteran digestive tract have been restricted to the larva (Lyonnet 1762, Peterson 1912, Larsson 1929, Teotia and Pathak 1957, Dreckträh et al. 1966) or adult (Burgess 1880, Bordas 1920, Pyle 1940). Only a few studies such as those by Minot and Burgess (1884) on *Aletia xylina* (L.) and Chauthani and Callahan (1967) and Standlea and Yonke (1968) on the bollworm, *Heliothis zea* (Boddie), include the larval, pupal, and adult stages.

This work provides a description of the morphological changes that occur in the alimentary canal of *H. cecropia* during metamorphosis. Subsequent papers will discuss the histology of these changes as well as the effects of juvenile hormone and molting hormone on the development of specific components of the digestive tract.

### MATERIALS AND METHODS

Larvae of *H. cecropia* were reared locally on wild black cherry trees; pupae were obtained commercially. Diapausing pupae were placed at 6°C for at least 6 weeks, then they were removed as needed and restored to 25°C to initiate adult development. The time sequence of adult development was followed by observing morphological changes through the pupal cuticle (Schneiderman and Williams 1954). Morphological studies were carried out by dissection under a stereo dissecting microscope. Most observations were made on freshly dissected specimens, but occasionally the insects were fixed in Kahle's fluid or aqueous Bouin's. Dilute methylene blue was frequently used to enhance the visibility of various structures.

### OBSERVATIONS

#### Fifth-Stage Larva

Fig. 1 reveals the gross anatomy of a feeding 5th-stage larva. The alimentary canal is divided into 3 regions according to embryonic origin: the foregut (stomodaeum) which includes the buccal cavity, pharynx, esophagus, crop, and stomodaeal valve; the midgut (ventriculus, mesenteron, midintestine); and the hindgut (proctodaeum) which includes the pylorus, ileum, colon, and rectum. Table 1 lists the average length of each of these gut regions in a larva 100 mm long.

*Foregut.*—The foregut (Fig. 2) originates at the mouth opening behind a pair of heavy mandibles and extends about 1 mm into the body as the buccal cavity (bc, Fig. 2). This chamber is supported by paired dorsolateral and ventrolateral dilator muscles,

[1] Supported by grant AM-02818 from the National Institutes of Health. Accepted for publication Mar. 27, 1969.

[2] Former predoctoral fellow of NIH training grant GM-903. Present address: Zoecon Corporation, 975 California Avenue, Palo Alto, Calif. 94304.

Reproduced with permission from *Annals of the Entomological Society of America*, 62:1438–1445, 1969. Published by the Entomological Society of America, College Park, Maryland.

Table 1.—Average lengths (in mm) of regions of the alimentary canal of *H. cecropia* during metamorphosis from 5th-stage larva to adult.

| Stage of development | Body | Foregut | | | Midgut | Hindgut | | | |
|---|---|---|---|---|---|---|---|---|---|
| | | Pharynx | Esophagus | Crop | | Pylorus | Ileum | Colon | Rectum |
| 5th-stage larva | 100 | 2 | 5 | 6 | 60 | 2 | 7 | 6 | 10 |
| Pupa | 35 | | 10[a] | | 18 | | | 7[a] | |
| Pharate adult | | | | | | | | | |
| Day 5 | 35 | 10 | | 1[b] | 18 | | 10[b] | | 5[b] |
| Day 10 | 35 | 10 | | 3 | 18 | | 18 | | 8 |
| Day 15 | 35 | 10 | | 4 | 7 | | 22 | | 10 |
| Newly emerged adult | 35 | 10 | | 6 | 7 | | 25 | | 15 |

[a] The fore- and hindgut of the pupa show no differentiated subdivisions.
[b] For a discussion of the adult crop, ileum, and rectal pouch, see text.

and is surrounded by 3 rings of circular muscle where it joins the pharynx (ph, Fig. 2). The pharynx begins just beneath the frontal ganglion, is surrounded by moderately heavy bands of circular muscle, and several paired dorsal and ventrolateral dilator muscles arise from its walls. Details of the musculature of the cephalic portion of the foregut are similar to those described by Henson (1931), Snodgrass (1935), and Drecktrah et al. (1966) in other insects.

Beneath the brain, the pharynx joins the esophagus (e, Fig. 2) which is a straight, thin-walled tube of the same diameter as the pharynx. Numerous dorsolateral and ventrolateral dilator muscles insert in its walls between bands of circular muscle. As the esophagus enters the thorax, it expands considerably to form the thin-walled crop (cr, Fig. 1). The walls of the crop are covered by thin, regularly spaced bands of longitudinal (inner) and circular (outer) muscles while a few branched muscle fibers also are present. The wall of the crop receives tracheal tubes directly from the 1st thoracic spiracle. In all except the last days of the 5th instar, the crop is filled with ingested leaf clippings.

The crop empties into the midgut (mg, Fig. 1) between the pro- and mesothoracic segments, and there is no well-developed proventriculus. The posterior portion of the crop extends into the lumen of the midgut for a short distance, forms the stomodaeal valve, and then folds back over itself to attach to the anterior edge of the midgut. This posterior portion of the crop is surrounded by slightly thicker bands of muscle than the anterior region. The junction of foregut and midgut occurs more cephalad in the ventral region than in the dorsal, so the crop appears to lie somewhat above the anterior midgut.

*Midgut.*—The midgut is the largest organ in the larval body and occupies most of the hemocoel from the 2nd thoracic segment to the 6th abdominal segment. It is a straight tube and lacks diverticula except for a ring of small caecalike protuberances around its anterior end (gc, Fig. 1).

The midgut is generally filled with food throughout the 5th stage but is emptied prior to the spinning of the cocoon. Its walls are characteristically covered by rings of circular muscle, external to which are a few fine, longitudinal muscle fibers. Prominent pairs of longitudinal muscle strands (lm, Fig. 1) run along both the dorsal and ventral midlines for the length of the midgut. Numerous tracheae serve the midgut and attach to it dorso- and ventrolaterally in each body segment along its length. Two dorsal and 1 ventral pair of malpighian tubules lie along the posterior ⅔ of the midgut and will be considered subsequently. A peritrophic membrane lines the lumen along the entire length of the midgut, and between the 6th and 7th abdominal segments the midgut tapers to join the hindgut.

*Hindgut.*—The hindgut (Fig. 3) may be divided into 4 distinct regions: the pylorus, ileum, colon, and rectum. The anterior pylorus has about the same diameter as the posterior midgut so their junction is not immediately obvious. The arrangement of circular and longitudinal muscles of the pylorus is the same as that of the midgut, but the muscle fibers of the two are not continuous. The muscles of the midgut terminate at the anterior pylorus where they insert against its walls. The longitudinal fibers of the pylorus originate at the junction of the 2 gut regions. Only the prominent middorsal and midventral longitudinal fibers of the midgut extend for a significant distance along the pylorus, but these fibers also terminate before reaching the pyloric valve (pv, Fig. 3). The longitudinal fibers of the anterior pylorus terminate at the pyloric valve, a region constricted sharply by heavy circular muscle bands. Still another set of longitudinal fibers arises just anterior to the pyloric valve. These fibers pass over the circular bands of the sphincter and either insert in the anterior walls of the ileum or join into 6 heavy bundles that pass over the surface of the ileum (il, Fig. 3). The inner wall of the anterior pylorus is thrown into 6 shallow, longitudinal folds. At the pyloric valve, 4 of these folds are much larger than the other two and are so opposed as to effectively occlude the lumen when the sphincter is constricted. A pair of common malpighian ducts inserts beneath the pyloric sphincter to empty into the hindgut.

Each of these common ducts drains an enlarged ampulla in the hemocoel lateral to the pyloric valve. These ampullae (am, Fig. 3) collect the secretion of the 3 pairs of malpighian tubules that comprise the

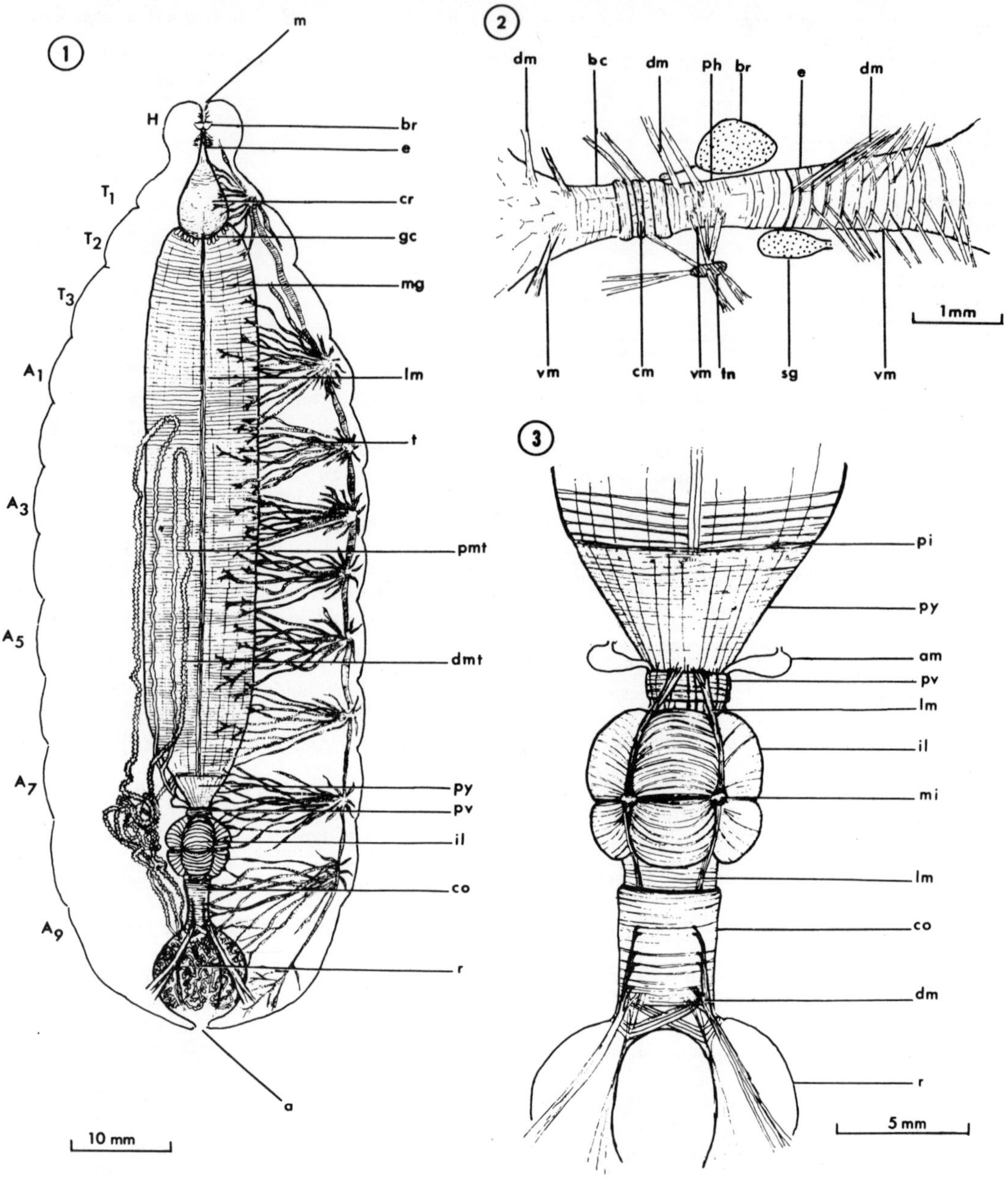


FIG. 1–3.—*H. cecropia,* 5th-stage larva. 1, Dorsal dissection to show the alimentary canal (malpighian tubules are shown for the left side only; tracheae are shown for the right side only). 2, Lateral view of the cephalic portion of the foregut (mouth opens on left side). 3, Dorsal view of the anterior hindgut (malpighian tubules are omitted).

insect's excretory system. From the collecting vesicles, these 3 pairs of tubules extend forward along the midgut, 2 pairs dorsal and 1 pair ventral, to a point about ⅔ of the way from the pylorus to the crop. Here the tubules turn sharply and continue back to the pyloric region where they become elaborately entwined with one another as well as with the tracheae and portions of fat body. The proximal tubules (those passing forward along the midgut, pmt, Fig. 1) possess smooth walls, while the distal tubules (those continuing posteriad from the midgut dmt, Fig. 1) have numerous protuberances in their walls. In the region of the rectum the coiled malpighian tubules separate and pass into the space between the rectal epithelium and the 2 sheaths of thin cellular membranes surrounding the rectum. This reassociation of the tubules with the alimentary canal comprises the cryptonephric excretory system typical of many larvae.

The malpighian tubules first pass beneath an outer single sheath and then continue posteriorly along the rectum (r, Fig. 1, 3). Near the anal duct, the tubules pass inward beneath a double-layered sheath and continue forward. These inner tubules are extremely convoluted and difficult to follow. The malpighian tubules terminate in this space between the rectal epithelium and the inner, double cryptonephric sheath.

Posterior to the pyloric valve, the hindgut enlarges to form the ileum. It is situated in the 7th abdominal segment and is supplied by numerous tracheal branches from the spiracle in that segment. The ileal wall is thin and divided by deep folds into 6 longitudinal bulges. These bulges are further divided by a circular groove around the middle into anterior and posterior portions. The walls of the ileum are covered by thin bands of circular muscle, and those covering each distended portion of the ileum focus to a common insertion at the junction of the longitudinal and circular grooves in the wall (mi, Fig. 3). At this point the fibers fuse into a junctional complex with the posterior ends of those longitudinal fibers previously described that arise near the pyloric valve and pass over the ileal wall in 6 bundles. These longitudinal muscle bundles lie in and are responsible for the deep, longitudinal grooves over the ileum. The circular and longitudinal muscles insert into thick, spherical protrusions of the wall of the ileum at the junctions of the circular and longitudinal grooves. These spherical protrusions have not been described previously. Another set of 6 longitudinal muscle bundles originates at this point and continues posteriorly. The ileum tapers sharply to form the next portion of the hindgut, the colon.

The colon (co, Fig. 1, 3) is a straight tube connecting the enlarged ileum with the rectum, and its wall is thrown into 6 longitudinal folds as is that of the rest of the hindgut. Its anterior end is often folded back over itself as a sleeve. The walls are covered by heavy bands of circular muscle overlying 6 major bundles of longitudinal muscle. At regular intervals along the colon, dorso- and ventrolateral dilator muscles (dm, Fig. 3) arise that extend into the body cavity and insert on the body wall. The most posterior of these muscles pass back at an acute angle and form 6 deep grooves in the normally distended wali of the rectum. At the junction of the colon and rectum several diagonal bands of muscle from the colon crisscross and insert in the wall of the rectum.

The rectum (r, Fig. 1, 3) is a large thin-walled sac covered by fine bands of circular and longitudinal muscle. Beneath this muscularis lie the cryptonephric sheaths and invested malpighian tubules described earlier. Numerous tracheoles also lie beneath the muscularis and arise from the profuse branching of the lateral tracheal trunks that terminate in the posterior abdomen. The inner wall of the rectum forms 6 thin, longitudinal folds into the lumen. In the most posterior abdominal segment, the rectum tapers to form a very short anal duct that exists at the extreme posterior end of the body between the anal prolegs.

## The Larva-Pupa Transformation

*Foregut.*—Metamorphic changes in the alimentary canal are initiated when the larva begins spinning the cocoon in which it will pupate. Before spinning, the insect ceases feeding and voids all its gut contents. Once emptied, the foregut becomes reduced to a narrow tube of nearly uniform diameter from mouth to midgut. This reduction in diameter is apparently caused by the contraction of the circular muscles that results in the formation of many longitudinal folds in the wall of the foregut. Most of the musculature of the foregut degenerates during pupation.

During the larva-pupa transformation the head is flexed ventrad, causing a ventral curvature of the anterior foregut to the mouth in the pupa. At its junction with the midgut the posterior foregut is slightly swollen as a result of the thickening of the epithelium in this region. The lumen of the foregut is somewhat occluded by the intima that has been sloughed at the larval-pupal molt.

*Midgut.*—The midgut which has been empty for a short period is reduced in length and diameter. However, it is soon refilled with waste material which persists throughout the pupal period. The small caecalike bulges at the anterior end become indistinct, but the musculature of the midgut does not undergo extensive degeneration.

*Hindgut.*—At the posterior end, the midgut tapers sharply to the pyloric valve. However, this pyloric valve is not the same sphincter that is present in the larva. This pupal sphincter is situated at the junction of the mid- and hindgut, a position somewhat anterior to that of the larval valve. Only a few fine, circular muscle fibers mark the position of the pupal valve and the larval sphincter degenerates completely.

Occlusion of the lumen at the pyloric valve is aided by the pyloric plug, a mass of intima membrane sloughed from the hindgut lining. The hindgut, like the foregut, becomes reduced to a narrow tube by a combination of longitudinal folding and circular muscle contraction. The anterior portion of the pupal hindgut forms the pupal ileum and is derived from the cells of the larval pylorus.

The common malpighian ducts enter the pupal ileum a short distance posterior to the pupal pyloric valve. The larval ileum, colon, and rectum become reduced in the pupa to a slightly swollen tube between the pupal ileum and anus. The swollen appearance of this region is largely a result of the mass of diffuse tissue that surrounds it. This tissue is formed from degenerating muscle fibers, blood cells, and the cells of the cryptonephric malpighian tubules that disintegrate during the larva-pupa transformation. At 2 points along this diffuse tissue mass, fibers which are remnants of the muscles of the larval ileum and the colon extend into the hemocoel for a short distance. In some specimens, numerous nodules containing a milky-white substance lie embedded in the tissue mass. These nodules are remnants of the cryptonephric malpighian tubules containing trapped waste fluids.

At the completion of pupal development, the posterior hindgut is flattened dorsoventrally, occluding the lumen, but the 6 longitudinal folds are still evident. The anal duct is occluded by an intimal mass, the anal plug. The hindgut bends ventrad to the anus, a condition brought about by the flexure of the posterior portion of the larva during pupation.

The body of the insect has been shortened considerably when pupation is completed (Fig. 4). The alimentary canal consists of a slender foregut (fg, Fig. 4). from the mouth to the junction with the midgut (mg, Fig. 4) in the last thoracic or 1st abdominal segment, a large smooth-walled midgut that is filled with waste material and extends into the 7th abdominal segment, and a relatively straight, narrow hindgut that is surrounded by a diffuse tissue mass at the posterior half. The arrangement of tracheae is the same as in the larva and except for the degenera-

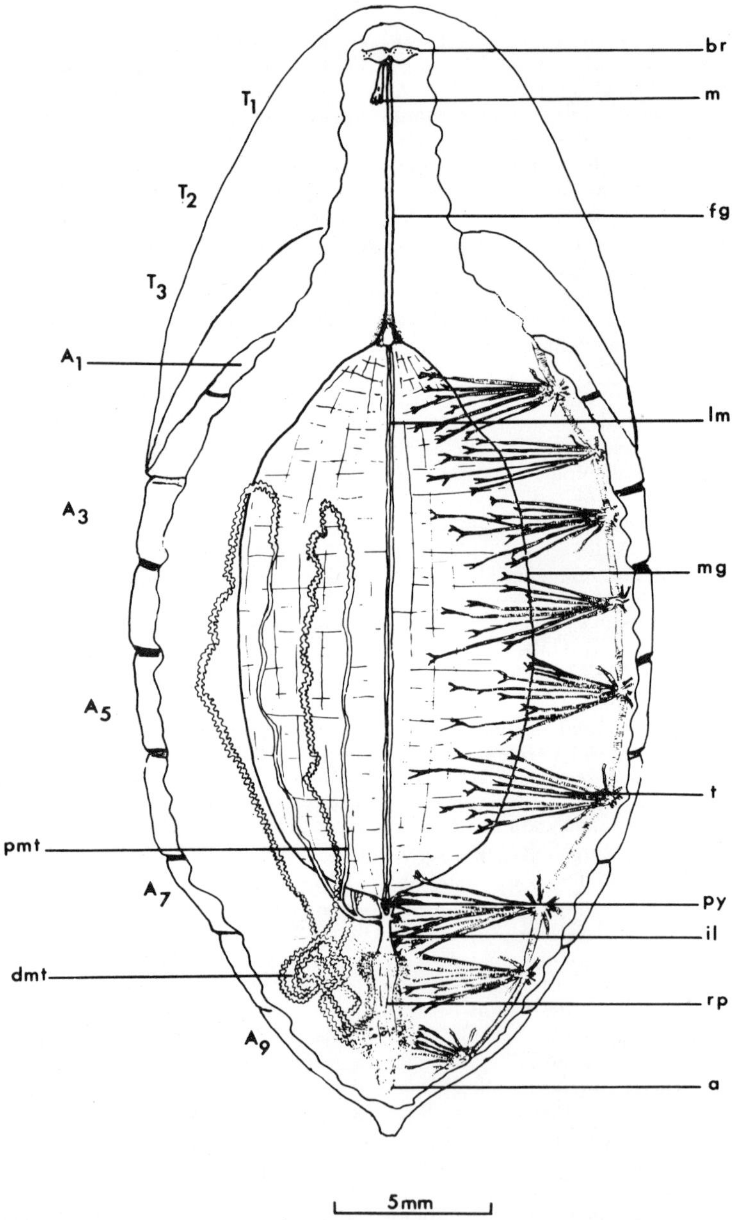


FIG. 4.—*H. cecropia,* pupa. Dorsal dissection of pupal alimentary canal (malpighian tubules are shown for left side only; tracheae are shown for right side only).

tion of the cryptonephric tubules, the malpighian tubules are situated as they are in the larva.

### The Pupa-Adult Transformation

The final stage in the metamorphosis of the alimentary canal (Fig. 5) involves the formation of several new organs, the retention of several pupal features, and the reduction of the midgut. It should be noted at this point that the elaboration of new gut structures in the adult *H. cecropia* moth is not correlated with different feeding habits since the adult moth does not feed.

*Foregut.*—The foregut (fg, Fig. 5) undergoes little change during pharate adult life. For the most part, this portion of the alimentary canal remains reduced throughout the remainder of the life cycle, the only exception being the posterior end of the foregut. By the 4th day of pharate adult life, a distinct swelling is visible in the dorsal portion of the slightly enlarged posterior foregut. The swelling increases during the 1st week and becomes an ovoid, fluid-filled sac. During the remainder of pharate adult life, this sac (the adult crop) enlarges as an uninflated, thin-walled bag overlying the anterior midgut. A wedge-shaped mass of tissue that resembles fat body is usually found adhering to the anterior wall of the developing crop.

At about the time of adult ecdysis the crop (cr, Fig. 5) is inflated with gas and has the appearance of a somewhat elongated, transparent balloon, free in the hemocoel except for its narrow ventral connection to the posterior foregut. The lumen of the crop is continuous with that of the foregut. A weak sphincter segregates the midgut from the foregut, but occasionally a small amount of the dark waste material in the midgut is found in the foregut or crop.

*Midgut.*—Little change occurs in the morphology of the midgut (mg, Fig. 5) during the first 2 weeks of pharate adult life. (In some cases, there is a slight reduction in the diameter of the anterior midgut.) The walls of this chamber are extremely delicate and easily ruptured. During the last week of adult development, the midgut empties its contents into the ileum and becomes reduced in length and diameter. Throughout adult life the midgut persists as a narrow tube ca. 7 mm long and 1.5 mm diam. It may be distinguished from the ileum by its location anterior to the insertion of the malpighian ducts (pmt, Fig. 5) and by the deep circular constrictions in its walls.

*Hindgut.*—The earliest observable change in the alimentary canal associated with adult development is the elongation of the anterior hindgut or adult ileum (il, Fig. 5). This elongation may be detected as early as 2 days after a chilled pupa has been placed at 25°C. The zone of elongation is situated between the point of attachment of the common malpighian tubules and that portion of the hindgut surrounded by diffuse tissue. After 2 days at 25°C, the ileum elongates to a smooth-walled tube ca. 2–4 mm long and by the 5th day it has extended to 10 mm. At this time the posterior hindgut becomes noticeably expanded with a clear fluid to form the rectal pouch. By the 7th day, the rectal pouch is expanded to a diameter of ca. 4 mm and the diffuse tissue surrounding it has become thinly spread. The anterior portion of the rectal pouch forms a diverticulum, the caecum (ca, Fig. 5), which projects forward over the insertion of the ileum. The walls of the rectum proper contain numerous small, round thickenings, the developing rectal pads (rpd, Fig. 5).

By the 10th day of pharate adult life the ileum may

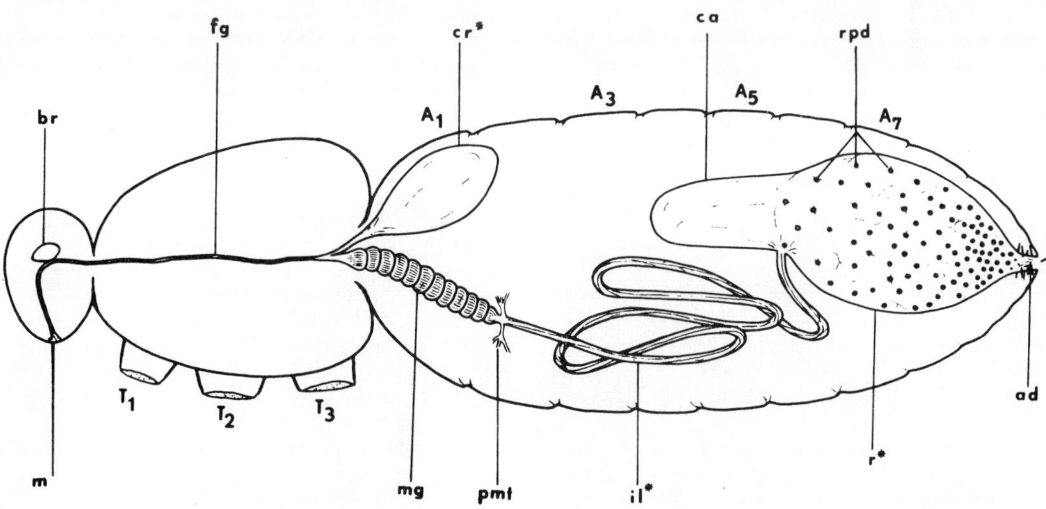


FIG. 5.—*H. cecropia,* adult. Lateral view of the alimentary canal (trachae and malpighian tubules are omitted: rectal pouch is shown distended with meconium).

be 20 mm long and is slightly coiled. It enters the rectal pouch ventrally between the caecum and rectum. The entire rectal pouch is distinctly pear-shaped with the caecum forming the narrower anterior portion and the enlarged rectum proper comprising the posterior portion. About 300 rectal pads dot the walls of the rectum, the majority being in the posterior portion near the anal duct (ad, Fig. 5). Throughout the remainder of adult development, the size of the rectal pouch increases as it becomes gradually filled with excretory fluid.

One week prior to ecdysis, the midgut empties its contents into the rectum as a result of peristaltic contractions of the ileum. The rectal pouch, receiving both the contents of the midgut and the excretory fluid produced by the malpighian tubules, now swells and becomes the largest organ of the alimentary canal. The caecal region projects well forward and often protrudes into the 3rd abdominal segment. The weight of the filled rectum now constitutes as much as 20% of the total body weight.

The ileum reaches its maximum length of ca. 25 mm in the adult moth. It is coiled several times in the center of the body, and where it is not dilated by its contents, the ileal wall is creased by 6 longitudinal grooves. Active peristaltic contractions occur in the ileum from the 2nd week of adult development throughout adult life. The muscularis of the ileum consists of a few longitudinal fibers covered by a thin layer of circular fibers.

The rectum is thin-walled and transparent, so solid particles in the meconium can be seen through the rectal wall. Two thin muscle layers surround the rectum with the circular fibers overlying the longitudinal. Some branching of fibers occurs. In general, the hindgut lacks external muscle attachments, lies free in the hemocoel, and is surrounded by masses of fat body. Fine tracheal branches supply the ileum and rectal pouch. The anal duct is regulated by a sphincter that normally allows the release of the accumulated meconium immediately following adult ecdysis. No morphological changes occur in the alimentary canal during adult life.

## DISCUSSION

*Larval Alimentary Canal.*—The gross morphology of the larval alimentary canal of *H. cecropia* is typical of Lepidoptera in general (Snodgrass 1935). The foregut in the 5th larval stadium is a relatively simple tube designed to function in the ingestion of food with no provision for the trituration of food beyond the mandibles. The crop region seemingly serves to temporarily accumulate food prior to its passage into the midgut. The stomodaeal valve is poorly suited to function as a sphincter and more likely serves to insure the entry of food particles within the peritrophic membrane as suggested by Wigglesworth (1965).

The large size of the midgut is expected both from the nature of the food consumed by the phytophagous larva and the rapid growth that occurs during the final larval stadium. The function of the large middorsal and midventral longitudinal muscle bundles remains obscure, and we cannot explain the close association of the malpighian tubules with the wall of the posterior ⅔ of the midgut.

The hindgut and malpighian tubules described for the *H. cecropia* larva are generally similar to those of other Lepidoptera as shown by Snodgrass (1935). Exceptions noted in this insect include absence of a heavy band of circular muscle at the junction of the mid- and hindgut, presence of a deep circular constriction around the ileum, presence of ovoid protrusions from the ileal wall for muscle attachment, and presence of an elongate colon between the ileum and the rectum.

The cryptonephric condition of the malpighian tubules over the rectum agrees with the descriptions of Ishimori (1924) and Saini (1964) for other Lepidoptera. It is generally accepted (Wigglesworth 1965) that the cryptonephric arrangement is an adaptation for water absorption from the rectal contents. Rectal pads have been described in the larvae of *Heliothis* (Chauthani and Callahan 1967), but they are not present in larvae of *H. cecropia.*

*The Larva-Pupa Transformation.*—The metamorphosis of the gut in *H. cecropia* may be viewed as a distinct 2-step process because of the intervention of a pupal diapause. The larval alimentary canal undergoes an extensive reduction at the time of pupation and maintains this condition through diapause. With the onset of adult development, certain gut regions differentiate to a form quite unlike that of the larval condition.

During the larva-pupa transformation, 2 types of changes occur in the alimentary canal that are related to the embryonic origins of the regions involved. The fore- and hindguts, both of ectodermal origin, empty and collapse to form narrow tubes. The midgut, an organ of endodermal origin, on the other hand, first empties and then fills again with waste material that is retained until adult development is nearly complete. The various regions of the fore- and hindgut that are readily distinguishable in the larva become obscured by reduction in the pupa.

Perhaps the most extensive alteration associated with the alimentary canal during pupation is the deterioration of the mass of cryptonephric tubules. The loss of this system at pupation and the development of the rectal pads in the pharate adult emphasize the divergence of these 2 stages in the life cycle.

In general, the changes from larva to pupa are of a degenerative nature. The alimentary canal apparently dedifferentiates morphologically to a condition that might be termed "a least common denominator" between the larva and adult.

*The Pupa-Adult Transformation.*—With the onset of adult development, a new series of alterations occurs in the alimentary canal. The major portion of the foregut remains reduced and unspecialized in the nonfeeding adult. However, an adult crop is formed from the posterior foregut and presumably fills with air at the time of adult emergence. The function of this region is clearly not digestive, and it has been generally accepted that this chamber inflates so as to distend the body at ecdysis (Wigglesworth 1965). We find this suggestion inadequate, at least in the case of *H. cecropia,* for 3 reasons. First, at ecdysis, the major split in the pupal cuticle from which the adult exits occurs over the thorax, and the adult crop lies chiefly in the anterior region of the abdomen. Second, observation of adults at the time of ecdysis reveals that very little force is required to split the partially digested pupal cuticle, and this force seems to be provided by the alary muscles. Third, numerous dissections of adults at the time of ecdysis repeatedly show the crop to be uninflated. We propose the following as the principal function for the adult crop.

Shortly after adult eclosion, the wings are inflated with hemolymph by periodic, strong contractions of the abdomen. Unless a compensatory mechanism is present, the hemolymph pumped into the wings would be drawn back as soon as the abdomen relaxed. If the return flow of hemolymph is merely prevented by a valve of some sort, then the insect would be forced to remain contracted, since relaxation would tend to produce a vacuum in the hemocoel. The crop may provide a compensatory mechanism for maintaining hemocoel volume. Such a function could be accomplished by the passive inflation of the crop at the time the abdomen is relaxed. At that time, a volume of air may be drawn in through the mouth roughly corresponding to the volume of blood pumped into the wings. In preliminary experiments we found that adults in which the esophagus was cut during the pupal stage were in fact unable to expand their wings. However, further and more conclusive experimentation is required.

The midgut serves no apparent function in the adult stage. Some of the material stored in it during pupal life may be absorbed and reused in adult development, but the amount involved is probably small. Based on its growth and differentiation during adult development, it appears that the hindgut may play a significant role in adult physiology. The formation of the long, narrow adult ileum, the enlargement of the rectal pouch, and the development of the rectal pads are unique to the adult. Although the rectal pads are considered to be organs of water reabsorption, no conclusive experiments support this supposition for the Lepidoptera (Wigglesworth 1965). The observed accumulation of caps of condensed meconium over the rectal pads of *H. cecropia* adults may be an indirect indication that they are functional, at least during adult development. Histological and physiological studies of the rectal pads of *H. cecropia* will be reported in subsequent publications.

## ABBREVIATIONS USED IN FIGURES

A, abdominal body segments
a, anus[4]
ad, anal duct
am, ampulla of common malpighian ducts
bc, buccal cavity
br, brain
ca, caecum
cm, circular muscle
co, colon
cr, crop[5]
dm, dorsal dilator muscle
dmt, distal malpighian tubule
e, esophagus
fg, foregut
gc, gastric caecae
H, head
il, ileum[5]
lm, longitudinal muscle
m, mouth
mg, midgut
mi, muscle insertion
ph, pharynx
pi, posterior interstitial ring[6]
pmt, proximal malpighian tubule
pv, pyloric valve
py, pylorus
r, rectum[5]
rp, rectal pouch
rpd, rectal pads in wall of rectum
sg, subesophageal ganglion
T, thoracic body segments
t, tracheae
tn, tentorium
vm, ventral dilator muscle

[4] In the pupa (Fig. 4) a = anal region since there is no opening to the outside.
[5] In Fig. 5 adult structures having the same name as certain larval regions are probably not formed from those larval tissues.
[6] Situated at the junction of the mid- and hindgut but can be demonstrated only histologically.

## REFERENCES CITED


**Bordas, L.** 1920. Étude anatomique et histologique de l'appareil digestif des lepidopteres adultes. Ann. Sci. Natur. Zool. (10e ser), iii: 175–250.

**Burgess, E.** 1880. Contributions to the anatomy of the milkweed butterfly *Danais archippus* Fabr. Anniv. Mem. Bost. Soc. Natur. Hist. 1–16.

**Chauthani, A. R., and P. S. Callahan.** 1967. Developmental morphology of the alimentary canal of *Heliothis zea* (Lepidoptera Noctuidae). Ann. Entomol. Soc. Amer. 60: 1136–40.

**Drecktrah, H. G., K. L. Knight, and T. A. Brindley.** 1966. Morphological investigations of the internal anatomy of the fifth larval instar of the European corn borer. Iowa State J. Sci. 40(3): 257–86.

**Henson, H.** 1931. The structure and post-embryonic development of *Vanessa urticae* (Lepid.) 1. The larval alimentary canal. Quart. J. Microscop. Sci. 70: 321–60.

**Ishimori, N.** 1924. Distribution of the malpighian vessels in the wall of the rectum of lepidopterous larvae. Ann. Entomol. Soc. Amer. 17: 75–84.

**Larsson, S. G.** 1929. The internal anatomy of the larva of *Pyrausta nubilalis* Hb. Int. Corn Borer Invest. Sci. Rep. 2: 146–59.

**Lyonnet, P.** 1762. Traité anatomique de la chenille, qui ronge le bois de saule. La Haye, Pierre Crosse Jr. and Daniel Pinet, Lib. de S. A. S. a la Haye, p. 616.

**Minot, C. S., and E. Burgess.** 1884. Report on the anatomy of *Aletia xylina* (Lepid., Noctuidiae). 4th Rep. U. S. Entomol. Comm. Washington, D. C., Chap. 5: 45–58.

**Peterson, A.** 1912. The anatomy of the tomato worm larva *Protoparce carolina.* Ann. Entomol. Soc. Amer. 5: 246–69.

**Pyle, R. W.** 1940. Anatomy of the gut of the supposedly nonfeeding adult of *Callosamia promethea* (Lepid., Saturn.). Entomol. News 51: 181–5, 211–5.

**Saini, R. S.** 1964. Histology and physiology of the cryptonephridial system of insects. Trans. Roy. Entomol. Soc. London 116: 347–92.

**Schneiderman, H. A., and C. M. Williams.** 1954. The physiology of insect diapause. IX. The cytochrome oxidase system in relation to the diapause and development of the Cecropia silkworm. Biol. Bull. 106(2): 238–52.

**Snodgrass, R. E.** 1935. Principles of Insect Morphology. McGraw Hill Book Company, Inc., New York. 667 p.

**Standlea, P. P., and T. R. Yonke.** 1968. Clarification of the description of the digestive system of *Heliothis zea*. Ann. Entomol. Soc. Amer. 61: 1478–81.

**Teotia, T. P. S., and M. D. Pathak.** 1957. The anatomy of the larva of *Enarmonia pseudonectis* Meyr. (Eucosmidae: Lepidoptera). Ann. Zool. (India) 2(6): 65–85.

**Wigglesworth, V. B.** 1965. The Principles of Insect Physiology. 6th ed., London: Methuen & Co. Ltd., New York; E. P. Dutton and Co., Inc., New York. 741 p.

# The Flight Mechanics and Evolution of the Wings of Ephemeroptera, with Notes on the Archetype Insect Wing [1]

George R. Edmunds, Jr.
University of Utah
and
Jay R. Traver,
University of Massachusetts

(Communicated by Herbert H. Ross)

The classification of the Ephemeroptera, or of any other group of organisms, should be based upon the phylogeny of the group or, to be more precise, upon a reconstruction of the probable phylogenetic relationship as indicated by all available evidence. This paper presents the results of one phase of an investigation into the problem of establishing a more natural classification of the order Ephemeroptera. The wings of mayflies have been used as a primary evidence of relationship among extant mayflies and have been the principal evidence available for the classification of fossil forms, many of which are known only as wing prints. To utilize this data to the full it is necessary to be exact in determining the homology of veins and other structures. Thus it was considered necessary to re-examine critically the problem of the wing venation and structure of the mayflies and their fossil allies.

As the study progressed it was realized that there was a correlation between wing structures and the flight pattern of the mayflies studied. This led to an extensive study of the mechanics of flight of the Ephemeroptera. As the Ephemeroptera are apparently the most primitive of all winged insects, it was subsequently necessary to enter into some of the basic problems concerning the origin and evolution of the insect wing.

## STRUCTURE AND MECHANICS OF THE MAYFLY WING

The wing[2] of a mayfly (Fig. 1) is somewhat triangular in outline and has a regular series of corrugations or fluting, which gives it a fanlike form. Each longitudinal vein follows along the crest of a ridge or the bottom of a furrow. Veins that follow ridges are called convex (indicated by a + sign in the figures); those that follow the bottom of the furrows are called concave (− sign). At the margin of the wing there is a complete alternation of concave and convex veins. These corrugations, much as in a fluted fan, greatly strengthen the wing and make it quite rigid. The convex veins and their attached crossveins and membranes act as braces to keep the wing rigid on the downstroke, and the concave veins and their attached crossveins and membranes act as braces to maintain rigidity on the

[1] This paper is modified from a thesis submitted to the University of Massachusetts in partial fulfillment for the requirements of the degree doctor of philosophy. Dr. Jay R. Traver collaborated extensively on this section of the thesis. The research was supported by financial assistance from the University of Utah Research Fund.

[2] Reference is always to the forewing, unless otherwise specified.

Reproduced with permission from the *Journal of the Washington Academy of Science,* 44:390–400, 1954. Published by the Washington Academy of Science, Washington, D.C.

upstroke.[3] The anterior margin of the wing is further strengthened by the close grouping of the costa, subcosta, and radius 1, which are held together at the base by the costal brace. This brace is attached to the convex costa and radius, while the concave subcosta lies between and below and is held firmly in place by the floor of a cup-shaped depression (Fig. 2).

As a result of the above-mentioned structure, the mayfly wing would seem to be rather rigid. This is not entirely so, however, for some of the main longitudinal concave veins, namely Sc, $R_{2+3}$, $MP_1$, and sometimes $MP_2$, have weakened spots called bullae (Fig. 3). In *Siphlonurus* and most of the mayflies, the last vein with a bulla is $MP_1$, but in some specimens of *Cinygmula* bullae were observed on $MP_2$. The bullae allow the concave veins to bend, and consequently, during the upstroke, the concave veins fail to support the distal half of the wing. From these facts it can be seen that the corrugations or fluting of the wing and the bending allowed by the bullae serve important flight functions. On the downstroke the wing remains rigid, and the insect gets a maximum amount of lift and propulsion from the stroke (Fig. 4). On the upstroke, however, the distal half of the wing bends downward under the pressure of the opposing air because of the failure of the concave veins, which are weakened by the bullae, to support the wing tip (Fig. 5). The opposing air thus slips away in such a manner as to offer lessened resistance to the upstroke of the wing. The wing tip is the most effective portion of the wing surface because the arc through which the wing travels and the moment of force both become greater as the distance of the wing surface from the point of attachment to the thorax increases.

When the upper and lower surfaces of the wing are separated by soaking in caustic potash it is seen that the costal vein and costal crossveins are equally represented on both surfaces of the wing. All other convex veins are developed almost entirely on the upper surface of the wing, and the concave veins are on the lower surface. Crossveins are found on the upper surface except at the end that attaches to the concave veins where a distinct short section is found on the lower surface (Fig. 6). The veins and crossveins are arranged on the wing surfaces in such a way as to give maximum support during flight. The greatest stress falls on the convex veins and they are noticeably larger than the concave ones. The crossveins are arranged so as to give support on both upstroke and downstroke but the major development of these veins is in that part that attaches to the convex veins.

In the higher winged insects the anterior margin of the wing is quite rigid, and the remainder of the wing membrane is comparatively flexible. These insects fly by means of a sculling action of the wing in which the rigid anterior margin leads the membrane in the direction of the wing stroke. As the wing moves downward, the membrane is inclined upward. This drives the air backward, giving forward thrust, as well as causing a reduced pressure area anteriorly into which the insect is drawn.

In the Ephemeroptera the folding along the line of the bullae (Figs. 1, 5) inhibits sculling action to a noticeable degree. It is probable, therefore, that the mayflies obtain their forward thrust primarily from the downstroke, but even then sculling action is not especially effective.

With no problems of obtaining food, the adults of Ephemeroptera have almost every structure adapted for reproductive activity. The wings appear to be no exception to this. The nuptial flight of most mayflies consists of flying upward and then passively coasting or leisurely flying downward. In most species the lines of flight are mainly vertical. The mechanics of flight thus seems to be well adapted for this particular behavior pattern. The corrugation or fluting of the wings also appears to give stability to the downward sailing motion with the wings motionless. Even though the Ephemeroptera have a very primitive wing type, the wing structure is specialized to the degree that it is closely correlated with the behavior of the nuptial flight.

An intensive study was conducted on the wings of *Siphlonurus*, but the other members of the Siphlonuridae, and representatives of the Ametropodidae, Heptageniidae, Baetidae, Leptophlebiidae, Ephemerellidae, Baetiscidae, Ephemeridae, and others, show a similar structure and

[3] The physical principles involved in strengthening a plane surface by fluting are widely used in engineering. The fact that corrugated steel is much more rigid than flat sheets depends on the same physical principles as does that of the wing of a mayfly. The fact that the support is primarily a one-way support is readily demonstrated by a common steel rule as used by carpenters, *et al.* When held with the concave surface up, such a rule is so rigid that it supports itself when extended to full 6-foot length, but is so flexible as to roll readily into a small circular case when bent in the opposite direction.

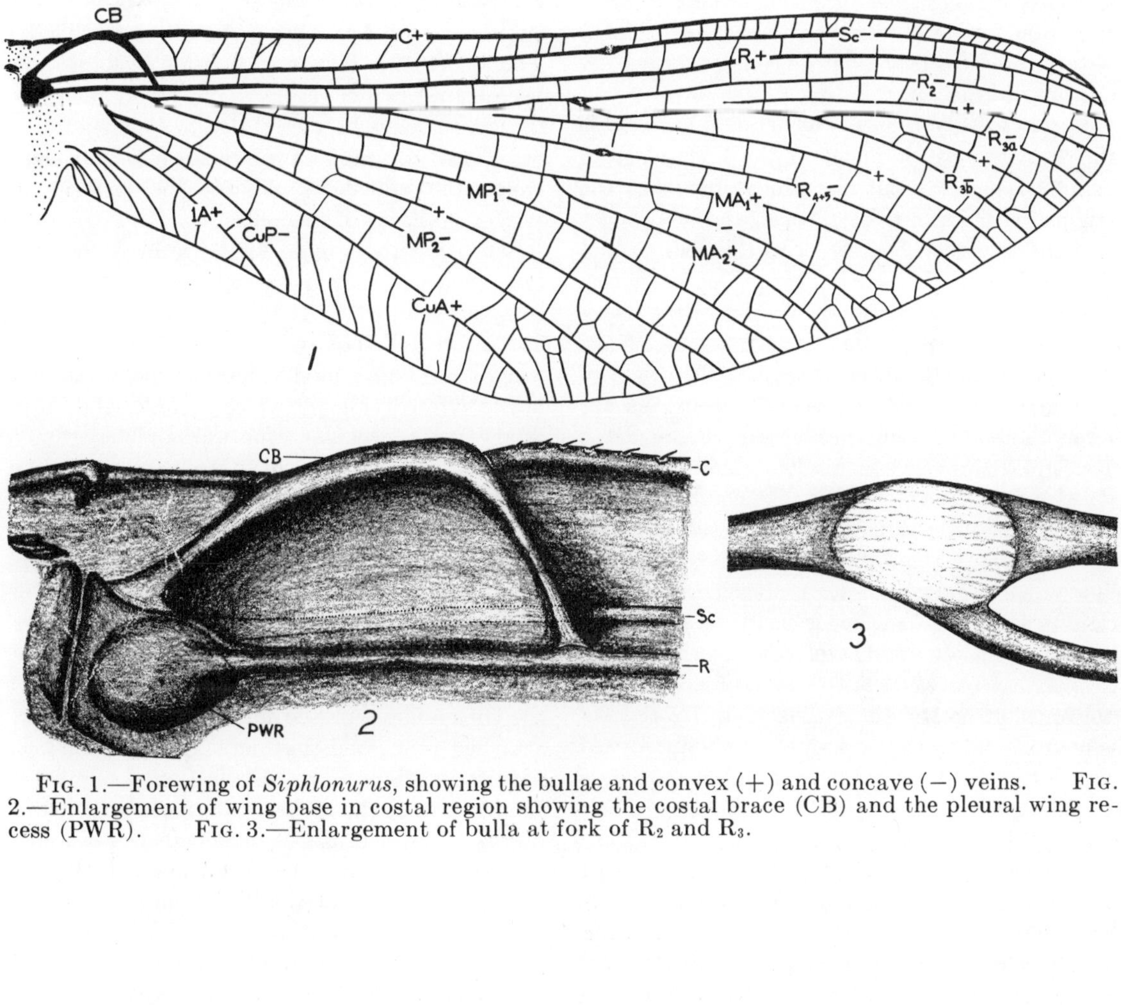


Fig. 1.—Forewing of *Siphlonurus*, showing the bullae and convex (+) and concave (−) veins. Fig. 2.—Enlargement of wing base in costal region showing the costal brace (CB) and the pleural wing recess (PWR). Fig. 3.—Enlargement of bulla at fork of $R_2$ and $R_3$.

presumably have similar flight mechanics. The theories of flight functions gained by a study of the wings were tested extensively by means of paper models which performed in the manner which had been already hypothesized on the basis of the study of the structure of the wings. Many direct observations upon the wing motion of the larger mayflies such as *Hexagenia* and *Ephoron* also corroborated these findings to some extent.

An examination of the wings of mayflies belonging to a number of genera indicated some correlation between the degree of development of the bullae and the wing shape. In the genera with long and narrow wings, the bullae are usually well developed on the anterior four, and sometimes five, longitudinal concave veins. Genera possessing shorter and broader wings usually have fewer bullae, feebly developed (probably vestigial) bullae, or both.

### THE ARCHETYPE WING

The oldest winged insects are found in the

upper Carboniferous, but it is evident from the great variety of wing types in these fossils that the first winged insects are of greater antiquity. Thus, on the basis of evidence available from the study of fossils and development, entomologists have been forced to speculate on the nature of the first insect wings.

Throughout the winged insects one finds evidence of concave and convex veins. The presence of such a condition led Adolph (1879), and later Lameere (1922), to propose nomenclatorial systems based on the concave or convex positions of the veins.

Forbes (1943) discusses in detail the probable stages in the development of the first wings in insects. He points out the probability that the original wing was a thick process, and he believes that the areas between the veins thinned out and the veins became the main strength of the wings. The present study of the mechanics of the wings of Ephemeroptera has led the writers to believe that the thinning of the wings and the origin of fluting in the wings were simultaneous. As the wings thinned out they needed support, and it is doubtful if the veins themselves could have provided the necessary rigidity. The combination of veins and fluting certainly could have provided better support than the veins alone. That the fluted surface would have provided greater rigidity than a flat surface is easily demonstrated by comparing the rigidity of a flat piece of paper to one that has been fluted by folding. The importance of the veins in establishing this rigidity is pointed out by the fact, as has been mentioned above, that the wing tip loses its support because of the desclerotized areas or bullae on a few principal longitudinal veins.

The actual origin of fluting of the wings is lost in antiquity, but some ideas concerning it have been offered. Needham (1935a) would attribute its origin to the gathering together of the basal connections of the veins as the base of the wing narrowed. Forbes (1943: 393) explains it by the articulation and muscle attachment of a few principal veins to the thorax.

However, once fluting actually started, from whatever cause, it would increase the rigidity of the wing. Once a wing membrane was made more rigid by the fluting, it would have been possible for the wing membrane to become thinner without deleterious effect on the flight powers of the insect.

On the basis of a study of the wings of all groups of insects, living and fossil, Comstock and Needham (1898), arrived at a concept of the hypothetical archetype wing venation (see Comstock 1918: 19). After Lameere (1922) pointed out that the media was composed of two elements, an anterior convex and a posterior concave portion, Bradley (1931) presented a modified hypothetical archetype which included MA. This archetype was modified by Snodgrass (1935: 216) to include the two branches to the anterior division of the cubitus, which had been pointed out earlier by Tillyard (1919). The drawing from Snodgrass has been copied herein (Fig. 7),[4] but the branches of the cubitus have been relabeled to conform to the venational terminology of this paper and signs have been added to indicate the concave or convex positions of the veins.

We herewith propose the hypothesis that the first winged insects possessed a wing that was completely fluted, i.e., had a complete alternation of concave and convex veins at the margin of the wing. Therefore, a modified archetype venation is herein offered (Fig. 8) which includes the intercalary veins necessary for a completely fluted wing. Of necessity, such an archetype venation must include the so-called intercalary veins that are typical of many fossil forms and of some modern orders, principally the Ephemeroptera. Comstock (1918) refers to intercalary veins as veins which arose without a basal attachment, and cites the ephemerid wing as an example. The fossil record clearly demonstrates, however, that the intercalaries of the mayflies were originally attached veins, and that detachment at the base has been a subsequent specialization. A number of the features of the archetype venation are highly questionable because of their great variability in primitive groups. There is reason to believe that subcosta was originally an unbranched vein and that the so-called $Sc_1$ is merely a realigned crossvein while $Sc_2$ is the true Sc. It is uncertain as to whether MP should still be considered to be four-branched or if it should be only a 2-branched vein. In what we consider to be the most primitive type of wing as seen in the Syntonopteridae it is only 2-branched, but even the Syntonopteridae have wings that have undergone considerable evolution. Those Palaeodictyoptera that have more than two branches to MP appear to have more specialized wings than do the Syntonopteri-

[4] From *Principles of insect morphology*, by R. E. Snodgrass, courtesy McGraw-Hill Book Co., Inc.

dae. It is, of course, possible that such a wing as illustrated in Fig. 8 could have been the precursor of the Syntonopteridae and those Palaeodictyoptera with more than two branches to MP. The evidence is not sufficient in our estimation to necessitate a relabeling of the branches of MP in the ephemeropteran wing to indicate the possible 4-branched condition in the ancestral stock.

In the foregoing discussion it has been pointed out that fluting actually is a severe detriment to the sculling action of flight. There seems to be little doubt that the sculling type of flight is superior to that of the mayfly type. It is our belief *that fluting of the wing surface was important to the first winged insects for the purpose of strengthening the wing surface and was a necessary requisite before the wing could thin out and form an effective surface for flight. Yet once the wing became thin, fluting actually is detrimental to sculling flight except in the few veins near the anterior margin where rigidity is still required. The trend in the evolution of insects is, in most cases, toward wings which have a rigid anterior margin with a flexible membrane behind.* This type of wing is well adapted for speed in flight, while the fluted wings are essentially lift-type wings. None of the remarks above are intended in the Lamarckian sense, but are meant only to imply that the mechanical fitness of any variation in the wing is the principal selective force determining the general direction of the evolutionary trend. Simpson (1950) has recently remarked that most major trends in evolution are adaptive, and it would appear that this is certainly true in the case of wing structure and venation of insects.

It may well be that the mayflies have maintained functional fluting only because of their unique vertical flight habits, which are dependent upon the bullae.

A number of changes in the venation of the wings of insects seems to bring about flexibility in the wing by destroying functional fluting. A few of these noted during the present study are: (1) the shifting of the position of the longitudinal veins so as to cause them to diverge to the rearward; (2) broadening and shortening of the wing (this often aids process 1, above); (3) migration of concave veins to the vicinity of, and subsequent fusion with, convex veins; (4) elimination or atrophy of intercalary veins; (5) staggering of the path of a vein; (6) desclerotization of veins; and (7) atrophy of crossveins.

The weakening of the wing membrane beyond a certain point is detrimental and those insects which are strong fliers continue to have a well developed system of veins. The problem is most easily explained by the analogous case of fanning air into a fire. If one selects a flat board and waves it much in the fashion of the action of an insect wing, he will find that it is quite ineffective. A small piece of rug which can be held stiff on one margin, however, has a fine sculling action and is much more effective. If for a final test one selects a piece of cloth or paper and holds one edge stiff, the material will prove so pliable as to be of little value in moving air. The wing operates in much the same manner. A wing membrane that is either too rigid or excessively pliant is ineffective; the most effective degree of pliancy is intermediate between the two extremes and undoubtedly varies according to the size and shape of the wing, loading, speed of flight, wing beat frequency, atmospheric pressure, and numerous other factors.

Another evolutionary trend that has been significant in providing an effective sculling surface for insect flight has been the inclusion of the hind wing surface with the forewing by means of coupling devices. Such wing coupling is well known in such insect orders as Lepidoptera and Hymenoptera, but it also occurs in the Ephemeroptera. We have noted such coupling devices in certain Palingeniidae, Polymitarcidae (Campsurinae), Oligoneuriidae, and Siphlonuridae, and a survey of the order may prove wing coupling to be more common than suspected.[5] The nuptial flight pattern of certain Heptageniidae suggests that there is a greater sculling action than the forewing alone would be expected to provide and that the two pairs of wings may be coupled.

Among the smallest insects where the surface area of the insect is large in relation to volume one finds a wide variety of wing types. Many have greatly reduced venation as in the Coccidae, various parasitic Hymenoptera and the tiny Diptera, but even when the wing venation is reduced to an apparent single vein, this is along or near the costal margin of the wing. With reduction of the size of an insect there is an accompanied reduction in the number of wing

[5] Dr. Georges Demoulin has informed us (*in litt.*) that all Palingeniidae, Polymitarcidae (including Campsurinae), Euthyplociidae, and Isonychiidae, which he has examined have the hind wings coupled to the forewings.

veins. Dr. H. H. Ross has informed one of us (personal conversation with G. F. E., Jr. December 1953) that the order in which the wing veins drop out shows a regular parallel development in five separate phyletic lines of caddisflies in which reduction of the size of the insect has taken place. Because of the reduced venation among smaller forms, small species are rarely of value in attempting to arrive at a generalized wing venation pattern of any particular taxon.

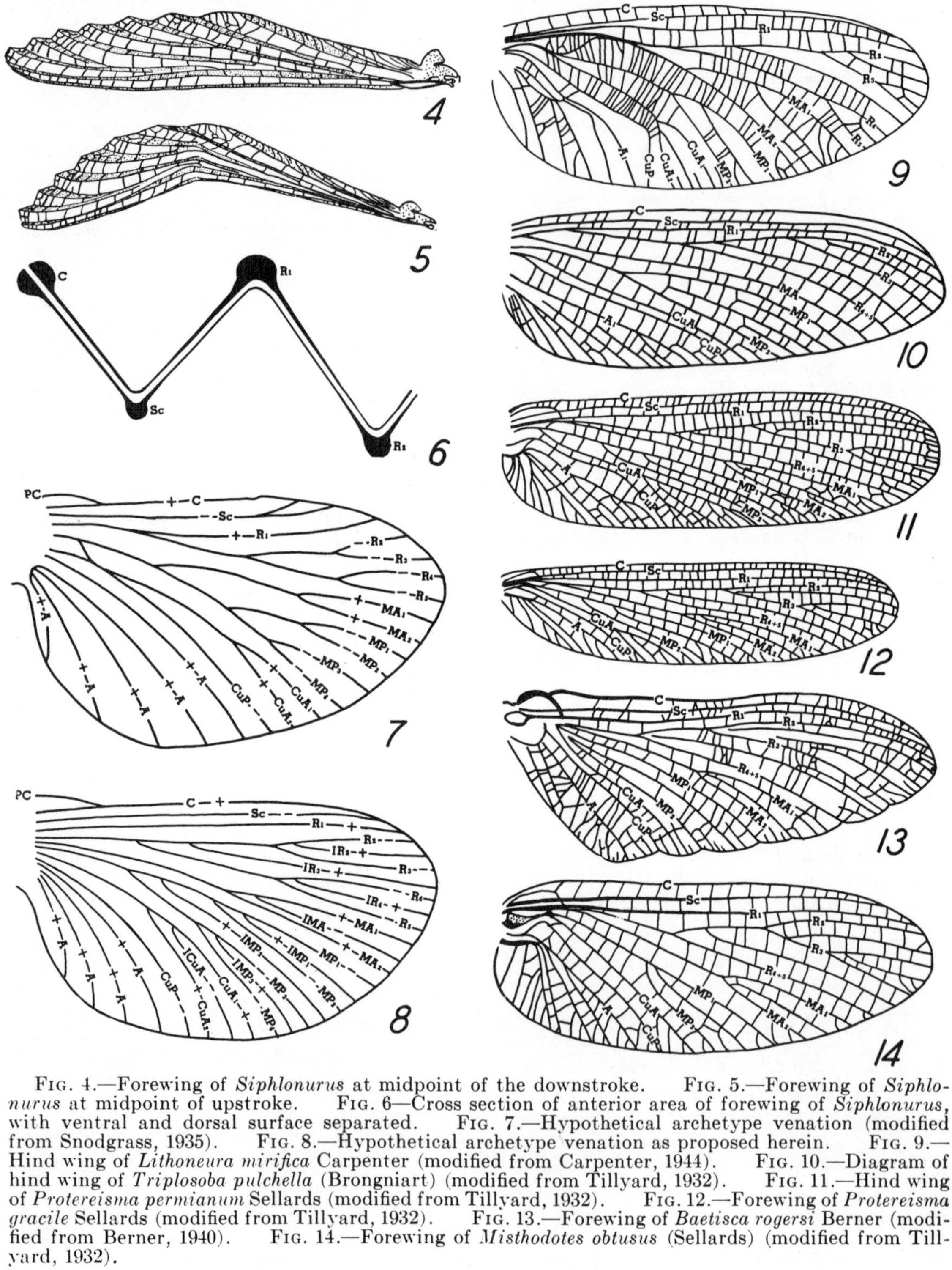


FIG. 4.—Forewing of *Siphlonurus* at midpoint of the downstroke. FIG. 5.—Forewing of *Siphlonurus* at midpoint of upstroke. FIG. 6—Cross section of anterior area of forewing of *Siphlonurus*, with ventral and dorsal surface separated. FIG. 7.—Hypothetical archetype venation (modified from Snodgrass, 1935). FIG. 8.—Hypothetical archetype venation as proposed herein. FIG. 9.—Hind wing of *Lithoneura mirifica* Carpenter (modified from Carpenter, 1944). FIG. 10.—Diagram of hind wing of *Triplosoba pulchella* (Brongniart) (modified from Tillyard, 1932). FIG. 11.—Hind wing of *Protereisma permianum* Sellards (modified from Tillyard, 1932). FIG. 12.—Forewing of *Protereisma gracile* Sellards (modified from Tillyard, 1932). FIG. 13.—Forewing of *Baetisca rogersi* Berner (modified from Berner, 1940). FIG. 14.—Forewing of *Misthodotes obtusus* (Sellards) (modified from Tillyard, 1932).

Among the smallest insects occur also the forms with fringed wings as in the Thysanoptera, Hymenoptera (i.e., Mymaridae), Coleoptera (Orthoperidae or Corylophidae, and Ptiliidae or Trichopterygidae), Diptera (Nymphomyidae), and some of the small Lepidoptera. Even among these fringed-wing insects the density of the fringe on the wing is directly correlated with size, the larger species having the denser fringe of setae. The independent aquisition of the fringed-wing type by several orders of insects suggests that this wing is particularly suited for flight by smaller insects.

Within the Ephemeroptera representatives of several families seem well on the way toward abandoning functional fluting, and some have completely abandoned fluting and present a well developed sculling-type wing. Members of the families Caenidae and Tricorythidae show a parallel development in that the wings of both are extremely broad and short, the cross venation is greatly reduced, bullae are not evident, and fluting is poorly developed. All of the Caenidae have lost the hind wing and it is much reduced or wanting in the Tricorythidae. The flight of both *Tricorythodes* and *Caenis* differs from the usual type in being rapid with a short vertical undulation.

In the Palingeniidae and Behingiidae the essential fluting has been altered in the radiomedial area by the migration of the concave veins to the vicinity of the convex veins. In the Oligoneuriidae there is a series of genera which show various degrees of fusion of the concave with the convex veins. In *Pseudoligoneuria*, whose wing venation is known only from a study of the nymphal wing pads, the venation apparently approximates the usual type although the concave veins are weak. The genera *Oligoneuriopsis* and *Oligoneuriella* have very weak concave veins which lie next to the convex veins. In *Lachlania* the concave veins lie in folds beneath the convex veins, while in *Homoeoneuria* and *Oligoneuria* the fusion is apparently even more complete. This family (except perhaps *Pseudoligoneuria*) no longer exhibits the conventional type of mayfly flight and has developed a sculling type of flight. The veins have no bullae, crossveins are greatly reduced, and the veins behind $R_1$ are divergent from the longitudinal axis of the wing. Also the wing membrane remains pliant because these mayflies fail to shed the subimaginal pellicle from the wings—it is shed only from the body, legs, and tails. The flight of *Lachlania* lacks the typical vertical pattern, and is extremely rapid and powerful. We have observed males of *Lachlania* making progress upwind against a breeze that would drive individuals of most other mayfly genera from the air. Mayflies of this genus have an excellent flying power that clearly demonstrates the superiority of the sculling type of flight for speed and power.

If functional fluting is maintained, as it is in most mayflies, the venational changes, except for modifications in the basal connections or path of a vein, are distinctly limited. Almost all changes in the main venational pattern fall into the categories of either (1) branching of existing veins with the formation of an intercalary vein, or (2) simple fusion of two branches of the same vein with the elimination of the intervening intercalary.

As functional fluting disappears changes in the venation are not as limited, and may include such possibilities as: (1) atrophy of a vein; (2) addition of a vein; and (3) fusion of veins of opposite sign, i.e., a concave and a convex vein. Even when one considers the great number of venational changes that are possible, it seems that many have come about primarily by simple fusion of the branches of a vein. It would cause much less upset in the mechanical action of a wing to reduce a vein by fusion of its branches than it would if the vein were to drop out completely. Students of wing venation should give serious thought to the possibilities of explaining present wing venation by fusion before assuming that a vein is lost.

Except for the veins along the anterior margin of the wing, fluting or alternation of concave and convex veins would seem to be of little use to insects that have a sculling type of flight. Also there is no apparent reason why convex veins should belong to the upper surface and concave veins to the lower surface in such forms. Yet, the fluting of wing veins is found in practically every order of insect and even when fluting is obscure, the separation of the wing surfaces often reveals which veins are concave and which are convex. The logical explanation appears to us to be that the fluted type of wing is the ancestral type and that the convex and concave vein positions are nonfunctional in most modern insects and are to be regarded as vestigial conditions. If this is true, then the concave or convex positions of veins should be valid evidence for establishing vein homologies. The present study, however, points out one possible pitfall in using this evidence alone. In the wing of *Lachlania* where the concave veins are now in the process of fusing with the convex ones, $R_3$ has lost its identity by having its distal half under $IR_3$. Thus, in a fundamentally concave area of the wing, it is evident that the sole surviving vein in the center of the radial sector will be a convex intercalary.

In many groups of insects a careful study of concave and convex vein positions will, when used in conjunction with other evidences, lead to a more clear exposition of wing vein homologies. Positional evidences are the most important single evidence in interpreting vein homologies in the Ephemeroptera; without this evidence determination of the venational homologies of the highly modified wings of the Oligoneuriidae would probably have been impossible. In other groups of insects in which the wings are greatly modified venational positions will probably yield a minimum of evidence. In some insects some of the wing veins are convex at the base and concave distally (or vice versa). In such cases the vein positions may be a result of secondary modifications but in others such evidence may aid in interpreting vein fusions. An example of the latter type is found in the dragonflies, where the concave subcosta has fused with the convex costa beyond the nodus and has imparted a concave nature to the combined costa and subcosta along the leading edge of the wing.

### ORIGIN AND EVOLUTION OF THE EPHEMEROPTERA

If the archetype venation proposed herein is accepted, the most primitive known winged types are found in the family Syntonopteridae of the Palaeodictyoptera. The main difference between syntonopterid venation and that of the archetype is that $R_3$ has formed a triad (although this may have been an archetype feature), and there have been some changes in the basal connections in the hind wing. In the genus *Syntonoptera*, MA is independent of Rs, but in *Lithoneura* (Fig. 9) MA has branched anteriorly near the base to fuse for a short distance with Rs. Carpenter (1938: 450) remarks that "It is very probable that the family Syntonopteridae occupied a position in the Palaeodictyoptera not far from the stock which produced the Plectoptera [= Ephemeroptera]." We would agree with Carpenter although we would note that the wings of the Syntonopteridae have no features which would preclude their inclusion among the Ephemeroptera.[6] The Palaeodictyoptera are characterized by having two cerci, while the mayflies have two cerci *plus* a median filament (although this is greatly reduced in about half the extant families); the number of caudal filaments of the Syntonopteridae is unknown. The point seems minor, but may be fundamental. If we accept the belief that Machiloid-like Thysanura were ancestral to the winged insects, then it seems likely that the first winged insects would agree with their thysanuran ancestors in having three caudal filaments. We suspect not only that the Syntonopteridae might have had three caudal filaments but that possibly they were true Ephemeroptera. We are also inclined to suspect that further discoveries of fossils of primitive winged insects might lead to the conclusion that the Palaeodictoptera themselves represent a specialized offshoot from the Ephemeroptera rather than being ancestral to them.

Although the mayflies are generally considered to be the most primitive of winged insects, most workers have considered the fluting and intercalaries of the mayfly wing to represent specializations rather than a primitive condition. Tillyard (1932: 105), however, is inclined to regard the wing venation of mayflies as the most primitive of all insects, both fossil and recent, but, of course, the wings of the fossil genus *Lithoneura* were unknown to him.

Tillyard (op. cit.: 102) seriously questioned, but did not completely reject, the view of Martynov (1923) that the Dictyoneuridae and the genus *Triplosoba* (Protephemeridea) (Fig. 10) were the ancestors of the mayflies. While *Triplosoba* has the general facies of a mayfly, has three caudal filaments, and has retained the complete alternation of concave and convex longitudinal veins, the simple condition of MA[7] and the backward slanting of all the branches of Rs behind $R_2$ so as to form a somewhat pectinate Rs show specializations toward the sculling type of flight. There are, in our opinion, no members of the modern mayfly fauna that give any indication of having arisen from the Protephemeridea as represented by *Triplosoba*. Tillyard (loc. cit.) suggests that the Prote-

[6] Our judgment of the phylogenetic position of the Syntonopteridae is based primarily upon the fossils of *Lithoneura* rather than upon the fragment of the wing of *Syntonoptera*.

[7] The original figure of the wings of *Triplosoba* by Brongniart (1893) shows MA arising from $R_1$ rather than from MP as in the subsequent figures published by Handlirsch (1908) and Tillyard (1932).

phemeridea represent a side branch not in the direct ancestral line of the true Ephemeroptera, and Needham (1935b: 207) and Martynov (1938) have indicated a similar position for this form. As *Triplosoba* has three tails present, the cerci plus a median filament, a condition shared with the true mayflies of the Permian, it would seem likely that if we are correct in assuming a common ancestry for *Triplosoba* and the Syntonopteridae, the latter may, as we have remarked above, prove to have three tails, when and if specimens are found with these structures preserved. The position of the Dictyoneuridae would seem to be even more remote from the ancestral stock of the mayflies.

Tillyard (op. cit.) has discussed at length the possibilities concerning the relationships of the Permain Protereismatidae and Misthodotidae to modern mayflies. While he draws no definite conclusions as to which family gave rise to Recent forms, he states that the evidence favors the Protereismatidae. He seems concerned about his failure to demonstrate the presence of the CuA triad of Protereismatidae (Figs. 11, 12) in extant mayflies. The CuA of the forewing of most Siphlonuridae (Fig. 1) is so similar to the same vein in the hind wings of some species of *Protereisma* (Fig. 11) that it is difficult to understand Tillyard's concern on this point. Indeed, the CuA triad can be detected in most modern genera of mayflies and is very well developed in some Ephemeroidea.

Tillyard noted that the otherwise primitive wing of *Baetisca* (Fig. 13) shows no sign of a triad on CuA. *Baetisca* and its somewhat aberrant relative, *Prosopistoma,* have long been recognized as being markedly distinct from the remainder of the mayflies. The wings of *Baetisca* are so fundamentally similar to those of Permian family Misthodotidae (Fig. 14), which likewise lacked the cubital triad, that it would seem most logical to account for the distinctness of *Baetisca* and *Prosopistoma* by assuming that they originated from the Misthodotidae or some misthodotid-like group. On the basis of their cubital veins all the remaining mayflies probably arose from the Protereismatidae or some protereismatid-like form, although it is possible that the Palingeniopsidae or some related form gave rise to the burrowing mayflies (Ephemeroidea).

### VENATIONAL NOMENCLATURE OF THE MAYFLIES

At the present time there is some dispute over the nomenclature and homologies of veins of the mayfly wing. The history of the venational nomenclature of the mayflies has been adequately reviewed in several publications (Tillyard, 1932: Spieth, 1933; and Needham, 1935b) so the treatment here will be limited to a discussion of the identity of the vein variously called MA or $R_4$ and $R_5$. In 1923, Tillyard proposed a system of nomenclature based on studies of fossil forms, and adults and nymphal wing pad development of some primitive modern genera. At this time he had not yet seen the paper by Lameere (1922) in which it was recognized for the first time that the media was composed of two parts, an anterior convex and a posterior concave element. In 1926, Tillyard modified his earlier system by recognizing the median anterior vein (MA) of Lameere. He again modified the system slightly in 1932 to make the venational symbols coincide with the Comstock-Needham system.

Needham (1935b) concludes that MA is not represented in the mayfly wing. He therefore proposed a nomenclature which is somewhat similar to Tillyard's (1923), but he made several small, yet significant, changes. Forbes (1943) reviewed the problem again and accepted the Needham nomenclature except that he refers to the $Cu_2$ of Needham as the plical vein.

At the present time both the Needham system of 1935 and the Tillyard nomenclature of 1932 are in common usage. There is but a single major point of disagreement between the two aforementioned systems, namely: Is the convex vein which lies between the concave radial sector and the concave media part of the radial sector or is it media anterior (MA)?

From the previous discussion it is obvious that the Tillyard system of 1932 has been accepted in this paper. His interpretations appear to be correct and are given support by two new evidences: (1) the discovery and description of the two fossil species of Syntonopteridae, *Lithoneura mirifica* and *L. lameeri,* and clarification of the position of *Syntonoptera* by Carpenter (1938, 1944); and (2) the present work on the mechanics of flight of mayflies.

The wings of the family Syntonopteridae are remarkably primitive and very similar to the archetype venation herein proposed. Radial sector is five-branched as the result of the formation of a triad on $R_3$. In the genus *Syntonoptera* MA is entirely free from Rs although it does arch toward it basally. In the wing of *Lithoneura mirifica* Carpenter (described from Mazon Creek, Illinois, Upper Carboniferous formation)

*MA branches anteriorly, fuses for a short distance with the base of Rs and then continues as a convex vein.*

In comparing members of the pre-mayfly family Syntonopteridae with the Permián mayflies of the family Protereismatidae (figs. 11 and 12) only a few differences are evident. In the latter a few veins have different courses, the forks are deeper, and $R_4$ and $R_5$ are fused. The main point of interest here is that MA is *still attached to MP basally but arches forward as in Lithoneura* to fuse with Rs for a short distance before continuing on as a convex vein. Tillyard (1932) pointed out the evidence of this fusion in *Protereisma*, but it is much more pronounced in *Lithoneura*.

The essential difference between the venation of the Permian *Protereisma* and the modern Siphlonuridae is that the basal connection of MA to MP has been lost and MA appears to arise from Rs. Actually, in the Recent genera studied in the present work, the bases of Rs and MA are independent and merely abut against one another.

The fact that the concave or convex positions of the vein serves an important function in flight most certainly would seem to preclude the possibility of complete loss of MA and the substitution in its place of parts of Rs which would apparently have had to change from concave to convex. The function of the wing would, however, be in no way impaired by a shifting of the base of MA so that it is associated with Rs rather than MP. There thus seems to be wholly adequate evidence to show that the vein in question was originally part of media and not a part of the radial sector.

The primary evidence for the belief that the vein in question is part of Rs rather than MA has stemmed from the fact that the vein is supplied by a trachea that arises from Rs. The tracheal connections in the developing mayfly wings vary considerably as has been demonstrated by Morgan (1912) and Tillyard (1923), yet the adult wing venation illustrates considerable conservatism. The relative stability of the venation has been imposed by the mechanical needs of the wing, each variation that interfered with the mechanics of flight being eliminated. While the tracheation of the mayfly wing apparently has played an important part in the phylogenetic development of venation, it most certainly does not have enough stability in nymphal mayflies to justify its use in maintaining a venational nomenclature contrary to the evidence from the fossil record. As Needham (1951) has so aptly pointed out, tracheal evidences are valuable only when used judiciously and with caution. It is possible that the small callus or sclerotized plate in the developing nymphal wing pad of Ephemeroptera may have played a role in uniting MA to the base of Rs, but this reasoning apparently would not apply to the other insect groups where MA has fused with Rs.

ACKNOWLEDGMENTS

We are deeply indebted to several persons who have aided us during the present study. Dr. C. P. Alexander (University of Massachusetts) made many helpful suggestions during the investigations and writing of this work. Dr. F. M. Carpenter (Harvard University) and Dr. H. T. Spieth (University of California, Riverside) critically reviewed early drafts of the manuscript but did not necessarily agree with all parts of the paper. Dr. Herbert H. Ross (Illinois Natural History Survey) and Dr. Georges Demoulin (Institut Royal des Sciences Naturelles de Belgique) have read the final manuscript and have made a number of valuable suggestions. Dr. Dan McLachlan, Jr. (University of Utah), has aided in the interpretation of the physics of flight mechanics. To these and many others who, through their constructive criticism, have aided this investigation, we offer our warmest thanks.

## LITERATURE CITED

Adolph, G. E. *Ueber Insektenflügel.* Nova Acta Leop.-Carol. deutschen Akad. Naturf. **41:** 215–291, pls. 27–32. 1879.

Berner, L. Baetisca rogersi, *a new mayfly from northern Florida.* Can. Ent. **72:** 156–160, 5 figs. 1940.

Bradley, J. C. *A laboratory guide to the study of the wings of insects:* 41 pp. Ithaca, N. Y., 1931.

Brongniart, C. *Recherches pour servir à l'histoire des insectes fossiles des temps primaires précédées d'une étude sur la nervation des ailes insectes.* Industrie Minérale, ser. 3, **7:** 1893.

Carpenter, F. M. *Two Carboniferous insects from the vicinity of Mazon Creek, Illinois.* Amer. Journ. Sci. **36:** 445–452, 3 figs. 1938.

———. *Carboniferous insects from the vicinity of Mazon Creek, Illinois.* Illinois State Mus. Sci. Papers 3: 20 pp., 5 figs, 4 pls. 1944.

Comstock, J. H. *The wings of insects.* Ithaca, N. Y., 1918.

Comstock, J. H., and J. G. Needham. *The wings of insects.* Amer. Nat. **33:** 117–126. 1898.

Forbes, W. T. M. *The origin of wings and venational types in insects.* Amer. Midl. Nat. **29:** 381–405, 10 figs. 1943.

HANDLIRSCH, A. *Die fossilen Insekten und die Phylogenie der rezenten Formen.* Leipzig, 1908.

LAMEERE, A. *Étude sur l'évolution des éphémères.* Bull. Soc. Zool. France **42**: 41–59, 61–81. 1917.

———. *Sur le nervation alaire des insectes.* Bull. Classe Sci. Acad. Roy. Belgique, sér. 5, **8**: 38–149. 1922.

MARTYNOV, A. B. *The interpretation of the wing venation and tracheation of the Odonata and Agnatha.* Rev. Russe Ent. **18**: 145–174, 3 figs., 1 pl. 1923. (Translation in Psyche **37**: 245–281. 1931.)

———. *Études sur l'histoire géologique et de phylogénie des ordres des insectes ptérygotes.* Trav. Inst. Paleontol., Acad. Sci. U.R.S.S., **7**(4): 1–150, 1 pl., 70 figs. 1938.

MORGAN, A. H. *Homologies in the wing veins of mayflies.* Ann. Ent. Soc. Amer. **5**: 89–106, pls. 5–9. 1912.

NEEDHAM, J. G. *Some basic principles of insect wing venation.* Journ. New York Ent. Soc. **43**: 113–129. 1935a.

———. *In* Needham, Traver, and Hsu. *The biology of mayflies.* Ithaca, N. Y., 1935b.

———. *Prodrome for manual of the dragonflies of North America, with extended comments on wing venation systems.* Trans. Amer. Ent. Soc. **77**: 21–62, 7 text figs., 5 pl. 1951.

SIMPSON, G. G. *The meaning of evolution.* New Haven, 1950.

SNODGRASS, R. E. *Principles of insect morphology.* New York and London, 1935.

SPIETH, H. T. *The phylogeny of some mayfly genera.* Journ. New York Ent. Soc. **41**: 55–86, 327–390, pls. 16–24. 1933.

TILLYARD, R. J. *The panorpoid complex. Part 3: The wing venation.* Proc. Linn. Soc. New South Wales **44**: 533–718, pls. 31–35, figs. 35–112. 1919.

———. *The wing venation of the order Plectoptera or mayflies.* Journ. Linn. Soc. London, Zool., **35**: 143–162, text figs. 1–10. 1923.

———. *The insects of Australia and New Zealand.* Sydney, 1926.

———. *Kansas Permian insects, Pt. 15. The order Plectoptera.* Amer. Journ. Sci. **23**: 97–134, 237–272, 22 text figs. 1932.

# Part III: The Physiology of Insects

**Physiology** is the study of the physical and chemical processes involved in the (1) transformation and utilization of energy, (2) interactions involved in the maintenance of the functional integrity and changes in development of tissues and organ systems, (3) duplication of essential chemicals, components, and organ systems, (4) reproduction of other individuals of the species, and (5) mechanisms involved in the interactions with the abiotic and biotic environment.

The metamorphosis (major change in form) of insects has great influence on their survival, evolution, and success. Yet, after many years of study, scientists still disagree as to the significance and the origins of the stages of metamorphic development. In *The Origin and Function of the Pupal Stage,* H. E. Hinton suggests an alternative theory to the two major theories on the origin and function of the pupa, i.e., those of Berlese and Poyarkoff. This theory results from comparative analysis of musculature and of premises and assumptions made by the previous two theories, but in a phylogenetic context. The other major aspects of metamorphosis are the biochemical processes occurring during these changes. *Control of Growth and Development in Insects* by H. A. Schneiderman and L. I. Gilbert reviews the hormones, such as brain hormone, ecdysone (the molting hormone), and juvenile hormone (which maintains larval development), and their significance as regulatory chemicals in the complex and interrelated processes known as molting and growth.

Growth in insects, as discussed by Schneiderman and Gilbert, is occasionally arrested. This *condition of arrested growth and development* is called **diapause.** K. Hasegawa and O. Yamashita *(Studies on the Mode of Action of the Diapause Hormone in the Silkworm, Bombyx mori L. VI. The Target Organ of the Diapause Hormone)* discuss the hormone (diapause hormone) secreted from a part of the nervous system, the suboesophageal ganglion, and its role in carbohydrate metabolism in pupal ovaries. As compared to the previous articles, this study provides detailed experimental results of the type extensively performed in physiology. Microsurgery (e.g., gonadectomy) and the process of extraction and substitution (e.g., extraction of suspected hormone sources followed by hormone injection or the transplantation of tissues and organs) are some of the techniques used by Hasegawa and Yamashita to determine the function of diapause hormone. These are common techniques in insect physiology.

As indicated in the definition of physiology, one of the aspects of physiology is the interaction between insect and environment. One of the best illustrations of this is *Defensive Secretions of Arthropods* by T. Eisner and J. Meinwald. This article, presented in Part V of this volume, points out all major aspects relating to arthropod secretion, its occurrence throughout the phylum, the morphology of the organs of secretion, the biochemistry of the defensive substances, and their efficiency in use.

S. D. Beck's *Physiology and Ecology of Photoperiodism* discusses the role of light in the survival and activity of insects. As in the article by Hasegawa and Yamashita, Beck discusses diapause, but at a more general level, i.e., the relationship between photoperiodism, diapause, and hormonal activities.

# A. Development

## The Origin and Function of the Pupal Stage

By H. E. Hinton

(*Department of Zoology, University of Bristol*)

### Introduction

The origin and functional significance of the pupal stage has been the subject of controversy almost since man first began to be interested in insects. By the second decade of this century the various views that had been expressed in the past were crystallised into two sharply opposed and largely irreconcilable theories, each of which has a considerable following. The first of these theories is known as the Berlese, or Berlese-Imms, or Berlese-Jeschikov theory. The second is known as the Poyarkoff or Poyarkoff-Hinton theory. The two theories are also concerned with the significance of the differences between the larval stages of hemi- and holometabolous insects and with homologies between the various stages of the two groups.

The Poyarkoff theory for the first time provided a very precise causal explanation of the functional necessity for the pupal stage. According to this theory, the pupa originated as a kind of mould required for the development of certain adult skeletal muscles. This view was accepted by many, *e.g.* Hinton (1948), Snodgrass (1954), Lower (1954), and DuPorte (1958), although what was claimed had not been demonstrated in detail with reference to any particular species. In order to do this, a study was made of the metamorphosis of the skeletal muscles of several species. The result of this work was unexpected; it showed that the theory was untenable.

### The Berlese Theory

According to this theory the eggs of the Hemimetabola (Exopterygota) are relatively large and contain greater reserves of food than the relatively small eggs of the Holometabola (Endopterygota). It is claimed that as a consequence of this difference, a difference that has been assumed but never demonstrated, the larvae of the Hemimetabola hatch in a post-oligopod stage of development, whereas the larvae of the Holometabola hatch in a polypod or oligopod stage of development. The larvae of the Holometabola are regarded as precocious embryos, and it is claimed that in the Hemimetabola the ontogenetic stages equivalent to the larvae of the Holometabola are passed within the egg. To support these views, holometabolous larvae, such as caterpillars, are cited as examples of larvae in a polypod stage of development. That is, the abdominal prolegs are either implicitly or explicitly considered to be homodynamous with the thoracic legs. The pupal stage of the Holometabola is considered to be the ontogenetic counterpart of the nymphal (larval) stage of the Hemimetabola in which several instars have been omitted so that there is now only one. In other words, it is claimed that the single pupal instar of recent Holometabola is the product of tachygenesis of the nymphal instars of the hemimetabolous ancestor of the Holometabola.

As early as 1651 William Harvey had stated that it was the lack of food reserves in insect eggs that caused what he called their premature birth, and the Berlese theory might be described as an elaboration of the thesis of Lubbock (1874, p. 80), "That the occurrence of metamorphosis arises from the immaturity of the condition in which some animals quit the eggs." The persistence of the belief that the eggs of the Hemimetabola are larger or have richer supplies of yolk than those of the Holometabola is most surprising in view of the number of times it has been shown that there is no evidence for it, *e.g.* Swammerdam (1737), Miall (1895), and Hinton (1955).

---

Reproduced with permission from the *Proceedings of the Royal Entomological Society of London*, 38:77–85, 1963. Published by the Royal Entomological Society of London, England.

Upon differences in the development of the segmental appendages during embryonic growth, it is possible to distinguish three successive phases through which insects normally pass. These are not sharply differentiated, and the embryo gradually passes from one to another as segmentation proceeds from the anterior to the posterior end. The appearance of the abdominal appendages marks the end of the

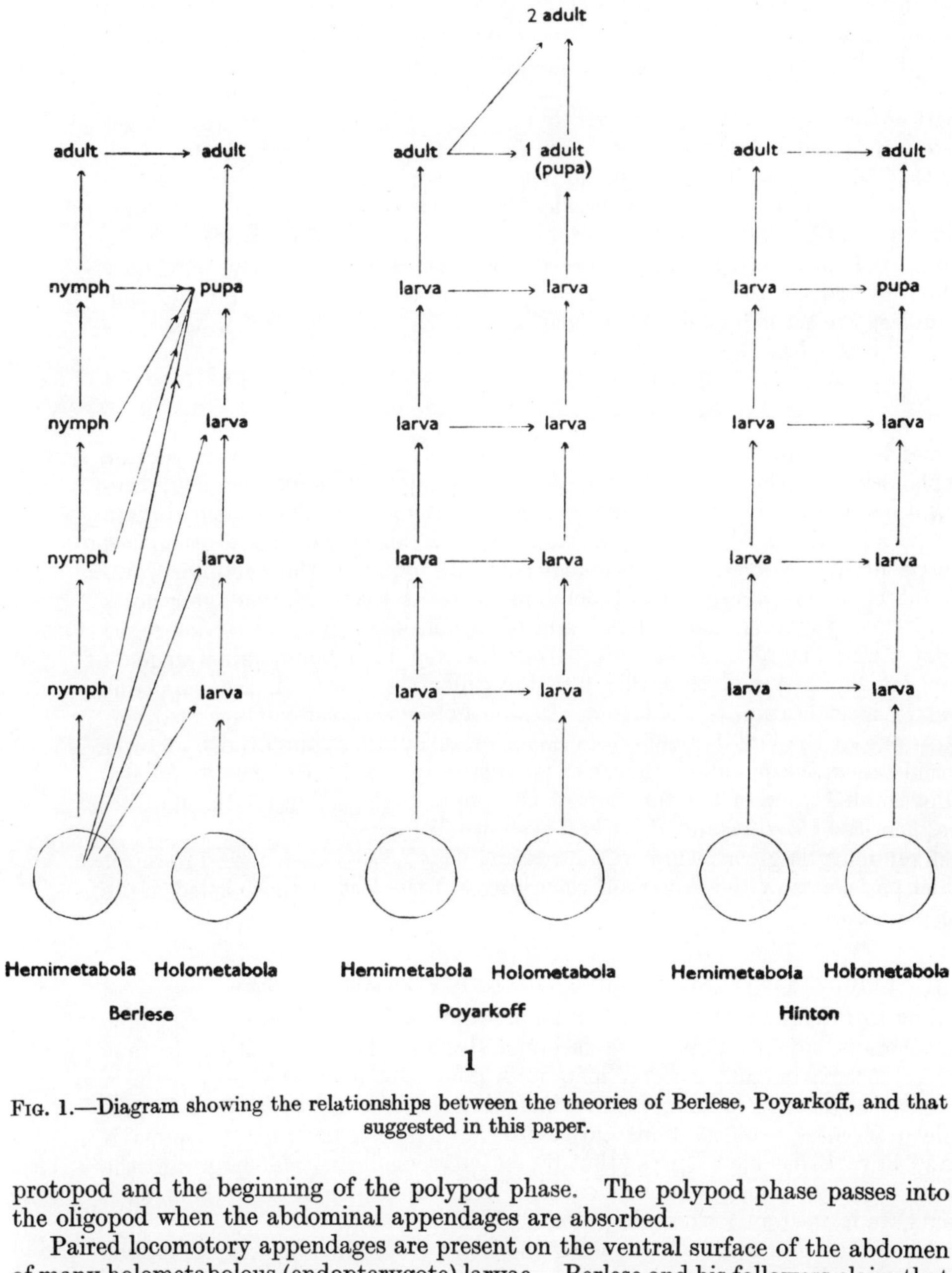


1

FIG. 1.—Diagram showing the relationships between the theories of Berlese, Poyarkoff, and that suggested in this paper.

protopod and the beginning of the polypod phase. The polypod phase passes into the oligopod when the abdominal appendages are absorbed.

Paired locomotory appendages are present on the ventral surface of the abdomen of many holometabolous (endopterygote) larvae. Berlese and his followers claim that larvae with abdominal prolegs like those of the Hymenoptera-Symphyta, Lepidoptera, and Mecoptera have emerged from the egg in the polypod phase, whereas beetle larvae and other endopterygotes that lack prolegs have emerged in a later embryonic phase, the oligopod. Apodous larvae, such as those of many Coleoptera, Hymenoptera, and Diptera, are said to be derived from the oligopod phase by reduction and loss of the thoracic appendages. The theory therefore postulates two primary categories of larvae: (1) the pre-oligopod or oligopod, and (2) the post-oligopod (nymph). It is also claimed that in the first category of larvae, the pre-oligopod or oligopod, there are two subsidiary but nevertheless very distinct kinds of larvae: (a) larvae differing according to the degree of embryonic development attained before hatching, and (b) larvae differing from the oligopod owing to adaptive modifications in the direction of simplicity, *e.g.* apodous larvae.

It seems evident that the assumption that in one form or another underlies the greater part of the Berlese (1913) theory is that the prolegs or locomotory appendages of the abdomen of many endopterygote larvae are homodynamous with thoracic legs. The view that these appendages are homodynamous with thoracic legs and are not secondary has been upheld by many besides the adherents of the Berlese theory, *e.g.* Snodgrass (1935). This view has been criticised by me (Hinton, 1955), and the reader is referred to that paper for a detailed discussion. More recently, Snodgrass (1957, 1961) altered his earlier views on this question and came to the conclusion that the prolegs are secondary or provisional larval organs that are not in the same series as the thoracic legs.

### The Poyarkoff Theory

The chief features of this theory are as follows: (1) it advances a very specific causal explanation for the origin and function of the pupal stage of the endopterygotes; (2) it advances the same explanation for the subimago of the Ephemeroptera and the "pupal" stages of some other exopterygotes, all of which are recognised to be independently evolved; (3) it homologises the pupa of the endopterygotes with the adult of the exopterygotes and therefore maintains that in the endopterygotes there are in effect two imaginal instars; and (4) it considers that the larvae of the exopterygotes and endopterygotes hatch from the egg in a comparable stage of embryonic development and that therefore the differences between them are due to the greater specialisation of the latter. It seems clear that the validity or otherwise of any one of these statements does not necessarily affect the others. Lower (1954), for instance, accepts all of the chief postulates of the Berlese theory and also accepts Poyarkoff's view of the function of the pupal stage. The relationship between the theories of Berlese and Poyarkoff is shown in figure 1.

A brief summary is given below of the reasons why I have been forced to reject all the chief postulates of the Poyarkoff theory except the last of those listed in the preceding paragraph.

### Functional Significance of the Endopterygote Pupa

According to Poyarkoff (1914*a*: 48), "La nymphe a la signification biologique d'une moule dans laquelle doivent se faconner des muscles imaginaux; la mue nymphale est nécessaire aux insectes métaboles pour leur permettre de prende la forme que doit avoir cette moule; la mue imaginale leur est nécessaire pour la production de différentes néoformations cuticulaires." The French term for pupa is nymph, and Poyarkoff makes it clear that by cuticular neoformations he means the attachments of the muscles to the new cuticle. According to him and others, such attachments are formed only when the cuticle is being secreted.

Thus, according to Poyarkoff, in order to develop certain adult muscles—the indirect flight muscles—a mould is required of the same shape as the adult with its cuticular parts having approximately the same spatial relations as those of the adult. This mould is the pupa. Since the secretion of the mould necessarily precedes the development of the adult muscles, the muscles cannot be attached to the mould. Once the necessary muscles have developed in the mould, they must be attached to the cuticle before they can function, and for this there must be another moult, the pupal-adult moult, and a new cuticle must be secreted.

The most important assumption of the Poyarkoff theory is undermined by the fact that even in some of the most highly advanced endopterygotes, *e.g.* nematocerous Diptera, the indirect flight muscles are present as fibre rudiments at the time that the larva hatches from the egg (Hinton, 1959*a*). These muscles subsequently develop in the larvae in precisely the same way as do both the direct and indirect flight muscles in the larvae of many primitive exopterygotes, *e.g.* Odonata, Dictyoptera,

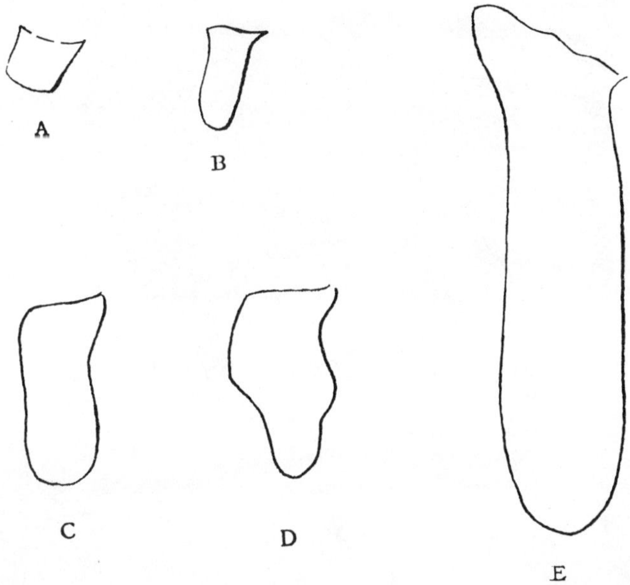


**2**

FIG. 2.—Front wing of pupa of *Myrmeleon europaeus* McLach.: (*A*) pharate pupa shortly after the larval-pupal moult; (*B–D*) pharate pupa 2, 6, and 14 days later, respectively; (*E*) pupa 1 day after the larval-pupal ecdysis (based on Sundermeier).

and Orthoptera (Hinton, 1959*b*). The development of both the direct and indirect flight muscles from fibre rudiments (discs) is widespread not only amongst primitive exopterygotes but also amongst specialised ones, *e.g.* Hemiptera. It therefore seems reasonable to regard this method of growth as the primitive method of growth of the flight muscles of insects. That is, departures from this way of growing, and there are a number both in exopterygotes and endopterygotes (summary in Hinton, 1959*b*), can be considered to be secondary specialisations.

Another important assumption of the Poyarkoff theory, namely, that tonofibrillae can only be formed at the time of the moult, now appears to be also incorrect. The myofibrils terminate at extensive Z bands of dense filaments. The opposed plasma membranes of muscle and epidermal cells are thickened and vesicular. Between them is material like the basement membrane. It seems probable that the cuticular fibres formed within the epidermal cells, *i.e.* the tonofibrillae, become firmly cemented to the basement membrane when the borders of the epidermal cells break down opposite the point of attachment of the muscle (Stuart and Edwards, 1958). In many insects the epidermal cells continue to secrete endocuticle long after the moult, and some are known to do so throughout the life of the instar. Tanning of the outer part of the endocuticle to form exocuticle is also known to continue in some insects long after the moult. Cuticularisation of the epidermal cells to form tonofibrillae may also continue for an appreciable period after the pupal-adult moult, *e.g.* in *Simulium* (Hinton, 1961).

Because of the space occupied in the larval thorax by muscles and other organ systems, there is insufficient room for the wing anlage to develop to the size required for the wings to be effective even allowing some growth after evagination. One moult only is required to evaginate the wings. At this first moult there is considerable

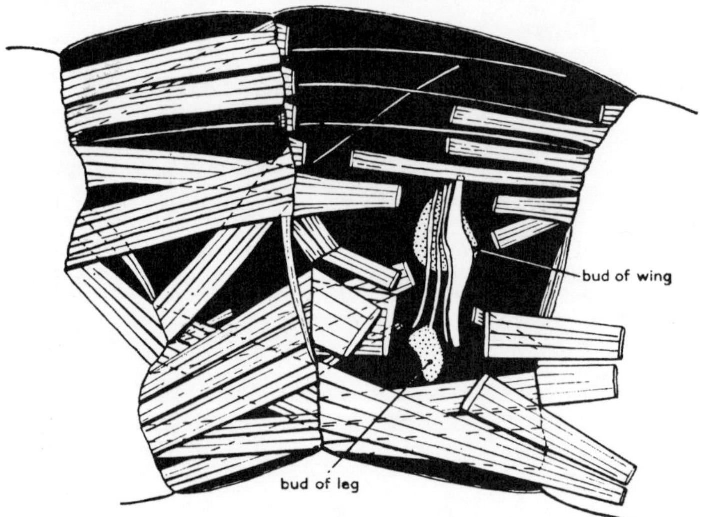


3

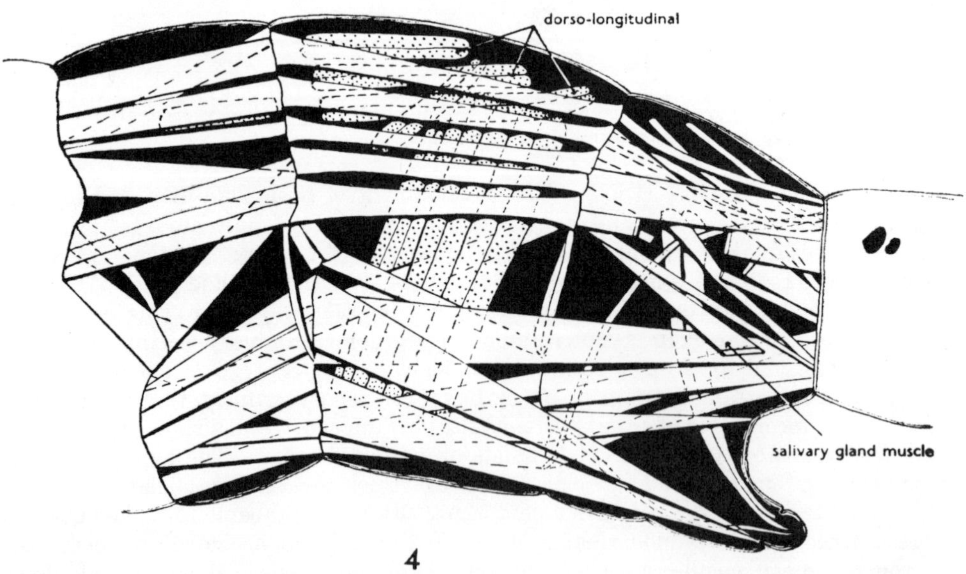


4

Figs. 3–4.—Meso- and metathorax of *Simulium ornatum* Meig. (3) Penultimate larval instar; the fibre rudiments of the indirect flight muscles and the tergal depressor of the trochanter are unshaded. (4) Pharate pupa; the fibre rudiments of the indirect flight muscles and the tergal depressor of the trochanter are stippled; most of the fibre rudiments of the imaginal muscles are now divided longitudinally.

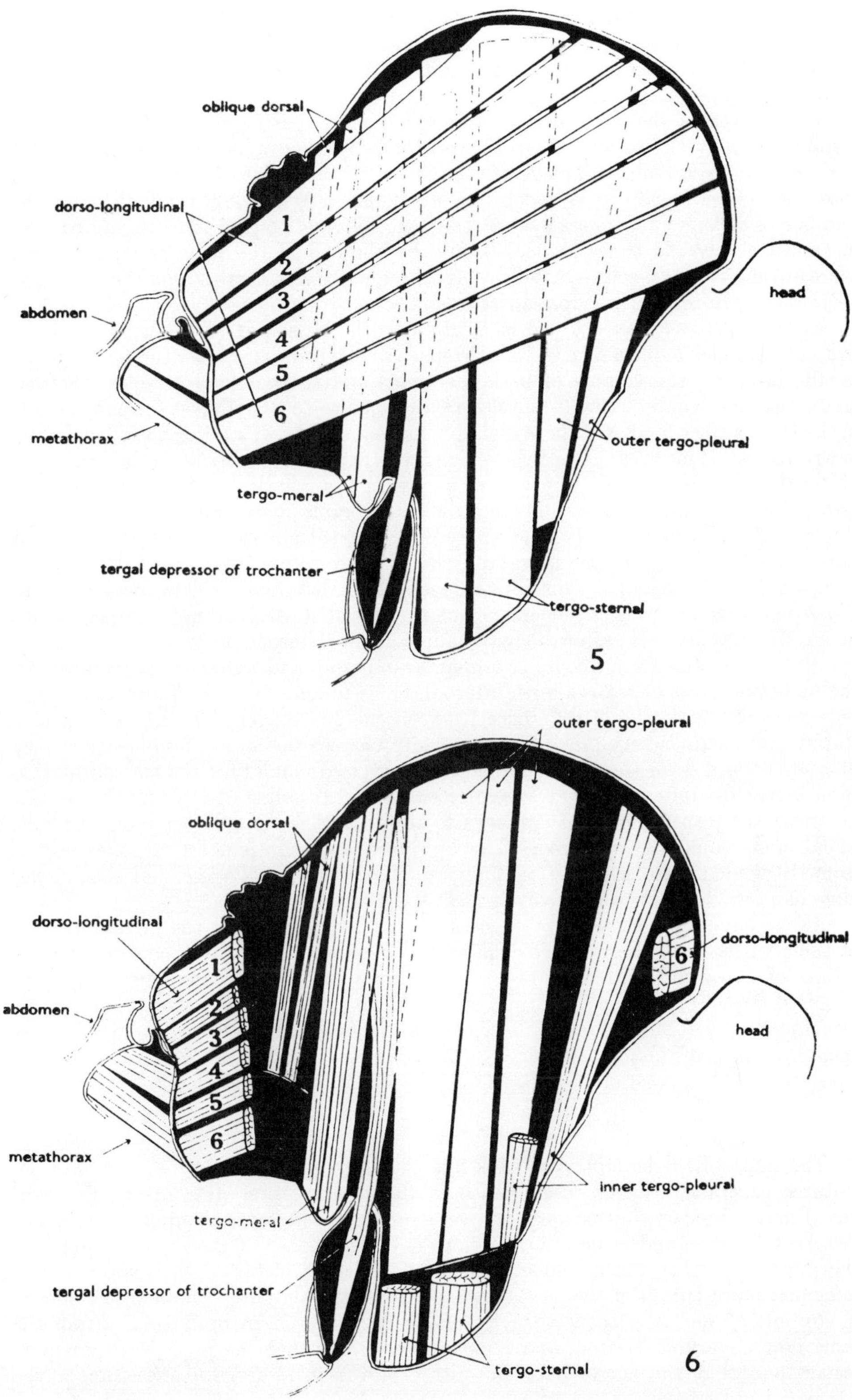


FIGS. 5–6.—Meso- and metathorax of the adult of *Simulium ornatum* Meig.

growth of the epidermis of the wings, and large pupal wing-cases are formed (fig. 2). A second moult is then required, during which the epidermis is freed from the pupal cuticle and the number of its cells increases greatly with the result that the definitive adult epidermis is often much folded within the pupal cuticle before secretion of the adult wing cuticle is completed.

The growth of the wing anlage after the larval-pupal moult is enormous. For instance, Pohley (1956) has shown that at the beginning of the last larval instar of *Ephestia kuhniella* Zell. the anlage of the hind wing has about 400 cells, not counting those of the epithelial sac in which it is contained. At the larval-pupal moult the number of cells increases greatly: at the time of wing evagination the number has increased nearly 100 times to 38,000, and by the end of the pharate pupal period there are about 70,000 cells. After the larval-pupal ecdysis, there is another prolonged outburst of mitosis at the pupal-adult moult.

In the larva, as may be seen in figure 3, both the anlage of the wings and the imaginal muscles require relatively little space in the thorax. After the larval-pupal moult, however, the volume of both the wings and the imaginal muscles increases enormously. The indirect flight muscles are destined to occupy almost the entire space in the thorax (figs. 5–6), and there is simply no room in the thorax for both developing wings and adult muscles. Once the wings are evaginated shortly after the beginning of the larval-pupal moult, the way is partly but not entirely cleared for the full growth of the adult muscles. These grow considerably in the pharate pupal period (fig. 4). At this time histolysis of many of the larval muscles begins, but it is not until the degeneration of the larval muscles is more or less complete that the space required for the imaginal muscles to reach their definitive size is available. In *Simulium ornatum* Meig. degeneration of the larval muscles is not complete until about 30 hours (at 8° C.) after the pupal-adult moult (Hinton, 1959*a*).

For the reasons given above, two moults—no more and no less—are required for the metamorphosis of a larva with internal wings to an adult with functional wings regardless of the degree of difference between the two stages: the larva of, say, a staphylinid beetle is very much like the adult, whereas the larva of a blowfly is very different indeed from the adult, but both require two moults for the transformation from larva to adult. It might be expected that if an adult endopterygote became apterous, the transformation from larva to adult could be done in one moult and the pupal stage could be discarded. The larviform apterous females of some beetles (some Stylopoidea and Lycidae) no longer go through a pupal stage, and the females have one less moult in their life-cycle than the males.

As we have seen, the endopterygotes require two moults for the metamorphosis of their wings from the larval to the adult condition. Therefore, contrary to the claims of Poyarkoff (1914*a*, 1914*b*) and Hinton (1948), the manner in which the imaginal muscles grow, and the time at which the epidermal cells strengthen muscle attachments by forming tonofibrillae, has no relevance to the origin or to the functional necessity of the pupal stage.

## Origin of the Pupal Stage

The pupa of the endopterygotes is like the adult in most respects: its external features generally bear no resemblance to those of the preceding instar, the final larval instar, but, on the contrary, are very similar to those of the adult. From this fact Lower (1954) and some other supporters of the theory of Berlese conclude that the pupa cannot be the homologue of the final larval instar of the exopterygotes but must represent all of the larval instars of the exopterygotes. Concrete evidence in support of such a process of tachygenesis has never been produced. From the same fact Poyarkoff, Hinton, and Snodgrass conclude that the pupa is an imaginal instar, and that the single imaginal instar of the exopterygote ancestors of the endopterygotes at some stage evolved the capacity for another moult. Here again what is claimed is inferred, and no evidence in support is provided. New developmental stages can arise only as the result of the modification or division of previously existing stages. The fact that the pupa resembles the adult so much more closely than the final larval instar has a simple explanation if in fact it originates as a specialisation of the last larval instar of the exopterygote ancestors of the endopterygotes, an explanation that does not involve the difficulties inherent in the Berlese theory of tachygenesis or in the Poyarkoff theory of a subdivision of the adult stage of endopterygotes. All that has to be assumed is that the last larval stage with external wings was retained in the endopterygotes to bridge the difference between a larva with internal wings and the adult. If in time such a stage became less and less mobile as the feeding larval stages became more and more specialised and therefore more and more different from the adult, we might well expect that in time the feeding larval stages would cease to resemble the last larval instar, now the pupa, which became relatively immobile and disrupted many of its former relations with the environment. In very many exopterygotes the last larval instar resembles the adult quite as much as the pupa of the endopterygotes resembles the adult endopterygote.

Thus, what is now suggested is in effect a functional subdivision of the larval stages into a series chiefly concerned with feeding and a final larval instar, now a pupa, chiefly concerned with bridging the gap between the specialised feeding stages and an adult specialised for reproduction and distribution. Pérez (1910) and earlier writers have held the view that the endopterygote pupa is the homologue of the last larval instar of the exopterygotes. Subdivision of the larval stages into two or more series specialised for very different modes of life and quite unlike each other in external form has occurred independently on a number of occasions in both primitive and highly advanced recent groups of Endopterygota, *e.g.* Neuroptera, Coleoptera, Hymenoptera, Lepidoptera, and Diptera (summary in Snodgrass, 1954).

### Subimago of Ephemeroptera and Pupal Stages of Exopterygotes

From what has already been said about the functional significance of the endopterygote pupa, it seems clear that the functional significance of the subimago of the Ephemeroptera and the so-called pupal stages of other exopterygotes must be different. Boas (1899) was perhaps the first to claim that the subimago of the Ephemeroptera was the homologue of the pupa of the endopterygotes. This claim can be maintained if it is considered that the subimaginal stage is a larval stage and therefore also homologous to the last larval instar of other exopterygotes. It cannot be maintained if the subimaginal stage is considered to be the first adult instar. It is now generally recognised (*e.g.* Martynova, 1961) that the endopterygotes are derived from some group near the Protoblattoidea, that is, from an exopterygote group that did not moult in the adult stage.

In some exopterygotes such as the Aleyrodidae and male Thysanoptera and Coccoidea, the general structure of the feeding larval instars has departed very widely from that of the adult. The structural differences between the feeding larval instars and the adult are normally bridged in the last larval instars. As the differences between the two stages became greater and involved a greater degree of re-organisation of the internal tissues, it would seem that the last or last two larval instars became more and more quiescent and eventually ceased to feed. It may be noted here that the structural re-organisation required to bridge the gap between the feeding larval stages and the adult of some exopterygotes is greater than in the primitive endopterygotes, *e.g.* some Megaloptera. No difficulty necessarily arises if these quiescent or semi-quiescent stages of exopterygotes are called pupae provided that it is recognised that their origin is quite independent from that of the endopterygote pupa and their initial functional significance is different.

## Summary

The theories of Berlese and Poyarkoff are briefly reviewed. Reasons are given for rejecting the theory of Berlese in its entirety. Poyarkoff's explanation of the origin and functional significance of the endopterygote pupa is rejected and an alternative view is suggested.


## References


BERLESE, A., 1913, Intorno alle metamorforsi degli insetti. *Redia* **9** : 121–36.

BOAS, J. E. V., 1899, Einige Bermerkungen über die Metamorphose der Insekten. *Zool. Jb.* (*Syst.*) **12** : 385–416.

DUPORTE, E. M., 1958, The origin and evolution of the pupa. *Canad. Ent.* **90** : 436–9.

HINTON, H. E., 1948, On the origin and function of the pupal stage. *Trans. R. ent. Soc. Lond.* **99** : 395–409.

—— 1955, On the structure, function and distribution of the prolegs of the Panorpoidea, with a criticism of the Berlese-Imms theory. *Ibid.* **106** : 455–545.

—— 1959*a*, Origin of indirect flight muscles in primitive flies. *Nature, Lond.* **183** : 557–8.

—— 1959*b*, How the indirect flight muscles of insects grow. *Sci. Progr.* **47** : 321–33.

—— 1961, The role of the epidermis in the disposition of tracheae and muscles. *Ibid.* **49** : 329–39.

LOWER, H. F., 1954, A morphological interpretation of post-embryonic insect development. *Arch. Zool. exp. gén.* **91** : 51–72.

LUBBOCK, J., 1874, *On the origin and metamorphosis of insects.* 2nd ed. London.

MARTYNOVA, O., 1961, Palaeoentomology. *Annu. Rev. Ent.* **6** : 285–94.

MIALL, L. C., 1895, The transformations of insects. *Nature, Lond.* **53** : 152–8.

PÉREZ, C., 1910, Signification phylétique de la nymphe chez les insectes métaboles. *Bull. sci. Fr. Belg.* **44** : 221–33.

POHLEY, H. J., 1956, Untersuchungen über das Zellteilungwachstum der Flügelanlagen von *Ephestia kühniella* Z. *Biol. Zbl.* **75** : 86–98.

POYARKOFF, E., 1914*a*, Essai d'une théorie de la nymphe des insectes holométaboles. *Horae Soc. ent. ross.* **41** : 1–51 (1913).

—— 1914*b*, Essai d'une théorie de la nymphe des insectes holométaboles. *Arch. Zool. exp. gén.* **54** : 221–65.

SNODGRASS, R. E., 1935, *Principles of insect morphology.* London.

—— 1954, Insect metamorphosis. *Smithson. misc. Coll.* **122** (9) : 1–124.

—— 1957, A revised interpretation of the external reproductive organs of male insects. *Ibid.* **135** (6) : 1–60.

—— 1961, The caterpillar and the butterfly. *Ibid.* **143** (6) : 1–51.

STUART, D. C. and EDWARDS, G. A., 1958, Intercellular bars at myochitin junctions. *Rep. N. Y. St. Dep. Hlth.* **1958** : 49–50.

SWAMMERDAM, J., 1737, *Biblia naturae.* Leiden.

# B. Hormonal Control

## Control of Growth and Development in Insects

Several growth hormones appear to be isoprenoid derivatives, and some may act upon the cell nucleus.

Howard A. Schneiderman and Lawrence I. Gilbert

The postembryonic growth of higher organisms is under hormonal control during most of the organism's life. The best-studied groups are the vertebrates and higher plants, but increasing attention is being paid to invertebrates, especially arthropods.

Insect endocrinology has made startling advances during the past decade. Three insect hormones have been crystallized: the molting hormone, by Butenandt and Karlson (*1*); the queen substance of bees, by Butler and his co-workers (*2*); and a substance with some of the properties of the brain hormone, by Kobayashi and his co-workers (*3*). Extracts of several other hormones have been prepared and partially purified: (i) a polypeptide or protein with the activities of the brain hormone, by Ichikawa and Ishizaki (*4*); (ii) the juvenile hormone, by Williams (*5*) and by others (*6, 7*); (iii) the diapause hormone, by Hasegawa (*8*); (iv) a hormone regulating carbohydrate metabolism, by Steele (*9*); and a hormone regulating hardening and darkening of blowflies, by Cotrell and by Fraenkel and his associates (*10*). A year ago we knew of no chemicals which could mimic the action of insect growth hormones. We now know of more than a dozen chemicals which can simulate the effects of these hormones. In this discussion we summarize a few recent developments in the field of insect growth hormones which seem to suggest new approaches to comparative endocrinology and perhaps to developmental biology as a whole (*11*).

Dr. Schneiderman is professor of biology at Western Reserve University, Cleveland, Ohio. Dr. Gilbert is associate professor of biological sciences at Northwestern University, Evanston, Ill. This article is based on a lecture presented 28 December 1962 at the Philadelphia meeting of the AAAS.

The distinguishing feature of insect growth and development is the process of molting. The epidermis of an insect is bound to an outer layer of cuticle, and it cannot grow until it becomes detached from this cuticle. Periodically the epidermal cells of immature insects become detached from their cuticle, grow by cell division or cell enlargement, or both, and secrete a new and extensible cuticle with folds. When this new cuticle is nearly complete, the epidermal cells produce and release enzymes which digest the inner layers of the old cuticle, which they then absorb. Thereupon the epidermal cells waterproof the new cuticle, and the old cuticle is shed. An insect, such as a young cockroach, that has a rigid body fills itself with air immediately after the molt, when its cuticle is soft; when the cuticle hardens the insect has an enlarged, rigid skin which provides room for its internal organs to grow. A soft-bodied insect such as a caterpillar remains quite wrinkled after it molts, but gradually increases in size during the intermolt period until its cuticle is stretched and smooth.

Because of the process of molting, it is commonly claimed that insect growth is discontinuous. This is not strictly true in many insects: caterpillars, as we have just noted, increase in body weight steadily during their life. However, if we consider only one aspect of growth—namely, cell division—then indeed it may be said that the growth of many insect tissues is discontinuous. For example, in caterpillars and in many other insects the increase in size of the skin and of many internal organs between molts commonly occurs without an increase in cell number. Here, cell multiplication is periodic and is restricted to the period just prior to or just after shedding of the skin. Thus, unlike immature mammals, whose growth is essentially continuous, most immature insects have discontinuous growth, the visible expression of which is molting.

The periodic cell multiplication and molting of immature insects is brought about by two hormones, one produced by secretory cells in the insect's brain and the other by glands in the prothorax, the prothoracic glands. The "brain hormone" (*12*) acts by stimulating the prothoracic glands. The prothoracic glands respond to this stimulus by releasing prothoracic gland hormone, "ecdysone," which in turn acts on various cells and causes them to grow. In the case of the insect's epidermal cells, the prothoracic gland hormone sometimes stimulates cell division, but it always causes these cells to deposit a new cuticle and to molt. If a few micrograms of ecdysone are injected into an isolated fragment of an insect—for example, an isolated abdomen—the fragment will molt. The prothoracic gland hormone is thus a true growth and molting hormone for insects and perhaps for all arthropods.

A third hormone, the juvenile hormone, is secreted by endocrine glands known as the corpora allata, which are located near the brain of the insect. Nearly 30 years ago Wigglesworth showed that this hormone promotes larval development but prevents metamorphosis (*13*). The presence of juvenile hormone in the immature insect insures that when the larva molts it will retain its larval characters and not differentiate into an adult. The juvenile hormone is a remarkable agent which permits growth but prevents maturation.

When larval cells are stimulated to grow and molt by the prothoracic gland hormone, the presence of a high concentration of juvenile hormone causes them to use their synthesizing machinery to secrete larval cuticle. In the presence of a small amount of juvenile hormone, the same cells secrete pupal cuticle; in the absence of juvenile hormone, these cells may secrete adult cuticle directly and by-pass the pupal molt. Through regulation of the release of brain hormone, molting is controlled; through regulation of the release of juvenile hormone, maturation is controlled. Control of maturation by the juvenile hormone is also evident in many adult insects, where the juvenile hormone, or at least a secretion of the corpora allata, stimulates the matura-

Reproduced with permission from *Science*, 143:325–333, 1964. Published by the American Association for the Advancement of Science, Washington, D.C.

tion of eggs in females and the development of accessory glands in males. These various endocrine events are summarized schematically in Fig. 1 (see *7, 14–16*). With these basic endocrine events in mind, let us consider the hormones themselves.

### "Brain Hormone"

The first extracts of "brain hormone" were prepared in 1958 by Kobayashi and Kirimura, who succeeded in extracting oily material from 8500 surgically isolated silkworm brains which had "brain hormone" activity (*3*). Injection of 0.1 milligram of extract into brainless silkworm pupae caused them to molt. The active principle appeared to be a lipid. Subsequently, water-soluble extracts with brain hormone activity were prepared by Ichikawa and Ishizaki (*4*), and these appear to be polypeptides.

Recently, Kobayashi and his co-workers reported the isolation and identification of an active principle in lipid extracts of insect brains (*3*). They painstakingly dissected out 220,000 brains from silkworm pupae, homogenized and extracted them with methanol, re-extracted the dried methanolic extract with ether, and then purified the extract, largely by countercurrent distribution in a variety of solvent systems. They reported that the crystallized brain hormone is probably identical with cholesterol. Brainless pupae molted when injected with the extract (0.02 $\mu g$ in 10-percent ethanol). Kobayashi and his associates also reported that commercial preparations of cholestanol and 7-dehydrocholesterol had "brain hormone" activity.

This astonishing result must have prompted a number of laboratories to check their findings. John Edwards and Krishnakumaran at Western Reserve University and Gilbert at Northwestern University have confirmed a key result. Commercial preparations of cholesterol and of several other sterols in microgram amounts activate prothoracic glands and cause brainless pupae to molt. Table 1 gives data for a typical experiment. Hundreds of control preparations of brainless pupae have been examined, and none of these has ever molted, nor have isolated abdomens injected with these same sterols.

The reader may wonder how entomologists, who are traditionally accused of bombarding insects with every known chemical, could have failed to try cholesterol. The answer is that many did try cholesterol but apparently did not use the proper solvent system or the right concentrations. For reasons which are not yet clear, 9- or 10-percent ethanol is the solvent of choice.

Although it has been possible to confirm the effectiveness of cholesterol and other sterols as mimics of the brain hormone, the results in Table 1 are difficult to understand, for there appears to be no clear correlation between sterol concentration and brain hormone activity. The data suggest that some factor other than chemical structure may be responsible for the brain hormone activity of these alcoholic preparations. Perhaps surface activity or some other properties of the sterol solutions and suspensions are involved. There is also other evidence that brain hormone and cholesterol are not identical. For example, Goodfellow and Gilbert (*17*) have shown that cholesterol levels in the blood are quite high at all stages: female pupae of the cynthia moth contain 520 micrograms of sterols (largely cholesterol) per milliliter, and male pupae contain 300 micrograms per milliliter of blood. The pupa possesses about 1 milliliter of blood. Thus the amount of cholesterol which appears to cause molting is about 0.0001 the amount normally present in the blood (*18*)! Notwithstanding these curious observations, it seems clear that substances with brain hormone activity derived from insect brains have properties similar to cholesterol, and that several sterols have brain hormone activity. If this work is further confirmed and the active principle in the extracts should prove to be a steroid, it would be most exciting. For the present, the work stands as the first report of a steroid with hormonal effects on an insect.

Table 1. Effect of injecting sterols into brainless pupae of *Samia cynthia*. The brains of the pupae were removed 3 months prior to injection, and the pupae were inspected prior to assay to make sure they were not developing. All injections were in 9-percent ethanol; 50 to 250 microliters were injected into each pupa. No effects were observed when injections were made into isolated abdomens, nor were effects observed with injections of up to 5.0 $\mu g/g$ (fresh weight) of stigmasterol, stigmasterol acetate, dehydrocholesterol, viosterol, or dihydroisoandrosterone. About nine pupae were tested at each concentration except when ethanol alone was used; in that case 50 pupae were tested.

| Amount injected ($\mu g/g$, fresh weight) | Percentage completing adult development |
|---|---|
| *Cholesterol* | |
| 0.005 | 40 |
| .05 | 60 |
| .5 | 20 |
| 5.0 | 15 |
| *Ergosterol* | |
| 0.005 | 0 |
| .05 | 33 |
| .5 | 0 |
| 5.0 | 48 |
| *Beta sitosterol* | |
| 0.005 | 33 |
| .05 | 67 |
| .5 | 0 |
| 5.0 | 0 |
| *Ethanol 9-percent only* | |
| | 0 |

How does the brain hormone act? The brain hormone has as its principal target the prothoracic glands. As Williams first showed in 1947, the brain hormone stimulates the prothoracic glands to secrete ecdysone (*19*). What is this "stimulation" in biochemical terms? To answer this question Oberlander (*20*) has studied, with autoradiographic techniques, the effects of brain hormone on the prothoracic glands of various wild silkworms. The first event of activation that he has observed is intense RNA synthesis within the nuclei of prothoracic gland cells, followed by the appearance of cytoplasmic RNA and by protein synthesis. Presumably these events reflect the synthesis of enzymes necessary for the production of ecdysone, and they are followed by the release of the hormone. From these observations it is clear that, as far as the prothoracic glands are concerned, brain hormone is a growth hormone and stimulates the prothoracic gland cells in much the same way that ecdysone stimulates other cells of the insect's body. Whether the brain hormone stimulates other cells besides those of the prothoracic gland remains to be demonstrated.

### Ecdysone

Ecdysone was first crystallized by Karlson and Butenandt in 1954 (*1*). Its partial structure has just been published. From an analysis of 250 milligrams of crystals isolated from 1000 kilograms of dried pupae of *Bombyx mori* (the silkworm of commerce), Karlson and his co-workers (*21*) concluded that ecdysone was a steroid with the empirical formula $C_{27}H_{44}O_6$. The key to its identification as a steroid was an x-ray diffraction analysis of the crystals by a new "diffuse scattering" method, which permitted determination of the shape of the molecule and the molecular weight without the introduction of heavy atoms. From ultraviolet, infrared, and proton resonance spectra it was concluded that ecdysone had a keto group at carbon atom 12 conjugated to a double bond between carbon atoms 9 and 11, a hydroxyl group on the side chain at carbon atom 25, and four other hydroxyl groups at undetermined locations. Its identification marks the culmination of two decades of painstaking work. Karlson is sensibly cautious in advancing his claim and states that "even the partial structure requires further confirmation" (*14*). However, it is clear that he is convinced that ecdysone is a steroid, a view which is shared by most organic chemists who have examined his data, and a view which we have adopted.

The discovery that ecdysone is a steroid is of uncommon interest. But equally remarkable are some recent reports on its mode of action, which appear to have important consequences for students of growth and suggest that this agent may act on the genes themselves or on the nuclear membrane. Now, it has long been assumed that the activity of genes varies in different tissues and at different developmental stages. We might expect to find different genic activity in an epidermal cell of a dormant pupa and in the same cell engaged in adult synthesis. Experimental support for this opinion comes from recent studies by Beermann and others, who have described characteristic puffs or swelling in certain segments of the giant chromosomes in the salivary glands of developing Diptera, as well as differences in giant chromosome morphology in different tissues (*22*). These giant chromosomes consist of hundreds of individual chromosomal threads and are found in several tissues. Chromosomal puffs indicate a local loosening of the bundles of threads and reflect enhanced metabolic activity in particular chromosomal regions which probably are the site of heightened gene activity (*23*). Studies with tritiated uridine and with labeled amino acids reveal that RNA synthesis and protein synthesis proceed at a rapid rate in puffed regions. These observations suggest that the changing biochemical activities of insect cells during development may be caused by the differential activation or suppression of different sets of genes (*24*). Since ecdysone initiates the postembryonic growth and development of insects, one is led to the view that ecdysone may have as a principal target the nucleus itself, where it activates particular chromosomal regions and brings about "the elaboration of specific substances (RNA?) that are destined to participate . . . in the cytoplasmic syntheses that characterize growth and molting" (*25*).

Recently, Clever and Karlson (*26*) have provided direct evidence to support this conjecture. They have shown that injection of pure ecdysone into *Chironomus* larvae causes prompt and characteristic changes in the puffing pattern of the chromosomes, changes identical to those that occur during pupation. In more recent experiments, Clever (*27*) has shown that injection of ecdysone causes new puffs to appear within 15 minutes of injection and that the size of the puffs is proportional

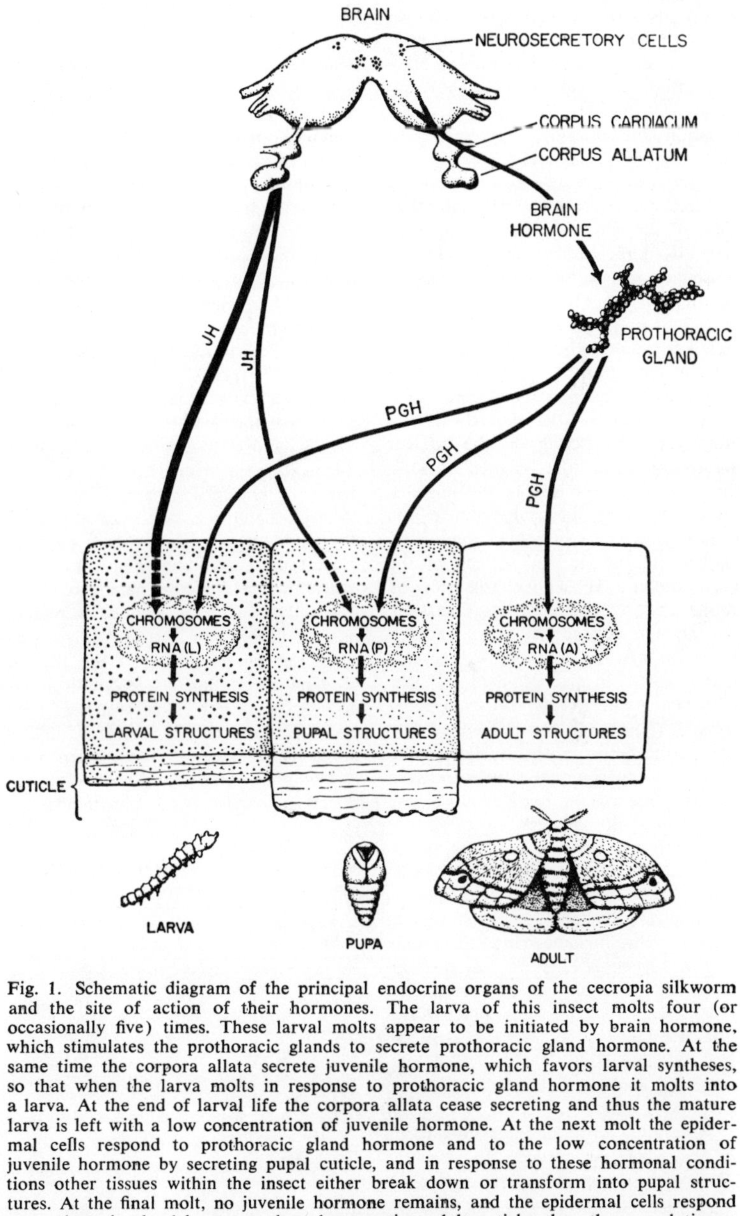


Fig. 1. Schematic diagram of the principal endocrine organs of the cecropia silkworm and the site of action of their hormones. The larva of this insect molts four (or occasionally five) times. These larval molts appear to be initiated by brain hormone, which stimulates the prothoracic glands to secrete prothoracic gland hormone. At the same time the corpora allata secrete juvenile hormone, which favors larval syntheses, so that when the larva molts in response to prothoracic gland hormone it molts into a larva. At the end of larval life the corpora allata cease secreting and thus the mature larva is left with a low concentration of juvenile hormone. At the next molt the epidermal cells respond to prothoracic gland hormone and to the low concentration of juvenile hormone by secreting pupal cuticle, and in response to these hormonal conditions other tissues within the insect either break down or transform into pupal structures. At the final molt, no juvenile hormone remains, and the epidermal cells respond to prothoracic gland hormone alone by secreting adult cuticle; the other pupal tissues either break down or develop into adult structures. The Lepidoptera, unlike many other insects, appear not to require the corpora allata for egg maturation.

to the dose. As little as $10^{-5}$ microgram is effective. One particular puff on one chromosome appears to be affected before any of the others. In these experiments ecdysone appears to act on the chromosomes and activates particular gene loci to bring about a variety of syntheses that culminate in molting.

In an equally impressive series of experiments, Kroeger (*28*) has demonstrated that a number of chemicals copy the effects of ecdysone on the puffing pattern of salivary chromosomes of *Chironomus*; zinc ions, cadmium ions, and narcotics cause a puffing pattern identical to that caused by ecdysone (although they do not cause pupation and molting). By surgically removing parts of the chromosome set, Kroeger was able to show that each puff responded independently of other parts of the genome. This finding suggests that puff activity patterns are not caused by interactions of various parts of the genome (for example, by one puff activating a second puff, and so on) but result from a changed intranuclear environment, perhaps from a change in the nuclear membrane (*28*). Kroeger believes (*29*) that these several agents, by acting on the nuclear membrane, affect the balance of sodium and potassium ions in the nucleus. He argues that it is changes in the sodium and potassium ions that cause changes in the puffing pattern of the chromosomes. To support this view he shows that by changing the potassium and sodium concentrations of the nucleus he can cause puffs to increase and decrease in size on demand. Presumably, in the normal situation ecdysone is responsible for the changes in the permeability of the nuclear membrane; in experimental situations other agents, such as zinc ions, produce similar permeability changes.

Perhaps the juvenile hormone, which can modify events initiated by ecdysone, has a similar site of action in the nucleus. Becker (*30*), in studies of salivary chromosomes in *Drosophila*, and Kroeger (*28*), in studies on *Chironomus*, have shown that wounding blocks and reverses puff patterns induced by ecdysone. Kroeger believes that this reversing of the puffing patterns is analogous to the action of juvenile hormone.

Recently Laufer and his co-workers (*31*) have succeeded in relating changes in the puffing pattern of chromosomes with changes in the synthesis of specific enzymes. They showed, for example, that actinomycin D, an antibiotic which inhibits DNA-dependent synthesis of RNA in many systems, inhibits both puffs and the synthesis of certain proteins in the salivary gland cells of *Chironomus thummi.*

Table 2. Incorporation of tritiated thymidine into pupae of *Hyalophora cecropia* at various times after pupation. The pupae were fixed 2 hours after a single injection (10 μc/g) (*66*).

| Tissue | Days after pupation | | | | |
|---|---|---|---|---|---|
| | 0 | 2 | 5 | 10 | 20 |
| Fat body | 0 | 0 | 0 | 0 | 0 |
| Epithelium | 0 | 0 | 0 | 0 | 0 |
| Muscle | 0 | 0 | 0 | 0 | 0 |
| Blood cells | + | + | + | 0 | 0 |
| Brain | + | + | + | 0 | 0 |
| Thoracic ganglion | 0 | 0 | 0 | 0 | 0 |
| Malpighian tubules | 0 | 0 | 0 | 0 | 0 |

These experiments on the puffing of giant chromosomes promise to become landmarks in biology. They have removed insect hormones from the parochial field of entomology to the center of contemporary biological thought. For a central problem of biology today is: How are genes activated and what is gene activity? (*32*). With this evidence that the activity of certain genes may be under the control of a specific chemical agent (ecdysone), and because the chromosomes of Diptera are so large that the biochemical events of gene activity can be analyzed by fine-resolution autoradiography, it seems likely that within the next few years the study of insect chromosome metabolism will become a most active area of biological research. For the student of genetic mechanisms in multicellular organisms, insect chromosomes may again become strategic research objects and regain some of the esteem they lost in 1947 when Max Delbrück introduced geneticists and biochemists to the elegance and simplicity of bacteriophage as an object of research.

Table 3. Rate of uptake of tritiated uridine into pupae of *Hyalophora cecropia* at various times after pupation. The pupae were fixed 2 hours after a single injection (10 μc/g) (*67*).

| Tissue | Days after pupation | | | | |
|---|---|---|---|---|---|
| | 0 | 2 | 5 | 10 | 20 |
| Epithelium | ++ | ++ | + | 0 | 0 |
| Muscle | + | + | 0 | 0 | 0 |
| Fat body | + | + | + | + | 0 |
| Blood cells | ++ | ++ | ++ | ++ | ++ |
| Malpighian tubules | +++ | +++ | +++ | ++ | ++ |
| Nerve tissue | + | + | + | + | + |
| Oenocytes | ++ | ++ | ++ | ++ | ++ |
| Connective tissue | + | + | + | 0 | 0 |
| Wing bud | + | + | + | + | 0 |

Thus, ecdysone may have the chromosomes as a primary target, but how are its effects manifested in different cells and tissues? Do all cells require ecdysone for DNA synthesis and cell division, for RNA synthesis and protein synthesis? What synthetic activities remain possible in cells which have been deprived of ecdysone? The newly molted insect provides a perfect experimental situation for answering this question, because, after a molt, ecdysone secretion ceases. Krishnakumaran attempted to answer the question by measuring with autoradiography the ability of various tissues to synthesize DNA, RNA, and proteins (*33*). We studied, first, the larval-pupal molt of the cecropia silkworm. This is a special molt, for the pupa that is formed enters a prolonged period of developmental arrest, or pupal diapause, during which ecdysone secretion ceases. At various times after pupation the pupae were injected with tritiated thymidine to determine the rate of DNA synthesis, or with tritiated uridine to determine the rate of RNA synthesis. The results of a typical experiment are seen in Table 2. When the larva molts into a pupa destined to enter diapause, DNA synthesis ceases except in the blood cells and in parts of the brain that continue to make DNA for nearly a week. Table 3 indicates the rate of RNA synthesis in different tissues. The different tissues "shut down" their RNA synthesis at different times after pupation. Thus, 3 weeks after pupation, epithelium, muscle, fat body, and imaginal discs are inactive, while blood cells, malpighian tubules, nerve tissue, and oenocytes continue to synthesize RNA. Apparently, almost all tissues of the pupa require ecdysone for DNA synthesis. By contrast, RNA synthesis continues in many tissues when little or no ecdysone is present, while other tissues require ecdysone for RNA synthesis. These findings are being pursued in a variety of ways and promise a simple and direct approach to some of the problems of ecdysone action in insects.

Ecdysone profoundly affects the development of insects, but it must not be supposed that developmental processes cannot proceed without it. In many insects, fragments deprived of prothoracic glands or brains are still capable of considerable biosynthesis. Thus, when a large piece of the imaginal disc which is destined to give rise to the future pupal and adult wing is removed from the larva of the flour moth, *Ephestia*, the insect will regenerate the imaginal disc completely in the absence of both brain and prothoracic glands (*34*). A fragment of a larval genital disc of *Drosophila* will also regenerate and grow to full size when transplanted into an adult that has no prothoracic glands (*35*). Similarly, isolated fragments of cecropia pupae can heal large wounds and secrete a modest cuticle. Even adult insects whose prothoracic glands have degenerated during the course of adult development are capable of growth, wound healing, and tissue regeneration. Adult cockroaches, for example, during their long adult life, continue to produce spermatogonia, oogonia, ovarial sheath cells, and probably many other cells throughout life; they can also regenerate nerve fibers that have been surgically removed (*36, 37*). In our preoccupation with the growth hormones of insects, we must not forget that the tissues of most insects, like those of most other metazoa, are capable of considerable regeneration whether or not hormonal conditions favor the continuation of development. A besetting problem in insect endocrinology is that of distinguishing the intrinsically controlled biosynthetic activities of wound healing and regeneration from the extrinsically controlled biosynthteic activities of molting and metamorphosis (*38*).

## Juvenile Hormone

In 1956 Williams prepared the first extract of juvenile hormone from cecropia moths (*5*). When these extracts were injected into pupae, the pupae molted into second pupae: the epidermal cells, instead of secreting an adult cuticle, synthesized a second pupal cuticle. Injection of varying amounts of these extracts produced graded effects, thus providing the basis for a bioassay of the juvenile hormone. Later, we showed that extracts of bovine adrenal glands possessed juvenile hormone activity (*39*). Subsequently, juvenile hormone activity was found in extracts of the tissues of many invertebrates and vertebrates and in plants and microorganisms (*40*). These results indicate that we are dealing with a group of substances which are widespread in nature.

Table 4. Some chemicals with juvenile hormone activity (*68*).

| Substance | Units of juvenile hormone activity per gram* |
|---|---|
| Cecropia oil | 1000 |
| Phytol | 32 |
| Isophytol | 0 |
| Farnesol (commercial) | 16 |
| All-*trans* farnesol | 140 |
| Farnesal | 32 |
| Farnesyl acetate | 5.4 |
| Farnesenic acid | 7.8 |
| Hexahydrofarnesol | 0 |
| Nerolidol | 8.9 |
| Linalool | 0.08 |
| Geraniol | 0 |
| Geranyl linalool | 0.14 |
| Solanesol | .05 |

*Different units are employed by different workers to assay juvenile hormone activity. The common feature of these assays is the use of male cecropia oil as a standard. The assay used in the experiments recorded in this table was the *Galleria* wax test (see text and *44*). We have arbitrarily established as 1 *Galleria* unit of juvenile hormone activity, the juvenile hormone activity shown in the wax test by 1 mg of oil extracted by ether from the abdomens of male cecropia moths which have been permitted to age for 3 days at 25°C. These cecropia oil preparations are remarkably uniform in activity, and we have kept standard extracts at -20°C for a number of years without detectable loss of activity.

Recently the active principle of one of these juvenilizing extracts was isolated by Schmialek (*41*) from mealworm feces and shown to consist of the open-chain terpene alcohol, farnesol, and its aldehyde, farnesal. Farnesol copies the action of the juvenile hormone and also copies the action of the gonadotropin secreted by the corpora allata in promoting egg development (*42*). It was the first substance of known structure to act like an insect hormone, and its isolation was a major achievement in comparative endocrinology. Unfortunately, farnesol does not appear to be the juvenile hormone, for in most test systems pure farnesol is required in milligram or, at best, tenths of a milligram amounts, while purified preparations of the juvenile hormone of insects are active in millimicrogram amounts.

Farnesol is a straight-chain terpene primary alcohol composed of three isoprene units, and, as a precursor of squalene and other polyisoprenoids and of sterols in yeast and liver, it has been of considerable interest. It has four possible steric isomers, the isomerism occurring at the $\Delta^2$ and $\Delta^6$ positions of the molecule. Recently Yamamoto and Jacobson (*43*) determined that only isomers with a *trans* configuration at the $\Delta^6$ position were active. In our laboratory we discovered that hexahydrofarnesol was inactive, a finding which suggests that some unsaturation is necessary for activity. The fact that farnesol had some juvenile hormone activity led us to explore a number of other terpenes. During the past 2 years we have examined several hundred compounds in a variety of test systems. In the type of assay we used most often, substances to be tested were dissolved in mixtures of paraffin wax and peanut oil and applied to wounds made with hypodermic needles on the thorax of pupae of the moths *Antheraea polyphemus* and *Galleria mellonella* (*44*). If the substance was inactive, the cells of the healed wound secreted a thin, scaly adult cuticle which revealed itself at the pupal-adult molt. If the substance had juvenile hormone activity, the cells in the region of the wound secreted an island of thick, rough pupal cuticle which was surrounded by normal adult cuticle. This pupal cuticle is wholly different from incompletely developed adult cuticle and has an unmistakable morphology (*45*).

The first terpene that came to mind was phytol, the $C_{20}$ mono-unsaturated alcohol derived from chlorophyll (Table 4). Like farnesol it is an allylic primary alcohol. It too had some juvenile hormone activity, but isophytol, a rearrangement of phytol with a tertiary hydroxyl group, had no activity. We then turned our attention to a variety of other, related compounds. As Table 4 shows, some had activity while others did not, and there appeared to be no clear pattern except that all were polyisoprenoid compounds (*46*). The smallest was a $C_{10}$ tertiary alcohol, linalool. At this juncture we began carefully to examine compounds other than terpenes. We tested a variety of straight-chain saturated alcohols, from ethanol on up through compounds with 20 or more carbon atoms, as well as other compounds. One alcohol—1-heptanol—had some juvenile hormone activity in the wax test. Similarly, commercial preparations of a sterol, beta sitosterol, had some juvenile hormone activity when injected into pupae of *Antheraea polyphemus* (*47*). Recent studies by Wigglesworth and Schmialek (*48*) and by Karlson (*14*) have also uncovered terpene derivatives with juvenile hormone activity, including two with activity greater than farnesol—farnesyl-*o*-methyl ether and farnesyl diethylamine. Karlson reports that in his test for juvenile hormone activity these two compounds have more than 200 times the activity of farnesol or of cecropia extracts. In our assay systems—the wax test and the polyphemus injection assay—these compounds appear to be only about ten times as active as purified all-*trans* farnesol (*49*).

The significance of the juvenile hormone activity of these diverse compounds deserves analysis. (i) None of the compounds seems to be the natural juvenile hormone; and in our assay none of the active pure compounds appears much more active than the crude oil extracted from adult cecropia moths, which has served as a starting material for most purifications of juvenile hormone, and none is nearly as active as partially purified preparations of juvenile hormone. In considering the data of Table 4, it is worth noting that the most highly purified preparations of juvenile hormone for which a description of purification was given contained 200,000 Galleria units per gram (*7*, and Table 4, footnote), while the most active compounds discovered thus far contained only 1000 Galleria units per gram. (ii) Since terpenes, sterols, and alcohols occur widely in nature, it seems likely that the juvenile hormone activity reported in extracts from vertebrates, invertebrates, plants, and microorganisms is due to compounds of this sort rather than to the juvenile hormone. (iii) The widespread occurrence of such juvenilizing compounds makes the isolation of the true active principle of the corpora allata somewhat more difficult (*50*), since several of these substances contaminate insect extracts. [For example, lipid extracts of male cecropia moths contain 5 to 15 $\mu g$ of farnesol per gram and three times this amount of phytol (*17*).] Thus, the insect endocrinologist faces the same problem that the "bar substance" posed for the *Drosophila* geneticist and the "inducer" posed for the experimental embryologist—namely, that substances other than the natural one cause developmental effects (*42*, *51*). The possible relation of these substances to the true juvenile hormone deserves consideration. There are many possibilities, but two are particularly attractive.

The first is the possibility that the substances resemble the juvenile hormone in some structurally significant way and act at the same site in the cell. The second is the possibility (which does not exclude the first) that the various compounds are precursors of the juvenile hormone and may be converted within the insect's body. Certainly insects can synthesize higher alcohols, and probably terpenes such as geraniol (*52*). However, if these substances are precursors of the juvenile hormone, they must be transformed by the cells they affect and not elsewhere, since, in the wax-test assay and other topical assays, their juvenilizing action can be restricted to the area of application.

Table 5. Effects of injecting farnesol into brainless pupae of *Antheraea polyphemus*. The brains of the pupae were removed 3 months prior to injection. All injections of farnesol were at a dosage of 5 mg/g (live weight). All suspensions of farnesol contained 1 part of farnesol to 3 parts of peanut oil or water and were made up in a vortex mixer (*67*).

| Solution injected | Percentage dead | Percentage completing adult development |
|---|---|---|
| Farnesol | 100 | 0 |
| Farnesol and peanut oil | 100 | 0 |
| Farnesol and water | 17 | 50 |
| Farnesol and 9-percent ethyl alcohol | 33 | 33 |
| Water | 0 | 0 |
| Ethyl alcohol (9-percent) | 0 | 0 |
| Uninjected controls | 0 | 0 |

Another point worth mentioning is the fact that purified juvenile hormone preparations from cecropia have recently been shown to have gonadotropic activity (*53*), as do most chemicals which copy the juvenile hormone (*48*).

In addition to possessing juvenile hormone and gonadotropic activity, crude and purified juvenile hormone extracts from cecropia also activate the prothoracic glands (*54*). Injection of 50 milligrams of crude juvenile hormone extract or 50 micrograms of purified juvenile hormone extract caused molting in brainless pupae of the giant silkmoths, *Samia cynthia* and *Antheraea polyphemus* (*54*, *55*). Similarly, implantation of active corpora allata also causes molting in brainless pupae. It has been argued that the corpora allata are able to activate the prothoracic glands because they contain neurosecretion from the brain (*56*). This may be true, but there is convincing evidence that the juvenile hormone itself is the agent. Oil extracted from allatectomized male cecropia moths lacks both juvenile hormone activity and brain hormone activity, so the factor that activates the prothoracic glands must come from the corpora allata.

Table 6. Some notable endocrine relationships.

1. All known insect endocrine organs involved in growth derive from embryonic ectoderm. Prothoracic glands and corpora allata orginate in adjacent segments.
2. Ecdysone, brain hormone, juvenile hormone, cholesterol, and farnesol activate the prothoracic glands.
3. Isolated abdomens may molt when brains or corpora allata are implanted, as well as when the abdomens are injected with ecdysone.
4. Juvenile hormone extracts and terpene derivatives, such as farnesol, that stimulate the juvenile hormone have gonadotropic activity.
5. Many terpenes (sterol precursors in many organisms) have juvenile hormone activity.
6. Ecdysone is probably a steroid.

A more convincing argument comes from the recent discovery by Krishnakumaran that farnesol and farnesol derivatives can activate the prothoracic glands of brainless diapausing pupae and cause molting. Results of a typical experiment are shown in Table 5. Alcoholic solutions of farnesol, or aqueous suspensions, activate the prothoracic glands of brainless pupae of *Antheraea polyphemus* and cause the pupae to develop. These agents also activate the prothoracic glands of brainless cecropia pupae, which are ordinarily quite resistant to activation. Thus, farnesol and its derivatives copy not only juvenile hormone and gonadotropin but also the brain hormone.

Why should chemicals with juvenile hormone activity and, presumably, the juvenile hormone itself be able to activate the prothoracic glands? There appear to be four main possibilities.

1) It may be that activation of the prothoracic glands by these diverse compounds is a pharmacological rather than a physiological phenomenon (*15*) and represents no more than the nonspecific effect of comparatively large doses of physiologically active molecules on a "poised" target organ.

2) Perhaps brain hormone and juvenile hormone are chemically related, so that there is some overlap in the responses they can induce in target tissues.

3) It may be that the insect uses the corpus allatum and the juvenile hormone to activate the prothoracic glands during larval life, and uses the brain hormone to activate the prothoracic glands only during the larval-pupal and pupal-adult transformations.

4) Another possibility which seems particularly attractive is that the insect uses two hormones to control the prothoracic glands during larval life, one produced by the brain and the other

by the corpora allata. The argument runs as follows. During larval life the prothoracic glands are continuously active at a low level and produce and release prothoracic gland hormone continuously. This continued low level of ecdysone is necessary for the growth of many tissues, such as the imaginal discs, which grow continuously during larval life. The prothoracic glands are kept continuously active at a low level by the juvenile hormone of the corpora allata. Periodically the brain hormone "superactivates" the prothoracic glands, which respond by releasing a large amount of ecdysone. Then molting occurs.

There is considerable experimental evidence to support this last hypothesis. Three observations are particularly pertinent. (i) Ligature, implantation, and histological studies have shown that larval imaginal discs require a continuous supply of ecdysone for growth (*57*). (ii) The juvenile hormone is necessary for maintaining the integrity of the prothoracic glands at all stages (*54*). If molting occurs in the absence of the juvenile hormone, the prothoracic glands will break down. This ability of small amounts of juvenile hormone to preserve the prothoracic glands may be related to the ability of larger amounts of juvenile hormone to activate the prothoracic glands. (iii) Autoradiographic studies have revealed that the prothoracic glands are active throughout larval life and only get completely "turned off" at pupation (*20*).

Hence, it appears possible that the prothoracic glands have two activation mechanisms—one via the juvenile hormone of the corpora allata, which can be copied by farnesol, and the other via the true brain hormone, which is produced by neurosecretory cells in the brain.

Another substance which appears to activate the prothoracic glands is ecdysone itself (*58*). Inasmuch as ecdysone is a steroid, it may be that the "cholesterol" preparations of Kobayashi discussed earlier activate the prothoracic glands because they resemble ecdysone rather than because they resemble the true brain hormone. Alternatively, inasmuch as farnesol-like compounds and sterols are chemically related, these "cholesterol" preparations may activate the prothoracic glands not because they resemble brain hormone but because they resemble juvenile hormone.

## A New Look at Insect Hormones

What do all these findings mean? Let us re-examine the basic endocrine picture of insects (Fig. 1). During the past few years a number of significant deviations from this pattern have emerged. In addition, there are a number of fundamental relationships that are not apparent in the picture. These are summarized in Table 6. It must be borne in mind that some of these relationships may be exceptional and do not hold true for all insects.

1) The three insect endocrine organs involved in growth all derive from embryonic ectoderm. The prothoracic glands and the corpora allata derive from adjacent body segments. Recent light- and electron-microscope studies of silk moths (*59*) have shown that the cells of pupal corpora allata and prothoracic glands are so similar in structure that they are virtually indistinguishable. Berta Scharrer (*60*) has reached a similar conclusion from her studies of the cockroach *Leucophaea* and reports that the endoplasmic reticulum in both glands is similar to that found in mammalian endocrine cells engaged in steroid metabolism. These observations are consistent with the view that the corpora allata and prothoracic glands and their hormone products are related.

2) The fact that ecdysone, brain hormone, juvenile hormone, farnesol, and cholesterol activate the prothoracic glands has already been discussed and supports the suggestion that the hormones may be related chemically.

3) Further evidence that the hormones are related comes from the remarkable finding of Ichikawa and Nishiitsutsuji-Uwo that, in certain insects, the brain hormone *alone* or the juvenile hormone *alone* may substitute for ecdysone and will cause molting in isolated abdomens which have no known endocrine organs (*61*). Similarly, Kobavashi and Burdette (*62*) have shown that cholesterol preparations with brain hormone activity act synergistically with ecdysone and cause molting in isolated abdomens of fly larvae which have received subthreshold injections of ecdysone. This observation suggests that in some situations the three hormones can replace one another. A similarity between brain hormone and ecdysone is also suggested by the observation of Dietrich Bodenstein (*63*) that brain hormone may substitute for ecdysone in cockroach nymphs which can be induced to molt by brain implantation after extirpation of their prothoracic glands.

4) The finding that purified extracts of juvenile hormone have gonadotropic activity and that farnesol, which mimics juvenile hormone, has gonadotropic effects has important implications. Apparently the juvenile hormone and the gonadotropin are so similar in structure that a single substance may act in either capacity or may serve as a precursor for both agents. Another implication is that in the adult insect, where there is little or no ecdysone, juvenile hormone or farnesol *alone* can act as a growth hormone for the ovaries. If juvenile hormone acts as a growth hormone at one stage, may it not act as a growth hormone at other stages?

Items 5 and 6 in Table 6 have already been discussed.

Taken as a whole these findings suggest the following: (i) ecdysone, juvenile hormone, gonadotropic hormone, and perhaps the brain hormone may be related chemically; (ii) these substances may all be polyisoprenoid derivatives; (iii) the juvenile hormone and the brain hormone may act directly on many tissues to cause growth *without* ecdysone; and (iv) the possibility remains that the mechanism whereby the juvenile hormone activates the prothoracic glands is totally different from the mechanism whereby the brain hormone activates the prothoracic glands; and the brain hormone may turn out to resemble the more conventional neurosecretory polypeptides (for example, oxytocin and vasopressin).

Should most of the insect growth hormones turn out to be polyisoprenoid compounds, it would be most exciting from a number of standpoints. First, insects cannot synthesize sterols from acetate, mevalonic acid, or any of the other nonsterolic compounds that have been tested (*64*). All insects that have been studied appear to have a dietary requirement for sterols (*65*). Since some insect hormones are steroids, then either they must be made by modifying a dietary sterol (*66*) or different pathways of sterol biosynthesis exist in insects than those reported in microorganisms and mammals.

More significant than this is the implication for comparative endocrinology of the presence of steroid hormones in insects. If groups as diverse as mammals and insects use steroids and polyisoprenoid compounds as hormones, may not their neighbors on the phylogenetic tree, the echinoderms and the annelids, use them too? It is our belief that this important class of biologically active compounds plays a central role in the humoral control of growth in invertebrates as well as vertebrates. On another occasion we speculated (*39*) that, "rather than being a recent innovation of the vertebrates, steroid hormones may prove to have a far more ancient lineage." Studies in many laboratories during the past 5 years appear to have confirmed this conjecture.

#### References and Notes


1. A. Butenandt and P. Karlson, *Z. Naturforsch.* **96b**, 389 (1954).
2. C. G. Butler, R. K. Callow, N. E. Johnston, *Proc. Roy Soc. London* **155B**, 417 (1962).
3. M. Kobayashi and J. Kirimura, *Nature* **181**, 1217 (1958); M. Kobayashi, J. Kirimura, M. Saito, *ibid.* **195**, 575 (1962); J. Kirimura, M. Saito, M. Kobayashi, *ibid.*, p. 729; ———, *Mushi* **36**, 85 (1962) (published by Fukuoka Entomological Society, Kyushu University).
4. M. Ichikawa and H. Ishizaki, *Nature* **191**, 933 (1961); **198**, 308 (1963).
5. C. M. Williams, *ibid.* **178**, 212 (1956); *Science* **133**. 1370 (1961).
6. H. A. Schneiderman and L. I. Gilbert, *Anat. Record* **128**, 618 (1957).
7. L. I. Gilbert and H. A. Schneiderman, *Am. Zoologist* **1**, 11 (1961).
8. K. Hasegawa, *Nature* **183**, 397 (1957).
9. J. E. Steele, *Gen. Comp. Endocrinol.* **3**, 46 (1963). In this important paper Steele reports the discovery that the corpora cardiaca of insects secrete a substance which regulates the blood-sugar level in cockroaches. Secretion or injection of this agent causes glycogen breakdown in the fat body and a rapid increase in the circulating levels of the insect's blood sugar, trehalose. Similar observations have also been reported by W. S. Bowers and S. Friedman [*Nature* **198**, 685 (1963)]. These exciting findings open up experimentally the entire field of homeostatic regulation in insects. One can expect more reports in the near future of hormones which affect salt and water metabolism, levels of circulating metabolites, and so on [see, for example, S. H. P. Madrell, *J. Exptl. Biol.* **40**, 247 (1963)].
10. C. B. Cotrell, *J. Exptl. Biol* **39**, 413 (1962); G. Fraenkel and Hsiao, *Science* **138**, 27 (1962).
11. We sincerely thank Dr. Andre Meyer, senior biochemist at the Schering Corporation, Bloomfield, N.J., for many helpful discussions and suggestions and for penetrating criticism. We also thank Miss Irene Colvin, Miss Barbara Keelor, and Mrs. Frances Parson, who conducted thousands of bioassays with fastidious care, and Professor Dietrich Bodenstein and Drs. John S. Edwards, Hans Laufer, and J. A. L. Watson for their helpful comments on typescript. Experiments from our laboratories were supported in part by research grants from the U.S. Public Health Service and the National Science Foundation.
12. Following a common convention of insect endocrinologists, we use the term *brain hormone* in this article to refer specifically to the humoral agent which activates the prothoracic glands. The brain is the source of other humoral agents besides this one.
13. An excellent synthesis and review of earlier work is found in V. B. Wigglesworth, *The Physiology of Insect Metamorphosis.* (Cambridge Univ. Press, Cambridge, England, 1954).
14. P. Karlson, *Angew. Chem. Intern. Ed. Engl.* **2**, 175 (1963).
15. W. G. Van der Kloot, *Ann. Rev. Physiol.* **24**, 491 (1962).
16. V. B. Wigglesworth, *The Control of Growth and Form* (Cornell Univ. Press, Ithaca, N.Y., 1959).
17. R. D. Goodfellow and L. I. Gilbert, unpublished observations.
18. P. Karlson (*14*) reports high concentrations of cholesterol in *Bombyx mori.* He expresses doubts that cholesterol is brain hormone and argues that regulating hormones must be formed by the organism itself and cannot be part of the diet. The argument has some general validity, although it does not apply to *Bombyx*, which feeds on leaves that do not contain cholesterol, and does in fact make its own cholesterol by modifying the sterols of its food plant.
19. C. M. Williams, *Biol. Bull.* **93**, 89 (1947).
20. H. Oberlander and H. A. Schneiderman, in *Proc. Intern. Congr. of Zool. 16th* **2**, 67 (1963).
21. P. Karlson, H. Hoffmeister, W. Hoppe, R. Huber, *Ann. Chem.* **662**, 1 (1963).
22. W. Beermann, *Chromosoma* **5**, 139 (1952); ———, *Z. Naturforsch* **76**, 237 (1952); ———, *Chromosoma* **12**, 1 (1961); M. E. Breuer and C. Pavan, *Caryologia, Suppl.* **6**, 758 (1954); ———, *Chromosoma* **7**, 371 (1955); H. Kroeger, *ibid.* **11**, 129 (1960).
23. W. Beermann, *Am. Zoologist* **3**, 23 (1963).
24. ———, in *Developmental Cytology* [16th symposium of the Society for the Study of Development and Growth], D. Rudnick, Ed. (Ronald Press, New York, 1959), pp. 83–103.
25. H. A. Schneiderman and L. I. Gilbert, in *Cell, Organism, and Milieu* [17th Symposium of the Society for the Study of Development and Growth], D. Rudnick, Ed. (Ronald Press, New York, 1959), pp. 157–184.
26. U. Clever and P. Karlson, *Exptl. Cell Res.* **20**, 623 (1960).
27. U. Clever, *Chromosoma* **13**, 385 (1962).
28. H. Kroeger, *J. Cellular Comp. Physiol.* **62** (suppl. 1), 45 (1963).
29. ———, in *Proc. Intern. Congr. Zool. 16th* **4**, 251 (1963); unpublished observations.
30. H. J. Becker, *Chromosoma* **13**, 341 (1962); see also U. Clever, *J. Insect Physiol* **8**, 357 (1962).
31. H. Laufer *et al.*, unpublished observations.
32. F. Jacob and J. Monod, *J. Mol. Biol.* **3**, 318 (1961).
33. A. Krishnakumaran and H. A. Schneiderman, unpublished observations.
34. H. J. Pohley, *Arch. Entwicklungsmech. Organ.* **153**, 443 (1961).
35. H. Ursprung, *Develop. Biol.* **4**, 22 (1962).
36. D. Bodenstein, *J. Exptl. Biol.* **136**, 89 (1957); personal communications.
37. D. M. Guthrie, *J. Insect Physiol.* **8**, 79 (1962).
38. The problem may be stated in another way. Most insect cells have a certain endogenous biosynthetic "noise level." The effects of hormones like ecdysone must be assessed against this "noise level." The insects of choice for analyzing the action of ecdysone are obviously insects with low endogenous biosynthesis. For this reason a great deal of effort has been spent on studying diapausing pupae of giant silkworms, which are about as biosynthetically "quiet" as organisms can be. See, for example, H. A. Schneiderman and C. M. Williams, *Biol. Bull.* **106**, 210 (1953); W. H. Telfer and C. M. Williams, *J. Insect Physiol.* **5**, 61 (1960); F. Stevenson and G. R. Wyatt, *Arch. Biochem. Biophys.* **99**, 65 (1962).
39. L. I. Gilbert and H. A. Schneiderman, *Science* **128**, 844 (1958).
40. H. A. Schneiderman and L. I. Gilbert, *Biol. Bull.* **115**, 530 (1958); H. A. Schneiderman, *Casopis Ceskoslovenské Spolecnosti Entomol.* **58**, 12 (1961); H. A. Schneiderman, L. I. Gilbert, M. Weinstein, *Nature* **188**, 1041 (1960); C. M. Williams, L. V. Moorhead, J. F. Pulis, *ibid.* **183**, 405 (1959).
41. P. Schmialek, *Z. Naturforsch.* **16b**, 461 (1961).
42. V. B. Wigglesworth, *J. Insect Physiol.* **7**, 73 (1961).
43. R. T. Yamamoto and M. Jacobson, *Nature* **196**, 908 (1962).
44. L. I. Gilbert and H. A. Schneiderman, *Trans. Am. Microscop. Soc.* **29**, 38 (1960).
45. Several authors have reported that a variety of toxic compounds such as colchicine produce "abnormal adults containing large areas of a thick rough cuticle without scales" which was "pupa-like" [K. Slama, *Casopis Ceskoslovenské Spolecnosti Entomol.* **58**, 117 (1961); **59**, 323 (1962)]. Such effects have been called "pseudojuvenilizing," and it has been claimed that the effects of juvenile hormone extracts are "pseudojuvenilizing" and not true juvenilizing effects. The available evidence hardly favors this view. An elegant study of the cuticle of *Galleria mellonella*, the insect upon which most of our juvenile hormone bioassays are based, has been conducted by Adelheid Heims [*Arch. Entwicklungsmech. Organ.* **148**, 538 (1956)]. Heims points out that the structure of adult cuticle is similar to certain types of pupal cuticle and to certain types of larval cuticle. Nonetheless, the cuticles can be distinguished by their surface architecture, by their thickness, and by the structures they bear (scales, setae, and so on). We have made approximately 40,000 wax tests in *Galleria* and in other Lepidoptera. In fewer than two cases per 1000 were we unable to decide whether a juvenile hormone effect was positive (and in all such cases the experiments were repeated). Thus, there is no difficulty in distinguishing "pupa-like" cuticle induced by colchicine and other toxic agents from true pupal cuticle. True pupal cuticle and the second pupal cuticle which is secreted in response to juvenile hormone compounds are identical. Pupal cuticle has a characteristic surface architecture and pigmentation and bears pupal setae of characteristic structure. Pupa-like cuticle is abnormal adult cuticle; it does not have this surface architecture and it does not bear pupal setae. Indeed, the reason we have chosen the pupal cuticle of Lepidoptera for our hormone test system is that it provides a clear, unequivocal, highly sensitive yet highly specific end point in the biological assay for juvenile hormone, an end point which cannot be confused and which is not subjective. Furthermore, we routinely run parallel assays in which the agents to be tested are injected as well as applied topically. In such assays many structures besides the cuticle (for example, the genitalia) are affected by juvenile hormone [see L. I. Gilbert and H. A. Schneiderman (*44*)].
46. The following compounds were inactive in all of our test systems—that is, they had less than $5 \times 10^{-5}$ times the activity of cecropia oil and less than $5 \times 10^{-4}$ times the activity of purified all-*trans* farnesol: abietic acid, absynthin, allyl alcohol, ambrettolide, aritason, ascaridole, batyl alcohol, bixin, camphene, carotene, D-carvone, L-carvone, cedrene, cedrol, chimyl alcohol, cholesterol, cholesteryl chloride, cineole, citral, citronellal, citronellol, 1-decanol, *n*-dibutyl phthalate, digitoxin, *n*-dodecyl alcohol, ergosterol, estradiol benzoate, fenchone, fenchyl alcohol, geraniol, glutathione, 1-heptadecanol, 1-hexacosanol, δ-hex-α-ene lectone, 1-hexanol, hydrocortisone alcohol, α-ionone, β-ionone, iridodial, isodibutyl phthalate, isoiridomyrmecin, lanosterol, lignoceryl alcohol, D-limonene, linalyl acetate, lumisterol, lycopene, menthone, monobutyl phthalate, myrocene, nerol, 1-nonadecanol, 1-nonanol, octadecanol, 1-octanol, oleic acid, phthalic anhydride, pinene, 2-propyn-1-ol, pulegone, pyrethin I, α:β-santalol, selachyl alcohol, stearyl alcohol, tetradecanol, α-thujone, β-thujone, undecanol, $\Delta^{10}$-undecenol, vitamin A, vitamin A acetate, vitamin A alcohol, vitamin D, vitamin D-3, and vitamin K.
47. A few commercial preparations of other compounds (for example, hexadecanol) showed slight juvenile hormone activity, but gas chromatographic analysis revealed that these preparations were contaminated with other compounds. When preparations of these compounds were obtained which appeared pure when analyzed by gas chromatography, they proved to be inactive. We continue to be puzzled by the effect of 1-heptanol; this is the only pure substance which is not a polyisoprenoid derivative in which we have been able to detect juvenile hormone activity. So far as we can determine, 1-heptanol has slight but real juvenile hormone activity.
48. V. B. Wigglesworth, *J. Insect. Physiol.* **9**, 105 (1963).
49. The mode of applying these hormones is crucial, and it is difficult to compare in detail the results of different bioassays. For example, when farnesol is injected either as a pure substance or in peanut oil, it is toxic to polyphemus pupae at doses of 5 mg per gram, live weight. By contrast, if it is suspended in distilled water or made up in a solution of 10 percent ethanol, injections of 5 mg per gram are tolerated and have

hormonal effects. Similar problems arise in the topical application of farnesol. In a typical experiment, solutions of farnesol were mixed with wax and applied to *Galleria* pupae in a wax test. In one series the farnesol was dissolved in peanut oil and mixed with wax; in a second series the farnesol was dissolved in an inactive oil derived from allatectomized male cecropia moths. The farnesol showed ten times as much activity when it was dissolved in inactive insect oil as when it was dissolved in peanut oil, although the inactive insect oil alone had no juvenile hormone activity. Formulation thus appears to be all-important in the bioassays. Possible reasons for these differences have been discussed elsewhere [see L. I. Gilbert and H. A. Schneiderman (*44*); V. B. Wigglesworth (*48*)]; they include differences in rates of inactivation or conversion of the active principle in different insects, the presence of inhibitors, the presence of substances which facilitate penetration of the active principle, and so on.

The lack of comparability of the several bioassays now used demands that extreme care be taken in making quantitative comparisons of activity when different bioassays are used. Indeed, different bioassays give different relative activities. Thus, in the *Tenebrio* assay of Karlson [Karlson and Nachtigall, *J. Insect Physiol.* **7**, 210 (1961)], farnesol is about as active as cecropia oil, and farnesyl methyl ether and farnesyl diethylamine are 100 times more active [see P. Karlson (*14*)]. In our wax test, even highly purified farnesol is less than 15 percent as active as active male cecropia oil, and in both the *Galleria* wax test and the polyphemus injection assay, farnesyl methyl ether and farnesyl diethylamine are about as active as cecropia oil.

A final problem in assaying terpene derivatives and related compounds is that many of them isomerize readily and are somewhat unstable. Indeed, there is considerable opportunity for change, both in the insect after injection or topical application and in the bottle prior to testing. The possiblity that the isomers or decomposition products may interfere with the action or metabolism of the active principle must also be reckoned with.

50. Because of the widespread distribution of these juvenile-hormone-like compounds, some workers have questioned whether there is any true juvenile hormone in the ether extracts which Williams (*5*) first prepared from male cecropia moths and which are now used in many laboratories as a starting material for juvenile hormone purification [see, for example, J. Beetsma, L. de Ruiter, J. deWilde, *J. Insect Physiol.* **8**, 251 (1962); H. E. Hinton, *Sci. Progr. London* **51**, 306 (1963); K. Slama, *Casopis Ceskoslovenské Spolecnosti Entomol.* **58**, 117 (1961); ———, *ibid.* **59**, 323 (1962)]. For example, they point out that these extracts do not affect lepidopterous larvae. Recently we have conducted experiments which strongly support the view that the material found in male cecropia moths is the juvenile hormone. In one experiment [H. A. Schneiderman and I. Colvin, unpublished] a group of ten male cecropia pupae were allatectomized. Special care was taken to insure that the corpora allata were completely removed and that no small fragments remained. These animals were allowed to develop into moths, and 3 days after they had emerged each was extracted with ether and the extracts were tested for juvenile hormone activity. We used the *Galleria* wax test, which is sensitive and permits detection of juvenile hormone activity in extracts which contain about $10^{-5}$ as much of the activity as ether extracts of male cecropia moths. None of these extracts from allatectomized animals had any juvenile hormone activity. Hence, it appears that the juvenilizing principle in the male cecropia extracts is produced by the adult corpora allata. In a second experiment we attempted to determine whether the adult corpora allata produce a substance which has juvenilizing effects on larvae—in other words, juvenile hormone (L. Fuchsman, J. S. Edwards, H. A. Schneiderman, unpublished). We implanted adult corpora allata into last-instar larvae of *Galleria*. The donors were *Samia cynthia* moths, close relatives of cecropia; from *S. cynthia* active juvenile hormone extracts are obtained which are very similar to cecropia extracts. In a number of cases implantation of three pairs of adult corpora allata caused the larvae to molt into larval-pupal intermediates in which the only pupal characteristics were a mid-dorsal suture of pupal cuticle in the prothoracic segment and the beginnings of pupal antennae. In a similar experiment, S. Fukuda [*Annotationes Zool. Japon.* **35**, 199 (1962)] demonstrated that adult corpora allata of the silkworm, *Bombyx mori*, produce a substance which causes larval molting in allatectomized fourth-stage larvae. Thus, the corpora allata of adult moths produce an agent which has all of the characteristics of the juvenile hormone, and can cause supernumerary larval molts in Lepidoptera and replace extirpated larval corpora allata. This agent is not found in allatectomized animals. For these reasons, and in the absence of contrary evidence, we believe that the active principle in male moths is the juvenile hormone, although we recognize that this has not been proved beyond the shadow of a doubt.
51. For references to "bar substances," see S. Kaji, *Mem. Coll. Sci. Univ. Kyoto*, **B25**, 161 (1958).
52. A. J. Clark and K. Bloch, *J. Biol. Chem.* **234**, 2578 (1959); R. Boch and D. A. Shearer, *Nature* **194**, 704 (1962).
53. D. H. Chen, W. E. Robbins, R. E. Monroe, *Experientia* **18**, 577 (1962).
54. L. I. Gilbert and H. A. Schneiderman, *Nature* **184**, 171 (1959); C. M. Williams, *Biol. Bull.* **116**, 323 (1959).
55. H. A. Schneiderman and L. I. Gilbert, unpublished observations.
56. M. Ichikawa and J. Nishiitsutsuji-Uwo, *Biol. Bull.* **116**, 88 (1959).
57. D. Bodenstein, in *Recent Advances in Invertebrate Physiology*, B. T. Sheer, Ed. (Univ. of Oregon Publ., Corvallis, 1957), p. 197.
58. C. M. Williams, *Biol. Bull.* **103**, 120 (1952).
59. L. I. Gilbert and N. Goodfellow, unpublished observations.
60. B. Scharrer, unpublished observations.
61. H. Ichikawa and J. Nitshiitsutsuji-Uwo, *Mem. Coll. Sci. Univ. Kyoto* **B27**, 9 (1960).
62. M. Kobayashi and W. J. Burdette, *Proc. Soc. Exptl. Biol. Med.* **107**, 240 (1961).
63. D. Bodenstein, unpublished observations.
64. See, for example, R. B. Clayton, *J. Biol. Chem.* **235**, 3421 (1960); D. Gilmour, *Biochemistry of Insects* (Academic Press, New York, 1961); Z. H. Levinson, *Proc. Intern. Congr. Entomol. 11th, Vienna, 1958* (1960), vol. 3B, pp. 145, 154; *J. Insect Physiol.* **8**, 191 (1962).
65. H. House, *Ann. Rev. Biochem.* **31**, 653 (1962). There have been two recent claims that certain insects are capable of sterol biosynthesis from $C^{14}$ acetate [R. B. Clayton, A. M. Edward, K. Bloch, *Nature* **195**, 1125 (1962); M. Saito, M. Yamazaki, M. Kobayashi, *ibid.* **198**, 1324 (1963)]. In both cases the sterol-synthesizing activity of microbial and fungal contaminants was not excluded, hence the results are not convincing. However, the possibility still exists that certain insect tissues are capable of some sterol synthesis.
66. On the basis of nutritional studies, a number of workers have been led to the view that insects use sterols for hormone synthesis. M. G. Horning [in *Cholesterol*, R. P. Cook, Ed. (Academic Press, New York, 1958), p. 451] wrote: "It seems likely that the sterol acts as a source of steroidal substances which probably fill hormonal functions in insects." Similarly, Z. H. Levinson [*Proc. Intern. Congr. Entomol. 11th, Vienna, 1958* (1960), vol. 3B] speculated that "cholesterol may act as a precursor of steroid hormones, in addition to its structural role. . . ." Very recently P. Karlson and H. Hoffmeister [*Z. Physiol. Chem.* **331**, 298 (1963)] have shown that tritium-labeled cholesterol is converted by fly larvae into ecdysone, and that cholesterol is a precursor of ecdysone.
67. A. Krishnakumaran and H. A. Schneiderman, unpublished observations.
68. Highly purified *trans-trans* farnesol used in these experiments was generously provided by Givaudan and Company, Geneva, Switzerland, and was prepared from all-*trans* nerolidol. Hexahydrofarnesol and farnesyl acetate were prepared by Dr. John Pollard and his associates at the California Biochemical Corporation and were assayed as 99.5 percent pure. Other preparations were obtained from diverse commercial sources and were said to be at least 90 percent pure. Different batches of these commercial substances varied somewhat in activity. The values given in Table 4 are typical. Hoffmann LaRoche, Inc., of Basel, Switzerland, generously provided samples of farnesyl methyl ether and samples of farnesyl diethylamine and gave us permission to publish the results of experiments made with these compounds.

# C. Diapause

## Physiology and Ecology of Photoperiodism[1]

*By* Stanley D. Beck[2]

*University of Wisconsin*

The invitation to present this year's Memorial Lecture has come to me as a very great but undeserved honor. Although a pleasant and extremely challenging assignment, it is one that I have approached with no little trepidation, in view of the outstanding quality of the lectures presented by my predecessors. I wish to express my deepest appreciation to the Society for placing this high trust in my hands.

The memory we honor this year is that of S. A. Forbes, a pioneer in insect ecology and a number of other aspects of modern entomology. It is a commentary on the speed at which scientific fields are being explored and masses of knowledge being accumulated within our times, to note that at the time of Professor Forbes' active career, the word *photoperiodism* and the subject of study denoted were completely unheard of. With a few minor exceptions, photoperiodism as a biological phenomenon was not recognized until the 1920's, following the work of two botanists (Garner and Allard 1920). And it has been only within the past two decades that there has been an active growing interest in the zoological aspects of the subject.

Stanley D. Beck

Photoperiodism is a relatively new field of biological research, but it is one that has matured rapidly, has developed its own concepts, methods, language, and biases. I do not wish to review the history of the subject. Our concern here is the current status, the principal concepts, and the future promise of photoperiodism as an area of biological interest.

Photoperiodism is the responses of organisms to the daily and seasonal environmental rhythm of daylight and darkness; that is, to photoperiod. Because the earth has been revolving since the beginning of its existence, life on earth arose and has pursued its evolutionary course under the constant influence of a 24-hour environmental rhythm of photoperiod. Although this is an obvious geophysical fact, we are only now beginning to appreciate the tremendous biological importance of the photoperiodic rhythm. We find its influence in the form, behavior, physiology, and ecology of nearly all forms of life, including man. Photoperiod appears to play an important role in the internal organization and external relationships of plants and animals.

Seasonal changes in the number of daylight hours per 24-hour day are also biologically important. It is, of course, well known that the number of hours elapsing between sunrise and sunset at any given locality depends on only two factors: the time of the year, and the latitude of the locality in question. The main body of the continental United States lies, roughly, between 50° N. and 30° N. latitudes. At 50° N. the summer days are longer than at 30° N., and the winter days are shorter, and the rate of change during spring and autumn is also greater. The annual cycle of daylengths has also played an important part in biological affairs on earth. These seasonal cycles are exactly the same from year to year, and living things have evolved the ability to use them as guide lines in their seasonal biology. As the days become shorter in late summer, insects and other forms of life begin to undergo the physiological changes necessary for winter survival.

The ability of organisms to use photoperiodic cycles as clues to impending seasonal changes implies that they possess the ability to distinguish between long and short daylengths. This would seem to require some time-measuring ability, and leads us to believe that organisms possess a physiological clock system. Animal clock mechanisms is the subject of an invitational address by Professor C. S. Pittendrigh later in the program, and I will not, therefore delve into this aspect of periodism.

Turning now to the specific effects of photoperiod on insect biology, one immediately thinks of the fact that some insects are nocturnal and others are diurnal in habit. In the case of a nocturnal form, such as a cockroach, the change from daylight to darkness was once thought to serve as the stimulus inciting locomotor and feeding behavior. The beginning of daylight would act as a suppressing stimulus, and the insect would remain relatively quiescent, unless disturbed, during daylight hours. On the other hand, the response to day-night changes would be just the opposite in a diurnal form, such as the house fly. The situation is not that simple, however, and a more fundamental periodism is to be found in insect behavior. This is schematically shown in Figure 1, which is based largely on the results of Harker (1956, 1960 a, b, c) with the American cockroach, *Periplaneta americana* (Linnaeus).

In Figure 1, the light and dark hours of the photoperiods are represented as white and black bars beneath the time scales. Locomotor activity periods are represented by triangles rising above the time scale baselines. A regular recurring photoperiod (left column) results in the cockroaches showing a maximum of physical activity shortly after the beginning of the dark phase of the cycle each day. After several hours of activity, the insects became relatively inactive until the next day at the same time. The center column shows the activity when the insects were kept in darkness after the first day. The activity rhythm continued; the period of the rhythm—that is, the number of hours between peaks of activity—continued to be approximately 24 hours. The period of the rhythm was a little less than 24 hours, as evidenced by the fact that activity began a little earlier

[1] The 1962 Founders Memorial Lecture. Presented on December 3, 1962, at the annual meeting of the Entomological Society of America in Phoenix, Arizona.

[2] The original research results presented in this paper are from projects supported in part by a grant (GM 07557) from the National Institutes of Health of the U. S. Public Health Service. The author was the recipient of the 1962 Founders Memorial Award.

Reproduced with permission from the *Bulletin of the Entomological Society of America,* 9:8–16, 1963. Published by the Entomological Society of America, College Park, Maryland.

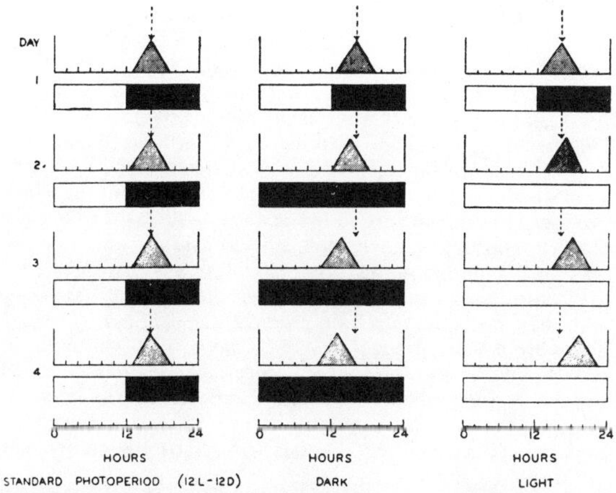


Fig. 1.—Locomotor activity rhythm of the American cockroach, *Periplaneta americana* (Linnaeus), under different photoperiodic conditions (schematic).

each day. When the cockroaches were maintained under continuous light (right column), the activity rhythm also continued, but with a periodicity of slightly more than 24 hours.

Such a pattern of activity rhythms is quite informative. First, we see that the nocturnal activity rhythm is not the simple result of the onset of darkness. The beginning of darkness does not, in itself, cause the activity. A more fundamental 24-hour periodicity is apparent, which does not depend on photoperiod for its expression. Under constant conditions—either light or dark—the activity rhythm appears to be endogenous. Also this tells us that the cockroach has the ability to measure time, as the rhythm shows a remarkable regularity. The fact that the period of the rhythm is not exactly 24 hours is also quite significant. Its period is only approximately 24 hours, and for this reason, such a rhythm has been named *circadian*—that is, *about daily.* It appears that the biological effect of photoperiod is to maintain the rhythm of activity in synchrony with the daily cycle of light and darkness.

The emergence of adult insects from pupae also has been found to be rhythmic. Since emergence involves considerable physical exertion, this is probably another example of an activity rhythm. Adult emergence rhythms have been demonstrated among moths of the genus *Ephestia* (Moriarty 1959) and flies of the genus *Drosophila* (Brett 1955, Pittendrigh 1954). Since each insect can emerge but once, emergence rhythms can be studied only on a population basis. In *Drosophila* cultures, for example, that have been reared in continuous dark from the egg stage, adult emergence is random in regard to the time of the day that the adult flies emerge. When reared under a photoperiod, however, a well-defined pattern of fly emergence becomes apparent. Most of the flies that are ready to emerge on a given day will do so shortly after the beginning of the lighted phase of the photoperiodic cycle. The photoperiod has, in such a case, served to synchronize the endogenous activity rhythms of the individual members of the population. Pittendrigh (1960) has shown that a *Drosophila* population reared in the dark can be synchronized by but a single brief light signal occurring during larval development. As the result of a single flash of light, an adult emergence rhythm becomes apparent when the flies begin to emerge several days later.

Such activities as feeding, flying, mating, and oviposition have been found to display daily periodism in many species. And in most cases, these behavior patterns cannot be accounted for solely on the basis of the insects' immediate responses to light and dark as "token stimuli." It is quite apparent that there is a more fundamental endogenous periodism involved, and that the environmental rhythm of photoperiod adjusts and regulates the manifestation of the endogenous rhythms (see Harker 1961 for a recent review).

To learn more about insect periodism in general and photoperiodism in particular, we must look beyond the outward behavior, and inquire into the physiological processes involved. The best study, to date, of the physiology of insect periodism is represented by the work of Harker (1956, 1960a, b, c) on the locomotor activity rhythm of the American cockroach. She found that if the insect's head were removed, the insect survived but became lethargic and no longer displayed an activity rhythm. If, however, the decapitated cockroach were put in parabiosis with a normal roach, the rhythm returned. The timing of the reestablished rhythm was determined by the insect possessing a head, and not by the headless member of the pair. Such an experiment tells us that the activity rhythm is caused by something produced in the head, carried by the blood, and inducing physical activity.

The immediate explanation of this effect is that hormones are involved, and that these hormones are produced in the head of the insect. Dr. Harker was able to demonstrate that such is the case. Through a series of ingenious experiments, she showed that the activity rhythm is induced by hormones produced by specialized neurosecretory cells located in the subesophageal ganglion. The ganglion displayed a daily rhythm of hormone production, and this rhythm was found to be the driving force underlying the daily activity rhythm. Rhythmic hormone production by the subesophageal ganglion continued even when it had been severed from the brain. But when separated from the brain, the activity rhythm was no longer sensitive to photoperiod. The rhythmic production of hormones by the subesophageal ganglion is apparently regulated by a control center in the brain, and it is this brain center that is sensitive to photoperiod.

And what is this brain system? It appears to be, at least in part, the complex of hormone-producing tissues—the neurosecretory cells of the so-called pars intercerebralis, the corpora allata, and the corpora cardiaca. There is a small nerve running from the cardiacum and allatum down to the subesophageal ganglion. Dr. Harker found that neurosecretory products from the brain reach the subesophageal ganglion via this nerve. If this small nerve is cut, the brain no longer controls the hormone production of the subesophageal ganglion.

This was a most significant discovery, because these hormone-producing organs are known to be involved in growth and differentiation processes. Their sensitivity to photoperiod is, therefore, of tremendous biological importance. For here, through this reacting system, the environmental rhythm of day and night and the seasonal cycles of daylength may exert profound effects on nearly all aspects of insect biology. If photoperiod can play a part in the timing and programming of endocrine events, we should expect to find instances where photoperiod influences metabolism, growth rate, diapause induction, diapause termination, form differentiation, and gametogenesis. In one species or several, all of these processes have been shown to be affected by the environmental rhythm of photoperiod (Ball 1958, Beck 1962, Corbet 1956, Dickson 1949, Fukaya and Mitsuhashi 1961, Grison 1949, Hodek and Cerkasov 1958, Lees 1952, Paris and Jenner 1959, Parker and Thompson 1927).

In my laboratory at the University of Wisconsin, we have recently begun an investigation of metabolic periodism. For this study, we are using the German cockroach, *Blattella germanica* (Linnaeus). Although the investigation is still in its preliminary stages, the results obtained appear promising. Reasoning that a daily rhythm of hormone production and physical activity should be accompanied by a daily rhythm in the rate of metabolism, we started by looking for a rhythm in the insect's rate of oxygen consumption. As shown in Figure 2, such a rhythm has been found. The rate of oxygen uptake was at its daily low near the beginning of the light phase of the photoperiodic cycle. The rate increased slowly, and then more rapidly to reach a peak at about one hour after the beginning of the dark phase. It then declined, rose again slightly toward the latter part of the dark phase, and finally dropped sharply to reach its minimum at the beginning of the next light phase. This rather complex respiratory rhythm continues day after day. We have been able to show that the respiratory rhythm is regulated through photoperiod, because we can adjust the positions of the peaks and troughs of the daily rhythm by making systematic changes in the photoperiodic schedule.

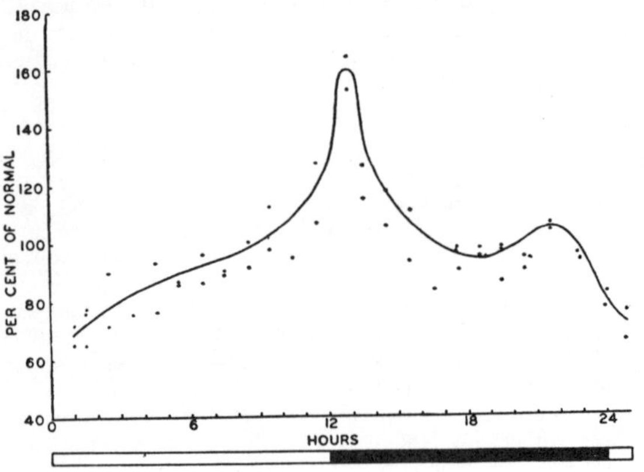


Fig. 2.—Photoperiodically regulated respiratory rhythm in adult male German cockroaches.

It seems very likely that the daily rhythm of metabolic rate is a result of the hormone production rhythm and is the energy source driving the daily rhythm of physical activity. However, measurements of crude oxygen consumption cannot tell us a great deal about the specific aspects of metabolism that might be involved in the insect's photoperiodism. For this reason, we are now beginning some more specific experiments. We thought that it might be of interest to test the possibility that the insects would show rhythmic changes in their sensitivity to toxicants and metabolic inhibitors. From this type of experiment, we hope to get leads on specific metabolic reactions involved in photoperiodism.

Male German cockroaches that had been well conditioned to a photoperiodic rhythm were injected with toxicants at different times during the 24-hour photoperiod. They were then held in continuous dim light, and the incidence of mortality was followed during the subsequent 24 hours. We found a sensitivity rhythm in their responses to DDT, Dimetelan, potassium cyanide, and other compounds. Figure 3 shows the response to a standard dose of potassium cyanide. Such a dose administered at different times produced quite different 24-hour mortality. Like the oxygen consumption rhythm, the sensitivity to cyanide was at a minimum at the beginning of the light, increased to a maximum shortly after the onset of darkness, and then declined to a minimum at the beginning of the next cycle. Sensitivity rhythms have been detected in the insect's response to the other chemicals mentioned above. It seems likely that the demonstration of rhythm sensitivity to toxicants might be of interest to those entomologists involved with bioassay systems and toxicology problems.

The aspect of insect biology that is the most profoundly affected by photoperiod is that of growth and metamorphosis. Growth processes are controlled by the endocrine organs concentrated in the insect's head, and since it is this endocrine center that is influenced by photoperiod, we might expect that photoperiod would affect growth patterns. One of the most spectacular of these effects is that of seasonal dimorphism. More and more evidence is accumulating in support of the concept that seasonal dimorphism is a photoperiodic response. Since the early work of Marcovitch (1924), it has been known that the appearance of sexual forms in some aphids is a response to short daylengths. The recent work of Lees (1959, 1960a, b) on the photoperiodism of a vetch aphid, *Megoura viciae* Buckton, has done much to elucidate some of the physiological processes involved in this type of dimorphism. Seasonal dimorphism manifested as different color patterns is known to occur among butterflies; the work of Muller (1960) and Pease (1962) has shown that such dimorphism is photoperiodically induced in at least some species. Muller (1957 and 1960) has also demonstrated photoperiodically induced anatomical dimorphism in some homopterous species.

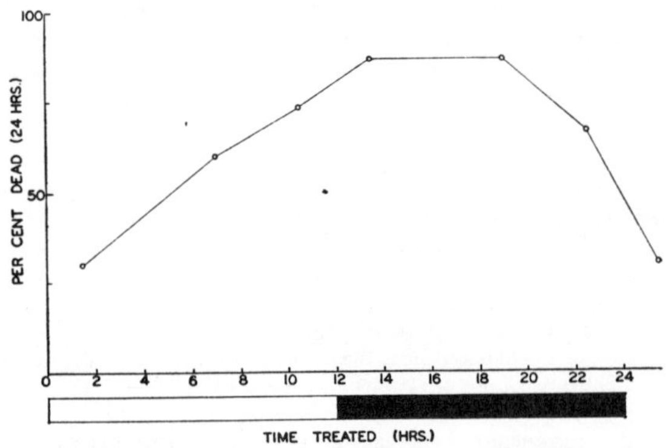


Fig. 3.—Photoperiodically regulated rhythm of sensitivity to potassium cyanide in the German cockroach (adult males).

The aspect of photoperiodic control of growth that has been studied most intensively is the phenomenon of diapause. Diapause is a state of arrested development in which the insect's respiratory metabolism has been suppressed (sometimes to an exceedingly low level); morphogenesis and gametogenesis have also been arrested. In the temperate parts of the world, most insect species survive the winter in a state of diapause. Diapause may occur in the egg, larval, pupal, nymphal, or adult stages, depending on the species. In some, diapause regularly occurs in every generation and is said to be obligatory. In other species, diapause is facultative, and there may be one or more nondiapausing generations within a growing season, followed by a generation that diapauses through the winter. From the standpoint of photoperiodism, species showing facultative diapause are of the greatest interest, because in many of these species, diapause occurs in response to photoperiod. As the days get short in late summer, the insects enter diapause and are ready for winter long before cold weather arrives (for reviews see Lees 1955, DeWilde 1962).

An insect is a highly specialized, complex organism, and its growth and morphogenesis must involve an intricate programming of physiological events. How photoperiod may act to modify the programming to cause diapause is a problem of considerable biological interest and one of no small magnitude. At Wisconsin, we have been studying this problem for the past several years. Our principal experimental insect has been the European corn borer, *Ostrinia nubialis* (Hübner) which is a lepidopterous species with a facultative larval diapause. The corn borer overwinters as a mature larva in diapause.

In our experiments, borer larvae are reared from hatching to maturity on a standard synthetic diet (Beck and Smissman 1960). They are reared in constant-temperature incubators that have been wired with lights and timers so that the photoperiodic conditions may be controlled (Beck and Hanec 1960, Beck 1962). When we rear experimental populations of borers under different photoperiods, we find that different proportions of them go into the diapause state upon reaching larval maturity. This we express as diapause incidence in percent.

When the diapause incidence is plotted against the photoperiods in which the larvae were reared, a response curve is obtained (Fig. 4). Under continuous darkness or continuous light, there is no appreciable incidence of diapause. Diapause is induced by a narrow range of photoperiods—those having from 10 to 14 hours of light per day. This response curve is not unique to the European corn borer; it is quite similar to the curves obtained by a number of workers using a variety of insect species.

It is of interest to note that diapause is induced only by a combination of light and dark—both are required. In other words, diapause is a response to an environmental photoperiodic rhythm. It is also of interest to observe that the borer population responds quite sharply to small differences in photoperiod. A one-hour difference in the number of hours of light and dark per day may induce an 80 percent difference in the incidence of diapause.

This very sharp response range would lead us to believe that the larvae are able to measure the duration of either light or dark periods, or both. A long series of experiments has shown that the duration of the dark phase is very critical in the induction of diapause (Beck 1962). Diapause induction was shown to be closely associated with a 12-hour period of darkness; the duration of the light was found to be much less critical.

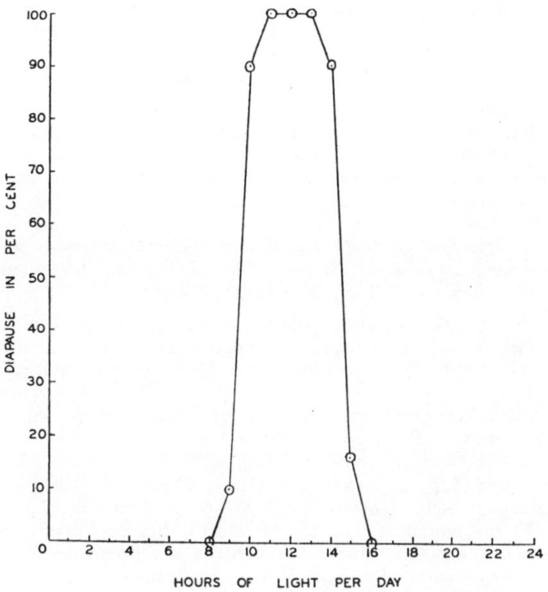


Fig. 4.—Effect of photoperiods on the incidence of diapause among larvae of the European corn borer (from Beck 1962).

We now come face-to-face with two fundamental questions concerning the insect's photoperiodism. How can the insect measure the length of the dark period? And once measured, how does photoperiod determine the developmental fate of the insect? The European corn borer larva must follow either of two developmental pathways. It may reach larval maturity and pupate, or it may reach larval maturity and enter diapause. Sometime during larval development, the insect is committed to one or the other developmental course. With some insect species, it has been found that the insect is sensitive to photoperiod only at one sharply defined time in its life cycle. With the European corn borer, however, we have not found this to be the case. Although the temperature and photoperiod experienced by the larvae during the latter third of their development seems to be a major factor in diapause determination, the final response is at least modified by the conditions experienced during the earlier growth stages (Beck and Hanec 1960).

Diapause is induced in the European corn borer by short day photoperiods (*ca* 12 hours of dark per 24-hour day). Diapause can be terminated also by photoperiod. Subjecting diapausing larvae to either long day photoperiods (16 or more hours of light per day) or to continuous light brings about the termination of the diapause state. In this species, termination of diapause does not require a period of exposure to low temperatures (McLeod and Beck 1963). Many species require a chilling of several weeks duration before diapause can be ended, but the corn borer has no such requirement. Whereas we found the duration of the dark phase of the photoperiod to be the major factor in the induction of diapause, the light phase is the more important in terminating diapause. It should be pointed out that under field conditions, diapause in the corn borer is not terminated in response to the lengthening days of the spring season. Diapause has been completed early in the winter, and by spring the borers are not responsive to photoperiods; they pupate in response to moisture and rising temperatures.

Diapause is a complex physiological event in its induction, maintenance, and termination. It involves an arrest of growth and a slow-down of metabolism, and appears to be the result of changes in the hormone balance within the insect. Before going into the details of the humoral aspects of diapause, let us review briefly the principal concepts concerning the endocrinology of insect growth and differentiation. These are presented in a schematic fashion in Figure 5.

Four endocrine structures are known to be involved in insect growth processes: (1) neurosecretory cells in the brain (P in Fig. 5), which occur in from two to three groups—medial, lateral, and posterior. These are specialized nerve cells that have secretory as well as neural functions; (2) corpora cardiaca (CC), a pair of neuroendocrine organs communicating with the brain via two nerves; (3) corpora allata (CA), a pair of endocrine glands closely associated with the corpora cardiaca; and (4) the prothoracic glands (PG), which are epidermally derived endocrine organs located in the prothorax and posterior parts of the head. In some forms—such as the European corn borer—the corpora cardiaca and corpora allata are fused to form a single pair of organs. In the higher Diptera, all four endocrine organs are fused to form the so-called ring gland.

The general sequence of humoral events associated with growth and metamorphosis is taken as follows: The medial and lateral neurosecretory cells of the brain elaborate substances that are transferred via their axons to the corpora cardiaca. The corpora cardiaca release the *brain hormone* into the blood stream, probably via nerve fibers to the dorsal vessel. This brain hormone stimulates somatic metabolism and also stimulates the prothoracic glands to elaborate and release the growth and molting hormone *ecdysone*. Ecdysone exerts its physiological effects by stimulating other tissues—such as epidermis and gonads—to initiate differentiation and molting processes. The corpora allata secrete the *juvenile hormone*, which plays a role in gonadal development and in determining whether the insect will molt to a larval, pupal, or adult stage. (For recent reviews of insect growth processes, see Schneiderman and Gilbert 1959, Wigglesworth 1959, Gilbert and Schneiderman 1961, Van der Kloot 1961).

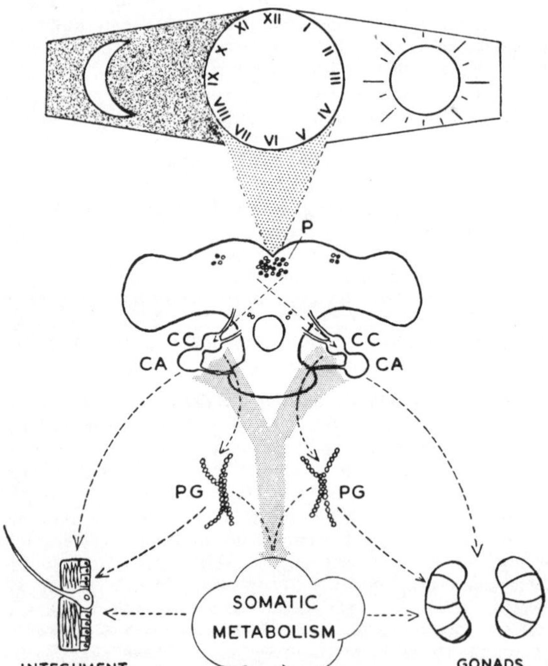


Fig. 5.—Schematic presentation of the endocrine structures and relationships involved in insect growth processes. (Symbols explained in text).

Diapause is currently thought to be caused by a failure in this series of events. Specifically, it is a failure of the brain to produce or to release the brain hormone. The absence of brain hormone results in inactive prothoracic glands and, therefore, no production of ecdysone. Without ecdysone, growth and differentiation come to a halt, metabolism is suppressed, and this arrested state is called diapause (Williams 1952, Van der Kloot 1955).

The insect cannot remain in diapause indefinitely; it must eventually resume developmental activity. During diapause, physiological changes must occur that lead to the termination of the state of diapause. These changes have been called "diapause development" by Andrewartha (1952). The completion of diapause is characterized by a renewal of brain activity (Van der Kloot 1955), production of brain hormone (Williams 1956), and a resumption of growth and differentiation.

As is the case with most generalized schemes, not all of the available experimental evidence is in agreement. The major features of the current concept of diapause were worked out for pupal diapause by C. M. Williams and his colleagues. Larval diapause in a number of dipterous and lepidopterous species appears to fit the concept reasonably well. Embryonic diapause, however, does not seem to have this sort of humoral basis, and adult diapause with gonadotrophic dissociation appears to involve the corpora allata as well as the other endocrine tissues (de Wilde and de Boer 1961). With minor reservations, larval diapause in the European corn borer is in agreement with the generalized picture, with diapause resulting from a failure of the brain to produce or to release the brain hormone.

Diapause in the borer, and in many other insects, is induced by shortday photoperiods; this might lead us to think that photoperiod exerts its effects directly on the neurosecretory cells of the brain. There is good evidence that the photoperiodic effect is on the insect's brain. Dr. Harker's work on the cockroach, discussed earlier, demonstrated that photoperiod influences a brain center, which in turn controls the rhythm of hormone production in the subesophageal ganglion. A brain-subesophageal ganglion system has been demonstrated also in the commercial silkworm, where it controls the production of diapause eggs (Fukuda 1951, Hasegawa 1957). Some very neat fiber-optic experiments by Lees (1960 a, b) have shown that the dorsal-central region of the aphid's brain is the center of the photoperiodic response system. Recent work by a Russian biologist (Shakhbazov 1961) has implicated the central portion of the brain of pupae of a Chinese oak silkworm, *Antheraea pernyi* (Linnaeus), as the site of photoperiodic sensitivity.

According to this concept, short day photoperiods exert a suppressing effect directly on the neurosecretory cells, switching off the neurosecretory activity. It does not seem likely, however, that the situation is quite that simple. The neurosecretory cells may well be the *effectors* of the photoperiodic response, but they must be merely following orders from elsewhere in the brain or central nervous system. Larval diapause in the corn borer occurs in the last larval instar. At each of the previous larval molts, the endocrine system was involved, and the brain cells functioned normally, in spite of the influence of a short-day photoperiodic cycle. We have found no evidence of even so much as a reduced growth rate that can be attributed to photoperiod (Beck and Hanec 1960). We also have found that the neurosecretory cells of the borer brain are potentially functional during diapause, but are held in an arrested state by some other factor (Cloutier et al. 1962). It seems fairly certain that the suppression of neurosecretory activity is the result of the action of photoperiod on an unidentified physiological system to which the neurosecretory system is coupled.

Since diapause comes as a photoperiodic response, there is good reason to suspect that diapause induction and maintenance is coupled to internal processes that show daily rhythms in their functioning. Under continuous light or darkness or under a long-day photoperiod, such rhythms would be free running, and development would be uninterrupted by diapause. Under the influence of short-day photoperiods, the phase relations of the rhythms would be altered, and the insect's developmental patterns would be directed down the path leading to diapause. Thus far, we have not been able to detect the existence of such rhythms in the European corn borer. Although we have found a respiratory rhythm not unlike that described earlier for the German cockroach (Beck et al. 1963), we have not been able to demonstrate a neurosecretory rhythm. Obviously, there is still much to be done in this field of interest. If we could identify the diapause-inducing system, we might also gain some information on how the insect's physiological system is able to measure time. But at this point we have little experimental evidence to lead us, and we must leave many questions unanswered.

Let us now turn to a consideration of the ecological aspects of photoperiodism. Photoperiod is an important environmental factor that influences the seasonal biology, ecological adaptations, and evolutionary trends in a vast multitude of insect species. We currently know very little about the ecology of photoperiodism, but what is known is enough to indicate that a meaningful and fruitful field of study awaits exploration. We must, of course, realize from the outset that photoperiod does not act in isolation; other ecological factors are always involved in the effects we measure. And of these other factors, moisture and temperature are probably the most important.

The most obvious ecological aspect of photoperiodism has to do with its role in phenology. The developmental stages and behavioral phases of an insect's biology must be maintained in some degree of synchrony with the changes occurring in the insect's environment. This is true not only in respect to the seasonal changes in the insect's physical climatic environment, but is also equally true in regard to the insect's biological environment. As insects emerge from their overwintering condition in the spring, it is important that they emerge into a tolerable environment in respect to temperature and moisture, and also in respect to available food, microhabitat, and the opportunity for mating and reproduction. In many instances, it has been found that the lengthening days of spring act as a factor helping to synchronize insect activity with the environment. Emergence from winter diapause has been shown to be a photoperiodic response in a few instances, most notably among midges and mosquitoes (Baker 1935, Palmen 1958, Paris and Jenner 1959). Undoubtedly, more such instances will be found as more species are investigated.

The role of photoperiod in the insect's preparation for adverse climatic conditions has been extensively investigated. The induction of diapause for survival of hot dry seasons—diapause for aestivation—and the induction of diapause for winter survival have been described for many species. The survival value of such diapause is quite obvious. The exact manner in which a particular species has become adapted varies, depending on the growth stage diapausing, and also depending on the other biological adaptations involved in the species' life history.

The ecological significance of some photoperiodic responses is not so immdiately apparent. Not all photoperiodism is involved in the insect's response to seasonal climatic cycles. A case in point concerns the photoperiodism of *Drosophila*. As discussed earlier, *Drosophila* adults tend to emerge from their puparia shortly after the beginning of the light phase of a photoperiod. Under natural conditions, this means that the flies tend to emerge an hour or so after dawn. This effect has been shown to be a photoperiodic response. This response has been found to be of considerable importance in the ecological adaptations of *Drosophila* species. Pittendrigh (1958) found that among *Drosophila* species living in relatively dry environments, the adult emergence rhythm is an important adaptation serving to minimize water loss. At about dawn, daily temperatures tend to be at their minimum and relative humidity tends to be at a maximum. Flies emerging at such a time are subjected to minimum water loss; emergence later in the day would expose the flies to much more severe conditions and the risk of fatal water loss. Pittendrigh and his co-workers have been able to demonstrate that the photoperiodism of a number of *Drosophila* species is an extremely important factor in their successful adaptation to the ecological niches they now occupy.

Finding that photoperiodism can involve adaptations important to water conservation, host plant phenology, feeding activity, reproduction, and diapause would lead us to believe that photoperiodism might be a factor in the differentiation of geographical populations and in speciation. From the currently available evidence, it would seem certain that such is the case. A great deal more research is needed, however.

To explore this problem, we shall again use the example of diapause in the European corn borer. In studying the photoperiodic induction of diapause, we use experimental populations of insects. Each individual insect either diapauses or does not, depending upon whether the photoperiod used fell above or below the insect's response threshold, When we speak of a critical photoperiod, we must be speaking of the mean response level of the population with which we are dealing. If increments of response (diapause) are plotted against the range of photoperiods employed, an approximately normal distribution is obtained (the objections of de Wilde (1962) notwithstanding). The peak of the distribution can be taken as a measure of the critical photoperiod. In the corn borers with which we have done most of our work, the critical photoperiod has been between 15 and 15.5 hours of light per day (Beck and Hanec 1960). That is, at 15 hours of light per day, about 60 per cent more of the larvae diapaused than was the case at 15.5 hours.

If such a borer population were transferred to field conditions in a northern area—50° N., for instance—the long daylengths of such a locality would be far greater than the critical daylength. Nearly all of the larvae would pupate; few would diapause. Such a northern area would, however, have an extremely short growing season. The new generation of borer larvae would perish, because there would be insufficient time to complete larval growth. The only survivors from the introduced population would be those few that represented the extreme tail of the distribution curve. They would be the larvae that had a higher than normal critical photoperiod and went into diapause despite the relatively long daylengths of the northern summer. A few seasons of such intensive selection would result in a borer population showing a high mean critical photoperiod.

Are there such geographical races of the European corn borer? Photoperiodic races have been described in a few other arthropod species. Bondarenko and Hai-Yuan (1958) demonstrated differences in the photoperiodic responses of spider mites from different latitudes Such racial differences have been demonstrated also in a few lepidopterous species (Danilyevsky 1957, Fukaya and Mitsuhashi 1957). Since the corn borer has a rather wide north-south distribution, it seemed likely that population differences might be found.

Following the work of Apple (1952), field development of the corn borer has been followed by a method of temperature accumulation. The seasonal development of the insect shows a temperature requirement, and has been calculated on the basis of the accumulation of day-degrees above a base of 50° F. Borers in southern Wisconsin require a temperature accumulation of about 1500 day-degrees before any of the early summer generation will pupate. Laboratory study of the photoperiodism of southern Wisconsin borers has shown that the critical photoperiod for diapause induction is about 15 hours of daylight per day. With this information at hand, M. W. Burley, Wisconsin State Climatologist, plotted long-term normal temperature accumulations up to the date at which daylengths fell to 14.75 hours; the isotherm map shown in Figure 6 was obtained. On this map we next plotted the locations where midsummer pupation of the borer had been reported during the past five years by the Wisconsin State Department of Agriculture. These localities are shown as black dots in Figure 6. If there were no population differences among the corn borers of the State, we would expect summertime pupation to occur only where the temperature accumulation was 1500 or more day-degrees prior to the critical daylength. As can be seen in the figure, some pupae were reported in areas where the normal temperature accumulation was well below 1400 day degrees.

These results might lead us to believe that the borers living in the northern parts of the State have either a higher critical photoperiod or a lower temperature requirement. Further investigation has shown that both types of differences are to be found (Beck and Apple 1961).

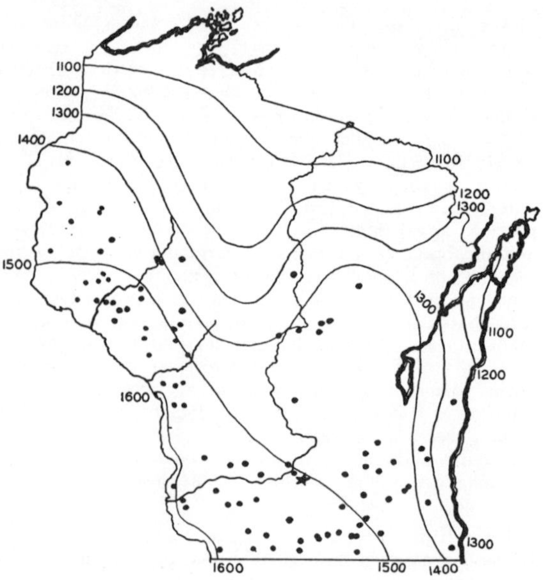


Fig. 6.—Midsummer pupation of the European corn borer and the normal temperature accumulation prior to a 14.75 hour daylength in the State of Wisconsin.

Table 1 shows a comparison of several borer populations in respect to critical photoperiod and day-degrees required for larval growth from the first instar to pupation. These determinations were made by relatively small numbers of laboratory rearings, and must be considered only approximations. More intensive study with greater replication and more extensive samplings needs to be carried out. But these preliminary data demonstrate some interesting population differences. First, we find that in regard to the two Wisconsin populations, both temperature and photoperiodic adaptations have been made. As predicted from Figure 6, the northern borers showed a higher critical photoperiod and lower temperature accumulation requirement than did the more southern borers.

The Ontario population we studied resembled the northern Wisconsin type of borer. The Massachusetts population displayed the lowest critical photoperiod and the lowest temperature requirement. This is an interesting finding, as it helps to explain the consistent two-generation performance of the borer under the cool conditions of the Massachusetts growing season. There

Table 1.—Comparison of different European corn borer populations in respect to critical photoperiod and temperature accumulation requirements.

| Source | Lat. (°N) | Critical Daylength (Hrs.) | Borer-Day-Deg. |
|---|---|---|---|
| Wis. | 45 00 | >15 | 540 |
| Wis. | 43 05 | 15 | 645 |
| Ont. | 42 45 | >15 | 555 |
| Mass. | 42 30 | <14.5 | 480 |
| Iowa | 42 00 | 15 | 620 |
| Kansas | 39 15 | 15 | 655 |
| Mo. | 36 50 | 14.5 | 580 |

seems little doubt that the corn borer has adapted or is still adapting to the different conditions existing in different areas of the country.

Some of the conditions that must be taken into account in predicting population trends of the European corn borer are shown in Figure 7. In this figure we have plotted the normal seasonal temperature accumulations against the daylength changes for three localities: Wausau, Wisconsin at 45° N, Madison, Wisconsin at 43° N, and Cairo, Illinois at 37° N. The shaded zone between 14 and 15 hours of daylight is to represent a general range of critical photoperiods. A temperature accumulation of about 1600 day-degrees is taken to be the requirement for the first generation to reach a pupation incidence of about 50 percent. The second and third generations each require only about 1300 day-degrees. The difference of 300 day-degrees is accounted for on the basis that the survivors of the first generation come from the latter half of the eggs laid; the early larvae show poor survival because of unfavorable host plant availability.

In any year that happened to be warmer than normal, the temperature accumulation curves would be shifted to the right, and this would improve the chances of a second generation in the Madison area and a third generation in southern Illinois. Any evolutionary change in the borer populations that resulted in a lower critical photoperiod and a lower temperature requirement would also result in a greater number of generations. These adaptive trends might be expected in the southern Wisconsin and southern Illinois populations. But they should not be expected in the northern Wisconsin population. This is predicted because the temperature accumulation in northern Wisconsin is so low that any borers that pupate will represent a loss from the species gene pool, because there will not be sufficient growing season available for their progeny to reach maturity. Therefore, the trend will be toward a higher critical photoperiod. But because the seasonal temperature accumulation is normally very little over 1600, there should be a tendency for natural selection to favor those borers having a lower than average temperature requirement. And these are, indeed, the characteristics we have found in the northern Wisconsin corn borer population (Beck and Apple 1961).

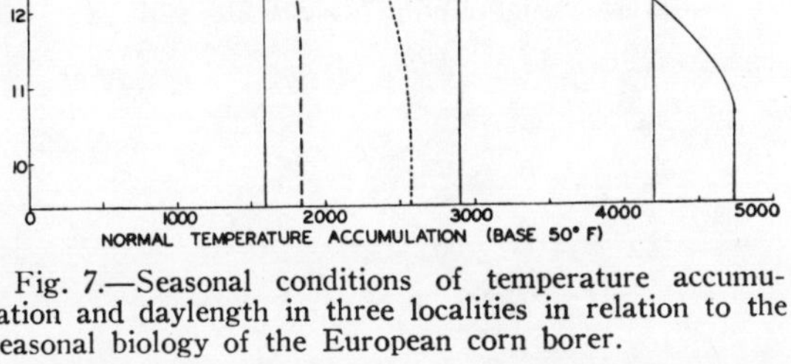


Fig. 7.—Seasonal conditions of temperature accumulation and daylength in three localities in relation to the seasonal biology of the European corn borer.

This has by necessity been a rather brief exploration of insect photoperiodism. From the knowledge now at hand, it is apparent that only a good beginning has been made. The subject holds much promise for the future, for through the study of the effect of photoperiod on insect biology, new insights are to be obtained concerning problems of the control of growth processes, behavior, ecology, and speciation.

## REFERENCES CITED

**Andrewartha, H. G.** 1952. Diapause in relation to the ecology of insects. Biol. Rev. 27: 50–107.

**Apple, J. W.** 1952. Corn borer development and control on canning corn in relation to temperature accumulation. Jour. Econ. Ent. *45:* 877–879.

**Baker, F. C.** 1935. The effect of photoperiodism on resting treehole mosquito larvae. Canad. Ent. *67:* 149–153.

**Ball, H. J.** 1958. The effect of visible spectrum irradiation on growth and development in several species of insects. Jour. Econ. Ent. *51:* 573–578.

**Beck, S. D.** 1962. Photoperiodic induction of diapause in an insect. Biol. Bull. *122:* 1–12.

**Beck, S. D.** and **J. W. Apple.** 1961. Effects of temperature and photoperiod on voltinism of geographical populations of the European corn borer, *Pyrausta nubilalis* (Hbn). Jour. Econ. Ent. *54:* 550–558.

**Beck, S. D., E. J. Cloutier,** and **D. G. R. McLeod.** 1963. Photoperiod and insect development. 23rd Biology Colloquium, Oregon State University (in press).

**Beck, S. D.** and **W. Hanec.** 1960. Diapause in the European corn borer, *Pyrausta nubilalis* (Hbn). Jour. Insect Physiol. *4:* 304–318.

**Beck, S. D.** and **E. E. Smissman.** 1960. The European corn borer, *Pyrausta nubilalis,* and its principal host plant. VIII. Laboratory evaluation of host resistance to larval growth and survival. Ann. Ent. Soc. Amer. *53:* 755–762.

**Bondarenko, N. V.** and **K. Hai-Yuan.** 1958. The peculiarities of the appearance of diapause in different geographical populations of the spider mite. Dok. Akad. Nauk SSSR *119:* (AIBS) 1247.

**Brett, W. J.** 1955. Persistent diurnal rhythmicity in *Drosophila* emergence. Ann. Ent. Soc. Amer. *48:* 119–131.

**Cloutier, E. J., S. D. Beck, D. G. R. McLeod,** and **D. L. Silhacek.** 1962. Neural transplants and insect diapause. Nature *195:* 1222–1224.

**Corbet, P. S.** 1956. Environmental factors influencing the induction and termination of diapause in the emperor dragonfly, *Anax imperator.* Jour. Exp. Biol. *33:* 1–14.

**Danilyevsky, A. S.** 1957. Photoperiodism as a factor in the formation of geographical races in insects. Ent. Oboz. *36:* 5–27.

**Dickson, R. C.** 1949. Factors governing the induction of diapause in the oriental fruit moth. Ann. Ent. Soc. Amer. *42:* 511–537.

**Fukaya, M.** and **J. Mitsuhashi.** 1957. The hormonal control of diapause in the rice stem borer, *Chilo suppressalis.* I. Some factors in the head maintaining larval diapause. Jap. Jour. Appl. Ent. and Zool. *1:* 145–154.

**Fukaya, M.** and **J. Mitsuhashi.** 1961. Larval diapause in the rice stem borer with special reference to its hormonal mechanism. Bull. Nat. Inst. Agric. Sci. (Jap.) C (13): 1–32.

**Fukuda, S.** 1951. The production of the diapause eggs by transplanting the subesophageal ganglion in the silkworm. Proc. Imper. Acad. Japan *27:* 672–677.

**Garner, W. W.** and **H. A. Allard.** 1920. Effect of relative length of day and night and other factors on growth and reproduction in plants. Jour. Agric. Res. *18:* 553–606.

**Gilbert, L. I.** and **H. A. Schneiderman.** 1961. Some biochemical aspects of insect metamorphosis. Amer. Zool. *1:* 11-51.

**Grison, P.** 1949. Effects d'implantation de cerveaux chez le Doryphore (*Leptinotarsa decemlineata* Say) en diapause. C. R. Acad. Sci. *228:* 428–430.

**Harker, J. E.** 1956. Factors controlling the diurnal rhythm of activity of *Periplaneta americana.* Jour. Exp. Biol. *33:* 224–234.

**Harker, J. E.** 1960a. The effect of perturbations in the environmental cycle of the diurnal rhythm of activity of *Periplaneta americana.* Jour. Exp. Biol. *37:* 154–163.

**Harker, J. E.** 1960b. Internal factors controlling the suboesophageal ganglion neurosecretory cycle in *Periplaneta americana* L. Jour. Exp. Biol. *37:* 164–170.

**Harker, J. E.** 1960c. Endocrine and nervous factors in insect circadian rhythms. Cold Spring Harbor Symposia on Quant. Biol. *25:* 279–286.

**Harker, J. E.** 1961. Diurnal rhythms. Ann. Rev. Ent. *6:* 131–146.

**Hasegawa, K.** 1957. The diapause hormone of the silkworm, *Bombyx mori.* Nature *179:* 1300–1301.

**Hodek, I.** and **J. Cerkasov.** 1958. A study of the imaginal hibernation of *Semiadalia undecimnotata* (Coccinellidae: Col.) in the open. Acta Soc. Zool. Bohem. *22:* 180–192.

**Lees, A. D.** 1952. The physiology of diapause in the fruit tree red spider mite. Trans. IXth Intern. Congr. Ent. *1:* 351–354.

**Lees, A. D.** 1955. The Physiology of Diapause in Arthropods. Cambridge University Press.

**Lees, A. D.** 1959. The role of photoperiod and temperature in the determination of parthenogenetic and sexual forms in the aphid *Megoura viciae* Buckton. I. The influence of these factors on apterous virginoparae and their progeny. Jour. Insect Physiol. *3:* 92–117.

**Lees, A. D.** 1960a. The role of photoperiod and temperature in the determination of parathenogenetic and sexual forms in the aphid *Megoura viciae* Buckton. II. The operation of the 'interval timer' in young clones. Jour. Insect Physiol. *4:* 154–175.

**Lees, A. D.** 1960b. Some aspects of animal photoperiodism. Cold Spring Harbor Symposia on Quant. Biol. *25:* 261–268.

**Marcovitch, S.** 1924. The migration of Aphididae and the appearance of the sexual forms as affected by the relative length of daily light exposure. Jour. Agric. Res. *27:* 513–522.

**McLeod, D. G. R.** and **S. D. Beck.** 1963. Photoperiodic termination of diapause in an insect. Biol. Bull. (in press).

**Moriarty, F.** 1959. The 24-hour rhythm of emergence of *Ephestia kuhniella* Zell. from the pupa. Jour. Insect Physiol. *3:* 357–366.

**Muller, H. J.** 1957. Die Wirkung exogener Faktoren auf die zyklische Formenbildung der Insekten, insbesondere der Gattung *Euscelis.* Zool. Jahrb. *85:* 317–430.

**Muller, H. J.** 1960. Die Bedeutung der Photoperiode im Lebensablauf der Insekten. Zeit. ang. Ent. *47:* 7–24.

**Palmen, E.** 1958. Diel periodicity of pupal emergence of some north European chironomids. Proc. Xth Inter. Congr. Ent. *2:* 219–224.

**Paris, O. H.** and **C. E. Jenner.** 1959. Photoperiodic control of diapause in the pitcher-plant midge, *Metriocnemus knabi.* A.A.A.S. Publ. no. 55 Photoperiodism, pp. 601–624.

**Parker, H. L.** and **W. R. Thompson.** 1927. A contribution to the study of hibernation in the larva of the European corn borer (*Pyrausta nubilalis* Hb.). Ann. Ent. Soc. Amer. *20:* 10–22.

**Pease, R. W.** 1962. Factors causing seasonal forms in *Ascia monuste* (Lepidoptera). Science *137:* 987–988.

**Pittendrigh, C. S.** 1954. On temperature independence in the clock system controlling emergence time in *Drosophila.* Proc. Nat. Acad. Sci. *40:* 1018–1029.

**Pittendrigh, C. S.** 1958. Adaptation, natural selection, and behavior. Pp. 390–416, in Behavior and Evolution, edited by Roe and Simpson. Yale University Press.

**Pittendrigh, C. S.** 1960. Circadian rhythms and circadian organization of living systems. Cold Spring Harbor Symposia on Quant. Biol. *25:* 159–182.

**Schneiderman, H. A.** and **L. I. Gilbert.** 1959. The chemistry and physiology of insect growth hormones. Pp. 157–187 in Cell, Organism, and Milieu, edited by Rudnick. Ronald Press, New York.

**Shakhbazov, V. G.** 1961. The reaction to the length of daylight and the light receptor of the pupa of the Chinese oak silkworm, *Antheraea pernyi* G. Dok. Akad. Nauk. S.S.S.R. *140:* (AIBS) 944–946.

**Van der Kloot, W. G.** 1955. The control of neurosecretion and diapause by physiological changes in the brain of the Cecropia silkworm. Biol. Bull. *109:* 276–294.

**Van der Kloot, W. G.** 1961. Insect metamorphosis and its endocrine control. Amer. Zool. *1:* 3–9.

**Wigglesworth, V. B.** 1959. The Control of Growth and Form. Cornell University Press. Ithaca, New York.

**de Wilde, J.** 1962. Photoperiodism in insects and mites. Ann. Rev. Ent. *7:* 1–26.

**de Wilde, J.** and **J. A. de Boer.** 1961. Physiology of diapause in the adult Colorado beetle—II. Diapause as a case of pseudo-all-atectomy. Jour. Insect Physiol. *6:* 152–161.

**Williams, C. M.** 1952. Physiology of insect diapause. IV. The brain and prothoracic glands as an endocrine system in the Cecropia silkworm. Biol. Bull. *103:* 120–138.

**Williams, C. M.** 1956. Physiology of insect diapause. X. An endocrine mechanism for the influence of temperature on the diapausing pupa of the Cecropia silkworm. Biol. Bull. *110*: 201–218.

# Studies in the Mode of Action of the Diapause Hormone in the Silkworm, *Bombyx Mori* L.

## VI. THE TARGET ORGAN OF THE DIAPAUSE HORMONE

By KINSAKU HASEGAWA and OKITSUGU YAMASHITA

*The Sericultural Laboratory, Faculty of Agriculture, Nagoya University, Anzyo, Aiti-ken, Japan*

(*Received* 1 *January* 1965)

### INTRODUCTION

In general, glycogen and crude fat contents in insects increase before the onset of the diapause. In silkworms it is known that diapause eggs and the ovaries of diapause egg producers (in short 'diapause' ovaries) have more glycogen than those of non-diapause type (Chino, 1957; Yamashita & Hasegawa, 1964*b*). As already reported, the diapause hormone released from the suboesophageal ganglion (SG) of the silkworm is responsible for the induction of diapause in this insect (Hasegawa, 1952, 1957).

Our recent experiments have shown that the removal of the SG from 'diapause' pupae just after pupation not only causes the pupae to produce non-diapause eggs but also raises the carbohydrate content of the blood and fat body and lowers that of the ovaries to the same level as in natural 'non-diapause' ovaries, without any detectable changes in the carbohydrate content in the Malpighian tubules, integument, midgut or the whole pupa. On the contrary, the injection of extracts of diapause hormone into 'non-diapause' pupae lowers the fat body glycogen and the blood sugar level, but raises the glycogen content in the ovaries proportionally to the amount of the hormone dosage. From these results, it has been inferred that the diapause hormone facilitates the penetration of blood sugars into pupal ovaries and/or glycogenesis therein, resulting in lessening the blood sugar level and raising ovary glycogen content, a hypoglycaemic effect which is similar to the action of insulin in mammals (Yamashita & Hasegawa, 1964*b*).

The SG in male pupae is scarcely concerned with carbohydrate metabolism, for SG-removal from male pupae induced no significant changes in carbohydrate content in the pupal tissues (Yamashita & Hasegawa, 1965). From this one may conjecture that the SG in males functions differently from that in females or that the SG does no secrete the diapause hormone. Nevertheless, male pupae are able to produce diapause eggs when ovaries have been transplanted into them (Hasegawa, 1952), and, in fact, diapause hormone can be extracted from the heads of male moths (Hasegawa, 1963, 1964).

The present experiments were undertaken to elucidate the role of pupal ovaries in the carbohydrate metabolism of silkworm pupae and the effect of the diapause hormone on this metabolism. For this purpose, ovariectomy and ovary transplantation into males were carried out during the larval stages, the SG being subsequently removed on the day of the pupation. From the results obtained it is clear that the SG in males, as well as in females, is concerned with carbohydrate metabolism so far as ovaries have been implanted and that, from the standpoint of carbohydrate metabolism also, the ovary is the target organ of the diapause hormone.

## MATERIALS AND METHODS

The silkworms used were diapause egg producers of the bivoltine race (Daizo) which had experienced a high temperature and long photoperiod during embryonic life. Ovariectomy, unilateral or bilateral, was carried out on the third day of the fifth larval stage. Ovary transplantation into male larvae was performed at the same stage in the same race; immediately after unilateral gonadectomy of the male larvae, each operated male received an ovary implanted through the incision of the hemi-orchotomy and the male pupae, each with a developing ovarial implant, were used for further experiments. SG-removal was carried out on the day of the pupation of both the operated and non-operated pupae of both sexes.

Of the carbohydrates in the pupal tissues, reducing sugars and trehalose were determined by the anthrone method (see Wyatt & Kalf, 1957), and glycogen by the method of Kemp & van Heijningen (1954). In each experiment about fifteen pupae were sacrificed. Blood collected in cold tubes was used after centrifugation at 2500 r.p.m. for one minute, and fat body, ovaries and whole pupae were used after homogenation with a glass homogenizer in an ice bath.

The earlier experiments, in which the effect of SG-removal and the effect of diapause hormone injections on the carbohydrate content of pupal tissues was investigated, showed that these procedures affected chiefly the major carbohydrate components in the fat body, blood and ovaries (Yamashita & Hasegawa, 1964*b*). In the present paper, therefore, only the major components in each tissue are indicated in the accompanying tables, i.e. glycogen in fat body and ovaries, and trehalose in blood, and these are expressed in terms of glucose equivalent.

## EXPERIMENTAL RESULTS

### 1. *Effects of SG-removal and ovariectomy on carbohydrate content of the tissues of female pupae*

For the determination of tissue carbohydrate content both normal female pupae and hemi-castrated or wholly castrated female pupae (with removal of SG) were sacrificed on the eighth day after pupation or 2 days before adult emergence. The major components in the fat body, blood and ovaries are tabulated in Table 1.

*Fat-body glycogen.* Although SG-removal from female pupae raised the glycogen level in the fat body as expected from the previous papers (Yamashita & Hasegawa, 1964*b*), the effect of SG-removal is different owing to the castration. Ovariectomy in the larval stage resulted in a high level of glycogen in the fat body; i.e. the glycogen content of hemi-castrated and wholly castrated pupae is about three times and more than seven times, respectively, that of pupae with the ovaries intact. Another interesting point is that, although SG-removal raised the fat-body glycogen content in both non-castrated and hemi-castrated pupae, this effect vanished after complete castration. In other words, when ovaries are absent the fat-body glycogen which has been accumulated during the early pupal stage (Yamashita & Hasegawa, 1965) is not mobilized so readily as it is in normal female pupae; it remains at the same high level irrespective of the presence or absence of the SG.

*Blood trehalose.* As already reported (Yamashita & Hasegawa, 1964*b*), blood sugars, especially trehalose, increase when the SG is excised from female pupae. It is of interest that unilateral castration causes the pupae to store more blood trehalose than normal pupae, but the removal of both ovaries maintains the blood trehalose at about the same level as in non-operated pupae. Combining SG-removal with castration is also interesting; removal in hemi-castration raises the blood trehalose level higher than that of pupae with two intact ovaries, whereas this effect does not appear in total castration.

Table 1. *Effect of SG-removal and castration on major components of carbohydrates in some tissues of female pupae*

| Presence of: SG | Presence of: Ovary | Glycogen in fat body (mg./g.) | Trehalose in blood (mg./100 ml.) | Glycogen in ovary (mg./g.) | Total carbohydrate in whole pupae (mg./g.) | Remarks |
|---|---|---|---|---|---|---|
| + | ++ | 5·1 | 461 | 20·1 | 14·2 | Non-operated or normal pupae |
| − | ++ | 8·2 | 627 | 14·5 | 13·4 | Pupae with SG removed |
| + | + | 17·7 | 590 | 19·8 | 13·6 | Pupae hemi-castrated |
| − | + | 24·5 | 696 | 15·3 | 12·8 | Pupae hemi-castrated but with SG removed |
| + | − | 38·5 | 483 | . | 13·4 | Pupae completely castrated |
| − | − | 37·8 | 493 | . | 12·5 | Pupae completely castrated but with SG removed |

Silkworms used, Daizo race (diapause type); SG-removal, on the day of pupation; hemi-castration and complete castration, on the third day of 5th-instar larval stage; pupal age at sacrifice, on the eighth day after pupation or 2 days before adult emergence.

*Ovary glycogen and total carbohydrate in whole pupae.* Even when pupae suffered hemi-castration at the larval stage the glycogen content (per wet weight) of the remaining ovary was identical with that in non-castrated pupae; and the content was decreased by SG-removal in the same way as observed on non-castrated pupae. Neither SG-removal nor castration could alter the total carbohydrate content of the whole pupae.

### 2. *Effects of SG-removal and ovary transplantation on the carbohydrate content of the tissues of male pupae*

The effect of SG function in male pupae seemed to be different from that in female pupae, for SG-removal from males on the day of the pupation, in contrast to its effect in female pupae, induced little change in the carbohydrate content of the pupal tissues (Yamashita & Hasegawa, 1965). Male pupae, however, are able to produce diapause eggs if ovaries have been transplanted into them (Hasegawa, 1952). For an analysis of the function of the SG in males each male larva was grafted with one ovary in lieu of a testis after hemi-castration. On the seventh and ninth day after pupation the male pupae were sacrificed and used as material for carbohydrate analysis of the tissues in those specimens in which the ovary implants had developed similarly to the ovaries in female pupae at the corresponding stage. The results obtained are summarized in Table 2.

In this table we can see that, as pupal development proceeds, fat-body glycogen and total carbohydrate in the whole male pupae decrease, and blood trehalose increases, which confirms the results of the previous experiments (Yamashita & Hasegawa, 1965). The following additional points are worth mentioning.

*Fat-body glycogen.* It is of interest that the transplantation of ovaries into males diminishes the glycogen content in the fat body. Where male pupae had not been grafted with ovaries, fat-body glycogen is not altered by SG-removal, as expected from the previous paper (Yamashita & Hasegawa, 1965). On the other hand, the glycogen content of pupae which had received ovary transplantation during the larval

Table 2. *Effect of SG-removal and ovary transplantation on major components of carbohydrates in some tissues of male pupae*

| Pupal age at sacrifice | Presence of: SG | Presence of: Ovary graft | Glycogen in fat body (mg./g.) | Trehalose in blood (mg./100 ml.) | Glycogen in ovary graft (mg./g.) | Total carbohydrate in whole pupae (mg./g.) | Remarks |
|---|---|---|---|---|---|---|---|
| Seventh day after pupation | + | − | 12·2 | 472 | . | 8·7 | Non-operated or normal pupae |
| | − | − | 11·9 | 494 | . | 8·8 | Pupae with SG removed |
| | + | + | 7·9 | 403 | 24·3 | 8·7 | Pupae with ovary graft |
| | − | + | 10·1 | 455 | 15·6 | 8·5 | Pupae with ovary graft but with SG removed |
| Ninth day after pupation | + | − | 6·5 | 547 | . | 4·9 | Non-operated or normal pupae |
| | − | − | 6·6 | 561 | . | 4·8 | Pupae with SG removed |
| | + | + | 3·1 | 460 | 27·0 | 5·1 | Pupae with ovary graft |
| | − | + | 4·7 | 529 | 19·3 | 5·5 | Pupae with ovary graft but with SG removed |

Silkworms used, the same race as in Table 1; SG-removal, on the day of pupation; ovary transplantation into males, on the third day of 5th-instar larval stage; donors of ovaries, the same race at the same larval stage.

stage varied according to the presence or absence of the SG; i.e. SG-removal from these pupac induced a rise in fat-body glycogen above the level in pupae without SG-removal. In other words, the SG of male pupae, into which ovaries had been implanted, functions as well as that of female pupae in decreasing the glycogen content of the fat body.

*Blood trehalose.* The trehalose level in male pupae is scarcely affected by SG-removal, but is markedly different when ovaries have been transplanted; ovary transplantation lowers blood trehalose, but if this is combined with SG-removal the blood trehalose tends to resume its original level.

*Glycogen content of transplanted ovaries and total carbohydrate content in whole pupae.* As expected, the glycogen level in ovary implants developing under the influence of the SG or the diapause hormone is higher than that of ovaries of pupae deprived of the SG. The total carbohydrate content in whole male pupae, however, is not affected by SG-removal, and it is of interest that the total carbohydrate content of male pupae is almost the same in those with or without ovary implants.

## DISCUSSION

### 1. *Pupal ovaries and carbohydrate metabolism in the silkworm*

As already reported, the glycogen content of pupal ovaries (glycogen is the major component of ovary carbohydrates) is different according to the nature of the ovaries—'diapause' or 'non-diapause'. The glycogen content in 'diapause' ovaries, induced by higher temperature and increased photoperiod during their embryonic life, or by injecting diapause hormone extracts into 'non-diapause' pupae, exceeds that of 'non-diapause' ovaries induced by lower temperature and darkness during their embryonic life, or by SG-removal from 'diapause' pupae. In the latter group the fat-body glycogen content and blood trehalose level (the major components of carbohydrate in each tissue) are raised. The diapause hormone has been thought to regulate carbohydrate metabolism in silkworm pupae by stimulating glycogen accumulation in the ovaries (Yamashita & Hasegawa, 1964*b*, 1965).

It has also been shown that, in 'diapause' pupae, ovary glycogen accumulating day by day after the onset of histogenesis is delayed by SG-removal, while fat-body glycogen and trehalose in the blood increase; whereas both trehalose and the level of reducing sugar in male blood become high after the middle pupal age with no relation to the SG (Yamashita & Hasegawa, 1964*b*, 1965).

As shown in the present experiments, the trehalose level in the blood of female pupae is increased by hemi-castration, but, after total castration, trehalose approximates to its original level. Fat-body glycogen, however, is always increased by castration; i.e. the glycogen content of pupae castrated unilaterally and bilaterally becomes, respectively, about three times and more than seven times as much as that of normal females (Table 1). On the other hand ovary transplantation into males decreases their fat-body glycogen and blood trehalose level (Table 2). Blood trehalose, however, is altered by these treatments only when ovaries are present in the body cavity, suggesting that the blood trehalose is the form in which fat-body glycogen is transferred to ovary glycogen. The carbohydrate content of whole pupae is not affected in either sex by ovariectomy and ovary transplantation (Tables 1, 2).

Considering these points it is highly probable that the increase in glycogen in the fat body after ovariectomy is the result of feedback control and that the accumulation or mobilization of fat-body glycogen during the early pupal period is dependent on the presence or absence of ovaries. This indicates that the pupal ovary plays an important part in carbohydrate metabolism in the silkworm; it synthesizes glycogen from carbohydrate which is carried via the blood from the fat body.

### 2. *The target organ of the diapause hormone*

In the first series of the studies on the mode of action of the diapause hormone (Hasegawa, 1963) it was suggested that the oocyte is the sole target organ of this hormone. This suggestion gained partial support from the accumulation of 3-hydroxykynurenine in 'diapause' ovaries (Yamashita & Hasegawa, 1964*a*). The experimental results of SG-removal and injections of diapause hormone, however, did not entirely support this suggestion, because of accompanying changes in the carbohydrate content of the fat body and blood as stated above. The possibility that the ovaries are the target organ of the diapause hormone therefore could not gain complete acceptance from the standpoint of carbohydrate metabolism. The uncertainty on this point, however, has been removed by the present experiments.

As indicated in Table 1 the effect of removal of the SG from female pupae, on the carbohydrate content of the fat body and blood, occurred in non-castrated and hemi-castrated females, but there was no effect after total castration, indicating that the SG-removal does not primarily affect the carbohydrate content of fat body and blood, but that the effect is secondary, and that, for its realization, the presence of ovaries is necessary.

That the presence of ovaries is required if the SG is to influence the carbohydrate content of the fat body and blood was further substantiated by another experiment (Table 2). In this experiment SG-removal from male pupae did not induce any significant increase of carbohydrate content in fat body and blood, but SG-removal from male pupae carrying ovary implants brought about just the same effect as manifested in female pupae with two intact ovaries or in hemi-castrated pupae.

From these results it is clearly demonstrated that the SG or the diapause hormone regulates the carbohydrate metabolism through the pupal ovaries and that, without ovaries, the effect does not appear, indicating that the target organ of the diapause hormone is the ovary itself.


## SUMMARY

1. Glycogen synthesis in the ovary of *Bombyx mori* is accelerated by the diapause hormone.

2. Such glycogen is derived, by way of the trehalose of the haemolymph, from carbohydrate in the fat body.

3. Extirpation of the suboesophageal ganglion has the converse effect: a fall in glycogen in the ovary and an increase in blood trehalose and fat-body glycogen.

4. Mobilization of fat-body glycogen is dependent on the presence of the ovaries, and of the suboesophageal ganglion.

5. The same results can be obtained in male pupae after the implantation of ovaries.

6. These observations on carbohydrate metabolism confirm the belief that the 'target organ' of the diapause hormone is the ovary itself. The changes in other tissues are due to feedback.


This work was aided, in part, by a grant from the Scientific Research Committee of the Department of Education.

## REFERENCES


CHINO, H. (1957). Carbohydrate metabolism in diapause egg of the silkworm, *Bombyx mori*. I. Diapause and the change of glycogen content. *Embryologia*, **3**, 295–316.

HASEGAWA, K. (1952). Studies on the voltinism in the silkworm, *Bombyx mori* L., with special reference to the organs concerning determination of voltinism. *J. Fac. Agric. Tottori Univ.* **1**, 83–124.

HASEGAWA, K. (1957). The diapause hormone of the silkworm, *Bombyx mori*. *Nature, Lond.*, **179**, 1300–1.

HASEGAWA, K. (1963). Studies on the mode of action of the diapause hormone in the silkworm, *Bombyx mori* L. I. The action of diapause hormone injected into pupae of different ages. *J. Exp. Biol.* **40**, 517–29.

HASEGAWA, K. (1964). Studies on the mode of action of the diapause hormone in the silkworm, *Bombyx mori* L. II. Content of diapause hormone in the suboesophageal ganglion. *J. Exp. Biol.* **41**, 855–63.

KEMP, A. & VAN HEIJNINGEN, J. M. K. (1954). A colorimetric micromethod for the determination of glycogen in tissues. *Biochem. J.* **56**, 646–8.

WYATT, G. R. & KALF, G. F. (1957). The chemistry of insect hemolymph. II. Trehalose and other carbohydrates. *J. Gen. Physiol.* **40**, 833–47.

YAMASHITA, O. & HASEGAWA, K. (1964*a*). Studies on the mode of action of the diapause hormone in the silkworm, *Bombyx mori* L. III. Effect of diapause hormone extract on 3-hydroxykynurenine content in ovaries of silkworm pupae. (In Japanese with English summary.) *J. Seric. Sci. Jap.* **33**, 115–23.

YAMASHITA, O. & HASEGAWA, K. (1964*b*). Studies on the mode of action of the diapause hormone in the silkworm, *Bombyx mori* L. IV. Effect of diapause hormone on the glycogen content in ovaries and the blood sugar level of silkworm pupae. (In Japanese with English summary.) *J. Seric. Sci. Jap.* **33**, 407–16.

YAMASHITA, O. & HASEGAWA, K. (1965). Studies on the mode of action of the diapause hormone in the silkworm, *Bombyx mori* L. V. Effect of diapause hormone on the carbohydrate metabolism during adult development. (In Japanese with English summary.) *J. Seric. Sci. Jap.* **34** (in the Press).

# Part IV: The Ecology of Insects

Ecology has recently been equated by the public with the problems of man and his environment, particularly as man affects himself through pollution of his oxygen and water supplies. However, **ecology** deals with *the relationships of all organisms to each other and to their environment,* and man is only one of these organisms. Many of the major principles operating in organismic relationships and interactions with the physical environment have been developed from investigations using insect populations. The following series of articles is intended to highlight some of these basic areas of ecology and ecological research on insects.

POPULATION ABUNDANCE – The topic of population abundance and the variety of mechanisms regulating the size of insect populations is of interest especially with economically important insects. The quantitative and qualitative aspects of **population dynamics** *(the form, growth and development of the unique characteristics and interactions of a population)* are highlighted by R. Glen in *Factors that Affect Insect Abundance.* With definitions of both general concepts (e.g., density-dependence, balance of nature, etc.) and economic concepts (e.g., outbreaks), a foundation is established for the other articles in this section.

In proposing more emphasis on ecology, Glen touches upon an important and rarely discussed controversy in the philosophy of research. This controversy involves the relevance of laboratory work to that done in the field. Glen points out that a combination of the two is perhaps the most productive.

*Density Dependence in the Australian Thrips* by F. E. Smith focuses attention on the effect of density on the severity of action of environmental factors. **Density dependence** *(whether the effect of any environmental factor is constant or changes with the number of individuals in a population)* is a key aspect in determining the effects of changes in population abundance. Its potential action as a regulatory mechanism of population size makes it a special and basic concept of insect ecology. The article also stresses the growing impact on experimentation of statistical analysis.

DISPERSAL AND MIGRATION – The **migration** *(periodic movement and return)* and **dispersal** *(movement into or out of an area)* of insect populations has been widely studied because it is a major part of the life cycles of different insects. Certainly, insect migration and dispersal has had a major impact in the competitive struggle for resources between man and the insects since the early days of man's history.

*Long-range Transport of the Harlequin bug and the Greenbug into Minnesota* by A. C. Hodson and E. F. Cook documents the dispersal activity of economic pests and illustrates the mechanisms and patterns of dispersal. The authors also emphasize the effects of weather factors in determining dispersal patterns and the potential value of monitoring dispersal of economic pests for predictions of pest outbreaks.

Migration of insects is not only exhibited by a variety of insect groups (Lepidoptera, Coleoptera, and Orthoptera), but it is a highly complex phenomenon differentially manifested in these groups. C. G. Johnson *(Physiological Factors in Insect Migration by Flight)* illustrates the effect of changes in the physiology of individuals on their migratory capacities, such as the roles played by ovarian development and the balance between environmental cues and neurosecretory activity.

PHYSICAL FACTORS—The influence of physical factors, such as temperature, light and moisture, on insect life is among the most studied areas of insect ecology. Most studies, whether on population dynamics, behavior, ecosystem structure, or physiology, must at some point have information on the effect of physical factors on individual insects.

Two major physical factors are investigated by W. A. Rowley and C. L. Graham in *The Effect of Temperature and Relative Humidity on the Flight Performance of Female Aedes aegypti.* These parameters, temperature and humidity, have a regulating and controlling influence on the biological activities (locomotion, feeding, mating, and others) of insects, since the body temperature of insects conforms generally to that of their environment and also because insects have a high surface-to-volume body ratio with its consequent effect on water loss. In this study the specific effects of temperature and humidity on flight distance, duration, and speed is assessed. Although humidity had little effect on flight, low temperatures relative to the optimal activity temperature, 27° C, were found to be preferable for flight in female *A. aegypti.* The flight behavior is of particular interest in this insect because of its medical importance in the transmission of yellow fever.

BIOLOGICAL INTERACTIONS—In ecology, **interaction** among organisms is *the mutual or reciprocal action or influence they have on each other.* This behavior in essence defines ecology, whether **synecology** *(dealing with associated groups of organisms)* or **autecology** *(dealing with the individual organism).* This reciprocity of influence between insects can take many forms and can have an effect on many phases of an insect's life. Interactions affect the development of an insect population and thus determine the significance of the role it plays in the **ecosystem** *(the biotic and abiotic parts of the environment).* They can similarly determine to a great extent the nature, development, and structure of the ecosystem itself, since the resultant ecosystem is the sum total of continuing or potentially successful interactions. The interactions between insects can be intraspecific or interspecific, or they can be between the individual insect and its host plant. All of these activities are generally categorized by more familiar nomenclature. These include phenomena such as **competition** *(the active seeking of factors in limited supply, which adversely affects both competitors),* host-parasite and host-plant relationships, disease transmission, **symbiosis** *(obligate association of two organisms which benefits both),* and predator-prey interactions.

The articles presented under the areas of host-plant relationships and relationships with other organisms illustrate many of the important ecological interactions in insects. *Mechanisms of Host-plant Selection* by J. S. Kennedy not only discusses how insects locate their hosts, but deals with the topics of **taxis** *(a directed movement)* and **kinesis** *(an undirected movement)* and with insect senses in general. This critical review and analysis of host-plant selection research highlights the variety of hypotheses that are currently proposed to explain host selection.

In the relationship between predator and prey a co-adaptation and balance may be particularly evident when both have evolved compatible survival mechanisms. *The Detection and Evasion of Bats by Moths* by K. D. Roeder and A. E. Treat discusses just such an example of mechanisms in both a vertebrate predator and its invertebrate prey. In addition, this article is an exceptional example of how laboratory experience, knowledge, and technology can be brought to the field and be combined into a very productive effort.

*Cyclic Fluctuations of Population Density Intrinsic to the Host-Parasite System* by S. Utida again deals with one of the more basic antagonistic relationships in insect ecology, i.e., host-parasite relationships. The emphasis of this article is on the mutual influence of host and parasite on the intrinsic numerical changes of each population. This article also stresses the importance of the long-range study, particularly when investigating the characteristics of population growth.

R. S. Miller's detailed study entitled *Larval Competition in Drosophila melanogaster and D. simulans* considers the role of competition in single species and two species populations. Included among the points is the importance of **compensatory responses** *(factors that tend to insure temporal survival of a portion of a stressed population)* due to overcrowding, such as their role as regulatory mechanisms in insect populations. These responses to stress can affect physiological and ecological characteristics of the population. Some of these changes in the individuals of stressed populations are exhibited in development, ultimate size, and survival.

Miller's approach to the study of competition has become a classic way of dealing with species interactions. Basically the approach entails the analysis and evaluation of single species populations, the prediction of possible outcomes of two species combinations, and the actual experimental mixing of the species. Such a procedure will help determine if the interaction of the species consists of **interference** *(any activity which directly or indirectly limits a competitor's access to a resource)* or **exploitation** *(the simple utilization of resources without adverse chemical or physical activity on the other competitor).*

The discussion of results brings out the controversy over Gause's principle of competitive exclusion, which states that complete competitors cannot coexist. This principle has been accused of being an example of circular reasoning in that it can be interpreted as being true whatever the result of species competition.

DISEASE TRANSMISSION – Insect transmission of diseases is as significant in plants as it is in animals. E. S. Sylvester's article, *Latent Period Phenomena in Transmission of Sugar Beet Yellow-net Virus by Green Peach Aphids,* demonstrates techniques used to study virus acquisition through feeding, subsequent serial transmissions to other host plants, and the latent period (the time after acquisition of the virus but before transmission can occur). The investigation of disease transmission of this type requires a rigorous scrutiny of a multidisciplinary nature.

ENERGETICS – The **energetics** approach *(energy flow and utilization through the ecosystem),* as described in *The Role of Soil Arthropods in Community Ener-*

*getics* by M. D. Engelmann, represents a viewpoint which stresses the importance of energetics in determining the role of arthropods in many ecosystems. This introduction to the trophic-dynamic approach, one which emphasizes **trophic feeding levels** *(the system of sequential levels consisting of organisms with common feeding habits, linked and dependent on a continuous energy flow),* to ecosystem analysis illustrates one of the ways in which insect ecologists and others are attempting to quantify interactions of the ecosystem. The greater reliability of an energetic framework in the evaluation of insect communities over the use of biomass or numbers of insects is stressed.

# A. Population Fluctuation

## Factors that Affect Insect Abundance[1]

ROBERT GLEN,[2] *Entomology Division, Ottawa, Canada*

Factors affecting insect abundance are the primary concern of every field entomologist as the best methods of utilizing beneficial species or of controlling pest species can rarely be devised without a knowledge of the principles that underlie the fluctuations in the populations concerned. A substantial body of fact and philosophy bearing on these principles has been amassed and we are better prepared than ever before to undertake their further clarification. The subject cannot be adequately reviewed in one short paper. However, some salient features are discussed under "Working Concepts," "Examples from Current Work," and "Lines of Emphasis in Future Research."

WORKING CONCEPTS.—Nature functions as an indivisible complex in which no species of plant or animal exists in total isolation and no factor, physical or biotic, operates in complete independence. We cannot, therefore, divorce the broad ecological relationships of insects from the discussion of factors affecting their abundance. These broad relationships are reviewed in reference to primary habitat and taxonomic units, factors, balance of nature, and outbreaks.

*Primary units.*—Ecologists have long recognized the existence of a wide variety of plant and animal associations, and the term ecosystem (Tansley 1935) has been given to the complex of organisms and physical factors that together form a more or less distinct natural community. Examples of major ecosystems are: a woods, a meadow, a lake, a sand dune area. Such systems, of course, are not mutually exclusive since, for example, a species of insect, in its different developmental stages, may inhabit different communities, and predators may range widely through several such areas. In general, however, interrelations are stronger within ecosystems than between them and it is concluded by Elton (1949) that the minor habitats of an ecosystem form the primary habitat units for animal communities. Such minor habitats are represented by a tree, a group of shrubs, a glade in a woods, or a grassy hummock.

This primary habitat unit can readily be linked in our thinking with the primary taxonomic unit. Modern systematists (Mayr 1942, Mayr *et al.* 1953) regard all taxonomic categories, including the species and the subspecies, as heterogeneous, each being an assemblage of partially isolated local populations that are distinct genetically and therefore frequently distinguishable physiologically and morphologically. The *population* and not the species is recognized today as the basic taxonomic unit. The spatial arrangement of these populations differs for different species, but in general a minor

[1] Contribution No. 3175, Entomology Division, Science Service, Department of Agriculture, Ottawa, Canada; presented as an invitation paper at the meeting of the Entomological Society of America, Los Angeles, Calif., Dec. 9, 1953.

[2] Chief of Division.

Reproduced with permission from the *Journal of Economic Entomology,* 47:398–405, 1954. Published by the Entomological Society of America, College Park, Maryland.

habitat is occupied by a single population of each species represented in it.

Each natural community, or ecosystem, is thus composed of a complex of smaller, interlocking, dynamic systems: biological systems (within the organisms) interacting with one another and with the multiphased physical system that surrounds them. It is this web of interrelations that controls insect abundance.

*Factors.*—The factors that comprise this co-ordinating web may include quantities of energy in the form of light, heat, moisture, food, disease organisms, parasites, competitor species, etc.; environmental niches or inherent development and behavior patterns that confer various degrees of protection against such forces; and even man-made components such as insecticides. The importance of any given factor varies with time, place, species, stage of development, etc., i.e., its importance varies in relation to other factors that are concurrently active. It is not the individual factor, therefore, but the whole complex of factors that controls populations; and equivalent changes in insect abundance may arise from several different expressions of this complex.

Even though the factors function as a complex, certain of them may have a predominating influence on a population at a given time or place. Physical factors such as climate, weather, and space exercise an all-pervading effect: (1) by determining, in the absolute sense, what organisms can exist in a habitat; and (2) by affecting thresholds of activity and other biological behavior and thus influencing, often decisively, the importance of biotic forces. Nevertheless, when climatic conditions favor population increase, such biotic factors as predators, disease, and food supply may determine the upper levels of density attained.

Since the physical and biotic factors (*sens. lat.*) together comprise the whole ecosystem, it follows that when one set of factors predominates the other is correspondingly reduced in importance. Howard & Fiske (1911), followed by Nicholson (1933) and others, stress biotic factors, more specifically insect parasites and predators, as the principal, if not the only, regulators of the density of insect populations; Bodenheimer (1928), and later Uvarov (1931), attribute fluctuations in insect numbers primarily to the influence of climatic factors. Thompson (1929) sums up the importance of physical in relation to biotic factors as greatest in polar regions, deserts, and other barren places where physical conditions are unfavorable to life; and least in tropical rain forests and places where physical conditions tend to be uniformly favorable and life most abundant. Following this line of thought we may add that in an area occupied by an insect species physical factors probably exert a greater influence in the periphery, where population density is low, than in the optimum portions, where density is high.

These observations obviously relate to processes in the ecosystem that vary with changes in the density of the population itself. Smith (1935) describes as *density-dependent mortality factors* those components that destroy an increasing proportion as the population density increases as, for example, in many host-parasite relationships; and he describes as *density-independent mortality factors* those whose effect is not related to population density as, for example, excessive heat or cold, drought or rain. In general, biotic factors are regarded as mostly density-dependent and physical factors as mostly density-independent; but various combinations occur, further emphasizing the essentially dynamic and interdependent nature of the ecosystem.

*Balance of nature.*—Ecologists disagree over the interpretation and even over the admissibility of the "balance of nature" theory. The argument centers on (1) the emphasis given to a theoretical mean density about which the population fluctuates (Nicholson 1933, Smith 1935, 1939) and (2) the importance attached to biotic factors, especially intraspecific competition for food, in restoring the so-called steady density, or equilibrium (Nicholson 1933, 1947).

The concept of regular fluctuations oversimplifies the sometimes gentle, sometimes violent changes that characterize many field populations. As Solomon (1949) observes, ". . . many . . . population curves do not give the least impression that the fluctuations are related to the mean value. Movements towards the mean are often reversed before it is reached." It appears that the mean value or steady state is a mathematical interpretation that does not fit the biological facts concerning the individual populations.

That competition for food is *the* mechanism that regulates abundance and restores the balance is not acceptable as a general concept. Undoubtedly density-dependent factors are frequently important and competition, in one form or another, may often predominate. This, however, should not obscure the facts that biotic factors always act in association with physical factors and that in such joint action the physical factors may also become density-dependent in their effects. One part of this effective complex should not be credited with the joint achievement. Furthermore, there are many situations in which competition for food does not appear to be a limiting feature of the habitat. Therefore, Nicholson's theory, as currently stated, fits only specific cases and fails as a general explanation.

Nevertheless, a broad type of balance or order in nature is obvious. It is not balance in the sense of equilibrium but rather as a changing state of *order* that results from the coexistence of interdependent forces that individually may vary, undergo modification, be added to, or be totally removed in the course of events. Hence, the expression "balance of nature" probably should be taken simply as a shorter way of saying: The ecosystem is whole; it functions as a complex of dynamic subsystems with an interdependence of parts that controls and directs all relationships; none of its components is constant, but their interrelations are such that excessive variation by any of them is ultimately checked by others, thus producing an elastic state of balance within which a state of flux prevails. In this sense, the ecosystem is always balanced though its constitutent populations may fluctuate widely.

*Outbreaks.*—An outbreak is simply an extreme upward fluctuation in the density of a single population living within a balanced habitat. During an outbreak the population is not out of control or unnatural or artificial, in the ecological sense, even when the outbreak may be attributed directly to man's intervention. Man is a natural creature who, like any other organism, lives in, utilizes, and modifies his environment with attendant natural consequences. That his effects are drastic does not justify excluding him from a consideration of community relationships. Natural laws must embrace all relevant natural phenomena. A discussion of insect outbreaks may, therefore, properly include reference to man-made epidemics.

de Gryse (1951), writing on forest entomology, states: "The final answer to the question 'What is the cause of insect outbreaks?' has not been forthcoming. It is more than likely that the solution of this problem will be found in a combination of causal relations and this will be different for every species of insect." With this we must agree. However, a few generaliza tions seem warranted.

Solomon (1949) concludes that "population variability tends to be inversely related to the degree of complexity of the ecosystem." This sums up the common observation that insect outbreaks occur more frequently in homogeneous than in mixed forests, and in large areas devoted to the production of one or a few related types of crop than in areas of diversified production. Since climate is a primary determinant of the nature of plant cover, these facts may be put on a geographic basis, complexity and stability of habitat decreasing generally from tropical through temperate to polar regions, with a corresponding increase in the probability of insect outbreaks. The accepted explanation is that an abundant or superabundant supply of food is assured to a limited number of phytophagous species and that their natural enemies suffer a reduction in efficiency through the loss of favorable relationships that exist only in the more varied habitats.

This may simply be another way of saying that any one condition affects different species in different ways and the more intensive or extensive the condition the more it is likely to promote an outbreak. The economic entomologist has witnessed this principle in action in innumerable instances as: (1) the widespread adoption of cultural practices such as strip-farming and combine-harvesting, which brought increased damage from the wheat stem sawfly, *Cephus cinctus* Nort.; (2) the occurrence of excessive quantities of flood-killed, wind-thrown, or fire-scorched timber, with consequent outbreaks of bark beetles and wood borers; and (3) the rise in prominence of phytophagous orchard mites following the general use of DDT.

The reference to DDT may more appropriately be included with a consideration of catastrophic factors. The term *catastrophic mortality factor* is applied by Ullyett (1947) to those forces that reduce the population of a host insect to insignificant proportions in a very short time. Bacteria, fungi, protozoa, and viruses that produce epidemics of insect disease are among the most important of the natural catastrophic factors. Drastic or sudden changes in weather or in other physical factors that result in a high percentage of deaths may similarly be classified. The application of chemicals and of sprays of disease organisms are examples from applied entomology. These factors do not normally exert a continuous effect, but a sudden and temporary one. They cause great changes and then virtually leave the scene. Parasites and predators may thus be reduced as drastically as phytophagous species, or even much more drastically; and in situations where natural enemies normally make an important contribution to the total host mortality, serious repercussions may follow. When the catastrophic factor has spent itself, the remnants of the host population begin to multiply and, in the absence of the density-dependent controlling factors, may rapidly reach destructive densities. In this way, catastrophic factors may be regarded as important contributory causes of outbreaks.

Weather has long been known as a primary influence holding in check many insect species with fantastically high reproductive potentials. This factor regularly eliminates enormous numbers of individuals under so-called normal conditions. Franz (1949) associates changes in population density with genetic variability in the population: weak physiological strains are normally destroyed, but under favorable conditions they survive and reproduce and eventually give rise to outbreaks; finally, lack of vigor permeates the whole population, which collapses from internal weakness especially when less favorable environmental conditions supervene. Wellington *et al.* (1950), on the other hand, lay the emphasis on the "relaxation of climatic control"; without this relaxation, destructive populations of the spruce budworm, *Choristoneura fumiferana* (Clem.), fail to develop even in the presence of an optimum food supply (Prebble & Morris 1951).

Finally, some outbreaks may be attributed to an evolutionary change that confers at least a temporary advantage on an insect species. For example, through natural selection a population can arise that is resistant or immune to the factor or factors that previously accounted for an important fraction of the total mortality in the species. The development of strains resistant to chemicals has been clearly demonstrated and the development of strains resistant to parasites and disease organisms is to be expected. Likewise, an important advantage may arise through the selection of a phytophagous strain with improved synchronization between its development and that of its host plant, as for example the remarkable phenological timing that exists between the emerging spruce budworm and the bursting of the male flowers of the balsam fir.

Examples from Current Work.—The more desirable trends in research may be indicated by brief reviews of three projects based in large part on field studies and illustrating (1) a small-habitat study without economic implications, (2) part of an extensive program with predictive and silvicultural objectives, and (3) a comprehensive investigation involving fungicides and insecticides as ecological factors in the community.

*Example 1.*—The primary habitat unit, or "minor habitat," as designated by Elton (1949) has already been mentioned. Varley (1947) throws some light on the inner workings of one such small community in his quantitative study of the insects living in the flower heads of the black knapweed, *Centaurea nigra* (L.) ssp. *nemoralis* (Jord.) Gugl. He sought an answer to the question: What factors control the population density of the Knapweed gall fly, *Urophora jaceana* (Hering)?

Three generations of galls were examined over a two-year period, a total of 92 square-meter samples being taken from approximately 500 sq. m. on which knapweed was abundant. In this small fragment of the ecosystem, Varley found 25 species of insects: 9 plant-feeding, 15 parasitic, one predaceous. He diagrammed a rather complex network of food relationships, but only five were really significant in affecting the numbers of the gall fly. By combining theory and measurement he reached two major conclusions: "First, that in the census area the chalcid parasite, *Eurytoma curta*, was the factor primarily responsible for the control of the population density of the knapweed gall-fly, although it caused only a small proportion of the whole mortality. Secondly, that most of the remaining mortality factors, which together destroyed the greater portion of the gall-flies, had the long-term effect of increasing the average population density of the gall-fly. This was because they killed *E. curta* and the gall-fly indiscriminately, and hence reduced the efficiency of *E. curta* as a controlling agent."

It would be interesting to have additional minor-habitat studies made in the same ecosystem and to observe the interrelations that would surely be revealed. Likewise, it would be profitable to have further investigation of the same minor habitat over a wider geographic and ecological range to determine the general significance of the findings from the present limited study.

*Example 2.*—The spruce budworm, *Choristoneura fumiferana* (Clem.), has been studied intensively in both field and laboratory, expecially in Canada during the past 7 years. Although this native pest is highly destructive in cut-over areas, some of the severest outbreaks have occurred in virgin forests. A reconstruction of tree stands and budworm infestations in the past century has just been completed for a portion of northwestern Ontario to provide comparison with the more recent outbreak in the same area that has been under careful observation since 1945 (Blais 1954). It is now possible to compare the origin and spread of the current infestation with that of the early 1860's. Complementary studies on genetics, physiology, mortality factors, bioclimatology, and habitat relationships have significantly advanced our knowledge of this pest (Prebble & Morris 1951). The species thrives on the opening male flowers of mature balsam fir trees and outbreaks invariably begin in areas of concentration of this host. However, some areas of heavy balsam fir have escaped serious infestations in outbreak periods. There are numerous parasitic, predaceous, and pathogenic enemies.

The development of destructive budworm populations is explained on the basis of a relaxation of climatic control in areas of susceptible stands; and though the biotic agents also increase as the pest multiplies they do not appear to bring it under control until starvation and specific meteorological factors have reduced the population and exposed the survivors to relatively greater attack (Wellington *et al.* 1950, Prebble & Morris 1951). These observations reveal in striking fashion the interrelations of physical and biotic factors: one combination promoting population increase, a different combination causing decrease of the same species in the same locality. This study has recently been extended to include the long-term effect of a large-scale chemical control program (Flieger 1953).

*Example 3.*—In 1943, a highly significant series of investigations was initiated in the Annapolis Valley, Nova Scotia, to determine the long-term effects of spray chemicals on insect populations in apple orchards. A group of experimental orchards was used to study the effects of individual fungicides and insecticides, additional records being obtained from commercial orchards whose owners collaborated by applying recommended spray schedules. The project was undertaken to find an alternative to the constantly increasing spray program as a means of obtaining marketable fruit of good quality.

The approach has been ecological, embracing the whole insect fauna of the orchard and its environs and combining intensive field observation with field and laboratory experimentation. Emphasis has been upon the identification, life-histories, and behavior of entomophagous insects and mites and upon the effects of orchard sprays on their abundance and their interrelations with the four major pest species: the oystershell scale, *Lepidosaphes ulmi* (L.); the European red mite, *Metatetranychus ulmi* (Koch); the eye-spotted bud moth, *Spilonota ocellana* (D. & S.); and the codling moth, *Carpocapsa pomonella* (L.). A beginning has been made on a study of the climatic factors involved and on the development of modified spray programs for trials by growers. Results and tentative conclusions have been presented in a series of papers (Pickett *et al.* 1946, Lord 1947, 1949, MacPhee 1953, Lord & MacPhee 1953, Pickett & Patterson 1953, MacPhee & Sanford 1954) and include the following:—

1. In general, parasites are most effective at densities of the pest species that are higher than can be tolerated economically, whereas general predators may be relatively influential at lower pest densities.

2. The predaceous thrips *Haplothrips faurei* Hood is a density-dependent factor in regulating phytophagous mite populations, but acts largely as a density-independent factor in regulating populations of the codling moth and the bud moth.

3. Phytophagous mites and aphids appear to form an important food reservoir for many predators and to be key factors in the interrelations that achieve natural control in orchards.

4. Natural control of the oystershell scale is accomplished by a parasite or a predator or both; but a physical factor, minimum winter temperature, determines which biotic agent exerts the greater influence in different parts of Canada.

5. Under conditions prevailing in Nova Scotia orchards, DDT, parathion, and sulphur drastically reduce the numbers of natural enemies; nicotine, lead arsenate, summer oil, *Ovotran*, *Phygon*,[3] and ferbam are lethal to some beneficial species and not to others, whereas cryolite, glyodin (Crag Fruit Fungicide 341), *Tag 331*,[4] and copper fungicides are relatively harmless to all important parasites and predators.

6. Many apple growers have adopted a modified spray program based on the results of these investigations and report general satisfaction even though in some cases insecticides have not been included in the spray schedule.

This comprehensive study illustrates well the complications that arise in the ecosystem from the addition of chemical factors. It also confirms the existence of two principles: the oneness of the factorial complex in the field; and the oneness of the science of entomology regardless of the epithets "fundamental" and "applied."

Lines of Emphasis in Future Research.—The study of factors that affect insect abundance will probably spearhead the return of ecology as the field of major emphasis in entomology during the next decade. The success with which we meet this challenge will depend on how wisely we utilize the research resources that may be turned to this purpose. In this brief section I can do no more than introduce this important aspect of the subject.

From the experiences of the past 30 years we know that the estimating of insect abundance, though highly perplexing, is not nearly enough in itself. Numerical data must be related to physiology and behavior, which in turn depend, as we have seen, upon the interactions of internal and external forces, physical and biotic. Ultimately, then, we shall be required to measure concurrently many factors and to do this under natural conditions, the host population being but one of them. Once the basic concepts are established and accepted, progress will depend largely on our skill in devising adequate techniques.

However, before we reach the quantitative stage certain qualitative information is required. We cannot measure the fluctuations in abundance of a species until we know its life-history and basic behavior patterns and can find and recognize it in all its forms and stages. Furthermore, the measurement of physical factors is meaningless until we have established what needs to be measured and what limits of accuracy are required and possible. When refinements are expensive, we must appraise their necessity, as for example the need for measuring to a tenth of a degree a factor that influences an abundance that we can only estimate as scarce, common, or numerous. Finally, all quantitative data should be statistically adequate whether they are recorded simply as meaningful categories or as refined individual measurements.

In our endeavors to understand natural phenomena, we should use every possible approach: field observation, field and laboratory experimentation, theoretical considerations, surveys, and incidental records. Each has virtue. However, we must go to nature to study nature; hence skillful field observation is a continuing requirement. It is an art that warrants further development and greater use. However, to unravel the interrelations that affect insect abundance, selected problems must be taken to the laboratory for analysis under controlled conditions. But what is undertaken in the laboratory should arise

[3] 2,3-Dichloro-1,4-naphthoquinone, a product of the U. S. Rubber Co.

[4] Phenyl mercury salts in mixture.

from what is observed in the field. Furthermore, we must recognize from the start that the laboratory is a novel habitat. In it populations that co-exist in the field are separated and interacting natural factors are isolated. Such an experiment will supply factual information, but it is factual only in reference to its own universe—the laboratory. The information thus obtained can be used to develop theories, but the theories require successful testing in the field before we can accept them as principles.

Yet theories and concepts are useful; they assist in orientation and planning. But theories derived from simple or restricted biological systems such as a culture of the confused flour beetle, *Tribolium confusum* Duv., have limited application to field ecology, as also have formulae that employ mathematical constants to represent inconstant biological phenomena. The relation of biomathematics to insect population problems has been discussed by Thompson (1939) and more recently in a comprehensive paper by Ullyett (1953), who concludes: "The theories of Lotka, Volterra and Nicholson and Bailey, being all based on the primary assumption that animals act at random in relation to the things which are important for their survival, must be rejected as adequate general theories of animal behaviour." Undoubtedly, we should proceed with caution in deciding what fraction of our resources should be devoted to experiments based on mathematical concepts that are derived largely from mathematics or physics instead of from adequate biological data.

Increased attention is warranted in field experimentation. In Schneirla's words (1950): "It should be possible to carry many of our laboratory methods into a direct-manipulative use in the field, with appropriate modifications." This was imaginatively executed by Brown (1952), who constructed "dummy men" for testing under controlled conditions in the field the validity of his laboratory findings on individual factors affecting the attractiveness of humans to mosquitoes. In general, field investigations must be thoroughly planned and the individual assignments carefully limited or the project becomes diffuse and unwieldy and, eventually, unprofitable. Furthermore, if the specific and limited assignments are to yield more than specific and limited results, they must be wisely selected and skillfully integrated into a combined picture of general significance.

Recapitulation.—In this paper I have sought to depict each natural community as a plastic but balanced system of variable factors and each insect population as one of the unstable components. The density of such a population at any given time is a product of the interactions of all concurrently active factors, each of which differs in time and space in its effectiveness. Fluctuations in insect abundance thus arise from different causes and tend, therefore, to be irregular in their occurrence. It is extremely difficult to appraise the individual significance of factors that never act independently in nature. But the need for such appraisal is upon us. Indeed, the effects of pesticides on the food-producing and fibre-producing ecosystems of the world seem to have already touched off a new era in field ecology. This may well be the great entomological challenge of the century.

## Literature Cited


**Blais, J. R.** 1954. The recurrence of spruce budworm infestations in the past century in the Lac Seul area of northwestern Ontario. Ecology **35**(1): 62–71.

**Bodenheimer, F. S.** 1928. Welche Factoren regulieren die Individuenzahl einer Insektenart in die Natur? Biol. Zbl. **48**(12): 714–39.

**Brown, A. W. A.** 1952. Factors in the attractiveness of bodies for mosquitoes. Trans. 9th Int. Congr. Ent. **1**: 895–9.

**de Gryse, J. J.** 1951. Forest entomology. Sci. News **21**: 94–109.

**Elton, C.** 1949. Population interspersion: an essay on animal community patterns. Jour. Anim. Ecol. **17**(1): 1–23.

**Flieger, B. W.** 1953. Spruce budworm spraying project in New Brunswick. *In* Forest Biology Bimonthly Progress Report **9**(4): 1–2. Canada Dept. Agr., Ottawa.

**Franz, J.** 1949. Über die genetischen Grundlagen des Zusammenbruchs einer Maassenvermehrung aus inneren Ursachen. Z. Angew. Ent. **31**(2): 228–60.

**Howard, L. O., and W. F. Fiske.** 1911. The importation into the United States of the parasites of the gypsy moth and the brown-tail moth. U. S. Bur. Ent. Bul. **91**.

**Lord, F. T.** 1947. The influence of spray programs on the fauna of apple orchards in Nova Scotia: II. Oystershell scale, *Lepidosaphes ulmi* (L.). Can. Ent. **79**(11, 12): 196–209.

**Lord, F. T.** 1949. The influence of spray programs on the fauna of apple orchards in Nova Scotia. III. Mites and their predators. Can. Ent. **81**(8, 9): 202–14, 217–30.
**Lord, F. T.,** and A. W. MacPhee. 1953. The influence of spray programs on the fauna of apple orchards in Nova Scotia. VI. Low temperatures and the natural control of the oystershell scale, *Lepidosaphes ulmi* (L.). Can. Ent. **85**(8): 282–91.
**MacPhee, A. W.** 1953. The influence of spray programs on the fauna of apple orchards in Nova Scotia. V. The predacious thrips *Haplothrips faurei* Hood. Can. Ent. **85**(1): 33–40.
**MacPhee, A. W.,** and K. H. Sanford. 1954. The influence of spray programs on the fauna of apple orchards in Nova Scotia. VII. Effects on some beneficial arthropods. Can. Ent. **86**(3): 128–35.
**Mayr, E.** 1942. Systematics and the origin of species from the viewpoint of a zoologist. Columbia University Press, New York.
**Mayr, E.,** E. G. Linsley, and R. L. Usinger. 1953. Methods and principles of systematic zoology. McGraw Hill Book Co., New York.
**Nicholson, A. J.** 1933. The balance of animal populations. Jour. Anim. Ecol. **2**(1): 132–78.
**Nicholson, A. J.** 1947. Fluctuation of animal populations. Rept. 26th Meeting Australian and New Zealand Assoc. Advanc. Sci., pp. 134–48.
**Pickett, A. D.,** N. A. Patterson, H. T. Stultz, and F. T. Lord. 1946. The influence of spray programs on the fauna of apple orchards in Nova Scotia: I. An appraisal of the problem and a method of approach. Sci. Agr. **26**(11): 590–600.
**Pickett, A. D.,** and N. A. Patterson. 1953. The influence of spray programs on the fauna of apple orchards in Nova Scotia. IV. A review. Can. Ent. **85**(12): 472–8.
**Prebble, M. L.,** and R. F. Morris. 1951. The spruce budworm problem. *In* Forest entomology in relation to silviculture in Canada. Part III. For. Chron. **27**(1): 14–22.
**Schneirla, T. C.** 1950. The relationship between observation and experimentation in the field study of behavior. Ann. New York Acad. Sci. **51**(6): 1022–44.
**Solomon, M. E.** 1949. The natural control of animal populations. Jour. Anim. Ecol. **18**(1): 1–35.
**Smith, H. S.** 1935. The role of biotic factors in the determination of population densities. Jour. Econ. Ent. **28**(6): 873–98.
**Smith, H. S.** 1939. Insect populations in relation to biological control. Ecol. Monogr. **9**(3): 311–20.
**Tansley, A. G.** 1935. The use and abuse of vegetational concepts and terms. Ecology **16**(3): 284–307.
**Thompson, W. R.** 1929. On natural control. Parasitology **21**(3): 269–81.
**Thompson, W. R.** 1939. Biological control and the theories of the interactions of populations. Parasitology **31**(3): 299–388.
**Ullyett, G. C.** 1947. Mortality factors in populations of **Plutella maculipennis** Curtis (Tineidae: Lep.), and their relation to the problem of control. South Africa Dept. Agr. For., Ent. Mem. **2**(6): 77–202.
**Ullyett, G. C.** 1953. Biomathematics and insect population problems. Mem Ent. Soc. Southern Africa **2**.
**Uvarov, B. P.** 1931. Insects and climate. Trans. Ent. Soc. London **79**(1): 1–247.
**Varley, G. C.** 1947. The natural control of population balance in the knapweed gall-fly (*Urophora jaceana*). Jour. Anim. Ecol. **16**(2): 139–87.
**Wellington, W. G.,** J. J. Fettes, K. B. Turner, and R. M. Belyea. 1950. Physical and biological indicators of the development of outbreaks of the spruce budworm, *Choristoneura fumiferana* (Clem.) (Lepidoptera: Tortricidae). Can. Jour. Res., D, **28**(6): 308–31.

# Density Dependence in the Australian Thrips

Frederick E. Smith
*Zoology Department, University of Michigan, Ann Arbor*

The study of *Thrips imaginis* by Davidson and Andrewartha (1948a, b) has been cited as an example of an animal population whose numbers are controlled by the weather. To quote from Andrewartha and Birch (1954): ". . . not only did we fail to find a 'density-dependent factor,' but we also showed that there was no room for one." A search for density-dependence in these data has never been published.

It is the purpose of this note to show that such a search supports the thesis of strong density-dependence in this population, and furthermore that the analytic model used by Davidson and Andrewartha is not appropriate to their theoretical interpretation, and is in fact appropriate to the concept of density-dependent regulation.

Density-dependence exists when population change depends upon population size. Usually an inverse relation is expected; when the population level is high, decreases tend to occur, and vice versa. Responses to other factors are not ruled out; changes in the weather, for example, can certainly influence population change. In a density-independent system, population changes are largely explained in such a way, with population size exerting no effect upon population change, while in the density-dependent system population size persists as a powerful factor acting together with other factors.

A search for density-dependence involves an analysis of population changes, which in turn require accurate censuses of population size. The data in the thrips study are among the most detailed and comprehensive that are published for a field population. Population estimates were made almost daily during the periods of observation, producing monthly means with very small sampling errors. Such means are published by Davidson and Andrewartha for 81 consecutive months (1948a) and during the late spring and early summer for 14 years (1948b).

Seasonal trends are similar from year to year (Figure 1), showing population lows in winter and again in late summer, with moderate rises in the autumn and sharp rises to annual maxima in the spring. Davidson and Andrewartha treated years as a form of replication, thus permitting analyses of population size at certain times of the year in terms of weather factors for those years. The same approach is used here to permit an analysis of population change.

Davidson and Andrewartha converted census data to logarithms, since rates of change are best expressed in this way. As stated by them (1948b): "Experience indicates that the thrips numbers are not likely to change by a constant quantity for each increment of the causal agent, but rather are they likely to change by a constant proportion." Happily, this transformation produced frequency distributions very close to normal distributions, and produced sampling errors that were independent of population size. Thus, the data become ideally suited to statistical analysis.

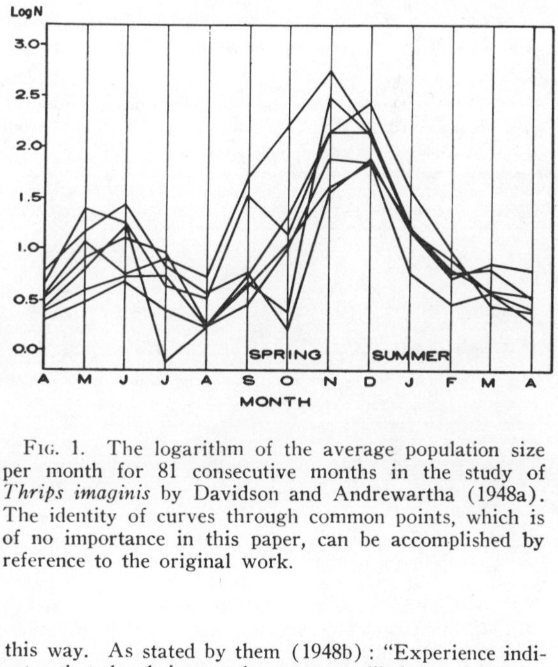


Fig. 1. The logarithm of the average population size per month for 81 consecutive months in the study of *Thrips imaginis* by Davidson and Andrewartha (1948a). The identity of curves through common points, which is of no importance in this paper, can be accomplished by reference to the original work.

Averages of the logarithms of population size for 3 calendar periods are published for 14 years (y's are the symbols used in the original papers; notations at the right will be used in this paper):

$y_4$ = average log N for the period 1-31 October = $\overline{\text{16 Oct}}$

$y'_6$ = average log N for 17 Oct. to 15 Nov. = $\overline{\text{1 Nov}}$

$y_5$ = average log N for the period 1-30 November = $\overline{\text{16 Nov}}$

Reproduced with permission from *Ecology*, 42:403–407, 1961. Published by the Ecological Society of America and the Duke University Press, Durham, North Carolina.

Each of these averages covers a period of about 30 days, and is based on censuses from as few as 22 to as many as 27 days, the missing days being mostly Sundays and holidays. Unpublished data exist for September and December which were utilized in other averages discussed below.

Estimates of the rate of population change can be obtained by differences from these average logarithms. Three such differences will be used here:

$y_5 - y_4$ = ave. 1-30 Nov. minus ave. 1-31 Oct. = Δ log N 16 Oct. to 16 Nov.

$y'_6 - y_4$ = ave. 1-15 Nov. minus ave. 1-16 Oct. = Δ log N 8 Oct. to 8 Nov.

$y_5 - y'_6$ = ave. 16-30 Nov. minus ave. 17-31 Oct. = Δ log N 23 Oct. to 23 Nov.

Note that in the last 2 the effect of subtraction is such that the period of overlap in the two 30-day means (17-31 Oct. in the second difference, 1-15 Nov. in the third) is virtually subtracted from itself and hence disappears. Thus all 3 differences span a period of about one month.

The correlations of these differences with appropriate estimates of population size (at or near the beginning of the period covered by the difference) are given on Table I. The consistency with which a negative correlation emerges is such that density-dependence is strongly suggested.

TABLE I. Correlations of population change (Δ log N) with population size (log N) using 14 years of data from Davidson and Andrewartha (1948b)

| Δ log N | log N | r | P |
|---|---|---|---|
| 16 Oct. to 16 Nov....... | $\overline{\text{16 Oct.}}$ | −0.80 | <.001 |
| 8 Oct. to 8 Nov....... | $\overline{\text{16 Oct.}}$ | −0.60 | .03 |
| 23 Oct. to 23 Nov....... | $\overline{\text{16 Oct.}}$ | −0.81 | <.001 |
| 23 Oct. to 23 Nov....... | $\overline{\text{1 Nov.}}$ | −0.73 | .004 |

Davidson and Andrewartha criticized the use of calendar periods as a basis for comparing different years on the grounds that spring arrives at different times in different years. They attempted to coordinate the seasons by using the time of the summer maximum of the population as a reference point. This time varied between 15 November and 13 December. After shifting each year's data to the left or right until the times of the maxima coincided, they computed new averages. Three such averages are published:

$y_1$ = ave. log N from 7 days before to 7 days after the maximum = $\overline{7-7}$

$y_2$ = same for the period 30-1 days before the maximum = $\overline{30-1}$

$y_3$ = same for the period 60-1 days before the maximum.

The last average covers a very long period. Estimates for the average of the period from 60 to 31 days (= $\overline{60\text{-}31}$) before the maximum can be obtained, however, by subtracting $y_2$ from $2y_3$. The estimates are inaccurate only to the extent that the number of days averaged in $y_3$ is not exactly twice the number averaged in $y_2$.

Differences can be taken as estimates of population change:

$y_2 - (2y_3 - y_2)$ = ave. 30-1 days minus ave. 60-31 days = Δ log N −45 to −15

$y_1 - (2y_3 - y_2)$ = ave. 7-7 minus ave. 60-31 days = Δ log N −45 to 0

$y_1 - y_2$ = approx. ½(7-1 before) plus 0-7 after the maximum minus ave. 30-8 days before = Δ log N −19 to +1

The last difference is especially awkward since the overlapping means are based on unequal time periods. Such irregularities do not, however, introduce bias into the analysis.

Correlations based on these time-shifted data are given on Table II. Obviously, attempting to coordinate the seasons has had little effect upon the inverse relation between population change and population size. It now remains to be seen whether this relation will persist if extrinsic factors are brought into the analysis.

TABLE II. Similar to Table I, but with the daily estimates shifted in time each year so that the summer maxima coincide. Times refer to days before or after the maximum

| Δ log N | log N | r | P |
|---|---|---|---|
| −45 to −15........... | $\overline{\text{60-31}}$ | −0.76 | .002 |
| −45 to 0.............. | $\overline{\text{60-31}}$ | −0.82 | <.001 |
| −19 to +1........... | $\overline{\text{60-31}}$ | −0.57 | .04 |
| −19 to +1........... | $\overline{\text{30-1}}$ | −0.49 | .08 |

Of the several weather factors studied by Davidson and Andrewartha, 2 were the most consistently significant, and accounted for the majority of the relation they found between population size and weather. These were the effective degree-days between the first fall rain and the end of winter ($x_2$ in their paper), and the total inches of rain during September and October ($x_3$).

The correlations of the 6 estimates of population change with these 2 factors are given on Table III. For data based on calendar periods, a general conclusion that the 2 factors are relevant is acceptable, while for the data based on time-adjusted periods no significance is evident.

TABLE III. Correlations of population change with effective degree-days between the first fall rain and Aug. 31 ($x_2$), and with inches of rain during Sept. and Oct. ($x_3$)

| Δ log N | CORRELATION WITH $x_2$ | | CORRELATION WITH $x_3$ | |
|---|---|---|---|---|
| | r | P | r | P |
| 16 Oct. to 16 Nov. | −0.57 | .04 | 0.54 | .05 |
| 8 Oct. to 8 Nov. | −0.42 | >.10 | 0.59 | .03 |
| 23 Oct. to 23 Nov. | −0.58 | .03 | 0.52 | .06 |
| −45 to −15..... | −0.29 | >.10 | 0.35 | >.10 |
| −45 to 0........ | −0.31 | >.10 | 0.23 | >.10 |
| −19 to +1...... | −0.22 | >.10 | −0.16 | >.10 |

The 2 weather factors themselves are correlated −0.63, and each is more or less correlated with the various estimates of population size. With so many interrelations present, statistical procedures are needed for clarification. Davidson and Andrewartha used multiple regressions. Where the prediction of a quantity is desired through the simultaneous use of several possible predictors, such a technique is indicated. In this case, however, the relative roles of possible factors are to be compared. For this, the related process of partial correlation is indicated.

Two of the kinds of estimates of population change were chosen for this analysis: Δ log N 16 Oct. to 16 Nov., and Δ log N −45 to −15. Each of these spans one month, and is based on 30-day means. Moreover, they afford an interesting comparison, since the average date of the population maximum was Dec. 1. For each, 3 factors are considered simultaneously: the initial 30-day mean, winter degree-days, and spring rain. Two of these were partialled out in the process of computing the partial correlation of the third with population change. Thus, each partial correlation shows the degree to which population change is related to a given factor, excluding that which can be attributed to the other factors.

The results of this analysis are given on Table IV. Of major interest is the observation that the inverse relation of population change with population size remains strong, and is much stronger than relations of population change with either of the 2 weather factors.

TABLES IV. Partial correlations of population change ($\Delta$ log N) with population size (log N), winter degree days ($x_2$), and spring rain ($x_3$). In each case 2 of the 3 factors have been partialled out. In the first row, log N is for the average of the period 1-31 Oct.; in the second row it is for the average of the period 60-31 days before the summer population maximum

| $\Delta$ log N | log N | P | $x_2$ | P | $x_3$ | P |
|---|---|---|---|---|---|---|
| 16 Oct. to 16 Nov...... | −0.70 | .02 | 0.36 | >.10 | 0.27 | >.10 |
| −45 to −15.......... | −0.84 | <.001 | 0.63 | .03 | 0.49 | .11 |

A comparison of the 2 sets of partial correlations supports the attempt by Davidson and Andrewartha to coordinate seasons. Using time-shifted data (2nd row), the partial correlations of population change and each of the weather factors are reasonably high, suggesting that with proper time-adjustments these effects can be revealed. In spite of this, however, the inverse relation of population change and population size is extremely high.

Although the method of correlation analysis is a powerful technique for discovering density-dependence, it cannot be used where the sampling errors of population size are large relative to the amount of change occurring in population size. Davidson and Andrewartha (1948a) reported a relative standard error in the daily estimates of about 15%. In terms of logarithms and of averages of many days, this error becomes negligible. Thus, the inverse relations obtained above cannot be ascribed to sampling error. Where errors are large, however, another technique is needed.

A related approach that escapes the problem of sampling error utilizes an analysis of the variance of population size at different times. This approach has broad applications, especially where a series of life tables or other population histories are available. Density-dependent control of population size produces a given level of variation among various sub-populations at any given time. Should this control be relaxed, the sub-populations being in slightly varying environments will steadily diversify—that is, the variance will increase. Variance will decrease again if and when regulatory mechanisms again operate.

In a truly density-independent model, of course, there is no method of reducing variance, since this can occur only through a negative correlation of population change and population size.

The 81 consecutive monthly means of thrips per flower (Davidson and Andrewartha 1948a), converted to logarithms, are shown on Fig. 1. Using years as replications, variances for each month can be computed, having 6 degrees of freedom for 9 months, and 5 for the other 3. These variances, together with an average curve of population size, are shown on Figure 2.

The most striking behavior of the variance is its sharp rise from August to October, during the early phase of spring growth, followed by an equally sharp decrease

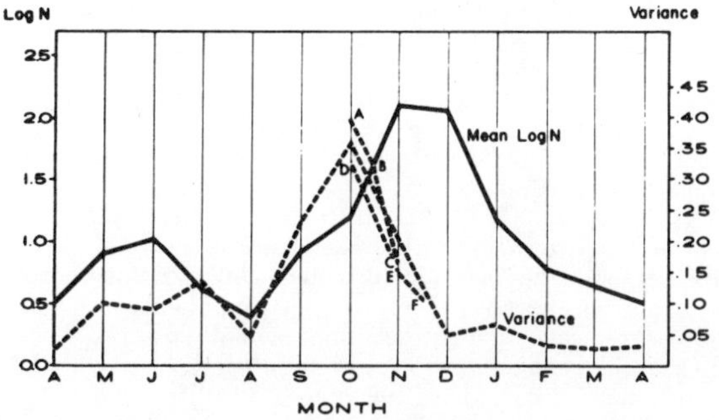


FIG. 2. The variance of the logarithm of the average population size each month, using the 6 and ¾ years of data as a form of replication. The average curve of population size from Fig. 1 is also shown. Additional points show the variance using 14 years of data from Davidson and Andrewartha (1948b).

from October to December, during the later phase of spring growth and into the beginning of population decrease. These variations in the variance are much too large to be attributed to chance. Furthermore, using the 14-year data, the variances of the 3 calendar periods (points A, B, C) and of the 3 time-shifted periods (points D, E, F), which cover the later phase of spring growth, show similar decreases.

The low variance through 9 of the months indicates that the thrips population is regulated by density-dependence most of the time. It should be emphasized, moreover, that this cannot be achieved through minor effects. Where the standard deviations of population size and population change are $s_a$ and $s_b$, respectively, a correlation of at least $-\frac{1}{2}(s_a/s_b)$ is necessary to prevent an increase in population variance. Only in September and October, when a marked increase in flower densities occurs, does the variance of population size behave according to the requirements of a density-independent system.

Reconciling these analyses with those of Davidson and Andrewartha is possible if their model is examined. Where density-independent interpretations are to be made, it is the relation of environmental factors to rates of change in population size that must be studied. Furthermore, as suggested by Davidson and Andrewartha, it is the relative change (percentage change in size, also measured by the change in the logarithm of population size) that should be related quantitatively to environmental factors. Hence, the appropriate multiple regression is:

$$\Delta \log N = a + bx_1 + cx_2 + dx_3 + \ldots.$$

where the x's represent various measures of the weather.

Population size, or its logarithm, will show a much poorer relation to weather factors if the system is density-independent. For example, although a population may always double in one month under given environmental conditions (a density-independent hypothesis), the size may in one case be very small and in another be very large. In a density-independent system, the approach is such that only population change can be accurately predicted; size cannot be predicted at all unless a known size is used for one point in time, and later sizes predicted through the successive application of predicted rates of change.

Since Davidson and Andrewartha used population size rather than change in their model:

$$\log N = a + bx_1 + cx_2 + dx_3 + \ldots$$

and obtained high degrees of predictibility, they have in fact demonstrated regulation in a natural population. Their multiple regression implies a predicted population level for specific environmental conditions, and the actual population follows this predicted level with remarkable fidelity. The predicted levels must, therefore, represent some kind of equilibrium.

The weather factors used are related frankly by Davidson and Andrewartha to the abundance of flowers, which are not only the food but also the habitat of the thrips. It is not unbelievable, therefore, that such factors as winter degree-days and spring rain should powerfully influence the regulatory system of the thrips.

It is not the purpose of this note to identify the density-dependent mechanism. The evidence presented by Davidson and Andrewartha that food is not limiting is convincing. So is their argument that density levels (thrips per rose) correlate with flower density. It is evident that some one or more factors hold the thrips to very low levels in the flowers. Predation, parasitism, and other density-dependent mortalities are likely to interact with the capacity of the thrips to increase, which would be a function of flower density, so as to produce a variable equilibrium level. Predation and parasitism would also suffer time lags which could explain the sharpness of the peak of the annual maximum. Whatever factors control the abundance of thrips, it is evident that they have so far been overlooked.

Throughout this paper no distinction has been made between unadjusted (y) and adjusted (y′) means from the original papers. Adjusted means are corrected for 'activity" through the use of a multiple regression with the momentary weather. Although this regression is significant, it accounts for only 10% of the variance of daily counts around the population trend. In the means this becomes a trivial correction, presumably because over a 15, 30, or 60 day period these corrections nullify each other. This is shown in the correlation coefficient of +0.997 between $y_2$ and $y'_2$, the only case where both adjusted and unadjusted figures were published. In Andrewartha and Birch (1954), pages 571 to 574 can be deleted without loss to the analysis.

### Summary

Direct analysis of the published data on the thrips population studied by Davidson and Andrewartha reveals a strong inverse correlation between population change and population size. A rapidly decreasing variance in population size is also shown to exist during the later portion of the spring population increase. Such evidence of density-dependence persists through the use of partial correlations using environmental factors known to be related to population size. The multiple regression model used by Davidson and Andrewartha is not appropriate for the interpretations they have made. They have in fact demonstrated the degree to which the population conforms to a predicted level, hence the degree to which the population is regulated. They have no evidence for a density-independent system. The regulatory system by which the thrips are held to levels well below that set by their food supply remains unknown.

### References

**Andrewartha, H. G., and L. C. Birch.** 1954. The Distribution and Abundance of Animals. Univ. Chicago Press.

**Davidson, J. and H. G. Andrewartha.** 1948a. Annual trends in a natural population of *Thrips imaginis* (Thysanoptera). J. Anim. Ecol. **17**: 193-199.

———. 1948b. The influence of rainfall, evaporation, and atmospheric temperature on fluctuations in the size of a natural population of *Thrips imaginis* (Thysanoptera). J. Anim. Ecol. **17**: 200-222.

# B. Dispersal and Migration

## Long-Range Aerial Transport of the Harlequin Bug and the Greenbug into Minnesota[1]

Alexander C. Hodson and Edwin F. Cook[2]

ABSTRACT

On May 4, 1959, the harlequin bug, *Murgantia histrionica* (Hahn), appeared in large numbers in a restricted locality in southern Minnesota. Fourteen days later an outbreak of the greenbug, *Toxoptera graminum* (Rond.), was discovered in the same general area. A study of weather records for the first 4 days of May showed that weather features particularly favorable for long-range aerial transport of insects could have enabled them to be carried to Minnesota from areas to the south and southwest, where they were known to be abundant. The possible prediction of future major outbreaks of the greenbug is discussed briefly.

During early May 1959, southwestern Minnesota was invaded by the greenbug, *Toxoptera graminum* (Rond.), and the harlequin bug, *Murgantia histrionica* (Hahn). The greenbug, while not a permanent resident of the State, is known to be a frequent immigrant from areas to the southwest, and occasionally it has increased to outbreak proportions after its arrival (Wadley 1931). In contrast, the discovery of the harlequin bug in the town of Slayton in southwestern Minnesota was a unique event. This species had never before been collected anywhere in the State. Because of this a brief account of its arrival is given here.

A greenhouse operator was covering cold frames in the late afternoon of May 4 to protect plants from a threatening storm, when he noticed very large numbers of flying insects in the yard. At first he thought they were honey bees, but unlike honey bees they soon settled on young cabbage plants in flats, where they immediately started to feed and oviposit. A careful survey made by personnel of the State Division of Plant Industry showed that the infestation was localized in an area about 1 square block to the east and including the place where they were first found. This site is designated by a large dot in Central Murray County in figure 1. Within a few days more harlequin bugs or reports of their being found were received from one locality to the east and several northeast of the original find. Much later in the summer harlequin bugs were collected north and northwest of what is presumed to have been their line of flight upon entering the State in May. The locations of these late reports, received in August, also are shown on the map in figure 1.

This account of the arrival of the harlequin bug in Slayton, Minnesota, suggests that we may have been fortunate enough to obtain a record of the day, and perhaps even the hour that the bugs terminated their flight from the southwest. We cannot be as sure about the time of arrival of its fellow travelers, the greenbugs. When they were first reported they already were present in abundance in several southwestern counties, (fig. 1), and were causing conspicuous damage to small grains. The date of their discovery was May 18, 14 days after the first harlequin bugs were seen. However, it is assumed that the principal greenbug invasion from the southwest probably occurred at the same time as that of the harlequin bug, because a large aphid population could have arisen from early established immigrants with the temperatures that prevailed during this 14-day period (Wadley 1931).

SOURCE OF INFESTATION.—As neither of these species is a permanent resident of Minnesota, and both are common pests south of lat. 40° N., the northern border of Kansas, it is pertinent to ask whence they came and by what means. Records obtained from the Cooperative Economic Insect Reports issued by the Plant Pest Control Division, Agricultural Research Service, U. S. Department of Agriculture show that the harlequin bug was present in abundance and was killing crucifers and causing considerable damage to ornamentals in Oklahoma and Texas during the week ending May 8. Damage continued to be reported from these States during the month of May. The records for the week ending May 29 indicate that unusually large numbers of harlequin bugs were present on weed hosts in Nebraska.

Dense populations of the greenbug also were reported early in the season from Oklahoma and Texas. Heavy infestations occurred in Tillman County, Oklahoma, during the week ending March 13, and heavy populations of winged forms were noted in Noble and Payne Counties during the week ending March 30. At the close of the week ending April 17, populations were declining in Oklahoma and the first reports of light to moderate populations were received from Kansas. Greenbugs continued to increase in abundance in Kansas, and by the last week in April they were increasing in abundance also in southwestern Missouri. During the following week ending May 8, populations were still increasing in Kansas, and a large proportion of the populations in both Kansas and Missouri had developed winged forms. Thus the stage was set in late April and early May with both the har-

[1] Paper No. 4314 Scientific Journal Series, Minnesota Agricultural Experiment Station, St. Paul 1, Minnesota. Accepted for publication January 28, 1960.

[2] Professor and Associate Professor, respectively, Department of Entomology and Economic Zoology, University of Minnesota, St. Paul 1, Minnesota.

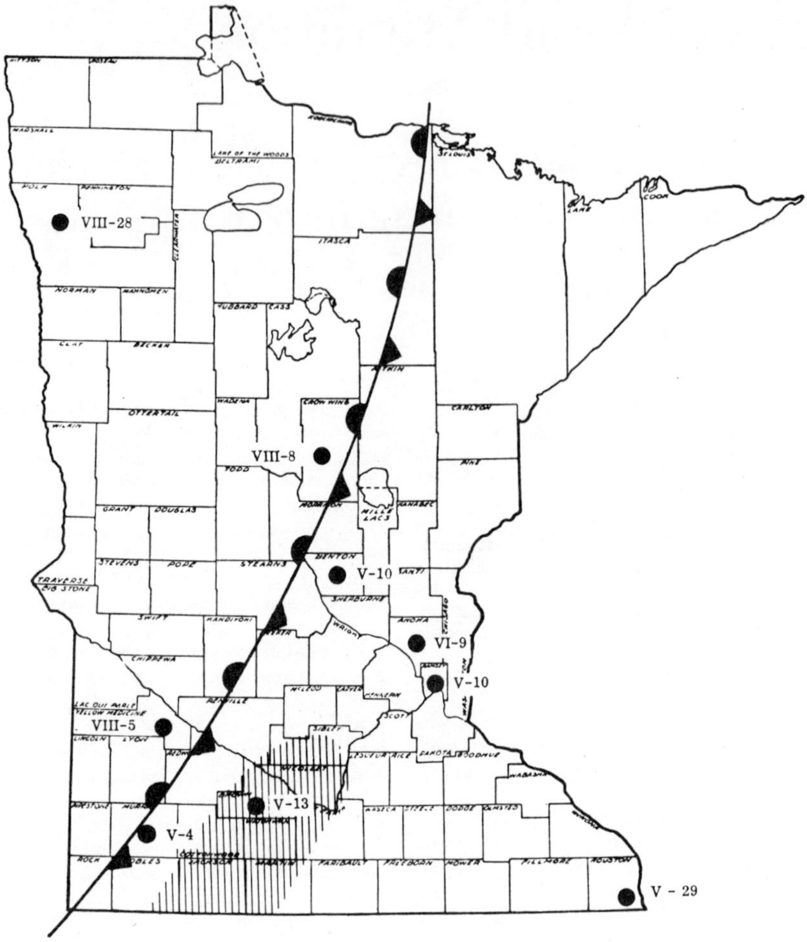


FIG. 1.—Position of front on the afternoon of May 4. Spots show localities and dates where harlequin bugs were collected. Shaded area shows distribution of greenbug when first detected.

lequin bug and the greenbug present in abundance in areas to the south and southwest of Minnesota.

METHOD OF MIGRATION.—The second part of the question stated earlier concerns the means of travel from the areas of known abundance, cited above, to southwestern Minnesota. Flight, aided by favorable wind circulation, is proposed as the simplest answer. This opinion is supported by the work of Wadley (1931). He states that prior to the discovery of the 1926 outbreak in Minnesota there were certain well-defined meteorological conditions that occur infrequently. These conditions were strong southwest winds that were sustained without interruption for 2- or 3-day periods. The winds were channeled from south of Oklahoma, across Kansas, and on northeastward into Minnesota. Wadley's description of weather conditions associated with the 1926 greenbug outbreak seems to satisfy the theoretical pattern required for long-range aerial transport of insects (Wellington 1945). According to Wellington there should be a high-pressure area which remains relatively stationary for several days, accompanied by strong winds that are maintained along a definite direction of circulation throughout the stationary period. Such conditions, he says, may occur in spring or autumn, and may well act to distribute numbers of a given species for hundreds of miles along a fixed track, thus insuring the establishment of the species in any suitable habitat along the wind path. With this information to serve as a guide the authors examined weather records assembled by the U. S. Weather Bureau for the first week in May 1959, the week during which it is assumed that the long range transport of the harlequin bug and the greenbug had occurred.

METEOROLOGICAL EVENTS.—The sequence of meteorological events leading up to the sighting of the harlequin bug on May 4 are illustrated in figures 2, 3, 4 and 5. According to Woffinden (1959), there was a temperature anomaly during the first week of May characterized by a strong east–west temperature gradient. Unseasonable cold weather prevailed over the western United States, while much above average temperature was reported in

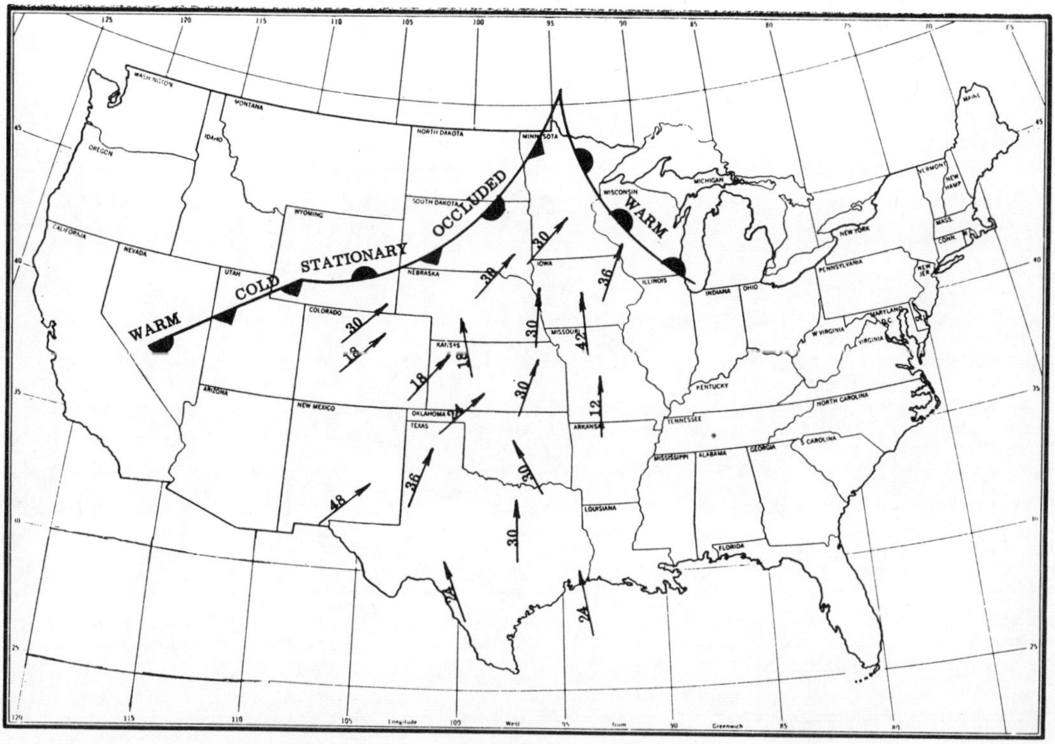


FIG. 2.—Position of front at noon of May 1. Arrows indicate wind direction and velocity in miles per hour at second standard level (between 200 and 1,200 feet above ground surface at weather station).

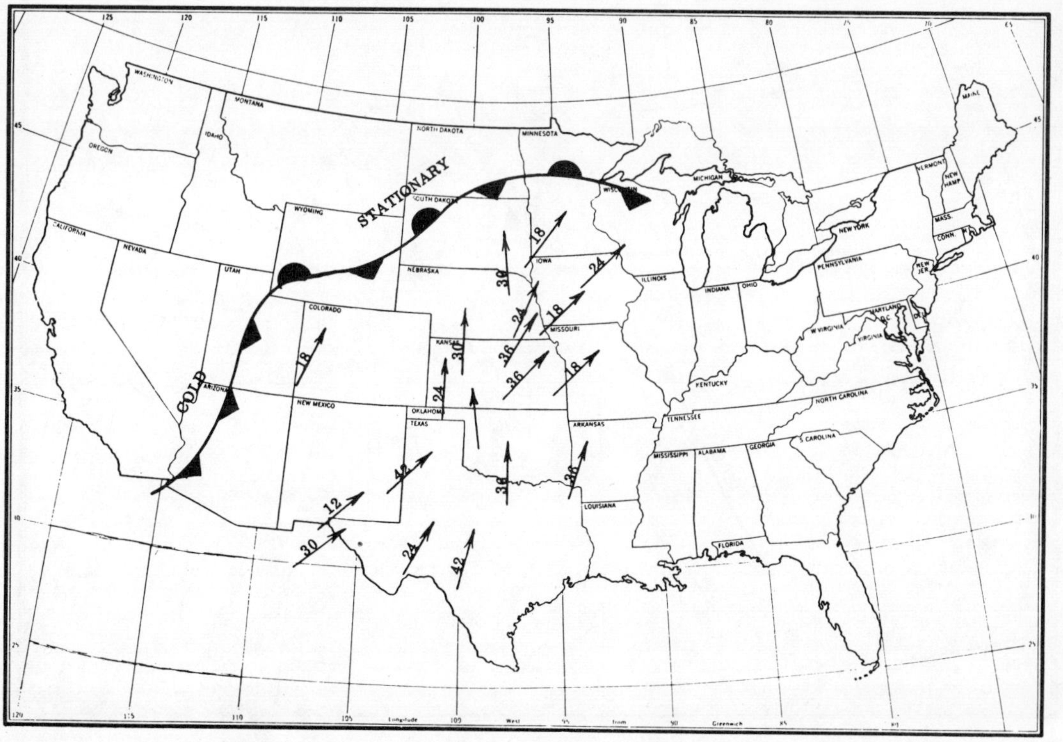


FIG. 3.—Position of front at noon of May 2.

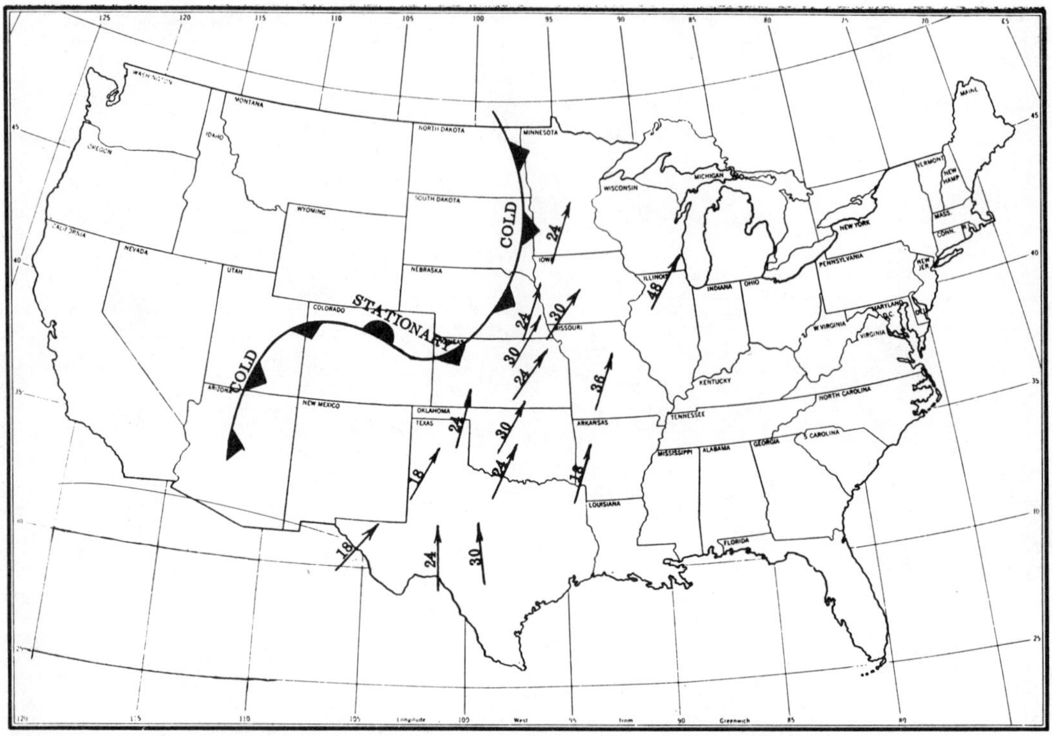


FIG. 4.—Position of front at noon of May 3.

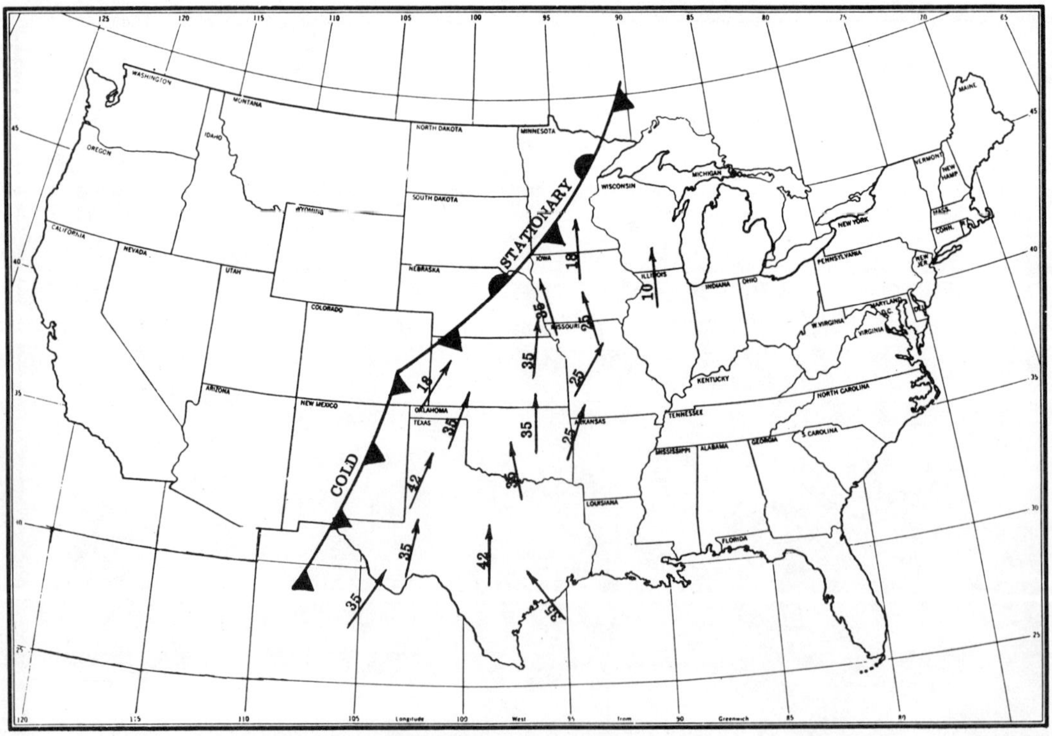


FIG. 5.—Position of front at noon of May 4.

the central Great Plains in advance of the frontal system shown in figures 2, 3, 4 and 5. A trough formed along the line of the frontal system, and a blocking effect amplified a ridge in front of the trough, thus advecting warm, moist air into the Mississippi Valley. Along the boundary between the contrasting air masses severe frontal weather developed with thunderstorms and considerable convective activity.

The frontal systems shown on the maps maintained a prolonged and relatively constant position throughout the 4-day period prior to the arrival of the harlequin bug in Slayton. A sustained circulation of strong southwest winds from Texas, across Oklahoma, Kansas, and eastern Nebraska was channeled along the frontal system at the same time. These winds were constant in direction from the surface to 18,000 feet. They were brisk from Texas to northern Iowa, but diminished in velocity in southwestern Minnesota by noon of May 4.

Thus the conditions for long-range transport of winged adult harlequin bugs and greenbugs, known to be abundant along the air-circulation path, fit the ideal pattern set forth by Wellington. Furthermore, the air turbulance and convective activity in advance of the frontal system could have been responsible for the insects' becoming airborne in the first place. Finally, the drop in wind velocity in southwestern Minnesota, which was most noticeable by noon of May 4 (fig. 5), may have permitted the airborne insects to descend into the area where they were seen first about 4:30 p.m. that afternoon. Harlequin bugs may have dropped out along the path of the winds before reaching Minnesota. This supposition seems evident from the fact that unusually large numbers were reported from Nebraska, and they also were found in Iowa (personal communication from H. M. Harris).

Mention was made earlier of reports that the harlequin bug had been found in August. The location of these finds seen in figure 1 shows them in positions behind the occluded front that lies diagonally across Minnesota. These bugs may have moved into western counties soon after they arrived in May and were not found until much later, or they may have been the adults of a second generation produced in Minnesota. The greenbug spread from the original heavily infested area (fig. 1) to be found later in nearly all counties in the southern two-thirds of the State.

Predicting Outbreaks.—The now well-established association of recognizable weather features with outbreaks of the greenbug in Minnesota suggests the possibility of predicting the development of major outbreaks in northern areas in the future. The first step would be to keep informed concerning the progress of greenbug development and numbers in areas in the South and Southwest. Then a daily scanning of weather maps during late April and May would make it possible to observe the development of the critical patterns of weather necessary for long-range aerial transport. Finally, whenever the required combination of weather features occurred, surveys could be conducted in small grain fields with a reasonable chance of detecting light populations in time to initiate control before extensive damage takes place.

## References Cited

**Wadley, F. M.** 1931. Ecology of *Toxoptera graminum*, especially as to factors affecting importance in the northern United States. Ann. Ent. Soc. America **24:** 325–95.

**Wellington, W. G.** 1945. Conditions governing the distribution of insects in the free atmosphere. IV. Distributive processes of economic significance. Canadian Ent. **77:** 69–74.

**Woffinden, C. M.** 1959. The weather and circulation of May, 1959. Monthly Weather Rev. **87:** 196–205.

# PHYSIOLOGICAL FACTORS IN INSECT MIGRATION BY FLIGHT

By Dr. C. G. JOHNSON
Rothamsted Experimental Station, Harpenden, Herts.

MOST migratory flight in insects is prereproductive, occurring at a particular time in adult life. This suggests a physiological and experimental approach hitherto lacking in migration investigations[1].

The modern view of migration sees the locomotory drive as of first and of general importance and the mechanisms of orientation secondary and specialized; many migrants are displaced by wind, and well-documented instances where the major direction of prolonged displacement is controlled by the insect are few. It is the drive, therefore, which is the main concern here, and particularly in females for reasons discussed later on.

## Migration in Females in Relation to Ovary Development

Most, if not all, flights generally recognized as migratory by their simultaneity, undistractedness and duration start at the breeding sites and usually begin either immediately after the teneral period, as with aphids, or relatively early in adult life[1]; the females nearly always have undeveloped or only partially developed ovaries—a fact known for many species, but not fully recognized as a generality. Examples are butterflies[2–5], moths[6–8], mosquitoes[9], syrphids[10], aphids[11], leafhoppers[12], aleyrodids[13], Heteroptera[14,15], coccinellids[16,17], scolytids[18–21], chrysomelids[22], thrips[23], locusts[24], dragonflies[25], ants and termites[26]. Exceptions are insects probably at the very end of migration or species, discussed later, that make prolonged flights in between bouts of oviposition.

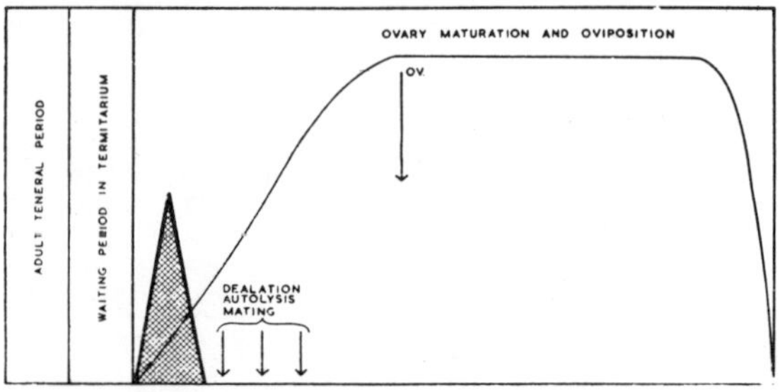


Fig. 1. Ants and termites. Migration flight shaded: *ov.*, oviposition.

*Primary migration from the breeding site.* Five examples are given: others can be obtained from the insects already listed.

The flight of termites has the behavioural and ecological characters of migration. Though short, it is not distracted by mating, feeding or oviposition, is made simultaneously by sexually immature females on the maiden flight and is followed by settling, mating, wing-muscle autolysis and ovary development. It is evidently adapted to transfer part of the population to another breeding site (Fig. 1*).

The almost irrepressible exodus flight of alate, parthenogenetic, female alienicolæ of *Aphis fabae* L. occurs, as with termites, on the maiden flight, but aphids contain partially developed embryos. After flight, which can range from a few seconds to many hours, aphids settle on a suitable host, feed, complete the development of the embryos and deposit larvæ. They may fly several times during the next three or four days before the flight muscles autolyse and flying ends. They spend the rest of their lives reproducing (Fig. 2)[11].

* All six figures in this article are diagrammatic and do not imply scales or quantities accurately.

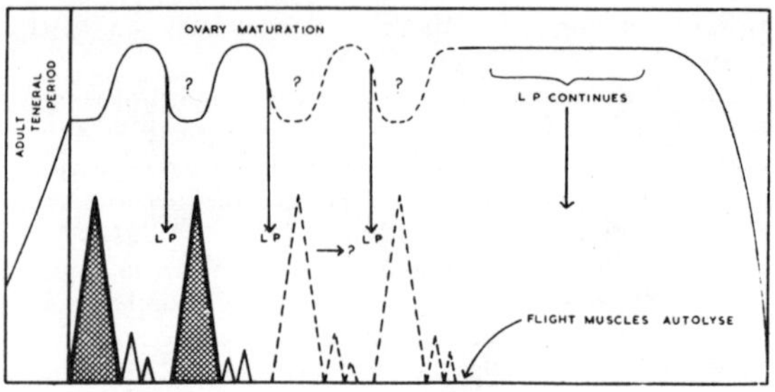


Fig. 2. Migratory aphids. Migratory flights may be repeated a variable number of times (dotted peaks). Small peaks represent more local flights. The course of ovary maturation between flights is uncertain. *LP*, larviposition

The butterfly *Ascia monuste* L. (Pieridae) in Florida (Fig. 3) migrates when females are sexually immature; migratory flights end on the first or second day of adult life. Thereafter, maturation and relatively short flights between feeding and oviposition sites are evident[2]. Many other butterflies begin to migrate before ovaries are fully developed; none is known to start otherwise.

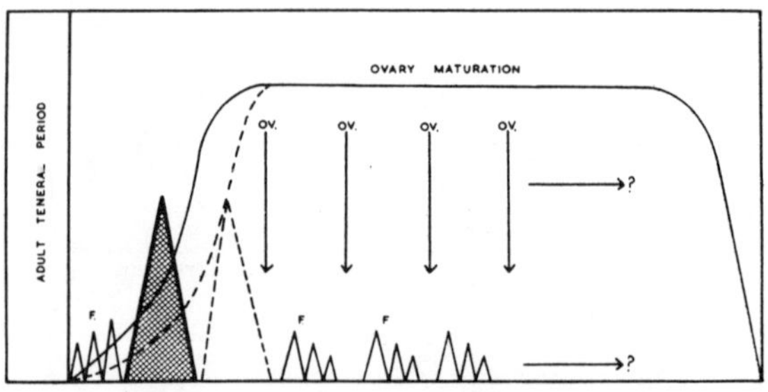


Fig. 3. *Ascia monuste* (Lepitopt: Pieridae). Normally with one migration flight, but with delayed ovary development an additional one occurs (dotted lines). *F*, flights at feeding sites; *ov.*, oviposition

Gregarious locusts begin to migrate while sexually immature and ovaries ripen during migration. Considerable migrations occur after initial oviposition, though how long migration and oviposition continue to alternate is controversial (Fig. 4). Three or four batches of eggs seem to be laid at intervals of a few days; but it is uncertain whether migration ends before all the eggs are laid[27–30].

Adults of some dragonflies[25] and chafers[31,32] leave their birth-place while sexually immature and fly straight to a feeding site a mile or two away, where the ovaries develop. They return later in the season, often to the original

Reproduced with permission from *Nature,* 198:423–427, 1963. Published by Macmillan (Journals), Limited, London, England.

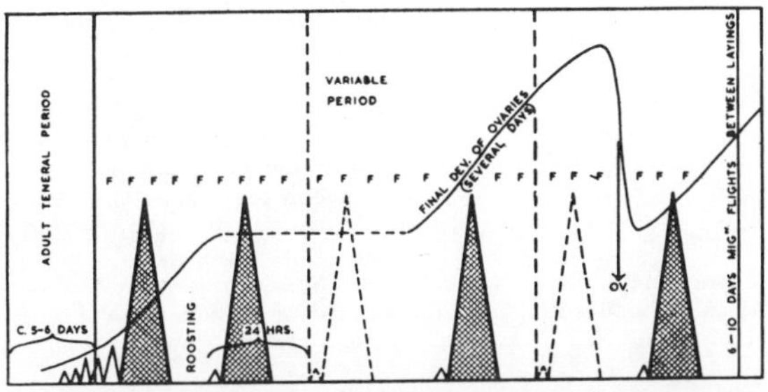


Fig. 4. *Schistoceria gregaria.* Migration flights occur often, during several weeks. Each is preceded by local 'milling and surging' flights (small peaks). *F*, feeding; *ov.*, oviposition. The number of successive ovipositions is variable and migration may occur in between

habitat, to oviposit (Fig. 5). This type of migration resembles ecologically those from breeding sites to hibernation or æstivation sites and the return to oviposition sites next season, as with classical migrants like the Monarch butterfly[33]. Perhaps migrations interrupted by diapause evolved from the non-diapausing type by winter intercepting ovary development.

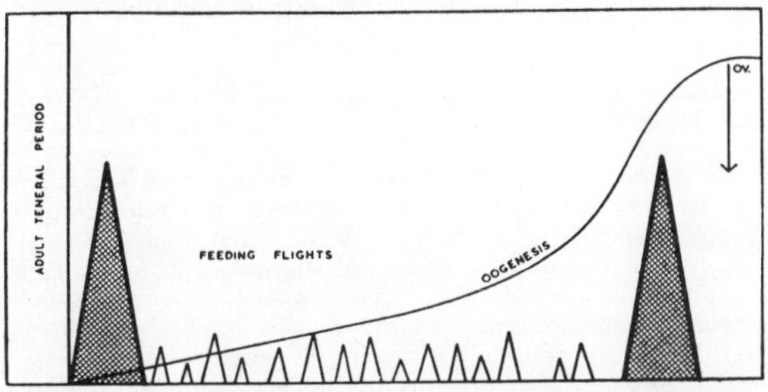


Fig. 5. Chafers, Dragonflies: initial migration from site of imaginal ecdysis (shaded), to a feeding site. Small peaks are flights within feeding site. Remigration to oviposition site (shaded); *ov.*, oviposition

*Migration in hibernating or æstivating insects.* Females of all migrants that hibernate (or æstivate) make the primary migration to hibernation quarters with the ovaries undeveloped, and they hibernate in this state. When diapause ends, the re-migration flight appears to end with oviposition (Fig. 6). Thus, in hibernating and æstivating insects migration is interrupted and temporarily suppressed[6,14,17,22,23].

## Physiological Factors in Migratory Flight

Neurophysiological processes cause the characteristic prolonged and undistracted migratory flight, but ecological factors determine when it occurs.

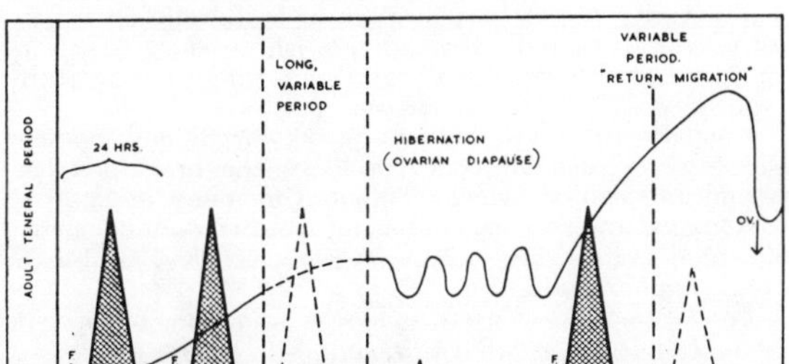


Fig. 6. Migrants with adult diapause: for example, *Danaus plexippus.* Migratory displacement flights shaded: *F*, feeding; *ov.*, oviposition

Because migratory species, or genotypical migrants, migrate with immature ovaries, factors that control ovary development probably control migration; a plausible hypothesis can be made on these lines explaining why some populations migrate at some times and not at others. It is more difficult to explain how migratory and non-migratory individuals differ physiologically and neurophysiologically. However, some of the factors involved are considered in the next section, before the evocation of migratory behaviour in nature is discussed.

*Factors affecting flight efficiency.* Sexually immature migrant females often have more fuel in the fat body, lighter bodies and relatively longer wings (making a lower wing-loading) than non-migratory individuals of the same species. These attributes may increase duration and perhaps efficiency of flight; but they do not cause the characteristic persistence and undistractedness of migratory flight, which is not only a matter of increased efficiency or lack of need to oviposit. The thresholds of response to stimuli that promote take-off and continuous flight appear to be lowered, or alternatively thresholds for settling are raised.

*Interrelation of flight and settling.* The antithesis of migration is a stable and prolonged rest: the physiological reasons for one illuminate those of the other. Flight and settling inevitably alternate in time, but they also have a functional relationship; for flight or any intense muscular activity is a physiological necessity for settling, feeding, mating or oviposition, at least with some, perhaps with all species.

Thus new, unflown alatæ of migratory aphids will not stay on the plant on which they were born and will not settle or feed until they have flown[11,35]. The Scolytid wood-boring beetle *Dendroctonus* behaves similarly[21]. Some mosquitoes[9,36] and *Hypoderma* (Diptera)[37] mate only after a short flight, and oviposition by noctuids[38] and muscids[39] alternates with bouts of flying; oviposition by the wax moth *Achroia* is preceded by intense walking[40].

The duration of settling and flight are related quantitatively. The longer an aphid flies the longer it later remains settled[11]; stability of the settling response in *Automeris* (Saturniidae) (as measured by an accompanying rocking motion) and the waggle dance of honey bees after settling are directly related to the length of the previous flight[41,42]. Incomplete settling induced by a succession of unsuitable hosts is reported to prolong flights of aphids[35].

This functional interplay is thought to be caused by reciprocating reflex systems which alternately release and suppress each other[35], and prolonged function of settling reflexes gradually lowers the flight thresholds so that flight replaces settling and vice versa; this is apparently a matter of reflex relationships, not of fatigue. With an insect in a migratory state flight exceeds settling; thus Lepidoptera feed hurriedly during migration flight[43], and settled immature locusts are said to join in a passing swarm more readily than those about to oviposit[28,29]; Weis-Fogh noticed that migratory locusts near to oviposition were reluctant to fly in the laboratory[44].

Thresholds of response to various general stimuli (such as sounds, movements, touch, wind, light) which initiate take-off and to others such as feed-back from the wind-sensitive organ on the head, needed to maintain continuous flight, appear to be abnormally low in migrants and evidently vary with the sexual state of females. This is perhaps why exceptionally vigorous individual migrants of *Ascia*[2] and *Aleyrodes*[13] produced in very crowded populations continue to fly at night at light-intensities that would normally inhibit flight.

That neural response thresholds are affected by endocrine balance was shown by Haskell and Moorhouse[45]. Locomotor activity in locust hoppers reaches its peak some days after moulting; the subsequent decline as the next moult approaches coincides with increased titre of the moulting hormone, ecdyson, from the prothoracic glands (or with some associated chemical process). That thresholds are so changed is shown by the lack of response by isolated hoppers to stimuli that cause orientated locomotory reactions in hoppers bred in crowds; and because injection of blood from crowded into isolated hoppers lowers their response thresholds so that they respond to stimuli which before were ineffective, and vice versa. Haskell and Moorhouse showed a still more intimate effect. When the metathoracic ganglion of mature adult male *Schistocerca* is bathed with ecdyson or with blood from moulting hoppers electrical activity in the motor neurones supplying the extensor tibialis muscles of the hind legs decreases. Thus the drive of migratory flight may well be caused by a lowering of the neural response thresholds by changes in endocrine balance acting selectively on the particular reflex systems involved in wing-beat, as it does in hind-leg movements of hoppers.

Migratory flight occurs frequently, though not always, immediately after metamorphosis; presumably the ecdyson titre is then decreasing, but it is not minimal and hence other associated factors may be involved, as Haskell and Moorhouse suggest for other reasons. Clearly, the analysis of flight response thresholds in relation to the sexual cycle and the endocrine system controlling it will be of great interest.

*Phototaxis and migratory flight.* Many migrants such as aphids, scolytids and frit flies[46] fly upwards at exodus with a strong, positive phototaxis, apparently an adaptation ensuring effective wind-borne travel. Eventually, however, the sign of phototaxis is reversed or the insects respond to the green-yellow of the earth rather than ultra-violet light of the sky[47]. So long as insects are high in the air this reversal may not end flight, for aphids in such a state will fly to exhaustion when free from visual distractions[47]. Descent on downward currents gives opportunities for settling, and, the insects being correspondingly receptive, their flight ends.

The change of phototaxis and the development of settling responses can be interpreted as a successive induction of reflex systems. However, the endocrine system is also involved for when the corpus allatum and corpus cardiacum are removed from the adult Colorado beetle a negative phototaxis and burrowing (which in nature precedes oviposition) is induced[48]. When prothoracic glands from *Locusta* (the moulting hormone, ecdyson) are implanted in photo-positive larvæ of the Eyed-Hawk-moth these become photonegative; neotinine, the juvenile hormone from the abdomen of *Cecropia*, causes a reversal back to a positive sign[49]. However, in migrants a negative phototaxis often develops long before ovaries are fully mature, when corpus allatum activity is presumably still small, as in scolytids after flight or when many insects hibernate with sexually immature ovaries. Phototactic responses also change in *Trypodendron* (Scolytidae) from positive to negative as air is swallowed during flight and pressure in the proventriculus increases. Artificial release of this air provokes positive phototactic flight once more[21]. Pressure on the abdominal ganglia in *Choristoneura* larvæ has a similar effect[50].

The end of migration often coincides with oviposition, and oviposition responses can be induced by hormones. Neurosecretory cells, or blood from ovipositing females of the heteropteron *Iphita* introduced into non-ovipositing females, induces them to make oviposition responses; blood from fertilized *Bombyx* females induces oviposition in non-fertilized females; the neurosecretory cells in ovipositing phasmids secrete actively[51].

Thus, while the short-term oscillations of flight and settling within the migratory period may depend on successive inductions of reflexes, the balance in favour of one response compared with another appears to be associated with endocrine balance.

*Differences between migrant and non-migrant species.* The internal factors that control migratory activity are partly humoral, partly neurophysiological, partly use of fuel, perhaps partly a lessened wing-load. All flying species possess some of these elements yet not all migrate. Thus the saturniid, *Cerodirphia speciosa*, has such a high threshold for settling that it never completely settles after flight and even slight stimuli cause it to fly again[52]. The insect is extremely restless, but not a migrant. Locomotory activity at the expense of settling goes a stage further without becoming migration when insects like chironomids and *Simulium* swarm for considerable periods over a stationary object.

Migrants in contrast possess a factor leading to displacement, such as a positive phototaxis which leads to wind-borne displacement, or an orientation which leads to a 'straightening out'[70] of flight, and they pass it from generation to generation; they are genotypical migrants. Nevertheless, individuals with the equipment to migrate often fail to do so. The evocation of this latent ability now needs considering.

## Evocation of Migratory Flight in Nature

*Variation in the pre-oviposition period and its effect on migration.* The length of the pre-oviposition period affects the duration of migratory activity, for it can be so brief as to obliterate it or prolonged to increase it. Information on the relation of migration to the speed of ovary development is scanty, but it suggests a quantitative relationship.

Observations with *Ascia monuste* indicate that lengthening the pre-oviposition period prolongs, and shortening it suppresses migratory flight. When the insects emerge too late in the day they are not flight-mature when it is time to migrate next day. The day afterwards, they are sexually mature and cannot migrate[2]. The migration flights of hundreds of miles to overwintering sites by *D. plexippus* is possible only because of prolonged ovarian immaturity, for individuals that mature quickly oviposit, die *en route* and never reach the hibernation sites[33].

Recent work by El Khidir[13] on *Aleyrodes brassicae* in England shows that during summer adults develop their ovaries quickly and fly only a few inches or feet to another and younger leaf before they settle and oviposit. In the autumn, ovaries develop slowly and sexually immature females accumulate as cold inhibits flight. On a warm day, however, they all fly off together and apparently unlike the summer adults fly for considerable distances *en masse* to overwintering sites.

The early generations of *D. plexippus* apparently do not migrate long distances presumably because of a short pre-oviposition period, but later generations do; they therefore resemble whiteflies except for distances travelled.

*Internal and external factors affecting the pre-oviposition period.* The speed of ovary development is controlled by the corpus allatum and the associated endocrines. The activity of the corpus allatum is diminished by environmental stimuli associated with a short day, insufficient or deficient food, crowding or high temperature, especially before metamorphosis.

Decrease in length of day during larval development slows ovary development, prolongs the preoviposition-period and causes ovarian diapause in many insects[53–57].

Removal of the corpus allatum from the adult has the effects of a short day during development, and inhibits or delays egg formation.

Food affects oogenesis, and the selection of specific feeding sites by syrphids, dragonflies, chafers and many others confirms this. Food supply is affected by crowding and the effects of one may be confused with the other, as in the literature on alate production in aphids.

Starvation or a diet of water or sugar solution delays oogenesis in *Oncopeltus* (Heteroptera) and *Calliphora* (Diptera) and removal of the corpus allatum has the same effect[58,59].

A most striking example of food affecting migration through ovary development, however, occurs with the spruce budworm *Choristoneura fumiferana*: first instars in spring feed on old leaves and young staminate flowers of balsam fir and move to new leaves, producing gravid adults too heavy to fly. Large populations, however, destroy new leaves and caterpillars then feed on old ones, so becoming small sexually immature adults, capable of upward phototactic flight. Here crowding appears to act by changing the food[8].

Either crowding or diminished food or both together also affect the bean beetle, *Callosobruchus maculatus* (F.), which has two alate forms. A bean with one larva produces an adult with rapidly ripening ovaries, a large wing-loading, unable to fly. With two larvæ in each bean the adults are sexually immature, have a small wing loading and can fly[60–62].

Crowding alone during larval stages lengthens the pre-oviposition period of some adults (*Locusta migratoriodes* (R. and F.)[63], *Leucania unipunctata* Haworth[64] and *Laphygma exigua*[7], *Pieris brassicae* L.,[65] *A. monuste*) and shortens it in others (*Schistocerca gregaria*[66], *Nomadacris septempunctata* Serville[67], *Plusia gamma*[65] and *Barethra brassica*[68]).

Low temperature during the adult stage slows metabolism and ovary development, but might stop flight. Such a retardation is not necessarily accompanied by a diminished corpus allatum activity which seems to produce migratory flight. But a relatively high temperature in pre-adult development decreases corpus allatum activity and produces an adult of abnormal sexual immaturity[56] A relatively high temperature even after metamorphosis might then permit migration before the ovaries fully mature.

*Displacement during the oviposition cycle.* Though migration is commonly pre-reproductive, some species (aphids, locusts, muscids) make considerable flights between bouts of oviposition. Spruce budworm adults fly on convective storms after first ovipositing[34]. Some chafers make additional migratory flights after ovipositing back to feeding sites where oogenesis is resumed[32]. There seems to be some doubt whether *Aedes taeniorhyncus* resumes migration after ovipositing[9], but other mosquitoes fly scores of miles after the blood-meal that follows post-teneral flight but precedes oogenesis[69].

The volume of the corpus allatum, at least in *Calliphora*, waxes and wanes as eggs develop periodically during adult life; some chafers, locusts and many other insects also take several days to mature successive egg batches; evidently the physiological condition associated with endocrine activity necessary for long migration occurs during these inter-oviposition periods, as in the pre-reproductive stage after imaginal ecdysis. Migration can be interreproductive as Kennedy points out[70].

*Environmental-endocrine system in relation to migration and habitat.* Southwood[71] showed that migratory species live in temporary habitats and migration can be regarded as an adaptation to ensure that such places are relinquished before they disappear or become untenable. But there is no evidence that migration is a reaction to current adversity and when this kind of behaviour has been claimed ('alimental' or 'climatic' migration as when locusts leave a drying habitat)[72,73] the ontogenetic hypothesis developed here seems more plausible. The factors associated with a habitat doomed to decay are the same as those that delay ovary development, namely, shortening day, senescing food, crowding and high temperatures, particularly with the later summer and autumn generations in temperate climates. These then may be supposed to produce an individual with the necessary sexual immaturity to evoke and prolong migratory flight.

In tropical countries day-length is more constant and food may be more ubiquitous; it may then be dryness[7] or heat during larval growth that delays sexual development. Many token stimuli may affect corpus allatum activity.

The operation of an eco-physiological system controlling ovary development and the migration associated with it in genotypical migrants will not be identical for all species, and many adaptations to ensure that migration is controlled ontogenetically might be expected. Sexual immaturity delayed for days or weeks is geared to long distance travel with locusts and the Monarch, when it is disadvantageous for oviposition to overtake migrants before they have reached their destination. With aphids the hosts of which are scattered but ubiquitous, and immense populations are broadcast among them, the migration period can afford to be shorter, and indeed migration occurs while embryos are nearly mature.

*Structural, physiological and behavioural 'polymorphism'* All the alate adults of migratory aphids are obligatory migrants and their non-migratory adults are wingless. It is easy to accept the ability or inability to migrate when it is rigidly associated with such gross structural differences as the presence or absence of wings[77]. But most migrant insects normally produce only winged adults some of which migrate and some do not. It is these that have made migration seem mysterious. Many of these species nevertheless have various degrees of alary polymorphism. Thus *gregaria* locusts have relatively longer wings than *solitaria*; with *Plusia gamma*, *Pieris brassicae*, *Callosobruchus quadrimaculatus*, crowding produces a smaller wing-loading. But although locust *gregaria* migrate in the classical swarms and so migration is connected with crowds and swarm development, *solitaria* also migrate. Indeed migration by many insects is popularly associated with dense populations and although this association may be real, and fits the hypothesis developed here, there is in fact little direct evidence for it[76]. There is an immense literature on structural polymorphism and its association with a short day, crowding and other factors affecting the endocrine system but differences in flight behaviour are rarely considered[75] Migration studies suggest that polymorphism does not stop short at structural differences but extends also to behaviour and that this, like form, is also a matter of ontogeny whether linked with morphological differences or not.

We can therefore expect a gradation from one extreme, of structural polymorphism and obligatory migration as in aphids and termites[77], to the other extreme where little or no visible morphological differences are apparent and the only differences are physiological, neurophysiological and behavioural, nevertheless controlled by the endocrine associated with development and metamorphosis and evoked by environmental factors.

## Migration in Males

Any mention in the literature of the state of the gonads in migrants almost always refers to the ovaries; therefore this article has dealt mostly with females. Females of all species have the same function, to ensure that offspring are deposited in suitable habitats; migration is a way of doing this in special circumstances. But the males are needed only to fertilize the females, and that can be done at the source without migrating, during migration, or at its end. Male migration is therefore much more variable from species to species than female migration.

At one extreme, as with the European pine shoot moth (*Rhyacionia buoliana* Schiff.), females fly readily only after mating and migrate without males accompanying them[78].

In *Aedes taeniorhyncus* mating often occurs at initial exodus and males accompany females probably only for a relatively short distance[9]. Some males accompany females to hibernation sites[14,17,33,43], but remain immature during hibernation[14]; others are mature soon after pupal emergence and migrate while sexually mature, with females, retaining live sperms in the seminal vesicles during hibernation, while the females are wholly immature[17,79]. With locusts, immature males and females migrate and mature together[24]. Thus the relation of endocrines and gonad development to thresholds of flight response does not show the obvious association in males as in females. Males probably have features peculiar to the sex and need separate investigation in each species.

The difference between the sexes, in general migration studies, has not been given the attention it deserves; usually both sexes have been considered as if belonging to the same causative system and this is not generally true.

## Conclusion

(1). Females of most migratory species migrate while the ovaries are immature. The migration from the breeding site usually occurs soon after imaginal ecdysis; re-migrations are also made before ovaries mature, often after imaginal diapause.

(2) This suggests two experimental approaches to migration: (*a*) an examination of the physiological and neurophysiological factors responsible for the migratory drive; (*b*) to investigate the hypothesis that migration is evoked, prolonged or suppressed in genotypical migrants by environmental factors affecting ovary development.

(3) Flight and settling thresholds are functionally independent and vary reciprocally over short periods. But an endocrine mechanism evidently controls the neural thresholds so that flight is emphasized while ovaries are immature and settling is emphasized when ovaries ripen. That increase in the moulting hormone raises the neural response threshold for locomotor activity in locusts hoppers supports this.

(4) The following hypothesis explains why genotypical migrants migrate at some times and places and not at others.

Migratory flight can be prolonged by a lengthened pre-oviposition period and obliterated by a short one. Therefore the factors which prolong sexual immaturity also probably evoke and prolong migration in females. These factors are crowding, too little food or food of the wrong kind, a short day and high temperature acting through the corpus allatum and associated endocrines, especially during pre-adult development. Extended photoperiod, lack of crowding and enough of the right food would have an opposite effect, and tend to shorten or suppress migratory flight.

(5) It is suggested that these factors, singly or together, cause adults in some generations of genotypical migrants to develop into obligatory migrants and to suppress migration in other generations. This could explain seasonal and 'climatic' and 'alimental' migration and why populations usually migrate before a habitat obviously deteriorates. The same eco-physiological system that produces morphometric differences associated with migratory behaviour may be expected also to produce behavioural differences alone.

(6) Males of some species migrate while sexually immature, accompanying immature females all the way; females of other species are fertilized before migration and the males do not migrate. Some males migrate while sexually mature. There is a gradation between these extremes. Migration of females is physiologically similar between species, but that of males is more variable and requires a separate investigation.

I thank Sir Boris Uvarov, Prof. V. B. Wigglesworth, Mr. F. C. Bawden, Dr. P. T. Haskell, Dr. Z. Waloff, Dr. T. R. E. Southwood and Mr. L. R. Taylor for their criticism of the manuscript.

1 Johnson, C. G., *Nature*, **186**, 348 (1960).
2 Nielsen, E. T., *Biol. Medd. Dan. Vid. Selsk.*, **23**, 1 (1961).
3 Tilden, J. W., *J. Res. Lepid.*, **1**, 43 (1962).
4 Williams, C. B., *Insect Migration* (Collins, London, 1958).
5 Roer, H., *Z. angew. Entomol.*, **44**, 272 (1959).
6 Common, I. F. B., *Austral. J. Zool.*, **2**, 223 (1954).
7 Faure, J. C., *Un. S. Afr. Dep. Agr. and For. Sci. Bull.*, 234 (1943).
8 Blais, J. R., *Canad. Entomol.*, **85**, 446 (1953).
9 Provost, M. W., *Mosq. News.*, **17**, 233 (1957).
10 Schneider, F., *Mitt. Schweiz. Entomol. Ges.*, **31**, 1 (1958).
11 Johnson, B., *Anim. Behav.*, **6**, 9 (1958).
12 Lawson, F. R., Chamberlain, J. C., and York, G. T., *U.S. Dept. Agri. Tech. Bull.*, No. 1030 (1951).
13 El Khidir, E., Ph.D. thesis, Univ. London (1963).
14 Feodotov, D. M., *C. R. Acad. Sci., U.R.S.S.*, **42**, 408 (1944).
15 Kelly, E. O. G. L., and Parks, T. H., *U.S. Dept. Agric. Bur. Entomol Bull.*, No. 95, 23 (1911).
16 Hagen, K. S., *Ann. Rev. Entomol.*, **7**, 289 (1962).
17 Hodek, J., and Cerkasov, J., *The Ontogeny of Insects* (Prague, 1959).
18 Henson, W. R., *Ann. Entomol. Soc. Amer.*, **55**, 524 (1962).
19 Atkins, M. S., and Farris, S. H., *Canad. Entomol.*, **94**, 25 (1962).
20 Chapman, J. A., and Kinghorn, J. M., *Canad. Entomol.*, **90**, 362 (1958).
21 Graham, K., *Nature*, **191**, 519 (1961).
22 Tower, W. L., *Carnegie Inst. Publ.*, 48, Washington, D.C. (1906).
23 Lewis, T., *Entomol. Exp. and App.*, **2**, 187 (1959).
24 Rainey, R. C., and Waloff, Z., *Anti-Locust Bull.* (London), No. 9 (1951).
25 Corbet, P. S., *A Biology of Dragonflies* (London, 1962).
26 Imms, A. D., *A General Textbook of Entomology* (Methuen, London, 1960).
27 Waloff, Z. (personal communication, and in the press).
28 Popov, G. B., *Entomol. Mon. Mag.*, **94**, 179 (1958).
29 Popov, G. B., *Trans. Entomol. Soc. London.*, **105**, 65 (1954).
30 Bodenheimer, F. S., *Govt. Iraq: Min. Econ. Bull.*, No. 29 (1944).
31 Schneider, F., *Mitt. Schweiz. Entomol. Ges.*, **25**, 111 (1952).
32 Couturier, A., and Robert, P., *Ann. Epiph.* (6 Année) No. 1, 52 (1955).
33 Urquhart, F. A., *The Monarch Butterfly* (Univ. Toronto Press, 1960).
34 Greenbank, D. O., *Canad. J. Zool.*, **35**, 385 (1957).
35 Kennedy, J. S., *Proc. Tenth Intern. Congr. Entomol., 1956*, **2**, 397 (1958).
36 Nielsen, E. T., and Haeger, J. S., *Miss. Publ. Entomol. Soc. Amer.*, **1**, 71 (1960).
37 Weintraub, J., *Canad. Entomol.*, **93**, 149 (1961).
38 Makings, P., Thesis for Dip. Imp. Coll. Lond. (1956).
39 Macleod, J. (personal communication: and in the press).
40 Makings, P., *Proc. Roy. Entomol. Soc. Lond.*, A, **33**, 136 (1958).
41 Bastock, M., and Blest, A. D., *Behaviour*, **12**, 243 (1958).
42 Blest, A. D., *Behaviour*, **16**, 188 (1960).
43 Williams, C. B., *The Migration of Butterflies* (Oliver and Boyd, London, 1930).
44 Weis-Fogh, T., *Phil. Trans. Roy. Soc.*, B, **237**, 1 (1952).
45 Haskell, P. T., and Moorhouse, J. E., *Nature*, **197**, 56 (1963).
46 Johnson, C. G., Taylor, L. R., and Southwood, T. R. E., *J. Anim. Ecol.*, **31**, 373 (1962).
47 Kennedy, J. S., and Booth, C. O., *Discovery*, 311 (August 1956).
48 De Wilde, J., in *The Ontogeny of Insects* (Prague, 1959).
49 Beetsma, J., de Ruiter, L., and de Wilde, J., *J. Insect Physiol.*, **8**, 251 (1962).
50 Wellington, W. G., *Canad. Entomol.*, **80**, 56 (1948).
51 Nayar, K. K., *Proc. Ind. Acad. Sci.*, B, **47**, 233 (1958).
52 Blest, A. D., *Zoologica*, **45**, 81 (1960).
53 Hodek, J., *Acta Soc. Entomol. Czechosl.*, **57**, 1 (1960).
54 De Wilde, J., *Ann. App. Biol.*, **50**, 606 (1962).
55 Bonnemaison, L., and Missonier, J., *Ann. Inst. Nat. Rech. agron.*, Ser. C: *Ann. Épiphyt.*, **6**, 457 (1955).
56 Wigglesworth, V. B., *The Physiology of Insect Metamorphosis* (Cambridge, 1954).
57 Lees, A. D., *The Physiology of Diapause on Arthropods* (Cambridge, 1955).
58 Strangeways-Dixon, T., in *The Ontogeny of Insects* (Prague, 1959).
59 Johansson, A. S., in *The Ontogeny of Insects* (Prague, 1959).
60 Caswell. G. H., *Rev. App. Entomol.*, A, **46**, 212 (1956).
61 Utida, S., *Rev. App. Entomol.*, A, **47**, 389 (1959 (1954).
62 Utida, S., *Contrib. Entomol. Lab. Kyoto Univ.*, No. 279, 93 (1956).
63 Norris, M. J., *Anti-Locust Bull.* (London), No. 6 (1950).
64 Iwao, S., *Contrib. Entomol. Lab. Kyoto Univ.*, No. 277, 60 (1956).
65 Zaher, M. A., and Long, D. B., *Proc. Roy. Entomol. Soc. London.*, A, **134**, 7 (1959).
66 Norris, M. J., *Anti-Locust Bull.* (London), No. 13 (1954).
67 Norris, M. J., *Anti-Locust Bull.* (London), No. 36 (1959).
68 Kirata, J., *Contrib. Entomol. Lab. Kyoto Univ.*, No. 278, 79 (1956).
69 Garrett-Jones, C., *WHO/MAL*/298 (1961).
70 Kennedy, J. S., *Nature*, **189**, 785 (1961).
71 Southwood, T. R. E., *Biol. Revs.*, **37**, 171 (1962).
72 Kennedy, J. S., *Biol. Revs.*, **31**, 349 (1956).
73 Volkonsky, M., *Arch. Inst. Pasteur. d'Alg.*, **20**, 236 (1942).
74 Southwood, T. R. E., *Proc. Roy. Entomol. Soc. Lond.*, A, **36**, 63 (1961).
75 Muller, H. J., *Z. angew. Ent.*, **47**, 7 (1960).
76 Key, K. H. L., *Quart. Rev. Biol.*, **25**, 363 (1950).
77 Lees, A. D., in *Insect Polymorphism*, Roy. Entomol. Soc. Lond., Symposium No. 1 (1961).
78 Green, G. W., and Ponting, P. J., *Canad. Entomol.*, **94**, 299 (1962).
79 Downes, J. A., *Trans. Roy. Entomol. Soc. Lond.*, **92**, 101 (1942).

# C. Physical Factors

## THE EFFECT OF TEMPERATURE AND RELATIVE HUMIDITY ON THE FLIGHT PERFORMANCE OF FEMALE *AEDES AEGYPTI*

WAYNE A. ROWLEY and CHARLES L. GRAHAM

Department of Zoology and Entomology, U.S. Army Biological Laboratories, Fort Detrick, Maryland

(*Received* 6 *April* 1968)

**Abstract**—The influence of temperature and r.h. on the flight performance of tethered virgin female *Aedes aegypti* was investigated. Mosquitoes of similar age were flown to exhaustion on flight mills at various temperatures and relative humidities. Parameters measured were distance flown, duration of flight, speed of flight, initial weight of mosquitoes, and live weight lost during exhaustive flight.

The temperature range at which sustained tethered flight occurred was 15 to 32°C. Temperature extremes at which flight was possible were 10 and 35°C; however, performance was minimal in both duration and distance at these temperatures. The optimal temperature for flight was 21°C. Mosquitoes flew farther at 15°C than at the commonly reported optimal temperature of 27°C. In general, flight performance was greater below 27°C.

Relative humidity within 30 to 90 per cent had no demonstrable influence on flight performance at any temperature studied except at 32°C, where 30 per cent had a marked limiting effect on mosquito flight.

### INTRODUCTION

Optimal temperature and humidity levels for adult mosquito longevity and biting activity have been well established, but no data are available on optimal conditions for flight activity. It is generally accepted that temperature and r.h. profoundly influence the behaviour and activity of adult mosquitoes, but authors disagree considerably as to how these environmental factors affect mosquito activity. The exact effects of either or both factors on mosquito flight can only be generally estimated because it is difficult, if not impossible, to separate the effects of one from the other when dealing with any biological entity. Christophers (1960) reports temperature effects as being 'rather complex' with regard to mosquitoes, affecting feeding (biting) primarily through a general influence upon the activity of the organism; he concludes that mosquitoes will attempt to bite at any temperature at which they are physically capable of doing so.

Most available literature concerning the effects of temperature and humidity on *Aedes aegypti* (L.) deals with the biting activity of the species at various temperatures. Connor (1924) found that *A. aegypti* was most active at 28°C, a temperature considered to be optimal for the species. It is commonly assumed that this species

---

Reproduced with permission from the *Journal of Insect Physiology*, 14:1251–1257, 1968. Published by Pergamon Press, London, England.

has an optimal temperature range between 27 and 32°C, and vital activities are retarded between 17 and 25°C.

CHRISTOPHERS (1960) states that, for routine testing, 70 to 90 per cent r.h. at 25°C will ensure maximal and standard biting results. However, he also indicates that an extremely high humidity has an inhibitory effect on feeding, especially at high temperatures. LUMSDEN (1947) reports that r.h. is of negligible importance with regard to biting by mosquitoes.

The present study was designed to determine the optimal temperature and humidity requirements for tethered flight activity in virgin female *A. aegypti.* A secondary objective was to investigate the range of both temperature and r.h. at which sustained flight was possible.

## MATERIALS AND METHODS

Female *A. aegypti* (Rockefeller strain) were used. Larval rearing, handling, and determination of pupal and adult age were identical to that reported by ROWLEY and GRAHAM (1968). Mosquitoes used in flight studies were held as adults in 1-pint ice-cream cartons at 27°C and 80 per cent r.h. and provided a 0·3 M sucrose solution in saturated cotton pads. These conditions had previously been demonstrated to be optimal for adult longevity (Rowley, unpublished data). Four to twelve mosquitoes were tested each day for 3 consecutive days at a prescribed set of conditions. Mosquitoes were removed from a carton and flown to exhaustion on flight mills at 10, 15, 18, 21, 27, 32, and 35°C. The r.h. at each temperature was regulated at 30, 50, or 90 per cent. Limitations in the regulation of our controlled environmental laboratory restricted flights at 30 per cent r.h. to 21, 27, and 32°C. Humidity could not be adequately regulated for flight studies at lower levels. The chamber was allowed ample time to reach a state of equilibrium at a designated set of conditions before flight tests were undertaken.

The effect of age on the flight ability of *A. aegypti* is striking. Flight performance is maximal in the first 2 weeks of adult life, but no differences in flight ability are detectable among mosquitoes that are within 1 week of age. Therefore, these tests were conducted using mosquitoes ±1 week of age. This afforded a comparison of the flight performances of similarly aged individuals under different environmental conditions. A detailed account of the effect age has on the flight ability of *A. aegypti* is presented in an earlier paper (ROWLEY and GRAHAM, 1968).

Constant surveillance of temperature and humidity was maintained during all flights. Humidity recorders* were employed to monitor r.h. variation throughout the test chamber and in each flight mill canopy. Once established, temperature remained at ±1·5°C and r.h. fluctuations were less than ±5 per cent, except at the 30 per cent level where ±5 to 10 per cent was maintained. Details of the flight mill system and techniques for flying mosquitoes were presented in a previous paper (ROWLEY *et al.*, 1968).

* Speedomax G. Recorder, Leeds and Northrup Co., Philadelphia, Pennsylvania, U.S.A.

## RESULTS

The mean preflight (live) weights of mosquitoes flown to exhaustion at the different temperatures and humidities are presented in Table 1. The initial live weight of mosquitoes had no observable effect on tethered flight ability, which was surprising since larger individuals might be expected to perform better. Weight loss during exhaustive flight was dependent on the length and duration of flights and was apparently independent of the environmental conditions (Table 2). Weight loss during flight was strikingly similar at each temperature tested regardless of the r.h. Above 18°C, weight loss increased slightly at 50 per cent r.h. even though the distance flown was slightly less than 90 per cent.

TABLE 1—THE INITIAL (LIVE) WEIGHT OF FEMALE *A. aegypti* MOSQUITOES FLOWN TO EXHAUSTION AT VARIOUS TEMPERATURES AT 50 AND 90 per cent r.h.*

| Temperature (°C) | 90% r.h. (mg ± S.E.) | 50% r.h. (mg ± S.E.) |
|---|---|---|
| 35 | 3·493 ± 0·125 | 3·306 ± 0·117 |
| 32 | 3·829 ± 0·119 | 3·591 ± 0·140 |
| 27 | 3·039 ± 0·103 | 3·886 ± 0·121 |
| 21 | 3·706 ± 0·099 | 3·637 ± 0·091 |
| 18 | 2·950 ± 0·068 | 3·302 ± 0·065 |
| 15 | 3·307 ± 0·166 | 3·726 ± 0·133 |
| 10 | 3·473 ± 0·100 | 3·679 ± 0·086 |

* $n = 12$.

TABLE 2—AMOUNT OF LIVE WEIGHT LOST IN EXHAUSTIVE FLIGHTS AT VARIOUS TEMPERATURES AT 50 AND 90 per cent r.h.*

| Temperature (°C) | 90% r.h. (mg ± S.E.) | 50% r.h. (mg ± S.E.) |
|---|---|---|
| 35 | 0·090 ± 0·011 | 0·181 ± 0·036 |
| 32 | 0·178 ± 0·042 | 0·225 ± 0·033 |
| 27 | 0·666 ± 0·073 | 0·724 ± 0·059 |
| 21 | 1·016 ± 0·076 | 1·116 ± 0·066 |
| 18 | 0·642 ± 0·054 | 0·651 ± 0·055 |
| 15 | 0·461 ± 0·070 | 0·424 ± 0·058 |
| 10 | 0·018 ± 0·005 | 0·011 ± 0·002 |

* $n = 12$.

A histographic representation of the flight performance (distance flown) by female *A. aegypti* at each temperature and the various relative humidities is presented in Fig. 1. Temperatures of 10 and 35°C clearly represent the extreme

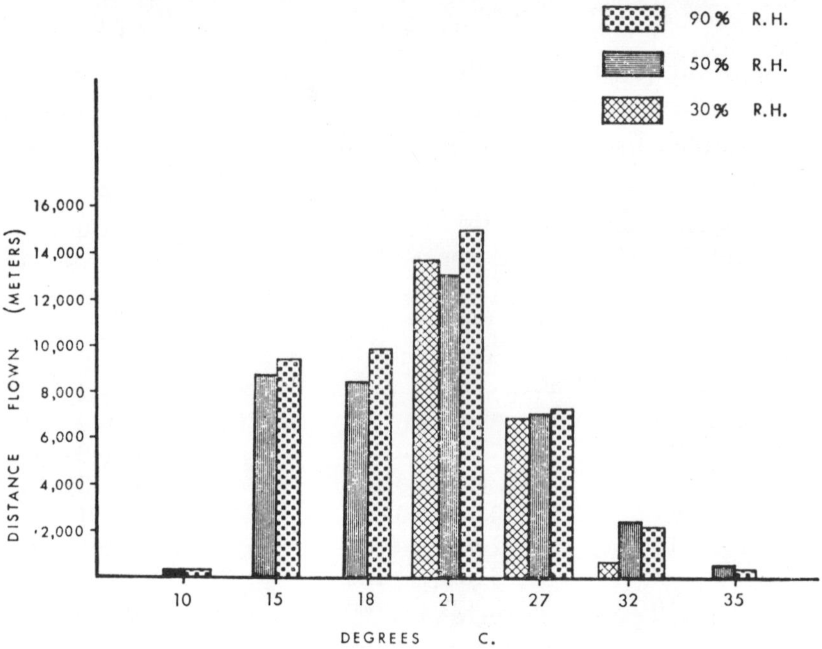


FIG. 1. Average distance flown by female *A. aegypti* at different temperatures and relative humidities.

limits at which tethered flight is possible in this species. The optimal temperature for flight was 21°C. Flights at 32 and 35°C were shorter than expected and indicated either a loss of flight ability or a 'reluctance' to fly at these temperatures. The performance of mosquitoes at 21 and 27°C was similar at all three humidity levels. At 32°C, a reduction in r.h. from 50 to 30 per cent resulted in a marked decrease in flight performance, thus indicating a definite negative response to low humidity at this and, probably, higher temperatures. However, this was the only case studied in which humidity had an obvious influence on flight.

Flight duration, like the other variables, was independent of the humidities tested. Maximal duration of flights at all three humidity levels occurred at 21°C. At 21°C and 30 per cent r.h., the average duration of flight was 494 min, which was 22 min longer than the mean duration of flights at the same temperature at 50 per cent r.h. and 60 min shorter than the mean duration of flights at 21 and 90 per cent r.h. At most temperatures, the duration of flight was slightly greater at the higher humidities, but in no case was the difference sufficient to be of statistical significance. Table 3 shows the mean duration or time of flight for the different temperatures at 50 and 90 per cent r.h.

Table 4 represents the average speed of flight (m/min) of mosquitoes flown at 50 and 90 per cent r.h. and various temperatures. Flight speed was similar regardless of the temperature from 15 to 32°C at 90 per cent r.h. However, at 50 per cent r.h. flight speed increased gradually as the temperature was raised to 32°C. At 32°C, the average flight speed was substantially higher at 50 per cent (34·1 m/min) than at 90 per cent (28·0 m/min). Speed decreased sharply when the temperature was elevated from 32 to 35°C. Flight speed was slightly higher at 35 than at 10°C regardless of r.h.

TABLE 3—MEAN DURATION OF FLIGHTS FLOWN BY FEMALE *A. aegypti* MOSQUITOES AT 50 AND 90 per cent r.h. AND AT DIFFERENT TEMPERATURES*

| Temperature (°C) | Duration of flight (min) | |
|---|---|---|
| | 90% r.h. ± S.E. | 50% r.h. ± S.E. |
| 35 | 17·6 ± 1·6 | 29·9 ± 4·2 |
| 32 | 80·8 ± 16·4 | 70·2 ± 8·8 |
| 28 | 258·3 ± 24·9 | 228·1 ± 18·9 |
| 21 | 554·2 ± 58·1 | 472·0 ± 30·3 |
| 18 | 361·9 ± 32·6 | 322·7 ± 32·1 |
| 15 | 414·1 ± 38·6 | 376·3 ± 47·9 |
| 10 | 30·0 ± 2·2 | 18·1 ± 1·9 |

* $n = 12$.

TABLE 4—MEAN FLIGHT SPEED OF FEMALE *A. aegypti* FLOWN AT 50 AND 90 per cent r.h. AND DIFFERENT TEMPERATURES*

| Temperature (°C) | Speed of flight (m/min ± S.E.) | |
|---|---|---|
| | 90% r.h. | 50% r.h. |
| 35 | 18·6 ± 1·11 | 18·2 ± 1·98 |
| 32 | 28·0 ± 1·61 | 34·1 ± 2·32 |
| 27 | 28·4 ± 0·80 | 31·1 ± 1·74 |
| 21 | 26·8 ± 1·04 | 28·1 ± 1·19 |
| 18 | 27·4 ± 0·86 | 26·2 ± 1·17 |
| 15 | 23·7 ± 0·85 | 23·5 ± 0·83 |
| 10 | 13·1 ± 0·53 | 8·9 ± 0·96 |

* $n = 12$.

## DISCUSSION

The optimal conditions for all vital activities of mosquitoes have been assumed by many workers to be approximately 27°C and 80 per cent r.h. Our studies indicate that such is not necessarily the case, at least if the tethered flight of adult female *A. aegypti* can be accepted as a vital activity. Apparently, flight is more readily undertaken and sustained at temperatures below the optimal level than at temperatures above it.

Published data (NEVILL and WEIS-FOGH, 1963; YURKIEWICZ and SMYTH, 1966) indicate that the temperature of the internal thorax (flight muscle) of flying insects increases from a few to several degrees above ambient. If this is true for mosquitoes, flight at lower temperatures would be enhanced, but at the higher levels it would probably be impeded. The unexpectedly strong flight abilities of this insect at 21°C and even at 18 and 15°C could possibly be the result of this phenomenon.

Relative humidity, within the range of 30 to 90 per cent had very little demonstrable effect on the flight activity of virgin female *A. aegypti,* except for a depression of flight ability at 32°C and 30 per cent r.h. If temperature conditions were suitable, mosquitoes flew well at all humidities, and performance at any given temperature was independent of the r.h. These results are in agreement with the findings of THOMSON (1938), who found that changes in humidity of as much as 40 per cent within the range of 30 to 85 per cent have little or no effect on the behaviour of the mosquito *Culex fatigans.*

YURKIEWICZ and SMYTH (1966) found that the wing-beat frequency of the blowfly *Phaenicia sericata* increased with increases in temperature between 15 and 30°C. This is somewhat in contrast to the tethered flight performance of our mosquitoes in that flight performance at the lower temperatures exceeded that at 27 and 32°C; however, wing-beat frequency was not measured in our studies. Below 21°C, mosquito flight performance was slightly better at 90 per cent r.h. than at 50 per cent r.h. However, at 27°C and above, flight performance was slightly higher at the lower humidities. The mean distances flown by 36 mosquitoes at 21°C were 14,419, 13,244, and 13,517 m at 90, 50, and 30 per cent r.h. respectively. This clearly indicates that at this temperature, flight is totally independent of r.h., at least, between 30 and 90 per cent.

Statistical analysis of data, using a two-error split plot with a completely randomized design for performance at the different temperatures and at 50 and 90 per cent r.h. indicated an absence of humidity effect. However, there was a strong indication of interaction between the two variables.

It was not possible to evaluate the effects of r.h. below 30 nor above 90 per cent. It seems reasonable to expect more pronounced effects of either or both conditions near the limits of tolerance. At the same time, however, extreme conditions are rarely, if ever, imposed in the laboratory and are of limited significance in nature.

PLATT *et al.* (1957) found relative humidities of 70 to 80 per cent to be optimal for *Anopheles quadrimaculatus*, but sharp avoidance reactions were noted at both low and high humidities. This could well be the case with *A. aegypti* flight ability; however, there is no apparent optimal humidity level.

Temperature, unlike humidity, appears to markedly influence flight ability in these mosquitoes. The wide temperature range from 15 to 32°C at which sustained tethered flight is possible validates this conclusion. In view of the results obtained, it appears that females of this species might be much better fliers in the cooler hours of the day, such as early morning or evening. This would be especially important in areas where extremely high day-time temperatures prevail.

## REFERENCES

CHRISTOPHERS R. S. (1960) Aedes aegypti (*L.*) *The Yellow Fever Mosquito.* Cambridge University Press, Cambridge.

CONNOR M. E. (1924) Suggestions for developing a campaign to control yellow fever. *Am. J. trop. Med.* **4**, 277–307.

LUMSDEN W. H. R. (1947) Observations on the effect of microclimate on the biting of *Aedes aegypti* (L.) (Diptera: Culicidae). *J. exp. Biol.* **24**, 361–373.

NEVILLE A. C. and WEIS-FOGH T. (1963) The effect of temperature on locust flight muscle. *J. exp. Biol.* **40**, 111–121.

PLATT R. B., COLLINS C. L., and WITHERSPOON J. P. (1957) Reactions of *Anopheles quadrimaculatus* Say to moisture, temperature and light. *Ecol. Monogr.* **27**, 303–324.

ROWLEY W. A. and GRAHAM C. L. (1968) The effect of age on the flight performance of female *Aedes aegypti* mosquitoes. *J. Insect Physiol.* **14**, 719–728.

ROWLEY W. A., GRAHAM C. L., and WILLIAMS R. E. (1968) A flight mill system for the laboratory study of mosquito flight. *Ann. ent. Soc. Am.* In press.

THOMSON R. C. M. (1938) The reactions of mosquitoes to temperature and humidity. *Bull. ent. Res.* **29**, 125–140.

YURKIEWICZ W. J. and SMYTH JR. T. (1966) Effect of temperature on flight speed of the sheep blowfly. *J. Insect Physiol.* **12**, 189–194.

# D. Host-Plant Relationships

## Mechanisms of Host Plant Selection

By J. S. Kennedy

*A.R.C. Unit of Insect Physiology, Entomological Field Station, 34A Storey's Way, Cambridge*

The recent revival of integrated control, Californian by inspiration, at first embraced chemical and entomological ('biological') control only (Smith, 1959). Extending the idea to include cultural, plant-breeding and other methods was an afterthought (van den Bosch & Stern, 1962; Rabb, 1962) and in inspiration more European (Franz, 1961). Van den Bosch & Stern (1962), for instance, spoke of integrated control as aimed at *not disturbing* the ecosystem. This seems to me an arbitrary limitation of the strategy of integrated control (Kennedy, 1953) and hardly compatible with the generally accepted thesis that integrated control is applied ecology. But, of course, there is a practical limitation of this kind at present, because we do not know enough. Rather than try to manipulate the delicate agricultural ecosystem, many people think we must always try and hit each pest directly with some Magic Bullet shot in from outside; an insecticide or pheromone, a parasite or predator, or a lot of sterilized males—leaving the rest well alone. And in the present state of our knowledge we cannot deny this, in most cases.

Nor, however, can anyone justly assert it, especially in deciding research policy. We are simply too ignorant to say much either way, in most cases. We lack the essential information on how the ecosystem itself is integrated. In particular, we are ignorant of how the insects and the plants are integrated, by the processes of host finding and host selection. Lipke & Fraenkel (1956) once described the problem of host preference in phytophagous insects as 'the very heart of agricultural entomology', and so it is. They went on to deplore the small attention it had received compared with other parts of the subject and it cannot be said that this situation has greatly changed in the ten years since they wrote.

A symptom of backwardness in any subject of great practical importance like this one, is a rash of resounding principles and grand generalizations: short-cuts to wisdom. Beck (1965) is quite scathing about these in his recent review of plant resistance to insects. 'The confusion of concepts that currently exists', he says, 'can, in part, be traced to premature and irresponsible generalisations.' It is true that early generalizations must be half-baked and can mislead. Nevertheless it is, after all, only by framing and testing generalizations that we progress. Let us then look at some of them, as a quick way of surveying the whole problem.

One hoary old principle in this subject says that 'the healthy plant is an immune plant'. Another is Hopkins's host selection principle which says that larval food conditioning determines adult oviposition preference. Neither of these will stand as a generalization, for both have been disproved in some cases, and confirmed in some others. The only safe conclusion for the present is that each case requires separate investigation. Five more principles may be listed for fuller consideration:

Reproduced with permission from the *Annals of Applied Biology,* 56:317–322, 1965. Published by the Cambridge University Press for the Association of Applied Biologists.

(1) G. Fraenkel's theory of host selection exclusively by means of 'odd' substances in plants.

(2) Our dual discrimination theory of selection by responses to nutrient substances or to tokens of them, as well as to odd substances.

(3) R. H. Painter's two principles of plant resistance to insects, non-preference and antibiosis.

(4) The traditional generalization that insects find their host plants by 'chemotropism', associated particularly with the name of N. E. McIndoo.

(5) The modern, revised one, that finding is by chance and selection by staying or leaving, as suggested by A. J. Thorsteinson.

The first two of these generalizations concern the classes of chemical compound to which insects respond in selecting their host plants, for it is generally agreed that chemical stimuli are pre-eminent here, although other senses are certainly important. At the Amsterdam Entomological Congress in 1951, Dethier (1953) and Fraenkel (1953) gave a new impetus to the idea pioneered by the Amsterdam botanist Verschaeffelt, that the reason why insects are confined to certain kinds of host plant is that they respond selectively to the secondary, 'odd' substances—essential oils, alkaloids, glucosides, etc.—of no nutritional importance, which different plants contain. In the following years Fraenkel continued to elaborate and press this proposition, and summed it up thus (Fraenkel, 1959): 'The food specificity of insects is based solely on the presence or absence of these odd compounds in plants, which serve as repellents to insects in general and as attractants to those few which feed on each plant species.'

If this were true it would certainly simplify our theory—although not necessarily our practical control problems (Kennedy, 1958). When Fraenkel put forward his proposition there already existed considerable evidence that the so-called 'botanical instinct' of insects really does depend on positive or negative olfactory and gustatory responses to plant-specific 'odd' substances. More and more evidence has since been forthcoming, notably from Fraenkel himself, his students and ex-students, and from Japanese workers. Verschaeffelt's original demonstration of the importance of mustard oil glucosides in the attachment of cabbage butterflies to Cruciferae, has been extended to at least three more Crucifer insects, a moth (Thorsteinson, 1953; Nayar & Thorsteinson, 1963), a weevil (Sugiyama & Matsumoto, 1959) and anaphid (Wensler, 1962). Many other examples could be quoted (see Beck, 1965), including substances which deter feeding by all but a few insects, such as the glycoalkaloids of Solanaceae.

Nevertheless, there is more to host selection than this. First, deterrent plant substances are not essential to inhibit feeding by phytophagous insects. When the maxillary palps are removed, some insects will readily eat not only plants that they ordinarily refuse, but also neutral materials like filter paper which they also ordinarily refuse (Torii & Morii, 1948; Dethier, 1953; Chin, 1950 in de Wilde, 1958; Waldbauer & Fraenkel, 1961; Ito, Horie & Fraenkel, 1959). The maxillary palps, then, seem to provide an inhibitory input, either spontaneously or perhaps in response to mechanical stimuli received when active palpation movements are started, and this inhibitory input normally prevents biting until it is disinhibited by positive stimuli from the proper host plant (Waldbauer, 1963). In that case, when there is, in addition, negative stimulation from deterrent substances, it may act via other receptors upon some other response in the chain, preceding or following biting. This problem well illustrates the need to analyse host selection into its several components, with due regard for inhibitory as well as excitatory effects.

Secondly, even intact, oligophagous insects will sometimes accept and grow quite well on diets without any secondary plant substances (*Acyrthosiphon pisum:* Auclair & Cartier, 1963; *Pieris brassicae:* W. A. L. David & B. O. C. Gardiner, personal communication). And thirdly, in agricultural entomology, we are not concerned solely with the 'botanical instinct', in other words with selection between species or higher plant taxa. We are particularly concerned with selection between varieties within one species, and between plants differing only phenotypically under different cultural treatments, in vigour, growth rate, development pattern, water status, etc. I do not think there is any definite evidence that odd substances determine preferences here.

Unfortunately, the odd-substance theory, valid as it clearly is up to a point, was stated so comprehensively by Fraenkel as to suggest that all plant chemical stimuli governing food preferences came from such substances. He pointed to the tastelessness for us of common nutrients such as proteins and fats, and claimed that leaf-feeding insects could develop equally well on any leaves provided only that they ate enough of them. This last idea has been 'laid to rest', as Beck (1965) put it, by Waldbauer (1962, 1964) using maxillectomized larvae of the tobacco hornworm *Protoparce*, which will eat some non-host leaves well enough, but still do not develop well on them.

More important is the ample evidence now available that phytophagous insects do, in fact, respond positively to moderate concentrations of various nutrient substances found in plants—to sugars above all, but also to amino acids and amides, phospholipids, sterols and even some vitamins (see Beck, 1965). Sugar often acts as a synergist or co-factor with other substances, e.g. with amino acids for grasshoppers (Thorsteinson, 1960) and for aphids (Mittler & Dadd, 1964), mulberry leaf fractions for silkworms (Nayar & Fraenkel, 1962), sinigrin for *Plutella* (Thorsteinson, 1960; Nayar & Thorsteinson, 1963)—substances which have less or no stimulating effect without sucrose. In the maize plant, varying concentrations of sucrose counteract in varying degree the deterrent effect of the 'Resistance factor A' isolated and identified by Beck (1965) and his colleagues.

To the extent that nutrients are often feeding stimuli, our dual discrimination theory (Kennedy, 1958) has been partially confirmed. But again, I do not think we have as yet any direct evidence that variation in the supply of stimulating nutrients does actually govern preferences between crop plants. There is now a great deal of circumstantial evidence which seems to point that way for aphids, from work in several countries (e.g. Auclair, Maltais & Cartier, 1957; Müller, 1961; Tambs-Lyche & Kennedy, 1958; van Emden, 1965). But some direct evidence one way or the other is an outstanding, urgent requirement.

The last three generalizations in my list concern the actual mechanisms and manoeuvres by which host selection is effected. Looking at these events the other way round, from the point of view of plant resistance to insects, Painter (1958) has separated the mechanisms into two classes: active *Non-preference*, and *Antibiosis*, in which the insect stays on the plant but does not thrive there. This division has obviously been a convenience for plant breeders but it refers strictly to resultant effects and not mechanisms. If the two categories are taken as mechanisms, then the exercise of distinguishing them can easily become a substitute for research. For it is now commonplace that host selection is a *catenary* process: a chain made up of different responses, usually to different stimuli—chemical, visual, tactile, etc.—each of which is received as a result of the insect making the previous response in the chain. The literature contains many examples (see Beck, 1965; Wigglesworth, 1953; Carthy, 1958, 1965; Yamamoto & Fraenkel, 1960; Hamamura, Hayashiya, Naito, Matsuura & Nishida, 1962; de Wilde, 1958; Thorsteinson, 1960); together with oft-revised schemes designed to generalize the sequences. These schemes, however, tend to concentrate on classifying the stimulating plant substances, under operational names like *attractant, repellent, arrestant, biting factor* and so on, including etymological chimaeras like *phagostimulant* and even semantic inversions like *rejectant*.

But the real weakness of such schemes is not verbal. It is their preoccupation with identifying stimuli, as if the response chain itself were self-evident. And they imply that each response has one self-sufficient kind of stimulus, as if each kind of stimulus acted on its own. This is to lose sight of the rich variety of interactions that are known. One kind of stimulus may induce a response to quite another kind of stimulus, being received simultaneously, as when a flower scent in the air makes a flying butterfly go and land on a coloured paper (Tinbergen, 1951). Again, one response not only brings in the external stimulus for the next following response in the chain, but may also prepare the insect for it by lowering the response threshold (Wigglesworth, 1953; Kennedy & Booth, 1963).

While the final links in such reaction chains—biting and feeding, or oviposition, once contact is made with the plant—are comparatively easy to see and to analyse, the earlier links, by which the plant is found and contacted, are not so. They constitute, I believe, the most backward part of the whole subject, for that reason; and yet the need to understand them increases with every step toward control methods more subtle than straight killing. The subject of host finding is hardly mentioned in our latest textbooks of agricultural entomology (Edwards & Heath, 1964; Jones & Jones, 1964); and it has always been beset by a confusion of ideas which is still not entirely cleared up in spite of the considerable clarification effected by Dethier, Browne & Smith (1960).

The confusion is between two kinds of host-finding movements: (i) movements that are truly directed, oriented toward the plant (*taxes*), and (ii) movements that are not so directed at all but, nevertheless, result in the insect finding itself at the host, rather than at non-hosts or elsewhere, either because stimuli from the host arrest the insect's random movements (*orthokinesis*), or, alternatively, because they make it change direction at random more often or less often (*klinokinesis*) and in one of those ways increase its chances of encountering the host.

Host finding by one or other type of mechanism or by both, directed and undirected, has been thought of as if it were always directed, using a variety of terms such as attraction, orientation and tropism which all bear that directive connotation. This is an old and extraordinarily deep-rooted idea. It can be found, for example, in every edition of Hubert Martin's standard work from 1928 to 1964, in Dethier's 1947 book, and echoed again this year in Beck's (1965) otherwise excellent review where *arrestants* are classed paradoxically as stimuli for *orientation*.

One may ask, is not this just a quibble? Does it matter in practice how the insect finds the plant, so long as it gets there? I think it does matter, and this was the main point of Dethier, Browne & Smith's well-known 1960 article. Suppose we wish to prevent the spread of viruses in potato fields. If the vector aphids were attracted to the plants from a distance by some characteristic odour, then any treatment that masked or suppressed this odour would reduce virus spread. In practice such a treatment would not work because host finding is in this case not by olfactory attraction, chemical stimuli from the host plant acting only as arrestants after landing (Kennedy, Booth & Kershaw, 1959), so that viruses would go on being spread. On the other hand, over the last inches aphids are directed, visually, to land on plants—and can be kept off visually, too—so that the chemokinesis may still not be the important part of the process from a control point of view.

The lumping of all host-finding mechanisms under the heading of attraction or orientation stems, probably, from the kind of laboratory apparatus commonly used to investigate olfactory responses. The classic instrument is of course the Y-tube olfactometer of McIndoo and many others. Any results obtained with this do indeed demonstrate directed responses (*tropotactic* or *klinotactic*) and probably only such. But, unfortunately, these responses depend on the extremely steep local odour gradient that the instrument provides and positive results do not justify any conclusions about longer-range attraction as envisaged in the field.

The second popular instrument is some kind of arena 'choice chamber' as used, for instance, in the 'screen tests' of Dethier (1947) and others, where leaves are placed under a perforated floor just out of reach of the insects, whose movements and distribution are then recorded. Arenas with two wide air streams, one scented and the other not, have also been used.

Positive results obtained with these devices may come from *un*directed, as well as from directed, reactions. The two cannot be distinguished by distribution counts; only detailed recording of activity and tracks will distinguish them. Thus Dethier's (1947) records from Monarch caterpillars in still-air screen tests show both increased random turning while over a host leaf (*klinokinesis*) and, on reaching its border, directed turning back over it after head-waving (*klinotaxis*). Only the undirected (*klinokinetic*) component could contribute to host finding from more than millimetres away. And the main effect of both types of response is not to attract but to hold the insects at the host, once they have arrived.

Laboratory experiments that have given good evidence of a capacity for directed movement toward a source of plant odour from some way off, are those in which the odour is borne on an air current and does not itself orient the insects, but rather stimulates them to respond to the air current by orienting into it and making headway against it and hence toward the odour source, as in Flügge's (1934) classical experiments with *Drosophila*. A case of what certainly appears to be such odour-conditioned anemotaxis was described recently by Haskell, Paskin & Moorhouse (1962), who observed a reversal from downwind to upwind walking by locust hoppers when grass odour was added to the air stream (the authors surprisingly interpret this as orientation to the odour itself—*tropotaxis*—through a misapprehension, as it seems to me). Otherwise, there has been little work on these lines with phytophagous insects; and the recent realization that other laboratory results had often been misinterpreted and that only very close-range orientation had in fact been demonstrated, seems to have discredited the whole idea. At any rate, neither Thorsteinson (1960) nor Beck (1965) in their recent reviews attach any importance to oriented host finding from a distance, except by visual stimuli which are often important but not usually very host specific. As far as chemical stimuli are concerned both authors plump for simple orthokinesis—the arrest of random movement.

This surely is a premature generalization, conditioned by the usual laboratory arrangements which do not permit the insects to display their full repertoire of host-finding manoeuvres. Take first the mechanism of odour-conditioned upwind orientation. The well-known examples of this *in flight* do not come from phytophagous insects proper but from host-finding mosquitoes (Kennedy, 1940; Wright, 1962), from the males of various insects stimulated by sexual pheromones (see Wilson & Bossert, 1963, Bossert & Wilson, 1963), and from dung-beetles (Steiner, 1953; Steiner & Wette, 1954; Danzer, 1956). One should not attach too much importance to this discrepancy because for these non-phytophagous insects the 'target' is small and temporary and the field observer can oversee the whole process. It is much easier to catch sight of the individual insects arriving at a dung-pat than at a whole field of crop plants.

For phytophagous insects the evidence is more anecdotal but, nevertheless, there are some striking reports of bark-beetles and butterflies flying specifically to their host plants or sources of host odour from due downwind, sometimes over considerable distances, up to half a mile or more (Beeson, 1930; Gardiner, 1957; Cripps, 1947; McMullen & Atkins, 1962). Admittedly, we tried hard and failed to find any evidence of this in aphids, but then aphids are at best rather limited behaviourally. With less degenerate phytophagous insects, the possibility of odour-conditioned anemotaxis now demands more field and laboratory experiments.

Secondly, the fact that the air is seldom still in the open means that quite sudden changes in odour concentration are a common experience for insects in nature, when they pass from downwind to just upwind of the source, or the other way. In this situation *klinokinesis* can play a most important part in host finding in the field. This term does not now refer to the only kind of klinokinesis that was recognized by Dethier, Browne & Smith (1960) (but not recognized by Fraenkel & Gunn, 1940) and found by Dethier in his caterpillars: where the rate of random turning increases as the smell gets stronger. It is just the reverse—the classical type of klinokinesis (once called phobotaxis) as described in the textbook of Fraenkel & Gunn (1940). Here the insect keeps going more or less *straight* so long as the smell stays the same or is getting stronger, but changes direction and begins 'casting about' as soon as the smell weakens. This gives an insect that 'overshoots' a chance of encountering the nearby host, or of re-entering the scented airstream and reacting anemotactically and thus making more 'runs' at the host from downwind. Again, there are only odd reports of this kind of thing in phytophagous insects in the field (e.g. Cripps, 1947; McMullen & Atkins, 1962), but more can reasonably be expected and it is stressed because reviewers have not mentioned it recently, neither Dethier, Browne & Smith (1960), nor Thorsteinson (1960), nor Beck (1965).

I have perhaps painted too gloomy a picture of the applicability of present knowledge to the development of integrated control. But my point is that the possibilities of frustrating would-be insect pests are much richer than anyone would guess from the crude principles on which we have been working so far and these should therefore not be allowed to blinker research policy aimed at more integrated control.

## REFERENCES

References in the text which are not included in the following list will all be found in the review by Beck (1965).


AUCLAIR, J. L., MALTAIS, J. B. & CARTIER, J. J. (1957). *Can. Ent.* **88**, 457–64.

BECK, S. D. (1965). *Annu. Rev. Ent.* **10**, 205–32.

BEESON, C. F. C. (1930). *Nature, Lond.*, **126**, 12.

VAN DEN BOSCH, R. & STERN, V. M. (1962). *Annu. Rev. Ent.* **7**, 367–86.

BOSSERT, W. H. & WILSON, E. O. (1963). *J. theoret. Biol.* **5**, 443–69.

CARTHY, J. D. (1958). *An introduction to the behaviour of Invertebrates.* London: Allen and Unwin.

CARTHY, J. D. (1965). *The behaviour of arthropods.* London: Oliver and Boyd.

CRIPPS, C. (1947). *Proc. R. ent. Soc. Lond.* **22**, 42–3.

DANZER, A. (1956). *Z. vergl. Physiol.* **39**, 76–83.

DETHIER, V. G. (1947). *Chemical insect attractants and repellents.* Philadelphia, Pa.: The Blakiston Co.

EDWARDS, C. A. & HEATH, G. W. (1964). *The principles of agricultural entomology.* London: Chapman and Hall.

VAN EMDEN, H. F. (1965). (This symposium.)

FLÜGGE, C. (1934). *Z. vergl. Physiol.* **20**, 463–500.

FRAENKEL, G. S. & GUNN, D. L. (1940). *The orientation of animals.* Oxford: Clarendon Press.

FRANZ, J. M. (1961). *Annu. Rev. Ent.* **6**, 183–200.

GARDINER, L. M. (1957). *Canada Dept. Agric. Forest Biol. Bi-monthly Prog. Rep.* **13** (1), 2.

HASKELL, P. T., PASKIN, M. W. J. & MOORHOUSE, J. E. (1962). *J. Ins. Physiol.* **8**, 53–78.

JONES, F. G. W. & JONES, MARGARET G. (1964). *Pests of field crops.* London: Edward Arnold.

KENNEDY, J. S. (1940). *Proc. zool. Soc. Lond.* **109**, 221–42.

KENNEDY, J. S. (1953). *Chem. & Indus.* 1953, 1329–32.

MCMULLEN, L. H. & ATKINS, M. D. (1962). *Can. Ent.* **94**, 1309–25.

MITTLER, T. E. & DADD, R. H. (1964). *Ent. exp. appl.* **7**, 315–28.

RABB, R. L. (1962). *Bull. ent. Soc. Amer.* **8**, 193–6.

SMITH, R. F. (1959). *J. econ. Ent.* **52**, 1106–8.

STEINER, G. (1953). *Naturwissenschaften*, **40**, 514–15.

STEINER, G. & WETTE, R. (1954). *Naturwissenschaften*, **41**, 172.

TAMBS-LYCHE, H. & KENNEDY, J. S. (1958). *Ent. exp. appl.* **1**, 225–39.

TINBERGEN, N. (1951). *The study of instinct.* Oxford: Clarendon Press.

WALDBAUER, G. P. (1964). *Ent. exp. appl.* **7**, 253–69.

WIGGLESWORTH, V. B. (1953). *The principles of insect physiology.* London: Methuen.

WILSON, E. O. & BOSSERT, W. H. (1963). *Rec. Progr. Hormone Res.* **19**, 673–716.

WRIGHT, R. H. (1962). *World Rev. Pest Control*, **1** (4), 12 pp.

# E. Relationships with Other Organisms

## THE DETECTION AND EVASION OF BATS BY MOTHS

By KENNETH D. ROEDER and ASHER E. TREAT[1]

A CENTRAL objective of a large segment of biological and psychological research is to provide a physiological basis for behavior. The first step toward this objective is analytic, and consists of determining the structure and function of neural components after they have been isolated from their connections with the rest of the nervous system. There has been much progress in this direction, and it is now possible to describe in terms of input and output performance the operation of many isolated sense cells, neurons, and muscle fibers, even though the principles of their internal operation are mostly not understood.

The next step, the synthetic process of assembling this information on isolated neural components and relating it to the behavior of the intact animal, is hampered by two kinds of difficulty. The first appears to be methodological, but is somewhat hard to define. When one regards the evergrowing literature on the unit performance of sense cells, nerve cells, and muscle fibers, it is to experience that sense of dismay first encountered at a tender age when the springs, gears, and screws of one's first watch were strewn upon the table. The *modus operandi* of analysis or taking apart seems to come naturally, and the problems encountered are essentially technical in nature. Synthesis or the derivation of a system from its components seems to lack the *a priori* logic of analysis.

The second general difficulty is technical, and stems from the fact that even the simplest behavior of the higher animals and man is accompanied by the simultaneous activity of millions of sense cells, nerve cells, muscle fibers, and glands. Even if it were possible to register the traffic of nervous and chemical information generated and received by each and all of these neural elements during the behavior, it is doubtful whether the record would provide a meaningful description of the action.

Even though these problems cannot be solved directly at the present time, they become less formidable if the behavior selected for study is simple and stereotyped, and only a small number of nerve cells are concerned in its execution. These conditions are partly fulfilled by the sensory mechanisms whereby certain nocturnal moths detect the approach of insectivorous bats.

[1] Much of the experimental work reported in this paper was made possible by Grant E-947 from the U. S. Public Health Service.

Reproduced with permission from *American Scientist*, 49:135–148, 1961. Published by the Society of the Sigma Xi, New Haven, Connecticut.

### *Echolocation and Countermeasure*

Bats detect obstacles in complete darkness by emitting a sequence of high-pitched cries or chirps and locating the source of the echoes. As Griffin (1958) and others have shown, this form of Sonar is unbelievably precise. By means of it, insectivorous bats locate and track flying moths, mosquitoes, and small flies (Griffin, *et al.*, 1960). North American bats, such as *Myotis lucifugus* and *Eptesicus fuscus*, emit chirps about 10 times a second when they are cruising in the open. Each chirp lasts from 10 to 15 milliseconds (msec) with an initial frequency of 80 kilocycles (kc) dropping about one octave in pitch toward its end (see Figure 5).

The frequencies in these chirps are ultrasonic, that is, inaudible to human ears, which cannot detect tones much above 15–18 kc. The higher frequencies used by bats make possible more discrete echoes from smaller objects. The chirps can be rendered audible by detecting them with a special microphone and rectifying the ultrasonic component. They then can be heard through headphones as a series of clicks. These clicks fuse into what Griffin has called a "buzz" when the bat is chasing an insect or avoiding an obstacle.

Several families of moths (in particular the owlet moths or *Noctuidae*) have evolved countermeasures enabling them to detect the chirps of bats. A pair of ultrasonic ears is found near the "waist" of the moth between thorax and abdomen (Figure 1). An extremely thin eardrum or tympanic membrane is directed obliquely backward and outward into the recess (dark area) found at this point.

FIG. 1. External opening of the right ear in *Agrotis ypsilon*. The external surface of the tympanic membrane faces obliquely backwards and outwards into the cavity below arrow. The body of the moth is about $^3/_4$ inch in length.

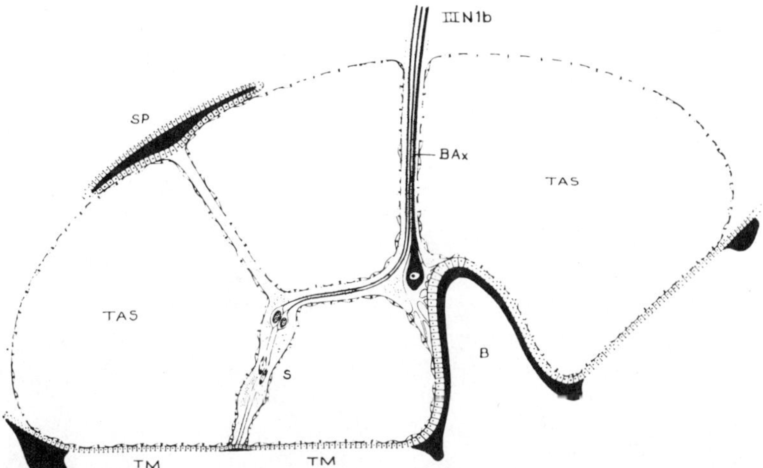


FIG. 2. Diagram of the tympanic organ of a noctuid moth. The sensillum (*S*) contains the pair of acoustic receptors or A cells. The A nerve fibers are joined by that of the B cell (*BAX*) to form the tympanic nerve (*IIINIB*). *TAS*, tympanic air sac; *B* and *SP*, skeletal supports; *TM*, tympanic membrane. (After Treat and Roeder, 1959.)

Internal to the eardrum is an air-filled cavity that is spanned by a thin strand of tissue running from the center of the eardrum to a skeletal support (Figure 2). This tissue contains the sound-detecting apparatus, consisting of two acoustic sense cells (A cells). A single nerve fiber arises from each A cell and passes close to the skeletal support, where the pair is joined by a third nerve fiber arising from a large cell (B cell) in the membranes covering the support. The three fibers continue their course to the central nervous system of the moth as the tympanic nerve.

The traffic of nerve impulses passing over the three fibers from A cells and B cell to the nervous system of the moth can be followed if a fine metal electrode is placed under the tympanic nerve. Another electrode is placed in inactive tissue nearby. As each impulse passes the site of the active electrode it can be detected as a small action potential lasting about 1 msec. Since the tympanic nerve contains only three nerve fibers, it is not difficult to distinguish and to read out the respective reports to the nervous system from the pair of A cells and the B cell. A similar experiment in a mammal is practically meaningless since the auditory nerve contains about 50,000 nerve fibers.

This method of detection shows that the A cells transmit organized patterns of impulses over their fibers only when the ear is exposed to sound (Roeder and Treat, 1957). The B cell transmits a regular and continuous succession of impulses that can usually be distinguished from the A impulses by their greater height. The B impulses are completely unaffected by acoustic stimulation, and change in frequency only when the skeletal framework and membranes lining the ear are subjected to steady mechanical distortion (Treat and Roeder, 1959). The B cell behaves in a manner similar to receptors found in other parts of the body that convey information about mechanical stress on joints, muscles, and skeleton. The role of such a receptor in the ear of a moth is unknown.

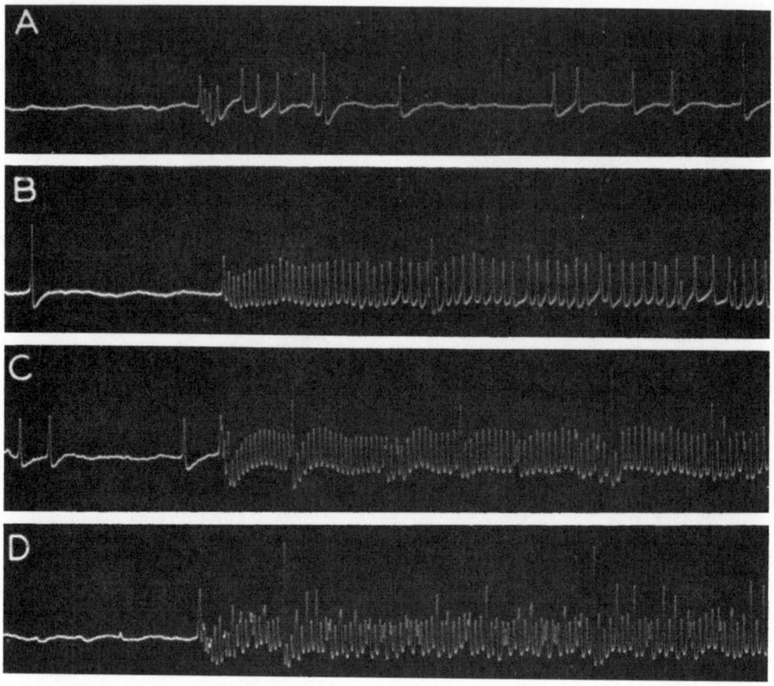


FIG. 3. Tympanic nerve response in *Prodenia eridania* to a pure tone of 40 kc. The occasional large spikes originate in the B cell. (*A*) Response to a sound intensity close to the threshold of the sensitive A cell. (*B*) Intensity 7 db above that in (*A*). (*C*) Intensity 15 db above that in (*A*). (*D*) Intensity 23 db above that in (*A*). The less sensitive A cell discharges occasionally in (*C*), and frequently in (*D*), as indicated by the double peaks. Time line 100 msec. (From Roeder, 1959.)

In the absence of sound, the A cells discharge irregularly spaced and relatively infrequent impulses (Figure 3A). A continuous pure tone of low intensity elicits a more regular succession of more frequent impulses in one of the A fibers (Figure 3B). The other fiber is not yet affected. Any slight increase in the intensity of the tone causes a corresponding increase in the impulse frequency of the active fiber. When the intensity of the tone is increased to about 10-fold that producing a detectable response in the more sensitive A fiber, the second A fiber begins to respond in like manner. Its action potentials are superimposed on those of the first (Figures 3C and D) by the method of recording, but actually reach the central nervous system over their own pathway. This experiment reveals two of the ways in which the moth ear codes sound intensity. It is like an instrument having a graded fine adjustment (the intensity-frequency relation) and a coarse adjustment of two steps (the pair of A cells). Other ways of coding intensity will appear later.

The moth ear responds in this manner to tones from 3 kc to well over 100 kc but there is no evidence that it is capable of discriminating between tones of different frequency. It is most sensitive near the middle of its range, that is, to frequencies such as those contained in bat chirps.

In Figure 3 it will be noticed that, in each of the recordings, the intervals between the successive impulses increase as the pure tone stimulus continues. In terms of the nerve code outlined above, the A cells report that the sound is declining in intensity with time, although in fact it was kept constant. This adaptation to a constant stimulus occurs in most receptors registering changes in the outside world. In terms of our own experience, the impact of our surroundings would be shocking and unbearable if it were not distorted in this manner by sense organs. The brilliance of a lighted room entered after dark would continue to be blinding and the noise of a jet engine would remain unbearable. However, the A cells of the moth's ear adapt very rapidly to a continuous tone, and their full effectiveness as pulse detectors is revealed only when they are exposed to short tone pulses similar to bat chirps.

In the experiment illustrated in Figure 4 a tone pulse of 3 msec duration was generated at regular intervals. It is similar to a bat chirp except for its regularity and the absence of frequency modulation. A microphone (upper trace) and moth ear (lower trace) were placed within range, and the intensity of the stimulus pulse was adjusted so that it just produced a detectable response in the most sensitive A fiber (0 db). The intensity was then increased by 5 decibel[1] (db) steps as each recording was made. It will be seen that the microphone begins to detect the sound pulse when it is about 10 db above the threshold of the most sensitive A cell in the moth's ear. As before, the increase in frequency of A impulses is evident if the 5 and 10 db records are compared, and a response of the less sensitive A cell appears first in the 25 db record where the extra peaks of its action potentials overlap those of the more sensitive A unit. In addition to these two ways of coding intensity, two more can now be recognized. If the interval between detection

[1] The decibel (db) notation expresses relative sound pressures. An intensity of 20 db is 10-fold that of the reference level (0 db), a 40 db sound is 100-fold the reference level.

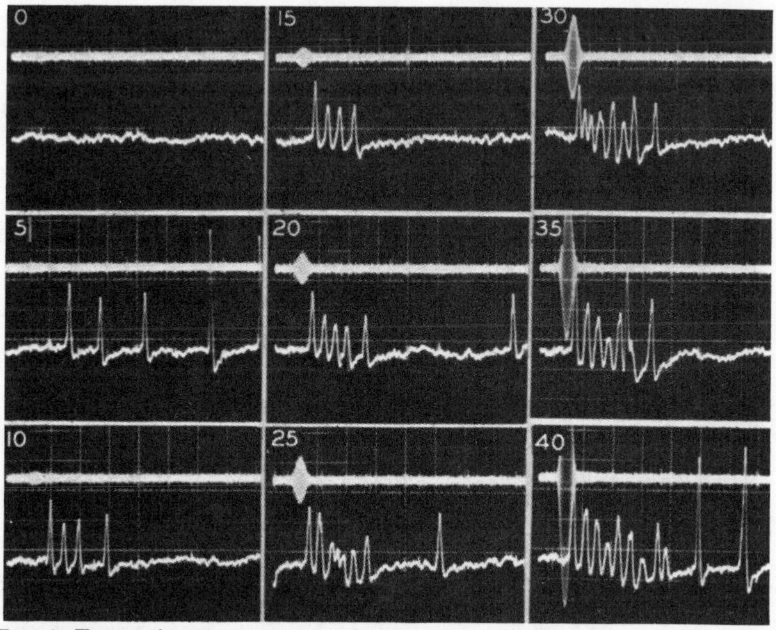


FIG. 4. Tympanic nerve responses (lower traces) of *Noctua* ( = *Amathes*) *c-nigrum* to a 70 kc. sound pulse recorded simultaneously by a Granith microphone (upper traces). The numbers indicate the intensity of the sound pulse in decibels above a reference level (0). The threshold of the sensitive A cell lies between 0 and 5 db. The large spikes appearing in some of the records are from the B cell. The less sensitive A cell responds in the 25 db recording. Vertical lines, 4 msec. apart.

of the sound by the microphone and by the moth ear is compared at different sound intensities, it will be noticed that the tympanic nerve response occurs earlier and earlier on the horizontal time axis. In other words, the latency of the response decreases with increasing loudness. Also, the sense cells are seen to discharge impulses for some time after the sound has ceased, and this after-discharge becomes longer with increasing sound intensity.

### *The Detection of Bats*

These experiments with artificial sounds suggest how the moth ear might be expected to respond to a bat cry. A few laboratory observations were made with captured bats. In one of these, in collaboration with Dr. Fred Webster, the cries of a flying bat were picked up simultaneously by a moth ear and a microphone, and recorded on high-speed magnetic tape (Figure 5). Interesting though they were, these experiments served mainly to show that the full potentialities of the moth ear as a bat detector could not be realized within the confines of a laboratory, and efforts were made to transport the necessary equipment to a spot where bats were flying and feeding under natural conditions.

Finally, about 300 pounds of equipment was uprooted from the laboratory and reassembled at dusk of a July evening on a quiet hillside in the Berkshires of western Massachusetts. Moths attracted to a light provided experimental material. The insect subject was pinned on cork so that one of its ears had an unrestricted sound field, and with the help of a microscope its tympanic nerve was exposed and placed on electrodes. After amplification, the action potentials were displayed on an oscilloscope. They were also made audible as a series of clicks by means of headphones connected to the amplifier, and were stored on magnetic tape for later study.

It was dark before all was ready, but bats immediately revealed their presence to the moth ear by short trains of nerve impulses that recurred about 10 times a second (Figure 6A). The approach of a cruising bat from maximum range was coded as a progressive increase in the number and frequency of impulses in each train, first from one and then from both A fibers. It was not long before we learned to read something of the movements of the bats from these neural signals. Long trains, sometimes with two frequency peaks, suggested the chirps of nearby bats that echoed from the wall of a neighboring house (Figure 6B). An increase in the repetition rate of the trains coupled with a decrease in the number of impulses in each train signified a "buzz" as the bat attacked some flying insect in the darkness (Figure 6C).

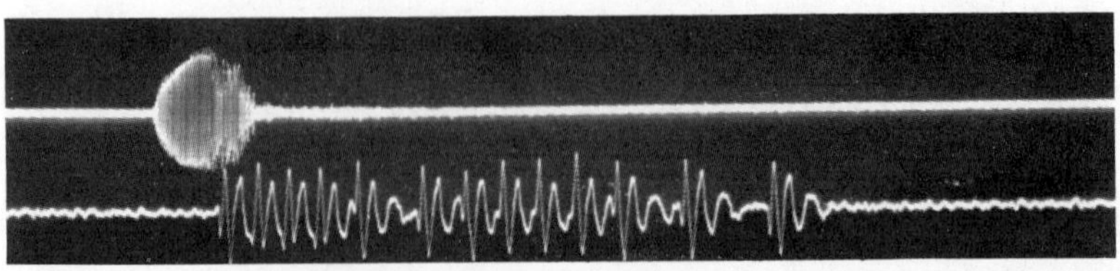

FIG. 5. The cry of a flying bat (*Myotis*) recorded by a Granith microphone (upper trace) and the A cells (lower trace) of a noctuid moth (*Agroperina dubitans*). The A spikes shown in the lower trace have been distorted in form by the recording technique. Time line, 10 msec. Made in collaboration with Dr. Fred Webster in his laboratory.

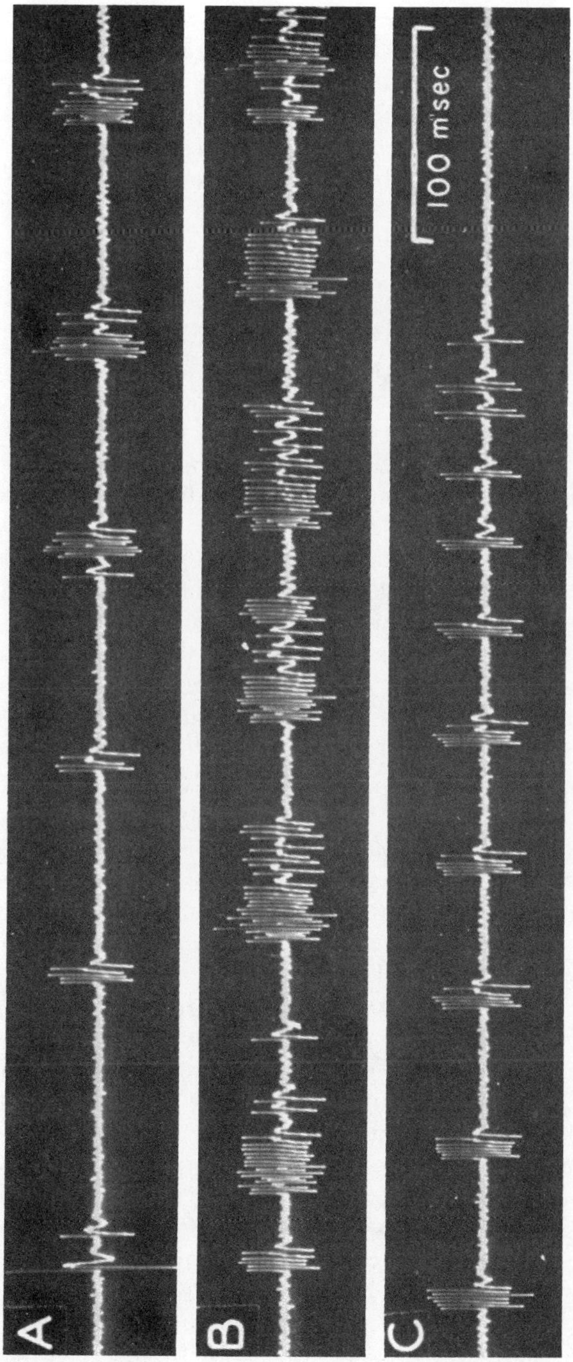


FIG. 6. Tympanic responses of *Noctua* (=*Amathes*) *c-nigrum* to the cries of bats flying in the field. (*A*) The approach of a cruising bat emitting pulses at about 10 per second. (*B*) Tympanic response to the original cry and its echo made by a bat cruising nearby. (*C*) A "buzz." Time line, 100 msec. (From Roeder and Treat, 1961.)

All of this was inaudible and invisible to our unaided senses. With a powerful floodlight near the nerve preparation we were able to see bats flying within a radius of 20 feet, and some attacks on flying insects could then be both seen and also "heard" through the "buzz" as coded by the moth's typanic nerve. However, most of the sounds detected by the moth ear were made by bats maneuvering well out of range of the light. A rough measure of the sensitivity of the moth ear to bat chirps was obtained at dusk on another occasion when the bats could still be seen. The A cells first detected an approaching bat flying at an altitude of more than 20 feet and at a horizontal distance of over 100 feet from the moth—a performance that betters that of the most sensitive microphones.

### *Direction*

Since differences in sound intensity are coded by the tympanic nerve in at least four different ways, the horizontal bearing of a bat might be derived from a comparison of the nerve responses to the same chirp in the right and left ears. A difference in right and left responses might be expected only if each ear had directional properties, that is, a lower threshold to sounds coming from a particular direction relative to the moth's axis.

Directional sensitivity was measured in an open area where echoes were minimal. A source of clicks of constant intensity was placed on radii to the moth at 45 degree intervals. The source was moved in and out on each radius until a standard tympanic nerve response was obtained, and the distance from the moth noted. Horizontal distances along 8 radii were combined to make a polar plot of sensitivity (Roeder and Treat, 1961). The plot showed that, although there was little difference in sensitivity fore and aft, a click on the side nearest the ear at about 90 degrees relative to the moth's longitudinal axis was audible at about twice the distance of a similarly placed click on the far side.

This led to further field experiments in the presence of flying bats. The tympanic nerve responses from both ears of a moth were recorded simultaneously on separate tracks of a stereophonic magnetic tape. The tape was subsequently re-played into a two-channel oscilloscope and the traces photographed (Figure 7). In the upper record (A) the increase in number of impulses in each succeeding train suggests the approach of a bat. When the signals from right and left ears are compared, it is evident that the greatest difference exists when the signal is faintest, the first response of the series occurring in one ear only. When both ears respond, the differential nature of the binaural response can be seen first as a difference in the number of spikes generated in right and left ears, second in the differential spike frequency, and third in the latency of the response, which is greater on that side generating fewer spikes. It is also evident that, as the sound intensity increases (presumably due to the approach of the bat), the differential becomes less until the responses of right and left ears become almost identical. In another experiment, it was found that the tympanic nerve response saturates, i.e., becomes maximal, when the sound intensity is about 40 db (100-fold) above threshold. From this it can be concluded that the moth's nervous system receives information that would enable it to determine whether a distant bat was to the right or left, but if the bat was at close quarters

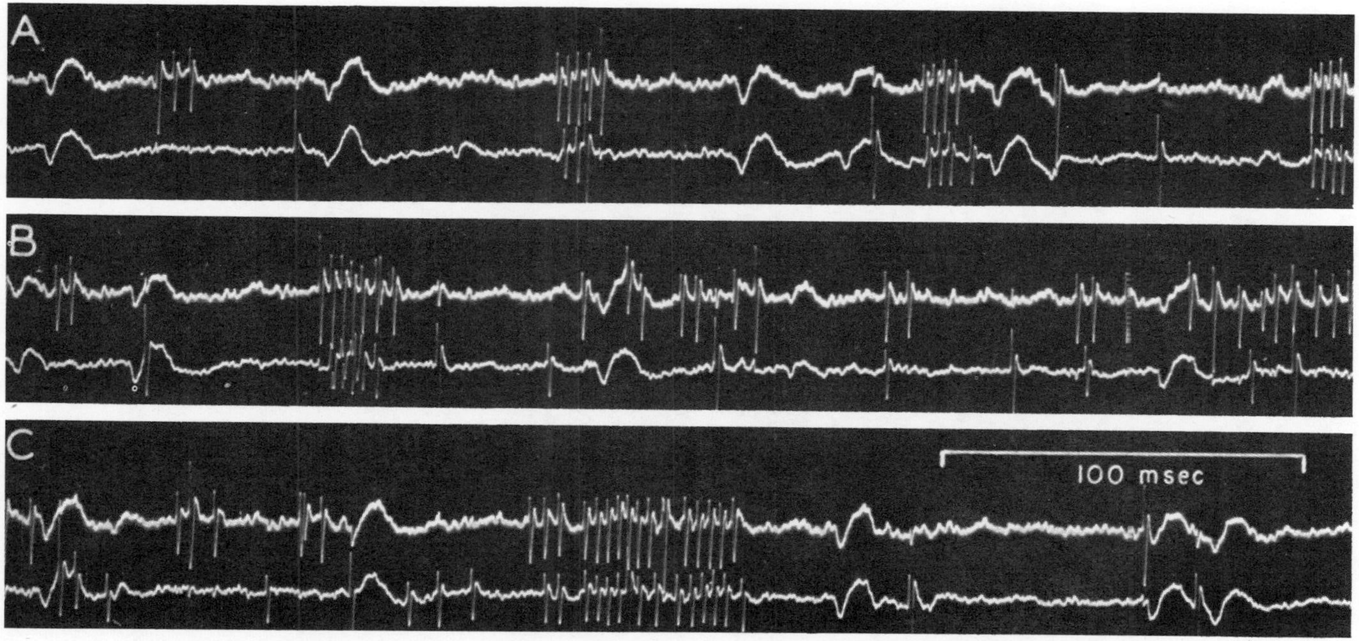


FIG. 7. Binaural tympanic responses of *Feltia sp.* to the cries of red bats flying in the field. The electrocardiogram of the moth also appears on both channels as slow waves. B impulses (large spikes) appear regularly in the records from both tympanic nerves. (*A*) An approaching bat. Differential response is marked at first (response latency, number of spikes) but has practically disappeared in the final train. (*B*) A "buzz" registered mainly by one ear. (*C*) A "buzz" registered by both ears. Time line, 100 msec.

this information would not be available. In Figure 7C the "buzz" was picked up by one ear only, presumably because during this part of its performance the chirps of a bat are much less intense.

It is tempting to estimate just how close the bat must be before the moth fails to get information on its location. If it is assumed that a bat is first detected at 100 feet and approaches on a straight path at right angles to the moth's course while making chirps of constant loudness, the differential tympanic nerve response would diminish throughout the approach and disappear completely when the bat was 15 to 20 feet away. However, we have not yet determined how much of the information that we are able to read out of its auditory mechanism is actually utilized by the moth in its normal behavior.

### *The Evasive Behavior of Moths*

Although the evasive behavior of moths in the presence of bats must have been witnessed hundreds of times, it is hard to find an adequate account of the maneuvers of either party. The contest normally takes place in darkness, and, even when it is illuminated by a floodlight, the action is too fast and complex to be appreciated by the eye. The flight path of the bat and its ability to intercept and capture its prey have been studied by Griffin (1958) and his students. More recently, Webster (in press) has shown by means of high-speed sound motion pictures that bats become adept at using echoes to plot an interception course with an object moving in a simple ballistic trajectory. Many people have noted the seemingly erratic dives and turns made by moths when bats are near, and similar behavior has been described when moths are exposed to artificial sources of ultrasound (Schaller and Timm, 1950; Treat, 1955).

In an effort to learn more about the behavior of moths under field conditions their flight was tracked photographically as they reacted to a series of ultrasonic pulses simulating bat cries. The sounds were generated by the equipment used in the experiment shown in Figure 4. The pulses were similar in form to those shown, although longer in duration (6 msec). Each pulse ranged from 50 to 70 kc with a rise and fall time of about 1 msec. Pulse sequences up to 50 per second could be released on closure of a switch. The sounds were emitted by a plane-surfaced condenser loud-speaker mounted so as to project a fairly directional beam over an open area of lawn and shrubs illuminated by a 250-watt floodlight.

The observer sat behind the sound generator and floodlight, holding in one hand the cable release of a 35 mm camera set on "bulb," and in the other the switch controlling the onset of the sound-pulse sequence. Many moths and other insects flew out of the darkness into this floodlight arena. A number were attracted directly to the light and were disregarded. Many others moved across the arena at various angles but without marked deviation toward the light. When one of these appeared to be in line with the loudspeaker the camera shutter was opened and the sound pulses turned on.

Some of the tracks registered by the camera as the illuminated moths moved against the night sky are shown in Figure 8. Many insects, including some moths, showed no change in flight pattern when they encountered the sound. In others, the changes in flight path were dramatic in their abruptness and bewildering in their variety. The

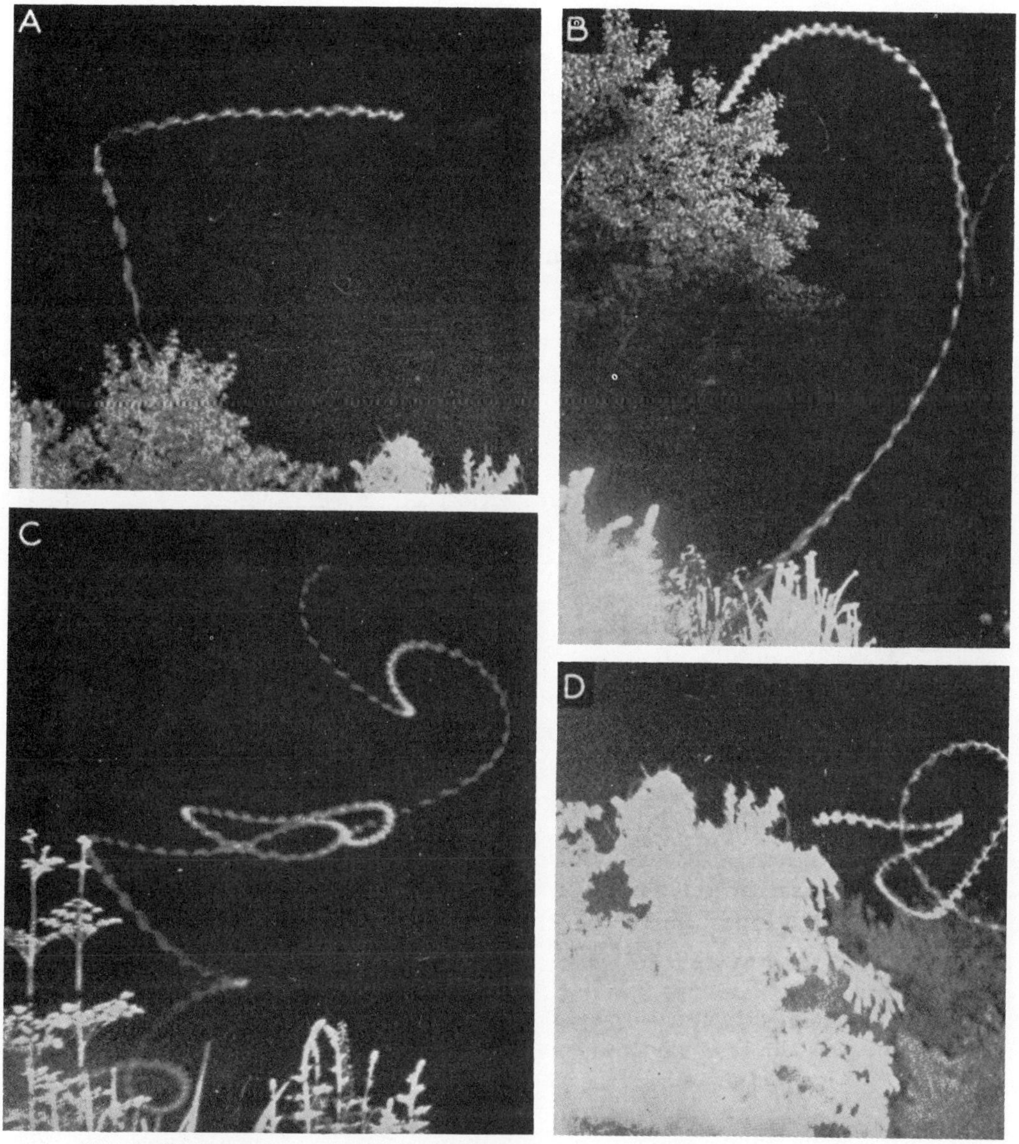


FIG. 8. Flight tracks registered by various moths just before, and immediately following, exposure to a series of simulated bat cries. The dotted appearance of the tracks is due to the individual wingbeats of the moth. The beginning of each track appears in each photograph, and the moth finally flies out of the field.

simplest, and also one of the commonest reactions was a sharp power dive into the grass (Figure 8A, B). Sometimes the dive was not completed and the insect flew off at high speed close to the ground. Almost as frequently the dive was prefaced or combined with a series of tight turns, climbs, and loops (Figure 8C, D).

It is not known whether these maneuvers are selected in some random manner from the repertoire of individual moths, or whether they are characteristics of different species. However, Webster (in press) has shown that bats soon learn to plot an interception course with food propelled through the air in a simple ballistic trajectory. The random behavior elicited by simulated bat cries in the natural moth population seems to be a natural answer to this predictive ability in bats, while the sharpness of the turns must certainly tax the maneuverability of the heavier predator.

The reacting moths shown in Figure 8 were mostly within 25 feet of the camera and sound source, and were exposed to an unknown but probably high sound intensity. Under these circumstances, the evasive behavior appeared to be completely unorientated relative to the sound source, as might be predicted from the binaural tympanic nerve recordings. In some instances, moths flying at a greater distance or only on the edge of the sound beam appeared to turn away from the area and fly off at high speed. This must be checked in future experiments.

### *The Survival Value of Evasion*

In spite of the evidence that the moth ear is an excellent bat detector, and that acoustic stimulation releases erratic flight patterns, one may well ask whether this behavior really protects moths from attack by bats.

This question has been answered (Roeder and Treat, in press) by observing with a floodlight 402 field encounters between moths and feeding bats. In each encounter we recorded the presence or absence of evasive maneuvers by the moth, and the outcome, that is, whether it was captured by the bat or managed to escape. From the pooled data we determined the ratio of the percentage of nonreactors surviving attack to the ratio of reactors surviving attack. Thus computed, the selective advantage of evasive action was 40 per cent, meaning that for every 100 reacting moths that survived, there were only 60 surviving nonreactors.

This figure is very high when compared with similar estimates of survival value for other biological characteristics. It seems more than adequate to account for the evolution of the moth's ear through natural selection even if the detection of bats turns out to be its only function.

### *Conclusion*

As with most investigations, this work raises more questions than it has answered. The role of the B cell remains completely obscure. There is no evidence to connect it with the auditory function even though it is located in the ear, and its regular impulse discharge is a characteristic feature of the tympanic nerve activity of many species of moth (Treat and Roeder, 1959. See also Figure 7). The manner in which the A cells transduce sound waves recurring 100,000 times a second into the much slower succession of nerve impulses remains a mystery, and the synaptic mechanisms whereby information from the A fibers is translated into action by the nervous system of the moth, await investigation.

During the field experiments it was noticed that many other natural sounds initiated impulses in the A fibers. These included the rustling of leaves, the chirp of tree and field crickets, and, in one instance, ultrasonic components in the wingbeat sounds made by another moth. Occasionally, the A fibers discharged regularly as if detecting a rhythmic sound, though none was audible to the observers and its source (if any) remains a mystery. There is no evidence that these identified and unidentified sounds are important in the life of a moth, yet it must be said that a moth can detect them, and a careful study of moth behavior in their presence would be of value.

Several families of moths lack ears and show no response to ultrasonic stimuli. Some of these, such as the sphinx or hawk moths and the larger saturniid moths, are probably too much of a mouthful for the average bat, and might find no survival advantage in a warning device. Others are of the same size and general habits as the noctuids and might be expected to suffer attacks by bats. Included in this group are some common pests such as the tent caterpillar. It will be interesting to learn whether these forms owe their success in survival to some structural or behavioral countermeasure that compensates for the lack of a tympanic organ.

In spite of these unanswered questions, we believe that some progress has been made in putting together the sensory information received by an animal, and relating this to what the animal does. That this has been possible in moths is only because of the small number of channels through which acoustic information reaches the nervous system in these insects. Further examples of this favorable situation have been described in other insects, and still others are waiting to be explored.

## LITERATURE

1. Griffin, Donald R. *Listening in the Dark.* Yale University Press, New Haven, 1958.
2. Griffin, D. R., F. A. Webster, and C. R. Michael. The Echolocation of Flying insects by Bats. *Animal Behaviour, 8,* 141–154, 1960.
3. Roeder K. D. A Physiological Approach to the Relation between Prey and Predator. *Smithsonian Miscellaneous Collections, 137,* 287–306, 1959.
4. Roeder, K. D. and A. E. Treat. Ultrasonic Reception by the Tympanic Organ of Noctuid Moths. *Journal of Experimental Zoology, 134,* 127–158, 1957.
5. Roeder, K. D. and A. E. Treat. The Detection of Bat Cries by Moths. *Sensory Communication* (ed. W. Rosenblith), M.I.T. Technology Press, 1961.
6. Roeder, K. D. and A. E. Treat. The Acoustic Detection of Bats by Moths. *Proceedings of the XI (1960) International Entomological Congress,* in press.
7. Schaller, F. and C. Timm. Das Hörvermögen der Nachtschmetterlinge. *Zeitschrift für Vergleichende Physiologie, 32,* 468–481, 1950.
8. Treat, A. E. The Response to Sound of Certain Lepidoptera. *Annals of the Entomological Society of America, 48,* 272–284, 1955.
9. Treat, A. E. and K. D. Roeder. A Nervous Element of Unknown Function in the Tympanic Organs of Moths. *Journal of Insect Physiology, 3,* 262–270, 1959.

# Cyclic Fluctuations of the Population Density Intrinsic to the Host-Parasite System*

Syunro Utida
*Entomological Laboratory, Kyoto University, Kyoto, Japan*

## Introduction

The cyclic fluctuations of animal population caused by the interaction between prey and predator or host and parasite have long been predicted mathematically by Lotka, Volterra and Nicholson and Bailey, and experimental work to verify this mathematical postulate has been done by Gause (1934, 1935), DeBach and Smith (1941) and the present author (Utida 1950, 1955). In experimental population of protozoans and mites, Gause demonstrated so-called relaxation oscillations and was able to produce cyclic fluctuations of the classic type only when periodic "immigrations" of prey were made. The experiment of DeBach and Smith, using the pupae of the house fly and its hymenopterous parasite, was devised very ingeniously, the host's power of increase being fixed at a constant value; the resulting fluctuations of population density obeyed the mathematical model of Nicholson and Bailey. Unfortunately, however, the experiment terminated at the seventh generation, before one complete wave of population fluctuation had been shown. The present author attempted to find cyclic fluctuations in the interacting population of the azuki bean weevil, *Callosobruchus chinensis,* and its larval parasite *Neocatolaccus mamezophagus,* avoiding artifical "immigrations" and without arbitrary limitation of the host's reproductive power, but no clear fluctuation having a definite amplitude and wave length could be found (Utida 1950). Thus experimental demonstration of theoretical cycles is not ordinarily achieved, and there remain many doubtful points in the results of the experiments. An exception may be made for some more recent work of the present author (Utida 1955), which gives a clear demonstration of cyclic fluctuations of population density, covering 25 generations of population interaction between the azuki bean weevil and the larval parasite *Heterospilus prosopidis* (Hymenoptera, Braconidae).

The present paper is a continuation of the previous work. Three of the six experimental populations were continued from the 25th generation until long after the termination of the experiments described in the previous paper. The main problem of the present paper is whether the populations continue to fluctuate with the same wave length and amplitude even after very many generations.

* Contribution from the Entomological Laboratory, Kyoto University, No. 261.

## Methods

Here only a brief outline of the method is given, since the full account has already been given elsewhere (Utida 1955). The mixed populations of the azuki bean weevil and its larval parasite *Heterospilus prosopidis* were run in Petri dishes 1.8 cm high and 8.5 cm in diameter. The temperature and the relative humidity were maintained constant at 30°C and 75 per cent respectively. Ten grams of azuki beans (the seed of *Phaseolus angularis*), with water content of about 15 per cent, were placed in each Petri dish every ten days to serve as food for the weevils.

Three mixed populations, called populations A, C, and E, were run in the Petri dishes, starting with different combinations of initial densities of host and parasite, *viz.* 64 weevils and 8 wasps, 8 weevils and 8 wasps, and 512 weevils and 128 wasps, respectively. The number of each sex of dead and living adults of both species was counted at ten-day intervals. Living adults were lightly anesthetized with ether, counted, and then put back in the culture dish. At each census a new supply of beans was provided and the most ruined part of the food was removed to another dish, where counts were continued until 20 days after the separation.

## Results

The results of the counts of both host and parasite populations show that the numbers of living and of dead adults increase and decrease in turn, with a period of about 20 days. This is apparently caused by the emergence and death of adults belonging to each generation. When the generations do not overlap, as in the present case, the number of individuals emerging in each generation is a better index to population fluctuation than is the number of living or dead animals observed at the times of census. To eliminate the intra-generational population fluctuation, the number of emerging individuals belonging to two consecutive counts should be summed, for the double census period of 20 days closely approximates the time required for a generation of hosts. In this way the total numbers of hosts and parasites that emerged in

Reproduced with permission from *Ecology,* 38:442–449, 1957. Published by the Ecological Society of America and the Duke University Press, Durham, North Carolina.

each generation can be obtained, and these values are treated as the population densities of each generation.

The changes of density in the host and parasite populations are shown in Figures 1 and 2, including the data for the first 25 generations as previously reported (Utida 1955).

Fluctuations of population density can be seen for both host and parasite in each of the three experimental populations. In population A, the densities of host and parasite fluctuated rhythmically and reciprocally up to the 112th generation, when the population was accidentally destroyed. Population C was also destroyed by careless handling at the 47th generation. By contrast, although population E was not subjected to accidental disturbance, complete extinction of the host population was observed as a consequence of eruptive increase of parasite density.

In each of these three experimental populations, it is evident that in the initial stages in their history the densities of host and parasite rose and fell reciprocally, as described in the previous paper (Utida 1955); with the progress of the experiment, however, the fluctuations diminished in violence and approached a relatively stable level or steady state. In other words, the oscillations of density are damped. This trend of convergence was not found in the data described in the previous paper, for the 25 generations provided too short a time in which to detect it. The damping is now easily seen to have occurred before the 25th generation in population C, but it was disregarded in the previous paper for lack of confirmation.

Damping of the oscillations is especially well illustrated when the densities of host and parasite in each successive generation are plotted against each other, as in Figures 3 and 4. The cyclic fluctuations of density are repeated one or several times with gradually diminishing amplitude, and finally change to a steady state, after which the cycles, if present, are too small to be seen on the graphs. The final level attained is the same as the singular point of Gause (1934).

A very interesting fact is that in the course of many generations the damping of fluctuations occurs repeatedly, as seen in Figures 1 and 2. After the populations passed through several tens of generations in a state of steady density, relatively violent oscillations suddenly recurred and were damped in their turn. In population A, the first series of damped fluctuations attained its steady state at about the 30th generation, after which the fluctuations increased in amplitude before approaching the next phase of damping. This second phase was relatively protracted and was not broken until after the 80th generation; a third phase of damping ensued. In population C two phases of damped fluctuations can be seen, the division between them coming at the 20th generation. In population E as many as 70 generations elapsed before a second phase began; the second phase ended when the hosts died out at the 82nd generation.

Although the host and parasite densities are essentially constant during any one phase of damping, a long-term trend seems to be superimposed on the damping cycle itself. That is, in each successive phase of damping the host populations seem to rise and the parasite populations seem to fall. This tendency is also clearly shown in Figures 3 and 4. In the first phase of damping the singular points (representing steady densities) are situated near the ordinate, while in the second or third cycle of damped oscillations they tend to move downward toward the abscissa.

When the more violent fluctuations are smoothed by moving averages of seven consecutive points, the long-term trend of population density is obtained, as shown in Figure 5. Seven consecutive points were taken because the mean length of the short-period fluctuations (obtained by statistical methods) was about seven generations, as mentioned in the previous paper (Utida 1955). In each experimental population the host density increases steadily while the parasite density decreases. Accordingly, the incidence of parasitism (the numerical ratio of parasite to host) decreases with the progress of the generations. Since the long-term trends were the same in the three experimental mixtures, despite the fact that the initial preparations of parasite and host were different, it would seem to follow that the equilibrium proportions are not given by the singular point of any one wave or of any one phase of damping, but by the mean proportions throughout the entire population history.

## Discussion

Some important points to discuss here are deduced from the results obtained.

The first point is whether the fluctuation of population density is intrinsic to the system of host-parasite interaction. Since internal oscillations have been clearly shown in the author's previous paper (Utida 1955), as well as in the present paper, their existence no longer affords any room for doubt. The relation between the cyclic fluctuations of host and parasite populations can be clearly understood by Gause's method of graphic representation, as shown in Figures 3 and 4. This representation of the interaction does not

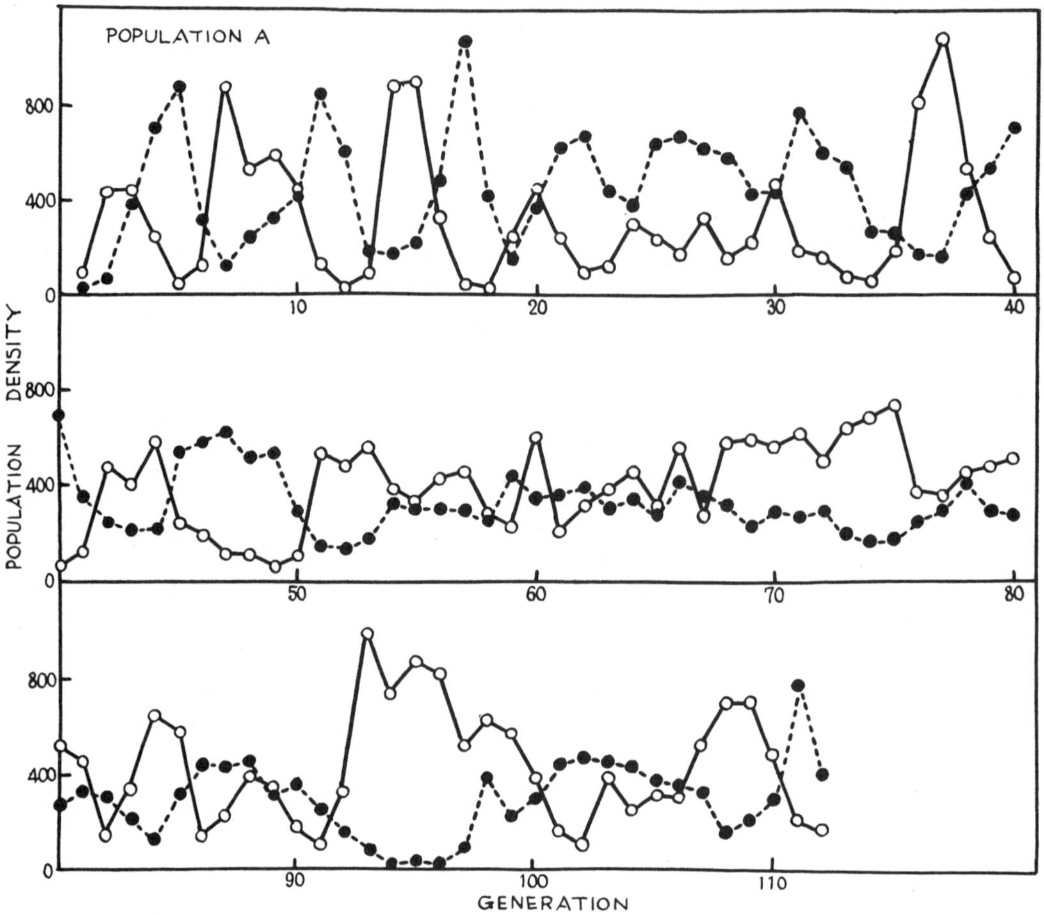


FIG. 1. Fluctuation of population density from generation to generation in population A. Host population:——o——, Parasite population: -----●-----

show the usual form of a circle or an ellipse, but the irregular form of an obliquely squashed polygon, which represents the delayed effect of the parasite upon the host population. One can conclude that lag effects are common in host-parasite or prey-predator interactions, as emphasized by Varley (1953).

Next, it must be remarked as an important finding that the fluctuation obtained is a damped one, decreasing in amplitude with the progress of generations, and finally converging to the level of steady density, the singular point of Gause (1934). It is noteworthy that the fluctuation obtained does not belong to the classical type of fluctuation in the mathematical theory of Lotka and Volterra.

Nicholson (1933) considered that this type of damped fluctuation should be found when a parasite attacks other species of host in addition to the host under consideration. It is also to be expected in two-species systems where the host population is closely controlled by the effect of its own density, as pointed out by the present author (Utida 1955). Experimentally, a similar finding was made in the interaction between *Paramecium* and *Schizosaccharomyces* (Gause 1934, 1935), but the problem requires further investigation. In the writer's opinion, the damped fluctuation is more important than the "conservative" or classical type of fluctuation, for the former type is generally characteristic when, as is usual, the host population is controlled both by attack of its parasite and by the influence of its own density.

Returning to the curves of Figures 1 and 2, we note that there is repetition of the cycle of damping when the interaction is observed for very many

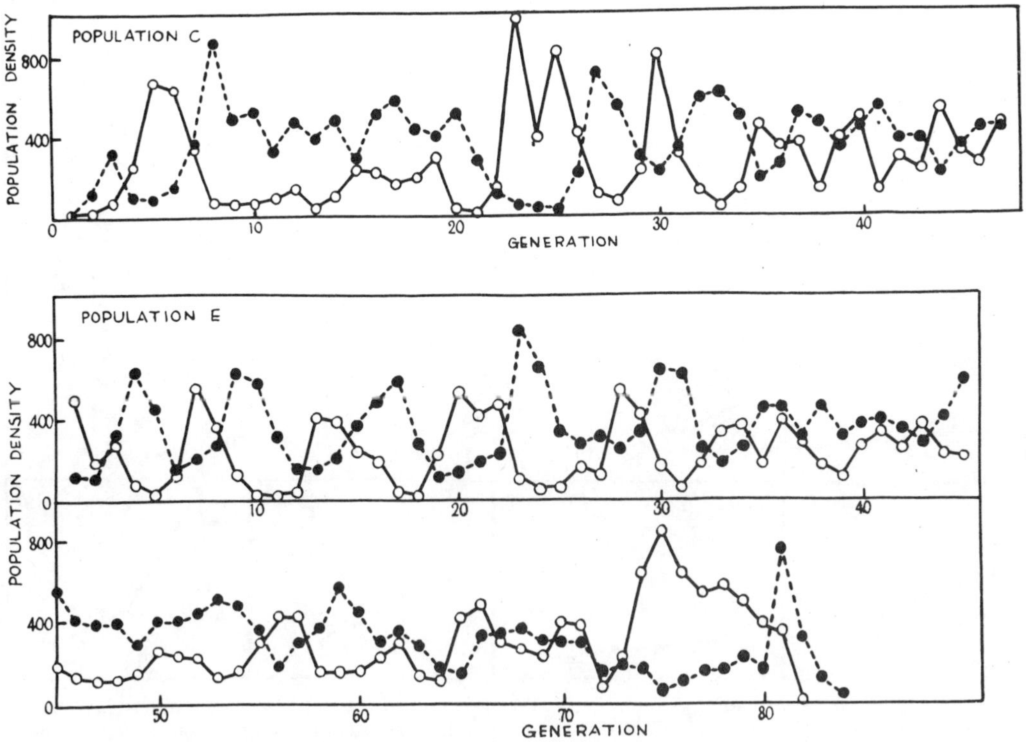


FIG. 2. Fluctuation of population density from generation to generation. Upper: population C, Lower: population E. Host population: ——o——, Parasite population: -----●-----

generations. There were three repetitions in population A, two in population C, and nearly two in population E.

Such cyclic repetitions of shorter-term cycles must be given special attention, for this may prove to be one of the most important points in population dynamics. In the present state of our knowledge, however, we cannot give any plausible explanation of the stimulus that forced the populations into a new phase of damped oscillations after they had previously attained steady states. In each experimental population the relative proportions of host and parasite populations changed with the onset of each successive phase of damped fluctuations. The change can be seen not only in the level at which the steady density of host and parasite is maintained, but in the relative form of the cyclic fluctuation, as clearly shown in Figures 3 and 4. In other words, the average percentage of parasitism decreases, as does the effect of the host's density on the parasite's power to keep the host population in check.

On the basis of mathematical analysis described in the previous paper (Utida 1955), it is considered that these changes in the values and form of the host-parasite interaction should reflect changes in the biotic potential and in the environmental resistance of host or parasite. In particular, the change in the biotic potential of the parasite may be the principal factor responsible. When we plot on a graph the relation between the density of the adult population and its rate of reproduction, as given by the ratio between the sizes of successive generations, the values obtained for the host species fall on a single curve; this is the curve for the equation describing the effect of population density on rate of reproduction:

$$R = \frac{N_2}{N_1}$$

$$= 2\left(\frac{1}{b + c N_1} - \sigma\right) \qquad (1)$$

where $N_1$ and $N_2$ are the population densities in two consecutive generations, $R$ is the rate of reproduction, and $b$, $c$, and $\sigma$ are specific constants (Fujita and Utida 1953). On the other hand, the values for the parasite's potential are scattered and cannot be described by a simple equation.

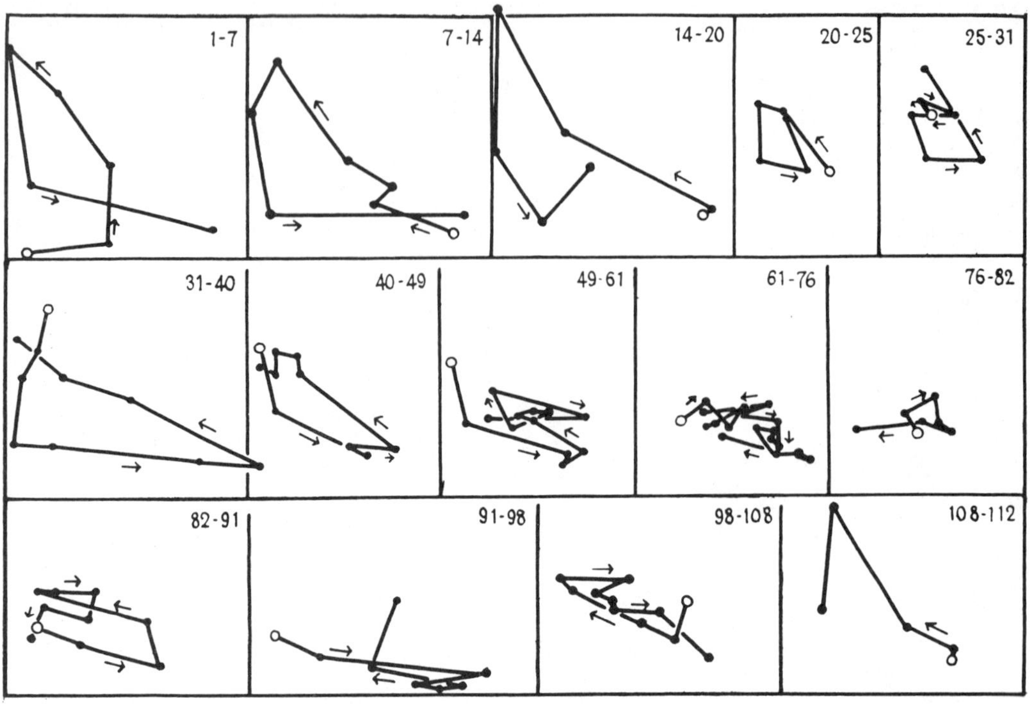


FIG. 3. Graphic representation of cyclic interaction between host and parasite populations in population A. Density of host population on abscissa, density of parasite population on ordinate. Each figure shows a complete cycle of fluctuation which starts and finishes at the generations whose numbers are given in the corner of each figure.

The relation is not improved when the parasite's adult density, $N_1$, is made more constant by taking the value $P_n / h_{n+1}$ (which reduces the parasite density to a constant proportion of hosts available for parasitisation). However, the plots obtained accord well with the above equation for the comparatively short span of the first 25 generations (Utida 1955).

This contradiction between short-term and long-term results can be resolved by breaking the graphs into several parts, each corresponding to a separate phase of damped oscillation. The separate graphs then yield straight lines, each of which is precisely described by the above equation, but with different numerical values for the constants, as shown in Table I. In particular, the constant $b$, the characteristic coefficient of the biotic potential, varies widely, while the constant $c$, an index of the environmental resistance, remains relatively stable. This result supports the suggestion that the changes in value and form of the cyclic interaction are mainly due to changes in the parasite's biotic potential.

TABLE I. Values of constants $b$ and $c$, representing parameters of the parasite population. Successive phases of damped oscillation are treated separately in calculating the constants from equation (1)

| Population | Range of generations | Numerical Values of Constants | |
|---|---|---|---|
| | | $b$ | $c$ |
| A | 1 - 31 | 0.35 | 0.162 |
| | 32 - 82 | 0.40 | 0.162 |
| | 83 - 112 | 0.45 | 0.153 |
| C | 1 - 19 | 0.30 | 0.213 |
| | 20 - 47 | 0.40 | 0.141 |
| E | 1 - 38 | 0.38 | 0.231 |
| | 39 - 62 | .... | ..... |
| | 63 - 82 | 0.30 | 0.231 |

The long-term trend of population change mentioned above can also be clearly represented by smoothing the data statistically; Figure 5 is the result of smoothing by moving averages of seven consecutive points. It reveals a rising trend in the host population, a declining trend in the parasite

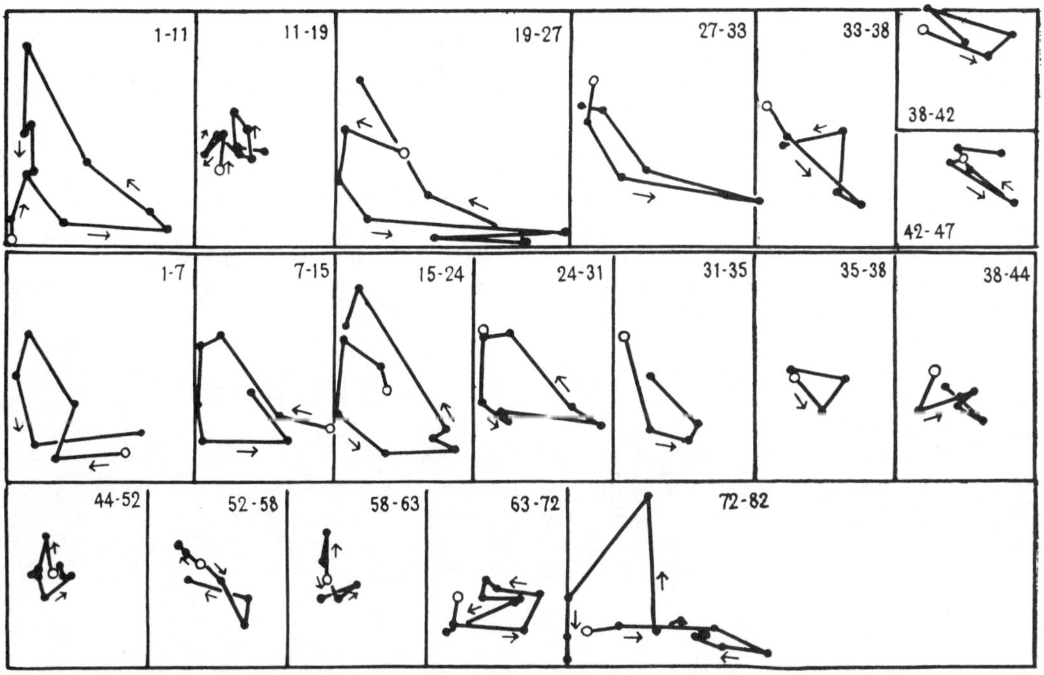


FIG. 4. Graphic representation of cyclic interaction between host and parasite populations. Upper figure: Population C. Lower figure: Population E.

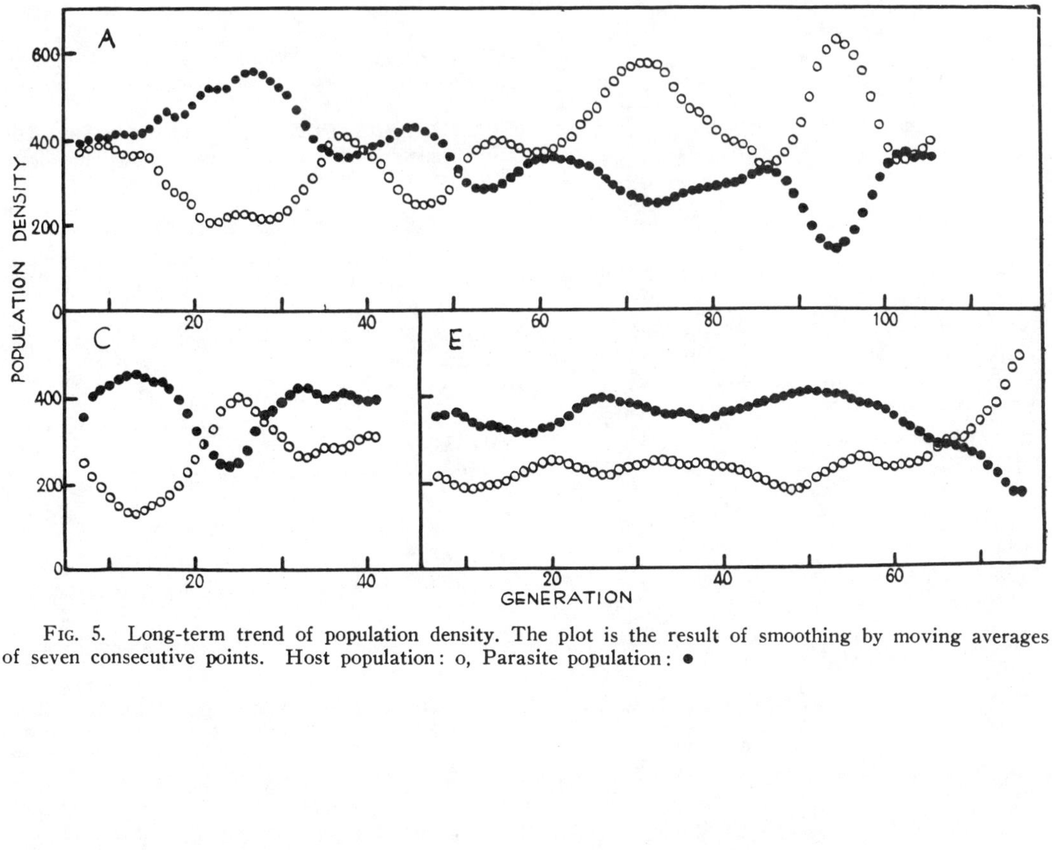


FIG. 5. Long-term trend of population density. The plot is the result of smoothing by moving averages of seven consecutive points. Host population: o, Parasite population: •

population, and also several slow undulations. The rising or declining trend is most conspicuous in population A, which was followed for the longest time, and is not at all clear during the short history of population C.

A similar tendency to long-term population fluctuation can be seen in the interaction between the azuki bean weevil and another larval parasite, *Neocatolaccus mamezophagus* (Utida 1950), when the data are similarly smoothed. Figure 6 is the result of this smoothing. In one of the two experimental populations the declining trend of the parasite population is particularly distinct. This additional example reinforces the soundness of the results reported here.

The problem remains to study further the stimulus that induces such changes in host and parasite populations. Natural selection provides the readiest explanation. That is, it may be presumed that the game of hide-and-seek between host and parasite intensifies the host's ability to conceal itself, or otherwise to avoid attack by the parasite. This selection pressure will be operative at all times, but should be most potent when the host population is in the trough of a cycle while the parasite population is at a peak. Increased success in concealment on the part of the host should

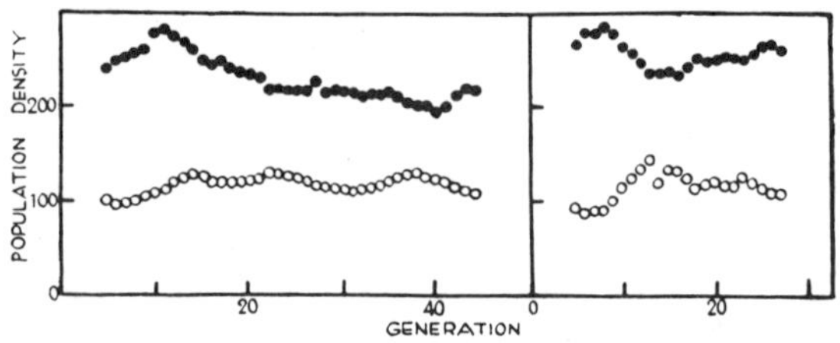


FIG. 6. Long-term trend of population density observed in the interaction between the azuki bean weevil and another larval parasite, *Neocatolaccus mamezophagus.* Host population: o, Parasite population: •

then cause the decline of the parasite's population. On the other hand, the change in the biotic potential of the parasite may be the primary cause of the parasite's own decrease, if some ecological advantage is linked to a lowered reproductive ability. In the present state of our knowledge we cannot judge which of the two explanations is more plausible.

## SUMMARY

Under experimentally controlled conditions, populations of the azuki bean weevil, *Callosobruchus chinensis,* were subjected to attack by the wasp *Heterospilus prosopidis,* which parasitizes the weevil's larvae. The interacting populations were maintained in the laboratory for as many as 112 successive generations.

Violent reciprocal fluctuations of population density were observed, the parasite's fluctuations being somewhat displaced in phase in relation to those of the host. The oscillations were gradually damped in amplitude, but episodes of renewed oscillation followed by damping were found to occur repeatedly.

The value of Gause's singular point, representing the steady proportion of parasite to host population, was different for each successive series of damped oscillations: that is, a long-term trend was superimposed on the cycle, such that the host population gradually increased and the parasite population gradually declined. This trend is inferred to be the result of a fall in the parasite's biotic potential.

## ACKNOWLEDGMENTS

The writer is indebted to the Scientific Research Expenditure of the Department of Education for financial assistance in carrying out this work. He also wishes to express his appreciation for the careful services of Miss H. Kakemi, who looked after much of the routine work, and to Dr. E. S. Deevey for assistance with the English style of the manuscript.

## REFERENCES

**DeBach, P., and H. S. Smith.** 1941. Are population oscillations inherent in the host-parasite relation? Ecology, **22**: 363-369.

**Fujita, H., and S. Utida.** 1953. The effect of population density on the growth of an animal population. Ecology, **34**: 488-498.

**Gause, G. F.** 1934. The struggle for existence. Baltimore: Williams & Wilkins. 1-163.

———. 1935. Vérifications expérimentales de la théorie mathematique de la lutte pour la vie. Paris: Hermann. 1-62.

**Nicholson, A. J.** 1933. The balance of animal populations. Jour. Anim. Ecol., **2**: 132-178.

**Utida, S.** 1950. On the equilibrium state of the interacting population of an insect and its parasite. Ecology, **31**: 165-175.

———. 1955. Population fluctuation in the system of host-parasite interaction. Mem. Coll. Agr., Kyoto Univ., **71**: 1-34.

**Varley, G. C.** 1953. Ecological aspects of population regulation. Trans. 9th Intern. Congr. Entom., **2**: 210-214.

# Larval Competition in *Drosophila melanogaster* and *D. simulans*

RICHARD S. MILLER
*Department of Biology, University of Saskatchewan, Saskatoon, Canada*

## INTRODUCTION

The Gause principle of interspecies competition states that if 2 species are forced to coexist in an undiversified environment, one inevitably becomes extinct, and if 2 species do coexist they must occupy different ecological niches. An extensive literature has been written on the development of this principle and the confusion that has arisen from the problem of competitive exclusion and identical niches (cf. Elton and Miller 1954; Andrewartha and Birch 1954; Hutchinson 1957). Gilbert *et al.* (1952) and Hardin (1960) have discussed the circularity that is usually involved in attempts to prove or disprove the Gause principle. If one species eliminates another we say the principle is "proved," but if they coexist we conclude that they differ ecologically and occupy different niches. Any meaningful concept of species or ecological niches recognizes that no 2 species are genetically or ecologically identical, so that the "axiom of inequality" provides automatic confirmation of the principle as it is stated. To this extent the principle can neither be proved nor disproved (Hardin 1960). The Gause principle is useful only when it can be applied to a conceptual model which can be tested with empirical data. When we find that a prediction is not verifiable we modify the model, but there is no procedural rule to tell us which element in the model we should abandon or change.

Elton and Miller (1954) attempted to resolve certain of the descriptive problems in the niche concept by considering "arenas" of potential competition between species. Hutchinson (1957) has more recently used set theory to provide a more precise definition of an ecological niche and the conditions of interspecies competition. According to Hutchinson's formulation, if we consider the array of environmental variables relative to a species $S_1$, we can define an $n$-dimensional hypervolume which corresponds to the state of environment which would permit species $S_1$ to exist indefinitely. For any species $S_1$ this hypervolume $N_1$ is its "fundamental niche." If $\mathbf{N}_1$ and $\mathbf{N}_2$ are 2 fundamental niches, they may have points in common which are said to intersect. Thus, $\mathbf{N}_1 \cdot \mathbf{N}_2$ is the subset of points common to both and is referred to as the "intersection subset." An intersection subset, according to Hutchinson's terminology, corresponds to what Elton and Miller (1954) referred to as an "arena of competition" in the overlap of 2 niches.

Identity of 2 fundamental niches would imply that $\mathbf{N}_1 = \mathbf{N}_2$, or every point of $\mathbf{N}_1$ is a member of $\mathbf{N}_2$ and every point of $\mathbf{N}_2$ is a member of $\mathbf{N}_1$. For the reasons mentioned above with regard to the "axiom of inequality," this condition is, as Hutchinson (1957) states, "so unlikely that the case is of no empirical interest." We frequently find among closely related species, however, that the intersection subset of fundamental niches provides a wide range of overlap in ecological requirements. The data provided by Park (1954, 1955) for interactions between the flour beetles *Tribolium confusum* and *T. castaneum* are an extremely well-documented example. A detailed analysis of these data by Neyman, Park, and Scott (1958) has shown that the outcome of competition between these 2 species is predictable in certain environments and "indeterminate" in others. In other words, within the intersection subset for *confusum* and *castaneum* there are points at which the ecological requirements and increase potentials of the 2 species are so similar that the outcome of competition is a statistical probability rather than a predictable certainty.

The genus *Drosophila* is a particularly good source of material for the study of competition among closely related species. The genus contains more than 700 recognized species and is well represented in all zoogeographical regions of the world (Patterson and Stone 1952); investigations of the genetics and ecology of *Drosophila* have shown that remarkable similarity exists among various sibling species which coexist in nature, in apparent contradiction to the Gause principle. Merrell (1951) studied competition in cage populations of *Drosophila funebris* and *D. melanogaster* which coexisted in the laboratory for almost 2 years. Analysis showed that a fresh yeast medium was more suitable for *melanogaster* whereas *funebris* was able to maintain itself in older food, so that periodic renewal of the food and medium induced environmental fluctuations which alternately favored one species and then the other. Moore (1952) obtained somewhat similar results with cage populations of the more closely related species *Drosophila melanogaster* and *D. simulans*. He concluded that *melanogaster* is superior to *simulans* in competition at 25°C and *simulans*

Reproduced with permission from *Ecology*, 45:132–148, 1964. Published by the Ecological Society of America and the Duke University Press, Durham, North Carolina.

superior at 15°C, but the progress of competition and its outcome was partly determined by the age and quality of the food. When the food cups were changed at a rate of 2 a week and the temperature was 15°C, *melanogaster* was reduced to 24% of the population after 200 days in one cage and to 45% after 402 days in another cage, but was not eliminated. The data also showed that these results could be reversed, suggesting that coexistence could have been maintained indefinitely by selecting an appropriate temperature and rate of renewal of the food cups.

These and other studies of competition in *Drosophila* populations have shown that a remarkable amount of overlap may exist in the intersection subsets of the niches of sibling species and that only minor variations in environment may be required to permit continuing coexistence. This paper reports the first in a series of investigations of the nature of competition in populations of the sibling species *Drosophila melanogaster* and *D. simulans* in an environment which seems to satisfy the needs of both species with almost equal success. These experiments were designed to test intraspecies and interspecies competition in populations of larvae in which density was artificially manipulated to produce crowding effects.

## Materials and Methods

### *Experimental stocks*

Moore (1952) has reviewed the pertinent similarities and differences in *Drosophila melanogaster* and *D. simulans.* The 2 species coexist throughout most of the temperate and tropic regions of the world, although *simulans* is generally regarded as the more southern species because of its abundance in the tropics (Burla 1948; Dobzhansky and Pavan 1950). Patterson (1943) observed seasonal changes in natural populations of *melanogaster* and *simulans* in the vicinity of Austin, Texas, and found that both species are relatively rare during the winter months, but that *melanogaster* responds quickly to temperature increases in the spring and summer. The *simulans* population remains at a low level until the hot dry months of summer, when the *melanogaster* population declines and *simulans* becomes the more abundant species.

These observations confirm expectations from laboratory studies. Sturtevant (1920) found that both species can be kept in mixed stocks for as long as 5 months with ordinary laboratory techniques, but that *melanogaster* is the more abundant species until the food becomes old and dry, after which *simulans* is the more successful. Moore (1952) also noted that *simulans* is more successful than *melanogaster* on dry medium.

Tantawy and Mallah (1961) measured the effect of temperature on the development of several strains of these 2 species from different localities in Lebanon, Egypt, and Uganda and found that *melanogaster* has an optimum developmental temperature of 25°C and *simulans* 22°C, on the basis of per cent emergence at different temperatures. Complete sterility of eggs occurred at 31.5°C in *melanogaster* and at 30.5°C in *simulans.* Moore (1952) found that *melanogaster* completes its development and emerges slightly earlier than *simulans* at 25°C but *simulans* develops considerably faster than *melanogaster* at 15°C. Sturtevant (1929), on the other hand, concluded that *simulans* develops somewhat more rapidly than *melanogaster,* although temperature and density were not carefully controlled when this observation was made.

The similarity in the external morphology of these 2 species caused them to be considered as one species until Sturtevant (1919) described *simulans.* Females of the 2 species cannot be distinguished by their external characters, but the characteristic male genitalia of *simulans* allow easy identification of the males.

The morphological, behavioral, and ecological similarities in *melanogaster* and *simulans* reflect their genetic affinities (Sturtevant 1919, 1920, 1929). Both species have 4 large linkage groups; there are striking similarities in the salivary chromosome bands of the 2 species; and many of the mutations that have appeared in *simulans* are allelomorphs of those in *melanogaster* (Patau 1935; Kerkis 1936, 1937; Horton 1939).

As Moore (1952) has noted, the general conclusion that one draws from the published work on *simulans* and *melanogaster* is that the species are strikingly similar in external appearance, genetics, chromosome morphology, geographic distribution, and habitat preferences. Unselected laboratory strains of wild-type *melanogaster* and *simulans* were used in these investigations. These were descendants of the same strains used by Moore (1952) in his studies of competition and were derived from single fertilized females collected at Brandenton, Florida, in 1949.

### *Culture methods*

Stocks of each species were kept in population bottles and raised on molasses medium seeded with dry Fleischman's yeast (Spassky 1943). Oviposition bottles were prepared by fitting 40 mm watch glasses, filled with a medium of agar, clear Karo, and water and covered with a 10% film of yeast solution, to ¼-pint milk bottles. A 3 × 5 in. file card with a small piece of plasticene in the center was attached to the convex surface of the watch glass so that the watch glass could be removed with the card, flies placed in the bottle, and the bottle and watch glass inverted on the card and placed on a table.

Four or 5 pairs of flies were placed in each oviposition bottle and the females allowed to lay eggs for 2 or 3 days before larvae were collected. Fresh medium and yeast were provided daily. When the number of eggs was sufficient to allow mass collections, fresh medium and yeast were provided, and early first-instar larvae were collected from the surface of the medium within 4 hours after hatching. The eggs and larvae were clearly visible against a dark background without the use of dyes in the medium. The larvae were transferred at a density of 5 per vial to 8-dram vials containing 5 cc of Kalmus medium (Kalmus 1943) to which had been added a 1/20 cc drop of a fresh 15% suspension of dry Fleischman's yeast and distilled water.

The medium was prepared in 1,000 ml batches to insure uniform quality, and water was added to restore the original volume after boiling. Kalmus medium cannot be autoclaved, but the vials were sterilized and Moldex was added. No molds were observed in the cultures. The liquid medium was delivered into the vials with a present automatic syringe and allowed to gel. The yeast solution was added with a microliter syringe.

The larvae were picked from the surface of the oviposition medium by allowing them to adhere to the tip of a sterilized teasing needle. When the needle with the attached larvae was inserted gently into the fresh yeast solution in a vial, the larvae readily freed themselves and dispersed into the yeast and medium. The vials containing these larval cultures were plugged with cotton and allowed to develop at 25°C (70-80% R.H.).

Pairs of $F_1$ adult flies obtained from these cultures were used as parent stocks for the larvae to be used in competition experiments. The adult flies were placed in oviposition bottles as before and $F_2$ first-instar larvae collected. During these collections the smallest larvae of uniform size were selected from the 0- to 4-hr age groups. These larvae were transferred at predetermined experimental densities to 8-dram vials containing 5 cc of Kalmus medium and 1/20 cc of yeast solution, and the cultures were allowed to develop under the same conditions as before. The $F_2$ adults which provided the data on intraspecies and interspecies competition were collected every morning from the first day of emergence until emergence had ceased. The numbers of males and females in each vial were recorded at the end of the period of emergence for each population.

*Measurements of body size*

Samples of adult flies from each series of replicates were etherized and weighed to the nearest 0.02 mg on a Mettler analytical balance. Flies from uncrowded cultures were usually quite uniform in size and appearance, and samples of 40 to 50 of each sex from each series of replicates gave an adequate representation of body weight. The weights and numbers of flies from crowded cultures were more variable, and it was often necessary to weigh the entire population of each series. Dry weight was determined by placing the flies in a drying oven at 60°C for 48 hr and weighing lots of 10 or more flies to the nearest 0.02 mg.

Measurements of wing length were made by removing one wing from each fly to be measured, mounting a series of wings of flies from the same series of replicates on a slide with balsam, and measuring with a 10-mm micrometer disc divided into 100 units. Wing length was measured from the base of the 2d basal cell to the distal end of vein IV, between the first and 2d posterior cells, as recommended by Robertson and Reeve (1952). Micrometer units were converted to mm for calculation.

## INTRASPECIES COMPETITION

*Rate of development*

*Drosophila melanogaster* larvae raised at 25°C hatch between 20 and 22 hr after the eggs have been fertilized, but freshly deposited eggs may be at any level of development from the period before pronuclear conjugation to just before emergence of the larvae (Sonnenblick 1950). To obtain material of uniform age and development it is, therefore, more practical to start cultures at the early first-instar larva stage, rather than the egg stage of the life history. Larvae collected within 4 hr of hatching have completed 20 to 26 hr of postembryonic development, and by further restricting the collections to only the smallest of these larvae even greater uniformity can be obtained.

Tables I and II show the effects of initial larval density on the time required to complete development from the early first-instar larva stage to imago. The figures in each row show the percentage of the total adult population that emerged on successive days after the larvae were transferred. The values in the far right column show the mean number of days required for the development of each population at each density.

With increased crowding the time required for development is prolonged, and there is a tendency for emergence to be spread throughout a greater period. At low densities emergence began on day 7 and was usually completed by day 9 or 10, but at high densities emergence was delayed until day

TABLE I. Rate of development of *D. melanogaster* in relation to larval density

| Larval density | Per cent emergence/day | | | | | | | | | | Mean days |
|---|---|---|---|---|---|---|---|---|---|---|---|
| | 7 | 8 | 9 | 10 | 11 | 12 | 13 | 14 | 15 | 16 | |
| **Males** | | | | | | | | | | | |
| 5 | 39.1 | 59.8 | 1.1 | | | | | | | | 7.6 |
| 10 | 27.8 | 72.2 | | | | | | | | | 7.7 |
| 20 | 36.6 | 58.4 | 4.7 | | | | | | | | 7.7 |
| 40 | 34.6 | 57.4 | 7.8 | 0.3 | | | | | | | 7.8 |
| 60 | 30.4 | 56.7 | 10.6 | 2.3 | | | | | | | 7.8 |
| 80 | 1.2 | 65.8 | 29.8 | 3.2 | | | | | | | 8.4 |
| 100 | 1.8 | 42.9 | 36.9 | 14.5 | 3.0 | 0.9 | | | | | 8.8 |
| 120 | | 53.4 | 35.3 | 6.4 | 4.3 | 0.5 | | | | | 8.6 |
| 160 | | 12.5 | 41.4 | 22.1 | 12.5 | 3.8 | 5.8 | 1.9 | | | 9.8 |
| 200 | | | 15.2 | 35.7 | 25.9 | 13.4 | 8.0 | 1.8 | | | 10.7 |
| 300 | | | 2.4 | 4.8 | 9.5 | 23.8 | 23.8 | 16.7 | 16.7 | 2.4 | 12.9 |
| **Females** | | | | | | | | | | | |
| 5 | 68.8 | 30.3 | 1.1 | | | | | | | | 7.3 |
| 10 | 66.7 | 33.3 | | | | | | | | | 7.3 |
| 20 | 73.2 | 25.5 | 1.3 | | | | | | | | 7.3 |
| 40 | 42.0 | 51.8 | 6.0 | 0.3 | | | | | | | 7.6 |
| 60 | 36.3 | 53.3 | 8.9 | 1.5 | | | | | | | 7.8 |
| 80 | 15.0 | 68.2 | 14.0 | 2.8 | | | | | | | 8.0 |
| 100 | 4.8 | 45.9 | 34.0 | 9.0 | 5.2 | 1.1 | | | | | 8.8 |
| 120 | 1.0 | 58.6 | 28.0 | 8.1 | 3.2 | 1.1 | | | | | 8.6 |
| 160 | | 13.1 | 32.1 | 35.7 | 7.1 | 2.4 | 7.1 | 2.4 | | | 9.8 |
| 200 | | | 22.4 | 37.3 | 13.4 | 16.4 | 6.0 | 4.5 | | | 10.6 |
| 300 | | | 4.3 | 4.2 | 8.3 | 41.7 | 12.5 | 16.7 | 8.3 | 4.2 | 12.6 |

TABLE II. Rate of development of *D. simulans* in relation to larval density

| Larval density | Per cent emergence/day | | | | | | | | | | Mean days |
|---|---|---|---|---|---|---|---|---|---|---|---|
| | 7 | 8 | 9 | 10 | 11 | 12 | 13 | 14 | 15 | 16 | |
| **Males** | | | | | | | | | | | |
| 5 | 4.2 | 89.6 | 6.2 | | | | | | | | 8.0 |
| 10 | | 88.1 | 11.9 | | | | | | | | 8.1 |
| 20 | 3.4 | 78.0 | 18.0 | 0.6 | | | | | | | 8.2 |
| 40 | 27.9 | 70.5 | 1.6 | | | | | | | | 7.7 |
| 60 | 13.5 | 60.5 | 21.2 | 4.8 | | | | | | | 8.2 |
| 80 | 5.2 | 49.2 | 27.6 | 13.4 | 4.5 | | | | | | 8.6 |
| 100 | 1.6 | 42.4 | 47.5 | 8.5 | | | | | | | 8.6 |
| 120 | | 8.0 | 17.5 | 45.3 | 15.3 | 8.8 | 3.6 | 1.5 | | | 10.2 |
| 160 | | | 16 7 | 42.9 | 21.4 | 11.9 | 4.8 | 2.4 | | | 10.5 |
| 200 | | | 11.1 | 34.7 | 41.7 | 6.9 | 5.6 | | | | 10.6 |
| 300 | | | | 50.0 | 12.5 | | 37.5 | | | | 11.2 |
| **Females** | | | | | | | | | | | |
| 5 | 12.0 | 87.4 | 0.6 | | | | | | | | 7.9 |
| 10 | | 95.8 | 4.2 | | | | | | | | 8.0 |
| 20 | 23.0 | 71.5 | 0.5 | | | | | | | | 7.4 |
| 40 | 59.5 | 38.7 | 1.8 | | | | | | | | 7.4 |
| 60 | 28.1 | 55.3 | 14.0 | 0.8 | 1.8 | | | | | | 7.9 |
| 80 | 6.7 | 61.1 | 20.1 | 8.7 | 3.4 | | | | | | 8.4 |
| 100 | | | 48.0 | 42.7 | 9.3 | | | | | | 9.6 |
| 120 | | 17.5 | 51.0 | 18.7 | 9.5 | 1.5 | 2.2 | | | | 9.3 |
| 160 | | | 15.9 | 38.6 | 22.7 | 15.9 | 4.6 | 2.3 | | | 10.6 |
| 200 | | | | 46.2 | 36.5 | 9.6 | 7.7 | | | | 10.8 |
| 300 | | | 20.0 | 60.0 | | 20.0 | | | | | 10.2 |

8 or 9 and required as many as 6 to 8 days to complete. Male flies usually required more time than females to complete development, but there were exceptions in some of the *simulans* populations.

The duration of the pupal period in *Drosophila melanogaster* is 4.3 days at 25°C and is relatively independent of nutrition or differences in body size (Sang 1956). Changes in the time required for development at any given temperature are consequently related to factors affecting larval growth and nutrition. Semistarvation or minor nutritional deficiencies will prolong larval life, and complete starvation at the beginning of the last instar will cause the larvae to stop growing and pupate prematurely (Bodenstein 1950). The general trend of a slower rate of development with increased density is undoubtedly due to various degrees of nutritional deficiency and semistarvation imposed by competition for food and space. The increased variability in the time required for each population to complete development and changes in pupal viability, which will be discussed in later sections,

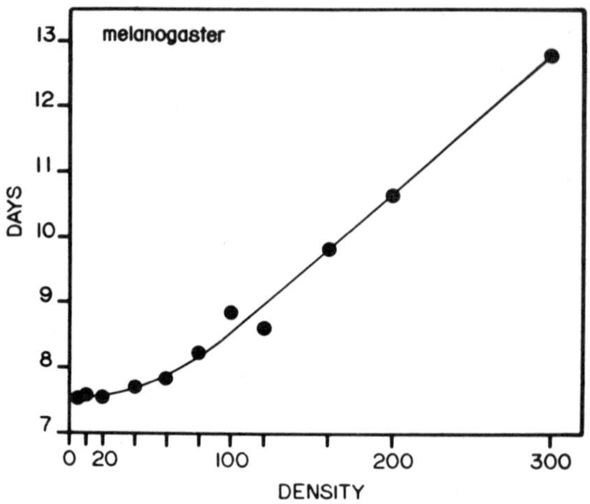


FIG. 1. Average number of days required for development of *melanogaster* from early first-instar larva to adult.

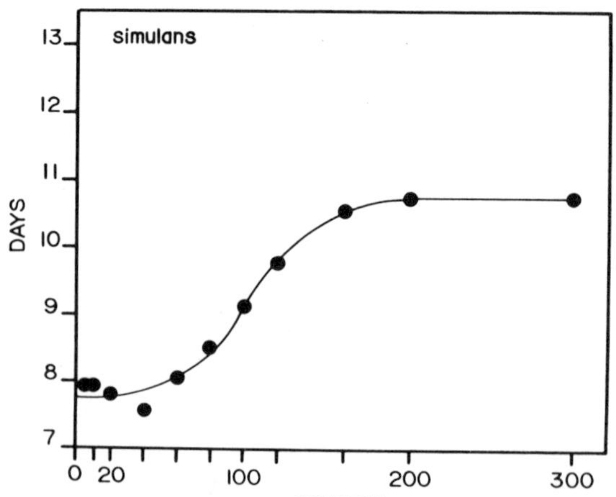


FIG. 2. Average number of days required for development of *simulans* from early first-instar larva to adult.

may, however, be due in part to premature pupation among some of the larvae.

Crowding affects the rates of development of *melanogaster* and *simulans* in different ways. This is illustrated in Figures 1 and 2, which show the average number of days required for the development of each species at each density. Both species showed gradually decreasing rates of development up to a density of 160 per vial, but further increases in density had relatively little effect on the rate of development of *simulans* although the rate of development of *melanogaster* continued to be prolonged at all densities tested.

The conflicting statements by Sturtevant (1929) and Moore (1952) with regard to developmental rates in these 2 species can, perhaps, be explained by the fact that larval density was not controlled during their observations or that different strains were used. At low densities in these experiments *melanogaster* had a shorter larval period than *simulans;* at intermediate densities their rates were similar; and at high densities *simulans* had the faster average rate of development.

*Adult body size*

The relationship between mean body weight and initial larval density is shown in Table III and Figure 3. Increases in larval density produced marked decreases in body weight, especially among female flies. Over the range of densities tested, the body weights of male *melanogaster* and *simulans* were reduced by 53 and 58% from their observed maxima of 0.84 and 0.86 mg, respectively. Female body weights were reduced by 64 and 66% from maxima of 1.29 and 1.27 mg for *melanogaster* and *simulans,* respectively. In the *melanogaster* populations the weights of female flies were reduced rapidly until, at a density of 120 per vial, their weights were only slightly greater than those

TABLE III. Adult body weights of newly emerged *melanogaster* and *simulans* in relation to density

| Larval density | Body weight (mg.) | | | |
|---|---|---|---|---|
| | *melanogaster* | | *simulans* | |
| | ♂♂ | ♀♀ | ♂♂ | ♀♀ |
| 5 | 0.83 | 1.29 | 0.86 | 1.27 |
| 10 | 0.84 | 1.27 | 0.83 | 1.22 |
| 20 | 0.81 | 1.25 | 0.81 | 1.21 |
| 40 | 0.74 | 1.16 | 0.70 | 0.98 |
| 60 | 0.62 | 0.87 | 0.67 | 0.99 |
| 80 | 0.61 | 0.85 | 0.55 | 0.76 |
| 100 | 0.58 | 0.82 | 0.48 | 0.75 |
| 120 | 0.57 | 0.58 | 0.47 | 0.65 |
| 160 | 0.46 | 0.54 | 0.39 | 0.47 |
| 200 | 0.39 | 0.47 | 0.38 | 0.46 |
| 300 | 0.44 | 0.55 | 0.36 | 0.43 |

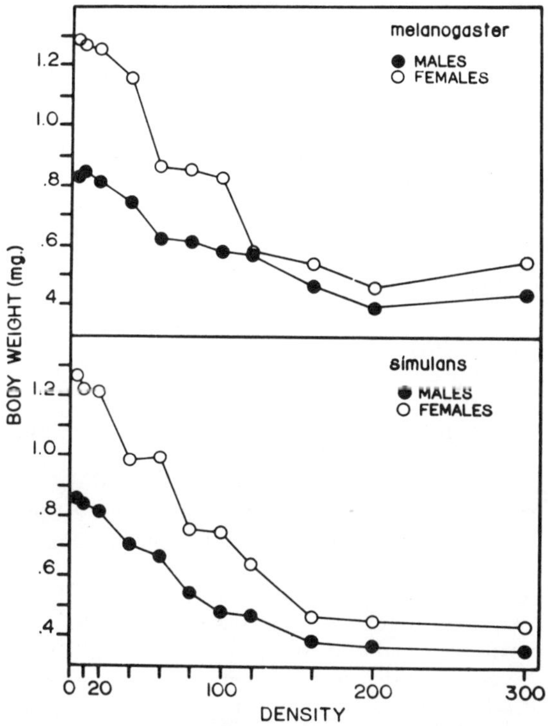


FIG. 3. Adult body weights of *melanogaster* and *simulans* in relation to initial larval density.

of male flies. Subsequent increases in density produced only minor changes in body weight in both sexes. A similar relationship existed in the *simulans* populations with maximum decreases in body weight occurring between 5 and 160 per vial.

The data for *melanogaster* show an increase in the mean body weights of both sexes with an increase in density from 200 to 300 per vial. This "rebound effect" has been observed previously in *Drosophila melanogaster* populations by Sang (1949a) and Chiang and Hodson (1950), in *Drosophila miranda* by Sokoloff (1955), and in *Musca domestica* and *phormia groenlandica* by Vladimorova and Smirnov (1938). Vladimorova and Smirnov (1938) attribute the increase in weight at high densities to scavenging by live larvae on dead larvae, and feel that the rebound effect is not attributable to density. Sang (1949a) concludes that it is due to quantitative or qualitative changes in the yeast population, and Chiang and Hodson (1950) suggest that it may be associated with extremely high density or extremely low food supply.

Contradictions in these hypotheses are probably more apparent than real. *Musca* and *Phormia* are flesh-eating flies which can undoubtedly use dead larvae as a source of food, and the high mortality that accompanies high densities would provide greater resources for necrophagous larvae.

To this extent the effect noted by Vladimorova and Smirnov (1938) could be attributed to density. Sokoloff (1955) observed *Drosophila* larvae feeding on dead adults but necrophagy of larvae upon larvae did not occur in his experiments. He suggests that necrophagy may have survival value in *Drosophila* by allowing larvae to survive until they find a new supply of yeast, but doubts that necrophagy will carry *Drosophila* larvae through to metamorphosis.

Tables I and II show that the rate of development within a population is more variable at high densities than at low densities, and it was noted earlier that the length of the larval period may be increased by minor nutritional deficiences or semi-starvation or decreased by complete starvation during the last larval instar. It will also be shown in later sections that the effect of larval density on the number of adults produced is reflected first in reduced pupal viability and later in a higher rate of larval mortality. At moderately high densities a relatively large proportion of the larval population may complete larval development and pupate, but many of the pupae will not be viable; at very high densities a much smaller proportion of larvae are able to pupate, even prematurely. These facts suggest that mortality occurring early in the larval population at high densities may eliminate a relatively large proportion of the competing population at a stage early enough to allow the survivors to obtain sufficient food and nutrients to reach normal size by the time they pupate. This possibility would be enhanced by the ability of *melanogaster* larvae to extend their period of development under unfavorable circumstances, thereby allowing the development of greater variability in size within the population.

A 2d aspect of this problem is that the number of flies involved in the "rebound effect" is usually quite small, owing to the high mortality that occurs at densities where the rebound effect is observed. It was noted earlier that flies raised at low densities are usually quite uniform in size and appearance but that one effect of crowding is to increase the variability in body weight, possibly through the mechanism outlined above. The data obtained by Sokoloff (1955) for *D. miranda* show an increase in the coefficient of variability for body weight with increased density and correlated reductions in mean body weight. At moderately high densities the range of body sizes in the population includes enough small flies to reduce the mean body weight of the population, even though a few larger flies are produced. At very high densities, where mortality rates are high and small flies are disproportionately eliminated, the few large flies that have produced may constitute a greater proportion of the total population, and the mean body weight for the surviving adults may therefore be increased.

Robertson (1960) has shown that *Drosophila melanogaster* have the ability to regulate body size in response to minor deficiencies in the diet by extending the larval period. To test for this mechanism in the competitive relationships in these experiments, measurements of body size and larval period were made in a series of *melanogaster* populations raised at densities of 10, 20, 40, 60, and 80 per vial. The index of body size used by Robertson is 3 times the natural logarithm of thorax length in hundredths of a mm, for unpublished tests have shown that a unit of change on this scale corresponds to a unit of change in the log weight of newly emerged flies. For tests in this study, wing length rather than thorax length was used, because of the convenience of preserving wings for later measurement. Robertson and Reeve (1952) have shown that there is a strong correlation between thorax length and wing length in *D. melanogaster*. The larval period was calculated in log days after subtracting the pupal period of 4.3 days from the observed development time from first-instar larva to adult.

Figure 4 shows the results of these tests expressed as the deviations in log units from the larval periods and body sizes of male and female *melanogaster* raised at a density of 10 per vial. It is immediately apparent that at these densities there was no tendency for regulation of body size in the direction shown by Robertson (1960). The per cent deviation in body size was in fact greater than that for larval period.

Evidence for the regulatory mechanism de-

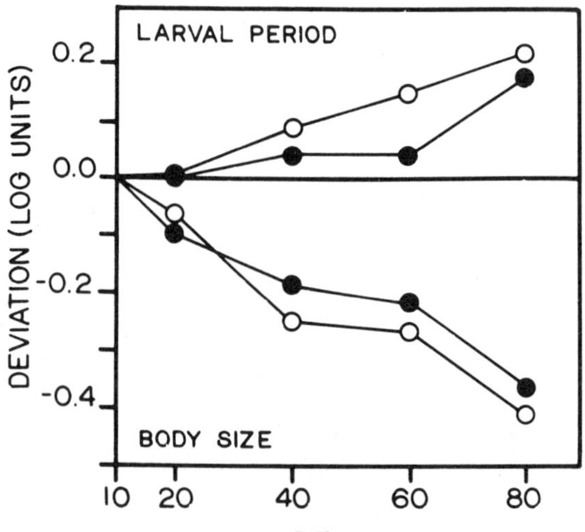


FIG. 4. Relationship between larval period and adult body size in *melanogaster*.

scribed by Robertson (1960) was obtained by comparing body sizes and larval periods of flies grown on live yeast medium with data from flies grown on synthetic media lacking or deficient in fructose, RNA, or protein. Reductions of 10, 20, and 30% in body size were accompanied by corresponding increase of 30, 50, and 70% in the length of the larval period in unselected strains of *melanogaster*. Robertson also noted that "when larvae are grown on increasingly poorer diets there is a distinct tendency for a given percentage reduction in final body size to be associated with an equivalent proportional lengthening of the larval period" (p. 300). If Robertson's observations are applied to the relationships shown in Figure 4 where the per cent change in body size with increasing density was invariably greater than the corresponding change in larval period, a severe nutritional deficiency is indicated. In a general way, but without comparable analysis, a similar relationship is evident in the data presented by Sokoloff (1955) for *persimilis, pseudoobscura,* and *miranda* and there is, in fact, no evidence that regulation of body size occurs in competition experiments.

TABLE IV. Pupation and adult emergence of *melanogaster* and *simulans* in relation to density

| Larval density | N | Pupation | | Adult emergence | |
|---|---|---|---|---|---|
| | | Pupae/vial ± S.E. | Per cent | Adults/vial ± S.E. | Per cent |
| | | ***melanogaster*** | | | |
| 5 | 630 | 4.4±0.07 | 88.6 | 4.1±0.08 | 82.9 |
| 10 | 570 | 8.2±0.20 | 82.1 | 7.8±0.21 | 77.9 |
| 20 | 860 | 17.9±0.28 | 89.5 | 17.0±0.29 | 84.9 |
| 40 | 920 | 36.6±0.48 | 91.6 | 33.6±0.58 | 84.0 |
| 60 | 960 | 52.3±0.66 | 87.2 | 45.7±2.32 | 76.1 |
| 80 | 1360 | 70.6±1.26 | 88.2 | 58.8±2.28 | 72.8 |
| 100 | 900 | 78.9±1.37 | 78.9 | 66.6±1.68 | 66.6 |
| 120 | 600 | 88.8±6.06 | 74.0 | 74.6±6.46 | 62.2 |
| 160 | 480 | 95.7±5.18 | 59.8 | 62.7±6.68 | 39.2 |
| 200 | 600 | 104.0±1.25 | 52.0 | 49.7±3.18 | 24.8 |
| 300 | 600 | 101.0±7.02 | 33.7 | 33.0±4.00 | 10.1 |
| | | *simulans* | | | |
| 5 | 440 | 4.5±0.08 | 90.0 | 4.2±.010 | 83.9 |
| 10 | 320 | 8.7±0.21 | 86.9 | 7.9±0.28 | 78.8 |
| 20 | 480 | 18.2±0.25 | 90.8 | 15.9±0.47 | 79.4 |
| 40 | 360 | 38.4±0.50 | 96.1 | 33.6±0.79 | 83.9 |
| 60 | 300 | 54.0±1.82 | 90.0 | 44.0±3.39 | 73.3 |
| 80 | 400 | 74.8±0.80 | 93.5 | 54.4±2.16 | 68.0 |
| 100 | 500 | 91.4±1.60 | 91.4 | 62.2±2.33 | 62.2 |
| 120 | 480 | 110.0±1.08 | 91.7 | 69.0±3.89 | 57.7 |
| 160 | 480 | 118.7±3.18 | 74.2 | 29.0±2.52 | 18.1 |
| 200 | 1000 | 115.6±2.95 | 57.8 | 14.8±2.94 | 12.4 |
| 300 | 900 | 86.0±1.25 | 28.7 | 4.3±0.88 | 1.4 |

Bodenheimer (1938), Chiang and Hodson (1950), and Sang (1949a, b, 1950) all have shown that the ecological events affecting larval nutrition during competition for food and space in a limited environment are often extremely complex; but the fact that a mechanism for regulation of body size does exist but is not evident during competition suggests that it is only possible under special circumstances and may be ecologically unimportant, or that the food and medium used in these experiments were deficient even at low larval densities.

*Larva-adult survival*

The effect of density on larval survival and pupal viability is shown in Table IV. Per cent pupation refers to the number of pupae formed in

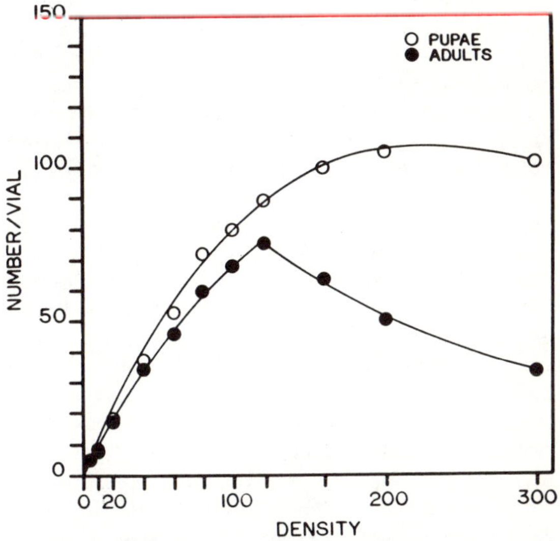


FIG. 5. Mean numbers of pupae and adult *simulans* in relation to initial larval density.

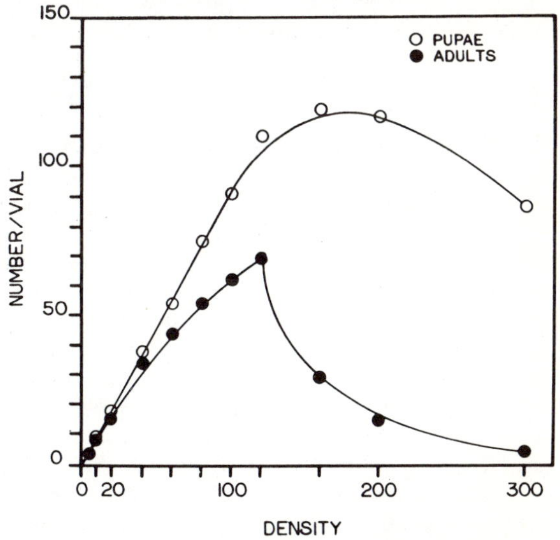


FIG. 6. Mean numbers of pupae and adult *melanogaster* in relation to initial larval density.

relation to initial number of larvae present and represents larval survival; per cent emergence of adults is based on the total number of adults produced from the initial numbers of larvae in the culture and is a function of larval survival and pupal viability.

Sang (1949a) refers to "minimal mortality" as the number of larvae and pupae which fail to survive even at low densities. Although mortality rates fluctuated somewhat between treatments, roughly 90% of the initial numbers of larvae reared at densities of 5 to 40 per vial were able to complete development and form pupae. In the *melanogaster* populations approximately 5% of these pupae failed to survive, and in the *simulans* populations about 9% were nonviable. The highest percentage emergence of adults observed for *melanogaster* was 84.9%, so that a normal mortality rate of about 15% from early first-instar larva to adult can be expected under optimum conditions with these techniques.

Figures 5 and 6 show a gradual trend of disproportionate decreases in pupal viability in relation to larval mortality with increasing density. When the initial larval density was increased from 5 to 120 per vial, there was a decrease of nearly 21% in total adult emergence though larval mortality increased by only 15% in the *melanogaster* populations. In the *simulans* populations there was a 26% decrease in adult emergence with no significant change in per cent pupation over the same range of densities. This relationship was even more noticeable when the larval density was increased from 120 to 160, 200, and 300 per vial. At these densities there was a significant decrease in per cent pupation but the effects of severe crowding were especially pronounced in relation to the viability of the pupae that were formed. This effect has also been noted by Sang (1949a), Chiang and Hodson (1950), and Sokoloff (1955) in similar studies of competition in *Drosophila*.

It was noted earlier that increases in density produced increases in the length of the larval period and decreases in the body weights of both species. At a density of 120 larvae per vial the males and females of both *melanogaster* and *simulans* were near their minimum body weights, or what Sang (1949a) refers to as their "smallest possible survival sizes," in relation to density. The most pronounced decreases in pupal viability and adult numbers occurred when density was increased to 160, 200, and 300 larvae per vial, which was beyond the point at which significant adjustments in body weight could be made. Between densities of about 40 and 120 per vial there was, however, a trend toward greater variability in the developmental rates and body sizes of individuals within a population, and it is likely that this increased variability was a function of individual nutritional deficiencies which led to the gradual decrease in pupal viability observed between densities of 5 and 120 per vial.

This relationship between larval period, body size, larval mortality, and pupal viability would appear, therefore, to reflect the effectiveness of changes in larval period and ultimate body size as compensatory mechanisms in competitive interactions. At low or moderate densities, increases in the length of the larval period and decreases in body size allow *Drosophila* to compensate for the adverse effects of competition without significant losses in adult numbers. At higher densities, as the limits of these compensatory mechanisms are approached, there is a gradual decrease in pupal viability, although larval mortality may remain at about the same rate or be only slightly affected. There is, however, a maximum critical density corresponding to the limits of effective compensation. Further increases in density may produce additional increases in the numbers of pupae formed, as shown in Figures 5 and 6, but the viability of these pupae is so drastically reduced that the total numbers of adults produced is less than at lower densities.

There were 2 notable differences in the responses of each species to these conditions. The per cent pupation in the *simulans* populations remained constant at about 90% in spite of a 26% decrease in pupal viability when the density was increased from 5 to 120 per vial, but a corresponding decrease of 21% in the adult emergence of *melanogaster* was accompanied by a 15% decrease in larval survival. This relationship was also evident at higher densities where *simulans* produced greater numbers of pupae than *melanogaster* but had a lower total survival rate. There was, in other words, no particular advantage in the fact that larval mortality was relatively lower in the *simulans* populations.

A second point of interest is that the *melanogaster* populations were able to extend their larval periods in response to each increase in density, even after further reductions in body size were no longer possible. The *simulans* populations lost this ability at a density of about 160 per vial, when an absolute minimum body size was reached. The greater numbers of adults produced by *melanogaster* at high densities may have been due to the additional compensation that was possible through further extensions of the larval period, allowing more time for individuals to reach a body size at which they could survive.

TABLE V. Sex ratios of adult *melanogaster* and *simulans* in single-species cultures

| Density | *melanogaster* ♀♀/100 ♂♂ | *melanogaster* $X^2$ | *simulans* ♀♀/100♂♂ | *simulans* $X^2$ |
|---|---|---|---|---|
| 5 | 100 | .000 | 90 | .978 |
| 10 | 90 | 1.297 | 88 | 1.016 |
| 20 | 102 | .088 | 115 | 1.913 |
| 40 | 100 | .000 | 134 | 6.411[a] |
| 60 | 85 | 4.762[a] | 110 | 1.818 |
| 80 | 100 | .000 | 111 | .795 |
| 100 | 81 | 6.626[a] | 127 | 4.402[a] |
| 120 | 100 | .000 | 100 | .000 |
| 160 | 81 | 2.128 | 102 | .001 |
| 200 | 54 | 13.590[a] | 72 | 3.226 |
| 300 | 57 | 4.908[a] | 62 | 2.769 |

[a] Significant at 5% level.

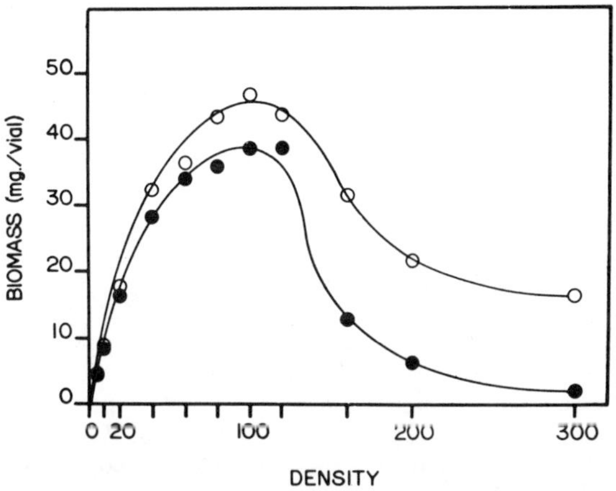


FIG. 7. Biomass of adult *melanogaster* (open circles) and *simulans* (dots) in relation to initial larval density.

## *Sex ratio*

The relationships between density and the observed sex ratios, expressed as numbers of females per 100 males, are shown in Table V. Values for the chi-square of the difference between the observed and expected ratios (1:1) are based on the total numbers of adults from all replicates at each density.

There were no consistent trends in sex ratio in relation to density, although significant chi-square values which could have been due to experimental error in the initial larval transfers were obtained at 4 densities of *melanogaster* and 2 of *simulans*. The 11 chi-square values for each species were tested for consistency, and significant interactions between the individual values and the chi-squares of the total ratios were observed in each case.

## *Biomass*

It is impossible to obtain accurate measurements of the food requirements of crowded populations of *Drosophila* in a live yeast system such as the one used in these experiments. The growing yeast provides a major source of protein (Begg and Robertson 1948), but a *Drosophila* culture is an interspecific association between primary producer (live yeast) and a primary consumer (*Drosophila*), with the added complication that some of the nonliving medium may also be consumed. As Sang (1950) points out, "we are concerned not with a situation involving only *intra-specific* competition, but with an *inter-specific* relationship of some complexity."

Chiang and Hodson (1950) made an effort to circumvent some of these complexities by confining the larvae entirely to yeast for their nourishment, and by minimizing access of the yeast to carbohydrates in the substrate. They thereby reduced the growth rate of the yeast so that reasonably accurate measurements could be made of the quantity of food available to the growing larvae. Their results showed that 1.6 mg dry weight of yeast is sufficient for the normal development of a single *melanogaster* larva. With as little as 0.56 mg of food the larvae are able to survive and pupate but produce adults of minimal size. Development cannot be completed on a ration of 0.42 mg of yeast.

The effect of initial larval density on total biomass in mg live weight per vial in these experiments is illustrated in Figure 7. Biomass was calculated from the mean live weights of each sex and the mean number of adults produced per vial at each density, assuming an equal sex ratio in each instance. The recorded changes in biomass were, therefore, the result of changes in body weight in response to crowding and nutritional deficiencies and the numbers of adults that were produced in relation to the initial larval density. There was an increase in biomass corresponding to increased numbers of adults of each species up to a density of 100 per vial. Since both *melanogaster* and *simulans* responded to changes in density with comparable changes in adult body weight, the greater biomass of *melanogaster* was due primarily to better survival of this species. At densities of 100 and 120 per vial *melanogaster* produced 46.6 and 43.3 mg live weight of adult flies respectively; the corresponding values for *simulans* at these densities were 38.6 and 38.6 mg. Subsequent increases in larval density from 120 to 160, 200, and 300 per vial caused decreases in total biomass which were also correlated with changes in adult numbers.

The chi-square of the difference between the observed biomass for each species at each density and the expected value (1:1) calculated from the average biomass of the 2 species, showed significant values at densities of 160, 200, and 300 but not at lower densities. We may conclude, therefore, that saturation densities, with values for biomass which were not significantly different, were reached by both species at the same initial larval densities. The test for heterogeneity of chi-square values with an expectation of 1 : 1 over 11 densities showed that the treatment fell into 2 groups, one with significant and one with non-significant chi-square values.

To obtain a crude index to the efficiency of utilization of food and medium in these cultures, the data for biomass and live weight were converted to dry weight by means of regression. Flies of both species were reared at selected densities which would provide a suitable range of body sizes. Newly emerged adults were etherized until dead and weighed in lots of 10 to 20. These flies were then dried and reweighed.

An analysis of variance gave no significant differences between the linear regressions of dry weight on live weight for each sex and species, and a highly significant value for the ratio between the mean square attributable to linear regression and the mean square of the residue ($F = 34.79$) showed that the average regression for both sexes and species is linear. The average regression of dry weight on live weight for male and female *melanogaster* and *simulans* is shown in Figure 8. Dry weight is not a constant percentage of live weight but can be calculated from the following equation:

$$\text{Per cent dry weight} = \frac{100a}{\bar{x}} \times 100b$$

where $x =$ observed live weight
and $a = \bar{y} - \bar{b}x$

The mean dry weight of 5 cc of Kalmus medium in a series of samples prepared in the usual manner was $511.5 \pm 3.16$ mg with a coefficient of variability of slightly less than 5%. The maximum biomass for each species in this amount of medium was observed when the initial larval density was 100; a mean of 66.6 adult *melanogaster* produced 14.1 mg dry weight of flies, and 62.2 *simulans* produced 12.8 mg of adult flies per vial. These represent ratios of 36 : 1 and 40 : 1 between the initial amount of medium and the final biomass of flies produced at saturation density. At this density the larvae penetrated the medium almost to the bottoms of the vials, and all available space seemed to be occupied. If both *melanogaster* and *simulans* larvae required at least 0.56 mg dry weight of yeast for each adult produced, the yeast production had to provide a total of 37.3 mg dry weight in the *melanogaster* cultures and 34.8 mg in the *simulans* cultures from an initial 1/20 cc of yeast solution in approximately 512 mg of medium.

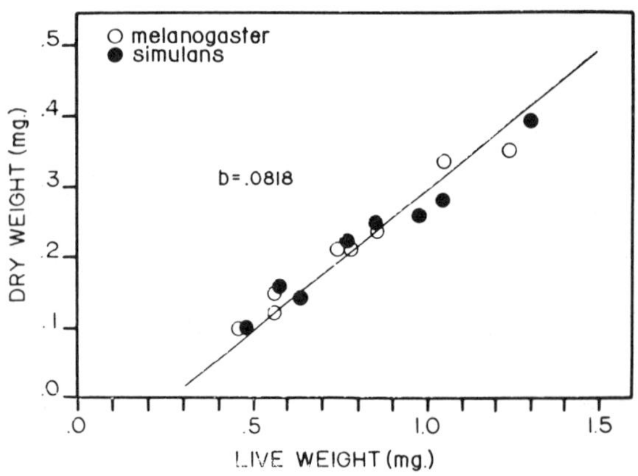


FIG. 8. Regression of dry weight on live weight in *melanogaster*.

Under normal conditions in the absence of larvae the growth of a yeast population is sigmoid, but the addition of larvae to the culture causes a lower and earlier peak in yeast growth, and the subsequent decline in the quantity of yeast in the cultures corresponds to the amount metabolized by larvae (Sang 1950). In fact, it can be shown that the growth form of the yeast population depends directly on the numbers of larvae and their stages of development. In terms of the conditions in these experiments and the evidence presented by Sang (1950), a large population of larvae of uniform age might so seriously retard yeast growth in its early stages that early larval mortality would be necessary before an increase phase could appear in the yeast-growth curve. The high-density populations in these experiments fit the requirements for such an interaction, and it is reasonable to suppose that the pronounced decreases in adult members and biomass that occurred when the initial larval density was increased beyond 120 per vial reflected a critical point at which the larval density became too great to allow significant yeast growth. At a density of 120 per vial 74% of the *melanogaster* and approximately 92% of the *simulans* larvae survived long enough to pupate, and the yeast production was sufficient to provide enough food for 62% of the *melanogaster* and 58% of the *simulans* to produce adults that were at or near the minimum survival size. At higher densities the impact of large numbers of larvae on the initial yeast population was so great that the adverse effects of competition were expressed earlier in the life cycle, as shown by the marked increases in larval mortality that occurred before pupation at densities of 160, 200, and 300 per vial. There may even have been some survival value in the fact that larval mortality was greater at lower densities in the *melanogaster* populations than in the *simulans* populations, thereby releasing greater amounts of food for the surviving larvae. It was noted earlier that the higher per cent pupation in the *simulans* populations did not confer a correspondingly higher total survival in terms of the numbers of adults produced, and it would seem that this lower rate of larval mortality may have had the disadvantage of intensifying competition in the prepupal stage. This would suggest that the inability of the *simulans* populations to extend their larval periods in conjunction with a relatively high rate of larval survival produced excessively severe competition effects which were reflected in the low numbers of adults produced at high densities.

## Interspecies Competition

The results of the experiments described in preceding sections define the specific responses of *melanogaster* and *simulans* to larval crowding in single-species cultures and form at least a partial basis for predicting the outcome of competition between the 2 species. In the range of densities between 5 and about 120 per vial, the population characteristics of the 2 species were remarkably similar. Their rates of development, adult body weights, numbers of adults, and total biomasses at each density were nearly equal, and changes in initial larval density initiated comparable ecological responses in each species. The performance of *melanogaster* was, in most respects, somewhat better than that of *simulans*, but the differences between them were not significant. At densities of 160, 200, and 300 per vial, *melanogaster* showed definite superiority. This species continued to extend its larval period in response to increased density, significantly larger numbers of adults were produced in comparison with *simulans*, and *melanogaster* seemed better able to withstand the adverse effects of crowding at high densities.

If interference, as defined by Elton and Miller (1954), is negligible in competitive interactions between these 2 species, and competition is limited to exploitation of the resources according to the specific ability of each to utilize the food and space available, it should be possible to predict the outcome of competition on the basis of the observed performances in single-species cultures. We would expect that the relative numbers of adults produced by each species at experimental densities of less than 120 per vial would not differ significantly from a 1:1 ratio. We would also expect that *melanogaster* would have a significant advantage in interspecies competition at total larval densities greater than approximately 120 per vial.

To test this hypothesis, combinations of equal numbers of early first-instar *melanogaster* and *simulans* larvae were cultured at combined total densities of 10 (5 *melanogaster* + 5 *simulans*), 20, 40, 80, 120, 160, 200, and 300 per vial, using the same techniques as those described for single-species cultures.

Table VI. Ratios of adult male *melanogaster* and *simulans* in 2-species cultures

| Density | *melanogaster* ♂♂ | *simulans* ♂♂ | $X^2$ | Per cent *melanogaster* |
|---|---|---|---|---|
| 5 x 5 | 54 | 52 | .036 | 50.9 |
| 10 x 10 | 68 | 61 | .380 | 52.7 |
| 20 x 20 | 57 | 52 | .229 | 52.3 |
| 40 x 40 | 71 | 71 | .000 | 50.0 |
| 60 x 60 | 67 | 78 | .834 | 46.2 |
| 80 x 80 | 98 | 79 | 2.782 | 56.3 |
| 100 x 100 | 33 | 18 | 4.412[a] | 64.7 |
| 150 x 150 | 23 | 11 | 4.235[a] | 67.6 |

[a]Significant at 5% level.

Table VII. Sex ratios of adult flies (*melanogaster* + *simulans*) in 2-species cultures

| Larval density | Total ♂♂ | Total ♀♀ | ♀♀ 100♂♂ | $X^2$ |
|---|---|---|---|---|
| 5 x 5 | 106 | 86 | 81 | 2.084 |
| 10 x 10 | 126 | 134 | 104 | .095 |
| 20 x 20 | 109 | 134 | 123 | 2.572 |
| 40 x 40 | 142 | 140 | 99 | .001 |
| 60 x 60 | 155 | 134 | 92 | .434 |
| 80 x 80 | 174 | 139 | 80 | 3.814[a] |
| 100 x 100 | 51 | 42 | 82 | .871 |
| 150 x 150 | 34 | 34 | 100 | .00 |

[a]Significant at 5% level.

### *Species ratios*

The relative numbers of adult male flies of each species that emerged from the 2-species cultures at each density are shown in Table VI. The number of replicates for each density can be calculated from the data in Table VIII. Since adult female *melanogaster* and *simulans* are indistinguishable, an estimate of the success of each species in interspecific competition must be based on the relative numbers of adult males, assuming that the sex ratios are equal. The combined sex ratios for both species are shown in Table VII. A significant chi-square value for the difference between the observed sex ratio and an expected 1:1 ratio occurred only at a density of 160 (80 + 80) when a ratio of 100 males: 80 females was observed.

The relative numbers of *melanogaster* and *simulans* at combined densities of 10 to 120 larvae per vial did not differ significantly from the expected 1:1 ratio. At a density of 160, *melanogaster* had a slight but not significant advantage, and at combined densities of 200 and 300 *melanogaster* performed according to expectation with a significant superiority in numbers of adults produced. Since these results agree with the predictions made on the basis of the ecological performances of each species in single-species cultures, they tend to confirm the assumption that competition between the larvae of these 2 species is essentially a numerical relationship in which the available resources are shared according to the specific ability of each species to survive at these levels of total density.

*Larva-adult survival*

The effect of density on larval survival and pupal viability in 2-species cultures is shown in Table VIII. These values have been calculated from the observed numbers of males of each species, the total numbers of females, and the species ratios shown in Table VII.

TABLE VIII. Pupation and adult emergence in 2-species cultures of *melanogaster* and *simulans* in relation to density

| Larval density (*mel* × *sim*) | N | Pupation | | Adult emergence | |
|---|---|---|---|---|---|
| | | Pupae/vial ± S.E. | Per cent | Adults/vial ± S.E. | Per cent |
| 5 x 5 ...... | 230 | 8.9±0.09 | 88.7 | 8.4±0.31 | 83.5 |
| 10 x 10 ...... | 320 | 18.2±0.47 | 90.9 | 16.4±0.46 | 82.2 |
| 20 x 20 ...... | 320 | 34.4±1.07 | 85.9 | 30.1±0.92 | 75.3 |
| 40 x 40 ...... | 400 | 68.8±2.22 | 86.0 | 56.4±2.36 | 70.5 |
| 60 x 60 ...... | 480 | 98.0±2.50 | 81.7 | 64.8±3.54 | 54.0 |
| 80 x 80 ...... | 800 | 117.6±1.92 | 73.5 | 62.6±8.06 | 39.1 |
| 100 x 100...... | 400 | 123.0±7.07 | 61.5 | 46.5±3.74 | 23.2 |
| 150 x 150...... | 900 | 67.0±2.84 | 22.3 | 23.0±1.88 | 7.7 |

Because the over-all outcome of competition between these 2 species is predictable in terms of species ratios, the species interactions can be examined in more detail by comparing the larva-adult survival of each species in single- and 2-species cultures. The factors for interactions between competing species are included in the Gause model for the simultaneous growth of populations, as represented in the following equations (Slobodkin 1961):

$$(1)\quad \frac{dN_1}{dt} = r_1N_1\left[\frac{K_1 - N_1 - \alpha N_2}{K_1}\right],$$

$$(2)\quad \frac{dN_2}{dt} = r_2N_2\left[\frac{K_2 - N_2 - \beta N_1}{K_2}\right]$$

in which the inhibitory effect of a member of species $N_1$ on its own species is $1/K_1$, where $K_1$ is the saturation density of $N_1$ alone. The inhibitory effect of species $N_1$ on another species $N_2$ is $\beta/K_2$ where $\beta$ represents the value of a species interaction which may be more or less than the effect of $N_2$ alone.

This system is at equilibrium when

$$\frac{dN_1}{dt} = \frac{dN_2}{dt} = 0,$$

in which case $(K_1 - N_1 - \alpha N_2) = 0$ and $(K_2 - N_2 - \beta N_1) = 0$. In the following analysis of the numbers of *melanogaster* ($N_1$) and *simulans* ($N_2$) it is assumed that the 2-species system is at equilibrium for each set of values at each experimental density. The values $K_1$ and $K_2$ therefore refer to the maximum numbers of each species in single-species cultures and $N_1$ and $N_2$ are the observed values for each species in 2-species cultures. Thus at each level of density the initial concentrations are equal but the maximum values of $K_1$ and $K_2$ and of $N_1$ and $N_2$ are set by the conditions of the experiment.

Table IX shows the mean numbers of adults produced by each species alone ($K_1$ and $K_2$) and during interspecies competition ($N_1$ and $N_2$) at total densities of 10, 20, 40, 80, and 120 per vial. In the preceding sections it was shown that the saturation density for numbers of each species occurred when the initial larval density was 120; the saturation density for biomass occurred at an initial density of 100. This analysis has therefore been restricted to values up to the maximum observed saturation density for numbers of adults.

TABLE IX. Numbers of adult *melanogaster* ($K_1$ and $N_1$) and *simulans* ($K_2$ and $N_2$) in single-species and 2-species cultures

| Initial density | Single-species | | 2-species | |
|---|---|---|---|---|
| | $K_1$ | $K_2$ | $N_1$ | $N_2$ |
| 10................. | 7.8 | 7.9 | 4.3 | 4.1 |
| 20................. | 17.0 | 15.9 | 8.7 | 7.7 |
| 40................. | 33.6 | 33.6 | 15.8 | 14.3 |
| 80................. | 58.8 | 54.4 | 28.2 | 28.2 |
| 120................. | 74.6 | 69.0 | 32.3 | 37.2 |

The results of an analysis of variance for the values of $K_1$, $K_2$, $N_1$, and $N_2$ at 5 levels of density are shown in Table X. The component of error variance from $K_1$ and $K_2$ was significantly higher than from $N_1$ and $N_2$, owing to the fact that the numbers of each species in 2-species cultures were about half what they were in single-species cultures. The ratio $F = 4.22$ (15, 15) of the variances of K's and N's was significant at the 2% level (2-tailed test). Each value for $N_1$ and $N_2$ was therefore doubled in order to analyze for variances of components for difference between columns (C) at various density levels (D).

Differences among densities were highly significant, as one would expect from the design of the experiments. The mean square values of

TABLE X. Analysis of variance for $K_1$, $K_2$, $N_1$, and $N_2$ at 5 levels of density

| Sources of variance | D.F. | S.S. | M.S. | F |
|---|---|---|---|---|
| COMPARISONS OF C | | | | |
| $K_1 \sim K_2$ | 1 | 12.1000 | | |
| $N_1 \sim N_2$ | 1 | 2.1160 | | |
| $K_1 - N_1 - N_2$ | 1 | 7.9207 | | |
| $K_2 - N_2 - N_1$ | 1 | 0.0007 | | |
| Interaction C x D | 12 | 79.9370 | 6.6614 | |
| COMPONENTS OF C × D | | | | |
| $(N_1 \sim N_2)$ x D | 4 | 54.4640 | 13.6160 | 3.93 (4.4) |
| $(K_1 \sim K_2)$ x D | 4 | 13.8700 | 3.4675 | |
| BASED ON LOG VALUES | | | | |
| $\alpha \sim \beta$ | 1 | 0.001649 | | |
| Error | 12 | 0.025886 | 0.002157 | |

deviations of the sums from their regression on density levels was very small. The regression covered almost all of the variance between densities. Similar tests showed that the residues from linear regression of $N_1$, $N_2$, and their total $(N_1 + N_2)$ were insignificant. Thus the larva-adult mortality rate was the same at all densities. Further, since the difference between coefficients of regression of $N_1$ and $N_2$ on density levels did not exist, it can be assumed that mortality was the same in both species.

Five comparisons among the column means shown in Table X were tested for significance. None was significant. We may therefore draw the following conclusions from the results of these tests:

(a) $K_1 \sim K_2$: the conditions of the experimental environment had the same survival value for both species at the density levels tested.

(b) $N_1 \sim N_2$: the total resources of the environment were shared equally between the 2 species.

(c) $(K_1 - N_1) \sim N_2$: the effect of $N_2$ on $N_1$ was the same as the effect of $N_1$ on members of its own species and, therefore, $\alpha = 1$.

(d) $(K_2 - N_2) \sim N_1$: the effect of $N_1$ on $N_2$ was the same as the effect of $N_2$ on members of its own species and, therefore, $\beta = 1$.

(e) $\alpha \sim \beta$: the effect of $N_2$ on $N_1$ was the same as the effect of $N_1$ on $N_2$.

In order to get a linear comparison to test for the differences in values of $\alpha$ and $\beta$, the logarithms of the values for $(K_1 - N_1)$, $(K_2 - N_2)$, $N_1$ and $N_2$ (observed values) were analyzed for variance. The comparison

$$\log \alpha \sim \log \beta$$

given by

$$[\log (K_1 - N_1) - \log N_2] \sim [\log (K_2 - N_2) - \log N_1]$$

was seen to have an insignificant mean square value. The values of $\alpha$ and $\beta$ may, therefore, be considered as equal.

## DISCUSSION

Population growth patterns are determined by interactions between the physiological properties of individuals, the interactions between individuals, and the physical, biological, and mechanical properties of the environment (Slobodkin 1961). We may investigate the nature of these interactions by allowing a population to grow from a low level to equilibrium density or, as was done in this study, by experimentally manipulating density in relation to food and space in a uniform physical environment. As Slobodkin (1961) has shown in a series of analyses of the results of selected competition experiments, the population growth patterns produced by the interactions of individuals in a controlled and limited environment have a definite relation to the normal selective pressures operating on a species in nature.

Two general categories of response are observed in *Drosophila* populations as numbers increase toward saturation densities in single-species cultures: (1) physiological and ecological adjustments to crowding effects, and (2) damping of the increase potential of the population. The rate at which the increase potential is damped or reduced as a function of intraspecific competition depends on the efficiency of the compensation mechanisms which allow a population to adjust to the adverse effects of crowding. As the initial larval density was increased toward saturation levels in the *melanogaster* and *simulans* populations in this study, there was a trend of decreased adult body size without substantial increases in total mortality. Thus, adult numbers were maintained at nearly normal levels, although the increase potential was decreased by the reductions in fecundity that accompany decreases in body size (Chiang and Hodson 1950). Miller and Thomas (1958) have shown that adult longevity may actually increase when body size is decreased, but it is doubtful that the nature and amount of change in survivorship under these conditions is a significant factor in the increase potential of the population. The 2d compensation mechanism that was noted was a tendency to prolong the larval period in response to crowding and nutritional deficiencies. This also seemed to be a means of avoiding excessive mortality and conserving adult numbers, but we may infer that the increase potential was reduced somewhat by extending the generation time (Slobodkin 1961). At saturation densities both species were at or near their minimum survival sizes; adjustments in body size were therefore no longer possible; and there was a marked increase in larva-adult mortality. However, *melanogaster* retained its ability to prolong the larval period even at extremely high densities, and this additional compensation, which was lost in the *simulans* populations when the density reached 160, allowed greater survival of *melanogaster* at these densities.

In his formulation of the fundamental niche, Hutchinson (1957) distinguishes 2 cases in which (1) $\mathbf{N}_2$ is a proper subset of $\mathbf{N}_1$($\mathbf{N}_2$ is "inside" $\mathbf{N}_1$), and (2) $\mathbf{N}_1 \cdot \mathbf{N}_2$ is a proper subset of both $\mathbf{N}_1$ and $\mathbf{N}_2$. In the latter case $S_1$ survives in the parts of the physical space of the biotope corresponding to $\mathbf{N}_1 = \mathbf{N}_2$) and $S_2$ in the parts corresponding to ($\mathbf{N}_2 = \mathbf{N}_1$) with the proviso that no point in $\mathbf{N}_1 \cdot \mathbf{N}_2$ can correspond to the survival of both species. The conditions of environment that existed for *melanogaster* and *simulans* in these experiments represented a particular example of case (2) in that amount of food per unit density and amount of space per unit density were the only environmental variables in the design, and other factors were assumed to be constant. Both species were able to survive in the part of the intersection subset represented by the experimental environment, but no point in the physical space of the environment would permit the simultaneous survival of both $S_1$ and $S_2$.

In the general case it is unlikely that points within $\mathbf{N}_1 \cdot \mathbf{N}_2$ will have equal *survival value* for both $S_1$ and $S_2$, for each point corresponds to the coordinates of all the environmental variables involved in the intersection and species responses to these variables will differ. The chance that identical survival values will occur would appear to depend on: (1) the number of points in $\mathbf{N}_1 \cdot \mathbf{N}_2$ which are common to $\mathbf{N}_1$ and $\mathbf{N}_2$, or amount of ecological overlap, and (2) the abilities of each species to adjust their survival rates in response to variations in their physical and biotic environments. This would imply that different species of homiotherms, with their greater ability to compensate for changes in environment, are more likely to have comparable survival values and to compete for common resources. The data from this study show that *melanogaster* and *simulans* larvae are capable of physiological adjustments which allow them to conserve numbers and to survive at nearly equal levels up to saturation density in this experimental environment. There were specific differences in the nature and limits of their compensation mechanisms, and mortality was not distributed in quite the same way among the populations of the 2 species, but the net effect was that the values for $K_1$ and $K_2$ were equal.

Interpretation of the results of interspecies competition depends on whether the problem is approached from the standpoint of (1) maximum exploitation of a unit of environment by one species, or (2) equitable utilization of the environment and possible coexistence in a mixed-species system. Considerable confusion has arisen through failure to distinguish between these 2 points of view. In the first and more conventional interpretation, emphasis is placed on the extent to which a population $N_1$ reduces the total resources available to another population $N_2$. This relationship is often described in terms of "degree of interference" or "intensity of competition" (cf Merrell 1951; Moore 1952; Sokoloff 1955). It is assumed that when the ecological requirements of $N_1$ and $N_2$ are similar, the intensity of competition will be high and coexistence impossible. When the requirements of the 2 species are dissimilar they will interfere less with each other and may coexist. Degree of similarity, or amount of overlap in fundamental niches, may define the extent to which 2 species have similar requirements and will, therefore, make similar demands on an environment, but it does not necessarily define the conditions of coexistence. In his study of competition between *pseudoobscura* (= A), *persimilis* (= B), and *miranda* (= C) Sokoloff (1955) observed the following order of competitive ability in species combinations: $B \cong A > C$ at low densities and $B > A >> C$ at high densities. The relationship between the sibling species *persimilis* and *pseudoobscura* was remarkably similar to that between *melanogaster* and *simulans;* the ecological requirements of the 2 species were nearly identical, and *persimilis* was significantly superior to *pseudoobscura* only at high densities in competition. *D. miranda* is more distantly related to *pseudoobscura* and *persimilis* than they are to each other and was relatively unsuccessful in competition with either. Sokoloff (1955) concluded that when closely related species (A and B) are involved in competition they interfere intensely with each other, although adult survival may be nearly equal for each species. When more distantly related species are in competition (A and C, or B and C) interference with the stronger competitor is not as great, even though the weaker competitor may be eliminated entirely. These, and the results of other studies of competition in *Drosophila*, pose an apparent contradiction to the Gause principle, in that coexistence has usually been more successful between closely related species than between more distant relatives.

The expression $(K_1 - N_1 - \alpha N_2)$ in the Gause equations implies that 2 species which compete with each other for food and space cannot survive at the same density levels that are possible in the absence of interspecies competition. In other words, $N_1$ cannot attain its maximum density $K_1$ in the presence of a competing species $N_2$. This statement refers to the exploitation of a common resource by 2 species, but does not exclude the possibility of coexistence, which depends on the relative values of $\alpha$ and $\beta$. By considering the relative values of $\alpha$ and $\beta$ with respect to $K_1$ and $K_2$ when $\frac{dN_1}{dt} = \frac{dN_2}{dt} = 0$,

Slobodkin (1961) has developed 4 cases for the Gause model in which the outcome of competition between 2 species in a limited environment can be predicted. Case 1, which corresponds to "indeterminate competition" in Park's (1955) studies, is particularly relevant to this discussion:

*Case 1:* Either $N_1$ or $N_2$ will be the sole survivor depending on initial concentration:

$$\alpha > \frac{K_1}{K_2} \quad \text{and} \quad \beta > \frac{K_2}{K_1}$$

This case is shown graphically in Figure 9. Neither species can increase above its saturation line $K_1, \frac{K_1}{\alpha}$ or $K_2, \frac{K_2}{\beta}$ and below its saturation line each species will increase. An unstable equilibrium is possible at point S, but it is expected that only one species will survive, depending on whether the initial mixture of species lies to the right or left of the line OS.

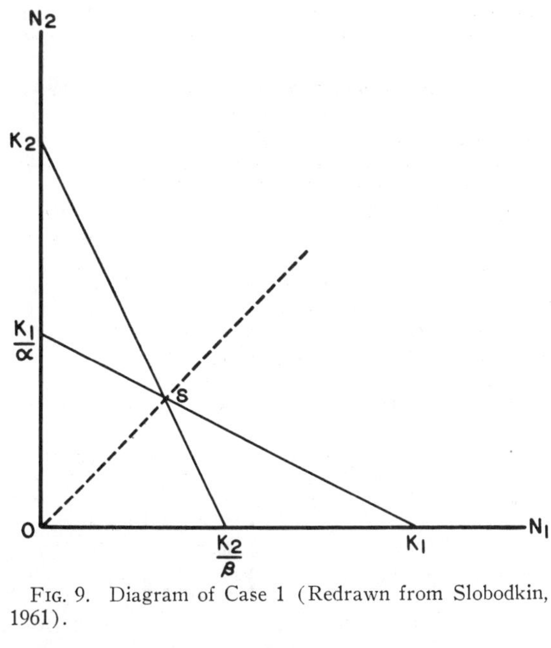


FIG. 9. Diagram of Case 1 (Redrawn from Slobodkin, 1961).

In the results for *melanogaster* and *simulans* the values of $K_1$ and $K_2$ at each density level up to saturation were not significantly different and we may assume that $K_1 = K_2$. It was also shown that $\alpha = 1$ and $\beta = 1$ and $\alpha$ and $\beta$ were not significantly different. We may therefore conclude that the competitive interactions within species were the same as those between species and $\alpha = \beta = 1$. Therefore $K_2 = K_1/\alpha$ and $K_1 = K_2/\beta$, which implies coincidence of the isoclines in Figure 9. These conditions fulfill the requirements for an unstable equilibrium whose position on the isocline $K_1 = K_2/\beta$, $K_2 = K_1/\alpha$ will depend on the initial concentration of each species. The general case for this relationship may be stated as follows:

*Case 5:* $N_1$ and $N_2$ will coexist at equilibrium densities which depend on initial concentration:

$$\alpha = K_1/K_2 \text{ and } \beta = K_2/K_1$$

As the initial concentrations of larvae were always equal in these experiments, the relationship $N_1 = N_2$ existed in each case and the predictability of this model with different concentrations of the 2 species was not tested. Equilibrium occurred at points along the line OS which corresponded to the density levels in the design of the experiment however, and to this extent the data for $N_1 = N_2$ confirm the prediction of the general case for a 2-species equilibrium (Case 5).

The ecological consequences of a 2-species equilibrium such as the one described in this study will depend on (1) whether food and space are limiting for extended periods in natural environments, and (2) the effect of environmental fluctuations on different parts of the competition process. As this study only considered competition during the larval stages of the *Drosophila* life cycle, there remains the possibility that factors affecting fecundity and fertility and the selection of oviposition sites alter the course of competition by varying the initial concentrations of each species in this system. Andrewartha and Birch (1954) have also stated a strong case for control of insect numbers by external factors such as weather. If they are correct in their conclusion that the increase potential of such populations is never fully realized and food and space are not limiting, coexistence may occur without serious consequences to the ultimate survival of either species. As Hutchinson (1948) points out, the rule of mutual exclusion from a niche need not apply when external factors act to rarefy the mixed population so that the environment is not heavily exploited, or when the values for $\alpha$ and $\beta$ are under environmental control and chance oscillations might continually reverse the direction of competition so that no equilibrium could be established. The significance of the results of this study is not that coexistence among larvae is a regular feature of *Drosophila* communities, but that variations in the relative abundance of species are probably introduced by factors other than larval competition.


### Summary

1. Intraspecies and interspecies competition among the sibling species *Drosophila melanogaster* and *D. simulans* was studied in larval populations of uniform age reared at experimental densities of 5 to 300 per 5 cc of medium.

2. In the range of densities up to saturation, the ecological performances of the 2 species were remarkably similar in single-species cultures; both species showed comparable changes in ultimate adult body weight, length of the larval period, pupal viability, larval mortality, numbers of adults produced, and biomass of adults in response to changes in initial larval density.

3. Saturation densities and maximum utilization of the environment occurred at the same densities in single-species cultures of the 2 species, and the numbers of adults and biomass were not significantly different.

4. At densities above saturation *melanogaster* was better able to survive than *simulans,* prolonging the larval period in response to crowding at these densities.

5. It was possible to predict the outcome of competition between the 2 species when it was assumed that the interactions between members of different species were no different from the interactions within species; the result of competition between these 2 species was an unstable equilibrium at densities up to saturation.

6. The results of these experiments are discussed with reference to recent developments in the niche concept and the Gause model of interspecies competition.



### Acknowledgments

This research was supported initially by a research grant from the American Academy of Arts and Sciences, and its later stages by N.S.F. Grant G-14290.

I am grateful to R. K. Misra who read the manuscript and advised on statistical procedures.


### Literature Cited


**Andrewartha, H. G., and L. C. Birch.** 1954. The distribution and abundance of animals. Chicago: University of Chicago Press. 782 pp.

**Begg, C. M. M., and F. W. Robertson.** 1948. Nutritional requirements of *Drosophila melanogaster.* Nature **161:** 769.

**Bodenheimer, F. S.** 1938. Problems of animal ecology. London: Oxford University Press. 183 pp.

**Bodenstein, D.** 1950. The postembryonic development of *Drosophila. In* M. Demerec, Biology of Drosophila. New York: John Wiley & Sons, Inc. 632 pp.

**Burla, H.** 1948. Die Gattung Drosophila in der Schweiz. Rev. Suisse Zool. **55:** 272-279.

**Chiang, H. C., and A. C. Hodson.** 1950. An analytical study of population growth in *Drosophila melanogaster.* Ecol. Monogr. **20:** 175-206.

**Dobzhansky, T., and C. Pavan.** 1950. Local and seasonal variations in relative frequencies of species of *Drosophila* in Brazil. Jour. Anim. Ecol. **19:** 1-14.

**Elton, C., and R. S. Miller.** 1954. The ecological survey of animal communities: with a practical system of classifying habitats by structural characters. Jour. Ecol. **42:** 460-496.

**Gause, G. F.** 1935. La théorie mathématique de la lutte pour la vie. Paris: Herman et Cie. 61 pp.

**Gilbert, O., T. B. Reynoldson, and J. Hobart.** 1952. Gause's hypothesis; an examination. Jour. Anim. Ecol. **21:** 310-312.

**Hardin, G.** 1960. The competitive exclusion principle. Science **131:** 1292-1297.

**Horton, I. H.** 1939. A comparison of salivary gland chromosomes of *Drosophila melanogaster* and *D. simulans.* Genetics **24:** 234-243.

**Hutchinson, G. E.** 1948. Circular causal systems in ecology. Ann. N. Y. Acad. Sci. **50:** 221-246.

———. 1957. Concluding remarks. Cold Spring Harbor Symposia on Quantitative Biology **22:** 414-427.

**Kalmus, H.** 1943. A factorial experiment on the mineral requirements of *Drosophila* cultures. Amer. Nat. **77:** 376-380.

**Kerkis, J.** 1936. Chromosome conjugation in hybrids between *Drosophila melanogaster* and *Drosophila simulans.* Amer. Nat. **70:** 81-86.

———. 1937. The cause of imperfect conjugation in chromosomes of hybrids of *D. simulans* and *D. melanogaster.* Bull. Acad. Sci. USSR 460-468.

**Merrell, D. J.** 1951. Interspecific competition between *Drosophila funebris* and *Drosophila melanogaster.* Amer. Nat. **85:** 159-169.

**Miller, R. S.** (unpublished). Competition between adult *Drosophila melanogaster* and *D. simulans.*

———, **and J. L. Thomas.** 1958. The effects of larval crowding and body size on the longevity of adult *Drosophila melanogaster.* Ecology **39:** 118-125.

**Moore, J. A.** 1952. Competition between *Drosophila melanogaster* and *Drosophila simulans.* I. Population cage experiments. Evolution **6:** 407-420.

**Neyman, J., T. Park, and E. L. Scott.** 1958. Struggle for existence; the *Tribolium* model: biological and statistical aspects. Gen. Systems **3:** 152-179.

**Park, T.** 1954. Experimental studies of interspecies competition. II. Temperature, humidity, and competition in two species of Tribolium. Physiol. Zool. **27:** 177-238.

———. 1955. Experimental competition in beetles, with some general implications. *In* J. B. Cragg and N. W. Pirie, The Numbers of Man and Animals. London: Oliver & Boyd. 152 pp.

**Patau, K.** 1935. Chromosomenmorphologie bei Drosophila melanogaster und Drosophila simulans and ihre genetische Bedeutung. Naturwiss **23:** 537-543.

**Patterson, J. T.** 1943. Studies in the genetics of Drosophila. III. The Drosophilidae of the Southwest. Univ. Texas Publ. No. 4313.

———, **and W. S. Stone.** 1952. Evolution in the genus Drosophila. New York: The Macmillan Company. 610 pp.

**Robertson, F. W.** 1960. The ecological genetics of growth in *Drosophila*. Body size and developmental time on different diets. Genet. Res. Camb. **1**: 288-304.

———, **and E. C. R. Reeve.** 1952. Studies in quantitative inheritance. I. The effects of selection for wing length and thorax length in *Drosophila melanogaster*. Jour. Genetics **50**: 414-448.

**Sang, J. H.** 1949a. The ecological determinants of population growth in a *Drosophila* culture. III. Larval and pupal survival. Physiol. Zool. **22**: 183-202.

———. 1949b. The ecological determinants of population growth in a *Drosophila* culture. IV. The significance of successive batches of larvae. Physiol. Zool. **22**: 202-210.

———. 1950. Population growth in Drosophila cultures. Biol. Rev., Cambridge Phil. Soc. **25**: 188-219.

———. 1956. The quantitative nutritional requirements of *Drosophila melanogaster*. Jour. Exp. Biol. **33**: 45-72.

**Slobodkin, L. B.** 1961. Growth and regulation of animal populations. New York: Holt, Rinehart & Winston. 184 pp.

**Sokoloff, A.** 1955. Competition between sibling species of the *Pseudoobscura* subgroup of *Drosophila*. Ecol. Monogr. **25**: 387-409.

**Sonnenblick, B. P.** 1950. The early embryology of *Drosophila melanogaster*. *In* M. Demerec, Biology of Drosophila. New York: John Wiley & Sons, Inc. 632 pp.

**Spassky, B.** 1943. Cream of wheat-molasses fly medium. D.I.S. **17**: 67-68.

**Sturtevant, A. H.** 1919. A new species closely resembling *Drosophila melanogaster*. Psyche **26**: 153-155.

———. 1920. Genetics studies on *Drosophila simulans*. I. Introduction. Hybrids with *Drosophila melanogaster*. Genetics **5**: 488-500.

———. 1929. The genetics of *Drosophila simulans*. Carnegie Inst. of Wash. Publ. 399, 62 pp.

**Tantawy, A. O., and G. S. Mallah.** 1961. Studies on natural populations of *Drosophila*. Heat resistance and geographical variations in *Drosophila melanogaster* and *D. simulans*. Evolution **15**: 1-14.

**Vladimorova, M. A., and E. S. Smirnov.** 1938. Intraspecific competition in and interspecific competition between *Musca domestica* L. and *Phormia groenlandica* Zeit. Meditzinskaya Parazitologiya i Parazitarnye Bolezni, 7: 755-777. Translation 245 F 1050 A, Bureau of Animal Population, Department of Zoological Field Studies, Oxford University, England.

# Latent Period Phenomena in Transmission of Sugar Beet Yellow-Net Virus by Green Peach Aphids[1]

Edward S. Sylvester[2]

ABSTRACT

Using the leaf cage technique in transmission tests of sugar beet yellow-net virus by green peach aphids, *Myzus persicae* (Sulz.), the following conclusions were reached: (1) A low percentage of the insects (1.1%) acquired virus if the cage was left in contact with a virus source leaf for 1 hour. Maximum acquisition (approximately 25% of the insects proving to be infective) occurred when the cage was left in place for a 48-hour period. (2) When leaf cages, containing two aphids which had been reared on virus source plants, were placed for 1 hour on healthy test plants, approximately 4% were infected. Increasing the test feeding access time increased the probability of transmission up to a maximum of approximately 25% with periods of 24 or 48 hours in duration. (3) Analysis of the results of serial transmission experiments with single aphids allowed a variable acquisition access period and subsequent sequential test feeding access periods each approximately 24 hours in length until death of the insects, lead to the conclusion that there was a measurable latent period averaging 3.5 days in length. In this analysis, the latent period was defined as the average time (in days) in excess of the acquisition threshold period required by the aphids to begin virus transmission.

In 1949 (Sylvester) experimental results were published which led to the conclusion that there was no latent period in the transmission of sugar beet yellow-net virus by green peach aphids, *Myzus persicae* (Sulzer). It is the intent of this paper to re-examine and modify this conclusion.

MATERIALS AND METHODS.—*General.*—The yellow-net virus used was a derivative of the original source (Sylvester 1949). Sugar beets (*Beta vulgaris* L.) used to maintain virus and at times aphids, and to provide test plants were of the variety U.S. 15. Test plants were sown as seed in pasteurized sand or vermiculite and later transplanted into pasteurized soil in 3-inch clay pots. Seedling test plants normally were used in the cotyledon stage, but occasionally 2 to 3 leaves developed before use.

Aphids were maintained as noninfectives on both smooth leaf mustard (*Brassica juncea* Coss) and sugar beet plants. Large groups of aphids were collected by aspiration, while smaller groups and individuals were sorted with a camel's-hair brush. Acquisition as well as test access times were done using leaf cages (Sylvester 1956). Routine spraying with nicotine and soap solution helped minimize unwanted aphid infestations.

Since the leaf cage technique lessened the danger of vector escapes, many experiments were done within the greenhouse where test plants were also incubated. In some cases if high temperatures were likely to occur, test feeding access periods were done in the laboratory.

*Inoculation Threshold Period.*—To establish the inoculation threshold period using the leaf cage technique, the following procedure was used successfully on five different occasions. Mature apterae or late instar nymphs, collected by aspiration from colonies which had been reared on virus source plants, were sorted into small leaf cages, two per cage. After sorting, the cages were attached to a series of 140 healthy beet plants, one cage per plant. At random, at each of the following intervals, 20 cages were removed: 1, 2, 4, 6, 8, 26, and 48 hours. In addition, 20 cages, none of which contained insects, were attached to 20 beet plants. In two of the trials, the insects which were removed at the end of 1 hour, were placed on another series of test plants and allowed a 2-hr test access time. The insects which initially had 2 hours of test access time were moved to a second series of test plants for a period of 1 hour. Four such reciprocal cycles were made, so that in each of the two trials, the same insects had access to three sets of 20 healthy plants for 1 hour and three sets of 20 plants for a 2-hour test access period.

*Acquisition threshold period and latent period.*—In general, the following procedure was used in the latent period and acquisition threshold period tests. A sample of apterous aphids (late instar nymphs or apterae) was obtained from a noninfective source by aspiration, and individually sorted into 30 leaf cages. Twenty of the cages contained 15 to 20 individuals, and 10 of the cages contained 10 individuals. The 10 cages of 10 aphids per cage were placed on healthy test plants, one cage per plant, to serve as checks. These cages were moved daily to new beet plants until all insects died. The remaining 20 cages of aphids (15 to 20 insects per cage) were placed upon 20 yellow-net virus source plants, one cage per plant. Three cages were removed from the virus source plants at the end of each of the following intervals: 1, 2, 4, 8, and 24 hours, and the remaining cages (usually five) were removed at the end of 48 hours. The extra cages were used in case of accidents due to breakage of the rubber band which held them in place, or excessive mortality in a particular group. The order of removal was at random. For any particular interval, the aphids from the removed cages were put into a container, and from the common pool, as rapidly as possible, a minimum of 20 aphids was picked and put into 20 cages, one aphid per cage. The leaf

[1] Accepted for publication June 2, 1958.

[2] Associate Entomologist, Department of Entomology and Parasitology, University of California, Berkeley.

Reproduced with permission from the *Journal of Economic Entomology,* 51:812–818, 1958.
Published by the Entomological Society of America, College Park, Maryland.

**Table 1.—Length (mean ± standard deviation) of the serial test feeding used in connection with the various acquisition access periods in the study of the latent period of the yellow-net virus transmission by the green peach aphid.**

| Acquisition Access Period (Hours) | Test Feeding Period | |
|---|---|---|
| | N[a] | Length[b] |
| 1 | 101 | 23.9±0.53 |
| 2 | 106 | 23.8±0.69 |
| 4 | 109 | 23.5±1.47[c] |
| 8 | 105 | 23.0±2.17[c] |
| 24 | 81 | 24.0±0.47 |
| 48 | 88 | 24.0±0.16 |

[a] N is the number of test feedings intervals recorded.
[b] The intended length of each test feeding was 24 hrs.
[c] The relative high variation found in this case was due to a gradual shortening the test feeding intervals in order to eventually allow all of the insects to be transferred at the same time.

**Table 2.—Results[a] of trials to determine the inoculation threshold period of beet yellow-net virus when transmitted by lots of two green peach aphids using variable test access times and the leaf cage technique.**

| | Test Access Time in Hours | | | | | | | |
|---|---|---|---|---|---|---|---|---|
| | 0 | 1 | 2 | 4 | 6 | 8 | 24 | 48 |
| Per cent | 0/96<br>0.00 | 10/257<br>3.8 | 18/255<br>7.0 | 9/100<br>9.0 | 15/98<br>15.3 | 19/99<br>19.1 | 25/100<br>25.0 | 25/99<br>25.2 |

[a] In ratios listed, numerator is number of plants infected; denominator, number tested. Variations in denominators (100 or 260 intended) due to plants or vector dying before termination of experiment.

cages were then attached to a series of 20 test plants and subsequently moved at approximately 24-hour intervals (table 1) until death of the insects. The number of aphids tested per interval per trial varied from 20 to 45, the larger number of individuals being tested for the 24- and 48-hour acquisition access times. Several such trials were run, with some variation in technique. For example, on a few occasions, the aphids were sorted in the late afternoon prior to being placed on the virus source plants the next morning. In these cases the aphids were kept in a refrigerator during the remainder of the afternoon and night before use. On one occasion, the 4-hour acquisition access interval was not included by mistake, and in this run, the number of insects tested at the 8-, 24-, and 48-hour intervals was increased to 25, 40, and 45, respectively. However, within practical limits, the conditions of any one trial were kept as uniform as possible, with the exception of some variation in the number of individuals tested per interval, and it was attempted to distribute the other sources of variation at random among the treatments.

Results.—*Inoculation threshold period.*—The results of all trials are given in table 2. Using two aphids per test plant, about 4% of the plants were infected when the vectors had 1-hour test access time. The chance of obtaining an infection gradually increased as the test access time was lengthened until 24 hours, after which interval no additional gain was noted. The difference between the results of the present work and the former (Sylvester 1949) where 15 minutes of test feeding was reported to be sufficient for inoculation, is not unreasonable if based upon the actual ability of certain insects to inoculate beet plants in relatively short test feeding times. In the former work, using 10 hours of access time on test plants, 7 of 25 insects tested (28%) were infective, whereas in the present work, using an 8-hour leaf cage test access time, an estimated 10% of the insects were infective. The general level of infectivity in the aphids in the former work was higher.

*Acquisition threshold period.*—During this work many of the aphids failed to survive for long series of transfers in the leaf cages, and there was variation in the death rate of the samples from trial to trial. Occasionally high temperatures occurred for brief periods in the greenhouse, and some of the later runs were made in the laboratory where the possible influence of radiant energy was considerably reduced. While the aphids lived longer under the laboratory conditions, there was no gain in the rate of transmission.

Originally it was thought that increasing the acquisition access time might increase the mortality rate in subsequent test feeding transfers. The summarized mortality data of table 3 indicates that the mortality rate in the 24- and 48-hour acquisition access periods was somewhat higher than in the other intervals. This is not so when the data are adjusted for the total time spent on beets. In general, 50% of the test insects had died by the time 72 hours of feeding on beets had been completed. The comparative mortality curves (fig. 1) for the total sample, for those aphids demonstrated to be infective, and for a sample of 98 aphids fasted under room temperature conditions, indicated that, aside from the fact that fasting increased the mortality rate, there was an apparent tendency for the infective aphids (214 cases) to die at a slower rate than the general population. This is most simply explained by the fact that since the probability of infecting a plant was less than 1, those insects which lived longest had more of a chance to demonstrate their infectivity. Presumably some infective aphids died before they could infect a plant.

From the results (table 4), in spite of a low percentage

**Table 3.—Cumulative mortality table for the aphids which were used in the yellow-net latent period trials, using single insects, the leaf cage technique, and daily transfers to seedling beets.**

| Acquisition Access Time in Hours | Total No. of Aphids Counted | Mean Per Cent of Original Aphids Dead on Day Interval | | | | | | | | | | | | |
|---|---|---|---|---|---|---|---|---|---|---|---|---|---|---|
| | | 0–1 | 1–2 | 2–3 | 3–4 | 4–5 | 5–6 | 6–7 | 7–8 | 8–9 | 9–10 | 10–11 | 11–12 | 12–13 |
| 1 | 337 | 5.6 | 36.6 | 59.9 | 80.1 | 87.8 | 91.4 | 94.6 | 97.3 | 98.8 | 100.0 | | | |
| 2 | 333 | 6.6 | 30.9 | 57.6 | 76.8 | 84.6 | 90.6 | 93.6 | 98.1 | 99.3 | 99.6 | 100.0 | | |
| 4 | 315 | 8.6 | 29.8 | 56.2 | 69.8 | 78.1 | 85.1 | 92.0 | 95.8 | 98.1 | 98.7 | 99.0 | 99.6 | 100.0 |
| 8 | 326 | 6.4 | 28.2 | 46.6 | 59.2 | 70.2 | 79.1 | 88.0 | 94.1 | 97.2 | 98.4 | 99.6 | 100.0 | |
| 24 | 352 | 15.3 | 50.0 | 72.7 | 85.8 | 93.7 | 96.0 | 98.0 | 99.4 | 99.7 | 100.0 | | | |
| 48 | 340 | 19.1 | 48.2 | 71.5 | 85.6 | 93.5 | 96.4 | 97.9 | 98.7 | 99.0 | 99.6 | 100.0 | | |

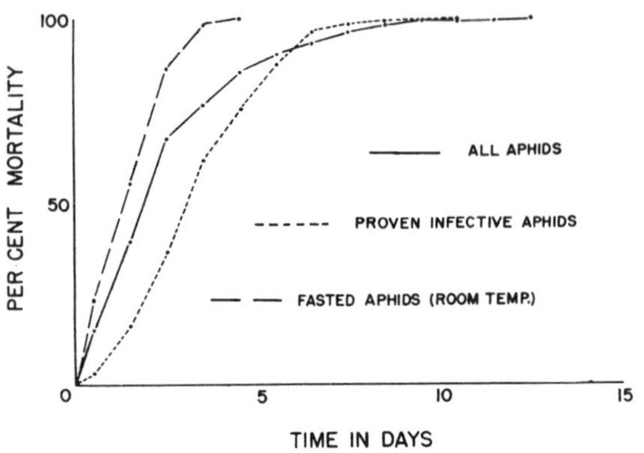


FIG. 1.—Mortality curves of various sub groups of the aphids used in trials made to determine the latent period of the yellow-net virus when transmitted by green peach aphids.

**Table 4.—Results[a] of trials to determine the acquisition threshold period of the beet yellow-net virus when transmitted by single green peach aphids using the leaf cage technique and a variable acquisition.**

| | Acquisition Access Time in Hours | | | | | | |
|---|---|---|---|---|---|---|---|
| | 0 | 1 | 2 | 4 | 8 | 24 | 48 |
| | 2/960[b] | 4/361 | 6/361 | 7/342 | 45/422 | 71/426 | 95/384 |
| Per cent | 0.2 | 1.1 | 1.7 | 2.0 | 14.0 | 16.7 | 24.7 |

[a] In ratios listed, numerator is number of insects infective; denominator, number tested.

[b] Check insects were tested in groups of 4, 5, or 10. One check plant was infective in each of two trials. Since lots of 5 to 10 insects were used, failure to get more than one plant infected in a series indicated that these infections were due to accidental spread in the greenhouse rather than infective source colonies of insects.

of contamination, it can be concluded that acquisition of virus increased most rapidly during the first 8 hours of acquisition access time on the virus source plant. The less rapid gain in infectivity during the 24- and 48-hour periods when compared with the 8-hour period, suggests that after the initial rapid rise there continued to be a slow increase in the number of individuals acquiring virus when access to a virus source was allowed to continue.

When these results are compared with those previously reported (Sylvester 1949) there is some difference. For example, in the former work transmission was obtained with an acquisition feeding of 5 minutes. Much of the difference was probably due to the lower level of infectivity in the present work. The proportion of plants infected using a 24-hour acquisition access and one 24-hour test access time in the present work was 5.3% (21/395), whereas in the former work, it was approximately 26% based upon 125 examples. In addition, in the previous work, a watched-timed acquisition feeding period was used for the intervals of 2, 4, 5, 10, 15, 20 and 30 minutes (weighted $\bar{x} = 12.3$ min.), while in the present work all acquisition times (minimum of 1 hour) were access periods in leaf cages. In the former work, combining all the results of the mentioned intervals gave a total of eight infections with 210 insects, while in the present work, only four infective aphids out of 301 tested were obtained with the 1-hour acquisition access interval, and this was only possible under conditions of serial transfer where most of the insects had more than one opportunity to demonstrate their infectivity. Thus the relatively low level of vector infectivity combined with the effects of the leaf cage technique probably accounts for the differences between this and the former work.

*Latent period.*—Table 5 gives the results considered in establishing the latent period, based upon the number of plants infected in the serial transfers of individual

**Table 5.—Results[a] of trials to determine the amount of transmission of the yellow-net virus to sugar beets by single green peach aphids, allowed a variable acquisition access time after which they were transferred at approximately 24 hr. intervals to a series of healthy test beets until death.**

| Trial No. | Acquisition Access Period in Hours | | | | | | | Total[c] |
|---|---|---|---|---|---|---|---|---|
| | 0[b] | 1 | 2 | 4 | 8 | 24 | 48 | |
| 4 | 0/8 | 0/42 | 0/40 | 0/41 | — | 6/52 | — | 6/175 |
| 5 | 0/21 | 0/54 | 0/65 | 2/63 | — | 6/90 | — | 8/272 |
| 6 | 0/22 | 0/78 | 0/85 | 5/101 | 15/116 | 12/90 | 25/79 | 57/549 |
| 7 | 1/55 | 2/72 | 0/77 | 1/77 | 6/71 | 9/63 | 13/94 | 31/454 |
| 8 | 1/40 | 0/71 | 0/96 | 0/71 | 7/91 | 5/79 | 2/55 | 14/463 |
| 9 | 0/35 | 0/48 | 1/60 | 1/53 | 0/57 | 2/31 | 3/62 | 7/311 |
| 10 | 0/35 | 2/59 | 5/67 | — | 27/141 | 55/137 | 84/146 | 173/550 |
| 11 | 0/52 | 0/83 | 1/84 | 0/87 | 1/95 | 0/71 | 5/84 | 7/504 |
| 13 | 0/29 | 0/57 | 0/57 | 0/53 | 2/67 | 0/49 | 2/41 | 4/324 |
| 14 | 0/32 | 0/50 | 0/58 | 0/55 | 1/58 | 4/59 | 5/65 | 10/345 |
| 15 | 0/33 | 0/69 | 0/58 | 0/69 | 1/63 | 1/48 | 8/76 | 10/383 |
| 16 | 0/33 | 0/64 | 0/64 | 0/70 | 1/71 | 5/61 | 8/83 | 14/413 |
| 17 | 0/37 | 0/76 | 0/64 | 0/78 | 3/98 | 3/61 | 2/59 | 8/436 |
| 18 | 0/84 | 0/140 | 0/143 | 0/146 | 1/132 | 0/103 | 1/78 | 2/742 |
| 19 | 0/68 | 0/130 | 0/105 | 0/123 | 2/187 | 0/217 | 1/98 | 3/770 |
| 20 | 0/68 | 0/132 | 0/124 | 0/160 | 4/179 | 1/92 | 5/99 | 10/786 |
| Total | 2/652 | 4/1225 | 7/1247 | 9/1247 | 71/1426 | 109/1213 | 164/1119 | 364/7477 |

[a] In ratios listed, numerator is number of plants infected; denominator, number tested.

[b] Insects, in lots of 4, 5, or 10 insects per lot, tested on number of plants indicated by denominator, with a total of 960 insects being used.

[c] Totals in this column do not include check (0 acquisition access time).

**Table 6.—Patterns[a] of yellow-net virus transmission by single green peach aphids given a variable acquisition access time, and then serially transferred at approximately 24-hour intervals to healthy beet seedlings until death.**

| Acquisition Access Time | Frequency |
|---|---|
| *1 Hour* | |
| − + − × | 1 |
| − + − − − − × | 1 |
| − − + − − × | 1 |
| − − − − − − + × | 1 |
| *2 hours* | |
| − − + × | 1 |
| − − + − − × | 1 |
| − − − + − × | 1 |
| − − − + − + − × | 1 |
| − − − − + − − × | 1 |
| − − − − − + × | 1 |
| *4 hours* | |
| + − − − + × | 1 |
| − + − − × | 1 |
| − − + − − × | 1 |
| − − + − − − − × | 1 |
| − − − + − × | 1 |
| − − − + − + − × | 1 |
| − − − − + − − × | 1 |
| *8 hours* | |
| + × | 1 |
| + − − × | 2 |
| + − + − × | 1 |
| + + − − − − 0 × | 1 |
| + + + + + + − × | 1 |
| − + − × | 1 |
| − + − − × | 1 |
| − + + − − × | 1 |
| − + − − − − − × | 1 |
| − + − − + + − − × | 1 |
| − + + − + − + − − × | 1 |
| − − + × | 1 |
| − − + − × | 1 |
| − − + − − × | 1 |
| − − + + − × | 1 |
| − − + + + × | 1 |
| − − + − + + × | 1 |
| − − + + + − × | 1 |
| − − + − − − − × | 1 |
| − − + − − − − − − × | 2 |
| − − + − − − − − − − × | 1 |
| − − − + × | 2 |
| − − − + − − × | 4 |
| − − − + + − × | 1 |
| − − − + − + × | 1 |
| − − − + − − + × | 2 |
| − − − − + − × | 1 |
| − − − − + − − × | 2 |
| − − − − + − − − × | 1 |
| − − − − + − + − × | 1 |
| − − − − − + − × | 4 |
| − − − − − + + × | 1 |
| − − − − − − + − × | 2 |
| *24 Hours* | |
| + × | 2 |
| + − × | 3 |
| + + × | 4 |
| + − − × | 4 |
| + − + × | 1 |
| + + − × | 5 |
| + + + × | 1 |
| + − + − × | 1 |
| + + − − × | 1 |
| + − + − − − × | 1 |
| − + × | 6 |
| − + 0 × | 1 |
| − + − × | 4 |
| − + + × | 1 |
| − + − − × | 1 |
| − + − + × | 1 |
| − + + − × | 5 |
| − + + + × | 1 |
| − + − + − × | 3 |
| − + + + − × | 2 |
| − + − − + − × | 1 |
| − − + − × | 10 |
| − − + + × | 3 |
| − − + + − × | 2 |
| − − + − − − × | 1 |
| − − − + × | 2 |
| − − − + − − × | 2 |
| − − − + − + − − × | 1 |
| − − − − + − × | 1 |
| *48 hours* | |
| + × | 4 |
| + − × | 15 |
| + + × | 1 |
| + − − × | 4 |
| + + − × | 2 |
| + + + × | 2 |
| + + − − × | 5 |
| + + + − × | 3 |
| + + + + × | 1 |
| + − − − − × | 1 |
| + − + − − × | 1 |
| + − − + − × | 2 |
| + − + + − × | 1 |
| + + + + − × | 3 |
| + − − − − − × | 1 |
| − + × | 2 |
| − + − × | 9 |
| − + + × | 4 |
| − + − − × | 5 |
| − + + − × | 3 |
| − + + + × | 2 |
| − + − − − × | 1 |
| − + + − − × | 1 |
| − + + + − × | 3 |
| − + + + + × | 1 |
| − + − + − − × | 2 |
| − + + + − − × | 1 |
| − + + + 0 − × | 1 |
| − + − − − − − × | 1 |
| − + + + + + + × | 1 |
| − − + × | 3 |
| − − + − × | 4 |
| − − + − − × | 1 |
| − − + + − × | 2 |
| − − − + × | 1 |
| − − − + − − × | 1 |

[a] Each symbol represents a plant fed upon. Minus (−) indicates no infection, plus (+) indicates infection, (×) indicates the day the insect died, and zero (0) indicates death of the test plant before it could be read.

**Table 7.—Results of analysis of data relevant to the determination of a latent period in the transmission of the sugar beet yellow-net virus by green peach aphids.[a]**

| Acquisition Access Period | Test Feeding Interval | Plants Infected: Observed | Plants Infected: Expected[b] | Plants Not Infected: Observed | Plants Not Infected: Expected |
|---|---|---|---|---|---|
| 48 hours | 1 | 46 | 50.6 | 49 | 44.4 |
| | 2 | 53 | 48.5 | 37 | 42.5 |
| | 3 | 38 | 38.9 | 35 | 34.1 |
| | $\chi^2=2.27$; d.f.$=2$; $p=0.50$ | | | | |
| 24 hours | 1 | 23 | 32.2 | 48 | 38.8 |
| | 2 | 37 | 31.3 | 32 | 37.7 |
| | 3 | 29 | 24.9 | 26 | 30.1 |
| | 4 | 17 | 17.7 | 22 | 21.3 |
| | $\chi^2=7.99$; d.f.$=3$; $p=0.05$ | | | | |
| 8 hours | 1 | 6 | 11.9 | 39 | 33.1 |
| | 2 | 8 | 11.6 | 36 | 32.4 |
| | 3 | 15 | 11.6 | 29 | 32.4 |
| | 4 | 14 | 10.6 | 26 | 29.4 |
| | 5 | 12 | 9.2 | 23 | 25.9 |
| | $\chi^2=9.48$; d.f.$=4$; $p=0.05$ | | | | |

[a] Single aphids were allowed a variable acquisition access period on a virus infected beet, and then were transferred until death to a series of healthy test sugar beet seedlings at 24-hour intervals.
[b] Expected based upon number of plants infected out of number tested.

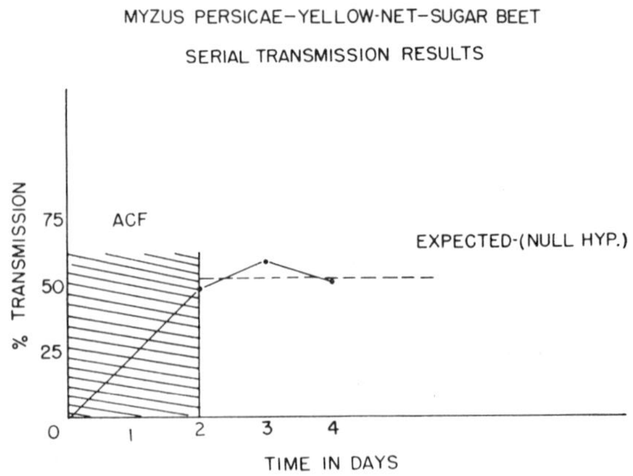


FIG. 2.—Graph of relevant results of trials made to determine the presence of a latent period in the transmission of the yellow-net virus by green peach aphids. Single aphids were allowed a 48-hour acquisition access feeding period after which they were serially transferred to test seedlings at daily intervals until death.

aphids, sorted as to length of the acquisition access time and as to the individual trials. The heterogeneity due to treatment and to trial is pronounced. It is evident that a considerable number of insects tested failed to inoculate plants in the test series. This could have been due to individuals not acquiring an infective charge of virus, or death occurring before they could demonstrate their infectivity. Which of these situations prevailed, or to what proportion both were true, is not known. Some approximations could be used to estimate the proportion of each, but it is not believed that it is essential to the interpretation of these data in connection with the latent period. The remainder of the discussion is limited to those individuals which were found to be infective sometime during their test feeding sequence. The individual patterns of transmission and the frequency of their occurrence are given in table 6.

For analysis, the following assumptions were made concerning the test populations.

1. If there is no latent period, the probability of transmission will be at a maximum at the completion of the acquisition access period.
2. The length of the acquisition access period determines the level of transmission, and measurement of the latent period can only be made after completion of the acquisition period.
3. The only deviations from the expected rate of transmission relevant to the question of a latent period are those which occur in the first few days of testing.

The data were tested in the following three cases. (1) acquisition access period 48 hours, (2) 24 hours, and (3) 8 hours. The results of the analysis are given in table 7, and graphically presented in figs. 2, 3, and 4.

Use of a 48-hour acquisition access period practically eliminated the possibility of demonstrating a latent period. The results obtained with both the 24-hour and the 8-hour acquisition access periods were of more value. In these cases the chi square value was at the 5% level of significance, indicating a rather strong tendency for the evidence not to support the null hypothesis, *viz.*, the variation that occurred was due to chance. Shortening the acquisition access time allowed more of the latent period to be measured.

Table 8 compares the time and frequency with which aphids given an 8-hour acquisition access period, first began virus transmission, with a distribution expected if no latent period was present. Again there is statistical support for the evidence of a latent period. Of interest is that the means of the two distributions are approximately the same. However, the observed resulted from a situation in which the probability of the event (the appearance of a new vector) was increasing (due to the latent period), while the other is the mean of a geometric series in which

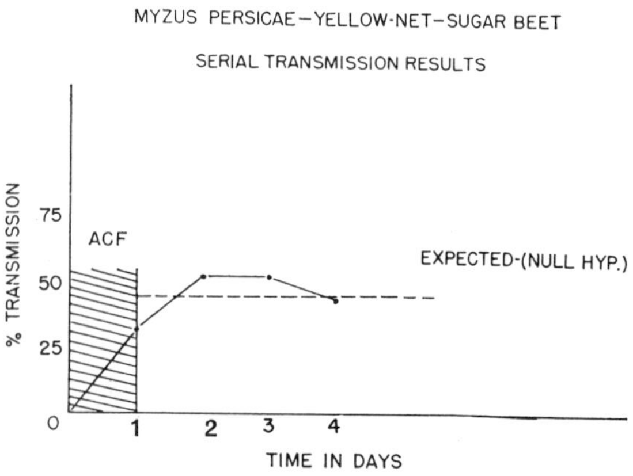


FIG. 3.—Graph of relevant results of trials made to determine the presence of a latent period in the transmission of the yellow-net virus by green peach aphids. Single aphids were allowed a 24-hour acquisition access feeding period after which they were serially transferred to test seedlings at daily intervals until death.

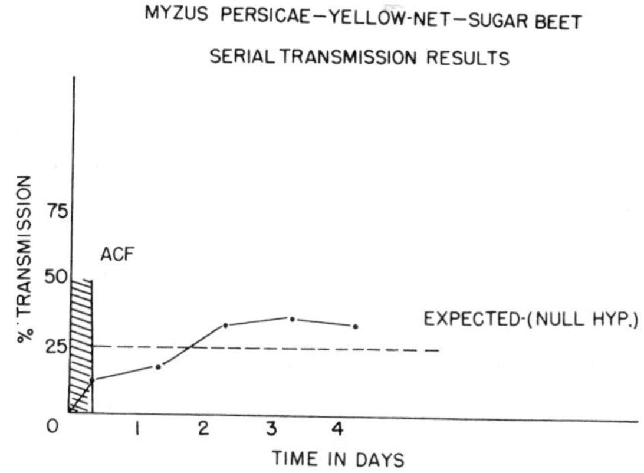


FIG. 4.—Graph of relevant results of trials made to determine the presence of a latent period in the transmission of the yellow-net virus by green peach aphids. Single aphids were allowed an 8-hour acquisition access feeding period after which they were serially transferred to test seedlings at daily intervals until death.

the probability of the event is decreasing.

DISCUSSION.—In contrast with leafhopper vector-virus relationships, in which latent period phenomena frequently have been well established, the question as to whether or not they exist in aphid vector-virus relationships is more pressing. With aphids there are limiting factors in many latent period investigations. Among these are (1) the relatively long period of feeding required to insure both adequate virus acquisition and adequate inoculation, (2) the relatively short life of aphids moved at frequent intervals, (3) the brevity of the latent period in many cases, and (4) the tendency for the probability of transmission to decrease as the test feeding sequence is prolonged.

**Table 8.—Comparison of results obtained in transmission of sugar-beet yellow-net virus by single green peach aphids with those expected under a "no latent period" hypothesis.**[a]

| TEST FEEDING PERIOD NUMBER | NUMBER OF APHIDS FIRST TRANSMITTING | |
|---|---|---|
| | Observed | Expected[b] |
| 1 | 6 | 11.7 |
| 2 | 6 | 8.6 |
| 3 | 11 | 6.4 |
| 4-7 | 22 | 18.3 |
| | Mean 3.6 | 3.9 |

$\chi^2 = 7.07$; d.f. $= 2$; $p = < 0.05$

[a] The acquisition access period was 8 hrs., and each serial test feeding access period was approximately 24 hours.
[b] Based upon 71 infections out of 274 tests.

Mistakingly emphasizing the importance of a minimum latent period, the author (Sylvester 1949) published the conclusion that in the transmission of sugar beet yellow-net virus by green peach aphids a definite latent period did not exist. The results indicated that there was no period of time beyond that occupied by the acquisition and inoculation threshold periods in which transmission could not be demonstrated, providing enough individuals were tested. This still may be true, but it does not mean that in properly designed serial transmission trials there will not be a delay in the development of infectivity within the test populations.

Thus the beet yellow-net virus seems to be another of the aphid-borne persistent plant viruses which possess a measurable latent period. The best known other examples of this group are potato leafroll (Smith 1931, Day 1955), pea-enation mosaic (Osborn 1935, Chaudhuri 1950, Simons 1954), filaree red-leaf (Anderson 1951), and strawberry virus 3 (Prentice & Woollcombe 1951). There are others which might be listed, but to date experimental work upon them is less detailed, or their status is somewhat uncertain, *e.g.*, carrot motley dwarf (Stubbs 1948).

## REFERENCES CITED

**Anderson, C. W.** 1951. The insect vector relationships of the filaree red-leaf virus, with special reference to a latent-period difference between nymphs and adults in *Macrosiphum geranicola* (Lambers). Phytopath. **41:** 699–708.

**Chaudhuri, R. P.** 1950. Studies on two aphid-transmitted viruses of leguminous crops. Ann. App. Biol. **37:** 342–54.

**Day, M. F.** 1955. The mechanism of the transmission of potato leafroll virus by aphids. Australian Jour. Biol. Sci. **8:** 498–513.

**Osborn, H. Y.** 1935. Incubation period of pea mosaic in the aphid *Macrosiphum pisi*. Phytopath. **25:** 160–77.

**Prentice, I. W.,** and T. M. Woollcombe. 1951. Resolution of strawberry virus complexes. IV. The latent period of virus 3 in the vector. Ann. Appl. Biol. **38:** 389–94.

**Simons, J. N.** 1954. Vector-virus relationships of pea-enation mosaic and the pea aphid *Macrosiphum pisi* (Kalt.). Phytopath. **44:** 283–9.

**Smith, K. M.** 1931. Studies on potato virus diseases. IX. Some further experiments on the insect transmission of potato leafroll. Ann. Appl. Biol. **18:** 141–56.

**Stubbs, L. L.** 1948. A new virus disease of carrots: its transmission, host range, and control. Australian Jour. Sci. Res. B **1:** 303–32.

**Sylvester, E. S.** 1949. Transmission of sugar beet yellow-net virus by the green peach aphid. Phytopath. **39:** 117–32.

**Sylvester, E. S.** 1956. Beet yellows virus transmission by the green peach aphid. Jour. Econ. Ent. **49:** 789–800.

# F Energetics

## The Role of Soil Arthropods in Community Energetics

Manfred D. Engelmann

*Department of Natural Science, Michigan State University, East Lansing, Michigan 48823*

Synopsis. Though the primary development of principles of energetics was pioneered in aquatic studies, considerable information has accrued from more recent investigations of terrestrial species. Studies of soil organisms—particularly arthropods—have given us some understanding of the applicability of principles of energetics to terrestrial communities. Data from soil communities have yielded an annual energy budget for oribatid mites, ecological efficiencies for two trophic levels, and a series of herbivore:carnivore ratios for a number of communities. Energy-flow analyses of the problems of species abundance in community organization, the relationship of population production and maintenance, and the energy relationships of the soil to above-ground communities have also been attempted. As data accumulate, however, the inadequacy of the Lindeman model becomes apparent, and there are prospects for change in this conceptual scheme.

Much of the earliest work on energy transfers in ecological systems was done in fresh-water lakes, and investigations of this nature culminated in the important theoretical formulations of Lindeman (1942). Viewed in this historical perspective, it is interesting to note that Bornebusch (1930) investigated the metabolism of forest arthropods more than 10 years before Lindeman's paper appeared. Bornebusch, a forest entomologist, was strongly influenced in his thinking by the contemporary physiologist, Krogh, who emphasized that the total metabolism of forest arthropods was as important a consideration as knowledge of total numbers. At Krogh's urging, Bornebusch studied soil animals of five different forest types, not only from the standpoint of composition and abundance of species, but also in terms of their metabolic activity. Although Bornebusch's work suffered from certain technical inadequacies, it continues to provide an important baseline for contemporary research. In fact, most of the studies in England and Europe which can be considered "energetic" are not patterned after Lindeman's model, but follow the Bornebusch tradition of emphasizing respiratory metabolism (*e.g.*, the work of Macfadyen (1963) in England, and Berthet (1963) in Belgium). There is, however, no reason why the soil community cannot be considered in terms of Lindeman's views.

Let us look just for a moment at the basic structure of the Lindeman system. In Figure 1 the trophic level is represented by the circular portion of the diagram. The trophic level consists of the energy within the bodies of the individual populations assigned to the level. The populations form a superstructure through which energy flows. The trophic level is not stable, for if its energy were cut off all the populations would die and the trophic level would disappear. And if the energy flow were to be resumed, a trophic level would reappear, but it might well have an entirely different species composition than before. In fact, populations do not make up trophic levels. Rather it is the *energy* that binds the elemental substances in the bodies of individuals. The mass of the body (*i.e.*, carbon, nitrogen, hydrogen, etc.) is not represented at all in the Lindeman trophic level. This point emphasizes the level of abstraction with which we are dealing. The arrows in Figure 1 indicate the flow of energy through the trophic level, and $\lambda_x$ represents ingestion. Of the total ingested energy, a portion enters the energy pool of the trophic level as assimilated energy, and another portion is egested. If one eats a handful of sawdust (with an energy content of about 5000 cals/g), it enters the digestive tract, but in 48 ± hours it will be defecated as sawdust and it will still contain 5000 cals/g of energy. The sawdust never entered another trophic level in spite of the fact that it was ingested. In other words, the lumen

Reproduced with permission from the *American Zoologist*, 8:61–69, 1968. Published by the American Society of Zoologists, Thousand Oaks, California.

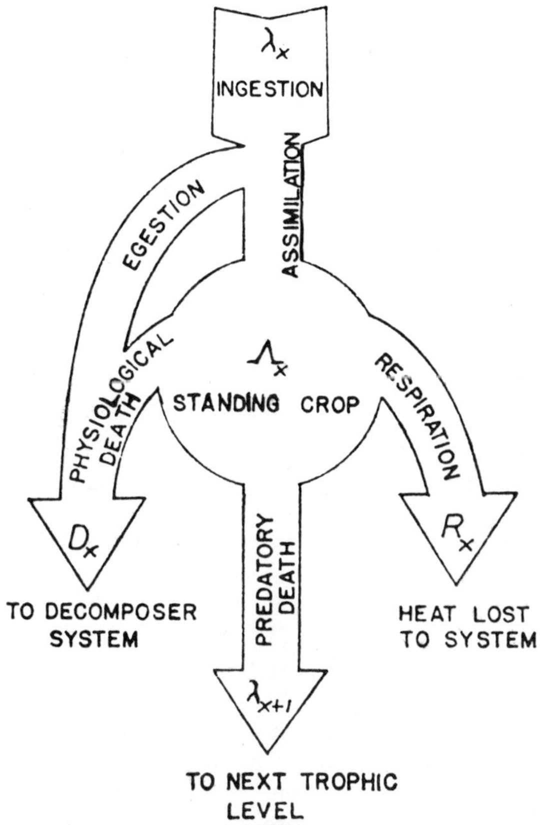


FIG. 1. Diagram of a single trophic level of the Lindeman system. Arrows indicate the direction of energy flow through the level. The symbols "Lambda," "D," and "R" are used in the manner of Lindeman (1942).

of the intestine is simply a tunnel of the external environment through the body as far as energetics are concerned. The other arrows in Figure 1 represent the different pathways by which assimilated energy may leave or be used in a given trophic level: respiration, physiological death, and death by predation. The simplest manner in which to visualize a community is to imagine three trophic levels: primary producers (plants), herbivores (plant eaters), and carnivores (eaters of other animals). Actually, of course, the community is much more complex.

Any useful model should not only accurately reflect nature, but also exhibit simplicity and order. The part of nature I choose to consider is the soil system—a portion of the ecosystem with several attributes useful in energy studies. One of the most intriguing features of the soil fauna is the enormous diversity of organisms, many of them very poorly understood. There is an almost endless variety of creatures—many no bigger than a speck of dust—living their clandestine lives among the rotting leaves and twigs of the forest floor: mites and collembolans, pseudoscorpions and centipedes. These tiny creatures are everywhere—just one stoop away in the nearest clump of grass or bunch of moss; in frozen winter earth, moist fall and spring soils, or dry summer sod. Animals can be obtained in great numbers from a lawn or any mulched area almost any time of the year (the poorest time is mid-summer). All that is necessary to bring them forth is a goose-neck lamp and coffee can (with a hole cut in the bottom and a screen inserted) set over a beaker of water. Twenty-four hours of heat from the lamp will drive the animals from the soil into the beaker. A few quick strokes of a camel's hair brush will remove them to a salve jar lined with moist cellucotton, paper towel, or filter paper; and observations can be made for hours or days. With proper feeding the animals will go through their life cycles, and cultures can be kept for months or even years.

Soil organisms are not only diverse, but also represent every trophic level—producers, herbivores, carnivores, and parasites. Populations are usually dense, and samples can be relatively small. A soil sample 2″ in diameter from glacial soils of a Michigan old-field have yielded as many as 450 individuals representing 13 species of oribatid mites alone. From 35 samples, Hairston and Byers (1954) counted a total of over 74,000 individuals of 177 different kinds of soil arthropods taken from this same old-field soil—an average of over 2000 individuals per sample. Soils over large areas are fairly uniform, and to that degree, observations can be repeated. There is more uniformity in species composition than in numbers. The soil system lends itself to manipulation. Clipping, fertilizing, shading, watering, and numerous other simple treatments can be used to alter the conditions in soil, thus exposing the effects of

TABLE 1. *A list of the major groups of organisms found in soils.*

| Group | Function and Food Source |
|---|---|
| Bacteria | |
| Autotrophic | Autotrophic; synthesis of carbon compounds. |
| Heterotrophic | Decay organisms; rapid decomposers. |
| Actinomycetes | Decay organisms; slow decomposers. |
| Fungi | Decay organisms; rapid decomposers (one predator). |
| Protozoa | Herbivorous on certain bacteria or predaceous. |
| Nematoda | Feed on bacteria; predatory. |
| Rotifera | Feed on bacteria. |
| Annelida | Feed on dead plant material. |
| Arthropoda | |
| Hexapoda | |
| Coleoptera (mostly larval forms) | Predatory or mycophagous. |
| Collembola | Feed on fungi (predatory on nematodes?). |
| Diptera (all larval forms) | Feed on bacteria and fungi; predatory. |
| Acarina | Feed on fungi and bacteria; predatory. |

such alterations on soil populations.

Studies of the soil system are not, however, without problems. The taxonomy of the U. S. soil arthropods is, in general, seriously incomplete. Also, because the animals may be extremely small, technical problems arise when equipment must be miniaturized and/or sensitized to deal with such species.

The major groups of soil organisms are summarized in Table 1. Many groups which are present in the soils are not represented in this table. Any one of three reasons may account for this situation: (1) the animals are a minor component in the soil (*e.g.*, algae, centipedes, millipedes, pseudoscorpions, moles); (2) the animals may burrow in the soil, but feed primarily upon components of the above-ground community (*e.g.*, ants, shrews, foxes, etc.); or (3) the animals, although abundant, may not yet be recognized as members of the soil community (*e.g.*, harvestmen, robins). This paper is concerned with the soil arthropods, which are second only to the nematodes in abundance in the soil. What have soil arthropods contributed to our knowledge of community energetics?

First, we have been able to make some estimates of the total productivity of some soil arthropods. Using the Oribatei (an abundant group of mites found in most soils), I estimated ingestion, egestion, respiration, and reproduction, using several laboratory approaches (Engelmann, 1961). By applying these estimates to counts of field populations, an energy-flow chart was constructed (Fig. 2). These values were estimated for animals found in an area one meter square and 12.5 cm deep. As can be seen, 75% of the ingested energy is egested, and of the 25% assimilated 95% is dissipated in respiration. Because the figures for each arrow were obtained separately, the total energy budget does not balance exactly. There is a 5% deficit in the ingestion-assimilation portion of the diagram and a 16% surplus in the figures for metabolism.

One of the major concepts of the Linde-

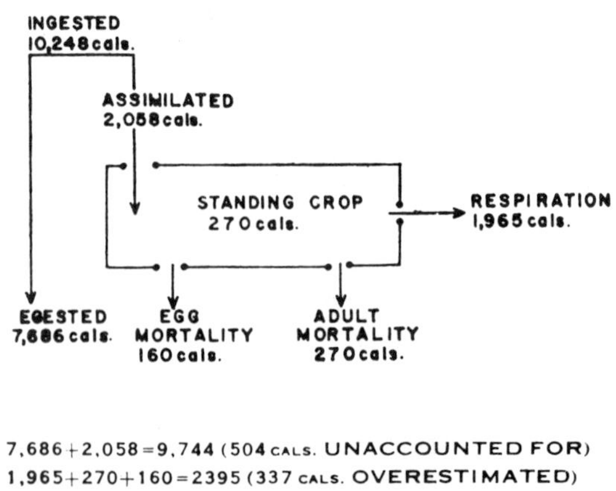


FIG. 2. Annual energy balance sheet for the oribatid mites found in the soils of an old-field in southeastern Michigan. Original data from Engelmann (1961).

TABLE 2. *Average numbers of individuals, biomass, and respiration of the soil arthropods found in a Michigan old-field. Values pertain to volume 1 m² and 20.3 cm deep. (Data from Engelmann, 1961).*

| Category | Average Number | Biomass (mg) | Standing Crop (cal) | Respiration (cal) | Food Habits |
|---|---|---|---|---|---|
| ACARINA | | | | | |
| Parasitoid | 12414.8 | 9.330 | 46.650 | 354.37 | Carn. |
| Oribatid | 87582.6 | 28.596 | 142.980 | 115.96 | Herb. |
| Sarcoptoid | 4721.8 | 3.466 | 17.330 | 138.66 | Herb. |
| Trombidoid | 1797.8 | 1.722 | 8.610 | 53.33 | Carn. |
| Eupodoid | 9191.0 | 3.262 | 16.310 | 136.42 | Herb. |
| | 20396.4 | 5.204 | 26.025 | 233.32 | Herb.? |
| | 2543.0 | 1.974 | 9.870 | 67.06 | Carn. |
| | 236.4 | 0.042 | 0.210 | 2.57 | Carn,? |
| Pauropoda | 679.9 | 0.244 | 1.220 | 10.73 | Herb. |
| Symphyla | 51.3 | 0.418 | 2.090 | 11.52 | Herb. |
| Protura | 662.1 | 2.387 | 11.935 | 66.53 | Herb. |
| INSECTA | | | | | |
| Japygida | 388.5 | 13.472 | 67.365 | 298.78 | ? |
| Podurid | 7764.2 | 7.534 | 37.670 | 274.81 | Herb. |
| Entomobryid | 243.2 | 2.095 | 10.475 | 53.60 | Herb. |
| Sminthurid | 175.7 | 0.121 | 0.605 | 3.64 | Herb. |

TABLE 3. *Estimated ecological efficiencies of the soil herbivores of a Michigan old-field. (Data from Engelmann, 1961).*

| Assimilation (%) | | Ecological efficiency (%) | | |
|---|---|---|---|---|
| Carnivores | Herbivores | Without *Japex* | *Japex* Carnivorous | *Japex* Herbivorous |
| | 32 | 18.9 | 27.5 | 16.9 |
| 32 | 25 | 15.4 | 22.9 | 13.7 |
| | 14 | 9.6 | 14.7 | 8.5 |
| | 32 | 23.0 | 32.7 | 20.7 |
| 25 | 25 | 18.9 | 27.5 | 16.9 |
| | 14 | 12.0 | 18.1 | 10.6 |
| | 32 | 32.9 | 45.4 | 30.1 |
| 14 | 25 | 28.1 | 39.4 | 25.5 |
| | 14 | 18.9 | 27.5 | 16.9 |

man model is the efficiency of transfer from one trophic level to another; sometimes called Ecological Efficiency (EE). For soil arthropods this can be given as:

$$EE = \frac{\text{cals ingested by carnivores}}{\text{cals ingested by herbivores}} \times 100.$$

The above formulation essentially asks "How effectively do the members of one trophic level use the members of another?"

Applying the information gained from populations of oribatid mites, energy estimates were made for the more extensive field data of Hairston and Byers (1954). These data are summarized in Table 2. The figures were used to estimate ecological efficiency. Two uncertainties arose in these last calculations. What is the efficiency of assimilation by animals in the field? An assimilation of 32% represents a starvation regime, while 14% represents a surfeit. The second uncertainty lay in the food habits of the thysanuran, *Japex,* which may be carnivorous or herbivorous. Because of these problems I have calculated a spectrum of possible efficiencies, rather than a single estimate, and this range of values is given in Table 3. The array of ecological efficiencies ranges from 8.5% to 45.4%, depending on the assumptions adopted, but most of them lie between 14 and 28%. These efficiencies fall within the range of values found for various aquatic communities and support the applicability of the Lindeman system to terrestrial as well as aquatic communities.

TABLE 4. *The comparison of percent or rank of annual caloric dissipation, and predator:herbivore ratio of different trophic groups found in nine natural communities on the British Isles and the European Continent. (Data from Macfadyen, 1963).*

| Community | 1 Grassland | 2 Limestone Grass | 3 *Juncus* Moor | 4 Oak Mull | 5 Beech R H | 6 Beech Mull | 7 Beech Mull | 8 Spruce Mull | 9 Spruce R H |
|---|---|---|---|---|---|---|---|---|---|
| Herbivore rank | 8 | 7 | 3 | 4 | 9 | 6 | 2 | 5 | 1 |
| Herbivore % | 22.29 | 25.65 | 31.77 | 30.85 | 20.36 | 27.25 | 32.08 | 28.36 | 32.91 |
| Large decomposer rank | 4 | 1 | 8 | 2 | 5 | 3 | 6 | 7 | 9 |
| L. decomposer % | 27.36 | 55.54 | 1.21 | 33.49 | 8.42 | 31.24 | 3.46 | 3.3 | 0.54 |
| Small decomposer rank | 4 | 9 | 1 | 6 | 2 | 8 | 3 | 5 | 7 |
| S. decomposer % | 39.32 | 17.38 | 64.18 | 29.87 | 46.06 | 23.92 | 45.41 | 36.48 | 29.12 |
| Predator rank | 6 | 9 | 8 | 7 | 3 | 5 | 4 | 2 | 1 |
| Predator % | 10.95 | 1.38 | 2.85 | 5.79 | 25.15 | 17.58 | 19.05 | 31.86 | 37.43 |
| Rank total Kcal/yr | 1 | 2 | 4 | 7 | 9 | 5 | 3 | 8 | 6 |
| Carnivore:herbivore Ratio | 0.1232 | 0.0140 | 0.0293 | 0.0614 | 0.3360 | 0.2133 | 0.2353 | 0.4675 | 0.5981 |

When we consider the uniformity of energy flow through communities, the data from soil arthropods prove more perplexing. Because nearly 80-90% of an animal's assimilated energy is normally respired, it is not surprising that early attempts to compare different communities relied upon measurements of metabolism. Macfadyen (1963) analyzed the animals found in the soil of several different communities in England and Europe. I have re-analyzed these data and estimated the calories consumed by each trophic group (Table 4). Then I have calculated the respiration of herbivores and carnivores as fractions of the total community respiration (Engelmann, 1966, p. 94). If there is functional uniformity in communities, the ratio of calories processed by the carnivores versus the herbivores should be about the same for all communities. As can be seen from Table 4, the data on soil arthropods do not support that hypothesis (the ratios vary from 0.01 to 0.59). The only trophic level which made up a relatively constant percentage for all nine communities was the Herbivore level (20-33%). The implications of these data are not yet clear.

In reviewing the productivity of animals in natural communities (Engelmann, 1966), I developed an empirical relationship relating net productivity to respiration (see Table 5 and Fig. 3). The relationship for

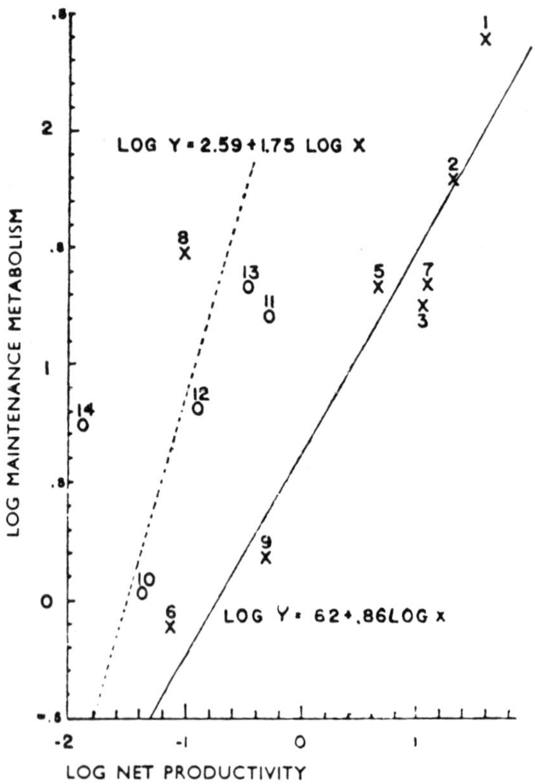


FIG. 3. Relationship between maintenance metabolism and net productivity. Open circles represent data from homeotherms, the "X's" data from heterotherms. The regression for the heterotherms (the solid line) is significant at the 5% level. Based on Engelmann (1966).

TABLE 5. *Productivity, maintenance-metabolism, percent utilization, and ecological efficiency of different animals found in various communities.*

| Taxon | Prod. in Kcal/m²/yr Net | Prod. in Kcal/m²/yr Gross | Maintenance Metabolism in Kcal/m²/yr | % Utilization of Net Production Food | Ecological Efficiency | Habitat | Reference |
|---|---|---|---|---|---|---|---|
| INVERTEBRATES | | | | | | | |
| 1. *Littorina* sp. | 40.6 | 290 | 249.4 | — | — | Salt marsh | Odum and Smalley (1959) |
| 2. Nematodes | 21.0 | 85 | 64.0 | — | 6.8 | Salt marsh | Teal (1962) |
| 3. *Orchelimum fidicinium* | 10.8 | 29.4 | 18.6 | 2 | — | Salt marsh | Smalley (1960) |
| 4. *Prokelisia* sp. | 70.0 | 275.0 | 205.0 | 4.6 | 6.8 | Salt marsh | Teal (1962) |
| 5. Orthoptera (3 spp.) | 4.0 | 25.6 | 21.6 | 2–7 | — | Old-field | Odum, *et al.* (1962) |
| 6. *Philaenus spumarius* | 0.048–0.096 | 0.582–1.167 | 0.778* | — | — | Old-field | Wiegert (1964) |
| 7. *P. spumarius* | 15.125 | 38.565 | 23.44 | — | 0.8–17.4 | Alfalfa field | Wiegert (1964) |
| 8. *Pogonomyrmex badius* | 0.09 | 14.29–47.79 | 31.03† | 64–213 | — | Old-field | Golley and Gentry (1964) |
| 9. Oribatei | 0.43 | 2.008 | 1.578 | — | 8–30 | Old-field | Engelmann (1961) |
| VERTEBRATES | | | | | | | |
| 10. *Passerculus* | 0.04 | 3.6 | 3.56 | 10–50 | — | Old-field | Odum, *et al.* (1962) |
| 11. *Microtus* | 0.517 | 17.517 | 17.00 | 1–6 | 2.3† | Old-field | Golley (1960) |
| 12. *Peromyscus* | 0.12 | 6.7 | 6.58 | 10–50 | — | Old-field | Odum, *et al.* (1962) |
| 13. *Loxodonta africana* | 0.34 | 23.34 | 23.00 | 9.6 | — | Savannah | Petrides and Swank (1965) |
| 14. *Mustela* | 0.013 | 5.564 | 5.551 | 31.0 | 2.3† | Old-field | Golley (1960) |

* Average.
† Single species.

heterothermic animals represented by the solid line is significant. The significance of the dotted line—pertaining to homeotherms—has not yet been established. Most of the points relating to heterotherms are based on populations of a single species. However, point #9 represents the 0.43 Kcal/m²/yr combined production of 22 species of oribatid mites in an old-field. If the relationship of maintenance metabolism to net production implied by the regression equation is meaningful, the aggregate production of 22 species should be calculable in terms of the estimated metabolism of the individual populations. This procedure is illustrated in Table 6. The

TABLE 6. *Net reproduction of oribatid mites calculated from the formula for maintenance-net reproduction. (Respiratory data from Engelmann, 1961.)*

| Species Number* | Species Name | Net Productivity in cal/m²/yr |
|---|---|---|
| 101 | *Tectocepheus velatus* | 86.0 |
| 102 | *Scheloribates pallidulus* | 39.3 |
| 103 | *Camisia* sp. | 25.0† |
| 108a | *Oppia nova* | 32.4 |
| 108c | *Cultroribula* sp. (*divergens*) | 9.5 |
| 108d | *Suctobelba* | 4.4 |
| 108e | *Oppia minutissima* | 0.6 |
| 109 | *Rhysotritia ardua* | 28.1 |
| 110 | *Peloribates curtipilus* | 27.4 |
| 111 | *Allogalumna alatum* | 17.5 |
| 113 | Adult Oppidae | 1.1 |
| 114 | *Trhypochthonius* sp. (*tectorum*) | 69.5 |
| 115 | *Belba* sp. | 0.5 |
| 117 | *Thyrisoma ovata* | 14.5 |
| 120 | *Liochthonius perpusillus and Brachychthonius jugatus* | 17.2† |
| 121 | *Zygoribatula rostrata* | 27.6 |
| 126 | Tarsonomid | 13.5† |
| 128 | Tarsonomid | 14.3† |
| TOTAL | | 428.4 |

* System of identification used by Engelmann (1961).
† Assumed 1/3 population productive.

resulting total (0.428 Kcal/m²/yr) is almost identical to the independently derived estimate of 0.430 Kcal. I will not make too much of the apparently remarkable accord because the agreement could be somewhat fortuitous. However, I do believe that the equation can give first approximations on the productivity of populations.

Energetics studies also shed light on other questions relating to community organization, *e.g.*, the problem of relative abun-

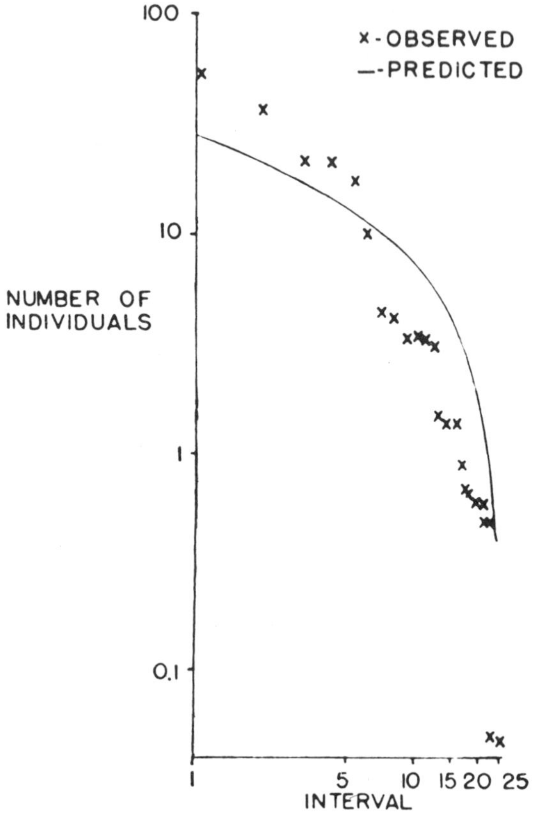


FIG. 4. Comparison of species abundance of oribatid mites from the soil of an old-field with the "broken-stick" model. From Engelmann (1961).

dance of species (see Hairston, 1959; Whittaker, 1965). Several mathematical models have been developed—mostly empirical and with no clear biological basis. However, MacArthur (1957) proposed three hypotheses with an underlying biological rationale. These models differed in the manner in which the environment was assumed to be partitioned among its occupants. One of these models, which visualized the environment as being subdivided into a number of contiguous but non-overlapping niches, has been tested with data from many kinds of animals (see King, 1964, for a review). In general, it has been found that the model underestimates the abundance of the most common species and overestimates the abundance of rare species. The problem is well illustrated by Figure 4, in which the estimated relative abundance of old-field mites is compared with predictions of the model. The point of this discussion is that some functional measure (rather than simply numerical abundance) may be a more meaningful way to consider community organization. Figure 5 shows total respiration of old-field mites plotted against the prediction of the model. Although there are still appreciable disparities, the general agreement seems improved.

Finally, studies on soil systems have given us a different view of the trophic-dynamic organization of the old-field community (Fig. 6). In addition to the above-ground levels, there are at least three soil levels which are dependent upon the energy falling to the soil in the form of dead bodies and feces. The soil trophic levels act more or less as does the Krebs cycle, eventually respiring all energy and dissipating it as heat. This concept is particularly pleasing, because it leaves no calories "sitting around" to be accounted for in some undefined population or trophic level.

These are the facts and concepts contributed by studies on soil arthropods, but

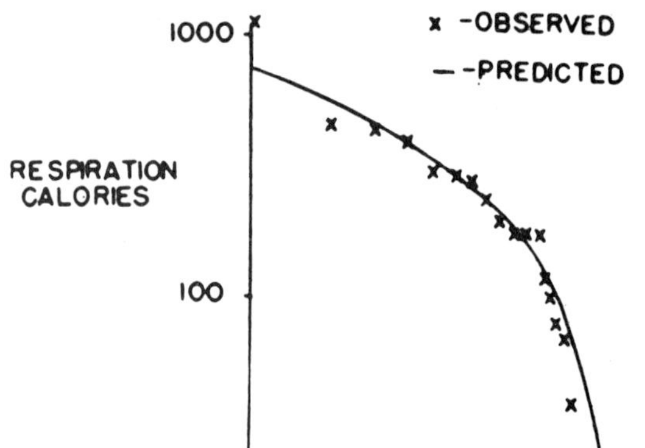


FIG. 5. Comparison of species abundance of mites as represented by respired calories with the MacArthur "broken-stick" model. From Engelmann (1961).

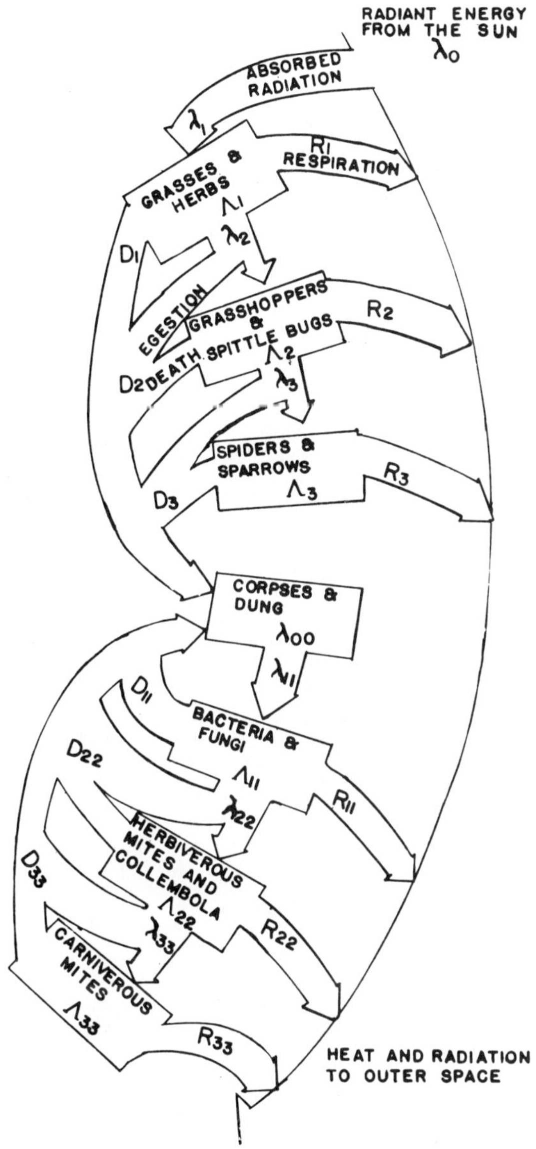


FIG 6. A schematic representation of the energy flowing through the various above-ground and soil trophic levels of an old-field in Michigan. From Engelmann (1961).

we must recognize that these ideas are not necessarily immutable. In energetics as in all endeavors of science, the facts hang on beautifully contrived constructs which we call models and theories. Too often we think of the theories and models as facts, when actually they are merely products of the human mind. Although theories and models provide the guidance and direction for research, these constructs of the human imagination are often useful for a limited time. Currently, the discipline of energetics is undergoing a change. Investigators are becoming disenchanted with the concept of the trophic level. How does one cope with the omnivore, or with the holometabolous insect whose larvae feed differently from the adults? How can one enumerate all of the populations in a given trophic level? These questions and others call for innovations in the Lindeman model as it is now conceived. I believe that these changes will be more accurate descriptively, will be more mathematically elegant, and will provide more precise direction to research on function and organization of the community.

## REFERENCES


Berthet, P. 1963. Mésure de la consommation d'oxygène des Oribatides (Acariens) de la litière des forêts, p. 18-31. *In* J. Doeksen and J. Van Der Drift, [ed.], Soil organisms. North Holland Publ. Co., Amsterdam.

Bornebusch, C. H. 1930. The fauna of the forest soil. Forstlige Forsøgsv. Danmark 11:1-224.

Doeksen, J., and J. Van Der Drift, [ed.]. 1963. Soil organisms. North Holland Publ. Co., Amsterdam.

Engelmann, M. D. 1961. The role of soil arthropods in the energetics of an old-field community. Ecol. Monographs 31:221-238.

Engelmann, M. D. 1966. Energetics, terrestrial field studies, and animal productivity, p. 73-115. *In* J. B. Cragg, [ed.], Advances in ecological research. Academic Press, London and New York.

Golley, F. B. 1960. Energy dynamics of a food chain of an old-field community. Ecol. Monographs 30:187-206.

Golley, F. B., and J. D. Gentry. 1964. Bioenergetics of the southern harvester ant, *Pogonomyrmex badius*. Ecology 45:217-225.

Hairston, N. G. 1959. Species abundance and community organization. Ecology 40:404-416.

Hairston, N. G., and G. W. Byers. 1954. The soil arthropods of a field in southern Michigan. A study in community ecology. Contrib. Lab. Vert. Biol. Univ. Mich. 64:1-37.

King, C. E. 1964. Relative abundance of species and MacArthur's model. Ecology 45:716-727.

Lindeman, R. L. 1942. The trophic-dynamic aspect of ecology. Ecology 23:399-417.

MacArthur, R. H. 1957. On the relative abundance of bird species. Proc. Natl. Acad. Sci. 43:293-295.

Macfadyen, A. 1963. The contribution of the microfauna to total soil metabolism, p. 3-17. *In* J. Doeksen and J. Van Der Drift, [eds.], Soil Organisms. North Holland Publ. Co., Amsterdam.

Odum, E. P., and A. E. Smalley. 1959. Comparison of population energy flow of a herbivorous and a deposit-feeding invertebrate in a salt marsh ecosystem. Proc. Natl. Acad. Sci. 45:617-622.

Odum, E. P., C. E. Connell, and L. B. Davenport. 1962. Population energy flow of three primary consumer components of old-field ecosystems. Ecology 43:88-96.

Petrides, G. A., and W. G. Swank. 1965. Estimating the productivity and energy relations of an African elephant population. Proc. 9th Intern. Grasslands Congr., São Paulo, Brazil.

Smalley, A. E. 1960. Energy flow of a salt marsh grasshopper population. Ecology 41:672-677.

Teal, J. M. 1962. Energy flow in the salt marsh ecosystem of Georgia. Ecology 43:614-624.

Whittaker, R. H. 1965. Dominance and diversity in land plant communities. Science 147:250-260.

Wiegert, R. G. 1964. Population energetics of meadow spittlebugs (*Philaenus spumarius* L.) as affected by migration and habitat. Ecol. Monographs 34:217-241.

# Part V: The Behavior of Insects

Insect behavior is the one aspect of entomology familiar to us all. The seemingly ceaseless activity of ants, the methodical visitations of bees from flower to flower, and the painful accuracy of a female mosquito in search of a blood meal essential for egg production are but a few examples. These casual observations are but fragments of larger behavioral patterns. Nevertheless, these are observations of **behavior** *(the ways in which an organism responds to stimuli from the physical environment or other organisms).*

The stimuli that evoke responses from an organism are of various types. Some are internal and reflect such changes as maturation of internal organs or whole changes in life stage. Among the important internal messengers are **hormones** *(regulating and controlling chemical substances produced and transmitted through the body).* Stimuli may also be received externally from parts of the environment, including other organisms. Stimuli from other organisms many times are transmitted through pheromones, which are much like hormones but act externally and affect the behavior of animals other than the producer.

Characteristics of the observer can play a significant role in the quality of an insect behavior study. The observer's originality and the ability to incorporate mechanical and electronic aids to investigate behavior are some of the manifestations of those characteristics. The more intangible but equally important qualities of the researcher include keenness of observation, patience, and objectivity.

A controversial area within the field of insect behavior is the question of instinct versus learned behavior. Today it is generally agreed that all behavior has a genetic base but that the modifications induced by abiotic and biotic factors play an important part in shaping behavioral patterns. However, the role of **instinct** *(behavior observable in all individuals of a species regardless of differences in habitat)* is still discussed on both theoretical and experimental levels (J. S. Kennedy, *The Experimental Analysis of Aphid Behaviour and its Bearing on Current Theories of Instinct*). Not only does Kennedy present a history of studies on instinct, but he delineates basic principles and concepts in this area such as reflex, priming, inhibition, and stimulation. These principles are presented in terms of the controversy that surrounds many tenets of behavior, but emphasizes theories of instinct. The general definition of instinct, behavior observable in all members of a species despite differences in their environment, is shown by Kennedy to be a drastic oversimplification.

Mating, feeding, communication, and defense have been investigated by behaviorists for many years. Advances in these areas occur from a sequence of investigations which ultimately allow for understanding of each behavior pattern. The initial step is a general description of the behavior pattern. This may be followed by a description of the sensory stimulus (some change that is detectable by the organism's sensory apparatus) that affects the organism and gives the observed behavior. Since the organism's response to stimuli is related to the structure and function of its sensory receptors, the receptors themselves are studied, as is the coding of sensory information by the nervous system.

All animals continually interact with their surroundings. It becomes important, therefore, to consider the role of past experience, heredity, life cycle, and age, on a behavioral response. This and other aspects of comparative behavioral studies can result in an analysis of how the particular behavior developed and of its survival value to the species. Evolutionary origins of the behavior must be included in this investigation.

The articles illustrate certain themes underlying research on insect behavior that apply regardless of the stage of investigation. Thus the orientation to and resultant pattern(s) of response to sensory input are major points of interest in behavioral studies. Secondly, the limitations and adaptive advantages reflected in the structure and function and in the threshold of response of sensory organs represent another overriding theme. Finally, analysis of data can attempt to reflect the relevance of the behavior pattern(s) to species survival.

There are many behavioral aspects of the development and survival of insects that have been studied, and most are relevant to the control of insect pests. Perhaps the most important of these is feeding behavior. Just as there are a number of feeding types (chewing, lapping, piercing-sucking, filtering, etc.), there are also a variety of feeding behaviors. A number of events occur in feeding behavior. The initial problem is host recognition, which may involve the use of any or all senses in a general determination or a specific nutritional selection. The process of feeding itself may not be a single event but may consist of a sequence involving attraction, selection of type, age or area of hosts, inducement to feed, and inducement to continue feeding.

V. G. Dethier *(Feeding Behavior)* details the patterns and sequences of feeding behaviors in various insects. An emphasis on sensory stimuli and their role in feeding leads to discussions of the roles of **token stimuli** *(secondary chemicals which have no nutritional value but may serve in the process of host selection)* and nutritional factors. This paper serves as an introduction to a very large and complex area of behavior.

The role of display is of major importance in most types of behavior. In general, it enables the performance of essential survival tasks and has been shown to be important in aspects such as communication, formation of new colonies, and mating. In mating, for example, one of the main functions of display is mate identification (species recognition). Display in an individual's defense behavior or in the defense of a territory is equally important. Such display may be of several types. M. H. Robinson *(The Defensive Behavior of the Javanese Stick Insect, Orxines macklotti de Haan with a Note on the Startle Display of Metriotes diocles Westw.)* describes two types that occur in walking sticks. Defensive display may entail aggressive postures such as increasing body size while presenting sudden color changes (e.g., the red and black hind wings of the walking stick). Defense may also result from a cryptic posture. The walking stick *Orxines macklotti* becomes immobile and undertakes rocking movements which conceal it from predators.

The role of display in both territorial protection and aggressive mating behavior is well illustrated in M. Jacobs' *Studies on Territorialism and Sexual Selection in Dragonflies.* The light colors of the upper abdominal surface of the males are used in aggressive display in protecting territories against other males. The display is not used when approaching females. This study also provides examples of other displays and an experimental analysis of its role in mating and territory defense.

Courting and mating are among the most complex areas of behavior. The importance of sequential events or chains of dependent behavioral patterns is paramount in courting and mating. One of these steps is mate recognition. This may involve color patterns, shape, movement (walking, flying), species-specific signals, chemical stimuli, etc. Territoriality may also affect mate pairing. A detailed discussion of mate pairing as an initial and essential phase of the continuation of a species is provided by L. Ehrman *(Courtship and Mating Behavior as a Reproductive Isolating Mechanism in Drosophila).* Sound can also play a major role in courtship. Not only do L. M. Roth and H. B. Hartman *(Sound Production and its Evolutionary Significance in the Blattaria)* survey sound production in the Blattaria, but they also discuss the possible evolution of the two basic mechanisms for sound production, stridulation (scraping the pronotum over the wing veins) and expulsion of air.

The behavior of social insects is perhaps one of the most interesting areas of insect behavior. The insect is the only arthropod that cares for its young to the adult stage. From this basic behavior has diverged a multitude of variations in social behavior. The many highlights of social insect behavior include the development of hierarchies, colony formation, and division of labor resulting in the development of **castes** *(morphologically and physiologically distinct groups of individuals which perform particular tasks of the colony).*

Survival and growth of the colony depends on communication. The most important aspects of the ability to communicate include the capacity to transfer food or food source information. Active transfer of food can occur by **trophallaxis** *(the mutual sharing of food).* The transfer of information about food sources can occur in a variety of ways. In ants this information is transferred chemically. The organization of mass-foraging and the communication of information with pheromones (chemical signals between organisms) in the fire ant is discussed by E. O. Wilson in *Chemical Communication among Workers of the Fire Ant. I. The Organization of Mass-Foraging.*

# A. Introduction

## The Experimental Analysis of Aphid Behaviour and its Bearing on Current Theories of Instinct

By J. S. KENNEDY

University of Cambridge — Cambridge, England

ABSTRACT

*The three types of hypothesis now current concerning the nature of instinctive behaviour have been tested on the conveniently simple example of the major changes of activity-pattern occurring during the early adult life of the winged aphid.*

*The first hypothesis is the new-old one advanced by the modern Ethologists, which draws a distinction of principle between reflex and instinctive action, the latter depending on an internal drive-generating mechanism which, even when it is set going by some external stimulus, has nothing in common with reflex mechanisms, and requires special new concepts borrowed from our subjective experience. The second hypothesis is a simple physiological one such as Loeb favoured, according to which reflex and instinctive action do not differ in principle and changes of behaviour pattern (i.e. changes of reflex thresholds) are due primarily to changes in the state of receptors and effectors. The third hypothesis is based on the work of the reflex physiologists, notably Sherrington, and also sees no difference of principle between reflex and instinctive action, but attributes changes of threshold in both cases primarily to the integrative activity of the central nervous system.*

*The experimental work gave no support to the first two hypotheses and entirely confirmed the third. The Ethological contention that there are central nervous factors governing responses, as well as immediate stimuli from outside, was confirmed, but there was every indication that these internal factors are themselves part of, and can be studied only through, reflex activity in the comprehensive physiological sense of that term. They do not belong to a separate category of fact, nor warrant a special vocabulary.*

### THEORIES ANCIENT AND MODERN

The ancient problem of Instinct is one in which entomologists have an almost proprietary interest. But it must be admitted that so far, the major developments in this field have come from other people, working mainly on other animals. We owe the latest development, which is commonly if improperly referred to as "Ethology", to Lorenz (1937, 1950) and Tinbergen (1942, 1950, 1951). This may possibly turn out to be the most important advance since Jacques Loeb's theory of Tropisms. One cannot yet be quite sure of that because from the position they have reached it would seem possible to go forwards or backwards; indeed some movement in both directions is discernible. So it is worth while looking back for a moment along the road we have travelled.

Loeb (1900), inspired by Ernst Mach, set himself the task of breaking down the very venerable animistic or anthropomorphic tradition in the study of animal behaviour. To his lasting credit, he succeeded; liberating research from that cramping anti-analytical tradition by establishing the idea that behaviour could be explained in material terms. And he did this with the help of his Tropism theory, with all its faults, because this theory did allot the primary causal role in behaviour to the environment, instead of to internal psychic forces beyond experimental control. "There is no sharp line of demarcation between reflexes and instincts", he declared, "in both cases we have to deal with reactions to external stimuli or conditions. . . . As long as such apparently complex things as the instincts are not analysed but treated as entities, it is easy to believe that they are based on very mysterious nervous structures." But his understandable anxiety to leave no dark recesses whatever in the nervous system for demons to lurk in, led him to oversimplify. He felt it necessary to deny any integrative activity to the central nervous system, asserting that "the central nervous system has nothing further to do with these phenomena than that it furnishes the protoplasmic connection between the skin and the muscles".

Reproduced with permission from *Proceedings Tenth International Congress of Entomology*, 2:397–404, (1956). Published by the Tenth International Congress of Entomology, Ottawa, Ontario, Canada.

It was entirely to changes there, in the state of peripheral structures, that he attributed the changes of responsiveness which are so prominent in instinctive behaviour. The organisation of responses to make up an instinct was not due in his view to any "centres" or hierarchical organisation in the central nervous system. An instinct was simply a chain of segmental reflexes. In other words he held to the crudest kind of "press-button" or "S-R" (stimulus-response) theory of behaviour.

The over-simplification of the Tropism theory was soon generally recognized, but its influence persisted in an experimental pre-occupation with immediate responses to purely excitatory stimuli, as if the nervous system really were no more than a transmitter of excitations. Analytical interest now centred on the organisation of some given response. The question of what governed the strength of the response, its appearance and disappearance, was neglected, as if —although no one believed it—behaviour were simply the sum of such fixed responses. That too was a cramping assumption and the modern Ethologists have now succeeded, to their lasting credit, in liberating research from it. Thanks to them we are not likely now to forget the important truth to which they have drawn attention, namely, that responses are not only elicited by specific immediate stimuli, but are also "primed" in varying degree; this being a central nervous function no less than conduction and not simply due to changes of state in the receptors or effectors as Loeb maintained.

That is a step forward. But the new theory of Instinct which the modern Ethologists have offered, to replace reflex theories such as Loeb's, is a very old one revived, a step backward. The neglected facts to which they have drawn attention have long been familiar to physiologists: mammalian and even insect physiologists. It reflects little credit on experimental students of instinctive behaviour that we have gone on for years virtually ignoring them. We may blame Loeb, but that is hardly good enough for it has been known since Loeb's own time that even the simplest reflexes do not at all fit the push-button, S-R model. Reflex action always depends on much more than the immediate excitatory stimulus and to study only the effects of that is to be satisfied with a peep-show view of behaviour. An elementary point is that every reflex action is inhibitory as well as excitatory. The knee-jerk occurs not only because the extensor muscles are excited to contract, but equally because the flexors are inhibited and relax. That is readily seen; but when the reactions under study are those of the entire animal, what we call its "behaviour", then one easily overlooks the concurrent inhibition. Behaviour students spend their time watching and therefore thinking about what an animal is "doing". What thanks to inhibition it is not doing is out of sight and out of mind. As far as one can see from outside, inhibition is just nervous inactivity, absence of excitation. It is hard to remember that "inhibition is coequally with excitation a nervous activity" (Sherrington, 1939).

The very idea of "Behaviour" as a separate study thus carries a certain anti-analytical flavour. It was not without reason that Pavlov preferred the ponderous but physiologically-insistent phrase "higher nervous activity". For what makes it doubly hard to remember that inhibition is a nervous activity, is that in ourselves, too, what we are aware of is only excitation. There is exactly the same kind of one-sidedness in what we see as "objective" observers of animal behaviour, and in what we know subjectively of our own. That is why despite the best of intentions behaviour students fall time and again into subjective analogizing. In ourselves, as in the animals we watch, we are conscious of what is being done, not of what is not being done. We are unconscious of inhibition as a nervous activity in ourselves—necessarily, because the function of inhibition is to permit us to get on with and "attend to" what we are doing. "The great psychical process of 'attention'", as Sherrington (1947) pointed out, absolutely requires inhibition, the shutting off of excitatory processes in whole regions of the nervous system; and naturally it is impossible to introspect that inhibitory activity by means of which our attention is directed elsewhere. Because inhibition is an alien, strictly objective factor in behaviour, a deliberate intellectual effort is required to keep it in mind during the experimental analysis of behaviour. For descriptive and ecological purposes it may be enough to know what the animal resultantly "does", that is, to know where, from moment to moment in the neuromuscular system, excitatory processes predominate. But for analytical purposes that is by no means enough. Inhibition, far from being the mere absence of excitation, is one of the causes of excitation: "A precurrent inhibition favours after its withdrawal the development of central excitation at the previously inhibited units" (Sherrington, 1939). In this way external stimuli serve to "prime" certain reactions as a inseparable concomitant of their directly eliciting other reactions. Thus the conventional restriction of the term "reflex" to more rigid cases where little priming is evident, may be verbally convenient but is physiologically quite misleading. The two functions of "priming" and of "release" are both contained in reflex action. The "extension-reflex predisposes to and may actually induce a flexion-reflex, and conversely the flexion-reflex predisposes to and may actually induce an extension-reflex" (Sherrington, 1947). This effect Sherrington discovered and called "successive induction"—around the turn of the century.

Reflex action, then, has long been known to include such things as the priming of responses, priming even to the point of "spontaneous" or "vacuum" activity. Yet these are the very things which Ethologists have presented as evidence that instinctive behaviour is not reflex. Lorenz (1937) actually cited the work of Sherrington and von Holst on "spinal contrast" (successive induction) in lower reflex action, as supporting evidence for his idea of "automatic stimulus-production" or "internal energy-accumulation" in instinctive behaviour. Yet his whole theory of instinct was founded on the claim that such internal energy-accumulation was an independent neuro-physiological process peculiar to instincts and absolutely different from reflex action. Thorpe (1948) sought to strengthen Lorenz's case by casting doubt on the importance of successive induction in lower reflex action itself. He dwelt particularly on an experimental example of it which Creed *et al.* (1932) thought was probably an artefact. It should therefore be underlined that physiologists still regard successive induction as an "important principle of reflex action" (Fulton, 1949) as Sherrington himself unmistakably did.

Reasoning like Loeb in reverse, the Ethologists evidently felt that in order to get rid of the rigid "S-R" conception of behaviour it was necessary to throw out the reflex theory altogether; for as Thorpe (1954) clearly states, what Lorenz resuscitated as "the essential idea of instinct", was "the concept of internal drive". Still apparently under the impression that reflex action means push-button action and that some bits of intact behaviour really are like that, they saw this was quite untrue of the rest and swung right over to denying that most behaviour was even in principle reflex—that is to say a reflexion of or adjustment to the environment. Thus they rejected that broad and far from rigid conception of behaviour (this was what Sherrington, 1947, and Pavlov, 1928, understood by the reflex theory), quite failing to distinguish it from the narrow mechanistic rigid kind of reflex theory to which Loeb and many subsequent behaviour students have in practice adhered. Thorpe (1954) called the distinction "a quibble"; Tinbergen (1951) similarly called the reflex theory as such, without discrimination, a "grotesque simplification".

The trouble is that, by rejecting the reflex theory altogether, even those Ethologists like Tinbergen who look to physiology for help in their attempt to overcome the deficiencies of mechanistic Behaviourism, deprive themselves of the help which reflex physiology can give in abundance, and have to import causal concepts borrowed from psychology—*energy*[1], *appetite*, *drive*, *mood*, etc. These concepts provide some sort of descriptive analogy for what is observed of the animal's behaviour but inevitably as we have seen they misrepresent its causes. Therefore they never are simply descriptive, analogical or "operational". The theory is at bottom a subjective one not merely removed from but opposed to physiology (Kennedy, 1954)[2]. If then this new-old theory were generally accepted by entomologists and other students of behaviour, we should soon find ourselves liberated from push-buttons only to lose ourselves in the old pre-Loebian anthropomorphic wilderness.

It is the practical importance of these theoretical considerations which I should like to bring out in this paper. For the difference between the Ethological theory and a broadly reflex one is a practical matter, not a mere matter of different words to describe the same things. It affects what one looks for, hence what experiments one does, and what one finds. On Ethological theory one will not expect to find and will not seek the causes of priming (threshold-lowering) in the very reflexes that make up the instinctive chain, and therefore in the results of external stimulation, but in some distinctive internal process; for "to say 'all behaviour is reflex' is stultifying for research" (Thorpe, 1954). This belief was put to the test in an investigation of the instinctive behaviour of the young winged adult of the Black Bean Aphid, *Aphis fabæ* Scopoli.

[1] The substituted terms *potential* (Thorpe, 1951) or better *potentiality* (Hinde, 1954) describe more objectively what is observed; but so long as they refer to internal processes different in principle from reflex mechanisms (as in Thorpe, 1954), little has changed.

[2] Work inspired by the theory has led to correction and re-correction of its earlier tenets (e.g. Hinde, 1954; Thorpe, 1954). Its distinctions between priming (motivation) and release, between appetitive behaviour and consummatory act, and even between reflex and instinct. have become so loose as to serve as difficult guides in experiment. Opposite movements are in progress, toward reflex physiology (Prechtl, 1956), and toward subjectivism and plain teleology (Thorpe, 1956). Attention is fixed in this paper on the continuing central idea of Ethology which has created these difficulties because it is flatly opposed to all reflex theories: the concept of the internal drive.

## AN EXPERIMENTAL TEST

The winged, parthenogenetic, viviparous adult of this insect has a conveniently simple instinctive repertoire made up of two main patterns: (i) most of its life is spent as an ectoparasite feeding on a host-plant, with intermittent excretion and parturition executed without even removing its beak from the plant tissues; but, before that phase begins, there is (ii) a preceding phase of intense locomotor activity in the course of which the young adult finds a new host plant. Moericke (1955 *et ante*) described the whole sequence of behaviour here in the Ethological way as a succession of "moods" (*Stimmungen*) controlled by internal factors. At the end of the teneral period (*Rühestimmung*) the threshold for host stimuli (which for brevity we may call the host threshold) is high: the insect will not settle down to feed on a plant however suitable the plant may be as a host. The threshold of its locomotor reponses on the other hand is low (*Flugstimmung*) and it moves in response to light to an exposed position on the plant and takes off for a considerable flight. After that its host threshold is found to have fallen: given a suitable host it does now settle down to feed and deposit young (*Befallsstimmung* and *Ansiedlungsstimmung*).

Now this kind of behaviour sequence, often more complicated than this example, is common, as the Ethologists have pointed out: wherein a phase of locomotor behaviour which they call "appetitive" leads up to some kind of "consummatory reaction" which they call the instinct *sensu stricto* and which can be defined following Sherrington as the stage when visceral effectors take over from skeletal muscle. The Ethological interpretation of such sequences is made clear in the following abridged version of Thorpe's (1948) "Diagram of instinctive behaviour in the wide sense showing the part played therein by instinct in the strict sense of Lorenz":

| DRIVE ——→ | APPETITIVE BEHAVIOUR ——→ | GOAL |
|---|---|---|
| The as yet unreleased *instinct* accumulates round itself a store of *energy* | Random or directed locomotion, being the expression of an urge or appetite for a particular direction or kind of activity | The *instinct* is released by the environmental *releaser*, with dissipation of the *energy*. |

Note that the locomotory activity which appears first, is called "appetitive" because it is attributed to the "drive" generated in the central nervous apparatus of the consummatory instinctive act proper, which appears last. The drive is *toward* that end. It is built up, on Ethological theory, either by an internal maturation process, or by external stimuli acting non-reflexly.

How far does the aphid's behaviour fit in with these causal assumptions? Taking the first alternative, is the lowering of the host threshold an internal maturation process? Is the time spent in flight simply the time needed for the maturation process to be completed, with the locomotion itself the first consequence of that process occurring in the centre for the consummatory host responses? That can hardly be so. Aphids can be kept on a suitable leaf for hours, even for a day or two, after they have become ready to fly, by keeping them in the dark—and they still take off promptly when exposed. Their host threshold is then as high or higher than ever; whereas, as Johnson (1954) first discovered, stimulating them by means of light to fly for as little as 20-30 seconds, as soon as they are ready to fly, produces a small but distinct drop in the host threshold: the aphid now wanders about and probes the leaf for a while instead of taking off at once. Indeed it has been found repeatedly that teneral alate aphids left unconfined on perfectly suitable host leaves (the still-furled leaves at the tip of a growing bean shoot), do not settle down but begin to walk off, even in the dark, from about the time they would fly off if exposed to light; and after about three days they have all left. If the aphids are confined on the leaf by a small flat cage so that they keep meeting the leaf again as they move round the cage, then they do eventually settle and deposit some young. But this cannot be said to indicate a strictly internal process of maturation of the host responses, since the aphids are exposed to repeated host stimuli which may summate, and the parturition is definitely delayed. Less than 10 per cent of the aphids have begun parturition 12 hours after the end of the teneral period, and about 70 per cent at 36 hours; whereas over 95 per cent have done so at ½–3 hours if they have flown during that time.

The first Ethological alternative seems therefore to be excluded. The lowering of the aphid's host threshold is not primarily an internal maturation process. It requires a prior flight: more precisely, it requires a sufficiency of locomotory or at any rate skeleto-muscular activity of some kind, for walking alone can lower the host threshold. Moericke (1941) was the first to note that flight is a pre-requisite for settling down on a host plant, and Johnson (1954) and then we confirmed it. Flight is of course a reflex action in the conventional push-button sense of an immediate response to external excitatory stimuli: light, in particular, for aphids. It is those external stimuli, therefore, which are responsible, indirectly, for the lowering of the host threshold. This is the kind of effect of external stimuli which, because it is not a direct push-button effect, Ethological theory regards as different in principle from reflex action. The external stimuli here set off an internal drive-generating mechanism supposedly peculiar to instincts. In order to test this second Ethological hypothesis to account for the lowering of the aphid's host threshold, we set out to examine the effect of flight on the host responses in some detail. It should be said that in doing so our own inclination was at first toward yet a third alternative hypothesis, not Ethological but mechanistically physiological. This hypothesis was in the Loebian tradition, attributing changes of responsiveness to peripheral changes in the body rather than to the integrative activity of the central nervous system, as many physiologists as well as behaviour students have tended to do. Müller and Unger's (1952) hypothesis was one of this type, when they attributed the fall in the aphid's host threshold to flight exhaustion. Johnson (1954) disproved that particular suggestion by showing that 20 seconds' flight was enough to reduce the host threshold appreciably, and he put forward the vaguer suggestion that "flight itself" was responsible. This still could be and generally was taken as an hypothesis of the same type, postulating some peripheral change which now had to be much quicker than exhaustion but was still presumably "physical", not nervous, in nature (such as accumulation of $CO_2$ in the tracheal system).

An interesting technical difficulty soon encountered was that seemingly simple and harmless experimental procedures, such as picking an aphid up with a capillary suction tube, themselves changed the aphid's host threshold. It was therefore necessary to use an apparatus in which the aphid could fly freely for set periods and then alight freely on a leaf, without any handling. This apparatus consists of a large black-walled chamber with a battery of lights above a central opening in the roof, and a fan above them blowing air down past them into the chamber. The aphid flies up toward the lights but is held in the air below them by adjusting the downward air speed to balance the aphid's rate of climb, which can thus be recorded continuously.

Using a highly suitable leaf (bean tip) as a take-off and alighting platform in the chamber, a few one-minute flights are found to be enough to change the aphid's response to the leaf from prompt taking off to settling down to feed and deposit young. If the aphid is kept flying for a long time so that it shows signs of fatigue in flight and finally appears quite exhausted, and only then is allowed to alight, then its response to the leaf is stronger, inasmuch as it deposits more young there in the ensuing hours. Using a less suitable leaf (mature bean), the aphid stays a while but takes off again without depositing any young after short flights, depositing young on the leaf only after a long flight. Using a quite unsuitable leaf, say potato, the aphid stays longer after a long than a short flight, but practically never stays long enough to deposit any young on the leaf. The priming effect of flight on the subsequent response to a leaf is thus a quantitative one: the longer the flight, the better the host response. But the direct external host stimulus has the last say in deciding whether the excited host reflexes will be carried through to completion or inhibited again.

On the face of it, this behaviour bears a close resemblance to the spinal phenomenon Sherrington called successive induction, and demands no new causal assumptions. External elicitation of the flight reflexes totally inhibits all host responses for the time being, but when the flight reflexes are finally inhibited in their turn (by contact with a host again, preferably after partial inhibition of flight by fatigue), then the threshold of the host reflexes is found to have fallen. The host reflexes now "rebound", or show "exaltation", in Sherrington's terms, as in the inductive priming of the leg extension-reflex by a precurrent flexion-reflex in the mammal. This hypothesis was then adopted for further work, since it was difficult to see any means of putting the Ethological hypothesis to direct experimental test; and experimental tests of the "Loebian" physiological hypothesis (of purely peripheral rather than central nervous changes), while perfectly practicable, had given confusing, inconclusive results.

## REFLEXES IN ANTAGONISM AND ALLIANCE

On the "Sherringtonian" hypothesis of successive induction between antagonistic reflex-groups, the experimental procedure was clear, namely, to study not merely the effects of flight on the host threshold, but the two-way interaction between these two types of response. On the Sherringtonian hypothesis, the observed quantitative excitatory influence of flight on the host responses implies that the central excitation of the host reflexes is cumulative, i.e. summates, during flight, concurrently with a summation of central inhibition of the flight reflexes. The growing central excitability of the host reflexes during flight is due to the temporary inhibition imposed on those reflexes by the external excitation of the antagonistic flight reflexes, which at the same time causes central inhibition to grow in the flight reflexes so that they become progressively easier to suppress by an external inhibitory (host) stimulus. If that interpretation were correct, then it should be possible to obtain by suitable external stimulation, reciprocal effects as well: cumulative central inhibition of the host reflexes, with concurrent summation of central exitation in the flight reflexes. Since the host responses are not apparently fatigable (aphids being parasites), in contrast with the flight responses, the demonstration of those reciprocal effects would provide strong evidence for the Sherringtonian hypothesis of central nervous integrative activity and against the Loebian type of physiological hypothesis giving pride of place to peripheral factors such as fatigue.

Inhibition of the host response is observed when the aphid makes contact with some solid but non-host surface, which first inhibits flight and excites the host response of probing the surface with the rostrum, but then causes retraction of the rostrum and re-take-off from the surface toward light. If this external inhibitory influence from a non-host surface is imposed repeatedly, it is found to have a cumulative effect. As the succession of short flights and alightments continues, the host response may weaken instead of strengthening as the flying time is added to. But the opposed effect—the strengthening of the host responses by the lengthening flight—of course continues as the same time. The result is that if this is done with a *moderately* unsuitable leaf (mature bean leaf), the host response eventually re-asserts itself and usually "wins" in the sense that parturition does sooner or later ensue. But before that final stage there is an irregular oscillation in the strength of the host response, now waxing, now waning, in just the way one would expect given two antagonistic processes, excitation and inhibition, which are both summating concurrently—a ding-dong battle between them. If an entirely unsuitable leaf is used instead, then alightment and re-take-off can be repeated many times, for the aphid will not settle down to feed here even when it is "finally" exhausted. The central inhibition of the host response can be built up by this means to the point where the insect will not settle long enough to deposit young even on the highly suitable leaf at the tip of a growing bean shoot. On such a leaf, as we have seen, an aphid that has not previously alighted on an unsuitable leaf will settle down and deposit young after very little flying indeed.

The other reciprocal effect which would be predicted from the Sherringtonian successive induction hypothesis, is the priming of flight by a pre-current host response. During a long series of short flights and alightments on an unsuitable leaf, there occurs not only the build up of inhibition of the host responses just described, but also a build up of excitation of the flight response. The aphid's rate of climb, instead of decreasing as in uninterrupted flight, increases. The rate often rises to well above the level that was reached at the first take-off when the aphid was perfectly fresh, so that the "exaltation" of flight here can hardly be attributed simply to recovery from fatigue (or mobilisation of reserves: Wigglesworth, 1949), during the aphid's brief rest on the leaf. This reciprocal inductive effect is also quantitative, i.e. summates, as analysis of the full set of records has shown. The better the leaf as a host, and hence the stronger the response to it and the longer-lasting the concurrently imposed inhibition of the flight reflexes, the more likely are the flight reflexes to show rebound "exaltation" when re-take-off does occur. Or again, the longer the flight before alightment is permitted, and hence the stronger the host response on alightment, the more likely are the flight reflexes again to show "exaltation" afterwards. Such "exaltation" of flight vigour occurs even when the aphid has already flown to the point of apparent exhaustion. Thus, *behaving* in an exhausted fashion is something decreed by the central nervous system long before the insect is exhausted in any "physical" sense. This kind of exhaustion can be completely reversed, many times in succession, if no suitable host is encountered, thus greatly prolonging the total period of flight. These reciprocal effects clearly show that we are dealing here with reflex interactions, and not with actions decided by any one, or any one-way, process inside or outside the central nervous system. "Priming" is not the special province of those particular centres which organize "consummatory" reactions like feeding and parturition in the aphid.

Other ways in which the results obtained with the intact aphid parallel those obtained in the spinal mammal can only be mentioned here. The generation of oscillations or crude rhythms by *continuous* stimuli (or by stimuli which are not correspondingly oscillatory) has already been touched on in one context (see above) but has been observed again and again in various contexts covering a wide range of time-scales and physiological amplitudes, from the oscillatory movements of the beating wings, and the oscillations in the rate of climb of the aphid when its alighting response is inhibited, right up to the grand oscillation from the 'vegetative' mode of life in the immature stages to the active locomotory existence of the young adult and back to the vegetative existence of the older adult when its flight muscles have been histolysed (Johnson, 1953). It is this generation of occillations by non-oscillatory stimuli which provides perhaps the most clinching evidence that we are dealing here with reciprocally-innervated, antagonistic reflex systems on the Sherrington model. And it is in his analysis of rhythmic reflex activities that he made the most direct contribution to the problems of the student of intact behaviour. That is the point from which we can proceed; and it is all the more regrettable that students of intact behaviour and neuro-physiologists alike have left his work largely on the shelf. The old idea of instincts as chain-reflexes recognized only one aspect of the chaining mechanism, the fact that the execution of each reflex brought in the appropriate external stimulus for the next reflex in the chain. The chain-reflex idea fell into disrepute because it contained the mistaken assumption that each reflex had a fixed threshold, while instinctive behaviour was marked above all by changes of threshold. But the chain-reflex conception of instincts regains its full force once the fact of successive induction between antagonists is conceded, since this means that a precurrent reflex lowers the threshold of, as well as bringing in the stimulus for, the next reflex in the chain.

Classical reflex physiology brought out another way in which reflexes are linked centrally so as to alter each other's thresholds. In addition to *antagonistic* reflexes where external excitation of one partner inhibits (and thereby primes) the other, and *vice versa* inhibition of one excites the other, there are also *allied* reflexes where excitation (or inhibition) of one irradiates to the other, exciting (or inhibiting) it simultaneously. The alighting response of the flying aphid, and its feeding response after alightment, appear to be allied reflexes in that sense. For flight increases the excitability of both at the same time, and what is more striking, external inhabition of one, the feeding response, by an unsuitable leaf, spreads centrally to the alighting response so that the aphid now "refuses to alight" although no external inhibitory action has been exerted on the alighting response itself.

Although the intact aphid's behaviour shows all these features which parallel the lower reflex co-ordination of the spinal mammal, it does of course show a number of new features as well. In the first place, the intact animal is a more complex and delicate system. The experimental advantage of the aphid here is that, as a parasite, its behavioural repertoire is limited. Its possible courses of action are few making the general laws governing the choice between them easier to discern; but they are still much less predictable in detail than spinal reflexes. In the intact aphid one is dealing with higher-order integrations, that is with the interaction not of simple reflexes but of compound reflex-groupings within which there is much room for variable subsidiary interactions. But there are more important differences. Events in the intact aphid also extend in time, over a scale which overlaps but goes well beyond the time scale of the classical mammalian experiments. Successive induction, for instance, can be observed in the aphid as quick or as long drawn out process. And whereas the antagonistic reflexes Sherrington studied were roughly equals, and consequently showed rather regularly rhythmic alternations (stepping, scratching, nystagmus), in the intact aphid the reflex-groups linked together in the same way are not equals. Thus flight is fatigable, but feeding is not. The antagonists do not either occur together in rhythmic alternation, or both cease together, for one partner can gain upon the other and finally suppress it altogether. And because these reflex groupings amount to the major activities of the animal, their interaction has much more radical consequences in the life of the animal than the interaction between single reflexes.

Nevertheless, the principles of the interaction between the reflex groups, as allies and antagonists, seem to be the same as between single reflexes. A set of single reflexes (acts) which are connected with one another as allies or antagonists, has now as a group similar relations of alliance and antagonism with other groups. The instinctive behaviour of the intact aphid when analysed experimentally as a reflex chain, does not show any features which require us to call in non-reflex concepts—certainly not psychological concepts such as appetite and drive. Such concepts put the cart before the horse. The aphid's flight, which brings it to a new host plant, is not appetitive or host seeking behaviour at all, for that would mean it was caused by a *feeding* drive. It is not the aphid's appetite for a host plant that makes it fly, but on the contrary, it is the flying that creates the appetite, makes it feed.

## REFERENCES


Creed, R. S., D. Denny-Brown, J. C. Eccles, E. G. T. Liddell, and C. S. Sherrington. 1932. Reflex activity of the spinal cord. *Oxford University Press*, London.

Denny-Brown, D. (Ed.) 1939. Selected writings of Sir Charles Sherrington. *Hamish Hamilton*, London.

Fulton, J. F. 1949. The physiology of the nervous system. 3rd Edn. *Oxford University Press*, New York.

Hinde, R. A. 1954. Changes in responsiveness to a constant stimulus. *Brit. J. anim. Behav.* 2: 41–55.

Johnson, B. 1953. Flight muscle autolysis and reproduction in aphids. *Nature* 172: 813.

Johnson, B. 1954. Effect of flight on behaviour of *Aphis fabæ* Scop. *Nature* 173: 831.

Kennedy, J. S. 1954. Is modern ethology objective? *Brit. J. anim, Behav.* 2: 12–19.

Loeb, J. 1900. Comparative physiology of the brain and comparative psychology. *G. P. Putnam's Sons*, New York.

Lorenz, K. Z. 1937. Uber der Begriff der Instinkthandlung. *Folia Biother.* 2: 18–50.

Lorenz, K. Z. 1950. The comparative method in studying innate behaviour patterns. *Symp. Soc. exp. Biol.* 4: 221–269.

Moericke, V. 1941. Zur Lebensweise der Pfirschlaus (*Myzodes persicæ* Sulz) auf der Kartoffel. Inaugural Dissertation, Bonn.

Moericke, V. 1955. Uber die Lebensgewohnheiten der geflügelten Blattlause (*Aphidina*) unter besonderer Berücksichtigung des Verhaltens bein Landen. *Z. angew. Ent.* 37: 29–91.

Müller, H. J., and K. Unger. 1952. Uber den Einfluss von Licht, Wind, Temperatur und Luftfeuchtigkeit auf den Befallsflug der Aphiden *Doralis fabæ* Scop. und *Myzodes persicæ* Sulz. sowie der Psyllide *Trioza nigricornis* Frst. *Züchter* 22: 206–228.

Pavlov, L. P. 1928. Lectures on conditioned reflexes. *International Publishers*, New York.

Prechtl, H. F. R. 1956. Neurophysiologische Mechanismen des formstarren Verhaltens. *Behaviour* 9: 243–319.

Sherrington, C. S. 1939. See Denny-Brown, 1939.

Sherrington, C. S. 1947. The integrative action of the nervous system. 2nd Edn. *University Press*, Cambridge.

Thorpe, W. H. 1948. The modern concept of instinctive behaviour. *Bull. anim. Behav.* 7: 12 pp.

Thorpe, W. H. 1951. The definition of some terms used in animal behaviour studies. *Bull. anim. Behav.* 9: 34–40.

Thorpe, W. H. 1954. Some concepts of ethology. *Nature* 174: 101.

Thorpe, W. H. 1956. Learning and instinct in animals. *Methuen & Co.*, London.

Tinbergen, N. 1942. An objectivistic study of the innate behaviour of animals. *Bibliotheca Biotheor.* 1: 40–98.

Tinbergen, N. 1950. The hierarchical organisation of nervous mechanisms underlying instinctive behaviour. *Symp. Soc. exp. Biol.* 4: 305–312.

Tinbergen, N. 1951. The study of instinct. *Clarendon Press.* Oxford.

Wigglesworth, V. B. 1949. The utilization of reserve substances in *Drosophila* during flight. *J. exp. Biol.* 26: 150–163.

# B. Feeding

## FEEDING BEHAVIOUR

V. G. DETHIER

*Department of Biology, University of Pennsylvania*

### *Introduction*

THE emphasis on description which in past decades has characterised studies of the feeding behaviour of insects has shifted in recent years to preoccupation with mechanisms. There is now less concern with the mechanics of eating and the cataloguing of menus and more with factors regulating patterns of eating and selection of diet. In the very broadest sense the basic nutritional requirements are the same for all insects, but different species satisfy these requirements with various kinds of food in accordance with their particular biochemical endowments and the ecological niche that they inhabit. Gordon (1961) has even suggested that all potential food sources are probably nutritionally unbalanced so that exploitation of a particular plant involves some metabolic adjustments (*cf.*, *e.g.*, Applebaum, 1964). In the absence of metabolic adjustments dietary imbalance could be counteracted by occasionally mixing the diet. There is evidence that some insects, as, for example, grasshoppers and locusts, fare better on a mixed diet than on a single plant (Dadd, 1963).

It is not surprising that different behaviour patterns have evolved in connection with the utilisation of different kinds of food, but in every case the behaviour involves a means of quality control and a regulation of quantity. In modern studies of feeding behaviour these two aspects have received most attention. They encompass mechanisms of discrimination and "on/off" mechanisms. This discussion will be concerned primarily with these two subjects.

### *Discrimination of quality*

Feeding behaviour has been conceived of as a complex and interacting sequence of responses to a variety of stimuli culminating in ingestion to repletion (Thorpe *et al.*, 1947; Dethier, 1947). While the sequence is less a chain than a continuous thread of varying character, understanding of the total pattern has been facilitated by recognition of a number of component parts. The nature of the components may vary from species to species (de Wilde, 1958; Thorsteinson, 1958; Hamamura *et al.*, 1961; Goodhue, 1963) and indeed may be interpreted quite differently by divers authors; nevertheless, it is generally true that the integrated behaviour consists of the following: locomotion bringing the insect to its food, cessation of locomotion on arrival, biting or its equivalent (probing, sucking, etc.), continued feeding, termination of feeding (satiation).

The discrimination of quality, which sets the pattern of feeding and determines the success or failure of ingestion, is accomplished by the whole procedure, but each successive step in the sequence constitutes a check on quality. Several examples will illustrate the generality of sequential responses leading to ingestion and will expose the variation in details.

The blowfly *Phormia regina* Meigen is guided by odour to its food and steps in it on arrival. Stimulation of chemoreceptors on the tarsi triggers proboscis extension; extension places the chemosensory aboral hairs of the labellum in contact with the food; stimulation of these hairs results in spreading of the labellar lobes which places oral taste papillae in contact with the food; stimulation of the papillae triggers and

Reproduced with permission from the *Royal Entomological Society of London Symposium on Insect Behavior*, 3:46–58, 1966. Published by the Royal Entomological Society of London, England.

drives ingestion (Dethier, Evans, and Rhoades, 1956). At each one of these steps quality is monitored. The locust *Schistocerca gregaria* Forskal approaches its food, lowers the tips of the antennae and palpi, then bites, and continues feeding accompanied only by palpal testing (Goodhue, 1963). Adult vegetable weevils (*Listroderes costirostris obliquus*) tap clover leaves with their antennae. If an appropriate signal is received, the maxillae are extended and tap in turn. If the correct stimulus is again received, biting commences (Matsumoto, 1962). The approach of leafhoppers to their food plant is guided by its colour, alightment is triggered by odour, colour induces probing, pH guides the mouthparts into the phloem (but see Day, Irzykiewicz, and McKinnon, 1952, who could not demonstrate response to a pH gradient and suggested that the insects feed on whatever tissue they encounter by chance probing, but feed *longer* on phloem because of gustatory factors there), taste drives continued feeding (Nuorteva, 1952). The feeding sequence of aphids consists of alightment, probing, insertion of stylets, location of phloem, maintenance of stylet tips in a sieve tube (Mittler and Dadd, 1965). Initiation of probing is hastened by leaf colour, a token stimulus characteristic of the plant (*e.g.*, sinigrin for the cabbage aphid *Brevicoryne brassicae* (L.)) initiates feeding after probing (Wensler, 1962), sucrose arrests further probing, amino acids plus sucrose determine the duration of sucking (Mittler and Dadd, 1965). The Colorado potato beetle (*Leptinotarsa decemlineata* Say) makes close range orientation to its food plant in response to colour and odour (which also arrest its locomotion), bites in response to olfactory stimuli, withdraws the antennae and continues eating as long as no adverse stimuli from within the plant are encountered in the process (de Wilde, 1958). Larvae of the spruce budworm (*Choristoneura fumiferana* (Clem.)) make their final orientation to feeding sites in response to humidity and thigmotactic stimuli, which also arrest locomotor activity, and continue their feeding on the basis of the taste stimuli received (Heron, 1965).

The significance as well as the mode of action of the stimuli operating at each level of feeding behaviour has engendered mountainous controversy. Part of the reason derives from the fact that most of the research on factors regulating quality discrimination (diet selection, diet choice, diet preference) has been conducted with phytophagous insects. Whether a study was termed "food-plant selection" or "host-plant selection", the prime consideration was usually one of nutrition whereas the plant may more properly be thought of as a place to live as well as something to eat (Kennedy, 1953; Dethier, 1953), and its selection as food may be based in part upon other considerations (at least in an evolutionary sense). Specificity, when it exists, may be not so much a matter of meeting the nutritional requirements of the individual as being a compromise between meeting these and various ecological optima as well (*e.g.*, habitat selection (Petersen, 1954), relation with predators (Brower, 1958) or parasites (Downey, 1962; Stride and Straatman, 1962)). Because of these complications it is perhaps unfortunate that the problem of diet selection was attacked first in phytophagous insects.

Leaving aside consideration of the factors leading to selection of the plant as "host", those underlying its selection as "food" are predominantly chemical. This is not to say that physical properties are entirely without effect. A number of interesting examples have been given by Merz (1959). One involves plants of families characterised by the abundant presence of raphides, bundles of needle-like crystals of calcium oxalate. Young larvae of the sphingid *Pterogon proserpina* Pallas could eat mature leaves of *Vitis* because the caterpillars' point-like bites fell between clusters of raphides. Older larvae, on the other hand, by reason of their large slashing bites could not avoid the crystals and refused the leaves. Dissolving the crystals in dilute hydrochloric acid made the leaves palatable to larvae of all stages (it is unlikely that this treatment caused significant chemical alterations). A more obvious example is that of the oak feeder *Lasiocampa quercus* (L.). These larvae would not eat holly (*Ilex*) until the sharp teeth were shaved from the edges of the leaves.

Before the importance of plant chemicals was fully realised the relationship between restricted feeders and their plants was described, in Fabre's term, as "botanical instinct." The classical work of Verschaffelt removed part of the mystery by demonstrating that specific chemicals provide clues for the insect. It was then proven experimentally that the senses of taste and smell were the means whereby the insect detected the chemicals in question (Dethier, 1937). The two primary questions that demanded answers then, as now, are: (1) what is the genetic basis of specific diet preferences? (2) how are the preferences implemented (Dethier, 1953)? The controversy mentioned earlier arose from three subordinate questions that have dominated the literature and dictated the direction of most research effort; namely: (a) are the effective stimuli token or nutrient? (b) is oligophagy based on unique acceptance of particular compounds or rather general rejection of many? (c) which is the more important sensory system employed by plant-feeding insects, gustatory or olfactory?

*Token stimuli.* Extrapolating from the work of Verschaffelt and from similar experiments with *Papilio* larvae, Dethier (1941, 1947, 1953) pointed out the paramount importance in plant selection of special compounds not necessarily related to the insect's nutritional requirements. These were conceived of as sign or token stimuli ("releasers" in ethological terminology). Fraenkel (1953) defined these tokens (secondary compounds, odd chemicals) as having no nutritional value *per se* and remarked (1959) that food specificity in insects is based solely on the presence or absence of tokens. Kennedy and Booth (1951) had already demonstrated by means of experiments with aphids that nutritional factors actually do play an important role. These authors suggested that "flavours" (equivalent to token stimuli) were important in the discrimination of different species of plants but that factors more closely related to nutrition mediated discrimination of leaves of different ages. They said that "nutrient" did not imply that the stimuli necessarily came from the same substances as were important nutritionally but that they were more closely related to nutrition than the flavours. ("Flavours" is too imprecise since it connotes a human sensation, the sensory basis of which is complex and not yet totally clear.) The value of this concept was that it brought the nutritional aspects of the plant back into a picture that had become unbalanced.

In the years that followed the so-called "token theory" and the "dual discrimination theory" were placed in juxtaposition and presented as mutually exclusive alternatives (Lipke and Fraenkel, 1956; Beck, 1965). The fact of the matter is that some of the concepts in each of the "theories" are correct. It is abundantly clear that compounds which are not nutritionally essential exercise an influence on food discrimination. This fact is well documented for Cruciferae-feeders. The odour of isothiocyanate is a weak attractant and arrestant for larvae of *Plutella maculipennis* (Curtis) (Gupta and Thorsteinson, 1960); mustard oil glycosides crystallised from Cruciferae and Tropaeolaceae (progoitrin, glucocheirolin, glucoconringiin, glucoerucin, glucotropaeolin, sinigrin, sinalbin, gluconasturtiin, gluconapin) are feeding stimulants even though two are toxic (Thorsteinson, 1953; Nayer and Thorsteinson, 1963). *Plutella* is said not to accept a nutrient unless one of the aforementioned phagostimulants is present (Thorsteinson, 1958). Another Cruciferae-feeder, the cabbage aphid (*Brevicoryne brassicae*) can be induced to feed on an abnormal host (*Vicia faba*) (and reproduce successfully) if the host leaves are permitted to absorb sinigrin via the stems (Wensler, 1962). For the mustard beetle, *Phaedon cochleariae* (Fabricius), allyl isothiocyanate elicits prompt biting, and sinigrin enhances feeding.

*Papilio polyxenes* Fabricius, which feeds primarily on Umbelliferae, is attracted from short distances (approximately 3 mm.) and induced to bite by compounds characteristic of the essential oils of Umbelliferae (*e.g.*, anethole, *l*-carvone, *d*-limonene, *l*-linalool, etc.) (Dethier, 1941). The vegetable weevil, *Listroderes costirostris obliquus* as a newly hatched larva is also attracted by the odours of these compounds (Sugiyama and Matsumoto, 1959). The adults are attracted from a distance of 3–5 mm. by the odour of coumarin, the odorous principle of sweet clover (*Melilotus alba*). The same compound terminates flight in *Sitona cylindricolis* (Fahraeus) but does not increase feeding (Thorsteinson, 1955, 1960). In addition to coumarin the following compounds attract the adult weevil: $\beta$-$\gamma$-hexenol, $\alpha$-$\beta$-hexenal (possessing fresh leaf odour and present in many plants), the homologous alcohols *n*-butyl to *n*-octyl inclusive, and $\beta$-phenylethyl alcohol (Matsumoto and Sugiyama, 1960). Silkworm larvae (*Bombyx mori* L.) are attracted by $\beta$-$\gamma$-hexenol and $\alpha$-$\beta$-hexenal from a distance of 3–4 mm. (Watanabe, 1958). The rice stem borer is attracted by oryzanone (Mumakata *et al.*, 1959).

Finally, with reference to token stimuli, the exhaustive evolutionary study of *Papilio* and its food plants conducted by Ehrlich and Raven (1964) should be studied because these authors concluded: "A systematic evaluation of the kinds of plants fed upon by larvae of certain subgroups of butterflies leads unambiguously to the conclusion that secondary plant substances play the leading role in determining patterns of utilisation."

*Nutritional factors.* Of the compounds that are clearly of nutritional importance to insects, sucrose is one of the most widespread. It is usually more stimulating than glucose, the other common sugar. It is an effective stimulus for the majority of insects, even those in which it forms no necessary part of the total feeding activity (Beck, 1956b; Dadd, 1960). It is noteworthy, however, that some species do not react to sucrose; most species of leafhoppers actually avoid it (Nuorteva, 1952). Its importance as a phagostimulant for phytophagous insects has been demonstrated in several cases. According to Beck (1956a), the choice by the European corn borer of sites on the corn plant is based on the highest sugar concentration and is not influenced by local attractants. For the Mexican bean beetle (*Epilachna varivestis* Mulsant) the principal phagostimulant is sucrose. For the aphid *Myzus persicae* (Sulzer) sucrose is important because it is quantitatively the major solute of phloem sap. As Mittler and Dadd (1963a, 1964) have shown, gustatory stimuli rapidly contribute toward acceptability or non-acceptability of a diet. The acceptability of sucrose increases with increasing concentration up to 40 per cent., beyond which there is a decrease.

In the phloem sap fed upon by aphids the next most abundant compounds are the amino acids. Singly or in combination they fail to stimulate and are not preferred to water. Generally speaking amino acids are poorly stimulating or totally ineffective to insects. The exceptions again are noteworthy. Amino acids cause aggregation and biting by wireworms (Thorpe *et al.*, 1947), some cause larvae of the European corn borer to aggregate and feed while others repel (Beck and Hanec, 1958), and the milkweed bug (*Oncopeltus fasciatus* (Dallas)) is stimulated to feed by a considerable number of amino acids (Feir and Beck, 1963).

Although amino acids are among the least stimulating of compounds, they do exhibit a curious synergism with some other compounds in so far as stimulation is concerned. For the European corn borer, serine and glucose are additive in their effect on feeding (Beck and Hanec, 1958). For the aphid *Myzus persicae*, leucine and methionine both enhance the acceptability of sucrose, and a mixture of six amino acids is even more effective when combined with sucrose (Mittler and Dadd, 1964). L-proline, hydroxy L-proline, and L-glutamic acid each synergise with sucrose for larvae of the spruce budworm (*Choristoneura fumiferana* (Clem.)) (Heron, 1965). Synergistic effects among other compounds have also been reported: sucrose plus sinigrin and other glycosides for *Plutella* (Thorsteinson, 1953; Nayar and Thorsteinson, 1963), glucose plus an unknown token for the tobacco hornworm (*Protoparce sexta* (Johan)) (Yamamoto and Fraenkel, 1960), sugar plus an unknown volatile constituent for *Epilachna* (Augustine *et al.*, 1964), sucrose plus glucose plus fructose for the spruce budworm (Heron, 1965), $KH_2PO_4$ plus sugar for *Camnula* (Thorsteinson, 1960), a mixture of amino acids plus sucrose for *Camnula* (Thorsteinson, 1960).

The nature of synergistic action in these cases is not at all clear. In the context of the experiments reported it usually means that the insect eats more when the two compounds are present than when either is present alone. Whether the effect is a sequential one (*i.e.*, one compound initiating biting and another driving continued feeding) or a simultaneous one (both compounds acting on the same phase of eating and possibly even on the same receptors) is not known. In cases of synergism among sugars (studied most thoroughly in the blowfly *Phormia regina* (Dethier *et al.*, 1956; Evans, 1961) and the honeybee (Wykes, 1952)), the best evidence is that there are at the molecular level multiple sites on the receptor surface, and unlike molecules, provided they are not competitive inhibitors, fill these sites. The suggestion of Thorsteinson (1960) that odd substances (*e.g.*, sinigrin) lower thresholds for sapid substances (*e.g.*, sucrose) merely restates the fact that there is synergism and would imply a sequential effect.

In addition to sugars and amino acids, a number of other nutrient and presumably nutrient substances are known to enhance feeding. Among them are the following: albumin for leafhoppers (Nuorteva, 1952), β-sitosterol for silkworm larvae (Hamamura *et al.*, 1961; but compare Niimura and Ito, 1964), unsaponifiable extracts of bran, olive and corn oil, and wheat germ oil for *Schistocerca gregaria* (Chauvin, 1951; Dadd, 1960), phospholipids, especially lecithins and phosphatidyl inositol, for older nymphs and adults of *Melanoplus bivittatus* (Say) and *Camnula pellucida* Scudder (Nayer and Thorsteinson, 1963), ascorbic acid and thiamine for *Chorthippus* (Thorsteinson, 1958), thiamine and α-tocopherol for *Tribolium confusum* (L.) (Thorsteinson, 1960), oxalo-acetic and citric acids for *Camnula* (Thorsteinson, 1960), asparagine, betaine, monosodium glutamate for *Chorthippus* (Thorsteinson, 1958), α-amino-butyric acid and alanine for the Colorado potato beetle (Thornsteinson, 1960).

It should be abundantly clear from an examination of the facts in the preceding sections that there are some compounds (*e.g.*, sucrose) that are almost universal in their stimulating action and that there are other compounds (*e.g.*, sinigrin) uniquely specific in their action. Some of the latter types of compound are themselves nutrients; others are only signs, that is, tokens, of a source of food and living accommodation. The finding of importance is not that many nutrients stimulate but that so many stimuli are not nutrients. Whether they are olfactory stimuli or gustatory stimuli hardly seems important in this connection. A logical conclusion is that discrimination (food selection) is based upon both.

Implicit in the foregoing conclusion is the additional conclusion that restricted feeding habits are based not merely on rejection of plants possessing deterrents or lacking nutrients but definitely upon specific acceptance as well. Jermy's (1958) original idea that oligophagy (by the Colorado potato beetle) is largely a matter of avoiding deterrents rather than accepting specific stimulants and Thorsteinson's (1958) emphasis on deterrents as one basis of oligophagy seem as restrictive as the original proposal of Dethier (1947) that the basis of oligophagy is specific attractance. Again the truth of the matter lies between, and a later statement of Jermy (1961) proposing a balance seems more in accord with the facts. Those instances cited above of enhancement of the stimulating effect of dietary substances by voltile components or glycosides exemplify very well the relationship. There are numerous other examples. Two will illustrate the point: the feeding behaviour of the milkweed bug and of larvae of the spruce budworm.

The milkweed bug orients to milkweed seeds from a distance of a few millimeters by attractants in the seed coat stimulating the antennae. Contact of the labium with chemical factors in the seed coat initiates stylet penetration, and maintenance of feeding is dependent on sensory input from chemoreceptors situated internal to the stylets. Sugars and amino acids, plus unidentified stimulants in the seed coat, drive feeding (Feir and Beck, 1963). Larvae of the spruce budworm show a preference in diminishing order for: staminate flowers, new shoots, old needles. Of the three tissues the flowers contain the most sugars. The free sugars are: glucose, fructose, and sucrose. The flowers also contain proline. As the needles grow older they get harder, the sugar content decreases, and the glycoside pungenin is produced. Of all of the sugars tested on the larvae the most acceptable are: L-fucose, D-fructose, maltose, sucrose and raffinose. D-arabinose, D-galactose, D-glucose, melibiose, trehalose, and melezitose are slightly acceptable. Only sugars approximate plant tissues in stimulating effectiveness. The amino acid proline is the only stimulating one, and it is less effective than sucrose. There are no responses to pungenin or to shikimic acid (present in new shoots and old needles). The conclusion is that differences in sugar content of different tissues influence feeding behaviour but that other unknown stimuli are important.

### *The sensory basis of discrimination*

The basis of food specificity is the specific interaction between an insect and the plant compounds it encounters, whether these be acceptable or unacceptable, nutrient or token. Specific interactions may differ from one another by involving different compounds or different combinations of identical compounds acting on different links of the feeding chain or in different ways on the same link. At the neural level there are differences either in the way the sense organs or the integrative centres of the central nervous system (via the sense organs), or both, interact with the chemicals present in plants selected as food and those rejected. To understand these interactions completely it is necessary to know not only what compounds are presented to an insect by a plant but what their effect on the sense organs may be. The ultimate realisation would be an understanding of food discrimination in terms of neural action. A small beginning toward this end has been made with lepidopterous larvae.

It was suggested many years ago that caterpillars discriminated their food-plant first on the basis of olfactory stimuli received through the antennae and maxillae and then, after biting, by stimuli perceived by organs of taste (Dethier, 1937). Several investigators subsequently discovered that the number of species of plants eaten by a caterpillar would be increased if the maxillae were amputated (Torii and Morii, 1948 for *Bombyx mori*; Dethier, 1953 for *Cingilia catenaria*, *Papilio polyxenes*, *Porthetria dispar*, *Pieris rapae*; Ito, Horie, and Fraenkel, 1959 for *Bombyx mori*; Waldbauer and Fraenkel, 1961 and Waldbauer, 1962 for *Protoparce sexta*). Chin (1950) produced a similar effect in larvae of *Leptinotarsa decemlineata*. The principal conclusion drawn from these experiments was that the maxillae normally receive stimulation from deterrents that prevent consumption of a leaf. This interpretation was enlisted in support of the idea of an oligophagy based principally on deterrence.

The first direct indication of the sensory function of the maxillae came from the electrophysiological studies of Ishikawa and his co-workers (Ishikawa, 1963; Ishikawa and Hirao, 1963; Ishikawa, Tazima, and Hirao, 1963), who proved that the two sensilla styloconica on the maxillary lobe of *Bombyx mori* are contact chemoreceptors. One contains a receptor for salt, a receptor for sucrose, one for glucose, and one for inositol. The other contains a receptor for water, two for salt, and one for repellent (*sic*) substances. The hypothesis was advanced that the preferred food, mulberry, stimulates principally the sugar and inositol receptors while unacceptable plants stimulate the repellent receptor. Recent unpublished studies by Schoonhoven and Dethier indicate that there are some differences in the response characteristics of the maxillary receptors of different species of caterpillars and that for *Protoparce* at least the chief difference between sensillum one and two relates to threshold levels. There is support for a generalisation of Ishikawa's idea that acceptable plants stimulate one set of receptors preferentially while unacceptable plants stimulate another set. These findings would, at least in so far as taste stimulation is concerned, argue against the contention that oligophagy is based primarily upon the absence of deterrents in the preferred plants. It also weakens the hypothesis that the principal, if not sole, function of the maxillae is to detect deterrents. Further light can be shed on this problem by refining the technique of maxillectomy so that either sensillum one or two alone can be amputated.

Important though the maxillae be, it is a mistake to attempt an explanation of food discrimination solely in terms of input from maxillary sensilla. The antennae certainly play a role, as originally suggested, but failure of antennectomy to cause alterations in feeding preferences discouraged for a while further experimentation with antennae. Recently, however, Morita and Yamashita (1961) confirmed electrophysiologically the olfactory function of the antennae and found that $\beta$-$\gamma$-hexanol, the compound that attracts larvae of *Bombyx mori*, caused an increase in firing of receptors of one of the large sensilla basiconica. This work has been extended by Schoonhoven and Dethier (unpublished) who have recorded responses to the odours of a number of plants by means of microelectrodes inserted into the antennae of *Protoparce sexta*. The data suggest that no single receptor is uniquely tuned to a single stimulus. Instead each receptor appears to respond to a spectrum of odours, but the maxima are not identical for each receptor. Thus, when the combined firing of all receptors is considered there is not only a difference of pattern between acceptable and unacceptable plants but also among acceptable plants. Comparable response patterns have been recorded from sensilla basiconica on the maxillae, which have been shown to be olfactory organs (Schoonhoven and Dethier). This additional finding further complicates interpretations of behavioural aberrations that occur after maxillectomy.

The rather broad spectra of sensitivity that seem to characterise the gustatory and olfactory receptors of caterpillars indicate that early ideas of the sense organs being highly specific filters permitting only one or a few key stimuli to place information in the nervous pathways were oversimplifications. Similar discoveries of broad spectra of sensitivity among receptors responsive to sex attractants indicate that the phenomenon is not a restricted one. The burden of discriminating between various diets lies with the central nervous system. An observation of Ishikawa *et al.* (1963) that a mutant of *Bombyx mori* having abnormal feeding habits received through the maxillae the same kind of information as did normal caterpillars is in agreement with this idea.

Assuming that the central nervous system can utilise information sent to it in the form of patterns from the antennal and maxillary olfactory organs, the caterpillar is in a position to discriminate among different plants by means of the odours emanating from them. If *Protoparce sexta* is at all typical, oligophagy is based upon acceptants as well as deterrents, olfactory stimuli as well as gustatory, and token stimuli as well as nutrients.

These conclusions are of further interest in connection with observations that feeding preferences of individuals may be modified by previous experience. Larvae of *Pieris brassicae* can distinguish among their preferred plants and tend to choose the species upon which they have previously fed, thus demonstrating an acquired taste (Johansson, 1951; Hovanitz and Chang, 1962). In *Galleria mellonella* (L.) "congenital food preferences may be considerably modified by temporary connections especially during the first days of post-embryonic development" (Balázs, 1958). Young larvae of the swallowtail *Papilio aegeus* Donovan exhibit a certain polyphagy and tend to become conditioned to the plant upon which they have fed whether it is a natural food or not (Stride and Straatman, 1962). Were the sense organs highly specific filters, modification of a response by an individual could not be easily accomplished because the sense organs probably would be silent to all messages except one. With multiple patterns available to the central nervous system modification is possible.

### *Regulation of quantity*

No insect eats uninterruptedly. Even the so-called continuous feeders, of which the larvae of mosquitoes and simulids are examples, alternate periods of feeding with periods of rest. The larvae of many phytophagous species, also said to feed continuously, rest more often than they feed. Prolonged observation shows that larvae of the moth *Philosamia cynthia* Drury, for example, feed for approximately 15 minutes at a time and rest for about 50. The antithesis of frequent feeding is to be found among blood-sucking insects, some of which feed only once in each intermoult period. The question in each case is: what initiates feeding and what terminates it? Contained in answers to these questions will be information relating to the determination of frequency patterns and the regulation of quantity.

The most extensive analysis of these matters has been made with the blowfly *Phormia regina* (for a complete summary see Dethier, 1966). As already explained, feeding is initiated by sensory input from tarsal and oral receptors. Once feeding has commenced continuous sensory input is supplied principally by the interpseudotracheal papillae (and to a lesser extent by tarsal and labellar hairs). Without this input feeding is neither begun nor maintained.

A single meal seldom exceeds two minutes in duration. During this period the rate of ingestion is related to the intensity of sensory input from receptors sensitive to sugars. The higher the sugar concentration the greater the frequency of sensory input and hence the faster the pumping. If the concentration is too high, however, the viscosity of the solution presents a mechanical impediment to sucking so that above 1 molar sucrose the rate of ingestion actually declines.

When the sense organs become adapted, ingestion ceases. Since a critical level of adaptation occurs more rapidly at low concentrations than at high, dilute solutions are fed upon for a shorter period of time than concentrated ones. The two phenomena, adaptation and intensity of sensory input, determine the volume of fluid ingested at a single meal. They operate independently of nutritional value. A fly will ingest a greater volume of a highly stimulating non-nutrient than of a moderately or poorly stimulating nutrient and in a two-choice situation will actually prefer the non-nutrient. In *Phormia*, then, the initiation of feeding, the termination of feeding, and the control of quantity is under the control of taste receptors.

The situation with regard to single meals appears to be simple and straightforward, but the long-term termination of feeding requires a different explanation. After a single meal the receptors disadapt within minutes yet the fly may not take another meal for one or more hours. The length of the inter-meal period is related to the quantity and nature of the meal just taken and the amount of energy expended in the interim. Additionally, when the meal is taken, the quantity ingested is related to the time since the last meal and its content. These interwoven relationships can best be expressed in behavioural terms by saying that proboscis extension, rate of sucking, and duration of feeding all depend in part on the acceptance threshold of the organism at the moment for the substance encountered.

In general, thresholds decrease with starvation; consequently, at the beginning of feeding, this threshold is at its lowest value. The failure to feed again after a meal is terminated results from an elevation of threshold in the central nervous system. The mechanism regulating threshold is located in the region of the fore-gut. The sugar that is imbibed goes initially both to the mid-gut and the crop. When the mid-gut is filled, the cardiac valve closes, and the continued influx of sugar is directed into the crop. Waves of peristalsis passing posteriorly in the crop duct constantly press the fluid back into the crop itself (Knight, 1962). When feeding has ceased, the crop valve closes. The region of the fore-gut now contains sugar residue from more anterior regions of the oesophagus and preoral cavity plus sugar from leaks in the crop valve as long as it is under pressure from a full crop (Green, 1964). When sugar in the mid-gut is utilised, the crop valve opens momentarily, peristalsis in the duct is reversed so that a slug of fluid is driven energetically into the fore-gut region, the crop valve closes, a wave of peristalsis in the fore-gut drives the fluid toward the proventriculus where the cardiac valve opens briefly to permit passage into the midgut. It has been postulated that there are receptors somewhere in the region of the fore-gut that are stimulated every time fluid passes, that this stimulation causes impulses to ascend the recurrent nerve, and that these impulses inhibit at some central point the sensory input from the now disadapted oral receptors, thus causing the behavioural threshold to be high and feeding to cease.

There is no direct evidence to support the idea of a hormonal link; indeed some of the evidence contra-indicates it. Evidence supporting the idea of a neural inhibitory feedback, while not conclusive, is derived from the observation that cutting the innervation supplying the fore-gut results in hyperphagia. It is entirely possible that more than one mechanism may be involved in threshold regulation. Experiments relating to central excitatory state seem to suggest this. In any case, the following can be eliminated as probable mechansims: the crop itself, the mid-gut, blood sugar, stored glycogen. In so far as the fly is concerned, therefore, sensory input is required to start feeding, the intensity of input controls the rate of ingestion, adaptation terminates it, and an elevation of central threshold prevents further feeding. When central threshold falls below a critical level, the fly is ready for another meal.

Many of the features of quantity control found in the fly must occur in one form or other in all insects. It is obvious from the discussion of quality discrimination that for many species a chemical stimulus is required to initiate feeding. It is clear, furthermore, that some sort of continuous sensory input is necessary for the maintenance of feeding, or influences it in such a way as to affect the quantity ingested. In *Periplaneta americana* L. there is a receptor in the pharynx that detects differences in the osmotic concentration of fluids being drunk (Davey and Treherne, 1963b). The crop valve is directed by signals from this receptor. The release of fluid from the crop is partially controlled by alterations in frequency of opening of the proventricular valve (Davey and Treherne, 1963a). The rate of removal of fluid from the mid-gut is limited by the rate at which the crop delivers fluid to this region (Treherne, 1957), that is, the crop never delivers fluid faster than the mid-gut can absorb it.

In many insects termination of feeding is probably brought by sensory adaptation as in the fly. This is the case with the milkweed bug (Feir and Beck, 1963). It is not likely to be the limiting factor in caterpillars and other insects whose meals last 15 minutes or more, but there is no information on the rate of sensory adaptation in these species. It is more likely that a rise in central threshold is the responsible factor.

There are undoubtedly major variations on the theme represented by the fly. For one thing, it appears that many insects do not require a specific stimulus from their food to start biting. The Colorado potato beetle is said to bite randomly at all sorts of objects (de Wilde, 1958). According to Dadd (1960), many grasshoppers need only contact with a "bitable" surface. Thorsteinson (1960), on the other hand, maintained that tactile stimuli do not normally cause feeding responses. The two views are not necessarily contradictory because there are obviously differences among species. This fact is brought out most forcefully by observing the behaviour of various insects in the throes of starvation. Some species will die without ever having essayed a single bite; others will become less fastidious than normally but still selective; still others will become polyphagous.

The tendency to bite in the absence of specific stimuli, to bite more frequently when deprived of food, and to become less discriminating as starvation progresses suggests that there may be in these species an endogenous mechanism driving feeding. If an insect does not require constant stimulation from its food to keep feeding going, as has been said of some species, this fact may also be construed as indicating an internal drive. Dadd (1960) suggested that such is indeed the case with *Schistocerca* whose feeding is initiated and sustained by "endogenous drives which are regulated by the utilisability of the ingested material". Yet Goodhue (1963) has said that *Schistocerca* appears to require a swallowing stimulus.

In fact we have little information on central manifestations of hunger (Dethier, 1964) and the extent of contributions made by active centres in the central nervous system to feeding behaviour. Analyses of the relative importance in feeding behaviour of central versus peripheral neural mechanisms and of neural versus hormonal control have only begun to yield answers.

## DISCUSSION

**J. D. Carthy.** The idea that it is input *patterns* from receptors which are distinguished by the insect is interesting in relation to the idea of learning as an important element in food preference. It seems to support Schneirla's suggestion that food preferences are not innate but learned.

**V. G. Dethier.** I agree that pattern input would permit learning by the insect whereas the older idea that there was sharp tuning of receptors to limited stimuli makes learning more or less impossible. It does not follow, however, that there is no innate contribution to feeding preferences.

**A. Manning.** Doesn't the observation that maxillectomy results in increasing amount and range of feeding suggest that endogenous activity drives feeding behaviour?

**V. G. Dethier.** I am not willing to go as far as that yet, since the antenna also must be involved.

**A. Manning.** What happens if the antennae are removed?

**V. G. Dethier.** No effect on feeding behaviour can be observed.

**A. Manning.** Surely that supports both of us?

**M. J. Parr.** Papilios are said to feed on Umbellifers and citrus because of token stimuli. If plants from other families be fed systemically on such token substances and the caterpillars then induced to feed on them, can they be gradually acclimatised to foreign plants, if the concentration of the token substances is gradually reduced?

**V. G. Dethier.** It has not been done, but feeding can be induced on some foreign plants by applying token stimuli to them, although again some foreign plants have repellent qualities which prevent this. Caterpillars may or may not survive on such a plant but no acclimatisation experiments of the sort you suggest have been tried. If they were I would speculate that the insect would gradually come to accept the new plant.

**E. T. Burtt.** Can you enlarge on the experiments in which hungry flies were fed by enema and had artificially raised blood sugar, but neither of these procedures satisfied them? It seems to indicate hunger as a central nervous phenomenon but you argued otherwise.

**V. G. Dethier.** Feeding is driven by the receptors on the proboscis and food then enters the crop and feeding stops because of the adaptation of the proboscis receptors. But in 15 minutes these sense organs are no longer adapted yet the animal will not feed for an hour, so that some further mechanism must exist. This is not the presence of food in the crop or in other parts of the mid-gut nor is it due to a blood-borne factor, and that is why the two examples I have quoted do not suggest a central nervous mechanism. However, the mechanism does involve nervous connections and probably inhibitory feedback from gut to brain, but the animal cannot feed in the absence of stimulation to proboscis receptors.

**E. T. Burtt.** But the central nervous system must be involved, or is it just that there is no urge to eat?

**V. G. Dethier.** The central nervous system receives input from oral receptors and inhibitory input from foregut receptors, integrates this information, and sends comments to motor centres. There seems to be no need at this time to postulate a C.N.S. centre whose endogenous activity drives feeding.

**E. T. Burtt.** Why did filling the gut by enema not stimulate the nervous pathways?

**V. G. Dethier.** Because we were unable to force the food into the foregut by this method, and only the foregut appears to have the necessary receptors.

**E. T. Burtt.** If you could get food through to the foregut by this method, would you say the animal would not go on feeding?

**V. G. Dethier.** Yes, probably it would not feed.

**W. A. L. David.** Would a fly, fully fed on carbohydrates, take protein if it were a female needing this for egg maturation?

**V. G. Dethier.** Yes, between day 4 and day 8 a female will take protein even if she is replete with carbohydrate. A specific protein mechanism operates here.

**H. C. Chiang.** Have you artificially emptied the fly's gut after a meal?

**V. G. Dethier.** Yes, we did this some time ago. The crop has been removed, or in some cases punctured. The fly then becomes hungry sooner, but not until all residual fluid near the fore mid-gut junction has gone.

**K. M. Mellanby.** Reverting to token stimuli—the human body louse will feed on rabbits which have been given a human odour and can then learn to feed on rabbits without the odour. Young lice, however, cannot learn or even take a meal.

**G. C. Varley.** Sensitivity threshold of the fly's sugar receptors must surely be below the concentration of blood sugar, yet blood sugar can be increased without inducing feeding responses—how is this?

**V. G. Dethier.** This is because the basement membrane forms a barrier to the blood sugar, preventing it from reaching the sense organ.

## *References*


APPLEBAUM, S. W., 1964, Physiological aspects of host specificity in the Bruchidae. I. General considerations of developmental compatibility. *J. Insect. Physiol.* **10** : 783–788.

AUGUSTINE, M. G., FISK, F. W., DAVIDSON, R. H., LAPIDUS, J. B., and CLEARY, R. W., 1964, Host plant selection by the Mexican bean beetle, *Epilachna varivestris*. *Ann. ent. Soc. Amer.* **57** : 127–134.

BALÁZS, A., 1958, Nutritional and nervous factors in the adaptation of *Galleria mellonella* to artificial diet. *Acta biol. hung.* **9** : 47–69.

BECK, S. D., 1956a, The European corn borer, *Pyrausta nubilalis* (Hübn.) and its principal host plant. I. Orientation and feeding behavior of the larva on the corn plant. *Ann. ent. Soc. Amer.* **49** : 552–558.

—— 1956b, Nutrition of the European corn borer, *Pyrausta nubilalis* (Hübn.). IV. Feeding reactions of first instar larvae. *Ibid.* **49** : 399–405.

—— 1965, Resistance of plants to insects. *Annu. Rev. Ent.* **10** : 207–232.

—— and HANEC, W., 1958, Effect of amino acids on feeding behaviour of the European corn borer, *Pyrausta nubilalis* (Hübn.). *J. Insect Physiol.* **2** : 85–96.

BROWER, L. P., 1958, Bird predation and food plant specificity in closely related procryptic insects. *Amer. Nat.* **92** : 183–187.

CHAUVIN, R., 1951, Sur les facteurs responsables de l'attraction que manifestent les Acridiens pour le son. *Bull. Off. nat. Anti-acrid., Paris* **1** : 15–18.

CHIN, CHUN-TEH, 1950, Studies on the physiological relations between the larvae of *Leptinotarsa decemlineata* Say and some solanaceous plants. Thesis. Amsterdam.

DADD, R. H., 1960, Observations on the palatability and utilisation of food by locusts, with particular reference to the interpretation of performances in growth trials using synthetic diets. *Ent. exp. appl.* **3** : 283–304.

—— 1963, Feeding behaviour and nutrition in grasshoppers and locusts. *Adv. Insect Physiol.* **1** : 47–109.

DAVEY, K. G. and TREHERNE, J. E., 1963a, Studies on crop function in the cockroach, *Periplaneta americana* L. I. The mechanism of crop emptying. *J. exp. Biol.* **40** : 763–773.

—— —— 1963b, Studies on crop function in the cockroach, *Periplaneta americana* L. II. The nervous control of crop emptying. *Ibid.* **40** : 775–780.

DAY, M. F., IRZYKIEWICZ, H., and MCKINNON, A., 1952, Observations on the feeding of the virus vector *Orosius argentatus* (Evans), and comparisons with certain other jassids. *Aust. J. sci. Res.* (B) **5** : 128–142.

DETHIER, V. G., 1937, Gustation and olfaction in lepidopterous larvae. *Biol. Bull., Wood's Hole,* **72** : 7–23.

—— 1941, Chemical factors determining the choice of food-plants by *Papilio* larvae. *Amer. Nat.* **75** : 61–73.

—— 1947, *Chemical insect attractants and repellents.* Philadelphia.

—— 1953, Host plant perception in phytophagous insects. *Int. Congr. Ent. IX* (Amsterdam, 1951) **2** : 81–88.

—— 1964, Microscopic brains. *Science* **143** : 1138–1145.

—— 1966, Feeding behavior of the blowfly (*in press*).

—— EVANS, D. R., and RHOADES, M. V., 1956, Some factors controlling the ingestion ot carbohydrates by the blowfly. *Biol. Bull.* **111** : 204–222.

DOWNEY, J. C., 1962, Host-plant relations as data for butterfly classification. *Syst. Zool.* **11** : 150–159.

EHRLICH, P. R. and RAVEN, P. H., 1964, Butterflies and plants: a study in coevolution. *Evolution* **18** : 586–608.

EVANS, D. R., 1961, Depression of taste sensitivity to specific sugars by their presence during development. *Science* **133** : 327–328.

FEIR, D. and BECK, S. D., 1963, Feeding behavior of the large milkweed bug, *Oncopeltus fasciatus. Ann. ent. Soc. Amer.* **56** : 224–229.

FRAENKEL, G. S., 1953, The nutritional value of green plants for insects. *Int. Congr. Ent. IX* (Amsterdam, 1951) **2** : 90–100.

—— 1959, The raison d'être of secondary plant substances. *Science* **129** : 1466–1470.

GOODHUE, D., 1963, Feeding stimulants required by a polyphagous insect, *Schistocerca gregaria. Nature, Lond.* **197** : 405–406.

GORDON, H. T., 1961, Nutritional factors in insect resistance to chemicals. *Annu. Rev. Ent.* **6** : 27–54.

GREEN, G. W., 1964, The control of spontaneous locomotor activity in *Phormia regina* Meigen. II. Experiments to determine the mechanism involved. *J. Insect Physiol.* **10** : 727–752.

GUPTA, P. D. and THORSTEINSON, A. J., 1960, Food plant relationships of the diamond-back moth (*Plutella maculipennis* (Curt.)). *Ent. exp. appl.* **3** : 241–250.

HAMAMURA, Y. and NAITO, K., 1961, Food selection by silkworm larvae, *Bombyx mori.* Citral, linalyl acetate, linalol, and terpinyl acetate as attractants of larvae. *Nature, Lond.* **190** : 879–880.

—— HAYASHIYA, K., and NAITO, K., 1961, Food selection by silkworm larvae, *Bombyx mori.* β-sitosterol as one of the biting factors. *Nature, Lond.* **190** : 880–881.

HERON, R. J., 1965, The role of chemotactic stimuli in the feeding behavior of spruce budworm larvae on white spruce. *Canad. J. Zool.* **43** : 247–269.

HOVANITZ, W., and CHANG, V. C. S., 1962, Three factors affecting larval choice of food plant. *J. Res. Lepidopt.* **1** : 51–61.

ISHIKAWA, S., 1963, Responses of maxillary chemoreceptors in the larva of the silkworm, *Bombyx mori,* to stimulation by carbohydrates. *J. cell. comp. Physiol.* **61** : 99–107.

—— and HIRAO, T., 1963, Electrophysiological studies of taste sensation in the larvae of the silkworm, *Bombyx mori. Bull. seric. Exp. Sta.* **18** : 297–357.

—— TAZIMA, Y., and HIRAO, T., 1963, Responses of the chemoreceptors of maxillary sensory hairs in a "non-preference" mutant of the silkworm. *J. seric. Sci., Tokyo,* **32** : 125–129.

ITO, T., HORIE, Y., and FRAENKEL, G., 1959, Feeding on cabbage and cherry leaves by maxillectomized silkworm larvae. *J. seric. Sci. Tokyo* **28** : 107–113.

JERMY, T., 1958, Untersuchungen über Auffinden und Wahl der Nahrung beim Kartoffelkäfer (*Leptinotarsa decemlineata* Say). *Ent. exp. appl.* **1** : 197–208.

—— 1961, On the nature of the oligophagy in *Leptinotarsa decemlineata* Say (Coleoptera : Chrysomelidae). *Acta zool. hung.* **7** : 119–132.

JOHANSSON, A. S., 1951, The food plant preference of the larvae of *Pieris brassicae* L. (Lepid., Pieridae). *Norsk ent. Tidsskr.* **8** : 187–195.

KENNEDY, J. S., 1953, Host plant selection in Aphididae. *Int. Congr. Ent. IX* (Amsterdam, 1951) **2** : 106–113.

—— and BOOTH, C. O., 1951, Host alternation in *Aphis fabae* Scop. I. Feeding preferences and fecundity in relation to the age and kind of leaves. *Ann. appl. Biol.* **38** : 25–64.

KNIGHT, SR. M. R., 1962, Rhythmic activities of the alimentary canal of the black blowfly, *Phormia regina. Ann. ent. Soc. Amer.* **55** : 380–382.

LIPKE, H. and FRAENKEL, G., 1956, Insect nutrition. *Annu. Rev. Ent.* **1** : 17–44.

MATSUMOTO, Y., 1962, A dual effect of coumarin, olfactory attraction and feeding inhibition, on the vegetable weevil adult, in relation to the uneatability of sweet clover leaves. *Jap. J. appl. Ent. Zool.* **6** : 141–149.

—— and SUGIYAMA, S., 1960, Attraction of leaf alcohol and some aliphatic alcohols to the adult and larva of the vegetable weevil. *Ber. Ohara Inst. landw. Biol.* **11** : 359–364.

MERZ, E., 1959, Pflanzen und Raupen. Uber einige Prinzipien der Futterwahl bei Gross-schmetterlingsraupen. *Biol. Zbl.* **78** : 152–188.

MITTLER, T. E. and DADD, R. H., 1963, Studies on the artificial feeding of the aphid *Myzus persicae* (Sulzer). I. Relative uptake of water and sucrose solutions. *J. Insect Physiol.* **9** : 623–645.

—— —— 1964, Gustatory discrimination between liquids by the aphid *Myzus persicae* (Sulzer). *Ent. exp. appl.* **7** : 315–328.

—— —— 1965, Differences in the probing responses of *Myzus persicae* (Sulzer) elicited by different feeding solutions behind a parafilm membrane. *Ibid.* **8** : 107–122.

MORITA, H. and YAMASHITA, S., 1961, Receptor potentials recorded from sensilla basiconica on the antenna of the silkworm larvae, *Bombyx mori*. *J. exp. Biol.* **38** : 851–861.

MUNAKATA, K., SAITO, T., OGAWA, S., and ISHII, S., 1959, Oryzanone an attractant of the rice stem borer. *Bull. agric. chem. Soc. Japan* **23** : 64–65.

NAYAR, J. K. and THORSTEINSON, A. J., 1963, Further investigations into the chemical basis of insect-host plant relationships in an oligophagous insect, *Plutella maculipennis* (Curtis) (Lepidoptera : Plutellidae). *Canad. J. Zool.* **41** : 923–929.

NIIMURA, M. and ITO, T., 1964, Feeding on filter paper by larvae of the silkworm, *Bombyx mori* L. *J. Insect Physiol.* **10** : 425–430.

NUORTEVA, P., 1952, Die Nahrungspflanzenwahl der Insekten im Lichte von Untersuchungen an Zikaden. *Ann. Acad. Sci. fenn.* A **19** : 1–90.

PETERSEN, B., 1954, Egg-laying and habitat selection in some *Pieris* species. *Ent. Tidskr.* **75** : 194–203.

SCHOONHOVEN, L. and DETHIER, V. G., 1966, Sensory aspects of host-plant selection by lepidopterous larvae (*in press*).

STRIDE, G. O. and STRAATMAN, R., 1962, The host plant relationship of an Australian swallowtail, *Papilio aegeus*, and its significance in the evolution of host plant selection. *Proc. Linn. Soc. N.S.W.* **87** : 69–78.

SUGIYAMA, S. and MATSUMOTO, Y., 1959, Attractiveness of umbelliferous plants to the vegetable weevil. *Nögaku Kenkyü* **47** : 141–148.

THORPE, W. H., CROMBIE, A. C., HILL, R., and DARRAH, J. H., 1947, The behaviour of wireworms in response to chemical stimulation. *J. exp. Biol.* **23** : 234–266.

THORSTEINSON, A. J., 1953, The chemotactic responses that determine host specificity in an oligophagous insect (*Plutella maculipennis* (Curt.), Lepidoptera). *Canad. J. Res.* **31** : 52–72.

—— 1955, The experimental study of the chemotactic basis of host specificity in phytophagous insects. *Canad. Ent.* **87** : 49–57.

—— 1958, The chemotactic influence of plant constituents on feeding by phytophagous insects. *Ent. exp. appl.* **1** : 23–27.

—— 1960, Host selection in phytophagous insects. *Annu. Rev. Ent.* **5** : 193–218.

TORII, K. and MORII, K., 1948, Studies on the feeding habit of silkworms. *Bul. Res. Inst. seric. Sci.* **2** : 3–12.

TREHERNE, J. E., 1957, Glucose absorption in the cockroach. *J. exp. Biol.* **34** : 478–485.

WALDBAUER, G. P., 1962, The growth and reproduction of maxillectomised tobacco hornworms feeding on normally rejected non-solanaceous plants. *Ent. exp. appl.* **5** : 147–158.

—— and FRAENKEL, G., 1961, Feeding on normally rejected plants by maxillectomised larvae of the tobacco hornworm, *Protoparce sexta* (Lepidoptera, Sphingidae). *Ann. ent. Soc. Amer.* **54** : 477–485.

WATANABE, T., 1958, Substances in mulberry leaves which attract silkworm larvae (*Bombyx mori*). *Nature, Lond.* **182** : 325–326.

WENSLER, R. J. D., 1962, Mode of host selection by an aphid. *Nature, Lond.* **195** : 830–831.

DE WILDE, J., 1958, Host plant selection in the Colorado beetle larva (*Leptinotarsa decemlineata* Say). *Ent. exp. appl.* **1** : 14–22.

WYKES, G. R., 1952, The preference of honeybees for solutions of various sugars which occur in nectar. *J. exp. Biol.* **29** : 511–519.

YAMAMOTO, R. T. and FRAENKEL, G., 1960, Assay of the principal gustatory stimulant for the tobacco hornworm, *Protoparce sexta*, from solanaceous plants. *Ann. ent. Soc. Amer.* **53** : 499–503.

# C. Defense

## The Defensive Behaviour of the Javanese Stick Insect, *Orxines Macklotti de Haan*, with a Note on the Startle Display of *Metriotes Diocles* WestW.

(Phasmatodea, Phasmidae)

By Michael H. Robinson

### Introduction

Some aspects of the general and reproductive biology of *Orxines macklotti* De Haan were described in an earlier paper (Robinson, 1964). The present contribution deals with those elements of the behaviour of *Orxines* for which an anti-predator function seems probable. The anti-predator adaptations of animals can usefully be regarded as belonging to two functional categories. There are those which operate to reduce the probability that the presence of the animal will be detected by a predator, and, in the second category, those which operate after the predator has found the animal. The latter function is to increase the probability that the animal will escape predation after it has been discovered by a predator. Crane (1952) listed the seven major types of defensive behaviour found in the animal kingdom, of these, six belong to the second category defined above. It is generally assumed that phasmids do not usually exhibit vigorous defensive behaviour (see for instance Bedford & Chinnick, 1966), but in fact *Orxines* possesses four of the seven types of defensive behaviour described by Crane (1952).

### Materials and Methods

The observations reported here were made on a laboratory culture of *Orxines* raised from stock kindly donated by Mr. G. J. Ashby, Overseer of the Aquarium and Invertebrates at the London Zoological Gardens. This culture was maintained at Oxford, under the conditions described in my earlier paper, for nearly two years. The maximum number of adult *Orxines*, of both sexes, reached nearly 200. Behavioural sequences were observed directly and, in addition, film records were prepared for subsequent analysis using a Bolex H 16 cine camera. Film records were analysed on a Specto 16 mm. movement analysis projector which permitted frame by frame analysis of movements. Some of the illustrations in this paper are based on such film records. Special techniques developed in connection with the investigation of particular problems are described in the account of defensive behaviour below.

### Elements of Defensive Behaviour

#### Category I. Concealment

*Orxines* is a highly cryptic lichenose insect which adopts an immobile diurnal resting position on tree bark. The resting attitude differs markedly from that of many more stick-like phasmids in that it does not involve protraction and apposition of the anterior limbs (stick-formation) and is assumed with the body appressed to the substrate. It can be argued that this resting attitude maximises the concealing effect of the lichenose coloration. The adoption of the diurnal resting position involves active background selection (Kuyten, 1961).

Reproduced with permission from the *Entomologist's Monthly Magazine*, 104:46–54, 1968. Published by the Entomologist's Monthly Magazine Ltd., Oxford, England.

In common with many other phasmid species and also with some stick- and leaf-mimicking forms of other insect orders, *Orxines* performs curious rocking movements when abandoning its diurnal resting position by slow locomotion. The question of the adaptive function of rocking

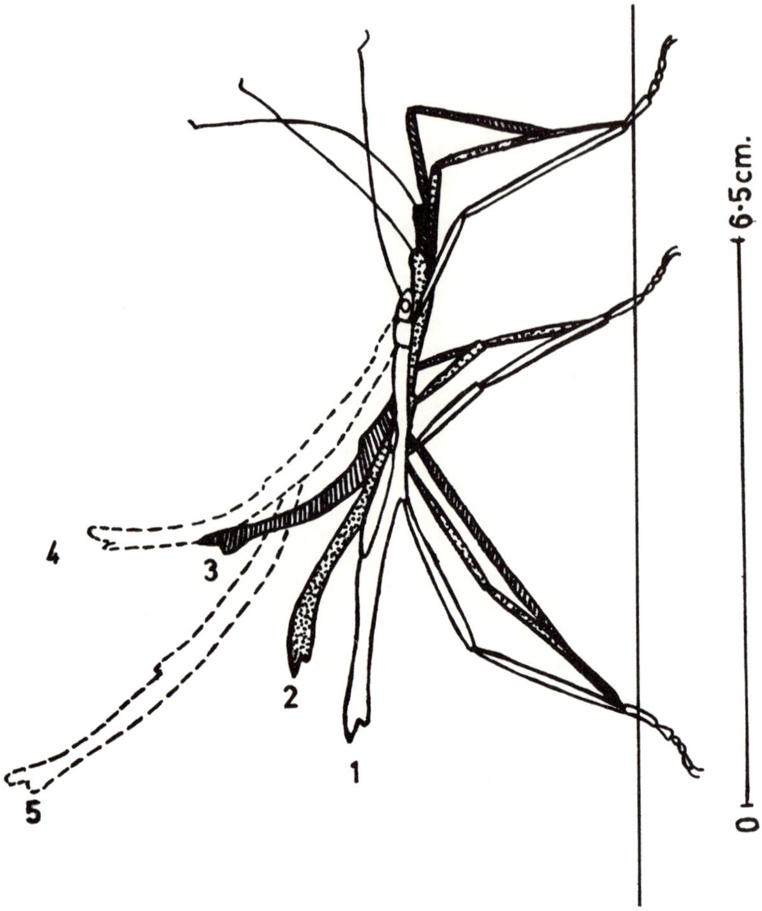


Fig. 1.—Male *Orxines macklotti* De Haan jumping from a vertical substrate. (Based on five successive frames of 16 mm ciné film, filmed at 24 frames per second.)

movements is to form the basis of a separate paper (in preparation) and is, accordingly, not dealt with here. However, it can be said that good evidence exists to support the view that rocking may function as a means of movement-concealment which can be regarded as an anti-predator adaptation belonging to Category I.

CATEGORY II. ACTIVE DEFENSIVE BEHAVIOUR

(a). *Active Escape.* Many insect species respond to the presence of predators by dropping or jumping off the substrate. This escape method utilises gravity to increase rapidly the distance between the predator and the prey. Dropping from the substrate following tactile stimulation has been described for several species of phasmid and is well known in the case of the Indian Stick Insect *Carausius morosus* Br. (Steiniger, 1933). In the case of *Orxines*, active escape, in both sexes, involves jumping from the substrate rather than just dropping. Figure 1 shows the sequence of movements involved in the act of jumping, following tactile stimulation. The horizontal distance travelled after jumping from a vertical substrate at the height of 4 ft. 0 in. is shown in fig. 2.

(Graphed results of tests on 69 insects.) From this it will be seen that the maximum horizontal distance travelled is not very great but this nevertheless may be sufficient to enable an insect resting on the bole of a tree to reach the ground clear of any lower obstructions, following jumping.

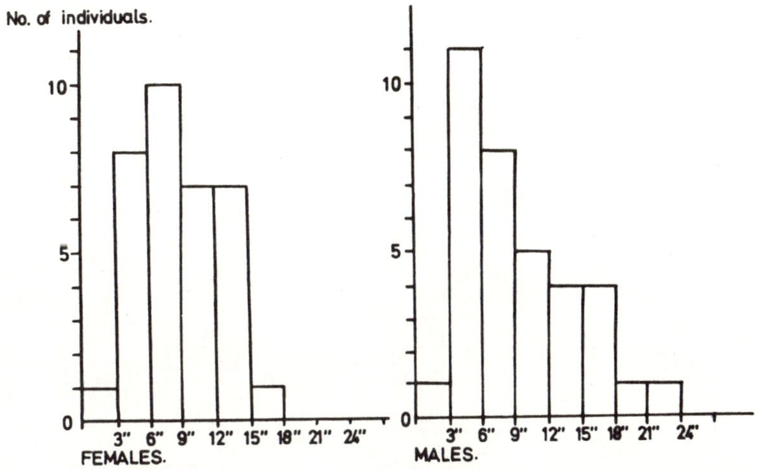


Fig. 2.—Histograms showing (in 3 inch intervals) the distance of the landing position of 69 adult *Orxines* after jumping from a vertical surface four feet high. The distances given being the horizontal distance from the take-off plane measured at right angles to this.

During falling the wings and tegmina are extended laterally and usually are immediately refolded on landing. The bright colouration of the hind wing which is visible during the fall is thus concealed when the insect lands. This rapid colour change may serve to confuse any predator which follows the falling insect. The lateral extension of the wings during falling is not simply induced by absence of tarsal contact with the substrate but seems to require the additional stimulus of moving air-currents.

(b). *Startle Display*. Displays involving a sudden apparent increase in size together with a change from inconspicuous to conspicuous coloration have been described for mantids by Crane (1952) and Varley (1939). Recently Bedford & Chinnick (1966) reported startle displays from two species of Australian stick insects. No experimental analysis has yet been made to determine whether startle displays depend for their efficacy on the sudden apparent size increase, the sudden colour change or a summation of both elements. Some, but not all, startle displays involve the revelation of typical aposematic colours and patterns and may gain effect from this. In certain circumstances (see below) *Orxines* produces a display which includes all the classic elements of a startle display. The tegmina and hind wings are snapped open suddenly into a vertically erect position thereby increasing the apparent size of the insect and revealing the very conspicuous red and black hind wings. In addition, the display, which may persist for up to one minute in the absence of further stimulation, is often accompanied by intense high amplitude rocking movements. Such movements accompany the display of several mantids (Crane, 1952) and may serve to heighten the startle effect. No ventrally directed defensive movements of the legs, such as those which accompany the display of the two Australian phasmids described by Bedford & Chinnick (1966), occur in the *Orxines* display. This difference can be correlated with the difference between the resting substrate of *Orxines* (normally tree bark) and the Australian species (twigs and branches). Attacks on *Orxines* from a ventral direction seem unlikely, if not impossible, at least when the insect is at rest.

The black borders of the hind wings of *Orxines* bear numerous small white spots which may function as simple eye-spot patterns and thereby contribute to the startle effect of the display. The work of Blest (1957)

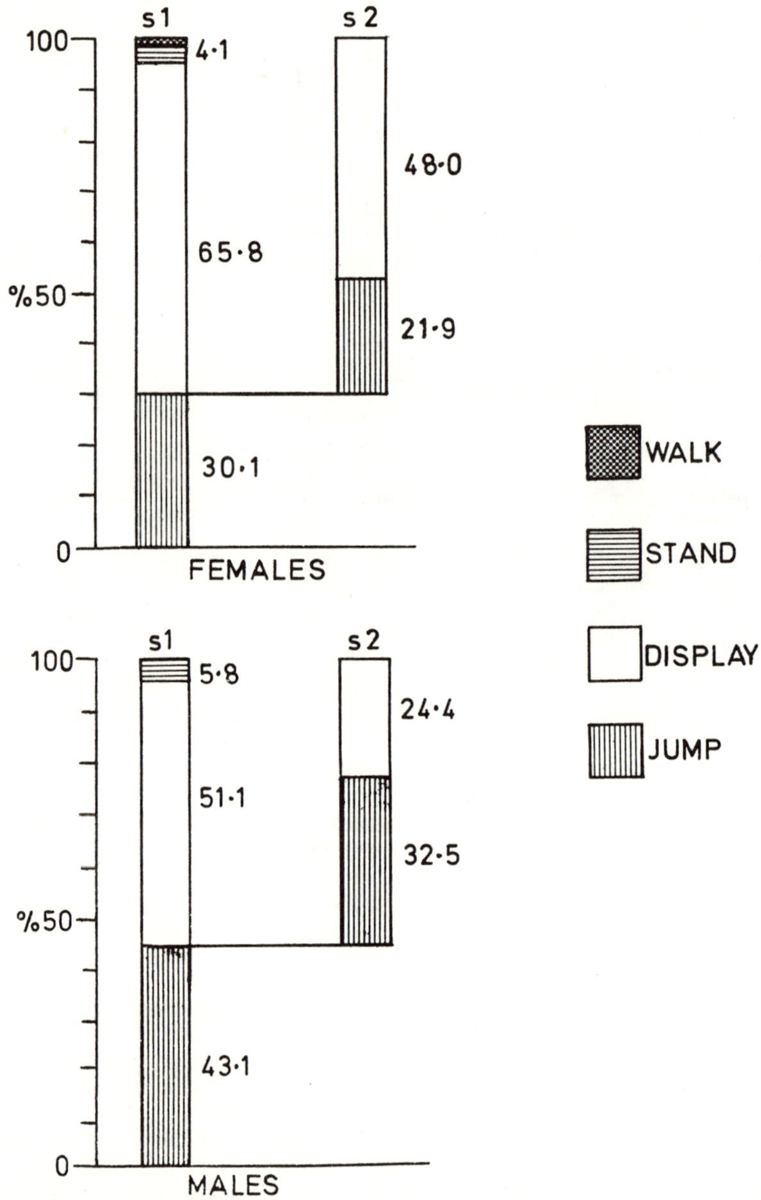


Fig. 3.—Histograms showing the response of male and female *Orxines* to mechanical stimulation. Explanation in text.

strongly suggests that even simple eye-spot patterns may have an antipredator function.

A similar wing-erection display occurs in the neotropical phasmid *Metriotes diocles* Westw.; this is described and figured in a note below.

TABLE 1.—JUMPING RESPONSES OF *ORXINES* (MALES AND FEMALES).

| | First stimulus | | Second stimulus | |
|---|---|---|---|---|
| | Jumps | Other responses | Jumps | Other responses |
| Males | 37 | 49 | 28 | 21 |
| Females | 22 | 51 | 16 | 35 |
| | $\chi^2=2.8$ | | $\chi^2=6.7$ | |
| | p is more than .05 (two-tailed) | | p is less than .02 (two-tailed) | |

*The Relationship of Escape Behaviour to the Startle Display.* Both the jumping response and the startle display are elicited by strong mechanical stimulation. Adult resting *Orxines* gave differing responses when subjected to such stimulation. Some insects moved into an erect standing position, others jumped off the substrate and some produced the startle display. These responses were investigated in tests involving 86 adult males and 73 adult females. Each insect was pinched on the terminal abdominal segments. Those which did not jump after the first stimulus were pinched again, one minute later. Table 1 shows the number of jumps against the number of other responses combined, for each stimulus. The responses of male and female insects are shown separately, expressed as percentages, in fig. 3. The left hand column in each histogram shows the responses to the first stimulus (s1) and the right hand column, the responses of those animals which were stimu lated a second time. The results can be summarised as follows:

(i) 51.1 % of the males displayed in response to the first stimulus and 43.1 % jumped. 5.8 % stood erect.
(ii) 65.8 % of the females displayed in response to the first stimulus and 30.1 % jumped. 4.1 % stood erect or walked.
(iii) 75.6 % of the males and 52 % of the females jumped after the first or second stimulus.
(iv) 24.4 % of the males and 48 % of the females responded to the second stimulus by displaying.

The responses of males and females were analysed separately because of an apparent greater tendency of males to jump rather than display when stimulated. The difference between the responses to the first stimulus of males and females is not significant, whereas the males show a greater tendency to jump in response to the second stimulus than do females (see Table 1).

Gwynne (1965) reported a staged response to disturbing stimuli by the East African saturnioid *Nudaurelia gueinzii* Karsch. This insect responded to the first disturbing stimulus by displaying its underwing eye-spot patterns and later, if stimulation persisted, the insect dropped off the substrate. If the initial stimulation was particularly strong, dropping occurred without further stimulation.

The response of *Orxines* to disturbance may be similarly staged. Thus the startle display could be regarded as the first stage and jumping as the second stage. More than half the insects, of both sexes, responded to the first stimulus by displaying. The greater tendency of males to respond to the second stimulus by jumping may be correlated with the fact that they are much smaller than the females and, in particular, their wings are less than a quarter of the area of those of the females. As a consequence the male startle display may be much less intimidating.

The fact that, in the tests, more than one type of response occurred to the same type of stimulus may be due to differences in stimulus intensity resulting from the somewhat crude method of stimulation. The further possibility that this was due to variations in sensitivity to mechanical stimuli cannot be ignored. The really critical test to determine whether the responses, display—jump, form a graded series, i.e. by applying graded stimuli, was not carried out because of the difficulties involved in standardising mechanical stimuli.

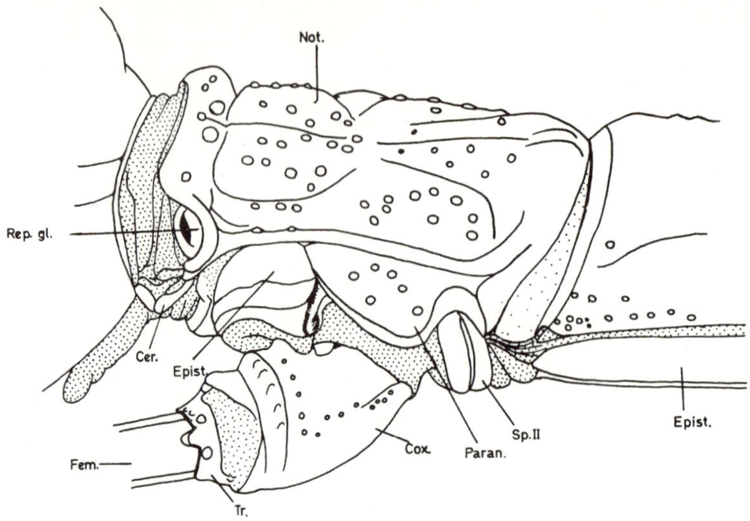


Fig. 4.—Anatomy of the anterior thorax of a female *Orxines macklotti* De Haan. Cer., cervical sclerite; Cox., coxa; Epist., episternum; Fem., femur; Not., notum; Paran., paranotum; Rep.gl., orifice of repugnatorial gland; Sp.II., spiracle; Tr., trochanter.

(c). *Repugnatorial Secretion.* Several species of phasmids possess glands situated in the prothorax for which a repugnatorial function is assumed (Beier, 1957). The secretion of these glands as a defensive spray by the phasmid *Anisomorpha buprestoides* proved effectively deterrent to ants, predaceous beetles, mice and blue jays in experiments carried out by Eisner (1965). In *Orxines* both sexes possess small paired glands situated in the prothorax and discharging through spiracle-like orifices in front of the dorso-lateral margin of the prothoracic tergum, shown in fig. 4. These glands are much smaller than those described for either *Anisomorpha* or *Paradoxomorpha* sp. (Moreno, 1940), being from 4–6 mm in length. The secretion is discharged when the insect is handled and is highly volatile. Chandless (1954) reported that it caused sneezing in dogs. I was able to obtain no evidence that it could be directed as a spray against potential predators. Nevertheless a similarly ephemeral secretion produced by the neotropical phasmid *Isagoras* sp. proved effectively deterrent to highly insectivorous Tamarins (*Saguinus geoffroyi*), in experiments which I carried out in Panama.

Preliminary histological investigations suggest that the glands are similar in structure to those of *Paradoxomorpha* sp. described by Moreno (1940). *Sipyloidea sipylus* Westw. a phasmid of the same sub-family as *Orxines*, the Necrosciinae, has similar glands in the prothorax although to the human observer the discharge has an entirely different odour.

Summary

*Orxines macklotti* has defensive behaviour at least as complex as that of many mantids. In addition to protective coloration and its associated patterns of behaviour (diurnal immobility, substrate selection and the adoption of a special resting attitude), *Orxines* has a well-marked startle display, active escape behaviour and a repugnatorial secretion.

Acknowledgments

This work was carried out during the tenure of a Research Studentship of the Science Research Council as part of a study involving more than 27 species of phasmids and mantids. Thanks are due to my supervisor Prof. N. Tinbergen, F.R.S., and to Prof. J. W. S. Pringle, F.R.S. for allowing me the use of the facilities of the Department of Zoology, University of Oxford.

NOTE ON THE STARTLE DISPLAY OF *METRIOTES DIOCLES* WESTW.

*Metriotes diocles* Westw., a winged flying phasmid of medium size (length of adult female: 75 mm) occurs fairly abundantly in lowland Panama feeding on the leaves of *Dieffenbachia oerstedii* Schott (Araceae). Adults of both sexes are pale fawn to deep straw coloured and have not been found resting on the food plant. It seems probable that they rest on a woody substrate during the day and fly down to the food plant at night. Both sexes erect the tegmina and hind wings vertically in response to mechanical stimulation. The tegmina and costal areas of the hind wings are concolorous with the body surfaces, but the main

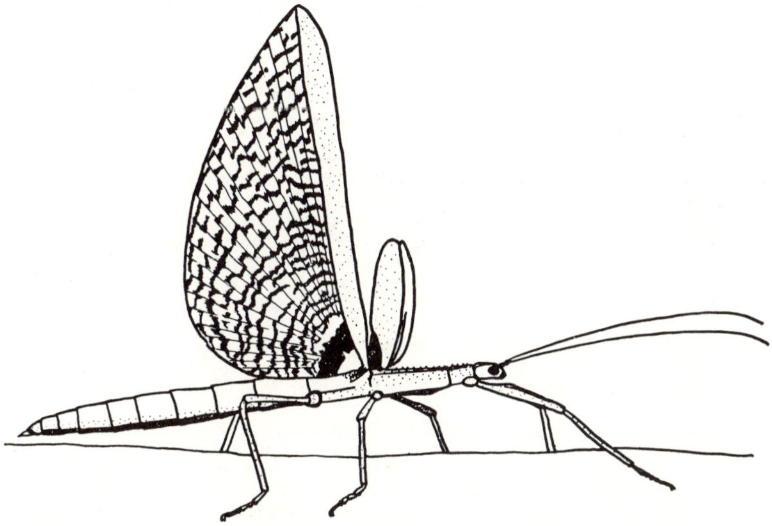

Fig. 5.—Startle display of *Metriotes diocles* Westw. (Based on a 16 mm film record.)

posterior element of the hind wing is tesselated with dark brown markings and basally has a bright peacock-blue spot which extends onto the metathorax and is ringed in dark brown or black. The wing-erection display greatly increases the apparent size of the insect and suddenly reveals the eye-spot. It is illustrated in fig. 5. In the most developed form of the display the wings are snapped up to the erect position and held there for up to 1½ minutes before being suddenly snapped down to the closed position. Further stimulation frequently results in the insect taking to flight or walking off with the wings raised. Rocking does not occur in conjunction with the startle display of this species.

REFERENCES

**Bedford, G. O., & L. J. Chinnick,** 1966, Conspicuous displays in two species of Australian stick insects, *Anim. Behav.*, **14**: 518–521. **Beier, M.,** 1957, Ordnung Cheleutoptera, in Bronns, *Klassen und Ordnungen des Tierreichs*, Leipzig. **Blest, A. D.,** 1957, The function of eye-spot patterns in the Lepidoptera, *Behaviour*, **11**: 209–256. **Chandless, R. C.,** 1954, *Orxines macklotti* de Haan (Orthoptera, Phasmidae), *Bull. amat. Ent. Soc.*, **13**: 88–89. **Crane J.,** 1952, A comparative study of innate defensive behavior in Trinidad mantids (Orthoptera, Mantoidea), *Zoologica, N.Y.*, **37**: 259–293. **Eisner, T.,** 1965, Defensive spray of a phasmid insect, *Science, N.Y.*, **148**: 966–968. **Gwynne, M. D.,** 1965, Protective display and underwing cryptic markings in an East African Saturnioid, *Nature, Lond.*, **206**: 957–958. **Kuyten, P.,** 1961, Verhaltens-beobachtungen an der Stabheuschrecke *Orxines macklotti* De Haan. 1. Ruheplatz und Ruhestellung, *Proc. K. ned. Akad. Wet.* (C) **64**: 305–314. **Moreno, A.,** 1940, Glandulas odoriferas en *Paradoxomorpha* (Orthopt. Phasm.), *Notas Mus. La Plata, Zool.*, **5**: 319–323. **Robinson, M. H.,** 1964, The Javanese Stick Insect *Orxines macklotti* De Haan (Phasmatodea, Phasmidae), *Entomologist's mon. Mag.*, **100**: 253–259. **Steiniger, F.,** 1933, Die Erscheinungen der Katalepsie bei Stabheuschrecke und Wasserläufern, *Z. Morph. Ökol. Tiere*, **26**: 591–708. **Varley, G. C.,** 1939, Frightening attitudes and floral simulation in praying mantids, *Proc. R. ent. Soc. Lond.*, **14**: 91–96.

Department of Zoology, Parks Road, Oxford.
Now Smithsonian Tropical Research Institute, P.O. Box 2072, Balboa, Canal Zone.
*March 23rd*, 1967.

*The Behaviour of Leopoldius coronatus (Rond.) (Dipt., Conopidae) towards its Hymenopterous hosts.* During the summer of 1965, on the Greek island of Skiathos, in the Sporades, I noted the behaviour of the Conopid *Leopoldius coronatus* (Rond.) which was common at the overflow of a spring, a muddy area 4–8 feet wide and 20 feet long, the workers of *Polistes gallicus* (L.), *Vespula vulgaris* (L.) and *V. germanica* (F.) also being in attendance. The spring was observed almost every day from 30th July to 24th August. *Leopoldius coronatus* was present on every occasion, frequently more than a dozen simultaneously although there were fewer towards the end of the month.

The flight of the Conopids was like that of small Muscids and the insects repeatedly landed for a few seconds. If a Conopid, on landing, sighted a wasp it immediately faced it and then took off, this time flying in the zig-zag manner typical of *Vespula* and alighted a little nearer the wasp. This behaviour was often repeated once or twice. If the wasp was a *Vespula* or *Polistes* worker the *Leopoldius* flew on to its back. Sometimes the victim flew up and sometimes it squirmed on the ground presumably with its wings pinned. The fly remained attached to its victim for 5–10 seconds.

*Vespula vulgaris* and *V. germanica* workers were most frequently attacked, workers of *Polistes gallicus* less often, but they were also less common. No queens were seen at the spring.

On a few occasions *Leopoldius* was attracted to *Sceliphron spirifex* (L.), also a common visitor to the spring. The *Leopoldius* flew in a *Vespula*-like manner towards the *Sceliphron*, but was never seen to attack it. Workers of *Vespa crabro* L. were found at the spring in small numbers, but were not attractive to the Conopid.

No *Leopoldius* was seen outside the immediate vicinity of the spring overflow. Two *Leopoldius coronatus* were seen at a smaller spring *circa* 2–3 miles from the one mentioned above. This was visited by very few *Vespula* workers. Unfortunately no examples of the parasitised wasps were preserved for dissection and so eggs were not recovered.

I thank Mr. K. G. V. Smith of the Department of Entomology, British Museum (Natural History), for identifying the flies and at whose suggestion I have written this note.—Anthony Raw, 4 Cliff Cottages, Burlam Road, Middlesborough: *January 25th*, 1968.

[There is little doubt that the *Leopoldius* females were ovipositing on the *Vespula* and *Polistes*. While *Leopoldius* has previously been associated with *Vespula* (see summary in Smith, 1966, *J.Zool.*, **149**: 266), only one rearing has been recorded (Meijere, 1912, *Tijdschr. Ent.*, **55**: 191), of *L. coronatus* from an abdomen of *V. germanica*. The above observation of the modified flight behaviour exhibited by *Leopoldius* when in pursuit of a potential host, is of special interest.—K. G. V. Smith: *January 29th*, 1968.]

*Conferment of Doctorate upon the Hon. Mrs. Miriam Lane.*—On February 24th, 1968, Oxford University conferred the Degree of Doctor of Science, *honoris causa*, upon the Hon. Mrs. Miriam Lane. We offer her our sincere congratulations on this well deserved recognition of her outstanding contributions to science, and also on her election to an Honorary Fellowship at St. Hugh's College, Oxford.—Eds.

# Defensive Secretions of Arthropods

Insects, millipedes, and some of their relatives, discharge noxious secretions that repel predators.

Thomas Eisner and Jerrold Meinwald

All organisms are chemosensitive, and all of them are also the source of substances to which others can potentially respond. In the course of evolution this potential for interaction has been thoroughly exploited, and organisms of the most diverse kinds have come to depend, in one adaptive context or another, on an exchange of chemical cues with other organisms in their environment. When the exchange is between members of the same species, the mediating substances are called pheromones (*1*). These interesting compounds, so important for the regulation of courtship and other social activities in animals, and including among others the sex attractants, alarm substances, aggregation-promoting substances, territorial markers, and trail substances, have recently been the subject of intensive research, and of excellent reviews (*2*).

Set apart from the pheromones, but certainly matching them in importance, is that vast array of substances—for which an adequate name has yet to be coined—that serve for the transmission of information between members of *different* species. Whether two species coexist in harmony (as in mutualistic associations) or in conflict (as in predator-prey or parasite-host relations), chances are that their interactions depend in some major way on chemical factors which they themselves produce. In fact, all living things—animals, plants, and microorganisms alike—are probably involved at one time or another in their existence with the emission or reception of extraspecific chemical messages. Surprisingly little is known about such cross-specific communication. Neither the chemistry of the "signals" themselves nor many of the factors relating to their reception and release are properly understood in most instances. This article deals with the chemical interactions that prevail between predators and certain of their prey. Defensive secretions, whether of arthropods or other organisms, are essentially chemical signals, since they represent the means by which predators and other potential enemies are "told" to desist or withdraw. Clearly, there is little room for ambiguity in the rather forceful message written in the stench of a skunk's spray. Among arthropods, chemical defenses comparable to that of the skunk are widespread. In fact, the vocabulary of protective chemical signals possessed by these animals may be one of the richest in the entire world of life.

The authors are, respectively, professor of biology and professor of chemistry at Cornell University, Ithaca, New York.

## The Glands and Their Discharge Mechanisms

Unlike pheromones, which are often produced in vanishingly small quantities and may require the development of special bioassays to demonstrate their presence (*1–3*), defensive secretions may be detected with relative ease since they are often strongly odorous and discharged in substantial amounts. The literature abounds with references to insects, millipedes, phalangids, and other terrestrial arthropods that eject an obnoxious liquid when disturbed (*4*). The glands responsible for this output are integumental organs, consisting essentially of infoldings of the body wall. Their outer openings may be small and relatively inconspicuous, but the invaginated portion is usually large and readily seen in dissection (Fig. 1, A and B). Typically, a gland is made up of a sac-like reservoir in which the secretion is stored, and of a glandular tissue that may be a part of the wall of the reservoir itself (Fig. 1D) or may lie apart from the reservoir and be joined to it by one or more special ducts (*4*) (Fig. 1C). The secretory cells are often strikingly elaborate in structure, but in only a few instances have they been examined with the electron microscope (*5, 6*). The glands are so variable in number, distribution, and morphological detail that there can be no doubt that they have arisen many times independently in the course of evolution. They may occur singly, in pairs, or in multiple pairs, and they may be present on head, thorax, or abdomen.

As regards their mode of discharge, the glands fall into three main categories. In one type, present commonly in millipedes (*7, 8*), the secretion simply oozes from the glands onto the animal's own surface (Fig. 2, A and B). In another, found in some caterpillars (Fig. 2E) and beetles, the secretion is "aired" by actual evagination of the gland as a whole (*9, 10*). The third and perhaps most spectacular category includes those glands that dis-

Reproduced with permission from *Science*, 153:1341–1350, 1966 Published by the American Association for the Advancement of Science, Washington, D.C.

charge their contents as a spray (Fig. 2, C and D), sometimes to a distance of several feet. Among the many arthropods known to spray are certain cockroaches (*11*), earwigs (*12*), stick insects (*13*), "stink bugs" (Hemiptera) (*14*), notodontid caterpillars (*15*), grasshoppers (*16*), carabid (*7, 17, 18*) and tenebrionid beetles (*19*), whip scorpions (*20*), and millipedes (*21*).

Special behavioral and other adaptations enable arthropods to discharge their secretion with maximum effectiveness and minimum wastage. Species that eject a spray usually possess the ability to aim it precisely toward that portion of their bodies subjected to disturbance (Fig. 2, C and D), thus insuring that an attacking predator is met by the full impact of a discharge. There are different ways in which aiming may be accomplished. The caterpillar *Schizura concinna*, whose single gland opens ventrally just behind the head, directs the spray by pointing its front end in the appropriate direction (*15*). In the whip scorpion, *Mastigoproctus giganteus*, the two glands open at the tip of a short revolvable knob that acts in the manner of a gun turret (*20*). Aiming may also involve postural adjustments of the abdomen (*12*) or of the body as a whole (*14*).

Animals that possess more than one gland may call them into action singly or in groups. This is strikingly illustrated by millipedes, many of which have their glands arranged segmentally, one pair per most body segments. A sharply localized stimulus may cause the nearest glands to respond, but no others (Fig. 2A). Only after persistent or generalized disturbance do the glands discharge in large numbers (Fig. 2B) (*7*). Eversible glands may also be employed in a specialized fashion. In the caterpillar of the swallowtail butterfly *Papilio machaon*, the two-pronged gland (Fig. 2E) may be partially or completely extruded, depending on where and how intensely the stimulus is applied. Moreover, the larva arches its front end in such a way that the "horns" are wiped against the instrument used for stimulation (*10*).

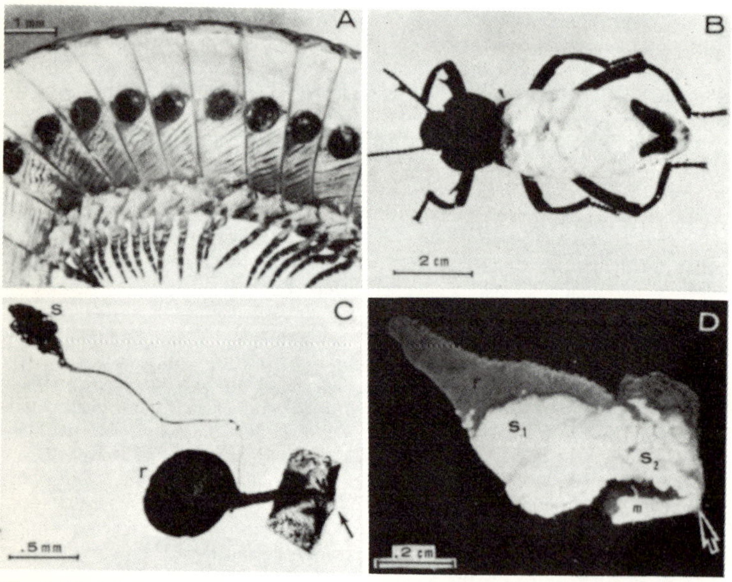


Fig. 1. (A) Midbody segments of a millipede (*Narceus gordanus*), sagittally dissected, showing spherical reservoirs of the defensive glands, amidst the intersegmental muscles of the body wall: the reservoirs are filled with dark-brown quinonoid secretion (see also Fig. 2, A and B). (B) A tenebrionid beetle (*Eleodes* sp.) with the dorsum of its abdomen dissected away to expose the two posterior defensive glands, turgid with quinonoid secretion (see also Fig. 2H). (C) Isolated gland of a carabid beetle (*Chlaenius cordicollis*); the secretory cells form a compact tissue (*s*), joined to the reservoir (*r*) of the gland by a narrow efferent tubule; a short duct (partly occluded by a rectangular piece of the body wall) leads from the reservoir to the outer orifice (arrow) of the gland. (D) Isolated defensive gland of the tenebrionid beetle *Eleodes longicollis*, showing two distinct secretory tissues ($s_1$, $s_2$) overlying the membranous reservoir (*r*) of the gland; the muscle (*m*) opens the valvular outer orifice (arrow) of the gland.

Although as a rule arthropods discharge their secretion only in response to direct contact-stimulation, exceptions are known. The walkingstick *Anisomorpha buprestoides* may eject its spray with accuracy toward a bird in its vicinity, even before any direct attack has taken place. Nothing is known about the particular sensory input that the insect relies upon for taking aim at the approaching predator (*13*).

The mechanisms whereby the discharge is effected vary. Sometimes muscular compression of the reservoir is involved (*7, 13, 22*). When intrinsic muscles are missing, the reservoir may be squeezed by blood pressure (*7, 16*). Blood pressure is also thought (*9, 10*) to force the extrusion of eversible glands, as in *Papilio* (Fig. 2E). Some glands are connected to respiratory tracheae and rely on air pressure for the discharge (Fig. 2G) (*4, 23*). In the remarkable bombardier beetles (*Brachinus* spp.), the secretion is expelled under pressure of gas generated in an explosive chemical reaction (*24*). Additional mechanisms have been postulated, some them perhaps unnecessarily elaborate (*21, 25*). An interesting discussion of the biophysics of the discharge mechanism of certain Hemiptera is provided by Remold (*14*).

### Chemistry of the Secretions

Although the earliest analytical work on arthropod secretions dates back to the 17th century (*26*), and occasional chemical studies have been made since then, the present widespread interest in this field was prompted largely by the recent studies of Pavan and his associates (*27*) on the toxins of ants and other insects. Several laboratories throughout the world are now actively concerned with elucidating the nature of the secretions, and dozens of active principles have been isolated and identified. Recent reviews have appeared in which these compounds are listed (*4, 28*), and in which some of the techniques that led to their isolation and identification are discussed (*29*). Most of the components identified to date are compounds of low molecular weight (about 30 to 200), belonging to a number of common chemical classes, including acids, aldehydes, ketones, esters, hydrocarbons, lactones, phenols, and *p*-benzoquinones. The group includes simple straight-chain compounds and both acyclic and cyclic monoterpenes (Fig. 3). With few exceptions they represent compounds previously known, including even common laboratory reagents. They are almost invariably strongly odorous, and they may occur within the glands at astoundingly high concentration [for example, acetic acid, 84 percent (*20*)]. Despite the simplicity of most of the compounds, their isolation and characterization have on occasion provided some challenge because of the small amounts of material available. In the few instances where new compounds have been discovered, and particularly in the case of the cyclopentanoid monoterpenes, their categorization and subsequent synthesis have been of considerable intrinsic chemical interest (*30*).

The secretions may be mixtures of some complexity. The odorous discharge of the pentatomid bug *Nezara viridula* contains 18 aliphatic components, including hydrocarbons, aldehydes, ketones, and esters (*31*). Quinonoid secretions usually contain two or three quinones, rather than a single one (*4, 28*), and the quinones may be accompanied by glucose, hydrocarbons, and traces of hydroquinones (*23, 32, 33*). The lesser components may offer clues to the biochemical mechanisms whereby the secretions are synthesized (*32*). In many mixtures only the dominant components have been identified, while minor ones remain unknown (*4, 28*). Secretions such as that of the cockroach *Eurycotis floridana* (*22*), which

Fig. 2 (right). (A) A millipede, *Narceus gordanus*, being tapped with a metal mallet, has begun to discharge its quinonoid secretion from two of its segmentally arranged glands (see Fig. 1A). (B) The same millipede discharging profusely from several glands, after persistent tapping. (C) The whip scorpion, *Mastigoproctus giganteus*, discharging its aimed spray in response to the pinching of one of its appendages with forceps (the arrow points to gland openings); the animal has been affixed to a rod, and the acid spray is rendered visible on a background of filter paper impregnated with alkaline phenolphthalein. (D) The same scorpion discharging a second time, in response to pinching of the right rear leg. (E) Caterpillar of *Papilio machaon*, being pinched with forceps, everting its two-pronged postcephalic gland. (F) An arctiid moth, *Utetheisa bella*, emitting froth from its right cervical gland. (G) The grasshopper *Romalea microptera* discharging a froth (arrow) from its anterior thoracic spiracle; the respiratory tracheal tubes leading inward from the spiracle are beset with glandular tissue and filled with secretion; the froth is a mixture of secretion and respiratory air. (H) Tenebrionid beetle *Eleodes longicollis* ejecting quinonoid secretion in response to the pinching of one of its legs with forceps (see also Fig. 1B); the secretion is rendered visible on a substrate of filter paper impregnated with an acidulated solution of potassium iodide and starch. (I) An oniscomorph millipede (*Glomeris marginata*) coiled into a tight sphere in response to disturbance, has discharged three droplets of secretion (arrows) from glands opening on the dorsal midline; the sticky secretion can be drawn into fine threads, as is here being done with the tip of a needle (see also Fig. 6E). (J) Unidentified onychophoran from Panama, discharging its sticky secretion in response to handling.

consists of a single, virtually pure component (*trans*-2-hexenal), appear to be exceptional.

Arthropods of very diverse types may produce secretions containing the same or closely similar components. For example, *trans*-2-hexenal is known from Hemiptera (*31, 34–36*), cockroaches (*22, 37*), and a myrmicine ant (*38*). Similarly, formic acid is secreted by ants (*39*), carabid beetles (*29, 40*), and notodontid caterpillars (*15, 41*). Particularly widespread are the *p*-benzoquinones, which have been identified from beetles, earwigs, millipedes, a cockroach, and a phalangid (Fig. 5) (*4, 28*). The glands of arthropods producing similar secretions may be very different structurally and obviously not homologous.

Instances of evolutionary diversification of the secretions are also known. In the beetles of the family Carabidae, the defensive (pygidial) glands (*42*), which are undoubtedly homologous throughout the group, have been shown in various species to produce compounds as different as formic acid (*29, 40*), *m*-cresol (*7*), *p*-benzoquinones (*24*), salicylaldehyde (*17*), and tiglic and methacrylic acid (*29, 43*) (Fig. 4). In millipedes, the orders Julida, Spirobolida, and Spirostreptida secrete *p*-benzoquinones (*29, 44, 45*), the Polydesmida produce cyanogenic secretions (*8, 46, 47*), and the single species of the Chordeumida that has been studied produces a phenol (*7*). Despite some differences in structure, the glands in these various orders are probably homologous.

Future studies may yet uncover secretions containing entirely different compounds, perhaps even some of interesting pharmacological activity. Existing evidence is already suggestive in this respect. A pharmacologically active choline ester closely resembling β, β-dimethylacrylylcholine is present at high concentration in the cervical glands of an arctiid moth (*48*). The grasshopper *Poekilocerus bufonius* sprays a secretion containing a digitalis-like compound (*49*). The salivary secretion of the reduviid bug *Platymeris radamantus*, which is ordinarily used as a lethal venom for injection into prey but can also serve for defense by being sprayed against predators, contains at least six proteins, three of which show trypsin-like proteolytic activity; it also has strong hyaluronidase and weak phospholipase activity. The saliva resembles snake venom both in number of proteins and in enzyme activity (*50*).

Although as a rule the secretions retain their fluid consistency after discharge, there are some that undergo rapid transformation on exposure to air and become viscous and sticky. The spray of Onychophora (*Peripatus* and its relatives) (Fig. 2J), as well as the discharge of certain millipedes (for example, *Glomeris marginata*) (Fig. 2I) and isopod crustaceans, are of this type (*51, 52*). None of these secretions has been categorized chemically.

### Defensive Effectiveness of the Secretions

Although encounters between predator and prey have only rarely been witnessed in nature, it is evident from laboratory tests in which arthropods were pitted against caged predators that the glands are weapons of extraordinary effectiveness (Fig. 6, A–F). Among the numerous predators that were shown to be repelled by the discharge of their intended prey are ants, praying mantids, carabid beetles, solpugids, spiders, toads, lizards, jays, armadillos, and mice (*4, 7, 10–14, 17–20, 35–37*). As a rule, arthropods eject their secretion during the initial phases of an attack, thwarting their assailants before receiving disabling injury. The ability to eject an aimed discharge, or to discharge from only those glands

| COMPOUNDS | | SOURCE |
|---|---|---|
| hydrogen cyanide | HCN | polydesmoid millipedes |
| acetic acid | $CH_3CO_2H$ | whipscorpion, coreid hemipteran |
| caprylic acid | $CH_3(CH_2)_6CO_2H$ | whipscorpion, tenebrionid beetle |
| 2-methylbutyric acid | $CH_3CH_2CH(CH_3)CO_2H$ | papilionid caterpillar |
| trans-2-dodecenal | $CH_3(CH_2)_8CH{=}CHCHO$ | spiroboloid millipede |
| p-cresol | 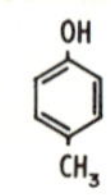 | chordeumoid millipede |
| p-benzoquinones | 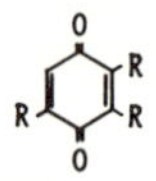 | see fig. 4 |
| dolichodial (anisomorphal) | 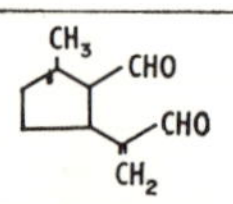 | dolichoderine ants walking stick |
| α-pinene | 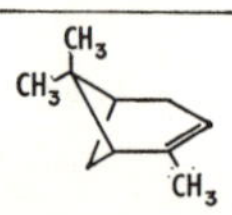 | termite |

| COMPONENT(S) | | SOURCE |
|---|---|---|
| formic acid | $HCO_2H$ | Pseudophonus spp., unidentified species of Calathus, Acinopus, Carterus |
| methacrylic acid (with or without) tiglic acid | $CH_2{=}C(CH_3)CO_2H$ <br> $CH_3CH{=}C(CH_3)CO_2H$ | Carabus spp., Abax spp., Pterostichus spp., Apotomopterus spp.. Damaster oxuroides, Cychrus rostratus |
| p-benzoquinone | 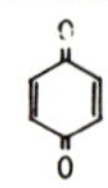 | Brachinus spp. |
| toluquinone | 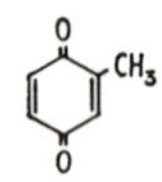 | |
| m-cresol | 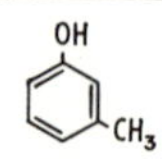 | Chlaenius cordicollis |
| salicylaldehyde | 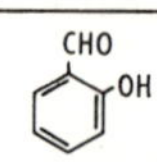 | Calosoma prominens |

Figs. 3 and 4. Fig. 3 (left). Representative compounds from defensive secretions of arthropods. [Compiled from *4, 7, 10, 28, 69, 87, 88* and references therein] Fig. 4 (right). Components in defensive secretions of beetles of the family Carabidae. [Compiled from *4, 7, 28, 29,* and references therein] that are closest to the site of attack, adds considerably to the efficiency of the weapon (Fig. 6D) (*11–13, 18, 20*).

The secretions act as topical irritants, causing the predators to desist from the attack and to perform more or less vigorous cleansing activities. Among vertebrates, the sensitive areas of the face—mouth, nose, and especially the eyes—appear to be the most vulnerable. Mice respond to the discharge by wiping their eyes and snout with their front feet, or by scurrying away, plowing their muzzle in the earth (*7, 17, 20*). Birds ruffle their plumage and rub their head against the feathers of the body; their eyes are cleansed by the wiping action of the nictitating membranes, which, in the manner of "windshield wipers," are drawn quickly back and forth over the eyeballs (*13, 17, 20*). Frogs and toads are sensitive over their entire body surface, and scratch affected regions with their feet (*12, 19*). Carabid beetles, ants, and other predaceous insects brush their sensitive antennae with their feet, or flee, dragging their mouthparts (*18, 52*). Small predators such as ants, whose entire bodies may become contaminated, literally attempt to "bathe" in the soil, rubbing their bodies against it while flailing their outstretched legs (*12, 20, 52*). Although prolonged confinement with the vapors of a secretion may eventually prove fatal to a predator (*53*), under conditions of adequate ventilation such as are likely to prevail in nature the secretions usually cause only transient effects. Predators may recover in a matter of seconds or minutes, depending on the species, the nature of the secretion, and other factors (*7, 11–13, 18, 20*). Future studies should attempt to establish quantitative criteria by which the repellent effectiveness of the secretions may be more precisely evaluated. Such studies should take into account the intriguing possibility that predators are more or less sensitive to the chemical defenses of their prey depending on the time of day. Rhythmic daily fluctuations in the susceptibility of animals to certain drugs is an established pharmacological fact (*54*).

No experimental studies have been made to determine how—that is, through which particular sensory receptors—the secretions exert their irritant effect on predators. In the case of vertebrates, it seems likely that the principal receptors are the peripheral

p-BENZOQUINONES

R = H, $CH_3$, $C_2H_5$ or $OCH_3$

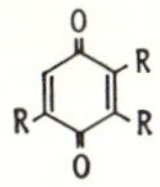


| CLASS | ORDER | FAMILY |
|---|---|---|
| Arachnida | Phalangida (harvest spiders) | Gonyleptidae 1/1 |
| Diplopoda (millipedes) | Spirobolida | Spirobolidae 2/3 |
| | | Rhinocricidae 1/1 |
| | | Trigoniulidae 2/2 |
| | | Floridobolidae 1/1 |
| | Spirostreptida | Spirostreptidae 3/6 |
| | | Harpagophoridae 1/1 |
| | | Cambalidae 1/1 |
| | Julida | Julidae 4/4 |
| Insecta | Coleoptera (beetles) | Tenebrionidae 12/21 |
| | | Alleculidae 1/1 |
| | | Carabidae 2/4 |
| | Dermaptera (earwigs) | Forficulidae 1/1 |
| | Blattaria (cockroaches) | Blaberidae 1/1 |

Fig. 5. List of arthropods producing secretions containing *p*-benzoquinone or derivatives thereof. In the fraction given after each family, the numerator indicates the number of genera examined; the denominator, the number of species examined. [Compiled from *29, 44*]

"free" nerve endings which, acting in the capacity of relatively insensitive and nonspecific chemoreceptors, mediate the so-called "common chemical" or "general chemical" sense (*55*). Highly vulnerable regions, such as the cornea, are indeed densely innervated by free nerve endings (*56*). Arthropods also possess a common chemical sense (*57*), but it is not clear whether free nerve endings—which they are known to possess (*58*)—are the receptors involved. As regards the precise mode of action of the secretions, it remains to be seen whether they act directly on the receptors or indirectly through induction of biochemical changes in nonnervous cells and in tissue fluids around the nerve endings. The growing literature on this general subject has recently been summarized (*59*). To some extent, the secretions may also act by way of true olfactory receptors. This is suggested by the fact that even the dilute vapors that linger after a discharge may be highly repellent. Such vapors are obnoxiously odorous, certainly to man, and have been shown to prevent some predators from closing in on their prey (*7, 18, 20, 35*). In the case of the odorless sticky secretions discharged by Onychophora and some millipedes (Fig. 2, I and J), the primary (but not necessarily exclusive) mode of action may be mechanical rather than toxicological (Fig. 6E). Small predators such as ants become trapped in the gummy discharge and may struggle for minutes or even hours before freeing themselves (*52, 60*). The secretion of certain termites has a similar immobilizing effect (*61*).

Although as a rule the defensive glands possess both the capability and sufficient secretion to discharge several times in succession (*7, 11, 12, 14, 18, 20*), they may ordinarily be called upon to discharge only once during a given attack. This is because residual secretion, remaining around the gland openings or elsewhere on the animal's body after the first discharge, acts to deter the predator from resuming its assault (*7, 18, 20, 35*). The beetle *Chlaenius* remains invulnerable to ants 8 to 13 minutes after a discharge. The estimated 100 meters it can cover during this period at its normal ambulatory pace must suffice for outdistancing even a dense swarm of ants (*7*). Since depleted glands may require several days to be completely replenished (*13, 18, 20*), it is obviously of adaptive value that the glands should ordinarily be required to expend no more than a minimum amount of secretion.

Vertebrates may quickly learn to discriminate against obnoxious prey. After having been sprayed by the walkingstick *Anisomorpha*, blue jays remained aloof from them even when the insects were presented to the birds no more often than every 2 to 3 weeks (*13*). Discrimination by learning on the part of predators has led to the evolution of mimicry, and some arthropods with defensive glands are known to have mimics. The tenebrionid beetle *Eleodes longicollis*, which when disturbed does a headstand and sprays a quinonoid secretion from its abdominal tip, is mimicked by another tenebrionid beetle (*Megasida obliterata*) that lives in the Arizona desert together with *Eleodes*. *Megasida* also lifts its rear when disturbed, but it lacks defensive glands (Fig. 6I) (*19*).

Some predators, although obviously sensitive to the secretions, overcome their prey through sheer persistence. In laboratory tests, grasshopper mice (*Onychomys torridus*) relentlessly pursued whip scorpions and, although temporarily deterred by each discharge, eventually subdued their quarry (*20*). Comparable results were obtained with

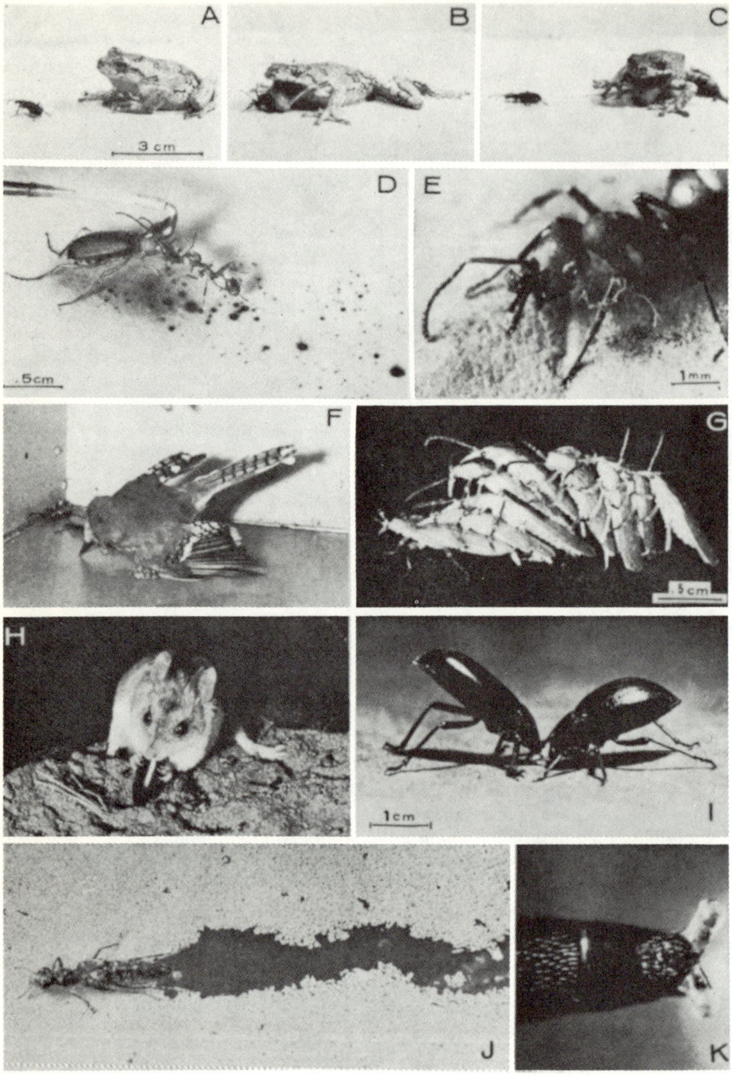


Fig. 6. (A–C) Toad (*Hyla versicolor*) being repelled by a bombardier beetle (*Brachinus* sp.). The toad (A) eyes the beetle, (B) strikes at it with its sticky tongue, and (C) rejects it. (D) Bombardier beetle (*B. ballistarius*), affixed to a rod, spraying in the direction of an attacking ant (*Pogonomyrmex badius*); the pattern of secretion is visible on a background of filter paper similar to that of Fig. 2H. (E) An ant (*Formica polyctena*), after having been contaminated with the sticky secretion of the millipede *Glomeris marginata* (see Fig. 2I); the tip of the left antenna is glued to the cranium; threads of dried secretion bind the left front leg to the head. (F) Blue jay (*Cyanocitta cristata*) at the moment of being sprayed by the stick insect *Anisomorpha buprestoides* (the prey is partly visible in the corner of the cage). (G) Group of caged male lycid beetles (*Lycus loripes*) drawn into a cluster by their own attractant pheromone. (H) Grasshopper mouse (*Onychomys torridus*) feeding on the beetle *Eleodes longicollis* (see also Figs. 1B, 2H); the beetle is held in such a way that its secretion, ejected from the tip of the abdomen, is discharged into the soil. (I) *Eleodes longicollis* (left) has defensive glands in the abdomen (see Figs. 1B, 2H) and performs a headstand when disturbed; its mimic, *Megasida obliterata* (right), also lifts its rear when annoyed, but lacks glands. (J) Staphylinid beetle (*Stenus* sp., about 5 millimeters long) propelling itself forward over water through touching the water surface with glands at the tip of its abdomen; the glands produce a secretion that depresses the surface tension of water; for experimental purposes the surface had been covered with *Lycopodium* powder; the clear trail behind the beetle marks the region of depressed tension where the powder has withdrawn. (K) Abdominal tip of *Stenus*, showing everted glands. [Photographs J and K, courtesy of K. E. Linsenmair (*66*)]

a rat opossum (*Marmosa demararae*) feeding on *Anisomorpha* (*13*). Other predators may attack their prey in such a way as to avoid exposure to the secretions. Grasshopper mice feed on certain beetles (*Eleodes, Chlaenius*) by holding them upright in their front paws, while jamming them butt-end down into the earth; the secretion, which is ejected from glands that open at the rear of the beetles' abdomen, is harmlessly expended in the soil (Fig. 6H) (*7, 19*). Toads, which characteristically catch their prey by a quick flip of their sticky tongue, may swallow millipedes (*Narceus*) or beetles (*Eleodes*) before the quinone-producing glands of the prey are discharged. The toads betray no immediate or long-range ill effects from secretion that might subsequently be ejected inside their stomach (*19, 52*). Bombardier beetles (*Brachinus*), which also produce quinones (*24*), avoid capture. They discharge the moment they are struck by the sensitive tongue of the toad, and are rejected instantly (Fig. 6, A–C).

The overall defensive effectiveness of a secretion may not be attributable solely to its chief constituents. Lesser components, even at concentrations of a few percent, may have important ancillary functions. The defensive spray of the whip scorpion, *Mastigoproctus giganteus*, consists of acetic acid (84 percent), water (11 percent), and caprylic acid (5 percent). Although the primary irritant in the mixture is acetic acid, caprylic acid adds to the effectiveness of the secretion as it is used against arthropod predators. The cuticular exoskeleton of terrestrial arthropods has a thin outermost layer, the epicuticle, which has the important function of waterproofing the cuticle, and which, by virtue of its high wax content, can act as a limiting barrier to the penetration of wax-insoluble compounds such as acetic acid. Caprylic acid is a wax solvent, and it apparently acts by disrupting the epicuticle, thereby facilitating the penetration of acetic acid. Caprylic acid also serves as a wetting agent that promotes the spread of the individual spray droplets over the cuticle of the enemy, thus increasing the area of contact of the poison. Actually, caprylic acid at a concentration of 5 percent can fulfill neither of these functions. However, the ejected droplets, during their aerial trajectory, undergo a differential evaporative loss of acetic acid, which insures that they arrive on target with an effective caprylic acid concentration of 10 to 20 percent (*20*). "Additives" other than caprylic acid are also known to occur. The aliphatic hydrocarbons in the secretion of Hemiptera promote not only the spreading and penetration of the mixture but also its seepage into the respiratory tracheae of insects (*14*).

The secretions may or may not have a noticeable effect on the arthropods that produce them (*22, 62*). In some cases there may be a high level of tolerance. The millipede *Apheloria corrugata*, which secretes a cyanogenic secretion (*47*), usually outlives other arthropods confined with it in a cyanide killing jar (*52*). Similarly, whip scorpions are apparently unaffected by the incidental dousing of acid spray that they inevitably receive when they eject their secretion (*20*). Whether insensitivity is attributable to an impervious integument or to the existence of special systemic detoxification mechanisms has not been determined. Some species can tolerate only limited contact with their own secretion, and they react, in much the same way that predators do, by cleansing sensitive regions (*17*). In certain Hemiptera the exoskeleton around the gland openings is elaborately sculptured to prevent the discharged secretion from spreading over the body (*14*).

### Other Functions of Defensive Secretions

Some secretions, although clearly defensive in function, may in addition serve as pheromones. The soldiers of the so-called nasute termites possess a pointed cephalic "nozzle" from which they eject a defensive spray. The secretion is used against insects or other small animals that venture into the termites' nest or foraging territory, and it has a rapid incapacitating effect. It also acts as an "alarm" substance, alerting additional soldiers to the state of emergency and inducing them to discharge spray toward the intruder (*61*). The ant *Acanthomyops claviger* produces a secretion in its mandibular glands that also serves both for defense and as an "alarm" pheromone (*63*).

Adults of the beetle *Lycus loripes*, an inedible, aposematically colored species, form dense and conspicuous aggregations on the plants upon which they mate and feed. The clustering behavior (Fig. 6G) is mediated by a volatile pheromone, produced by the males alone, capable of attracting both males and females. Aggregation in *Lycus* is more than just a means for bringing the sexes together. It is an important defensive behavior that enables the beetles to "pool" their aposematic resources and advertise themselves *en masse*, thereby more effectively warning visually oriented predators of their presence. The pheromone thus fulfills both a reproductive and a protective role (*64*).

Defensive secretions may also possess entirely different, nonpheromonal functions. Mention has already been made of the salivary secretion of *Platymeris*, which is sprayed in defense against predators, while also serving as a lethal offensive venom when injected into prey (*50*). Some defensive secretions have antimicrobial activity and may conceivably provide protection against both predators and microorganisms (*4, 36*). A mutant of the flour beetle *Tribolium confusum* has recently been discovered whose defensive glands contain far less quinone than those of the wild type. The nutrient medium in crowded populations of the mutant tends to become moldy, suggesting that the glands ordinarily serve to inhibit fungal and bacterial growth in the beetle's environment (*65*). A truly anomalous function is possessed by the eversible defensive glands in the abdominal tip of certain staphylinid beetles (*Stenus* spp.) (Fig. 6K). These small insects, which are normally terrestrial but may on occasion forage on water or be blown by wind onto water, rely on the surface-tension-depressant properties of their secretion to propel themselves across the surface of the liquid. By touching their everted glands to the water they weaken the surface tension behind them, and thus are carried forward by the "contracting" surface as it withdraws before them (Fig. 6J). The beetles employ this type of locomotion, appropriately called *Entspannungsschwimmen*, only when they are threatened by an emergency. Ordinarily they simply paddle along by walking (*66*).

### A Remarkable Parallel

Many plants are characteristically odorous, and their fragrance does not always originate from the flowers alone. Leaves, stems, and roots may also be scented, as is often especially apparent after they have been crushed. Although the attractant properties of floral perfumes have long been recognized, the adaptive significance of the odorous compounds of vegetative structures has remained a matter of some controversy. Organic chemists have long been interested in these natural products, many of which have been isolated and identified. Since the odorous principles appear not to play a major role in the fundamental biochemical processes of the living plant, they are usually classed—together with such other "irrelevant" compounds as saponins, alkaloids, and tannins—among the so-called "secondary" plant substances (*67*). An excellent catalog has recently appeared, in which these substances (except for alkaloids) are listed by chemical structure and source (*68*).

It is a striking fact that the active components in the defensive secretions of arthropods also occur, almost without exception, as secondary substances of plants. Some of the compounds, such as 2-hexenal, produced in the defensive spray of several insects (*4, 28*), and $\alpha$-pinene, discharged by certain termites (*69*), are extraordinarily widespread in plants (*68*). Others, such as isobutyric acid and 2-dodecenal, found in the secretion of a caterpillar (*10*) and a millepede (*45*), are more restricted in distribution (*68*). Sometimes the parallel even extends to the mechanism whereby the odorous principles are released. In cyanogenic plants, hydrogen cyanide is generated by hydrolysis of cyanohydrin glycosides (*70*). In the polydesmoid millipede *Apheloria*, cyanogenesis is effected by dissociation of mandelonitrile, the cyanohydrin of benzaldehyde (*47*).

The fact that the active principles in the defensive secretions of arthropods are effectively repellent to predators raises the possibility that these same principles, as secondary substances of plants, serve to protect plants against herbivores. This possibility has been advanced before, notably by Fraenkel (*67*). Irrespective of how one may feel about the hypothesis, it can lead to useful predictions. Among the more interesting compounds recently isolated from defensive glands of insects are certain cyclopentanoid monoterpenes, including iridodial and iridomyrmecin from ants, and dolichodial (= anisomorphal) from ants and a walkingstick (*30, 71*). Similar terpenes have been found in plants (*72*). One of these, nepetalactone or catnip, had long been of interest because of its peculiar excitatory effect on cats, a property that

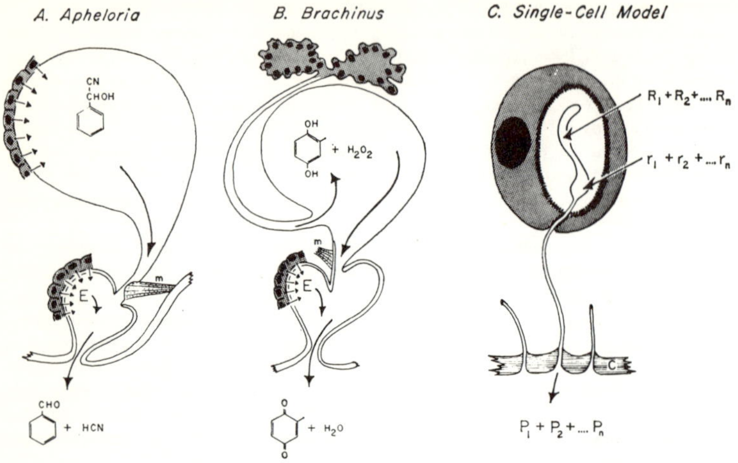


Fig. 7. (A–B) Diagrammatic representation of the known mechanism of operation of the defensive gland of a millipede, *Apheloria* [after Eisner *et al.* (*47*)], and a bombardier beetle, *Brachinus* [after Schildknecht and Holoubek (*24*)]. (C) Postulated mechanism of toxicant synthesis, such as may prevail in a single secretory cell; within the vesicle of the cell lies the hollow cuticular "reaction chamber," into which the cell passes sets of reactants (*R, r*) that interact to form the toxic products (*P*); a slender cuticular duct conveys the products from the reaction chamber to the wall (*C*) of the gland (see text).

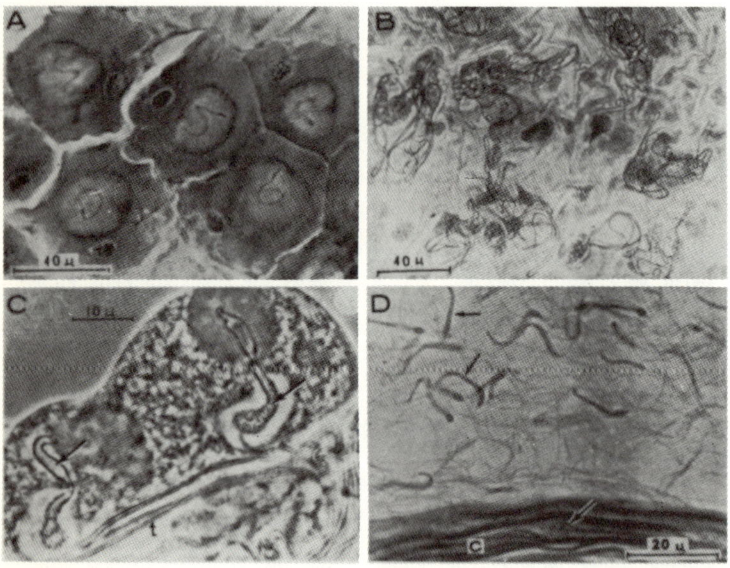


Fig. 8. Tubular cuticular organelles associated with secretory cells of defensive glands. (A) Secretory cells (from tissue $s_1$, Fig. 1D) of glands of *Eleodes* sp.; the cells have an eccentric nucleus, a large central vesicle, and within the vesicle a twisted cuticular organelle that, presumably, functions as a "reaction chamber" for synthesis of the poisons. (B) Surface view of a portion of the secretory epithelium of the quinone-producing gland of the cockroach *Deropeltis erythrocephala*; the long cuticular tubules lead from the cells to the lumen of the gland. (C) Cuticular "reaction chambers" (arrows) in secretory cells of the abdominal quinone-producing glands of the flour beetle *Tribolium castaneum;* the chambers are drained by cuticular tubules (*t*) that lead to the reservoir of the glands. (D) Isolated cuticular organelles from secretory cells of the quinone-producing tracheal gland of the cockroach *Diploptera punctata*; the organelles were isolated by dissolving all soft cellular components in hot aqueous potassium hydroxide; the "reaction chambers" (small arrows) are drained by long thread-like tubules that open (large arrow) on the cuticular wall (*c*) of the gland. [Photograph D, courtesy of D. W. Alsop]

is surely irrelevant to the actual adaptive function of the terpene in the mint plant that produces it. Since the cyclopentanoid terpenes in the secretions of insects were known to be repellent to other insects, it seemed likely that nepetalactone might be similarly repugnatorial and that it might protect the mint plant against herbivorous insects. Actual tests showed nepetalactone to be an insect repellent of considerable effectiveness (*73*). Clearly, new insect repellents of potential practical significance may yet be discovered among the secondary substances of plants. Some of the oldest repellents known to man are, in fact, derived from plants. One of the most familiar, citronellal, is now known to be produced also in the glands of an ant (*63*).

## Biosynthesis of the Secretions

Most of the relatively meagre evidence available to date suggests that the active components in the defensive secretions of arthropods are synthesized by the animals themselves from readily available metabolic precursors. Thus, the ant *Acanthomyops*, which produces citral and citronellal, synthesizes these acyclic monoterpenes from acetate units, by way of mevalonate, according to the usual terpenoid biosynthetic pathway (*74*). The cyclopentanoid monoterpene anisomorphal, secreted by a stick insect (*71*), is also synthesized from acetate and mevalonate (*75*). The results of such biosynthetic studies are sometimes unexpected and of intrinsic biochemical interest. In the beetle *Eleodes longicollis*, which secrets three closely similar quinones—*p*-benzoquinone, toluquinone, and ethylquinone (*76*)—the three compounds are apparently produced by two different routes. Whereas *p*-benzoquinone is synthesized from preformed aromatic precursors (phenylalanine, tyrosine), the two alkylated quinones are synthesized from acetate (*77*).

Herbivorous insects might conceivably be expected to make use of secondary plant substances obtained in their diet by incorporating such substances unchanged into their secretion. This does not appear to occur in pentatomid Hemiptera, one of which (*Nezara viridula*) was shown to be capable of synthesizing virtually all its many secretory products from acetate (*78*). And it evidently does not take place in the caterpillar of *Papilio machaon*, whose secretion had been thought to contain only the essential oils of the caterpillar's food plant but in fact contains two acids (2-methylbutyric and isobutyric acid) that are apparently absent from the umbelliferous species on which the caterpillar feeds (*10*). However, in the grasshopper *Poekilocerus bufonius*, the digitalis-like material isolated from its secretion (*49*) may stem from the diet. The insect feeds on Asclepiadaceae (milkweeds), which are known to be rich in cardiac glycosides (*79*). In view of the limited ability of insects to synthesize steroids from nonsteroidal precursors (*80*), this particular instance of a possible utilization by one of them of a preformed steroidal compound of plant origin takes on special significance.

The active principles of defensive secretions are, for the most part, relatively small, highly reactive molecules potentially capable of interfering with any number of vital chemical processes. It is therefore a matter of some interest that there should exist animals capable of producing such general toxicants, let alone capable of storing them at high concentrations. How do living cells make such poisons without poisoning themselves? There is as yet no final answer to this question, although suggestive clues have emerged from work done recently on millipedes and bombardier beetles. Paradoxical as it may seem, the gland cells responsible for the production of the poisons may themselves never really be exposed to them.

In the millipede *Apheloria*, which produces a cyanogenic secretion (*8, 47*), each gland is a two-compartmented organ (Fig. 7A). In the larger compartment is stored a compound, mandelonitrile, that consists of a stable combination of benzaldehyde and hydrogen cyanide; the smaller compartment contains a factor E (presumably an enzyme), capable of catalyzing the dissociation of mandelonitrile. A tight valve ordinarily blocks the tubular connection between the compartments, but at the moment of discharge—which occurs by compression of the larger compartment—the valve is opened by a special muscle, *m*, and the contents of the two compartments are mixed and ejected through the single outer orifice of the gland. Hydrogen cyanide emanates as a gas from the discharged mixture (*8, 47*). Aside from the intrinsic elegance of the mechanism, the point worth noting is that the final step in the production of the poison occurs, not inside the living gland cells, but outside them.

A remarkably similar mechanism occurs in bombardier beetles (*Brachinus* spp.), which eject a secretion containing quinones. The discharge is accompanied by an audible detonation, and the spray is hot to the touch. In *Brachinus*, each gland is also two-chambered (Fig. 7B). The large inner chamber is the source of phenolic precursors of the quinones, and of hydrogen peroxide, while the smaller, outer chamber produces the enzyme catalase (*E*). When the contents of the two chambers are mixed, catalase promotes the liberation of oxygen from hydrogen peroxide and the phenols are oxidized to quinones in an exergonic reaction. Under pressure of the liberated oxygen, the hot mixture suddenly "pops out" (*24*). To all intents and purposes this is an explosive mechanism. Unique in its own way, it resembles the mechanism in *Apheloria* in that the toxic end products are synthesized outside rather than inside the cells.

The majority of defensive glands differ from those of bombardier beetles and cyanogenic millipedes in that they consist of single instead of double compartments and the poisons are stored within them as finished products rather than as precursors. It is surprising in itself that the animals should be able to store the concentrated poisons without detriment to themselves, but the lumen of the glands is always lined with a cuticular membrane (*4*), which apparently acts as an effective insulating barrier (*14*). However, this does not explain how the individual secretory cells—which evidently must manufacture the finished toxins before pouring them into the lumen of the gland—withstand exposure to the poisons. The answer may eventually be found in the function of certain peculiar tubular organelles with which these cells are characteristically endowed (Fig. 8) (*22, 23, 81*). Cuticular in nature, the tubules are essentially links between the cells and the wall of the gland and, offhand, appear to be more than efferent ducts that drain the cells of their secretion. However, it is conceivable that the tubules fulfill a more specialized role, and that their lumen is actually the site where the final synthesis of the poisons takes place. At the free end of the tubule there is frequently to be found an expanded (Fig. 8D), sometimes elaborately specialized (Fig. 8C), bulbar portion, and this might possibly function as a "reaction chamber." One might imagine a mechanism whereby the cell manufactures a series of harmless precursors, which are prevented from interacting to form the final poisons until they are passed into the lumen of the organelle. The cuticular wall of the organelle could be of such nature as to permit the inward passage of the precursors, but not the outward diffusion of the poisons. Trapped within the organelle, the poisons would then be channeled along the tubule to the lumen of the gland (Fig. 7C). If this mechanism is correct —and the evidence for it is only circumstantial (*5*)—then it would be essentially similar to that occurring in bombardier beetles and cyanogenic millipedes. The difference is that, whereas in the latter animals the "reaction chamber" is the impervious two-compartmented lumen of the gland itself, in the other forms the reaction chamber consists of hundreds of tiny individual units, distributed singly among the secretory cells. It may be possible, by histochemical techniques, to reconstruct the sequence of biosynthetic events taking place in the cells and their cuticular organelles, and some efforts in this direction have already been made (*22, 23*).

## Nonexocrine Chemical Defenses

Most of what we know about the chemical defenses of arthropods stems from studies of such integumental glands as are discussed here. But arthropods also possess repellent substances in the blood (*82*), and possibly in other parts of their bodies, and only a very few of these systemic factors have been investigated to date. Although such factors are likely to be more difficult to extract and isolate than exocrine products, the efforts should be well worth while. Substances such as cantharidin and pederin, present in the body fluids of meloid and some staphylinid beetles, are not only unusual chemically but also of some medical and pharmacological significance (*83*). The compounds responsible for distastefulness in the many insects that figure as "Müllerian" elements of mimetic complexes would appear to be of particular interest. The pharmacological properties of a genin-like substance in the Monarch butterfly has recently been investigated (*84*), as well as cyanogenesis in the tissues of a moth (*85*), but literally hundreds of other species remain to be studied, including many butterflies, moths, and beetles. Interesting ecological and evolutionary publications have appeared recently that are likely to provide fruitful leads for experimental work in this general area (*86*).

There is also a good deal yet to be learned about the use that insects may make of secondary plant substances in their diet. Although, as pointed out above, such substances may not necessarily be appropriated intact for incorporation into defensive secretions, they may exert their protective action in other ways. When an insect feeds on a cyanogenic plant, the very act of chewing is likely to trigger cyanide emission at the sites of injury. Might the insect derive some protection from the emanating vapors, perhaps against small predators such as ants? The digestive tract of herbivorous insects often bulges with contained plant matter. Is it possible that intact secondary substances remaining in the gut act to deter appropriately sensitive predators? Many insects respond to disturbance by regurgitating some of their fluid intestinal contents. Is the regurgitate repellent to enemies? Is repellency attributable to factors in the diet or to endogenous substances supplied by the insect itself?

Evidently, the subject of chemical defenses, even as it pertains to the arthropods alone, is still far from exhausted. Like so many other areas of research concerned with the chemical interactions of organisms, it is likely to blossom for some years to come, if for no other reason than that chemists and biologists are now jointly involved in its exploration.

#### References and Notes


1. P. Karlson and A. Butenandt, *Ann. Rev. Entomol.* **4**, 39 (1959).
2. W. H. Bossert and E. O. Wilson, *J. Theoret. Biol.* **5**, 443 (1963); A. S. Parkes and H. M. Bruce, *Science* **134**, 1049 (1961); E. O. Wilson and W. H. Bossert, *Recent Progr. Hormone Res.* **19**, 673 (1963).
3. M. Jacobson, *Insect Sex Attractants* (Interscience, New York, 1965).
4. L. M. Roth and T. Eisner, *Ann. Rev. Entomol.* **7**, 107 (1962), and references therein.
5. T. Eisner, F. McHenry, M. M. Salpeter, *J. Morphol.* **115**, 355 (1964).
6. H. W. Beams and E. Anderson, *ibid.* **109**, 159 (1961).
7. T. Eisner, J. J. Hurst, J. Meinwald, *Psyche* **70**, 94 (1963).
8. T. Eisner, *Nat. Hist.* **74**, No. 3, 30 (1965).
9. G. Garb, *J. Entomol. Zool.* **7**, 88 (1915); L. M. Roth, *Ann. Entomol. Soc. Amer.* **38**, 77 (1945).
10. T. Eisner and Y. C. Meinwald, *Science* **150**, 1733 (1965).
11. T. Eisner, *ibid.* **128**, 148 (1958).
12. ———, *Psyche* **67**, 62 (1960).
13. ———, *Science* **148**, 966 (1965).
14. H. Remold, *Z. Vergleich. Physiol.* **45**, 636 (1962); *Nature* **198**, 764 (1963).
15. G. W. Herrick and J. D. Detweiler, *Ann. Entomol. Soc. Amer.* **12**, 44 (1919).
16. L. Fishelson, *Eos* **36**, 41 (1960).
17. T. Eisner, C. Swithenbank, J. Meinwald, *Ann. Entomol. Soc. Amer.* **56**, 37 (1963).
18. T. Eisner, *J. Insect Physiol.* **2**, 215 (1958).
19. ———, *Nat. Hist.* **75**, No. 2, 42 (1966).
20. ———, J. Meinwald, A. Monro, R. Ghent, *J. Insect Physiol.* **6**, 272 (1961).
21. J. P. Woodring and M. S. Blum, *J. Morphol.* **116**, 99 (1965).
22. L. M. Roth, W. D. Niegisch, W. H. Stahl, *Science* **123**, 670 (1956); B. Stay, *Ann. Entomol. Soc. Amer.* **50**, 514 (1957).
23. L. M. Roth and B. Stay, *J. Insect Physiol.* **1**, 305 (1958).
24. H. Schildknecht and K. Holoubek, *Angew. Chem.* **73**, 1 (1961).
25. J. P. Woodring and M. S. Blum, *Ann. Entomol. Soc. Amer.* **56**, 448 (1963).
26. J. Wray, *Phil. Trans. Roy. Soc. London Ser. B* **5**, 2063 (1670).
27. M. Pavan, in *Proc. Intern. Congr. Biochem. 4th, Vienna, 1958* (1959), vol. 12, p. 15 (references); *Atti Accad. Naz. Ital. Entomol.* **8**, 228 (1960).
28. M. Jacobson, *Ann. Rev. Entomol.* **11**, 403 (1966), and references therein.
29. H. Schildknecht, K. Holoubek, K. H. Weis, H. Krämer, *Angew. Chem.* **3**, 73 (1964).
30. G. W. K. Cavill, L. D. Ford, H. D. Locksley, *Australian J. Chem.* **9**, 288 (1956); G. W. K. Cavill and H. Hinterberger, *ibid.* **13**, 514 (1960); ———, *ibid.* **14**, 143 (1961); R. Fusco, R. Trave, A. Vercellone, *Chim. Ind. Milan* **37**, 251 (1955); R. Trave and M. Pavan, *ibid.* **38**, 1015 (1956).
31. A. R. Gilby and D. F. Waterhouse, *Proc. Roy. Soc. London Ser. B* **162**, 105 (1965).
32. J. J. Hurst, J. Meinwald, T. Eisner, *Ann. Entomol. Soc. Amer.* **57**, 44 (1964).
33. H. Schildknecht and H. Krämer, *Z. Naturforsch.* **17b**, 701 (1962).
34. M. S. Blum, R. D. Crain, J. B. Chidester, *Nature* **189**, 245 (1961); H. Schildknecht, K. H. Weis, H. Vetter, *Z. Naturforsch.* **17b**, 350 (1962); D. F. Waterhouse, D. A. Forss, R. H. Hackman, *J. Insect Physiol.* **6**, 113 (1961).
35. M. S. Blum, *Ann. Entomol. Soc. Amer.* **54**, 410 (1961).
36. L. M. Roth, *ibid.*, p. 900.
37. M. S. Blum, *ibid.* **57**, 600 (1964); M. S. Chadha, T. Eisner, J. Meinwald, *ibid.* **54**, 642 (1961).
38. C. W. L. Bevan, A. J. Birch, H. Caswell, *J. Chem. Soc.* **1961**, 488 (1961).
39. A. Quilico, P. Grunanger, M. Pavan, in *Intern. Congr. Entomol. Proc. 11th, Vienna* (1960), vol. 3, p. 66; R. Stumper, *Compt. Rend.* **233**, 1144 (1951); ———, *ibid.* **234**, 149 (1952), and references therein.
40. H. Schildknecht and K. H. Weis, *Z. Naturforsch.* **16b**, 361 (1961).
41. E. B. Poulton, *Brit. Assoc. Advan. Sci. Meeting Rep.* **5**, 765 (1887); H. Schildknecht and H. Schmidt, *Z. Naturforsch.* **18b**, 585 (1963).
42. F. Dierckx, *La Cellule* **16**, 61 (1899).
43. H. Schildknecht and K. H. Weis, *Z. Naturforsch.* **17b**, 439 (1962).
44. M. Barbier and E. Lederer, *Biochimia* **22**, 236 (1957); A. Béhal and M. Phisalix, *Bull. Museum Paris* **6**, 388 (1900); T. Eisner, J. J. Hurst, W. T. Keeton, Y. Meinwald, *Ann. Enotmol. Soc. Amer.* **58**, 247 (1965); A. Monro, M. S. Chadha, J. Meinwald, T. Eisner, *ibid.* **55**, 261 (1962); R. Trave, L. Garanti, M. Pavan, *Chim. Ind. Milan* **41**, 19 (1959).
45. J. W. Wheeler, J. Meinwald, J. J. Hurst, T. Eisner, *Science* **144**, 540 (1964).
46. M. S. Blum and J. P. Woodring, *ibid.* **138**, 512 (1962); G. Casnati, G. Nancini, A. Quilico, M. Pavan, T. Salvatori, *Experientia* **19**, 409 (1963); H. E. Eisner, T. Eisner, J. J. Hurst, *Chem. Ind. London* **1963**, 124 (1963); C. Guldensteeden-Egeling, *Pfluger's Arch. Ges. Physiol.* **28**, 576 (1882); H. J. Monteiro, *Anais Assoc. Brasil. Quim.* **20**, 29 (1961); W. M. Wheeler, *Psyche* **5**, 442 (1890).
47. T. Eisner, H. E. Eisner, J. J. Hurst, F. C. Kafatos, J. Meinwald, *Science* **139**, 1218 (1963).
48. G. W. Bisset, J. F. D. Frazer, M. Rothschild, M. Schachter, *Proc. Roy. Soc. London Ser. B* **152**, 255 (1960).
49. J. A. Parsons, *J. Physiol. London* **169**, 80P (1963).
50. J. S. Edwards, in *Intern. Congr. Entomol. Proc. 11th, Vienna* (1960), vol. 3, p. 259; ———, *J. Exp. Biol.* **38**, 61 (1961).
51. A. J. Alexander, *Ann. Natal Museum* **14**, 35 (1957); H. Dewitz, *Biol. Zentr.* **4**, 202 (1884); H. Gorvett, *Quart. J. Microscop. Sci.* **92**, 275 (1951); ———, *Proc. Zool. Soc. London* **126**, 291 (1956).
52. T. Eisner, unpublished observations.
53. We have found that ants and lizards confined in closed quarters with the vapors of the quinonoid secretions of millipedes may eventually die.
54. F. Halberg and R. B. Howard, *Postgrad. Med.* **24**, 349 (1958); F. Halberg and A. N. Stephens, *Proc. Minn. Acad. Sci.* **27**, 139 (1959); E. Haus and F. Halberg, *J. Appl. Physiol.* **14**, 878 (1959); J. W. Nowosielski, R. L. Patton, J. A. Naegele, *J. Cellular Comp. Physiol.* **24**, 199 (1962).
55. R. W. Moncrieff, *The Chemical Senses* (Wiley, London, 1944); G. H. Parker, *Smell, Taste and Allied Senses in Vertebrates* (Lippincott, Philadelphia, 1922).
56. G. Weddell and S. Miller, *Ann. Rev. Physiol.* **24**, 199 (1962).
57. E. S. Hodgson, in *The Physiology of Insecta*, M. Rockstein, Ed. (Academic Press, London, 1964), vol. 1, p. 363.
58. G. F. Mayer, *Zool. Jahrb. Abt. Anat. Ontog. Tiere* **74**, 339 (1955).
59. C. A. Keele and D. Armstrong, *Substances Producing Pain and Itch* (Arnold, London, 1964).
60. R. A. Holliday, *Ann. Natal Museum* **10**, 237 (1942).
61. E. Ernst, *Revue Suisse Zool.* **66**, 289 (1959).
62. L. M. Roth and R. B. Howland, *Ann. Entomol. Soc. Amer.* **34**, 151 (1941).
63. M. S. Chadha, T. Eisner, A. Monro, J. Meinwald, *J. Insect Physiol.* **8**, 175 (1962); R. L. Ghent, thesis, Cornell University (1961).
64. T. Eisner and F. C. Kafatos, *Psyche* **69**, 53 (1962).
65. M. Engelhardt, H. Rapoport, A. Sokoloff, *Science* **150**, 632 (1965).
66. K. E. Linsenmair, *Kosmos* **59**, 331 (1963).
67. G. S. Fraenkel, *Science* **129**, 1466 (1959).
68. W. Karrer, *Konstitution und Vorkommen der organischen Pflanzenstoffe* (Birkhauser, Basel, 1958).
69. B. P. Moore, *J. Insect Physiol.* **10**, 371 (1964).
70. J. M. Kingsbury, *Poisonous Plants of the United States and Canada* (Prentice-Hall, Englewood Cliffs, N.J., 1964).
71. J. Meinwald, M. S. Chadha, J. J. Hurst, T. Eisner, *Tetrahedron Letters* **1962**, 29 (1962).
72. J. Meinwald, *J. Amer. Chem. Soc.* **76**, 4571 (1954); T. Sakan, F. Murai, Y. Butsugan, A. Suzui, *Bull. Chem. Soc. Japan* **32**, 315 (1959); T. Sakan, A. Fujino, F Murai, *Chem. Abstr.* **56**, 11,644 (1962).
73. T. Eisner, *Science* **146**, 1318 (1964).
74. G. M. Happ and J. Meinwald, *J. Amer. Chem. Soc.* **87**, 2507 (1965).
75. J. Meinwald, G. M. Happ, J. Labows, T. Eisner, *Science* **151**, 79 (1966).
76. M. S. Chadha, T. Eisner, J. Meinwald, *J. Insect Physiol.* **7**, 46 (1961).
77. J. Meinwald, J. E. Rogers, T. Eisner, *Abstr. Intern. Symp. Chem. Nat. Prod. Kyoto* **1964**, 138 (1964); J. Meinwald, K. F. Koch, J. E. Rogers, Jr., T. Eisner, *J. Amer. Chem. Soc.*, in press.
78. H. T. Gordon, D. F. Waterhouse, A. R. Gilby, *Nature* **197**, 818 (1963).
79. R. Paris, in *Chemical Plant Taxonomy*, T. Swain, Ed. (Academic Press, London, 1963), p. 337.
80. R. B. Clayton, *J. Lipid Res.* **5**, 3 (1964).
81. P. Deegener, *Handbuch der Entomol.* **1**, 1 (1912); G. Gilson, *La Cellule* **5**, 1 (1889); L. M. Roth, *Ann. Entomol. Soc. Amer.* **36**, 397 (1943).
82. V. B. Wigglesworth, *The Life of Insects* (World, Cleveland, 1964).
83. C. Cardani, D. Ghiringhelli, R. Mondelli, A. Quilico, *Tetrahendron Letters* **29**, 2537 (1965); M. Pavan, *Ricerche Biologiche E Mediche Su Pederina E Su Estratti Purificati Di Paederus fuscipes Curt. (Coleoptera: Staphylinidae)* (Industrie Lito-Tipografiche Mario Ponzio, Pavia, 1963).
84. J. A. Parson, *J. Physiol. London* **178**, 290 (1965).
85. D. A. Jones, J. Parsons, M. Rothschild, *Nature* **193**, 52 (1962).
86. L. P. Brower and J. Brower, *Zoologica* **49**, 137 (1964); P. R. Ehrlich and P. H. Raven, *Evolution* **18**, 586 (1965); E. B. Ford, *Ecological Genetics* (Wiley, New York, 1964), and references therein; M. Rothschild, *Trans. Roy. Entomol. Soc. London* **113**, 101 (1961); ———, *Entomologist* **97**, 276 (1964).
87. D. F. Waterhouse and A. R. Gilby, *J. Insect Physiol.* **10**, 977 (1964).
88. Y. C. Meinwald and T. Eisner, *Ann. Entomol. Soc. Amer.* **57**, 513 (1964).


89. Our research on this subject has been supported by grant AI-02908 from the National Institutes of Health, travel stipends from the Sigma Xi RESA Research Foundation and the Bache Fund of the National Academy of Sciences, and unrestricted funds from the Upjohn Company. We thank Mrs. R. Alsop for excellent technical assistance.

# D. Communication

## CHEMICAL COMMUNICATION AMONG WORKERS OF THE FIRE ANT *Solenopsis saevissima* (Fr. Smith)

## 1. THE ORGANIZATION OF MASS-FORAGING*

By EDWARD O. WILSON
*Biological Laboratories, Harvard University, Cambridge, Mass., U.S.A.*

### Introduction

In preliminary studies of the social behaviour of the fire ant, nine categories of communication among workers have been recognized. These are listed in Table I, in the approximate order of their frequency of occurrence. It is the purpose of the articles in this series to take up a preliminary analysis of several of the message categories, with the aim of reconstructing in part the complex events of mass behaviour and social organization in the normal life of the fire ant colony.

It is commonplace supposition that communication in ants, especially in small-eyed species such as *Solenopsis saevissima*, must be partly chemical. Yet the idea has received little critical examination, and detailed studies of chemosensory behaviour are relatively scarce in the literature. Two classes of chemical releasers, or "olfactorily acting pheromones" (Karlson & Butenandt, 1959), have been the object of physiological research in recent years: trail substances (MacGregor, 1948; Carthy, 1951; Wilson, 1959) and alarm substances (Sudd, 1957; Wilson, 1958; Butenandt, Linzen & Lindauer, 1959; Brown, 1959; Wilson & Pavan, 1959). The specific releasers of necrophoric behaviour have been preliminarily characterized (Wilson, Durlach & Roth, 1959). On the basis of this new information, chemosensory physiology deserves

**Table 1. The Known Categories of Communication Among Fire Ant Workers.**

| Stimulus | Transmission | Response |
|---|---|---|
| Nest odour | Chemical | Nil, if odour is undisturbed. |
| Casual antennal or bodily contact | Tactual | Turning-toward movement or increased undirected movement |
| Body surface attractants | Chemical | Oral grooming; clustering |
| Oral grooming | Tactual | Decreased general movement, occasionally receptive posture |
| Ingluvial food solicitation | ? | Regurgitation |
| Regurgitation | At least partly chemical | Feeding |
| Emission of Dufour's gland secretion as trail | Chemical | Attraction, followed by movement along trail |
| Emission of Dufour's gland secretion during attack | Chemical | Attraction to disturbed worker |
| Emission of cephalic substance | Chemical | Alarm behaviour |

*Several persons provided essential assistance during the course of this study. Prof. Murray S. Blum generously contributed chemical fractions of fire ant workers used in his own biochemical studies of exocrine secretions. The technique of behavioural assays in the analysis of this material has been an unexpected and fruitful by-product of our collaboration. Prof. George A. Miller and Dr. Rudolf Jander followed the progress of the work, read the manuscript, and provided both new ideas and chastening criticisms. The work would be much less complete without the support of these colleagues, but the author alone must accept responsibility for any technical errors or misinterpretations that may remain lodged in the final report.

The research programme has been supported by a grant from the United States National Science Foundation.

Reproduced with permission from *Animal Behaviour*, 10:134–147, 1962. Published by Bailliere, Tindall and Cassell, London.

to become an increasingly important aspect of social behaviour studies.

For the present at least, it is useful to explore the hypothesis that most of the social life of ants can be explained as the product of the interplay of stereotyped patterns, each of which is induced by one of a limited series of chemical releasers. If this assumption is correct, it should be possible to separate the releasers and employ them in the experimental analysis of individual patterns. It is further useful for purposes of investigation to suppose that at least some of the releasers are produced as glandular secretions and tend to be accumulated in gland reservoirs. This first paper is concerned with several aspects of behaviour underlying mass-foraging.

### Materials and Methods

Colonies of *Solenopsis saevissima* were housed in circular artificial nests of the design shown in Fig. 4. A note concerning the construction of the nests may be of interest here. They are made entirely of plexiglas, the walls and ceiling being transparent and the base white opaque. Following a technique introduced by Brian (1951), a water channel leads into the base from the side to a central well stuffed with cotton wool, and water is allowed to pass in a continuous column from an outside reservoir. Four entrance spouts are set symmetrically in the wall, and are used to connect the nest, by means of clear plastic tubing, to additional nest units. The living area can thus be expanded indefinitely to accommodate the growing colony. The nests are placed on glass platforms supported on glass vials or bottles which in turn are set in dishes filled with mineral oil. The platforms serve as foraging fields for the ants. The nest interiors are constantly illuminated, so that within a year most of the living worker population has been reared in the light. As a result the entire colony can be examined conveniently at any moment without further disturbance.

The colonies used in the present study were reared from single recently fecundated light-phase queens collected at Baton Rouge, Lousiana. *Solenopsis saevissima* is an unusually fast-breeding species; under laboratory conditions healthy colonies raised from single queens reach "maturity", i.e. grow to include several thousands of workers, males, and virgin queens, in less than a year.

### The Organization of Mass Foraging

In an earlier paper (Wilson, 1959) it was shown that the odour trail of *Solenopsis saevissima* is secreted by Dufour's gland* and released through the extruded sting. The pheromone component can be extracted in petroleum ether and concentrated by steam distillation and chromatographic separation. It has proven to be a powerful attractment, serving not only as the trail orientator but also as a true "releaser" of hunting behaviour on the part of worker ants. In this and in subsequent reports special attention will be devoted to the behavioural mechanisms by which fire ants use the Dufour's gland secretion in multiple roles: to communicate the location of new food finds, to organize efficiently the whole of mass food-retrieving, and to alarm sister workers.

### Hunting by Solitary Workers

Foraging away from the nest is conducted preponderantly by workers in the small- and middle-sized classes. Only mature, fully-coloured individuals participate. Foraging is directed from the nest or way-stations by solitary workers, who move in irregular, looping paths.

When the periods of time spent away from the nest perimeter by 250 workers from a colony of about 3000 individuals were measured, these data formed the frequency curve shown in Fig. 1. When the measurements are transformed into number of workers left in the foraging field as a function of successive minute intervals following the departure from the nest of all workers simultaneously, a reasonably straight line is obtained in a double logarithmic plot. This shows that the function is moderately complex. In particular the probability of workers returning to the nest does not, as might be intuitively supposed, remain steady with time. Instead, the number remaining on the foraging field is best described in the double logarithmic form,

$$\log N = \log b - {}^1/_a \log t,$$

or

$$N = bt^{-1/a}$$

Where N is the number of workers still foraging at any given time, *b* is the number of workers foraging at the end of the first second interval, and *t* is the number of expired second intervals. The data conform specifically to the equation,

*Dufour's gland has also been referred to by past authors as the "accessory" or "alkaline" gland of the sting. Its morphology has recently been described in detail in the admirable study of the fire ant poison apparatus by Callahan, Blum & Walker (1959).

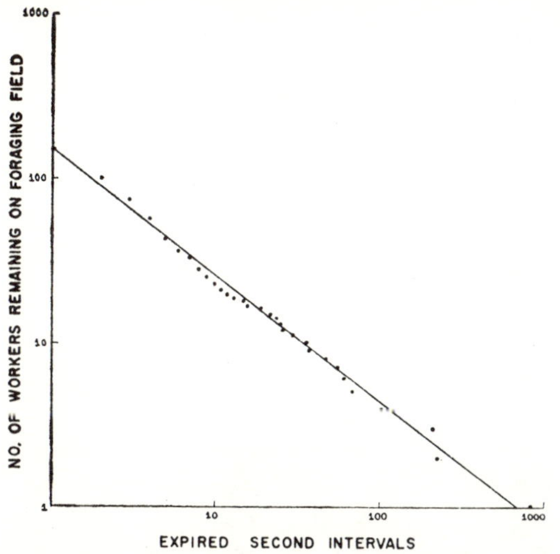


Fig. 1. Dispersion of foraging trip durations of solitary workers expressed as the decline of the number remaining on the foraging field with time. Colony size: about 3,000. Air temperature: 28±1°C.

$$N=100\ t^{-0.76};$$

which differentiated yields

$$\frac{dN}{dt} = -76t^{-1.76}$$

Thus the number of workers remaining foraging are decreasing at a steadily lessening rate with the passage of time. The longer a worker remains afield, the less becomes the probability that it will return home in the next second interval. This function can be at least partly explained by the common sense deduction that the longer a worker remains in the field, the farther it wanders from the nest, and hence the longer will be the journey home when it turns back. Nevertheless, the foraging time cannot be explained simply as the result of random movements out from the nest perimeter. The workers make frequent long straight runs and ordinarily follow a pattern of variable looping movements away from the nest perimeter. Evidence will be given later to show that in the laboratory hunting workers orient visually.

### Rapid Exploration of New Territory

When larger laboratory colonies are given sudden access to a new foraging field, such as a clean glass platform or blank piece of poster card, the workers respond in a manner which is not predictable from observation of hunting behaviour on the old foraging field. During the first several hours the workers move onto the new area in large numbers. New foragers are recruited by the laying of odour trails. The emission of the trail substance under these circumstances is remarkable, since no reward in the usual sense is available; the only stimulation that can be inferred is the unfamiliarity of the new field. Workers do not concentrate at the end of the trails, and no specific attraction points can be observed. The majority of workers entering the new field follow the odour trails to the interior, then diverge individually to perform exploratory looping movements. As a result, new fields only a few square metres or less are thoroughly explored and patrolled during the first hour or two. After the initial "land-rush", the number of workers on the field gradually declines and finally stabilizes, in most instances at less than half the peak number. This unusual behaviour seems indicative of an "exploratory" or "manipulative" drive analogous to that described in some mammals (see Thorpe, 1956: 10).

### Behaviour at the Food Find

When a solitary foraging worker is presented with a rich new food source, such as a freshly killed mealworm (*Tenebrio*) or roach (*Naupheta*), it abruptly mounts the object, and excitedly palpates it with its antennae while remaining motionless or creeping slowly and deliberately over the surface. If the food source is a single solid object and small enough, the ant soon picks it up and carries it forward or drags it backward in a homeward direction. If the source is an immovable solid object, the worker typically inspects it for a period ranging between 10 and 30 seconds without attempting to feed, and then commences laying a trail homeward. If the food source is a sugary fluid, the ant feeds to partial or whole repletion and returns homeward, laying a trail. Evidently there is nothing peculiar in the nature of liquid food that requires feeding before trail-laying, for if the liquid bait is taken away from the worker before it is able to feed, it nevertheless lays a trail.

### Trail-laying Behaviour

The trail-laying worker is easy to distinguish in the throngs of foraging workers through which it is passing. It moves at a slower pace, crouching closer to the ground, and the forward part of its body periodically swings to the right or left of the main line of motion but usually back again to follow the same approximate direction as before.

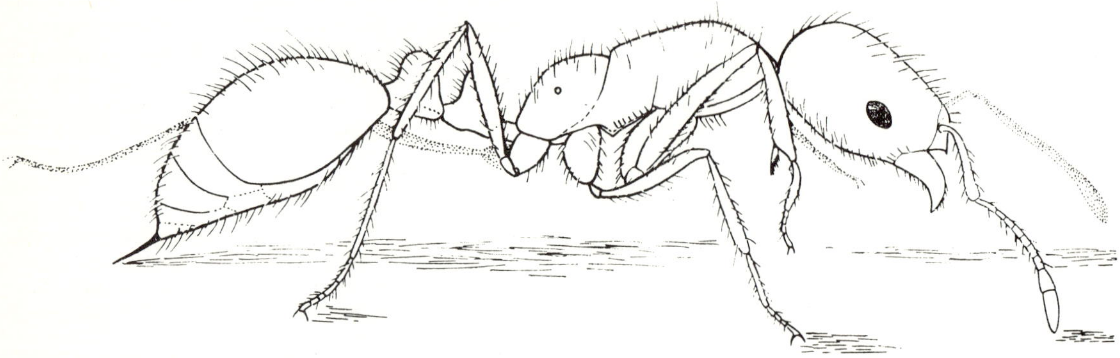

Fig. 2. Worker fire ant laying an odour trail, from left to right. Drawn from a photograph taken at 1/1000 second.

The gaster (abdomen) is bent downward slightly and the sting extruded. Examination of a series of photographs, taken with an electronic flash at 1/1000 sec. and showing lateral views of homing workers, reveal that only the tip of the sting touches the ground (Fig. 2). Furthermore, it is periodically lifted clear for short distances, or withdrawn altogether, breaking the trail into a series of streaks.

The chemical trail, proceeding from Dufour's gland and passing along the valves of the sting, is applied to the ground in a manner similar to a thin line being inked with a pen. It is nevertheless invisible to the naked eye and cannot be detected visually even with a dissecting microscope at magnifications of 50X. All efforts to render the trail visible, including dusting with fine powder (see Carthy, 1951), examination with near-ultraviolet light, and tracing with radioactive isotopes applied externally, have failed. Even strong trails, laid by hundreds of workers over a period of hours, are not visible.

While heading in a homeward direction, the trail-laying worker sometimes loops back in the direction of the bait, but only for short distances, before turning nestward again. During the backtracking movements, trail-laying continues. When another worker is contacted, the homing worker turns toward it. It may do no more than rush against the encountered worker for a fraction of a second before moving on again, but sometimes the reaction is stronger: it climbs partly on top of the worker and, in some instances, shakes its body lightly but vigorously, chiefly in a vertical plane. The precipitousness of the movement toward the encountered worker and the tendency to mount it recall strongly the behaviour shown when foragers first encounter food-finds (q.v.). The vibrating movement, however, is unique to these individual encounters. The behaviour at the encounter appears to function solely to bring the trail substance to the attention of the sister worker. It does not appear to communicate any important information about the food-find, because contacted workers do not seem to exhibit trail-following behaviour different from those not contacted. Moreover, as will be shown in a later section, the chemical trail is by itself sufficient to induce full and immediate trail-following behaviour when laid by artificial means. Trail-laying workers encountering other workers pause with them for only a fraction of a second, even when partially mounting them. Following contact they show a strong tendency to swing about and return in the direction of the food-find. At its weakest the fast-encounter movement involves at the least a partial deflection away from the homeward direction. Frequently the trail-layer turns a full 180°, heads for a short distance in the direction of the food-find, and then turns homeward again. As the ant approaches the nest perimeter, and contacts become more frequent, the looping movements become more pronounced. After several encounters (rarely, after only one) the worker now heads persistently in the direction of the food source. As it runs back, it continues laying a trail, so that a double trail is formed. It follows its own trail outward with no more precision than other workers are following it, wide deviations often occur, and hence an alternate trail is provided. If the target is sufficiently close for the trail-layer successfully to reach it, the ant usually turns homeward again, laying a third trail. Few workers get this far; in a single extreme case one was seen to lay a total of four trails, the final one ending in the vicinity of the target. When the target is so far away from the nest (on glass, about 50 cm. or more) that the outer stretch of the homeward-laid trail evaporates before the trail-layer can return along it, the trail-layer becomes confused. It continues for a while in an outward direction away from the nest, presumably either through kinesthetic momentum or menotactic orientation, but in the great majority of cases it is too far off to make an early second contact with the target.

**Recruitment of New Workers**

Most workers encountering a freshly laid trail respond by at once following it in an outward direction. They are able to detect it olfactorily over distances as great as 10 millimetres. Workers coming upon the trail by accident within the first half-minute respond with the same alacrity as those encountering the trail-layer directly. As noted before, there is no evidence that direct contact with the trail-layer confers any information beyond the attractive effect of the trail substance. No direct experimental test has yet been devised to determine whether or not the trail has an intrinsic orientation. However, several lines of indirect information virtually prove that it does not: (1) the trail is seemingly deposited in such a manner as to preclude the possibility of polarization by tapering or coding through varying intervals of segmentation; (2) the trail-layer commonly doubles back in looping motions while laying the trail, but followers continue along these inverted segments in an outbound direction without confusion; (3) in following trails workers are in direct contact with them only for short distances, often deflecting to one side and crossing them in an irregular zig-zag fashion; (4) workers that encounter nestward-bound trail-laying workers well away from the nest commonly follow closely behind them, sometimes most of the way back to the nest, rather than move toward the bait.*

Direct experimental evidence has been obtained to show that trails laid by multiple workers to a single food find contain no intrinsic orientation. In a series of tests, workers were forced to cross a bridge of white poster card 1·5 cm. wide and 30 cm. long in order to reach a bait. Approximately 10 minutes following maximum buildup, the bridge was removed, tapped free of ants, and placed sidewise on the near (nestward) side of the chasm in a position such that its centre lay athwart the oncoming stream of ants. Thus the ants approaching the bridge were all outbound and in order to proceed along the odour trail were required to choose between making right-angle turns to the right or left on it. If they were to choose predominantly the right-angle turn and thus follow the former outer half of the bridge, it could be assumed that they were orienting by some topographic or olfactory feature of the trail itself. If the stream divided evenly in both directions, it could be concluded that such features, if they exist, were not used. Counts were made of the numbers of ants on both outermost 10-cm. segment of the bridge 20 seconds after the bridge was turned. The direction of the bridge with respect to the lighting was alternated in successive trials to eliminate any special visual bias. In five replications a total of 59 workers travelled to the end of the bridge that previously would have led to the bait, and 62 travelled to the end that would have led back to the nest. In both the individual test scores, and the summed scores just cited, no significant difference was found in the direction taken by the ant stream. That the ants were truly following odour trails and not just randomly exploring was proven by control experiments in which newly constructed bridges were offered them in a sidewise position. The ants did not travel deliberately down the length of these as they had the test bridges but instead milled about in confusion near the centre. It was concluded that under the conditions of the experiment the odour trail offers no intrinsic features that allow polarity choices.

This negative conclusion is supported by the results of an additional experiment. The procedure just described was modified by simply removing the bridge, tapping it lightly to knock the ants from it, turning it around 180°, and replacing it across the gap. Under these conditions, both the outward- and inward-bound ant streams remounted the bridge and crossed it without evidence of confusion. Their transit was as swift and orderly as in other tests in which the bridge was replaced without reversal but notably faster than across fresh bridges lacking odour trails.

When the odour trail leading from the bait to the nest (or way-station) is short enough to allow workers to run its length before its outer segment evaporates, the build-up of workers at the bait is rapid. The build-up, in fact, is sigmoidal, accelerating at first and then stabilizing at an

*Macgregor (1948) and Carthy (1951) have demonstrated the lack of polarization communication in the odour trails of *Myrmica ruginodis* (=*rubra div. auct.*) and *Lasius fuliginosus*.

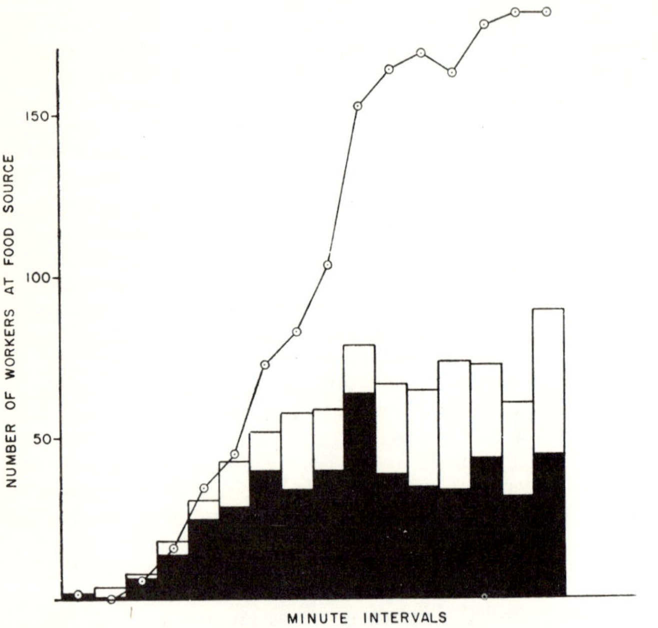


Fig. 3. The build-up of workers at a bait (adult *Naupheta* roach) placed near the edge of an artificial nest containing a large colony. Further explanation in the text.

asymptotic upper level (Fig. 3). The first newcomers to arrive at the bait repeat the behaviour of the discoverer worker: after a brief investigation they begin to lay trails homeward. The larger the number of trails, i.e. the greater quantity of Dufour's gland secretion, being laid at a given instant to the nest perimeter, the larger the number of workers that respond by going out to the bait. Since multiple workers are attracted by a single trail, and each of these in turn lay trails, the build-up at the bait is exponential.

The direct relation between the quantity of Dufour's gl--d deposit and the number of responding workers was demonstrated experimentally in the following manner. A highly concentrated preparation of Dufour's gland secretion was used.* To simulate the natural variation that obtains from the addition of multiple trials, the quantity of pheromone was varied by presenting differing areas of concentrate on the tips of glass rods. Rods of several diameters were dipped to measured distances into the concentrate and the adhering droplets removed to leave a smooth, flat liquid layer. By this means the quantity of pheromone released could be roughly measured in units of area of surface film exposed. Following treatment the glass rods were immediately placed at an entrance hole of a nest containing a small resting colony, as shown in Fig. 4. The evaporated pheromone was drawn over the colony by a steady air current being pulled by a vacuum apparatus through the nest. The number of workers attracted outside of the nest was then counted at the end of a ten-minute period. The results are shown in Fig. 5.

The data suggest that within the limits employed, the response increases approximately linearly with an increase in area of pheromone. The wide disparity in the slope of the curves obtained on two series of tests shows the influence of additional parameters of colony response not yet understood. Variation in trophic

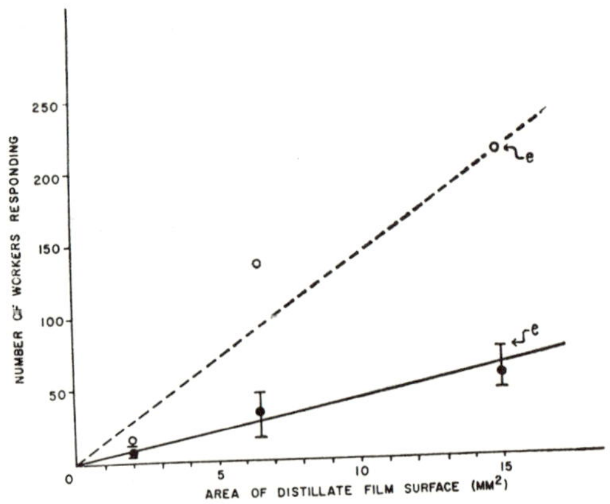


Fig. 5. Varying mass response of workers to different areas of Dufour's gland concentrate presented in the manner illustrated in Fig. 4. The results of two series of experiments on the same colony, containing 300-500 workers, are shown. In the lower series (joined by a solid line), three replications of each quantity of distillate were used. The range and mean are shown. In the upper series (joined by a broken line, only a single experiment was performed. The letter *e* designates the replications at which incipient emigration was induced. Further explanation in text.

*The concentrate used in this experiment was supplied by Dr. M. S. Blum, who obtained the pheromone by successively extracting wholeants in petroleum ether, steam-distilling the extract, and finally fractionating the distillate by column chromatography. The ethopotent chromatographic fraction was identified by the use of an artificial-trail assay (Wilson, 1959).

# PLATE XII

Fig. 4. Response of workers in an artificial nest to evaporated trail substance. *Above:* before the start of the experiment, air is being drawn into the nest (by suction tubing inserted to the left) from the direction of the untreated glass rod; only a few workers are foraging outside the nest. *Below:* within a short time after the glass rod has been dipped into Dufour's gland concentrate and replaced, a large fraction of the worker force leaves the nest and moves in the direction of the rod.

condition was probably not sufficient to cause the difference, since the colony was given a uniform diet throughout the tests, and population structure did not change significantly. Unfortunately, not enough concentrate was available to extend these tests beyond the few measurements shown.

Under natural conditions the number of workers at the bait approaches an asymptotic limit, evidently for three reasons: (1) as workers begin to crowd on the bait, the number reaching the bait during a given period of time reaches a limit; (2) only workers that have contacted and examined the bait lay trails when they return homeward; (3) the odour trail is impermanent, individual and compound trails evaporating to below threshold level within two minutes on a clean glass surface and within ten to twenty minutes on highly absorbent paper.

To measure the duration of single odour trails, the following experiment was performed. A single worker was allowed to discover a bait (freshly killed *Naupheta*) mounted on a glass slide. As the worker investigated the bait, a fresh glass plate was inserted between it and the nest, so that it was forced to lay its trail over this fresh surface. As soon as it reached the nest, the worker was diverted so it could not lay a second trail back over the glass. Further, the bait was removed, so that workers reaching the end of the trail were not reinforced and did not lay trails of their own. The period during which other workers followed the trail, or segments of it, was timed. In five replications, the trail remained active between 85 and 125 seconds, with an average period of 104 seconds. The air temperature during the tests was 28°C. No attempt was made to determine whether the passage of the workers in some way reinforced the trail and extended its life; it is at least certain that these workers did not lay trails of their own by the conventional means of sting extrusion. In any case, the measurements given were collected under the more significant condition, since in natural circumstances trails die out while being followed by unrewarded workers.

Glass of course has an unnaturally smooth surface, from which substances can be expected to evaporate rapidly. The duration of three trails laid under similar conditions to the above but on highly absorbent ink-blotter paper lasted 450 seconds, 750 seconds, and approximately 20 minutes respectively. The build-up at baits on blotter-paper is correspondingly faster. In a series of tests using a 1M sucrose bait at 20 cm., the interval was measured between the departure of the first trail-layer and the moment the build-up of trail-followers at the bait reached ten. On glass this interval ranged between 253 and 409 seconds, with an average of 329·6 seconds, in five tests; on blotter paper the interval ranged between 78 and 202 seconds and averaged only 133·8 seconds in five tests. It is scarcely necessary to add that the mean difference is highly significant.

Finally, artificial trails made by smearing accessory glands, containing far more of the pheromone than single natural trails and drawing many times more workers, remained active on a glass surface between 3·5 and 7 minutes (mean: 5·9 minutes) in ten trials. The densest natural compound trails laid by columns of workers simultaneously on glass usually declined below threshold value within twenty minutes.

In summary, it has been concluded that the failure of trail emission in frustrated workers and the ephemeral nature of the trail itself combine to provide an effective negative-feedback device that prevents the excessive build-up of workers. As the bait becomes crowded, the number of workers laying trails (hence, rate of emission of trail substance) stabilizes. Since a single worker's contribution vanishes within a few minutes, the number of new workers coming to the bait also stabilizes. In order to increase the number of workers, it is only necessary to increase the area of the bait (see Fig. 6).

### The Control of Build-up Rate and Limit

Starved colonies respond more swiftly to baits, and achieve a higher maximum density of attendance, than do well-fed ones. Similarly, the same colony responds more strongly to rich food sources than to poor ones; for example, the build-up at a 1M sucrose source is many times faster than at an adjacent pure water source, and the maximum attendance (build-up limit) is always greater at the former.

The basis of this variation in social response has been investigated by examining worker behaviour in detail at adjacent water and sucrose baits. Several key measurements are described in the sections to follow and summarized in Table II.

*a-b.* Workers approaching the water drop usually inspected it or fed from it in a brief and desultory manner, while those contacting the sugar solution invariably stopped at the point of

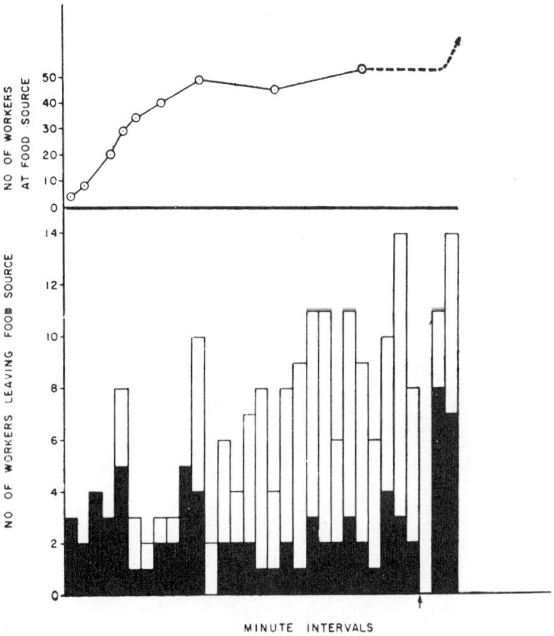


Fig. 6. A typical experiment demonstrating the relation between the area of the food source and the number of responding workers. *Lower:* number of laying workers (solid columns) and non-laying workers (open columns) leaving a bait at successive time intervals. At one point, designated by the arrow, the bait area was doubled. Following the disturbance, during which the total number of returning workers dropped off momentarily, the number of trail-laying workers increased significantly. *Upper:* the number of workers at the bait during the same experiment.

seconds and became semi-replete. In contrast, the abdomens of all ten of the workers at the sugar solution distended noticeably, 8 becoming replete and 2 semi-replete. If the ants in each sample are classified as feeders vs. non-feeders, the difference in proportion of feeding between the two samples is statistically highly significant (probability of randomness less than 0·1 per cent.). Moreover, in other trials using sucrose and water baits simultaneously, the wide disparity in response was invariably noted.

*c.* In other random samples of 20 workers at each bait, a significantly lower percentage among those leaving water laid trails (30 per cent.) than among those leaving sugar (90 per cent.). The probability that this difference in proportion is due to chance is less than 0·1 per cent. Of those laying trails from the water, all had fed to partial or whole repletion. But 6 others (30 per cent. of total sample) fed to some degree of noticeable repletion without laying trails; this fraction is of course separate from the 30 per cent. that laid trails. All of the 20 workers at the sugar solution fed to some degree of repletion, and of these 18 (90 per cent.) laid trails. The 2 that failed to lay trails were nevertheless fully replete; it is noteworthy that these and other observations suggest that after the ant has fed no correlation exists between the degree of repletion and the tendency to lay trails.

*d-e.* As the forager (from sample *c*) left the bait it was followed for 30 seconds and the total amount of time spent in trail-laying measured with a stop-watch. Only data pertaining to

**Table II. Responses of Foraging Workers to Water and Sucrose Baits** (Explanation in text).

| | a. Time spent at bait (range in secs.) | b. Time spent at bait (mean in secs.) | c. % extruding sting | d. Sting-extrusion period (range in secs.) | e. Sting-extrusion period (M±s.e.) |
|---|---|---|---|---|---|
| Pure water | 2—196 | 31·8 | 30 | 18—29 | 23·8±2·1 |
| 1M sucrose | 50—339 | 144·1 | 90 | 6—26 | 20·4±1·2 |

contact and fed lengthily. The first two columns give the periods of time spent in inspections and feeding by ten workers selected randomly as they first approached the water and by ten others selected as they approached the sugar solution. Only two of the workers drank enough water to distend their abdomens noticeably; one of these remained for 196 seconds and reached a replete condition, while the other remained for 42

workers laying trails are given. Actually, it is impossible under ordinary conditions to determine at each instant whether trail substance is being emitted, and it is furthermore very difficult to see whether at all times the tip of the sting touches the ground. Hence the measurement taken was the summed period during which the sting was extruded, on the assumption that this period is closely correlated with the duration of actual trail-laying and, similarly, quantity of trail substance emitted. The data in *d-e* show that trail laying is essentially an all-or-none response, since the average sting-extrusion periods differs very little. In fact the probability that the mean difference cited (3·4 seconds) is due to chance alone is 7 per cent., a magnitude greater than that conventionally accepted as indicating a significant difference. Workers that have fed on the bait either lay no trail at all or else engage in trail-laying (sting extrusion) more than half of the first thirty seconds. Other studies have shown that the all-or-none nature of the response holds until the ant returns to the nest. Note that a single worker leaving the sugar bait extruded its sting during only 6 seconds in the first 30-second period. The behaviour of this individual was exceptional in another, significant way. It was evidently disorientated for a short while after leaving the bait, wandering in irregular loops away from the homeward odour trail being followed by other foragers. During this time it did not extrude its sting. Only toward the end of the first 30-second period did it find the trail, start homeward, and commence extruding its sting. Thereafter it engaged in normal trail-laying, conforming to the general pattern. If this single individual is excluded, the periods of sting-extrusion by the remaining 17 workers ranged 14-26 seconds, and averaged 21·2 seconds with a standard error of 0·8 seconds. If these latter estimates are accepted, the probability that the new mean difference (2·6 seconds) is due to chance alone is 23 per cent.

Trail-laying by individual workers can therefore fairly be called an all-or-none response which is elicited, at least in the case of liquid foods, in a certain percentage of workers that successfully feed. The build-up at a poor source, such as pure water, is less than that at a rich source, chiefly because more foragers turn away without feeding at all. It may also be true that the poor source releases the trail-laying response in a smaller percentage of those workers that feed, but this control, if it exists, is clearly of secondary magnitude. The causes of failure of the response in a minority of feeding workers have not been further considered.

### Properties of Overcompensation in the Mass Response

The sum of the experimental evidence points clearly to the trail pheromone being the paramount, and probably sole, mode of communication in the mass response.* Further, and of equal importance, the individual emission of the pheromone is an all-or-none response, remaining relatively constant when released regardless of the quantity or quality of the food find.

Together these two properties of individual behaviour make it unlikely that the single worker is able to communicate information about the quality or quantity of the food find. In fact, it appears able only to command sister workers to proceed outward, orienting them by means of the trail, without varying the number of recruits in any directed manner. By means of individual choice at the food site, newly arriving workers then "vote", i.e. independently choose whether to lay trails of their own. As shown in the preceding section, control of build-up is mass-communicated; i.e. the number of workers responding is controlled by the sum of workers participating in trail-laying.

Assuming that these simple properties of the mass response are not significantly complicated by other behavioural phenomena, it should be possible to make predictions of the outcome of experimental modifications of the response without undue difficulty. One prediction convenient to test is the following: if the bait is removed in the midst of the build-up, the ants should not be able to communicate this fact at once but should "overshoot", i.e. fresh workers should continue to pour into the now-empty bait area for at least several minutes. Furthermore, the "maximum overshoot", defined as the largest number of workers present in the bait area following bait removal, should appear after an interval which is proportional to the distance of the bait area from the nest. Finally, and most importantly, the maximum overshoot should be independent of (1) quantity of the bait, (2) quality of the bait, (3) stage of the build-up at which the bait is removed, and (4) distance of the bait area from the nest.

Most of these properties were demonstrated in actual experiments (Table III, Fig. 7, and below).

The following procedure was used to measure maximum overshoot. The bait was placed in the centre of a 4-cm.$^2$ cover slip, which was in turn placed in the centre of a rectangular area 3 × 5 inches (7·6 × 12·7 cm.) marked on the lower

*That is, the mass response leading to the normal build-up at the food source. Later, extensive regurgitation occurs (Wilson & Eisner, 1957), with probable long-range social consequences that have not been adequately investigated.

**Table III. Maximum Overshoot Obtained Under Various Experimental Conditions.**

| Colony size | Bait | Bait (target) distance | Initial build-up | Maximum overshoot (range) | Maximum overshoot (mean ± s.e.) | Time of maximum overshoot (mins.) |
|---|---|---|---|---|---|---|
| 300±10 | 0·5 ml 1 M sucrose | 10 cm. | 10 | 1—8 | 5·0±1·2 | 1/2 |
| 13,000 ±2,000 | ,, | ,, | ,, | 17—29 | 20·0±2·6 | 1 1/2 |
| 26,000 ±4,000 | ,, | ,, | ,, | 7—33 | 19·0±3·6 | 1 1/2 |
| ,, | ,, | 20 cm. | ,, | 8—22 | 17·8±2·0 | 1 1/2 |
| ,, | ,, | 50 cm. | ,, | 10—33 | 19·0±3·6 | 3 1/2 |
| ,, | ,, | 80 cm. | ,, | 5—25 | 13·5±2·7 | 7 |
| ,, | ,, | 100 cm. | ,, | 10—17 | 13·0±1·2 | 7 |
| ,, | ,, | 150 cm. | ,, | 3—14 | 8·1±1·8 | 4·2 |
| ,, | ,, | 10 cm. | 1 | 3—7 | 5·4±0·7 | 2 |
| ,, | ,, | ,, | 50± 5 | 14—28 | 18·8±3·0 | 1 1/2 |
| ,, | ,, | ,, | 100±10 | 10—37 | 20·9±3·5 | 2 |
| ,, | distilled water | ,, | 10 | 13,21* | 17 | 2 |
| ,, | *Naupheta* | ,, | ,, | 11—34 | 19·9±3·2 | 1 1/2 |

*Based on two trials only; see text.

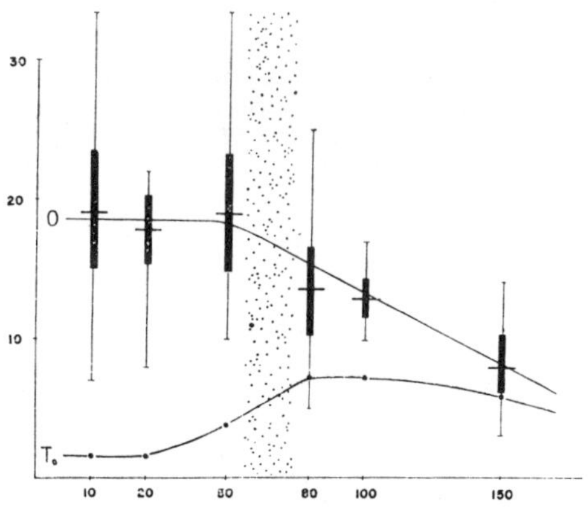


Fig. 7. Variation in maximum overshoot (O) and time of maximum shoot (To) as a function of distance of bait from the nest based in 8 replications for each distance. For the former, extreme limits, mean, and standard deviation are shown; for the latter only the mean is shown. The stippled area represents the target distance range in which the efficiency of the trail system of a larger colony begins a significant decline (see Fig. 8). Further explanation in the text.

surface of the glass platform. Workers were allowed to lay trails to the bait, and build up their numbers to a previously decided "initial build-up" level. Then the cover slip was lifted away, along with all the ants adhering to it. Thus at the outset, the ants feeding at the bait and those immediately arriving or leaving were removed. The number of ants remaining in the rectangle at this moment was noted. The increase in number of workers over the residue was then recorded at 30-second intervals. The excess is referred to as the "overshoot" (in the mass response) and the highest excess recorded at any half-minute interval the "maximum overshoot." The experiment was repeated eight times for each row given in Table III. One exception was made in the case of distilled water. Response to water was so slow that during repeated trials, an initial build-up of ten was attained only twice. The air temperature during the measurements was 28±1°C.

It will be seen from the data given in Table III and Fig. 7 that the maximum overshoot remains remarkably constant under a very wide variety of conditions. It is approximately the same, averaging between 17 and 21, for colonies of different sizes, for baits of different quality, for baits placed at varying distances from the nest, and for widely different initial build-ups allowed in the experiments. Specifically, the probability that the mean differences between the overshoot values are due to chance alone in most cases exceeds 95 per cent. The several statistically significant deviations from the constant maximum-overshoot value are not unexpected and do not detract from the significance of the general constancy. The very small colony (300 workers) overshot less because of the much smaller worker force at the nest periphery. The overshoot following an initial build-up of only one worker was significantly less because only a single odour trail was laid. Overshoot quickly reaches the expected level (by an initial build-up of 10) and does not exceed it in much higher build-ups (at least to 100). The reason for this can be deduced from the measurements given in Fig. 6. The upper limit of trail-laying activity, and hence the quantity of trail pheromone at any given instant, is quickly reached when the bait is relatively small. Finally, as seen in Fig. 7, the maximum overshoot significantly declines when the bait is greater than 50 cm. from the nest because by the time of an initial build-up of 10 the trail is still incomplete and recruitment less accurate.

### Trail Formation Over Great Distances

Beyond the distance of approximately 50 cm. on a glass surface, the trail-laying worker is not able to bring back other workers with it before the outermost (first-laid) part of the trail evaporates to below threshold level. Under these conditions the ants are unable to proceed directly to the bait. Under laboratory conditions, nevertheless, baits placed at this and greater distances are soon rediscovered and a solid system of multiple trails laid down to connect them with the nest. This is accomplished in the following manner. Ants reaching the shrinking trail terminus continue searching in its vicinity; a large proportion continue moving outward in the general direction that the trail led them. The result is that the area around the bait, providing it is not excessively distant, is soon being patrolled by an increased number of hunting workers. The bait will probably be encountered by one of these at the close of shorter interval than was required for the initial discovery. When this occurs another trail is laid, more workers enter the area, and the intervals between discoveries and the laying of individual trails grow ever shorter. Soon multiple trails are being laid simultaneously at various distances between the bait. These form links that ultimately sum into a continuous trail.

Although continuous compound trails are manufactured with ease for some distance beyond the maximum continuous distance of a single trail, this distance does have a measureable upper limit. The limit is affected by at least two parameters, physical structure of the foraging surface and colony size. The former has not been well analysed, glass being consistently used as the standard surface for comparative experiments. The influence of colony size is illustrated in Fig. 8.

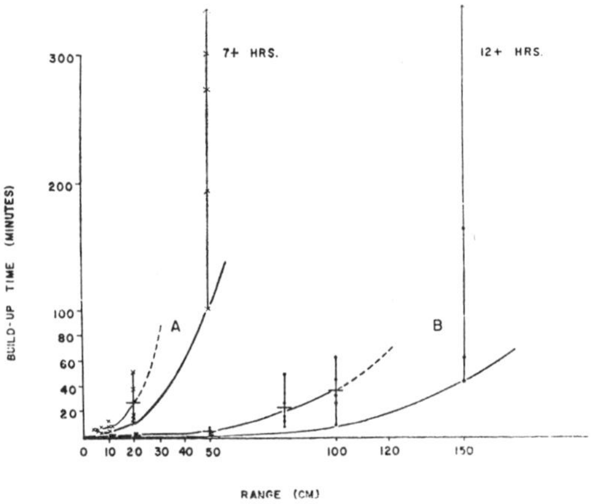


Fig. 8. Time required for colonies to build up forces of 10 workers at sucrose baits at various distances from the nest. All data are shown, with mean values for responses at each distance. A small colony (A), containing approximately 300 workers, is compared with a medium-sized colony (B), containing 26,000±4,000 workers. Two curves have been fitted by eye for the data from each colony, pertaining respectively to the lowermost and the mean values.

In nature, fire ant colonies commonly maintain odour trails leading to old food finds for long periods of time. Excavation often occurs along the trails, resulting in irregular roofs over the trail and gallery networks in the soil that serve as way-stations for foraging workers. These major trails, reminiscent of the "trunk routes" described in *Monomorium pharaonis* (L.) by Sudd (1960) and others, greatly enlarge the foraging fields of the fire ant colonies.

### Specificity of the Trail Substance

Colony fragments of the fire ant species *Solenopsis geminata* (Fabricius), collected at

**Table IV. Intra- and Inter-specific Responses to Dufour's Gland Secretion in the Artificial Trail Test.**

| Source species → test species | No. of replications | No. of workers responding (range) | Duration of response (range in mins.) |
|---|---|---|---|
| *Geminata* → *geminata* | 4 | 34—40 | 4—7 |
| *Geminata* → *saevissima* | 4 | 0 | — |
| *Geminata* → *xyloni* | 2 | 5 | 5 |
| *Saevissima* → *saevissima* | 4 | 50—200+ | 5—15+ |
| *Saevissima* → *geminata* | 4 | 0—3 | 1/2—3 |
| *Saevissima* → *xyloni* | 0 | — | — |
| *Xyloni* → *xyloni* | 2 | 16—51 | 12—19 |
| *Xyloni* → *geminata* | 1 | 52 | 20 |
| *Xyloni* → *saevissima* | 3 | 100—200+ | 3—5 |

Palmar, Costa Rica, and *S. xyloni* (McCook), collected at Loma Linda, California, were used in tests with *S. saevissima* to determine specificity of the Dufour's gland secretions among these closely related species. Bioassays were based on the artificial trail test described elsewhere (Wilson, 1959). The results, presented in Table IV, show conclusively that the secretion of *geminata* is chemically different from those of *saevissima* and *xyloni.* Furthermore, the secretions of *saevissima* and *xyloni* differ in their effect on *geminata* in the limited trials made.

The Dufour's gland secretions of other ant species have also been tested in *Solenopsis saevissima* and vice versa. The secretion of *saevissima* caused no reaction in the distantly related "thief ant", *Solenopsis* (*Diplorhoptrum*) *molesta* (Say), and similar negative responses were obtained in the myrmicines *Myrmica rubra* (L.), *Pogonomyrmex badius* (Latr.), *Crematogaster lineolata* (Say), and *Xenomymex floridanus* Emery, and dolichoderines *Iridomyrmex humilis* Mayr, *Monacis bispinosa* (Olivier), and *Tapinoma sessile* (Say). The Dufour's gland secretions of the myrmicine species listed above were tested on *Solenopsis saevissima*, also with negative effect. However, a most surprising discovery was made when the secretion of *Monacis bispinosa* was tested on *S. saevissima.* In repeated tests it was found that a single *Monacis* Dufour's gland smeared out into an artificial trail attracted approximately as many *S. saevissima* workers as a single *S. saevissima* gland, over the same period of time. However, the *saevissima* gland does not cause a similar response in *Monacis* workers. The latter are known to emit the trail substance from another organ, the ventral scent gland (Wilson & Pavan, 1961). The specificity experiments have not been extended beyond the species listed here.

## The Artificial Induction of Mass-Foraging Behaviour

If recruitment communication in *Solenopsis saevissima* is exclusively by means of a chemical releaser deposited in the trail, as the accumulated evidence seems to suggest, then it should be possible to evoke a full mass-foraging response from a laboratory colony solely by means of artificial trails made from Dufour's glands. That this is indeed the case is shown in the following experiment.

A freshly killed *Naupheta* roach was placed at a distance of a metre from the nest in a spot not being patrolled at the moment by workers. A thin artificial trail was then laid from the edge of the roach to the nest entrance, without allowing the applicator to touch workers during the inscription. In five replications of this procedure, the response was always the same. The great majority of workers within several millimetres distance of the freshly laid trail turned toward it and followed it at once in a direction away from the nest. Their initial response and movement along the trail differed in no detectable way from normal trail-following. When they reached the roach, they responded in the typical manner shown following the discovery of new food finds (q.v.), and when they departed they returned along the artificial trail, reinforcing it with their own secretions. In other tests workers led along artificial trails lacking rewards milled about in confusion at the end of the trail or continued on past the end a short distance before turning homeward, in movement patterns typical for natural dead-end trails. Workers could be led along tortuous paths in this manner, and even directed to form complete circles.

## Summary

1. Recruitment in mass foraging is organized almost exclusively by the use of the trail substance, which is secreted by Dufour's gland and released through the extruded sting. The secretion is a powerful attractant, which excites workers and draws them out of the nest in the direction indicated by the trail. No other stimuli are required to induce this behaviour. The initial build-up of workers at food masses is exponential. The build-up decelerates towards a limit as workers become crowded on the food mass because (1) workers unable to reach the mass turn back without laying trails, and (2) the trail deposits of single workers evaporate within a few minutes. As a result, the number of workers at food masses tends to stabilize at a level which is a function of the area of the food mass. But the nature of the communication requires that the feed-back system have a lag of up to several minutes. The lag is measurable in the "overshoot" of numbers of approaching workers and is a function of the distance of the food mass from the nest.

2. Trail-laying, as measured by the extrusion of the sting, is an all-or-none response of individual workers; hence individual trails do not communicate quantity or quality of the food find. Quantity of food is communicated in the mass response by means of the negative feed-back control of build-up just described. Quality is communicated by means of an "electorate" mass response, in which individuals choose whether to lay trails after inspecting the food find. The more desirable the food find, the higher the percentage of positive responses, the more the trail substance presented to the colony, and hence the more the newcomer ants that emerge from the nest.

3. The trail substance is highly species-specific. In transposition experiments utilizing the fire ant species *Solenopsis saevissima*, *S. geminata*, and *S. xyloni*, it was found that the *geminata* secretion is certainly different from that of the other two species. The secretions of *saevissima* and *xyloni* are more interchangeable but appear to differ somewhat. The thief-ant *S. molesta* and members of other myrmicine genera tested also do not share the *saevissima* trail substances. Curiously, the Dufour's gland secretion of the dolichoderine species *Monacis bispinosa* causes a full response in *saevissima* workers, but *M. bispinosa* uses another substance for its trails.

4. The foraging and homing behaviour of individual workers is partially analysed. An unexpected response, in which workers are attracted and lay trails to unexplored areas near the nest, is described.

## REFERENCES

Brian, M. V. (1951). Ant culture for laboratory experiment. *Entomologist's Monthly Mag.*, **87,** 134-136.

Brown, W. L. (1959). Releasers of alarm behavior in army ants. *Psyche*, **66,** 25-27.

Butenandt, A., Linzen, B. & Lindauer, M. (1959). Über einen Duftstoff aus der Mandibeldrüse der Blattschneiderameise *Atta sexdens rubropilosa* Forel. *Arch. anat. microscop. morph. Exp.*, **48,** 13-19.

Callahan, P. S., Blum, M. S. & Walker, J. R. (1959). Morphology and histology of the poison glands and sting of the imported fire ant (*Solenopsis saevissima* v. *richteri* Forel). *Ann. ent. Soc. Amer.*, **52,** 573-590.

Carthy, J. D. (1951). The orientation of two allied species of British ants. II. Odour trail laying and following in *Acanthomyops* (*Lasius*) *fuliginosus*. *Behaviour*, **3,** 304-318.

Karlson, P. & Butenandt, A. (1959). Pheromones (ectohormones) in insects. *Ann. Rev. Ent.*, **4,** 39-58.

Macgregor, E. C. (1948). Odour as a basis for oriented movement in ants. *Behaviour*, **1,** 267-296.

Sudd. J. H. (1957). A response of worker ants to dead ants of their own species. *Nature*, **179,** 431-432.

Sudd. J. H. (1960). The foraging method of Pharaoh's ant, *Monomorium pharaonis* (L.). *Animal Behaviour*, **8,** 67-76.

Thorpe, W. H. (1956). *Instinct and learning in animals.* Methuen.

Wilson, E. O. (1958). A chemical releaser of alarm and digging behavior in the ant *Pogonomyrmex badius* (Latreille). *Psyche*, **65,** 41-51.

Wilson, E. O. (1959). Source and possible nature of the odor trail of the fire ant *Solenopsis saevissima* (Fr. Smith). *Science*, **129,** 643-644.

Wilson, E. O. & Eisner, T. (1957). Quantitative studies of liquid food transmission in ants. *Insectes Sociaux*, **4,** 157-166.

Wilson, E. O., Durlach, N. & Roth L. M. (1959). Chemical releasers of necrophoric behavior in ants. *Psyche*, **65,** 108-114.

Wilson, E. O. & Pavan, M. (1959). Source and specificity of chemical releasers of social behavior in the dolichoderine ants. *Psyche*, **66,** 70-76.

(*Accepted for publication* 1*st January*, 1961).

# Sound Production and Its Evolutionary Significance in the Blattaria[1]

LOUIS M. ROTH AND H. BERNARD HARTMAN[2]

Pioneering Research Division, U.S. Army Natick Laboratories, Massachusetts 01760

ABSTRACT

Cockroaches produce sounds by rubbing the abdomen against the wings, tapping the substrate, striking the wings against the abdomen, expelling air through the spiracles, and scraping the pronotum over the costal veins (stridulation). Some of these sounds are made by both sexes when the insects are "disturbed" or by males when they court. Two methods of sound production, namely, stridulation and expulsion of air through the second abdominal spiracles, have evolved in the Oxyhaloinae. Stridulating structures have been found in several genera of the ovoviviparous Oxyhaloinae (*Nauphoeta, Leucophaea, Henschoudetenia, Oxyhaloa, Jagrehnia*) and 1 genus of Panchlorinae (*Panchlora*) of the Blaberidae. The hypothesis is presented that stridulating structures in the Oxyhaloinae first evolved, in both sexes, on those parts of the pronotum and tegmina that were rubbed together during struggling movements made when the insects were seized by a predator; originally the sound produced, after stridulating structures evolved, may have had a defensive function. Stridulation by males during courtship has so far been observed only in *Nauphoeta cinerea* (Olivier). Various behavior patterns of males of different blaberids suggest that vibratory and pumping movements, possibly employed to disseminate a pheromone, led to the production of rhythmical sounds in *N. cinerea* which may have sexual meaning to the female.

Relatively little has been done on sound production in cockroaches (Chopard 1938; Frings 1955, 1956; Kevan 1954, 1955; Dumortier 1963). Recently, Dumortier (1965) studied the hissing apparatus in *Gromphadorhina,* and Hartman and Roth (1967a, b) reported on stridulation during courtship behavior in *Nauphoeta cinerea* (Olivier). In this paper we report the results of a study of sound produced by several species of ovoviviparous cockroaches (Blaberidae).

## MATERIALS AND METHODS

The insects were reared on Purina laboratory chow at 25°–27°C and 45%–55% RH. The methods of recording and determining sound intensity are given in Hartman and Roth (1967a, b). When recording, no effort was made to control the temperature, which ranged from 25° to 29°C. Whole mounts of pronota and tegmina were prepared to study the stridulating structures. Darkly pigmented specimens were first treated with 10% KOH, dehydrated, cleared, and mounted in Permount. The number of striae indicated in the descriptions refers to 1 side of the insect only, unless otherwise stated.

## RESULTS

Many cockroaches produce sounds as part of their normal behavior. In general, sounds are made by one or more of the following means: (1) rubbing the abdomen against the wings, (2) tapping the substrate, (3) striking the wings against the abdomen, (4) expelling air through the spiracles, or (5) scraping the pronotum over the costal veins (stridulation). Each of these sound-production methods will be discussed separately.

### I. *Sounds not Associated with Specialized Structures*

A. *Rubbing the Abdomen Against the Wings.*—Main (1932) noted that a species of *Blaberus* would make "a distinct short hissing sound" when its tegmina were stroked. Main stated that the sound apparently was produced when the abdominal segments were simultaneously contracted toward the thorax. Main was correct in his description of the abdominal movement when *Blaberus* produces sound. We have observed that several species of blaberid cockroaches produce a sound, when restrained, by extending the abdomen posteriorly and then rubbing it anteriorly against the wings. These species are *Blaberus giganteus* (L.), *B. craniifer* Burmeister, *B. discoidalis* Serville, and *Archimandrita tessellata* Rehn. None have a pronotal-tegminal stridulating apparatus and all belong to the Blaberinae (Table 1). Contrary to Chopard (1938, p. 286), *Blaberus* evidently does not have a pronotal-tegminal stridulating apparatus.

The rhythmic disturbance sound made by *Archimandrita* is shown in Fig. 1. This method of sound production is not characteristic of all members of the Blaberinae, since *Eublaberus posticus* (Erichson), *E. distanti* (Kirby), and *Byrsotria fumigata* (Guérin) (males), when restrained, did not rub the abdomen against the wings to produce the sound. Other members of the Blaberidae which apparently do not produce sound in this manner are *Nauphoeta cinerea, Leucophaea maderae* (F.), ***Diploptera punctata*** **(Esch-**

[1] Accepted for publication September 29, 1966.

[2] Present address: Department of Zoology, University of Iowa, Iowa City, Iowa 52240.

scholtz), *Capucina patula* (Walker), *Phortioeca phoraspoides* (Walker), *Pycnoscelus surinamensis* (L.), and *P. indicus* (F.) (formerly considered the bisexual form of *P. surinamensis;* see Roth 1967).

When courting an unreceptive female, *Blaberus craniifer* performs trembling or quivering movements for brief periods after contact. These movements occur in short bursts lasting 1 or 2 sec, with a somewhat shorter rest period between successive bursts. The tegmina are held slightly off the abdomen, while the wings and/or abdomen tremble vigorously. During bursts of trembling, a soft, audible "buzzing" sound is produced (Barth 1961). It is possible that this sound is produced by rubbing the abdomen against wings.

During courtship, the male of *Eublaberus posticus* pulses the abdomen back and forth and lifts the wings off the abdomen. Both the tegmina and abdomen vibrate and the male may butt the female with his head. The abdomen is curved downward slightly (wings horizontal) and is not much extended during the pumping motions; the tegmina vibrate at about the same frequency. A not very rhythmical rustling sound (Fig. 2) is produced. We are not certain how this sound is produced.

B. *Tapping the Substrate.*—Some species of blaberids produce sounds by striking the abdomen against the substrate during courtship. The sound produced by males of *Leucophaea maderae* during courtship is probably the oldest and best documented example of a cockroach sound made by tapping (see Dumortier 1965). Details of this behavior are given by Barth (1961) and Roth and Barth (1967). Often during antennal fencing with the female, the male of *L. maderae* performs repeated high-frequency, small-amplitude, vertical jerks of the body; the abdomen moves down as the head and prothorax move upward. An audible tapping sound may be associated with this movement. Vertical vibration occurs in bursts lasting about 5 sec, though this apparently may vary with the receptivity of the female. The bursts are interspersed with brief rest periods, each lasting a few seconds. The wings may be raised up to 30° off the abdomen during later stages of vertical vibration. If the female is receptive, the male's vibrating movement is usually omitted, and he raises his wings to an angle of 60°–80° above the horizontal. Unreceptive females may be induced to mount the male after a prolonged courtship involving vertical vibration on the part of the male. However, mating does not occur even though the female may feed

Table 1.—Species of Blattaria[a] which lack stridulatory structures on the pronotum and tegmina.[b]

BLATTOIDEA

Blattidae

Blattinae.—*Blatta orientalis* L.; *Deropeltis erythrocephala* (F.) (♂); *Periplaneta americana* (L.); *P. australasiae* (F.); *P. brunnea* Burmeister; *P. fuliginosa* (Serville); *P. japonica* Karny (♂)

Polyzosteriinae.—*Eurycotis decipiens* (Kirby); *E. floridana* (Walker); *Methana convexa* (Walker); *Pelmatosilpha coriacea* Rehn; *Temnelytra truncata* (Brunner)

BLABEROIDEA

Polyphagidae

*Arenivaga* (*Arenivaga*) sp.; *Arenivaga* (*Psammoblatta*) *africana* L.; *A.* (*P.*) *cerverae* (Bolívar); *compsodes schwarzi* (Caudell); *Polyphaga aegyptiaca* (L.); *Therea petiveriana* (L.)

Blattellidae

*Aneurita viridis* (Hebard) (♀); *Arawakina frontalis* Hebard (♂); *Balta patula* (Walker) (♀); *B. similis* (Saussure) (♀); *Burchellia circumcincta* (Reiche and Fairmaire) (♂); *Ceratinoptera picta* Brunner (♀); *Euandroblatta propera* Rehn (♀); *Eunyctibora crassicornis* (Burmeister) (♂); *Maretina marquesana* Hebard (♂); *M. uahuka* Hebard (♂); *Megamareta phaneropyga* (Chopard) (♀); *Moluchia castanea* (Blanchard) (♀); *Nahublattella nahua* (Saussure) (♀); *Paratemnopteryx australis* Saussure (♀); *Pseudothyrsocera rectangularivittata* (Brunner) (♂); *Symplocodes ridleyi* (Shelford) (♂); *Xosablatta natalensis* (Rehn) (♀)

Anaplectinae.—*Anaplecta* sp.

Plectopterinae.—*Amazonina* sp. nov.; *Cariblatta lutea* (Saussure and Zehntner); *C. minima* Hebard; *C.* sp. nov.; *Chorisoneura apolinari* Hebard (♀); *Dendroblatta sobrina* Rehn; *Euphyllodromia angustata* (Latreille) (♂); *Euthlastoblatta compsa* Hebard (♀); *Latiblattella* sp. nov. (♀); *Lophoblatta arawaka* Hebard (♀); *L.* sp. nov.; *Neoblattella dryas* Rehn and Hebard (♀); *Onychostylus notulatus* (Stål); *Supella longipalpa* (F.) (formerly *S. supellectilium* (Serville))

Blattellinae.—*Blattella germanica* (L.); *B. humbertiana* (Saussure); *B. unicolor* (Brunner); *Ischnoptera deropeltiformis* (Brunner); *I. panamae* Hebard; *I. rufa rufa* (Degeer); *Parcoblatta bolliana* (Saussure and Zehntner) (♂); *P. pensylvanica* (Degeer); *P. uhleriana* (Saussure) (♂); *P. virginica* Brunner (♂); *Pseudomops intercepta* (Burmeister) (♂); *P. septentrionalis* Hebard (♂); *Shawella couloniana* (Saussure); *Symploce hospes* (Perkins); *Xestoblatta immaculata* Hebard

Ectobiinae.—*Ectobius lapponicus* (L.) (♀); *E. pallidus* (Olivier) (♀); *Phyllodromica* (*Luridiblatta*) *trivittata* (Serville) (♂)

Nyctiborinae. — *Megaloblatta blaberoides* (Walker) (♂); *M. regina* (Saussure); *Nyctibora azteca* Saussure and Zehntner (♂); *N. laevigata* (Beauvois) (♂); *N. tetrasticta* Hebard (♂); *N. tomentosa* Serville (♂)

Table 1.—Species of Blattaria[a] which lack stridulatory structure on the pronotum and tegmina.[b] (Continued.)

Blaberidae

Zetoborinae.—*Phortioeca phoraspoides* (Walker)

Blaberinae.—*Archimandrita tessellata* Rehn; *Blaberus craniifer* Burmeister; *B. discoidalis* Serville; *B. giganteus* (L.); *Byrsotria fumigata* (Guérin); *Eublaberus distanti* (Kirby); *E. posticus* (Erichson)

Panesthiinae.—*Panesthia australis* Brunner (♂); *P. laevicollis* Saussure (♀)

Pycnoscelinae.—*Pycnoscelus indicus* (F.)

Diplopterinae.—*Diploptera punctata* (Eschscholtz)

Panchlorinae.—*Capucina patula* (Walker)

Oxyhaloinae. — *Heminauphoeta hova* Saussure and Zehntner (♀); *Oxyhaloa deusta* (Thunberg) (♀)

Epilamprinae.—*Ataxigamia tatei* Tepper (♂); *Audreia jamaicana* Rehn and Hebard (♀); *Laxta granicollis* (Saussure) (♂); *Litopeltis bispinosa* (Saussure) (♂)

Perisphaeriinae.—*Gyna capucina* Gerstaecker (♂); *G. sculpturata* Shelford (♂)

[a] Classification principally according to McKittrick (1964, and personal communication) and Princis (1960).

[b] Females of the Polyphagidae listed (except *T. petiveriana*) lack tegmina and therefore only males were examined. Otherwise, where no sex symbol appears, both sexes were examined and found to lack stridulatory structures.

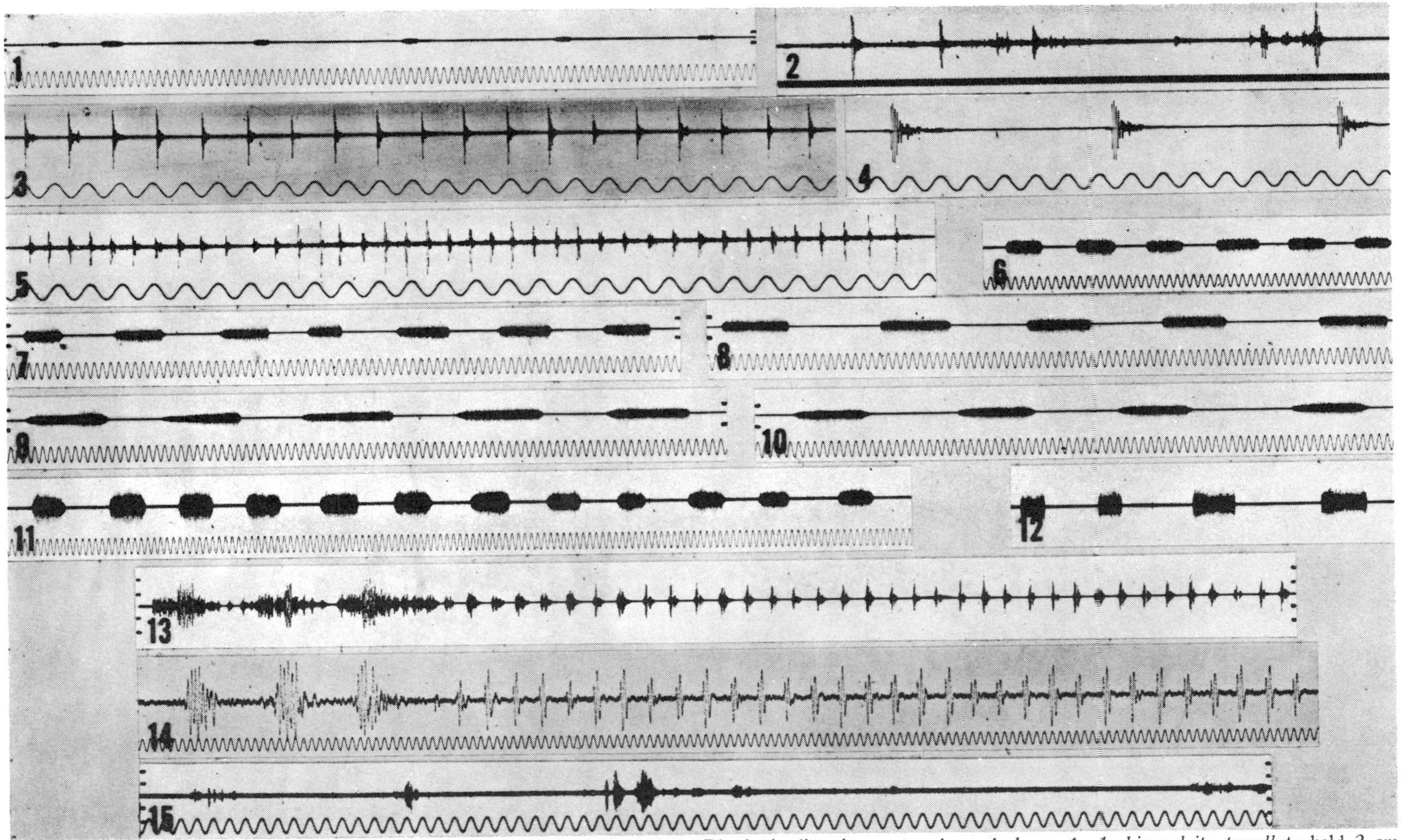


PLATE 1.—Oscilloscope records of sounds produced by cockroaches. FIG. 1.—Rhythmic disturbance sounds made by a ♂ *Archimandrita tessellata* held 2 cm from the microphone. FIG. 2.—Arhythmic rustling sound made by a courting ♂ *Eublaberus posticus.* FIG. 3.—Sound produced by a ♂ *Leucophaea maderae* tapping upon the substrate, which, in this case, was a ♀ upon which the ♂ was standing. FIG. 4.—Same record as in Fig. 3, filmed at higher speed. FIG. 5.—Courting sounds produced by a ♂ *Diploptera punctata* by striking the wings against the abdomen.

FIG. 6–12.—Disturbance hisses made by species of *Gromphadorhina.* 6, *G. javanica* ♂. 7–8, *G. portentosa* ♂ and ♀, respectively, held about 15 cm from microphone. 9–10, *G. brunneri* ♂ and ♀, respectively, held about 15 cm from microphone. 11–12, *G. chopardi* ♂ and ♀, respectively.

FIG. 13–15.—*Nauphoeta cinerea.* 13, phrase produced by stridulation during courtship by a ♂ at 1 cm from microphone. 14, substrate recording of a phrase produced by a ♂ during courtship stridulation. 15, disturbance sound produced by the ♂ while being handled 3–5 cm from microphone.

Sound-level reference marks are 65 db in Fig. 1 and 15; 70 db in Fig. 7–10; and 60 and 65 db in Fig. 13. Reference signals are 20 Hz for Fig. 1 and 5–14; 100 Hz for Fig. 2; 10 Hz for Fig. 3 and 15; and 50 Hz for Fig. 4.

briefly on the tergal gland secretion (Barth 1961, Roth and Barth 1967). The tapping sound is very rhythmical (Fig. 3, 4) and is made when the abdomen strikes the substrate. However, the pumping rhythm is still audible even though the abdomen fails to contact the substrate. In the example illustrated, the male was standing atop the female and used her as the substrate. The tap rate in this case was 7/sec. It is of interest that *isolated* males may perform vibratory movements and tap the substrate.

The males of *E. posticus* also may make a tapping sound during courtship, apparently when the end of the abdomen strikes the substrate. However, the vibrating motion is not as marked as in *Leucophaea*. Barth (1964) observed males of *Byrsotria fumigata* tapping during courtship.

C. *Striking the Wings Against the Abdomen.*—During courtship, the male of *Diploptera punctata* raises its tegmina and may flutter its wings (Roth and Willis 1955, Stay and Roth 1958). According to Barth (1961), the tegmina are pumped up and down "while the wings are fluttered, producing a barely audible whirring sound. This display may last several seconds and may be repeated a number of times, particularly if the female is relatively unresponsive." We have observed this behavior under a microscope using stroboscopic lighting and we have also recorded the sound, which is quite rhythmic (Fig. 5). The tegmina are raised off the abdomen and held slightly apart. The wings are not moved in a scissorlike motion; rather, they remain folded and move up and down, producing the sound principally by striking the abdomen. The fluttering rate depicted (Fig. 5) is about 29/sec. Courting males whose tegmina have been removed can still produce the sound.

### II. *Sounds Produced by Specialized Morphological Structures*

A. *Spiracular Sound-Producing Organs.*—The cockroaches known to produce sound by expelling air through the second abdominal spiracles are wingless forms of the genus *Gromphadorhina*. They are endemic in Madagascar. One species, *G. javanica* Hanitsch, was erroneously reported to have come from Java (Hanitsch 1930), but it also is found in Madagascar (Princis 1965). Because of the intense, characteristic hissing sound produced by these insects, they are called "blowers" by the Europeans of Madagascar, "Sanda" in Veza and Mahafaly dialects (Wintrebert *in* Dumortier 1965), and "Kofokofoka" in the high-plateau dialect (Chopard 1950). The morphology of the tracheal sound-producing device has recently been described and is present in *G. coquereliana* Saussure, *G. portentosa* (Schaum), and *G. brunneri* Butler (Dumortier 1965). We have found that *G. javanica* and *G. chopardi* Lefeuvre also produce a hissing sound when handled.

In *G. brunneri* (Dumortier 1965) and *G. portentosa* (Barth 1961), males hiss in defense of territory, and the latter hisses when it courts the female (Barth 1961). The female of *G. brunneri* hisses only when disturbed (Dumortier 1965), and *G. portentosa* does so particularly if disturbed when young are present (Roth and Willis 1960).

Fig. 6–12 are records of hisses produced by 4 species of *Gromphadorhina* when disturbed. These insects were held about 15 cm from the microphone, and were encouraged to hiss by the operator's pulling a leg with a pair of forceps. The *G. javanica* male in Fig. 6 hissed at the rate of 2.5/sec, with a hiss duration of 240 msec. The females of *G. javanica* do not hiss readily.

Both sexes of *G. portentosa* produce disturbance sounds readily. Their sounds, shown in Fig. 7 and 8, are produced at a rate of about 2/sec by the male and 2.5/sec by the female. The hiss durations are 320 msec and 400 msec, respectively, and the intensity in both cases exceeds 65 db.

Disturbance sounds by *G. brunneri* are shown in Fig. 9 and 10. The male hisses at the rate of 1.3/sec, the hiss having a 520-msec duration. The female produces hisses at the rate of 1/sec, and the hisses have a 480-msec duration. The intensity of these sounds also exceeds 65 db.

The *G. chopardi* male disturbance sounds are delivered at the rate of 2.5/sec, with each hiss having a 220-msec duration. However, the female does not hiss readily, and when she does, the hisses are not so uniformly spaced as in the male (cf. Fig. 11 and 12).

B. *Sound Production by Distinctive Stridulating Structures.*—Stridulation is known to occur in both sexes of *L. maderae, N. cinerea* and *Henschoudetenia epilamproides* (Shelford). The stridulating apparatuses in *Leucophaea* and *Nauphoeta* are very similar, and consist of parallel striae on the ventral lateroposterior margins of the pronotum and on the dorsoproximal regions of the costal veins of the tegmina.

Vosseler (1907), who first described the stridulating apparatus of *L. maderae* (Fig. 21–24), claimed that *Leucophaea* stridulated only when captured or "frightened." He concluded (after studying dead specimens) that the sound is produced when the insect rubs the pronotum sidewards against the proximal region of the costal veins. In the male, there are about 50 striae on the pronotum (2 specimens) spaced about 4$\mu$ apart, and between 700 and 900 striae on the costal vein (3 specimens). Nothing is known about stridulation in this species other than that they stridulate when disturbed. We have not heard them stridulate during courtship.

Hartman and Roth (1967a, b) reported on courtship stridulation by *N. cinerea*. The male possesses about 40 striae on the pronotum (Fig. 16–18, 20) and about 400 on the costal vein of the tegmen (Fig. 19). The striae on both structures are about 4$\mu$ apart. Both sexes produce disturbance sounds (Fig. 15). The arhythmic disturbance sounds illustrated were made by a male held 3–5 cm from the microphone. These noises were quite loud, exceeding 70 db, and are the result of relatively uncoordinated lateral pivoting of the pronotum across the costal veins. However, the male produces a patterned call (Fig. 13) by coordinated movements when it courts an unreceptive female (Fig. 33, 34). This sound is

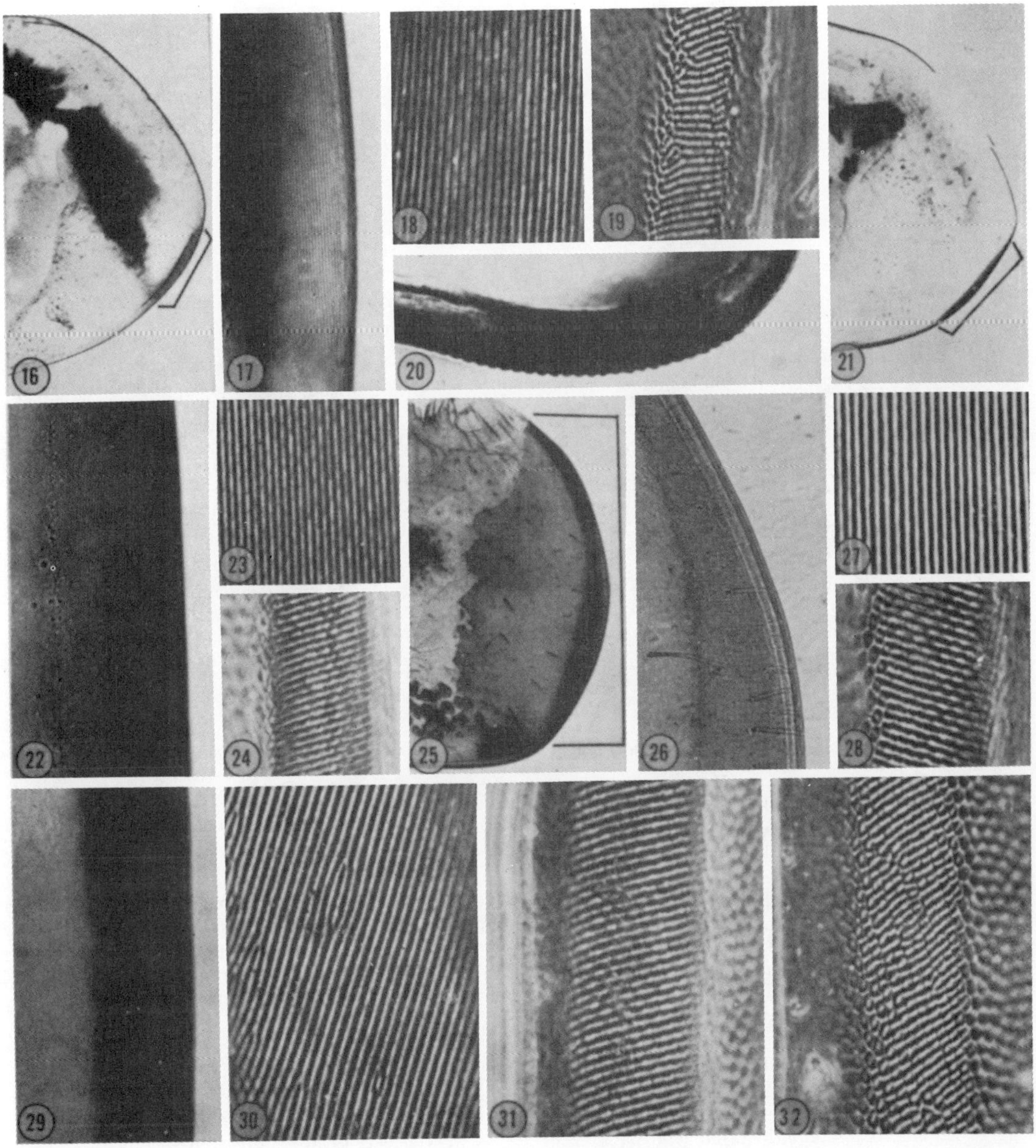

FIG. 16–32.—Stridulating structures on the pronota and tegmina of *Nauphoeta, Leucophaea, Henschoudetenia* and *Jagrehnia*.

FIG. 16–20.—*Nauphoeta cinerea* ♂, from Natick culture. 16, right side of pronotum; the striae are located in bracketed pigmented area, ×8.2. 17, portion of striae on pronotum, ×102. 18, portion of striae on pronotum, ×291. 19, portion of striae on tegmen, ×291. 20, cross section of the striae on pronotum, ×371.

FIG. 21–24.—*Leucophaea maderae*, from Natick culture. 21, right side of ♀ pronotum; striae are situated in bracketed pigmented area. 22, portion of striae on ♀ pronotum, ×102. 23, portion of striae on ♀ pronotum, ×291. 24, portion of striae on ♂ tegmina, ×291.

FIG. 25–28.—*Henschoudetenia flexivitta* ♀, from Natick culture. 25, portion of right side of pronotum, ×15; striae are found on bracketed pigmented area. This specimen was damaged so that anterior limit of area covered by striae is not shown. 26, portion of striae on pronotum, ×60. 27, portion of striae on pronotum, ×291. 28, portion of striae on tegmen, ×291.

FIG. 29–32.—*Jagrehnia gestroiana* ♀, from Somalia. 29, portion of striae on pronotum, ×102. 30, portion of striae on pronotum, ×291. 31, portion of striae on tegmen, ×291. 32, apical portion of striae on tegmen, near region where striae blend into wing sculpturing, ×291.

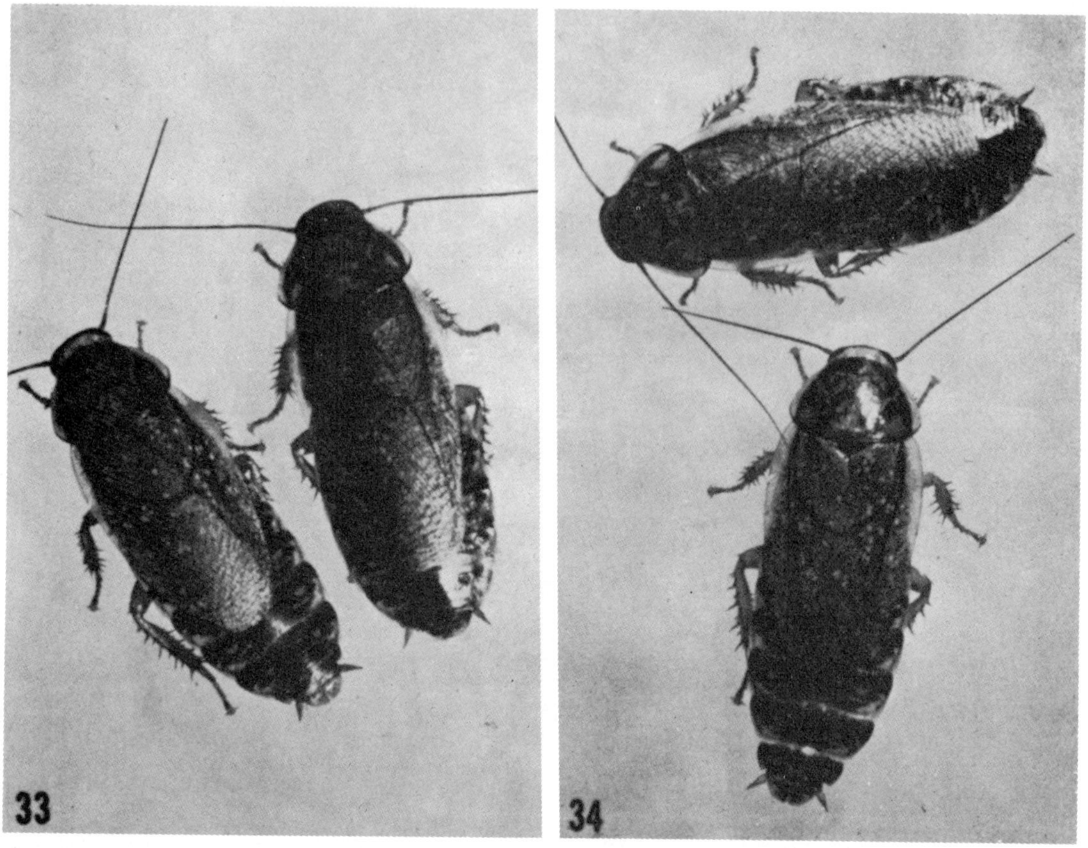


FIG. 33–34.—*Nauphoeta cinerea.* 33, male (left) prior to stridulating. 34, male (below) stridulating.

produced by posterior, anterior, and side-to-side displacement of the pronotum rubbing against the costal veins.

Guthrie (1966) described the sound produced by *H. epilamproides,* a species of cockroach which stridulates when it is disturbed. The stridulatory apparatus is similar to that found in *Leucophaea, Nauphoeta,* and *Henschoudetenia flexivitta* (Walker) (see following). According to Guthrie, the striae are about 5$\mu$ apart. The sound intensity rises to about 70 db and there is a carrier wave form of about 5 kc/s, amplitude-modulated to form pulses at 1–2 kc/s. The pulses form chirps of 100 msec or longer, and are simple or complex in structure. He noted that the recorded cockroach sound caused wild bank voles to run away.

Both sexes of *Panchlora nivea* (L.) possess a stridulating apparatus. The striae on the pronotum are often difficult to see, and many do not run the full length of the posterior lateropronotal margins (Fig. 38). One male had 14 striae on one side and 15 on the other, the striations being about 3$\mu$ apart. The rodlike structures (Fig. 35, 36) on the tegmina are arranged just below the costal vein. One female of *P. nivea* had 50 striae, and a male had 35, the spacing in both individuals being about 19$\mu$–21$\mu$.

We have observed *P. nivea* males stand and vibrate their bodies near a female while courting. However, we have been unable to hear or record any sound (Hartman and Roth 1967b) even though the movements are very similar to those of a stridulating male of *N. cinerea. Panchlora irrorata* Hebard performs similar vibratory motions when courting the female (Willis 1966) as a preliminary to successful mating. Both sexes of *P. irrorata* possess stridulating structures (Fig. 37, 39). A male had 26 striae on each tegmen and a female had about 30. Very fine striae are also present on the pronotum (Fig. 39). The striae on the tegmina are not so well differentiated as in *P. nivea* (cf. Fig. 35, 36).

A survey was made of museum specimens of Blattoidea and other Blaberoidea for the presence of stridulating structures on the pronotum and tegmina. They were absent in all Blattidae, Polyphagidae, and Blattellidae examined, as well as 21 Blaberidae (Table 1), but they were present in 3 genera belonging to the Oxyhaloinae.

The striae on the pronotum and tegmina of the *Jagrehnia gestroiana* (Saussure) (Fig. 29–32) female are very similar to those found in *Nauphoeta, Henschoudetenia,* and *Leucophaea.* The striae on the pronotum of *Jagrehnia* differ from those of the othe-

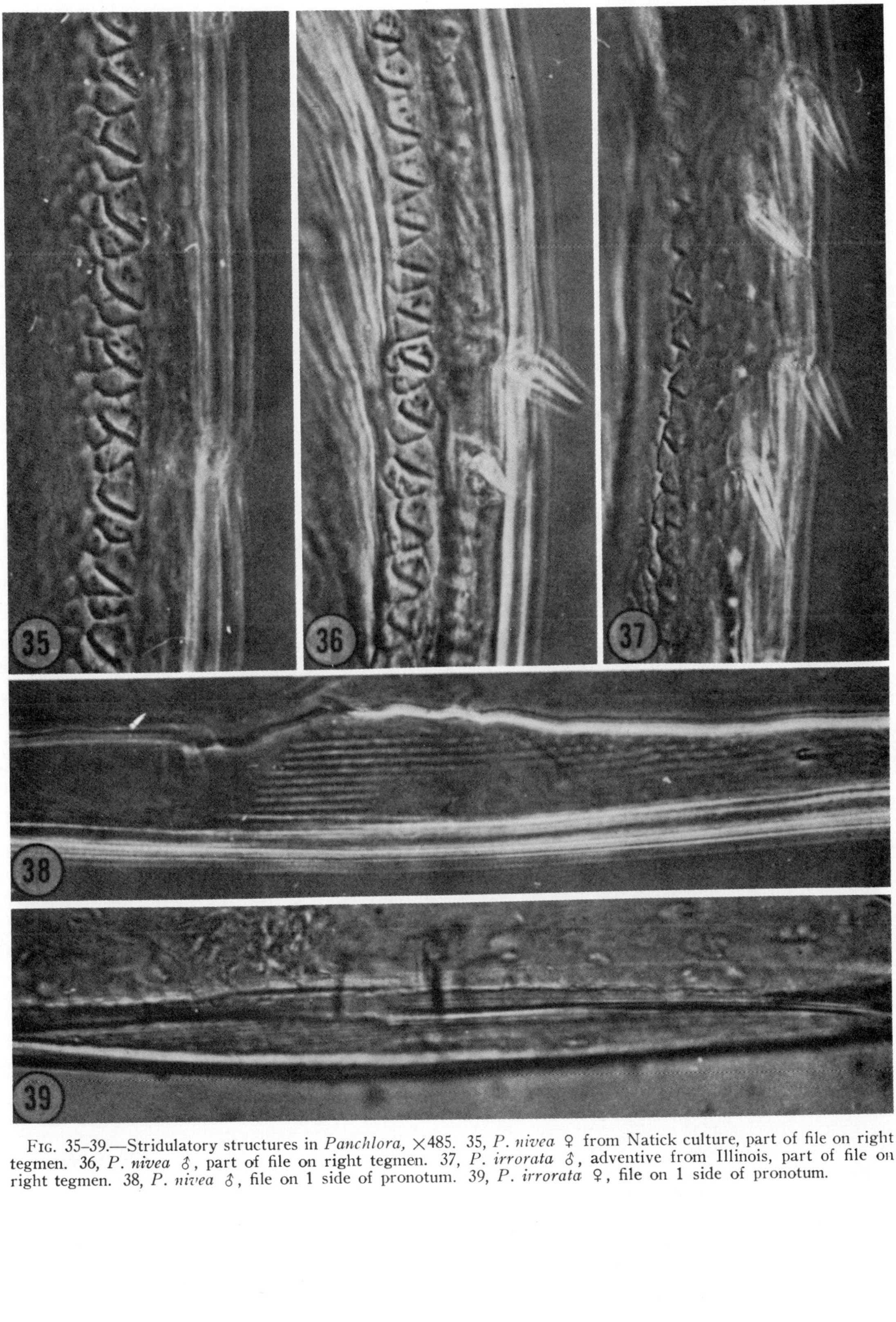

FIG. 35–39.—Stridulatory structures in *Panchlora*, ×485. 35, *P. nivea* ♀ from Natick culture, part of file on right tegmen. 36, *P. nivea* ♂, part of file on right tegmen. 37, *P. irrorata* ♂, adventive from Illinois, part of file on right tegmen. 38, *P. nivea* ♂, file on 1 side of pronotum. 39, *P. irrorata* ♀, file on 1 side of pronotum.

genera in that they are shorter, run obliquely (cf. Fig. 29 with Fig. 17, 22, and 26), and are much more numerous.

The pronotal-tegminal structures (Fig. 25–28) of the female of *H. flexivitta* are typical of those previously described, except that the striae cover a great part of the lateral margins of the pronotum; these striae extend over the anterior as well as posterior-lateral pronotal margins (Fig. 25). The female that was examined had about 65 striae on the pronotum, spaced approximately 4.7$\mu$ apart, and about 500 striae on the tegmen.

The stridulating apparatus on the tegmina of the *Oxyhaloa* species examined (Fig. 40–51) differs markedly from those previously described. The striae occur in distinct, separated patches of tubercles or crests along side the proximal region of the costal veins.

The single male specimen of *Oxyhaloa buprestoides* (Saussure) had 4 distinct patches on the left side (Fig. 40) and 5 on the right side (Fig. 41). The size of each group of striae is variable. Two females of this species were examined. One had 2 groups of striae on the left side (Fig. 42, 44), the second group smaller than the first. The same number of groups of striae are found on the right side (Fig. 43, 45) but the posterior group is relatively poorly developed. Another female also had 2 groups of striae on each tegmen (Fig. 46, 47).

Another *Oxyhaloa* species (probably *ferreti* (Reiche and Fairmaire)) had stridulating structures. On the male there are 2 well-developed groups of tubercles on the left tegmen (Fig. 48) and 3 groups on the right (Fig. 49). The right tegmen of the female has 2 groups of tubercles (Fig. 52) and, as in *O. buprestoides,* the second group is smaller than the first. There are no distinctive tubercles on the left tegmen of this female, but several peculiar aberrant patches (Fig. 50, 51) are present.

A female of *Oxyhaloa deusta* (Thunberg) lacked stridulating structures on the tegmina.

In both species of *Oxyhaloa* that had tegminal striae, the pronotum apparently lacks a stridulating structure, but it is conceivable that the edge of the pronotum may be used to rub against the tegmen. In *N. cinerea* the male can produce its courting sound, although the pronotal striae are cut off; in this experimental situation, the edge of the cut surface of the pronotum rubs against the tegmen (Hartman and Roth 1967b). Unfortunately, no observations of sound production have been made on living *Oxyhaloa.*

Of the large number of species and genera examined that lacked stridulating structures (Table 1), only 2 species had modifications of the tegmina (but not on the pronota) which were suggestive of striae. In a male of *Catara rugosicollis* (Brunner) (Blattidae) the tegminal sculpturing on the basal region of the costal vein was arranged to form faint parallel lines (reminiscent of *Nauphoeta,* etc.). These were difficult to see because the surface appeared smooth and the lines may simply be lightly pigmented areas (Fig. 53). In both sexes of *Gislenia australica* (Brunner) (Blattellidae: Blattellinae) the surface of the tegmina alongside the basal region of the costa is wrinkled and some areas show distinct, ridgelike elevations reminiscent of, but not so uniform as the file in *Panchlora* spp. These modifications are more distinct in the female than in the male, but in both sexes the size and number of "files" vary from 1 specimen to the next (cf. Fig. 54, 55) and also vary on the right and left tegmen of the same specimen (Fig. 54). The tegminal modification of *Gislenia* may represent a very early stage in the evolution of striae on the tegmina of some Blaberidae.

### III. *Sound Production in Other Blattaria*

With the exception of *Catara* and *Gislenia,* all the cockroaches discussed previously in this paper belong to the highly evolved Blaberidae and are ovoviviparous or viviparous (*Diploptera*). Almost nothing is known of sound production in oviparous cockroaches. Baier (1930) suspended a microphone among a group of *Blattella germanica* (L.) and could not detect any sound production.

Kleine (1918) was awakened by a sound like that produced by a locust, but not so pure. The only insect in the room was *Blatta orientalis* L. (Blattidae: Blattinae). On another occasion, he and an entomologist visiting him heard the same sound, and again the only insect in the room was the *B. orientalis.*

Rudow (1870, p. 322) claimed that *Ectobius lapponicus* (L.) (Blattellidae: Ectobiinae), when disturbed, made a short, abrupt, clicking sound before taking to flight. He assumed that the same sound could be made by rubbing the tegmina together. Karny (in Shelford 1909) suggested that the ulnar rami of the wings in *Gyna caffrorum* (Stål) (Blaberidae: Perisphaeriinae) can produce a rattling noise when the insect flies. Karny described this species as *G. stridulans.*

According to Skaife (1954) when the wingless female of the African species *Aptera fusca* (Thunberg) (Blaberidae) is handled, it makes a squeaking sound by rubbing the roughened edge of 1 segment of the abdomen against the surface of the next segment.

Karny (1924) described a stridulating apparatus on the hind wing rudiment and first abdominal tergite of females of *Archiblatta hoeveni* Vollenhoven (Blattidae). Since the males lacked this structure and small colonies consist of "mothers" and young, Karny suggested that the sound is used to keep the "brood" together. However, Karny produced the sound in dead females only by flexing the anterior and hind ends of the abdomen ventrally, and his conclusion is purely speculative.

We have heard a female of the wingless *Euzosteria nobilis* (Brunner) (Blattidae: Polyzosteriinae) produce a faint, squeaking sound when held. Slides were made of the entire abdomen and thorax, but no stridulating structures were found. The site of sound production in this species is unknown.

## DISCUSSION

Two distinct mechanisms for sound production—

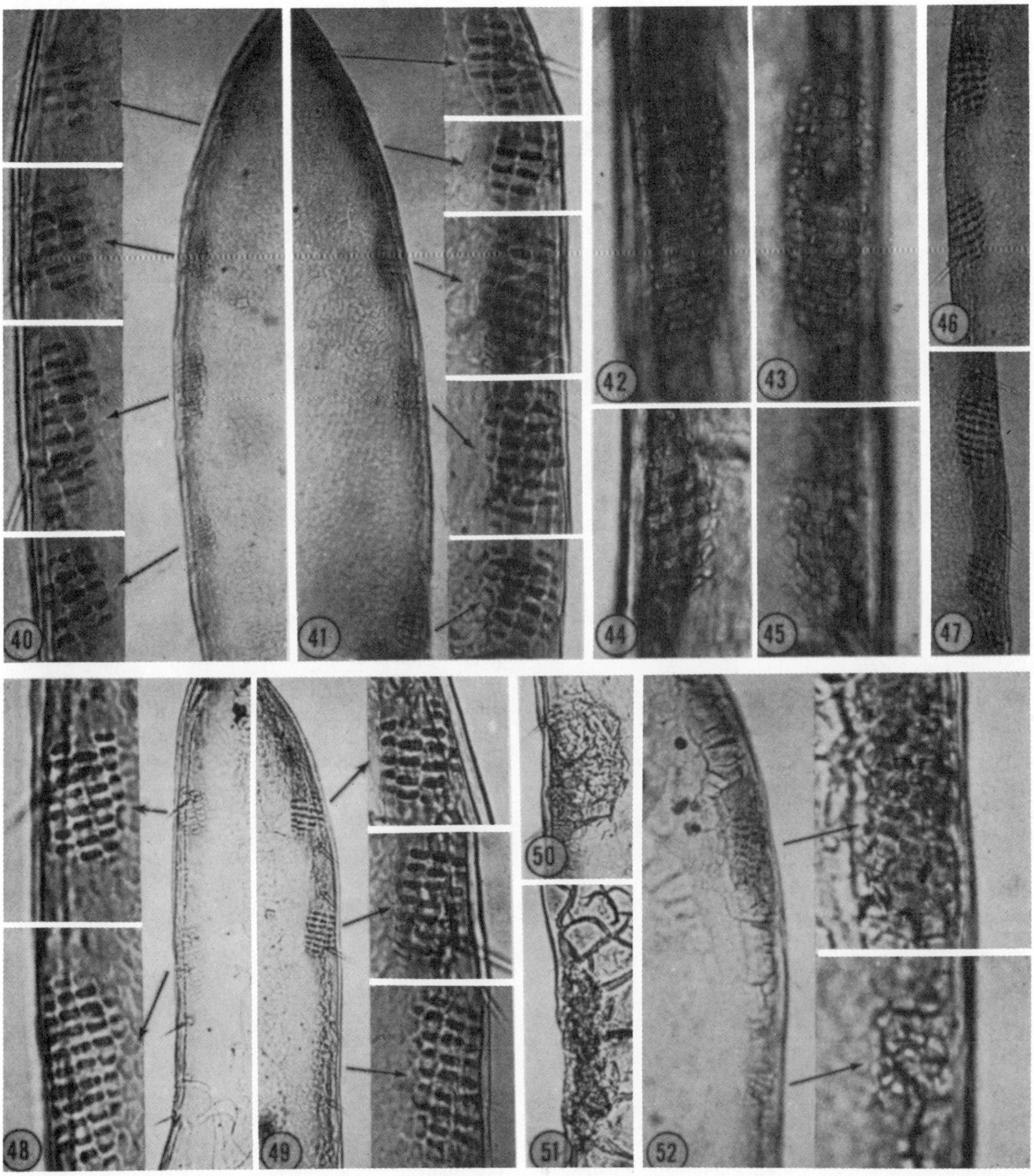

FIG. 40–47.—Stridulatory structures on the tegmina of *Oxyhaloa buprestoides*. 40, ♂ from Ruanda: Nguda (Cheff.) Terr. Nyanza; part of left tegmen, showing separated patches of striae, ×122 (arrows point to enlarged views, ×317, of respective groups of striae. 41, same as Fig. 40; right tegmen. 42–43, ♀ from Victoria, Njansa; patch 1 on left (42) and right (43) tegmina, ×317. 44–45, same as 42–43; patch 2 on left (44) and right (45) tegmina, ×317. 46–47, ♀ from Ruanda, Terr. Shangugu, Dendezi; patches on left (46) and right (47) tegmina, ×122.

FIG. 48–52.—Stridulatory structures on tegmina of *Oxyhaloa* sp. (probably *O. ferreti*). 48, ♂ from Ethiopia; part of left tegmen, showing separated patches of striae, ×123 (arrows point to enlarged views, ×317, of respective patches). 49, same as Fig. 48, right tegmen, 50–51, ♀ from Ethiopia; 2 of several aberrant patches on left tegmen. 52, same as Fig. 50–51; right tegmen, showing 2 patches, ×123 (arrows point to enlarged views, ×317, of striae).

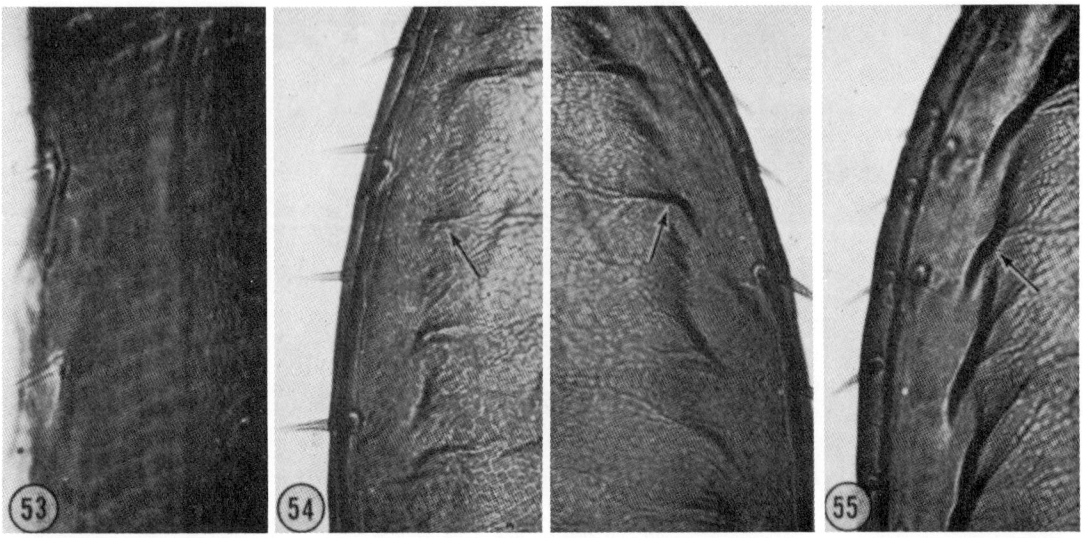

FIG. 53–55.—Portions of the tegmina of *Catara* and *Gislenia*. 53, *Catara rugosicollis* ♂; small area of proximal part of left tegmen, showing fine, light lines that appear to be striae, ×354. 54–55, *Gislenia australica*, 54, proximal regions of right and left tegmina of ♀, showing some of the ridges (arrows) below the costal veins, ×137. 55, proximal region of left tegmen of ♂, showing several thick, raised ridges, ×137.

stridulation and expulsion of air through abdominal spiracles—have evolved in the Panchloroid complex of the Blattaria. Within this complex, these mechanisms are represented only in the Oxyhaloinae and Panchlorinae. *Nauphoeta, Leucophaea, Oxyhaloa*, and *Gromphadorhina* are all members of the Oxyhaloinae (McKittrick 1964). *Henschoutedenia* and *Jagrehnia* were originally split off from the genus *Nauphoeta* (Princis 1954), and for this reason we place the former 2 genera in the Oxyhaloinae also.

The striae forming the stridulating structures appear to have evolved as thickenings or modifications of the reticulate sculpturing of the surface on which they are found. This condition is especially noticeable in areas where the striations gradually merge into the sculpturing of the surfaces of the tegmina (Fig. 32). Although the tegminal striae of *Oxyhaloa* (Oxyhaloinae) and *Panchlora* (Panchlorinae) look quite different from those found in other genera of Oxyhaloinae (i.e., *Nauphoeta, Leucophaea*, and *Henschoutedenia*), they also appear to have evolved as modifications of the tegminal surface sculpturing.

Locomotor, feeding, and cleaning movements have been suggested as having led to frictional stridulatory mechanisms in insects (see review by Haskell, 1964). We suggest that in certain species of blaberid cockroaches, struggling movements made by the insect when it was seized by a predator led to the formation of a specialized sound-producing apparatus on the pronotum and tegmina. In the Hemiptera "the chance production of sounds by the movement of one part of the body over another could have developed on the one hand into a deterrent to predators, and on the other into a means of ensuring the meeting of the sexes." (Leston and Pringle 1963.)

When highly developed stridulating structures are present in adult cockroaches, they occur in both sexes. Any hypothesis concerning the evolution of stridulation in the Blattaria must account for the presence of equally developed stridulating structures in both male and female. Although it has been suggested that the females of some Blaberidae employ sound to maintain the mother-young group (Dumortier 1965), there is no experimental evidence for this hypothesis (Hartman and Roth 1967b). It is probable that these devices for sound production originally served a nonsexual function—possibly they played a role in defense against predators. Only a relatively small part of the stridulating area is used by *Leucophaea* when "protesting" (Vosseler 1907) and *Nauphoeta* probably uses even less when stridulating during courtship. When *Nauphoeta* is physically restrained, its struggles and pronotal movements are vigorous and more stridulating surface is used in producing sound than during courtship; this probably accounts for the greater intensity of the disturbance sound, as compared with that made during courtship. The area bearing pronotal stria in *Henschoudetenia* (Fig. 25) is considerably greater than that found in the other Oxyhaloinae examined; this may represent a more primitive condition and suggests that in *Leucophaea* and *Nauphoeta* there has been a reduction in the stridulating surface area on the pronotum.

*Gromphadorhina* nymphs and adults produce sound, and it is likely that the tracheal sound apparatus also evolved as a defensive mechanism. The expulsion of air through the second abdominal spiracles could have been associated with ejection of defensive compounds. Barth (1961) claimed that an odor was associated with sound production in *G. portentosa* but Dumortier (1965) could not confirm this hypothesis in *G. brunneri*. In *D. punctata*, quinones used in defense are ejected from the second abdominal spiracles (Roth and Stay 1958, Eisner 1958). When *Diploptera* is anesthetized, it ejects its quinones and we have heard a weak squeaking sound produced (apparently originating from the second abdominal spiracles) when this occurs. When seized, *Leucophaea* gives off an offensive odor which originates from glands emptying into the second abdominal spiracle (Roth and Stay 1958); this cockroach may also stridulate at this time. Pumphrey (1955, in Dumortier 1963) suggested that the combination of sound production with the simultaneous release of an irritating chemical may cause the predator to associate a sound with an unpleasant experience; this could result in sound alone eventually becoming an effective predator deterrent. Dumortier (1965) questions the value of hissing by *Gromphadorhina* as a means of protection from natural predators. Although a dog retreated from a hissing *G. portentosa* (Roth, in Dumortier 1965) Dumortier dismisses this observation because the dog does not normally feed on this cockroach. He observed that a kinkajou and some jays ate *Gromphadorhina* in spite of hissing. However, neither the kinkajou (South American) nor jays, though both are carnivorous, could be considered natural predators of this cockroach since they do not occur in Madagascar. In this connection, it is known that even chemical defensive mechanisms may be effective against some predators but not others (Roth and Willis 1960). A study of the ecology of *Gromphadorhina* may reveal the presence of predators in its environment which are deterred by the hissing sound made by members of this genus.

How has stridulation evolved as a behavioral mechanism for courtship in the males of *N. cinerea?* Trembling or vibratory movements are characteristic of stridulation in *Nauphoeta*. This type of behavior is not uncommon in species of blaberids that lack stridulatory structures, particularly when courting males are confronted with an unreceptive female. The vibratory movements of males of *Blaberus giganteus* and *B. craniifer* become continuous and vigorous if the female is unreceptive (Barth 1961). Males of *Byrsotria fumigata* pump the abdomen and wings prior to full wing raising during courtship. Should the female be unreceptive, the wing pumping may be continued for several minutes (Barth 1964). Comparison of the behavior patterns made by different species suggests that the various wing and body movements made during courtship could be associated with dissemination of a pheromone used to induce the female to mount the male, and thus maneuver her into the proper position for copulation. The movement of the abdomen against the wings may first serve to transfer male pheromone from the terga to the wings and when the wings are finally raised during courtship or are fluttered, there is a greater dispersion of the pheromone. Males of *Diploptera punctata*, which raise the tegmina and *beat* the wings against the abdomen, would appear to have the most effective means of disseminating a pheromone.

In *N. cinerea*, the male stridulates during courtship, usually if the female is unreceptive. Stridulation apparently has evolved in *Nauphoeta* males as a means of promoting receptivity in the female. Yet we have been unable to show that male stridulation influences a nonreceptive female's behavior. Perhaps the female responds to the male's stridulation only when it is on the verge of becoming receptive. Skinner (1966) stated that "Useless structures with associated useless functions are as inevitable as superstitious ['ritualistic'] behavior." Perhaps male courtship stridulation by *Nauphoeta* is a useless function now and it may represent a stage of evolution prior to its becoming sexually meaningful to the female.

Our hypothesis of how courtship stridulation has evolved in the male of *N. cinerea* may be summarized as follows:

| Males and Females | Males |
|---|---|
| Uncoordinated struggling movements made when seized by a predator; defensive secretion ejected; stridulating apparatus absent | Vibratory movements of the body, wing and abdominal pumping associated with dispersal of a pheromone |
| ↓ | ↓ |
| [Pronotum rubs against tegmina. Striae develop on those parts of pronotum and tegmina that are rubbed together] | Coordinated rhythmic movements of the pronotum against the tegmina during courtship |
| ↓ | ↓ (also from the bracketed box at left) |
| "Disturbance" sound produced in conjunction with a defensive secretion | Stridulation when courting nonreceptive females |
| ↓ | |
| "Disturbance" sound produced for defense | |

In *Panchlora nivea* there may be no discernible preliminary courtship by the male if the female is receptive. However, if the female is unreceptive, the male behaves differently and performs vibratory movements (Roth and Willis 1958). Interestingly, the male of *P. irrorata* performs vibratory movements *prior* to successful mating with a receptive female (Willis 1966). This fact suggests that sounds or vibration of the substrate (see following) may be involved in releasing the female's sexual behavior. In both *P. nivea* and *P. irrorata*, the female does not feed on the male's terga prior to copulation; the male backs into the female to make connection, and there is no male tergal secretion. *Panchlora* probably evolved from a stock in which the males produced a tergal secretion and this genus may represent a group in which the pheromone has been lost, but the males have retained the vibratory movements which may have sexual meaning to the female.

In *Leucophaea maderae* the female is attracted to the male by odor (Smyth 1963, Roth and Barth 1967), but males may vibrate the body and tap the substrate even when isolated. This type of sound production (as well as stridulation in *Nauphoeta*) could have been the forerunner of acoustic "calling" devices to attract the female.

The fact that *Nauphoeta cinerea,* stridulates during courtship tends to support Alexander's (1956) hypothesis regarding the evolution of stridulation. In discussing the evolution of air-transmitted vibrations in the Ensifera (Tettigoniidae, Gryllidae, Gryllacridae, and Prophalangopsidae), Alexander suggested that substrate-transmitted vibrations produced by stridulation or body-jerking was an intermediate stage between proprioception and perception of air-borne sound. He stated that "some of the structural and functional precursors of tegminal stridulation must have been present in the common ancestor of the Ensifera ever since divergence from cockroaches." He also concluded that "tegminal stridulation became acoustically effective as a side effect of communicative vibrations and established the selective context for the later appearance of auditory function in proprioceptive tibial organs." Stridulating structures and stridulation in the Blattaria have been found definitely only in the highly evolved ovoviviparous Blaberidae (Oxyhaloinae and Panchlorinae) and they are probably late acquisitions. In Ensifera tegminal stridulation, both tegmina rub against each other to produce sound. In cockroaches, the mechanism differs in that the pronotum rubs against the tegmina.

Haskell (1955) showed that some of the energy resulting from stridulation by a grasshopper passes through the substrate either as acoustic energy or energy resulting from vibration of the body at the time of stridulation. Considerable energy is directed toward the substrate by a stridulating *N. cinerea* male (Fig. 14). No known auditory organs, like those found in the Ensifera, have been described in cockroaches. *Blattella germanica* has subgenual organs in the tibia of each leg (Graber 1882), and Autrum (1963) showed that similar organs in *Periplaneta americana* (L.) are extremely sensitive to substrate vibration. It is likely that *N. cinerea* and other stridulating cockroaches can detect sounds in this manner. Pumphrey (1940) suggested that tibial chordotonal organs developed as receptors for vibratory stimuli transmitted through the substrate, before the evolution of a tympanic membrane rendered them sensitive to air-borne sounds.

Our ideas concerning the functional significance of sound production by cockroaches are based upon laboratory observations. Field observations are unknown and are sorely needed to complete our understanding of this phenomenon.

ACKNOWLEDGMENTS

We thank Drs. Ashley Gurney, K. Princis, E. R. Willis and the late H. Grant for the loan of museum specimens, Mr. W. B. Hahn for the photographs, and Drs. B. Dumortier and R. D. Alexander for critically reading the manuscript.

REFERENCES CITED

**Alexander, R. D.** 1956. A comparative study of sound production in insects, with special reference to the singing Orthoptera and Cicadidae of the eastern United States. Ph.D. dissertation, Ohio State Univ., Columbus.

**Autrum, H.** 1963. Anatomy and physiology of sound receptors in invertebrates, p. 412–33. *In* R. G. Busnel [ed.] Acoustic Behaviour of Animals, Elsevier Publ. Co., New York.

**Baier, L. J.** 1930. Contribution to the physiology of the stridulation and hearing of insects. Zool. Jahrb. (Zool. Physiol. Tiere) 47: 151–248.

**Barth, R.** 1961. Comparative and experimental studies on mating behavior in cockroaches. Ph.D. thesis, Harvard Univ., Cambridge, Mass.

1964. The mating behavior of *Byrsotria fumigata* (Guérin) (Blattidae, Blaberinae). Behaviour 23: 1–30.

**Chopard, L.** 1938. La biologie des Orthoptères. Encycl. Entomol. Paris, sér. A, 20: 1–541.

1950. Sur l'anatomie et le developpement d'une blatte vivipare. Proc. 8th Int. Congr. Entomol., Stockholm 1948: 218–22.

**Dumortier, B.** 1963. Ethological and physiological study of sound emissions in Arthropoda, p. 583–654. *In* R. G. Busnel [ed.] Acoustic Behaviour of Animals. Elsevier Publ. Co., New York.

1965. L'emission sonore dans le genre *Gromphadorhina* Brunner (Blattodea, Perisphaeriidae). Étude morphologique et biologique. Bull. Soc. Zool. France 90: 89–101.

**Eisner, T.** 1958. Spray mechanism of the cockroach *Diploptera punctata.* Science 128: 148–9.

**Frings, M.** 1955. Bibliographie générale sur l'acoustique des insectes, p. 401–46. *In* R. G. Busnel [ed.] Colloque sur l'acoustique des Orthoptères, Ann. Épiphyt. (C) 5, Fasc. spéc. 448 p.

1956. Bibliography on sound production and reception by insects. First supplement. The Pennsylvania State University, State College, Penn. (mimeogr.).

**Graber, V.** 1882. Die chordotonal Sinnesorgane und das Gehör der Insecten. Arch. Mikroskop. Anat. 20: 506–640.

**Guthrie, D. M.** 1966. Sound production and reception in a cockroach. J. Exp. Zool. 45: 321–8.

**Hanitsch, R.** 1930. Über eine Sammlung malayischer Blattiden des Dresdner Museums. Stettin. Entomol. Zeitung 91: 177–95.

**Hartman, H. B., and L. M. Roth.** 1967a. Stridulation by a cockroach during courtship behaviour. Nature (London) 213: 1243.

1967b. Stridulation by the cockroach *Nauphoeta cinerea* (Olivier) during courtship behaviour. J. Insect Physiol. 13: 579–86.

**Haskell, P. T.** 1955. Vibrations of the substrate and stridulation in a grasshopper. Nature (London) 175: 639–40.

1964. Sound production, p. 563–608. *In* M. Rockstein [ed.] The Physiology of Insecta, I. Academic Press, Inc., New York.

**Karny, H. H.** 1924. Bemerkungen über einige Blattoiden, p. 3–19. *In* Beiträge zur malayischen Orthopteren-Fauna, V. Treubia 5.

**Kevan, D. K. McE.** 1954. "Unorthodox" methods of sound-production in Orthoptera. Spec. Pap. Univ. Nottn. Sch. Agr., Zool. Sect. no. 2. 22 p.

1955. Méthodes inhabituelles de production de son chez les Orthoptères, p. 103–42. *In* R. G. Busnel [ed.] Colloque sur l'acoustique des Orthoptères. Ann. Épiphyt. (C) 5, Fasc. spéc., 448 p.

**Kleine, R.** 1918. Hat *Periplaneta orientalis* einen Stridulationsapparat? Neue Beitr. Syst. Insekten 1(11): 85–86.

**Leston, D., and J. W. S. Pringle.** 1963. Acoustical behaviour of Hemiptera, p. 391–411. *In* R. G. Busnel [ed.] Acoustic Behaviour of Animals. Elsevier Publ. Co., New York.

**Main, H.** 1932. Recent insect immigrants. Proc. Entomol. Soc. London 7: 24.

**McKittrick, F. A.** 1964. Evolutionary studies of cockroaches. Cornell Univ. Agr. Exp. Sta. Mem. 389. 197 p.

**Princis, K.** 1954. Zur Kenntnis einiger afrikanischer Blattariengattungen. Ann. Mus. Congo Belge, Zool. I (Miscellanea Zool. H. Schouteden), Tervuren, p. 511–2.

1960. Zur Systematik der Blattarien. Eos 36: 429–49.

1965. Orthopterorum Catalogus. Pars 7 (p. 284–400). 's Gravenhage.

**Pumphrey, R. J.** 1940. Hearing in insects. Cambridge Phil. Soc. Biol. Rev. 15: 107–32.

1955. Rapports entre la réception des sons et le comportement, p. 320. *In* R. G. Busnel [ed.] Colloque sur l'acoustique des Orthoptères. Ann. Épiphyt. (C) 5, Fasc. spéc., 448 p. (Not seen; cited by Dumortier 1963.)

**Roth, L. M.** 1967. Sexual isolation in parthenogenetic *Pycnoscelus surinamensis* and application of the name *Pycnoscelus indicus* to its bisexual relative (Dictyoptera: Blattaria: Blaberidae: Pycnoscelinae). Ann. Entomol. Soc. Amer. 60: 774–9.

**Roth, L. M., and R. Barth.** 1967. The sense organs employed by cockroaches in mating behaviour. Behaviour 28: 58–94.

**Roth, L. M., and B. Stay.** 1958. The occurrence of paraquinones in Arthropods, with special emphasis on the tracheal glands of *Diploptera punctata* (Blattaria, Diplopteridae). J. Insect Physiol. 1: 305–18.

**Roth, L. M., and E. R. Willis.** 1955. Intra-uterine nutrition of the "beetle-roach" *Diploptera dytiscoides* (Serv.) during embryogenesis, with notes on its biology in the laboratory (Blattaria: Diplopteridae). Psyche 62: 55–68.

1958. The biology of *Panchlora nivea* with observations on the eggs of other Blattaria. Trans. Amer. Entomol. Soc. 83: 195–207.

1960. The biotic associations of cockroaches. Smithsonian Misc. Coll. 141: 1–439.

**Rudow, F.** 1870. Einige Beobachtungen über die Lebensweise der Heuschrecken. Z. Ges. Naturwiss. 36: 306–30.

**Shelford, R.** 1909. Studies of the Blattidae. Trans. Entomol. Soc. London 1909(2): 253–327.

**Skaife, S. H.** 1954. African Insect Life. Longmans, Green & Co., New York. 387 p.

**Skinner, B. F.** 1966. The phylogeny and ontogeny of behavior. Science 153: 1205–13.

**Smyth, T., Jr.** 1963. Mating behavior of the Madeira cockroach. 35th Annu. Meeting, Eastern Branch Entomol. Soc. Amer. (mimeogr. abstr.).

**Stay, B., and L. M. Roth.** 1958. The reproductive behavior of *Diploptera punctata* (Blattaria, Diplopteridae). 10th Int. Congr. Entomol. (1956) 2: 547–52.

**Vosseler, J.** 1907. Einige Beobachtungen an ostafrikanischen Orthopteren. II. Deut. Entomol. Z. 5: 527–30.

**Willis, E. R.** 1966. Biology and behavior of *Panchlora irrorata*, a cockroach adventive on bananas (Blattaria: Blaberidae). Ann. Entomol. Soc. Amer. 59: 514–6.

# E. Sexual Behavior

## Studies on Territorialism and Sexual Selection in Dragonflies[1]

Merle E. Jacobs
*Department of Zoology, Duke University, Durham, North Carolina*

In recent years territorialism has been studied extensively in vertebrates, but the invertebrates in general have received little attention. It is the writer's opinion that territorialism is related to sexual selection, and the insects, such as dragonflies, whose mating activities are highly localized and overt, are well suited for such studies.

There has been considerable controversy about the functions of territories and, consequently, about definitions. Noble (1939) has presented a general definition of territory as "any defended area" thereby leaving function to be described in relation to a particular case. Noble's definition will be used throughout the following discussion with an attempt at an analysis of function.

Darwin (1871) included dragonflies among insects illustrating his concept of sexual selection, and other naturalists have reported apparent displays of bright colors. Tillyard (1912) observed that males of *Libellago caligata* wave their long white inferior appendages toward females before copulation while the females replied to these signals by showing the white underside of the tip of the abdomen. Williamson (1906) says, "In *Rhinocypha* and *Calopteryx* the sexes are attracted to each other by movements and displays of color." Williamson was also of the opinion that the specific colors of male dragonflies are important in bringing about conspecific mating. He supported this view by pointing out the uniformly similar abdominal appendages of species with highly colored wings, in contrast to the specifically distinct appendages of colorless-winged species. He postulated that the influence of the male colors operated as an isolating mechanism among some genera, whereas the complex genitalia isolated the species in others.

Some indications that dragonflies show territorial behavior have been recorded. Williamson (1889) says that males of *Celithemis elisa* "make sudden dashes at passing diptera and smaller dragonflies, often returning to the same sedge time and time again. Each is the proprietor of a particular locality. When one encroaches on the

[1] Contribution no. 534 from the Zoological Laboratories of Indiana University, Bloomington, Indiana.

hunting territory of another he is quickly hustled away by the rightful and irate owner." Byers (1930) indicates defence of territory on the part of males of *Tetragoneura semiaquia.* Lieftinck (1934) describes males of *Crocothemis servilia,* "struggling for the mastery of their domain in the pursuit of other dragonflies." Wiliamson (1923) states that males of the genus Erythemis protect their own females against attack during oviposition. Whitehouse (1941) observed a male of *Plathemis lydia* poise on the wing above an ovipositing female with which he had mated and drive away other males.

Saint-Quentin (1934) made an attempt to study territorialism among dragonflies by means of marked individuals. He observed that dragonflies roam in restricted areas in which they tolerate no similar dragonfly. This behavior was interpreted as the eviction of intruders from the hunting ground, termed "ein Jagdrevier." Since none of the marked dragonflies was recovered, Saint-Quentin concluded that the Jagdrevier was of a transitory nature. He also noted that dead aeschnid males suspended by strings were attacked by other males.

Moore (1952a) repeated the work of Saint-Quentin using live male dragonflies suspended from strings and found that attacking males would sometimes couple with the suspended ones. Moore was successful in recovering some of his individually marked specimens, but failed to find any constancy in localization. He concluded that the inability of dragonflies to recognize sex results in inter-male clashes bringing about temporary dispersal throughout the breeding area to simulate territorial behavior.

Opinion has been divided in regard to whether there is inter-male competition among dragonflies. Williamson (1906) reports males of Libellula fighting for females. Walker (1912) in his studies of the genus Aeschna says there is no contest between males for females nor are the males apparently able to distinguish females of their own species from those of other species of the genus. Needham (1929) observed males of *Pachydiplax longipennis* meeting in combat. Hingston (1933) in interpreting animal adornment as an instrument of hostility includes the dragonflies and other insects in theoretical support.

The present investigation on the dragonflies *Plathemis lydia* (Drury) and *Perithemis tenera* (Say) was conducted in an attempt to obtain evidence on the following questions: (1) Do the bright colors of the males induce differential male mating success and, if so, is this because of the effect of these colors on females? (2) What sexual isolating mechanisms are operative? (3) Are territories maintained and, if so, what are the functions of these territories?

I wish to thank Dr. F. N. Young for his active interest and advice and for his help in the preparation of the manuscript, Dr. W. H. Thorpe for having reviewed the manuscript, Dr. T. W. Torrey for his care in seeing that photographic and other materials were provided, Dr. B. Saila for giving statistical advice, and Mr. R. H. Marlin for allowing free access to the pasture pond where part of the studies were conducted.

### Species and Localities

A series of observations was made on the behavior of dragonflies at ponds near Bloomington, Indiana. The two species selected for concentrated study differed greatly in behavior and were abundant in the area. A brief description of the species is as follows: *Plathemis lydia,* herein simply designated as Plathemis, was selected to represent a sexually dimorphic dragonfly in which male "display" to the female was not observed. The mature male abdomen is white dorsally, and the wings are clear with a dark median band. The female abdomen is dark, and the wings bear a median and terminal dark spot. *Perithemis tenera,* herein designated as Perithemis, was selected as a sexually dimorphic dragonfly in which males appear to "display" to the females. The male wings are amber in color except for a small dark spot near the triangle in some individuals. The female wings are clear with two dark spots on each, the degree of spotting being highly variable.

One of the situations at which observations were made was a small solution pond about 10 km. north of Bloomington, Indiana. It is circular, about 128 m. in circumference at the water's edge, and during the period of observation was about 0.6 m. deep except during heavy rains when it became as much as 0.9 m. deep. The sink in which the pond is located was first plugged in 1946 following ploughing of the field and has contained standing water since that time (*fide* Mr. R. H. Marlin, property owner). The pond contained little floating or emergent vegetation except around the edges where Potamogeton, sedges, grasses, and algae were found.

Observations were also made at an excavated pond about 1.6 km. north of the solution pond. This pond is rectangular measuring about 161 m. by 27 m. by 0.9 m. deep. The northeast and south banks are forested and partly shaded. The west bank has been cleared, and numerous stumps project into the water. The long axis of the pond lies north and south.

Other observations were made at an elongated reservoir, Wiley Lake, about 1.6 km. north of the excavated pond. Since the lake is surrounded by steep heavily wooded hills, most of the shore is partially shaded. In these shaded areas there is little aquatic vegetation. Along the sunny parts of the shore, the western and southeastern parts, patches of Potamogeton partly covered with algae are found.

## Methods

In catching dragonflies for marking, a net held open by means of wires was used to prevent injury to the specimens. When tenerals began to emerge, in May, as many as feasible were caught and marked with spots of paint in front of the thorax. Different color combinations were used on each day. When adults arrived at the pond after maturation, all the males and some of the females were individually marked with paint shot at them with a syringe or, as was usually possible, they were captured and marks of paint were placed on the wings and sometimes on the dorsal surface of the abdomen. Enamel paint was used exclusively.

The wing markings consisted of a dot about 4 mm. in diameter placed on the center of the median patch of the wing, a vertical line through the patch, a streak filling the costal cell of the patch, or a combination of the streak and vertical line. Any or all of the wings were marked in various combinations following a constant order. Yellow and aluminum were used exclusively for wing marking. Specimens thus marked could be identified even in flight.

Egg samples were obtained by dipping the abdomen of captive females into vials of pond water. The vials were shaken to scatter the eggs and were then laid on their sides until the eggs stuck to the glass. Eggs thus separated hatched readily at room temperature if the water was changed daily. If the water was not changed, or if the eggs were clustered, many failed to hatch.

In 1952 at the solution pond, Plathemis was observed from May 1 to June 17, and Perithemis from June 17 to July 31. In the Plathemis studies the pond was circled continuously each day when the species was active. The activities of Perithemis were observed from a vantage point overlooking all available oviposition sites. (Sites out of observation range were removed.) Further observations on Perithemis were made at Wiley Lake from August 1 to September 14, 1952 and at the excavated pond from June 6 to July 17, 1953.

A pond under study was visited daily from early morning until late evening when the last dragonfly had left (except during cool cloudy weather when dragonflies failed to arrive). Each time a specimen was observed its activities and location at the pond were recorded. Records of air temperature and cloudiness were made at regular intervals. At the end of the day, observations were tabulated. From time to time the results were analyzed statistically. In the following discussion "±" refers to standard error of the mean.

## Studies on Plathemis

### *Observations on Behavior*

*Emergence of Adults.* In 1952 the first teneral Plathemis was observed on May 1. Emergence continued through August. The naiads usually began creeping from the pond shortly after sunrise and continued until about 11:00 A.M. (CDT). They crawled into the grass usually 1-5 m. from the pond, but sometimes as much as 45 m. Solitary sandpipers captured large numbers of naiads on the shore, and kingbirds preyed upon tenerals flying toward the woods nearby.

*Teneral Adults.* The sex ratio of marked tenerals was 129:209 or 1 male: 1.62 females. This possibly does not indicate the true sex ratio, since males are more active than females and more readily escape capture. Of the marked tenerals, only 9 males and 6 females returned to the pond after maturation. Males returned a minimum of 8 and a maximum of 14 days after marking, and females a minimum of 13 and a maximum of 24 days. Although this may indicate that males return earlier than females, it could be only a sampling error. It is interesting, however, that some males show signs of immaturity upon return to the pond. Borror (1934) in his studies of *Argia moesta* notes that when adults are about 14 days old they return to the river to mate, the males by this time being almost or entirely pruinose.

Teneral males closely resemble females in color, but upon returning to the pond after the maturation period, males differ strikingly from the fe-white, or white abdominal pruinosity. Among seven males which were still bluish-brown after returning, and for which complete color notes were obtained, the pruinosity changed to bluish-white during the first day at the pond, and to white after 2-6 additional days.

*Male Behavior.* Single males appearing at the

pond cruise over about one quarter of the margin while flying close to sites at which females are later seen ovipositing. A lone male usually spends considerable time near one site which he repeatedly approaches very closely, sometimes stimulating oviposition by dipping the tip of the abdomen into the water after the manner of a female. The oviposition sites consisted of clumps of mud, matted vegetation, or sticks, logs and similar objects floating in water near shore. Sites with glistening surfaces seemed especially attractive to males, which would usually stay near them. Often the male would perch near the site.

When a female appears, the male flies quickly toward her with his abdomen lowered, thus apparently hiding the white upper surface. In some cases, especially when the temperature was below 20° C. and the male was sluggish, he would apparently hide behind objects while flying toward the female. As the female slows down near a site, or begins oviposition motions, the male seizes her with his feet and immediately fastens his abdominal claspers to her head.

Copulation occurs in the air usually close to the site where the male had been localized. In copulation, the female allows her wings to rest and wraps her feet around the male's abdomen. Timing of 200 undisturbed copulations showed the duration to be about 3 seconds and remarkably constant. After copulation the male usually places the female on the site. The female may oviposit where she has been placed, or may flit around the pond to another site. In any event, the male follows her and hovers above her when she begins to oviposit, as if guarding her.

In no case was a male observed to guard a single female without first having copulated with her or with another female in the same area shortly before. If the abdomen of the male was stiffened by a heavy layer of paint, thus preventing him from bending his abdomen to fasten the abdominal claspers to the head of the female, he would repeatedly seize ovipositing females with his feet sometimes as often as 25 times in 20 min. In no case, however, was such a male observed guarding an ovipositing female.

After oviposition, when the female flies from the pond, the male follows her closely until she rests in the field. The male then returns to the area where oviposition had occurred. No male was observed attempting to seize a female which was resting in the field or on an oviposition site.

If a male is guarding an ovipositing female and another female flies to the site, the male may catch the ariving female and, after copulation, place her near the other and guard both. If the arriving female dives to the site and begins to oviposit while the male is guarding, the male will guard both ovipositing females without having copulated with the second. If the second female takes flight, however, the male may catch her and, after copulation, replace her with the first and continue guarding. When one of the females stops ovipositing and flies, the male follows her then returns and guards the other.

Usually many males are active at the pond. When two males encounter one another, one male dashes at the other and pursues him. The pursuing male raises the white upper surface of the abdomen toward the new arrival, while the latter flies away with abdomen lowered. This appears to be a joint display of the conspicuous upper surface of the abdomen. After the second male has been "chased" 8-16 m. from the site over which the first had been active, the first usually turns and flies back with abdomen lowered while the second now pursues with abdomen raised. This pursuit display, alternating in direction, continues until the new arrival restricts his movements to the vicinity of another site.

If the newly arrived male stays for some time around the site of the first, and especially if he simulates oviposition thereon, a dual flight may follow. This flight is often preceded by the two males facing each other immediately over the site with abdomens raised (Fig. 1). They then fly quickly upward, usually to a height of about 3 m., level off, and fly side by side, or one a little behind the other, back and forth across the pond and adjacent fields. The speed of such flights, timed with a stop watch, was found to approximate 7.5-10.7 m./sec. in still air. It is obvious

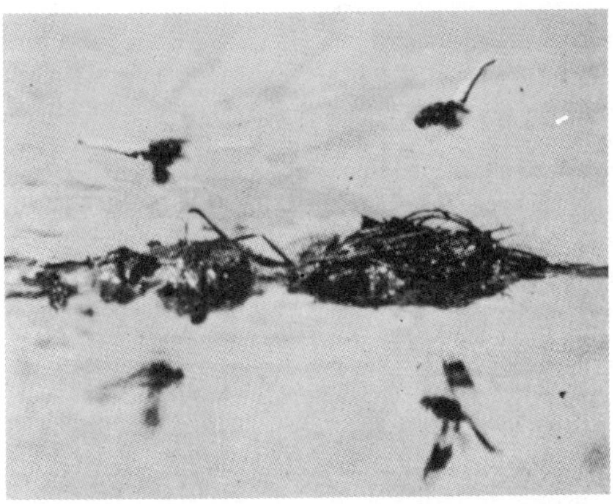

FIG. 1. Two Plathemis males mutually displaying the white upper surfaces of the abdomen near an oviposition site.

that the flight is not a speed test, for when one of the males is slowed by means of varnish placed on the wings, the other keeps pace with him. The flight may be interrupted after as little as one second or may continue for as long as 123 seconds, the mean of 37 samples being 40.5 seconds. One of the males seems to initiate the flight by suddenly flying up near the other. In a few cases males seemed consistently to refuse what appeared to be challenges to dual flight on the part of newly arrived males. After the flight, both males return to the original area and, after a little more pursuit display, one retires from the area.

As more males arrive, the interactions increase in complexity. The edge of the pond soon becomes populated with males each of which flies back and forth more or less parallel to the shore over a distance of about 9 m., about 4 m. on either side of one or more oviposition sites. It was found that the 128 m. circumference of the pond was thus normally divided into 12 sectors with lateral dimensions of about 11 m. The males in these sectors appeared to defend them against the intrusion of other males. When two males meet near the boundaries of their sectors they may first execute pursuit display, but gradually begin to avoid each other and sometimes perch in their sectors. Since there are usually more males present than sectors, pursuit displays and dual flights take place almost constantly in favorable weather. This activity appears to be partly responsible for continuous displacement of males throughout the day.

There was considerable disparity in number of matings (copulations and accessory events) which occurred in the sectors (Table I). There appeared to be more male interaction in those sectors in which most mating occurred. It also appeared that females mated and oviposited more readily at certain oviposition sites in much-used sectors. The eggs are laid in a glistening mass on the sides of projecting objects. Such glistening surfaces appear to attract both males and females and prodigious numbers of eggs may be laid at one spot.

TABLE I. *Distribution of 6,243 matings of Plathemis in the twelve sectors at the pond. Sectors are identified by means of compass directions*

| Sector | NNE | N | NNW | SSW | S | SSE |
|---|---|---|---|---|---|---|
| Matings | 3% | 6% | 10% | 12% | 12% | 12% |
| Sector | ESE | E | ENE | WNW | W | WSW |
| Matings | 2% | 13% | 8% | 7% | 9% | 6% |

Each sector had distinct characteristics; thus, the SSW sector contained numerous clumps of matted vegetation in addition to abundant emergent vegetation. The S and SSE sectors contained little emergent vegetation and few clumps of matted vegetation. The WNW sector was the only one with cattails. The ESE, E, and ENE sectors contained mostly grasses, and the NNE and N sectors mostly sedges. All the sectors changed aspect during the season so that only a general description can be given.

The excess of males over available sectors leads to many deviations from the pattern of lone male activity. In many cases a very active guarding male fails to ward off numerous others which are dashing at the ovipositing female. A male catching a female under these conditions flies with her in tandem over the pond in a jerky flight deflecting pursuing males from the course and finally returns to the site where copulation occurs. When other males disturb the copulating pair, copulation may last up to 15 sec. instead of the usual 3 sec. At such times the male appears to drive off rivals while he holds the female in a rather safe copulation position, then he releases the female when in a suitable position to dash at rivals (Fig. 2).

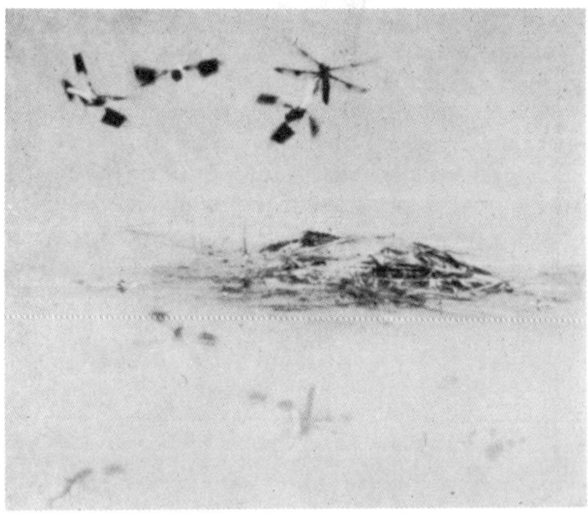

FIG. 2. Plathemis male releasing a female at an oviposition site while interacting with other males.

In many cases the female is chased by other males or is seized while ovipositing. When the female is seized, the guarding male becomes intensely active, flying toward other males and warding them from the site. In the meantime the male which seized the female copulates with her, then after replacing her on the site, guards her. The original male then dashes at the female and often catches her, then after copulation replaces

FIG. 3. Plathemis male seizing an ovipositing female as other males rush in.

her on the site. Many males may be involved in this process at one site leading to a melee in which they dash at one another swiftly (Fig. 3). Although males seldom contact one another, occasional collisions occur, after which one may drop into the water. During one of these melees the female may perch motionless on the site whereupon males will not seize her. (It appears they cannot see her, for if a female stops suddenly on the site, a pursuing male may assume the coupling position on the female's shadow or wing spot.) In these cases the female, being unable to oviposit, may fly from the pond followed by many males, some of which have come from adjacent areas. Kingbirds may fly into these flocks and catch males or, more frequently, the slower-flying females.

When two males simultaneously begin seizure of a female she may be injured and drop into the water. Wolf spiders (Dolomedes) often catch such females. With loss of the female, the males continue pursuit display and dual flights until many leave.

When a female does not oviposit on the site at which the male places her, but flies to another site, the guarding male follows her closely but does not seize her unless she dashes away suddenly. In such cases the female is frequently chased by other males. When a guarded female arrives at the site where another guarded female is ovipositing, the two guarding males do not interact strongly, but the visiting male soon retires while the other guards both females.

Males could be classed according to whether they arrived at the pond daily in the forenoon or afternoon: 13% of the marked males arrived mostly in the forenoon, and 60% mostly in the afternoon. The other 27% were too irregular to classify. The sequence of arrival of some of the males from day to day was quite constant.

Immature males fly in a slower, more fluttering manner than older males and show greater tendency to perch. They were frequently observed chasing males of *Anax junius* as if mistaking them for females of their own species. This reaction soon disappears as if they had learned to ignore Anax males. When a Plathemis female arrives, the immature males dash toward her immediately. They often made abortive copulations while flying for an extended period with the female in tandem. Although the female was usually soon released, one pair rested in the grass about five minutes before separating.

Immature males are easily driven away by the older males, but return from time to time. Such males, without a sector, if successful in capturing a female and copulating, may release her where no oviposition site is available. In such cases the female flies away without ovipositing. Immature males, if successful in holding a sector and mating, are less active in guarding than older males.

Some males showed definite localization to specific sectors. Many returned immediately to sectors previously occupied after having been absent from the pond for as long as five days during cool cloudy weather, indicating that memory may be involved. Many males showed affinity to specific oviposition sites in their sectors.

When many males are present, immatures show little localization the first days at the pond. After localization is established, perhaps on the fifth day, it may be maintained for perhaps ten additional days before declining. This analysis is based on complete records of only 20 males. This small sample makes quantitative statements tentative. However, the general statements are supported by shorter observations of many more males. Of the 20 immatures, 8 showed almost absolute localization during the 10-day middle period, but averaging all 20 males for this period showed that they were observed in their respective sectors a mean of only 42.5% of the times observed.

The mean daily number of matings of the 20 immatures increased during the period of increasing localization, while those of mature males marked on the same dates did not show the increasing localization nor an increase in mating number. The mating scores of all males showed decline with decline in localization. These data apply only when there is intermale competition and seem to indicate increasingly aggressive behavior with maturation.

The maximum period any of the immatures was present at the pond was 36 days, with a mean of 14.25 ± 2.00 days. Examination of the testes of 60 blue-brown, 88 blue-white, and 140 white males showed progressive degrees of encystment and spermiogenesis.

Males marked when blue-brown were less likely to repeat at the pond than were those marked when white on the date of their first arrival. Thus, of 137 males marked when blue-brown only 70, or 51% repeated one or more days, but of 127 marked when white 105, or 83% repeated. Though this may indicate heavier predation of immatures, it may also indicate a tendency for them to wander more than matures. In Perithemis there was a marked tendency for both sexes arriving early in the season to fly high and away from the pond after only brief examination. Perhaps a tendency to wander on the part of immature dragonflies is important in dispersal even in the absence of reaction between individuals.

There was an increase in the number of males present at the pond as well as in the number of matings from May 1 to June 17 when observations were ended. The maximum number of marked males recorded on any one day was 148: the daily mean for June was 104, and for May, 15. (Thirteen cool cloudy days in May were not counted.) During the first two weeks of July, Plathemis males were removed because studies of Perithemis were being started. Plathemis males dash at male and, especially, female Perithemis.) All males were taken from one sector by netting, or shooting with a BB air rifle while they were perched. In the two-week period 388 males were removed. The maximum removed on any one day was 130. When many mature males were removed, immatures gradually took their places.

Most matings occurred between 2-3 P.M. (generally the warmest part of the days). Males most in evidence at this time of day had higher mean daily mating scores than did those most in evidence during the first and last hours of observation. Perhaps males customarily arriving at the pond in the afternoon fresh from the woods, where they may have been feeding or resting, would be in a condition to compete with greater vigor and mate oftener than those arriving when few females are present, but when energy must still be expended in maintaining a site.

*Female Behavior.* On each day the first individual to arrive at the pond was a male, followed by a female after a mean of 41.0 ± 1.5 min. with a minimum of 1 min. in one case and a maximum of 110 in another. An arriving female flits about oviposition sites around the margins of the pond and may make oviposition motions on several sites. At such times a male may seize her. If, after copulation, the female is prevented from ovipositing, as when chased by other males or when no sites are available, she darts from males chasing her, except when near an oviposition site where she may slow down and be captured.

During oviposition the abdomen is struck into the water, thus washing off eggs against the site where they stick in a slimy mass. Undisturbed oviposition lasts 1-2 min. in which time 100-175 dips are made. .Ovipositing females are easily approached. An ovipositing female seized by a male and carried to a site nearby, or chased from a site by the observer, usually returns to the original site to complete oviposition.

Of 60 females marked after capture while ovipositing, 44 returned to the pond and oviposited after one or more days. The maximum number of times a female was observed to return was 7 (within 22 days). Returning females showed a tendency to oviposit in the sector of previous oviposition.

Females are attracted to other flying females or to other dragonflies such as *Anax junius* or *Libellula basalis.* They are also attracted to ovipositing females of their own species, the tendency leading to communal oviposition. The attraction of an ovipositing female seems greater than that of an oviposition site alone. It is interesting that Perithemis females follow not only one another but also males which fly slowly to the site during events leading to courtship. Plathemis females were never observed to follow mature males of their own species, but in two instances females followed, for short distances, immature males with which they had copulated. Since immature males fly in a less jerky fashion than mature ones, flight pattern may be partly responsible for inducing following by the female. No dragonflies which females were observed to follow move in the jerky manner of mature Plathemis males. Females would also follow other Plathemis females with abdomens painted white so as to resemble males, but which do not fly like males.

### Experimental Removal of Males from the Pond

To determine whether older males prevented newly arrived immatures from localizing, all males but a few immature ones were continuously removed, May 20-29, 1953. With 140 males removed, the few remaining immatures showed almost absolute localization to a sector from the time they first arrived. The sector taken did not always seem suitable to females. Thus, May 21, most matings were taking place in a certain sector at submerged logs. At 1 P.M. all males were removed. At 2 P.M. an immature male arrived (for the first time) at a sector where no oviposition sites were present. In spite of the fact that the submerged log sector was vacant and remained so, the immature male did not visit that sector, but stayed in the sector first encountered. Females arriving flitted around the submerged logs but, no male being present, they flew around the pond until they encountered the male in the sector where no sites were available. The male there copulated with all six females arriving that afternoon. These females left without ovipositing after flitting around sedges in the sector.

With most of the males removed, the remaining males mated far more frequently than usual. It had been found that the mean daily mating scores, computed by dividing the total number of matings of a male by the number of days that male was present at the pond, did not exceed four under ordinary conditions. The males remaining after most of the others had been removed, however, averaged 80 matings per day. A male would apparently mate with as many females as would enter his area. After having mated 20 or more times, males became less active, perched much, and were easily driven from the pond by newly arrived males. In egg samples obtained from ten females which had mated in succession with one immature male, hatching occurred in all but the seventh and ninth. Of 33 samples from females which had copulated with the few remaining males, hatching occurred in 23. Of 50 samples taken June 1, after removal of males had been discontinued, hatching occurred in 46.

With all males removed, females still came to the oviposition sites and some oviposited alone. Females arriving for the first time (perhaps virgins) sometimes performed oviposition motions at sites, but soon flitted around the pond and gradually became very active in chasing other dragonflies for short distances. If two such females were present at the pond they were not only attracted to one another, but performed dual flights much as do males, after which one would leave the pond. Only a few females were therefore present at the pond at one time when males were absent. No difference in this behavior was noted among females when some of their abdomens were painted white so as to resemble males. No female was observed to remain at the pond more than ten minutes when no males were present. In many cases they rested in the field, returning occasionally, then flying to the woods.

*Experimental Darkening of the Male Abdomen*

Since it appeared that the white upper surface of the abdomen was employed by males as a threat, experiments were performed to test this hypothesis as well as to determine whether the white color induced differential male mating success. In addition to the regular identification marks the upper surface of the abdomen of one of each three males marked during the study was painted black, another white, and a third left unpainted, 164 of each type being marked. Enamel paint was used and was confined to the individual abdominal segments, since it had been found earlier that painting the intersegmental areas prevents the males from bending their abdomens to clasp the females.

The males were observed closely to determine whether the darkening induced any differences in behavior, especially in regard to mating success. Reactions of females to marked males were also studied. The study was continued for fifteen days.

In compiling data, repeating males having blackened abdomens were compared with an equal number having whitened abdomens marked on the same dates and in the same sectors. Males having whitened abdomens were similarly compared with unpainted ones. In finding the mean daily mating score of a male, the total number of observed matings of that male was divided by the number of days he arrived at the pond. The mean daily mating score of a group was the average of these means. It was thought that perhaps 90% of all matings that occurred were observed.

*Results.* The mean mating score of the blackened males was significantly below that of the whitened ones (Table II). The lower mating success of the blackened males appeared to result from their lowered ability to ward off opponents. It was frequently observed that an intruding male flying toward a site would depart when the guarding male displayed his white abdomen at him even though the guarding male made no dash at the intruder. The blackened males displayed their abdomens at rivals, but the display seemed less effective. Those blackened males that were most successful in warding off opponents appeared to do so by greatly increased activity in dashing at intruders and performing dual flights with them. In one case while a white male was guarding a female, a male with his abdomen stiffened with an excess of black paint pounced on the female repeatedly, but the guarding male did not react appreciably to the blackened one.

TABLE II. *Mean daily matings of Plathemis males with painted abdomens*

| Treatment | Number of Males | Mean Mating Score | Sum of Squares |
|---|---|---|---|
| Whitened........ | 68 | 0.886 | 10.23 |
| Blackened....... | 68 | 0.731 | 9.62 |

"t"=2.34.

When there was an "aggregation" of white males, the blackened ones were often absent. Blackened males had a tendency to "aggregate" in the NNE sector where relatively few matings occurred. When it became cloudy and many white males departed, blackened ones often appeared in their places. Only 27% of the white males returned to the pond during the last hour of observation during the study, but 33% of the blackened ones did so. Perhaps blackened males lingered longer at the pond as a result of being inhibited by whitened males when most mating was occurring in mid afternoon.

Decreased mating success of the blackened males did not appear due to failure in attracting females: rather, males seize females often from behind and lower the abdomen to such an extent that it sometimes drags in the water. The white upper surface is thus not displayed to females as it is to other males.

Information concerning the influence on the male of white color was obtained by painting the upper abdominal surface white on 30 females. In a few cases males performed pursuit displays toward these females as they do toward other males. Usually, however, males seized whitened females as if they were ordinary. After copulation, a guarding male would fly toward the whitened ovipositing female and, when she flew, seize her and copulate again. This was repeated by the same male until the female flew away. A guarding male would also dash at spots of white paint on logs, dead males, or, especially, white feathers moved by the wind near the site.

Comparison of the mating scores of whitened and unpainted males showed no significant difference. The mean mating score of 60 whitened males was 0.925 as compared with 0.950 on the part of 60 unpainted males marked on the same dates and in the same sectors. The observed "t" value was 0.206.

## STUDIES ON PERITHEMIS

### *Observations on Behavior*

*Male Behavior.* On first arrival at the pond, males fly slowly along the margin and closely approach floating patches of Potamogeton, algae, roots, sticks, or logs. When thus engaged, the males make short dips at the site as does an ovipositing female, but the male does not bend down his abdomen as does a female. This activity will be referred to as examination of the site (Fig. 4). At this time the male is easily approached, but if an attempt is made to capture him he will fly away particularly readily. Having established himself at a site, the male hovers over the water within a radius of about 30 m. but returns occasionally to fly over the site. He may then perch near the site at a position commanding a view over the open water.

FIG. 4. Perithemis male performing examination behavior at an oviposition site.

When a female flies over the pond the male dashes toward her and follows her while swaying from side to side. He then turns and flies slowly to the site while the female follows, or if she does not follow flies after her again. Upon reaching the site, the male hovers intensely on it while fluttering his wings and sometimes bending up his abdomen. This activity will subsequently be referred to as courtship (Fig. 5). After the first courtship, or after several, the female's wings beat more slowly or stop while she perches on the site nearby. The male catches her as she flies slowly or perches, and fastens his abdominal claspers to her head. A male may apparently lose a perching female though she is only a few inches away. A male may assume a coupling position on the dark wing spots of a perching female as if mistaking them for a head.

In initiating copulation the male flies while supporting the female whose wings are at rest: he then perches, and copulation continues. Perhaps actual copulation does not begin until the female perches, for copulation time including the period

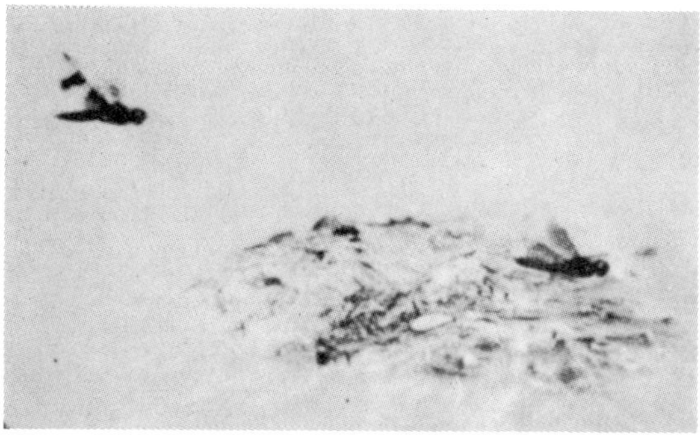

FIG. 5. Perithemis female following a male to the oviposition site.

when the pair was in flight in copulating position was highly variable, while if only the period of perching during copulation was considered, the time was often quite constant: such timing showed one male to copulate almost exactly 14 sec. with each of five females in succession.

On the whole, copulation time was variable; constancy prevailed only for short periods. One male copulated 17 sec. with each of four successive females from 11:00 to 11:40 A.M. June 29, 1953. Four other copulations for that male on the same day ranged from 15 to 20 sec. On July 1 he copulated 23 sec. with each of four successive females from 10:50 A.M. to 1:50 P.M., the only matings that day.

Copulation time seemed to depend on the male, for the same female would copulate long with a long-copulating male or briefly with a short-copulating male. The duration was often prolonged in shade or when the air temperature was below 25° C. Stop-watch timing of 623 undisturbed copulations of 25 males in sunlight at temperatures of 30-32° C. showed the mean duration to be 17.5 sec. If the copulating pair was disturbed by the observer, copulation sometimes lasted as long as 90 sec. Perhaps the male held the female until safe conditions prevailed.

After copulation the pair separates and the male flies slowly back to the site while the female closely follows. In some cases the female seemed to lose the male, whereupon he returned and led her to the site. Upon returning, the male hovers low over the site while the female slowly approaches. When the female begins to oviposit on the site, the male flies about her, dashing at any male intruders. If the ovipositing female begins to leave or is chased, the male follows her, whereupon she may follow him back to the site where he hovers until she resumes oviposition.

In many cases the female repeatedly avoids capture by a male at a site and finally fails to follow him, whereupon the male performs the activity described above as examination of the site.

It seems to require five minutes or more of examination before a male establishes himself at a site. If a female arrives while a male is still in the early stages of examination, the male may wander instead of leading her to a specific spot: the female quickly leaves in these cases.

If, while a pair is copulating, the oviposition site is removed, the returning male hovers near where the site had been, while the female follows. The female then flies to another site and oviposits alone, or the second male may court and catch her as usual.

If not ousted by other males, the male stays at one site throughout the day and returns to it on succeeding days with little or no re-examination. If the site is moved slowly to a new position, the male will follow and on subsequent days return to the new position. However, if a site is blown about the pond by the wind, it is soon vacated. Males would also vacate sites if the nearby water level lowered to less than about eight inches. Sites near steep banks in deep water were most used.

A male was rarely present at the pond alone for more than ten minutes. It was remarkable that regardless of the time the first male arrived, be it 8 A.M. or 3 P.M. he was succeeded by many others in about 15 minutes. When a second male arrives and encounters the first male, the latter dashes at him. If the second male had been at the pond previously, he would fly quickly to the site of the previous day and only occasionally encounter the first male. If the sites are at least 3 m. apart, the males gradually avoid encounters and perch near their sites. When many sites are available, the males come to be spaced about the pond at regular intervals of about 3-5 m. in areas including a site and sufficient water space to allow for flights over the water without extra-male molestation.

Males arriving at the pond for the first time may fly from site to site performing examination behavior. The localized males fly about them wildly and make roundabout flights over the pond. If the newcomer persists, wing fluttering may occur, the males hovering a few inches to a foot or more apart and making quick up and down movements over the site (Fig. 6). The duration of this activity may last a few minutes to an hour. Departure of one of the males is often preceded by the two facing each other in the air. Sometimes departure does not occur until after a pursuit flight, one male following the other at relatively great speed over the water in low roundabout

FIG. 6. Two Perithemis males fluttering their wings near each other.

flights. Such flights may last to 30 minutes. A departing male may leave the pond or fly to another site.

The initial spacing of the day requires little time, for the previous residents take their places quickly. As newcomers arrive, there is considerable wing fluttering at sites, the appearance being strongly reminiscent of flocks of "dancing" male Diptera. During such periods of violent activity many males leave the pond, and the organization becomes more tranquil. This tranquillity is again disturbed throughout the day as males return to the pond or as newcomers arrive. Usually there was complete displacement of localized males in the forenoon and another in the afternoon. The daily time of arrival of some males was remarkably constant and differed from male to male. Some males arrived daily in the forenoon and others in the afternoon.

Since the number of sites was limited, males sometimes took sites that were evidently unsuitable, for females consistently flew from courting males at these sites, then copulated at other sites. Males at these unsuitable sites frequently performed examination behavior and flew to other sites from time to time only to be driven away by male occupants of these other sites. Males repeatedly unsuccessful in obtaining sites flee readily when dashed at by other males, the flight being straight, like that of a female. This straight flight seems to induce male pursuit.

Instead of leading a female directly to a site, a male may deflect this course from another male's area as though aware that the occupant may interfere. Males from other areas often do pursue a female being led by another male, and a mass chase ensues. In these cases the female may fly until the number of males diminishes, whereupon she may turn and follow one of them. Extra males often make it impossible for a female to mate or oviposit, whereupon she may leave the pond, perch, or fly to another site.

Males examining sites were sometimes preyed upon. Wolf spiders (Dolomedes) were observed to catch six males, water bugs (*Belostoma lutarium*) four, fish two, and frogs (*Rana catesbeiana*) two. When a number of males are available, vacancies left after predation are quickly filled. If the male has been caught after the organization has become stable, however, the vacancy may not be filled for hours. Attraction of females to these sites by males is thus temporarily halted.

With bright sun and temperatures approaching 37° C. the males skew their wings causing a shadow to fall on the body, while they point their abdomens toward or away from the sun preventing the rays from striking the abdomen perpendicularly. The ordinary horizontal position is resumed when clouds obscure the sun.

*Female Behavior.* Each day, the first female arrived at the pond 10-30 min. after the first male. The organization was, therefore, relatively stable by the time the females began to arrive. During courtship at the site, the female may perch almost immediately, or fly and avoid capture by the male until she has been led back a number of times. If the female, after courtship, flies too rapidly to a perch and the male loses her, she may return to the site and, after further courtship, fly more slowly toward the perch where the male catches her. If an open net is placed over a courting male and moved slowly from the site, the female may follow the net in which the male hovers.

During copulation, the female wraps her feet around the male's abdomen. If, after copulation, the male flies back to the site too rapidly the female may apparently lose him, whereupon she may fly about and finally locate the site alone, but usually the male returns and leads her directly to the site. Both sexes showed a marked tendency to return to the exact spot where courtship had occurred and where the female had dipped her abdomen during courtship. In one case while a male was courting a female at one spot, the female repeatedly left and performed oviposition motions at a spot nearby. The pair copulated and oviposition occurred at the new spot. Thereafter, the male courted females only at the new spot.

If a female flies away after courtship the male attracts her again, whereupon she may follow or go to another site and stop for a second male or continue to yet another. During oviposition a female may be chased by an extra male whereupon she may return repeatedly to the spot to oviposit before being chased again. At such times the male which has copulated dashes at the extra male, then leads the female back to the spot. If extra-male interference makes oviposition impossible the female may go to another site where she is usually courted as a newcomer and may mate as usual, or may evade the male while ovipositing intermittently. In this case the male may pounce on the female and catch her; then copulation and oviposition occur in normal sequence. In this case the female may not copulate, but the pair flies in tandem about the site for an extended period while the male repeatedly bends his abdomen as in copulation. This behavior most nearly approximates what has been described as oviposition attended by the male. Needham (1929) says of this species, "The female is held by the male while ovipositing, but I have seen her oftener unattended, dropping her eggs on bits of floating dead pond skum by many successive dips at very nearly the same spot." Williamson (1899) says, "As in other related species (Celithemis and Sympetrum), the male retains his hold of the head of the female as she flies along near the surface of the water, occasionally striking the surface of the water with her abdomen to release the eggs." It seems unlikely, however, that Needham and Williamson were describing this rare behavior. In any case the writer has never seen the male hold the female during oviposition.

If an ovipositing female is netted and another female flies to the site and begins to oviposit, the guarding male does not usually court the newcomer but continues guarding. If the newcomer ceases ovipositing and begins to leave, or flies toward the male, the male may court her and copulation occur. Attraction of females to one another leads to communal oviposition. If a female arrives at a site where another female is ovipositing, the male may court or dash at either of the females. This may lead to re-copulation with the original female, copulation with the second, or chasing of one or both females from the site.

When all males are removed from the pond, the females fly to the oviposition sites, then either leave or oviposit alone. All of the egg samples (123) taken from females which had begun oviposition alone contained fertile eggs.

When ovipositing, the female dips her abdomen repeatedly toward the site and is easily approached when so engaged. If an attempt is made to capture her, she flies slightly from the site after each dip. In such cases she will continue oviposition more readily if the male is nearby, especially if the male has hovered on the site several times as he does after copulation. This behavior may have great survival value, for the male may here serve as an advertisement of conditions safe for the ovipositing female.

The egg masses "explode" in the water as observed by Montgomery (1937). This scattering may be important, for eggs in vials fail to hatch when clustered. Some females seem to lay their entire complement of eggs when their abdomens are dipped into vials of water: a few of the last eggs are sometimes small and presumably unripe. Counts of egg samples of 13 females which seemed to have laid unusually large numbers of eggs showed an average of 150 eggs with a maximum of 206.

Upon returning to the pond after the first oviposition, females showed a marked tendency to revisit the previous site. At Wiley Lake, where fourteen sites were being used, 150/394 repeats were at the previous site.

Females showed a peak of repeating after an interval of one day following the first oviposition (Table III). One female lived at least 34 days after the first oviposition.

*Experimental Removal of Perithemis Males*

From June 5, 1953 (when the first male of the season arrived) until July 19, only two males were permitted to remain at any one time in the vicinity of six sites on the east shore of the excavated pond. All other males were caught immediately

TABLE III. *Observed interval between first oviposition and repeated oviposition of 603 Perithemis females*

| Days After First Oviposition | Number Repeating | Days After First Oviposition | Number Repeating |
|---|---|---|---|
| 1 | 74 | 18 | 3 |
| 2 | 102 | 19 | 1 |
| 3 | 80 | 20 | 2 |
| 4 | 67 | 21 | 2 |
| 5 | 48 | 22 | 3 |
| 6 | 28 | 23 | 2 |
| 7 | 35 | 24 | 1 |
| 8 | 34 | 25 | 0 |
| 9 | 23 | 26 | 0 |
| 10 | 22 | 27 | 1 |
| 11 | 13 | 28 | 0 |
| 12 | 13 | 29 | 1 |
| 13 | 14 | 30 | 1 |
| 14 | 10 | 31 | 0 |
| 15 | 11 | 32 | 0 |
| 16 | 5 | 33 | 0 |
| 17 | 6 | 34 | 1 |

as they arrived. Egg samples were obtained from all females mating with these remaining males. This study gave further insight into the behavior of the species.

The remaining males showed almost absolute localization to particular sites throughout their duration at the pond and mated with the numerous females coming to these sites. After frequent copulation, the males often made abortive copulations and appeared weak. The time during which such males remained coupled to females was often prolonged. Under these conditions, the females were often observed to fly toward perching males and, in a few cases, pounced on them. After several days of abnormally high numbers of copulations, the males failed to return to the pond or, in two cases, left soon after returning. In one of these cases, after the male had mated 127 times in six successive days after first arriving at the pond, he returned the seventh day, mated one time, then left and did not return.

Egg samples from females with which the total of seven males had mated were checked for the approximate proportion of eggs hatching. Evidence of infertility of some or all of the eggs was found in the records of the three males for which most extensive records were obtained. No eggs hatched in a sample taken after the 43rd copulation of a male on the 8th successive day he had mated at the pond, and only about half of the eggs hatched in samples taken after the 45th and 46th matings of the same male on that day. The male returned next day, however, and copulated with four females whose eggs were fertile (perhaps they had been previously inseminated). The male returned on the 10th day, but left after about 15 min. without having mated and did not return.

The records of another male showed only about 10% hatching in a sample taken at the 15th (last) copulation on the first day at the pond, but a high percentage of the eggs hatched in samples taken after each of 85 subsequent matings of the male on seven additional days.

One male mated 8 consecutive days with a total of 101 females, but eggs hatched in high proportions in all samples. On the 9th day, however, the 2nd sample showed no hatching. On the 10th day the male returned, appeared inactive, and mated with only one female 25 min. after arriving: the eggs were fertile. After 45 min. he left and did not return.

*Correlation of Weather and Behavior*

In spite of wide fluctuations in air temperature, the mean number of daily matings (at least at temperatures 23-32° C.) was remarkably constant

TABLE IV. *Maximum air temperature and daily matings of Perithemis at Wiley Lake, 1952 and the excavated pond, 1953*

| Maximum Air Temperature | Matings | Days | Matings/Day |
|---|---|---|---|
| 23-24° C. | 106 | 5 | 21 |
| 25-26° C. | 244 | 9 | 27 |
| 27-28° C. | 235 | 12 | 20 |
| 29-30° C. | 570 | 25 | 23 |
| 31-32° C. | 319 | 15 | 21 |
| 33-35° C. | 347 | 11 | 32 |
| 36-37° C. | 102 | 2 | 51 |

(Table IV). Activity was slowed at temperatures below 25° C. and appeared then to depend on warmth derived from sunshine. At temperatures above 30° C. activity was about as great in cloudy weather as it was in sunshine. A great increase in mating was often noted just before a storm, during hot humid cloudy conditions. At maximum daily temperatures of 23-28° C. only 12% of the matings were before noon, but at temperatures of 29-37° C. 32% were before noon. On days with a maximum temperature of 33° C. or higher, there was a lull in mating from 1-3 P.M. After unfavorable weather there was often an abnormally large number of matings.

*Tests of the Function of the Amber Color of the Wings*

Observations of female behavior indicated attraction to males. It also appeared that females were influenced by courtship, since they would stop flying during such activity. If, while a female was being courted, the male was caught by a water bug and slowly pierced to death, the female would perch on the site while the male's wings were in motion, but when his wings finally stopped she no longer paused, but continued oviposition.

Following Darwin's views, one might suppose that the amber color influenced the female. Experiments were designed to test this hypotheses as well as the hypothesis that the color is employed in inter-male activity as had been found in Plathemis.

Live males and females were glued by their feet to sticks and placed near oviposition sites. Also, the wings of 200 males were blackened with shoe dye, while those of 200 controls were weighted with varnish thinned with turpentine to correspond roughly with the weighting effect of the shoe dye. The reactions of both sexes to the glued specimens and treated males were studied.

*Interactions of Blackened-winged Males and Normal Ones.* When a male with blackened wings flies to the pond, a localized male reacts to him as he does to a female: he flies toward the blackened male then turns and flies slowly to the oviposition site. The blackened male may follow, much as does a female. The normal male may court the blackened one at the site and, if the blackened male examines the site or flies slowly, the normal male may pounce on him, couple, and bend his abdomen into the precopulatory position. The blackened male is soon released, and after a few repetitions of this behavior the normal male fails to react to the blackened one. The blackened male may begin wing-fluttering near the normal male, but the latter fails to react and perches. The blackened male pounces repeatedly on the perching male, thus driving him away temporarily, but the localized male returns and perches again. After a time the blackened male may dash at and pursue the normal male over the water, but this flight usually is brief, and the blackened male leaves the area.

If a female arrives while the two males are in the area, she may follow either of the males to the site and be courted. Although it sometimes appeared that the female followed the normal male more readily than she did the blackened one, especially while she was being courted by a normal male or had copulated with him, this was not definitely determined. During courtship the female may stop and be captured by either male in the temporary absence of the other. It was not determined whether she would stop for a normal male more readily than for a blackened one.

When most males are removed and vacant sites are available, a blackened male leaving the area of a normal male may localize in a vacant area. A lone blackened male often seemed as successful in obtaining mates as the normal ones. Quantitative information on this point would be difficult to obtain, since mating success seems dependent on the nature and location of the site. Females may avoid capture by a blackened male at a site and copulate with a normal male at the same or another site, but the converse was found about as frequently.

When a blackened male enters the area of another blackened male, he is treated as a female. Upon reaching the oviposition site, the two blackened males may court each other simultaneously and may couple. They may later fail to react mutually and fly about the same area or perch as little as a foot apart. As many as six blackened males were observed "aggregating" in one area. Mating under these conditions was unsuccessful, for the females would be chased while following a male to the site.

Two or more blackened males may chase a normal male from an area by chasing him in turns, but a lone blackened male failed to do so except in a few cases after the normal male had mated a considerable number of times.

The males with wings weighted with varnish had a tendency to behave like blackened males, but the aberration (from normal) was less marked. These controls were sometimes treated as females by other males, so sex recognition may be partly based on flight pattern.

*Reactions of Males and Females Toward Specimens Glued to Sticks.* When male and dark female specimens (live or dead) are glued by their feet to sticks and placed near the site, the reactions of males toward female specimens differ from their reactions toward male specimens. Males bend their abdomens into the coupling position on the wing spots of female specimens or successfully couple with the head. On the other hand, they pounce repeatedly on the wings of male specimens without bending their abdomens. Coupling was most often observed among males which had just lost a female during courtship. Under these conditions a male would sometimes couple with a male specimen if the female specimen had been removed. With the female present, however, coupling was with that sex in all cases.

If a light female (with amber wings) was placed with a dark female and male specimen, the males reacted toward the light female sometimes as toward a male and sometimes as toward a female. Likewise, males, while wing-fluttering with other males, sometimes reacted to flying light females as if they were males: instead of courting the light female they would perform wing-fluttering with her and make roundabout flights over the pond. In such cases, males courted the light females after oviposition motions began.

### *Studies on Site Selection*

Some males at the pond had much higher mating scores than did others. This disparity of mating seemed to depend, at least in part, on female choice. Some observations suggested that females were influenced by the attraction of the oviposition site in their apparent choice of males. Males localized at certain sites were consistently deserted by females courted there, while other males at the same time were mating repeatedly at other sites. One male held such a site for seven days while courting many females all of which avoided capture. This male did not differ appreciably in vigor or appearance from other males. Furthermore, a female would sometimes desert an ordinary male at one site where she was being courted, and mate with a blackened male at another site. At one site where most of the mating of the day was occurring, the male had battered wings (males may clash their wings with females during capture). With males removed, females continued oviposition mostly at sites where most mating had occurred when males were present.

To determine whether females differentially selected sites, a variety of sites were placed in the solution pond June 7, 1952. Two lines of small logs were tied end to end and floated parallel in the water about 5-9 m. from the shore. Patches of Potamogeton were fastened to the logs at about 5 m. intervals. All other sites were removed. At a position on the shore commanding a view of the sites, an observation tower was built.

Marked males were classed into three groups according to size of wing spot as correlated with density of body pigmentation. Males with a spot one-half mm. or more in any dimension were classed as dark, those with a spot less than one-half mm. as medium, and those with no spot, as light. Females were classed three ways according to degree of pigmentation of wings and body. Specimens which could not be classified were not considered.

Upon arriving at the pond for the first time, a male would examine a number of points along the logs and patches. Finally, apparently having selected a spot, he would perch nearby. The site might be a stick or log projecting from the water or a patch of Potamogeton. If a male was displaced from a stick or log and selected another site that day or on following days, he would usually select another stick or log, while other males, would as consistently select patches of Potamogeton. The females, returning to oviposit, behaved similarly. Of all repeaters, 91% of the males and 92% of the females could be classified according to whether they selected sticks or patches.

A comparison of the records at the end of July, 1952 suggested that dark individuals of both

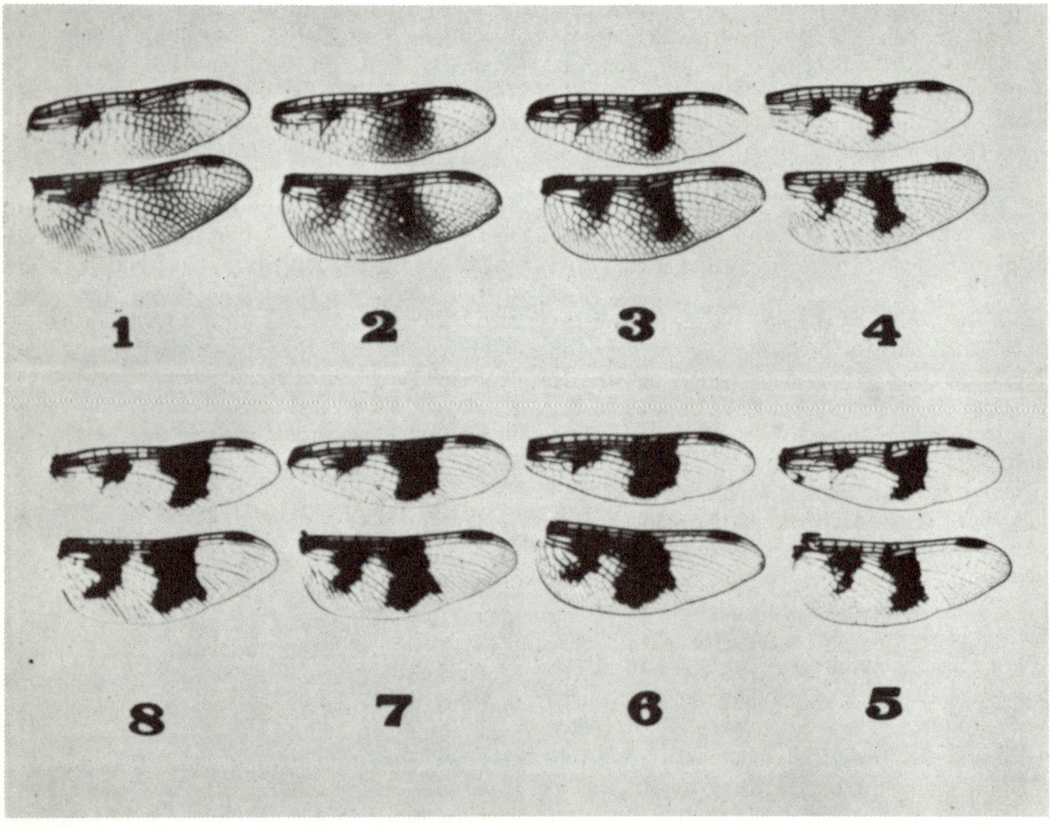


FIG. 7. Wings of Perithemis females classed according to degree of pigmentation. Classes 1-3 are light, class 4 is intermediate, and classes 5-8 are dark. The frequency distribution of color classes is as follows:

| Classes: | 1 | 2 | 3 | 4 | 5 | 6 | 7 | 8 |
|---|---|---|---|---|---|---|---|---|
| Number: | 24 | 157 | 327 | 344 | 287 | 259 | 79 | 20 |

sexes tended to select log or stick sites, while light ones tended to select patches of Potamogeton. On August 3, further observations were begun at Wiley Lake. Fourteen sites were placed at 5 m. intervals along the east bank. The sites consisted of logs charred by fire (which seemed to be especially attractive) alternating with patches of Potamogeton floated on small boards and almost sunk. All other oviposition sites were removed.

Standards were prepared by placing the wings of females between glass plates. Females were classed eight ways (Fig. 7) according to extent and density of wing pigmentation as correlated with density of body pigmentation. Males were classed three ways as before. Observations were continued until September 14. It was again found that dark forms of both sexes selected mostly logs and sticks for mating and oviposition, light forms mostly patches, and intermediate ones showed intermediacy in selection.

Further studies were conducted at the excavated pond from June 5 to July 19, 1953. Six sites similar to those studied at Wiley Lake were placed along the east bank. Only two males were permitted to remain at the studied sites at one time. Other males were caught each day as they began examination of the sites and were confined in jars containing fresh grass as perches and were kept in bags in shade. Males survived well under these conditions and were released in the evenings. In this way most of the males were prevented from localizing at a site. It was thought that this procedure might allow fuller expression of any tendency toward localization at a specific site. Females were studied according to methods employed at Wiley Lake.

The results agreed with those obtained at Wiley Lake in that a large proportion of the dark forms and a smaller proportion of the light forms selected logs, and the intermediate ones showed intermediacy in selection. Tables V and VI give a summary of the Wiley Lake and excavated-pond studies. The differences in proportions of color forms of both sexes showing differential selection of sites are highly significant.

Among those males which were captured when they arrived at the sites, only 63% of the returns were at sites visited previously, while non-captured males showed almost absolute localization to a specific site at the pond. Although the uncaptured males had much more opportunity to linger about the sites than did captured ones, both groups showed about equal tendency to select log or patch sites.

## Discussion and Supplementary Observations

*Competition and Sex Recognition among Males.* Dashes of males at one another, pursuit displays, dual flights in Plathemis, and wing-fluttering and pursuit flights in Perithemis seem to be forms of male competition. The white upper surface of the abdomen of Plathemis and the amber color of the wings of Perithemis appear to be employed in this competition. The psychological motivation behind the competition remained obscure. Association of aggressive and sexual motivations seemed very close, for when the upper surface of the abdomen of Plathemis females is painted white so as to resemble that of a male, males dash at them as they do at other males, but this may end in copulation when the female flies.

It appeared obvious that males of both species were able to distinguish sex, for males often made only short dashes at passing males, but undertook full pursuit when a female passed at about the same distance. Plathemis males often remained motionless when other males flew overhead in dual flights, but would intercept a female flying through the air at about the same distance. Groups of many males sometimes pursued a single

TABLE V. *Results of site selection studies of Perithemis females. Classes 1-3 are light, class 4 is intermediate, and classes 5-8 are dark*

| Classes | Log Selectors | Non-log Selectors | Total |
|---|---|---|---|
| **Observed Values** | | | |
| 1-3 | 200 | 308 | 508 |
| 4 | 185 | 159 | 344 |
| 5-8 | 376 | 269 | 645 |
| Total | 761 | 736 | 1497 |
| **Expected Values** | | | |
| 1-3 | 258.27 | 249.73 | 508.00 |
| 4 | 174.89 | 169.11 | 344.00 |
| 5-8 | 327.92 | 317.08 | 645.00 |
| Total | 761.08 | 735.92 | 1497.00 |
| **Chi-square** | | | |
| 1-3 | 13.15 | 13.60 | 26.75 |
| 4 | 0.58 | 0.60 | 1.18 |
| 5-8 | 7.05 | 7.29 | 14.34 |
| Total | .... | .... | 42.27 |

TABLE VI. *Results of site selection studies of Perithemis males. Class 1 is light, class 2 is intermediate, and class 3 is dark*

| Classes | Log Selectors | Non-log Selectors | Total |
|---|---|---|---|
| **Observed Values** | | | |
| 1 | 297 | 375 | 672 |
| 2 | 179 | 186 | 365 |
| 3 | 211 | 114 | 325 |
| Total | 687 | 675 | 1362 |
| **Expected Values** | | | |
| 1 | 339.36 | 332.64 | 672.00 |
| 2 | 184.33 | 180.67 | 365.00 |
| 3 | 164.13 | 160.87 | 325.00 |
| Total | 687.81 | 674.19 | 1362.00 |
| **Chi-square** | | | |
| 1 | 5.29 | 5.54 | 10.83 |
| 2 | 1.54 | 1.57 | 3.11 |
| 3 | 13.35 | 13.67 | 27.02 |
| Total | .... | .... | 40.96 |

female. Flight pattern may be partly responsible for sex recognition. In addition, the experiments with male and female Perithemis specimens glued to sticks made it appear that that dragonfly can recognize sex of even a dead specimen on the basis of wing-color pattern.

Sex discrimination sometimes appeared obscure during periods of intense mating activity, for male-by-male coupling might then result. Perithemis males often pounce on perching males without making coupling motions: in intense competition over females, however, male-by-male coupling sometimes resulted. Similar results were obtained with glued specimens when a male would couple with a glued male after a female which he had been courting had escaped.

Moore (1952a), in his studies of British dragonflies, postulates that the species he studied were unable to distinguish sex, and he was disinclined to view the inter-male reactions of dragonflies as forms of aggressive behavior, but considered them to be results of sexual attraction between males. This opinion was based on results of experiments with males tied to strings and held in the presence of free males, which sometimes coupled with the captive. Male guarding of the ovipositing female was interpreted as follows: The ovipositing female provides the male with an attractive stimulus. Passing males also present an attractive stimulus causing the male to dash at the intruder with sexual motive. If the intruder flies away, the male returns to the stimulus provided by the ovipositing female.

In the present study it was found that when male competition was very intense, the guarding male stayed unusually close to the female. This action can hardly be explained in terms of attraction to other males.

*Localization.* Under ordinary conditions, some males show nearly absolute localization to one sector of the pond throughout their adult life; and since they eliminated other males from their areas (whether the motive was sexual or aggressive) the species studied show conventional territorial behavior.

Speculations on the so-called territories of dragonflies have led to three interpretations: (1) The inter-male combat may lead to sexual isolation. (2) A territory may provide a male with a hunting ground. (3) Spacing males may insure dispersal.

Williamson (1906) considered male combat to be important in sexual isolation. He postulated that males show little discrimination for females of other species, but inter-species mating is prevented by "fighting" of males for females. In the present study, male interaction was almost entirely intraspecific. Plathemis males discriminate between their females and those of other species. Females of *Libellula pulchella,* which appear similar to Plathemis females, were common at the pond, but only one coupling (without copulation) between a Plathemis male and an *L. pulchella* female was observed. Plathemis males frequently chased *L. pulchella* females, but gave up the chase on close approach. Perithemis males showed almost no attraction to females of other species, but occasionally would pounce briefly on ovipositing Plathemis females, which are much larger than females of their own species.

Saint-Quentin (1934) considered the territory or "Jagdrevier" of the dragonfly to be a hunting ground. In the present study dragonflies were only rarely observed to feed at the pond, and mating seemed to be the primary, if not only, reason for their presence at the pond.

Moore (1952a) considered dispersal to be the function of the "so-called territories" of dragonflies. He postulated that sexual attraction, resulting in clashes, caused wandering of males to bring about dispersal. Moore maintains that dispersal is not significant in preventing male interference with ovipositing females since dispersal "does not prevent interference, and therefore cannot have this function."

Although male interaction may help distribute males, prevention of male interference with ovipositing females seemed to the present writer to be an obvious function of the dispersal. This view is similar to that stressed by Armstrong (1947) in connection with arena display in birds. Armstrong maintains that this form of territorialism is an adaptation to meet the problem of a differential sex ratio. With dragonflies, this differential sex ratio is occasioned by the tendency of males more than females to aggregate at the pond. The presence of several males seriously interferes with oviposition, sometimes preventing it entirely. Spacing and elimination of males greatly alleviates this condition.

Evidence in support of this last statement was obtained from studies of the damselfly, *Enallagma aspersum* Hagen, a species in which the male retains his hold on the female during oviposition. In this species, localization at the pond was only transitory among 488 individually marked males. These males showed little interaction, but flocked close together at times. It seemed that, since the male retained his hold on the female, thereby preventing clasping by another male, territorialism is not of great advantage. Males were often observed assuming the coupling position on an ovipositing female held by another male, but oviposition seemed little disturbed. At certain times, however, the females descend stems of Potamogeton and oviposit farther beneath the water. The male then releases the female while his wings are still above the water, perches or hovers over the spot where the female descended, and shows territorial behavior in dashing at other males entering the area. When the female arises after 5-25 min. she may get caught in the surface film. The waiting male seizes her, lifting her from the water, and the pair flies away and may continue oviposition.

*Sex Ratio.* If the apparent excess of males at the pond lowers the biotic potential of the species by inhibiting ovipositon, the question arises as to why the ratio is not otherwise adjusted. Indications are that emerging Plathemis females already outnumber the males by about 1.62:1. Borror (1934) found among 830 individuals of *Argia moesta* that females outnumbered males by 2:1. Garman (1917) notes that female dragonflies are more easily reared than males; they may therefore be more viable. Breeding experiments with other insects would make it appear that sex ratio can be altered by genetical means..

But it may be disadvantageous for too few males to be produced, for the following advantages for an excess of males in the breeding area may be postulated: (1) Inhibition of oviposition by excess males may prevent crowding of eggs or young. (Eggs clustered in vials often failed to hatch.) Competing males may drive females from attractive sites to cause them to lay eggs on sites less attractive but in the long run more favorable. (2) With many males present, defective males may be prevented from leaving progeny. Males with defective wing tips such as result from improper emergence had very low mating scores and appeared unable to defend a territory for an extended period. This defect is probably not genetical, however. (3) Competition may preserve genetical variability by inducing frequent changing of males. (4) Frequent replacement of males at sites may be necessary for complete insemination of females. There was lowered hatching of eggs in samples taken from females when many males had been removed from the pond.

*Simulation of Oviposition.* Males often simulate oviposition when guarding or examining a site or following the seizure of the guarded female by another male. This activity may have the following significance: (1) It may indicate to other males that a specific site is being guarded. Males in the same area reacted toward one another less violently if each were guarding a different site than they did when both were guarding the same site. (2) It may prevent predation of the female. Perithemis males were sometimes caught by predators while simulating oviposition at the site. If other males are not numerous, his place may not be taken for hours. Females will then not likely suffer the same predation, for remaining males will attract the females to other sites. (3) The activity may be involved in the examination of a site. This is supported by the fact that simulation of oviposition occurs mostly when a male first arrives at a site.

Moore (1952b) observed simulation of oviposition among male dragonflies and interpreted the activity as a displacement reaction, that is a substitutional action of inappropriate nature following frustration of a normal action.

*Differential Selection of Sites.* If the tendency of dark forms of Perithemis to select logs for mating and oviposition while light ones select mostly patches of Potamogeton covered with algae has a genetic basis, it may serve as a genetical isolating mechanism. It may be important, in view of the temporary nature of some of the situations in which Perithemis breeds, that forms selecting mostly Potamogeton (perhaps open pond forms) not be strictly isolated from stick- or log-selecting forms, which may normally occupy shaded habitats where Potamogeton and algae are scarce. In the words of Dobzhansky (1951) "Too early and too rigid isolation of favorable gene combinations . . . would mean too much specialization of the organism to environmental conditions that may be only temporary." Differential site selection may bring together males and females of similar type, while inter-male competition induces mixed matings. Courtship may be related to site selection, for the male's "waiting" during courtship may allow the female opportunity to examine the site without interference.

*Comparison of Color Forms of Perithemis.* Comparative study suggests that the amber color in the wings of males and light females results from diffusion through the wing of dark pigment characteristic of the two patches on the wings of dark females. Light Indiana females and some males of the dark species *P. seminole* from Florida show remnants of the outer patch characteristic of dark females. The outer patch seems to disappear first, then the inner one, and finally a small spot of the inner patch, a spot found surmounting the triangle in some males. On one individual the wings of the right side were like those of a light male, whereas those on the left side showed remnants of female-like patches. This specimen was several times caught by another male before it was netted. The genitalia did not appear different from those of an ordinary male, and many sperm were being produced.

Brimley (1903) reports that tenerals of Perithemis fly in upland fields, often resting on flowers of ox-eye daisies. In this connection it is interesting that light females are inconspicuous against a background of daisies, while dark forms stand out boldly. In view of the fact that birds prey on tenerals, light forms may have a selective advantage in a pasture-pond environment.

## Summary

The problems of territorialism and sexual selection were studied among marked individuals of *Plathemis lydia* and *Perithemis tenera,* at ponds near Bloomington, Indiana.

The results with Plathemis were as follows:

1. The sex ratio of marked tenerals was 129 males: 209 females. Teneral males closely resemble females in color, but upon returning to the pond after the maturation period, males differ strikingly from females in having developed a bluish-brown, bluish-white, or white upper abdominal color.

2. Males examine oviposition sites and establish themselves in restricted areas near these sites which they defend against other males.

3. Males were not observed to display to females, but they appear to hide the white upper abdominal surface by lowering the abdomen while approaching females for capture.

4. Copulation time was about 3 sec. and remarkably constant, but the time was prolonged if the pair was molested by other males.

5. After copulation, the male usually places the female near the oviposition site and guards her by hovering over her while making dashes at intruding males as the female oviposits. A male may mate with and guard several ovipositing females. In all observed cases, guarding behavior was preceded by copulation with the female being guarded, or with another female at about the same time.

6. Male defence activities consisted in dashing at other males, while displaying the white upper surface of the abdomen, and in performing dual flights with other males back and forth across the pond and adjacent fields. This activity appeared partly responsible for continuous displacement of males throughout the day, and led to the establishing of territories with lateral dimensions of about 11 m.

7. There was considerable disparity in number of matings occurring at various sites, and male interaction was greater at sites much used for oviposition. Males most in evidence at 2-3 P.M. (CDT), when most mating occurred, had higher mating scores than other males.

8. When males seize ovipositing females, the guarding male becomes intensely active in dashing at nearby males and at the female placed at the site (after copulation) by the male which seized her. A mass chase of the female may result to make oviposition temporarily difficult or impossible. The female may avoid repeated seizure by remaining motionless on the site. (Males appeared to be unable to see the female in these cases and sometimes made coupling motions on the female's shadow or wing spots.)

9. Immature males fly in a slower, more fluttering manner than older males and may temporarily chase males of *Anax junius* as if mistaking them for females of their own species. They quickly seize females of their own species, but sometimes make abortive copulations. They are less active in guarding ovipositing females than are older males and are easily driven from the pond by the latter.

10. Upon return to the pond, males usually re-

turned to the areas they had occupied previously, thus showing localization. Immature males showed little localization when numerous males were present, but localization was usually firmly established by the fifth day at the pond. Localization is correlated with the development of whiteness of the upper surface of the abdomen and increasing encystment of the testes and spermiogenesis.

11. Females are captured in the air, or in the vicinity of oviposition sites when they flit around the sites or perform oviposition motions thereon. They are more easily captured by males near some sites than they are at others. Returing females showed a tendency to oviposit at sites at which they had previously oviposited even after several days' absence from the pond.

12. Females are attracted to other flying or ovipositing females, the tendency leading to communal oviposition. They never followed mature Plathemis males, but, rarely, followed immature ones.

13. When many males were removed from the pond, the remaining ones mated much more frequently than usual, but became less active after having mated twenty or more times and were easily driven away by other males. Egg samples taken at these times showed lowered hatching.

14. When all males were removed from the pond, females flitted around the sites and performed oviposition motions thereon, then became very active in chasing other dragonflies. In these cases, females performed dual flights with one another much as do males, after which one or other female would leave the pond.

15. Males with the white upper surface of the abdomen painted black were less able to defend territories and had lower mating scores than males with the surface painted white or left unpainted. The blackened males "aggregated" in sectors where relatively few matings occurred and showed a tendency to remain longer at the pond in the evenings than other males.

16. When the upper abdominal surface of females was painted white, males sometimes displayed to them as they do to other males, but they usually captured the females and mated, then, after copulation, the guarding male would fly toward the whitened ovipositing female and, when she flew, would seize her and copulate again. Guarding males would also dash at inanimate white objects.

The results with Perithemis were as follows:

1. Males precede females at the pond and closely approach projecting sticks and logs and patches of Potamogeton partly covered with algae. They then fly about a particular site and defend it against other males: finally, having eliminated male competition, they perch near the site. When a female flies over the pond, the male approaches her, then he flies slowly back to the site while the female follows. The male then hovers closely over the site as the female approaches. If the female slows down or stops, the male seizes her.

2. Copulation time was usually about 17 sec., but was greatly prolonged in shade or when the copulating pair was molested.

3. After copulation, the male flies slowly to the site while the female follows. Oviposition usually occurs at the spot where courting had occurred and particularly where the female had dipped her abdomen during courtship. If the female is chased from the site, the male follows her and leads her back.

4. If not ousted by other males, the male stays at one site throughout the day and returns to it on succeeding days with little or no re-examination. Daily arrival time of some males was quite constant and differed from male to male.

5. Male defence activity consisted in "dancing" up and down over the site, wing-fluttering near other males, and pursuit flights over the pond. A group of males "dancing" over the site resembled a swarm of "dancing" male Diptera. Defense activity led to the establishment of territories with lateral dimensions of about 6 m. There was usually complete displacement of localized males in the forenoon and another in the afternoon.

6. Males sometimes held sites where females consistently deserted them during courtship. Such males examined the sites repeatedly and flew to other sites from time to time only to be driven away by the male occupants. Males repeatedly unsuccessful in obtaining a site fly readily when dashed at by other males; the flight, being straight, like that of a female, seemed to induce male pursuit.

7. Male competition often made oviposition impossible, whereupon the female would leave the pond, perch, or fly to another site.

8. Males examining sites were sometimes preyed upon, whereupon the site was left temporarily vacant. In these cases, females are not likely to suffer the same predation, for remaining males attract them to other sites.

9. Attraction of females to other females leads to communal oviposition. With all males removed, the females may leave the pond after visiting the sites, or may oviposit alone. Of 123 egg samples obtained after females had begun oviposition alone, all contained fertile eggs.

10. Upon returning to the pond after the first oviposition, females showed a marked tendency to re-visit the previous site. Females showed a peak of repeating after an interval of one day following the first oviposition.

11. With most males removed, the remaining males mated repeatedly, then often made abortive copulations and appeared weak. Under these conditions, females were observed to fly toward perching males and pounce on them. Egg samples under these conditions showed lowered fertility.

12. In spite of wide fluctations in air temperature, the mean number of daily matings was relatively constant. On days with maximum temperatures of 33° C. or higher, there was a mating lull from 1-3 P.M. Peaks in mating often preceded a storm, during hot, humid, cloudy weather, and after unfavorable weather.

13. Males with the amber color of the wings obscured with black shoe dye were unsuccessful in competing for sites with normal males and males with the wings weighted with thinned varnish. Blackened and, to a lesser extent, varnished males were courted as females by other males.

14. When live or dead male and female specimens were glued by their feet to sticks and placed near the sites, free males pounced on the wings of glued males without bending their abdomen into the coupling position, but they quickly coupled with glued females.

15. Some males at the pond had much higher mating scores than others. The disparity of mating seemed to depend on female choice of oviposition sites. Quantitative data showed dark forms of both sexes tended to select mostly sticks and logs for mating and oviposition, light forms mostly patches of Potamogeton covered with algae, while intermediate forms showed intermediacy in selection.

## References


**Armstrong, E. A.** 1947. Bird display and behavior. London: Drummond.

**Borror, D. J.** 1934. Ecological studies of *Argia moesta* by means of marked individuals. Ohio J. Sci., **34**: 97-108.

**Brimley, C. S.** 1903. List of dragonflies from North Carolina, especially from the vicinity of Raleigh. Ent News, **29**: 150-157.

**Byers, C. F.** 1930. A contribution to the knowledge of Florida Odonata. Univ. Fla. Publ., Biol. Sci. Ser., **1**: 1-327.

**Darwin, C.** 1871. The descent of man and selection in relation to sex. New York: Collier.

**Dobzhansky, Th.** 1951. Genetics and the origin of species. 3rd. ed. New York: Columbia University Press.

**Garman, P.** 1917. The Zygoptera, or damselflies of Illinois. Bull. Ill. St. Lab. Nat. Hist., **12**: 410-457.

**Hingston, R. W. G.** 1933. The meaning of animal colour and adornment. London.

**Lieftinck, M. A.** 1934. An annotated list of the Odonata of Java. Treubia, **14**: 4.

**Montgomery, B. E.** 1937. Oviposition of perithemis (Odonata, Libellulidae). Ent. News, **48**: 61-63.

**Moore, N. W.** 1952a. On the so-called territories of dragonflies (Odonata-Anisoptera). Behavior, **4**: 85-99. 1952b. Notes on the oviposition behavior of the dragonfly *Sympetrum striolatum* Charpentier. Behavior, **4**: 101.

**Needham, J. G.** 1929. A handbook of dragonflies of North America. Baltimore: Thomas.

**Noble, G. K.** 1939. Symposium on the individual vs. the species. IV. The role of dominance in the social life of birds. Auk, **56**: 263-273.

**Saint-Quentin, D.** 1934. Beobachtungen und Versuche an Libellen in ihren Jagdrevierin. Konowia, **13**: 275-282.

**Tillyard, R. J.** 1912. On some new and rare Australian Agrionidae. Proc. Linn. Soc. N. S. W., **37**: 463.

**Walker, E. M.** 1912. The North American dragonflies of the genus *Aeschna*. Univ. Toronto Studies Biol. Ser., no. 11.

**Whitehouse, F. C.** 1941. British Columbia dragonflies (Odonata) with notes on distribution and habits. Amer. Mid. Nat., **26**: 488-557.

**Williamson, E. B.** 1899. The dragonflies of Indiana. Ind. Dept. Geol. & Nat. Res., 24th. Ann. Rept.: 299-333.

———. 1906. The copulation of Odonata. Ent. News, **17**: 143.

———. 1923. Notes on the genus *Erythemis*. Univ. Mich. Mus. Zool., Misc. Publ., no. 11.

# COURTSHIP AND MATING BEHAVIOR AS A REPRODUCTIVE ISOLATING MECHANISM IN *DROSOPHILA*

Lee Ehrman

*The Rockefeller Institute, New York*

Species of sexually reproducing organisms are genetically closed systems. They are closed systems because they do not exchange genes or do so rarely enough so that the species differences are not swamped. Races are, on the contrary, genetically open systems. They do exchange genes by peripheral gene flow, unless they are isolated by extrinsic causes such as spatial separation. The biological meaning of the closure of a genetic system is simple but important—it is evolutionary independence. Consider these four species—man, chimpanzee, gorilla, orangutan. No mutation and no gene combination arising in any one of them, no matter how favorable, can benefit any of the others. It cannot do so for the simple reason that no gene can be transferred from the gene pool of one species to that of another. On the contrary, races composing a species are not independent in their evolution; a favorable genetic change arising in one race is, at least potentially, capable of becoming a genetic characteristic of the species as a whole.

Species are genetically closed systems because the gene exchange between them is impeded or prevented by reproductive isolating mechanisms. The term "isolating mechanism" was proposed by Dobzhansky in 1937 as a common name for all genetically conditioned barriers to gene exchange between sexually reproducing populations. According to Mayr (1963), isolating mechanisms are ". . . perhaps the most important set of attributes a species has. . . ." It is a remarkable fact that isolating mechanisms are physiologically and ecologically a most heterogenous collection of phenomena. It is another remarkable fact that the isolating mechanisms which maintain the genetic separateness of species are quite different not only in different groups of organisms but even between different pairs of species in the same genus.

Several classifications of the reproductive isolating mechanisms have been proposed. That of Mayr is a simple and convenient one. The two major groups are the premating barriers which prevent the formation of hybrid zygotes, and the postmating barriers which impede the survival or reproduction of these zygotes.

Three of the premating mechanisms are:

1) Potential mates do not meet (seasonal and habitat isolation).
2) Potential mates meet but do not mate (ethological or sexual isolation).
3) Copulation attempted but no transfer of gametes takes place (mechanical isolation).

Four of the postmating barriers are:

4) Gametes transferred but no fertilization, and hence no zygote formation takes place (gamete mortality).
5) Death of the zygotes (hybrid inviability).
6) The $F_1$ zygotes are viable but partly or completely sterile (hybrid sterility).
7) $F_1$ hybrids are fertile but the fitness of the $F_2$ or backcross hybrid is reduced (hybrid breakdown).

Since our primary interest today centers on the relations between genes and behavior, ethological isolation, sometimes also termed sexual or psychological isolation, should be discussed here in more detail. The phenomenon observed is usually that the mutual attraction between conspecific females and males is greater than the attraction between males and females of different species. Successful analysis of this form of isolation requires

The work reported here has been carried out under Contract No. AT-(30-1)-3096, U. S. Atomic Energy Commission.

Reproduced with permission from the *American Zoologist*, 4:147–153, 1964. Published by the American Society of Zoologists, Thousand Oaks, California.

a detailed description of the courtship rituals and the mating behavior in the pure species concerned. Species of the genus *Drosophila* offer abundant and favorable material for such analysis; their courtship and mating can be observed easily at any season of the year and under reasonably precisely controlled laboratory conditions.

My favorite materials are the incipient species of the *Drosophila paulistorum* complex or superspecies. As the name "incipient species" implies, these forms are very closely related and can be regarded either as very similar sibling species or as very distinct races standing on the brink of full species separation. This close relationship is, of course, very favorable for analytical purposes. What we wish to study is speciation in the process and the closeness permits a genetic analysis to be carried further than would be possible with full species.

The six incipient species of *Drosophila paulistorum* inhabit a part of the Neotropical zoogeographic region, from Guatemala and Trinidad in the north to southern Brazil in the south. Each incipient species has a distribution area of its own, but these are in part overlapping, so that in some places two, three, or even four incipient species live together, sympatrically, and apparently without producing any hybrids.

Thanks to the work of many authors, especially H. T. Spieth (1952) the sexual behavior in *Drosophila* is fairly well known. Sexual recognition is a trial and error affair among drosophilids. Males will generally attempt to court females of any *Drosophila* species, even distantly related ones.

At least four distinct elements can be distinguished in the courtship process in *Drosophila paulistorum.* The first element or stage is that of circling. The male approaches a female, attracts her attention, and may limit her movements by running around her. He never completes a full circle however, turns around, reversing the direction of his movement about every 330°.

Next he begins a tapping action—an important aspect of the courtship in this species (Fig. 1A). The male lightly touches the legs or the abdomen of the female with his own legs. At first, only one leg of each fly is involved. Since many taps may be necessary before it is established that the male has located a female of his own species, the male continues to circle between contacts. Here, the question of the female's receptivity is settled (this is probably mediated by samples taken by the great number of chemoreceptive hairs on the body of both insects); she will either remain still so that circling is no longer necessary, and then spread her wings to receive the male between them; or she will vigorously kick the courting male and do her best to depart.

Licking and wing vibrating occur next (Fig. 1B), both as a prelude to mounting. As Spieth (1952) has described it, the male "goes to the rear of the female and assumes a slightly crouched position with the tip of his abdomen slightly curled. Having positioned himself, he extends one wing 70° to 90° and vibrates it periodically, twisting his body on the longitudinal axis as he vibrates, taps (uppercuts), and occasionally licks or attempts to lick the female." In licking, the male proboscis contacts the female genitalia—and this is a reliable sign that the male is about to rush in for the mount. In *D. paulistorum* wing vibrations are not so important a part of courtship as in some of the other species. The one wing that is vibrated is used for leverage as the male raises himself on the body of the female.

Mounting and insertion seem to be accomplished simultaneously. A portion of the male reproductive organ is used in a clasping manner and, after mounting, the male secures his position on the female by placing his forelegs on top of her slightly-spread wings. This additional support is imperative because copulation lasts a fairly long time in this species (an average of 17 minutes and 12 seconds). During the copulatory period, the female may turn about or walk around; and she may even find herself having to fend off other males (Fig. 2).

When copulation is nearly over, the female attempts to dislodge the male by vigorously kicking and swinging her body from side to side. The male loses his hold on her wings when she snaps them together, and seconds later he falls off backwards. Thereafter, the female repels all sexually excited males by raising the tip of her abdomen, rendering her vaginal orifice inaccessible.

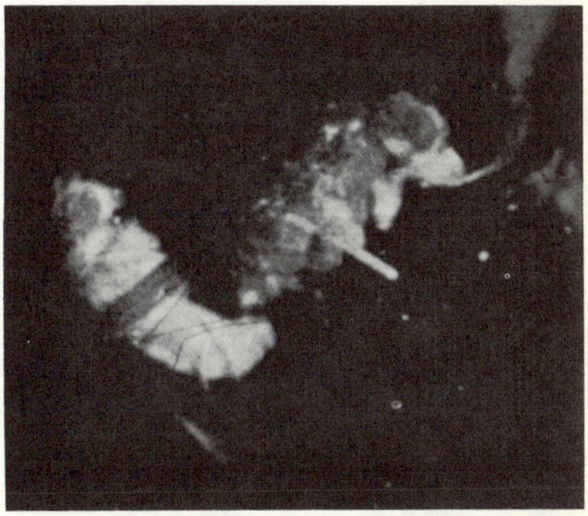

FIG. 2. Photograph of the full mount during which time the female may move about freely and even feed.

Notice that throughout this entire procedure, it is the female that at all stages, is "discriminatingly passive" while the male is "indiscriminately eager" (Bateman, 1948). *D. paulistorum* males are very active and will court almost anything: a dead fly, a mired fly, lumps of food, and often other males. It is not unusual to observe the formation of a chain initiated by a male courting a female, and in turn being himself courted by another male who is being approached by yet another male. These chains, of course, are of short duration (Fig. 1C).

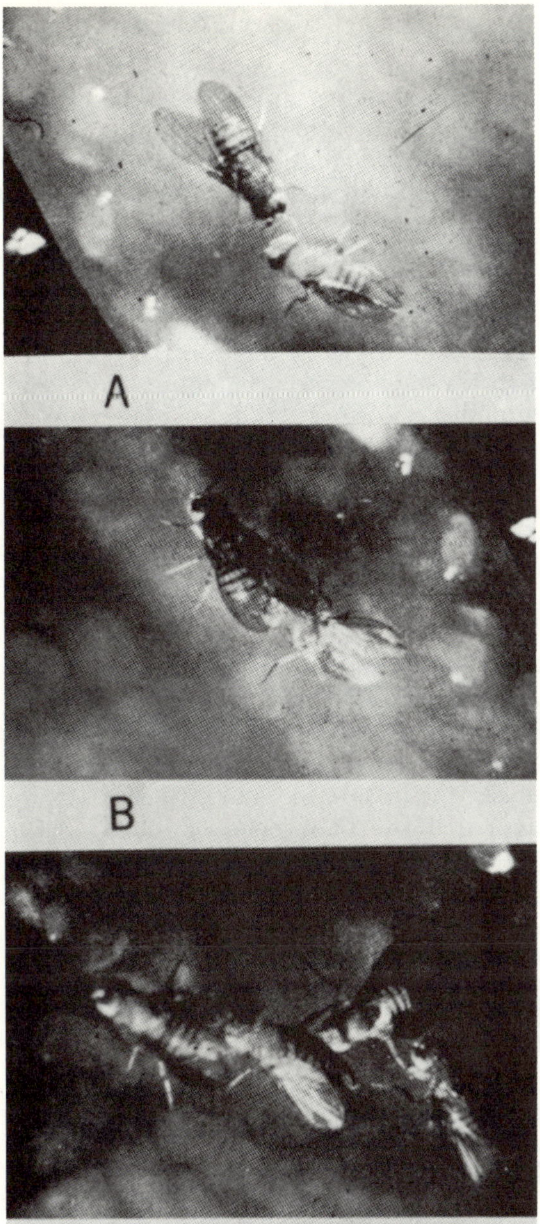


FIG. 1. These "stills" were taken from a 16 mm color film prepared by Mr. Richard F. Carter of the Rockefeller Institute and by the author; the project was supported by a grant from The Society of the Sigma Xi. This study of courtship be havior in *Drosophila paulistorum* was shown, in part, when this paper was delivered.

A. Tapping (the male is the smaller, active individual with the rumpled wings)—an initial stage in courtship.

B. Licking—often seen just before mounting.

C. The formation of a chain wherein one rejecting female (with extruded ovipositor) is being courted by a male who is in turn being courted by a male, etc., for a total of one female and three males—these chains are necessarily of short duration but do emphasize the fact that *D. paulistorum* males will certainly court the other males. See text for detailed explanation.

Now that the normal courtship and mating behavior of the species-complex has been recorded, we are ready to consider the evolution of this superspecies *Drosophila paulistorum* (Dobzhansky et al., 1964).

One of the first of the many interesting evolutionary phenomena exhibited by the *D. paulistorum* complex of seven known races or incipient species to be analyzed genetically was the complete hybrid male sterility discovered when crosses between the races were successful. The male sterility was found to be genic in nature (Ehrman, 1960a), and to be expressed via the genotype of hybrid mothers. This kind of hybrid sterility, unique in the genetic literature, operates in the following manner. Suppose that we cross two races, A and B, one or both of them having one or more chromosomes marked with suitable mutant genes. The distribution in the progeny of the chromosomes of different racial origin can then be followed by inspection of the external morphology of the flies. The interracial hybrid males are sterile, but the females are fertile and can be backcrossed to either A or to B males. After several backcrosses, flies are obtained which carry all but one of the chromosomes from race A, or all but one from race B. Females of this sort are, of course, fertile and can again be backcrossed to males of the recurrent race. Half of this progeny, of either sex, will carry all chromosomes of the same race, and half will contain one race-foreign chromosome. (Crossingover is suppressed in these hybrids.) Now, the striking fact is that males of both kinds are completely sterile. The sterility of the males identical in chromosomal constitution with males of one of the "pure" races can only mean that the sterility is in this case determined not by the chromosomal constitution of the individual itself but by that of his mother. The presence of at least one race-foreign chromosome in the developing egg cell before meiosis somehow modifies the cytoplasm or some other property of the egg, and makes a male individual developing from this egg sterile.

Once the above facts were established, the question logically arose whether the ethological isolation observed in races of *D. paulistorum* has the same peculiar genetic basis as the sterility of the hybrid males, that is, a genic-maternal effect. The ethological isolation is in all probability the mechanism which keeps the gene pools of these races separate in nature where the races share the same territory. Indeed, race hybrids have never been found in nature, and the cytological study of their chromosomes by Dobzhansky and Pavlovsky (1962) shows that genetically effective hybridization of the races is rare, if it occurs at all.

The genetic basis of the ethological (sexual) isolation was studied by Ehrman (1961). The method used was in principle the same as that applied for the analysis of the hybrid sterility, i.e., making crosses and backcrosses of strains of different races having some of their chromosomes marked with suitable mutant genes. It was here that the knowledge of the courtship and mating behavior of *D. paulistorum* became important.

Hybrid males of *D. paulistorum* transfer no sperm into the body of the female with which they copulate. $F_1$ hybrid males produce no motile spermatozoa, because a restitution nucleus is formed presumably after the first meiotic division, and then all the gametic material degenerates. Backcross hybrid males are even more profoundly sterile, since they frequently have no testes at all, or only one testis, or no gonial cells within the abnormally *thick* testicular sheath. Yet these males are normal in external morphology and have normal genitalia and internal reproductive organs other than testes. Their patterns of courtship and mating behavior are also normal. Since dissections of the female reproductive tract for the storage of transferred sperm (as is usually done in analyses of this sort where fertile males are involved) would be out of the question here; the entire analysis of the genetic architecture of sexual isolation as a reproductive isolating mechanism was made by using the simple but informative, direct-observation technique.

One may conclude from these direct-observation studies that the sexual isolation, which makes matings between the females and males of the *D. paulistorum* races much less likely to occur than matings within the races, is due to polygenes in every one of the three pairs of chromosomes which the species possesses. These polygenes control the sexual preferences of their carriers. Their effects seem to be simply additive. A female of hybrid origin which carries a majority of the chromosomal material derived from a given race is most likely to accept a male of that race. And conversely, a hybrid male is most likely to be accepted by females whose chromosomal constitution is closest to his. The source of the cytoplasm or the genetic constitution of the mother do not seem to matter. This is clearly not at all comparable with the genetic basis of the hybrid sterility, where the properties of an egg are determined, as far as the sterility of the backcross males is concerned, by the genotype present in it before meiosis, and not by that formed following fertilization.

We may surmise that this peculiar chromosomal-cytoplasmic mechanism of sterility arose first in the evolutionary history of these races, and that it arose in allopatric races becoming adapted to different environments in their respective geographic areas. Sexual isolation might then be built by natural selection, as a much less wasteful and more efficient reinforcement of the bar to gene exchange between the races. When this sexual isolation became strong or complete, the races became able to coexist sympatrically, as they have been found to do in some localities in the northern part of the South American continent (Dobzhansky et al, 1964) where as many as four races have been found to live in the same places.

How effective can sexual isolation be, what with omnipresent variations in environmental influences and the undeniable susceptibility of behavior patterns to these influences? Consider the case of the rare hybrids obtained in the laboratory between the races of *Drosophila paulistorum* inhabiting northern and southern Brazil, respectively. These hybrid individuals are most difficult to obtain and possess a genetic constitution discordant enough so that the hybrid females repel the courtship of all males, and will mate with none; their sterile hybrid brothers will court and will be rejected by almost all females, including their own hybrid siblings. There is no question of gene flow through the hybrid males, because they are absolutely sterile. Their sisters, however, are potentially fertile. This has been verified not only by dissection of their internal and external reproductive apparatus, but more conclusively by a direct experiment. Some etherized hybrid females are exposed to many mature, unanesthetized males; the males approach, mount and inseminate the females in question; the hybrid females subsequently recover and deposit fertilized eggs which develop normally (Ehrman, 1960b).

More recently, Carmody et al. (1962) investigated a possible correlation between the occurrence of hybrid sterility and sexual isolation within the *D. paulistorum* complex. In this massive study, more than sixteen thousand females from all of the then known *D. paulistorum* races were dissected and their sperm-storing organs checked. Each *Drosophila paulistorum* female, hybrid or not, has three sperm-storing organs. The "male choice" experimental method was used to test mating preferences of different strains comprising the races. Briefly, this involves groups of ten virgin females of each of two races ( a total of twenty females), aged for three to four days after hatching, marked for recognition by clipping a very small part of one of the wings of one of two types of females, and then confined with males of one of the two races for twenty-four to forty-eight hours, i.e., A ♂♂ × A ♀♀ + B ♀♀ or B ♂♂ × B ♀♀ + A ♀♀. All the females are then dissected in a physiological saline, and their sperm receptacles are examined for the presence or absence of sperm.

Strains of the same race but of different geographic origin often show significant preferences for homogamic matings, but strains of different races show such preferences to a much greater extent. On the average, sexual isolation is lower in interracial matings involving the transitional race, the bridge to fertile hybrids between all the other races. The degree of sexual isolation shows only a weak positive correlation with the fertility or sterility of the hybrids between the strains crossed.

More study is needed to unravel the interrelations of the sexual isolation and hybrid sterility. As indicated above, sexual isolation is most effective in limiting or preventing the appearance of interracial hybrids with reduced reproductive fitness. A problem of considerable interest is whether isolating mechanisms which decrease the chances of production of hybrid offspring are stronger between sympatric than between allopatric populations of the *same* pairs of races or subspecies or incipient species. When two or more Mendelian populations of sexually reproducing and cross-fertilizing organisms share the same territory, these populations are exposed to the risk of hybridization and gene exchange. If such a gene exchange leads to production of adaptively inferior genotypes, natural selection may favor genetic constitutions which hinder or prevent hybridization. On the other hand, the gene exchange and introgression may weaken the reproductive isolation, and may eventually lead to fusion of the previously separate populations. Experiments were made to test whether the sexual isolation between sympatric strains of a given pair of races is, on the average, greater or smaller than that between allopatric strains of the same races. They showed (Dobzhansky et al., 1964) that natural selection had encouraged the spread and maintenance, in sympatric populations of incipient species, of those genes which limit or prevent the reproductive wastage resulting from gene flow between these populations.

However, one may suppose as Muller (1942) does that reproductive isolation arises as an incidental by-product of genetic divergence. When populations become adapted to different environments they are likely to become different in progressively more and more genes. Reproductive isolation then might arise because the action of many genes is pleiotropic. Some gene differences selected for different reasons, or resulting from random genetic drift, may thus have isolating side-effects.

That selection can indeed produce, or at least strengthen, reproductive isolation has been demonstrated experimentally by Koopman (1950) and by Knight, Robertson, and Waddington (1956). Koopman set up experimental populations in laboratory population cages consisting of two species, *D. pseudoobscura* and *D. persimilis.* Each species was homozygous for a different recessive mutant gene with easily visible external effects. The pure species and the hybrids were thus made easily distinguishable. In every generation the adult flies were taken from the cages, etherized, classified, and counted. The hybrids were then discarded and new population cages were started with nonhybrid progenies. By these means Koopman was selecting the progenies of intraspecific, and excluding those of interspecific, matings. In a surprisingly small number of generations he obtained strains of *D. pseudoobscura* and *D. persimilis* which showed a more nearly complete sexual isolation than did the original strains. The results of Knight, Robertson, and Waddington are, in a way, even more dramatic, since they obtained by selection a significant, though of course incomplete, sexual isolation of strains of *D. melanogaster* which originally showed no such isolation.

The incipient species of *D. paulistorum* seem to furnish a good illustration of the two processes postulated above. It is not easy to imagine the sterility of male hybrids between these incipient species conferring any adaptive advantage upon them. But once this sterility has arisen as *a by-*

*product of their genetic divergence,* it is probable that natural selection would favor genetic constitutions which make the sterile hybrids rare. The sexual isolation between the incipient species may be a reproductive barrier contrived as such by the action of *natural selection.*

Dobzhansky and Spassky (1959) first suggested that *Drosophila paulistorum* is a cluster of species in *statu nascendi,* a borderline case of uncompleted speciation. This suggestion is borne out by the work since then. *Drosophila paulistorum* is a superspecies composed of six races or incipient species. The interest of the situation lies precisely in that these six may be considered about equally legitimately as very distinct races or as very closely related species. Each race inhabits a geographic area different from the others, but the areas of some of the races overlap. Where two or more "races" share a common territory they apparently do not interbreed, and thus behave like full-fledged species. The Transitional race and transitional strains yield, however, fertile hybrids with some other races. The possibility of gene flow between the incipient species is, therefore not excluded, although it is questionable whether it is actually taking place.

Last year, in his Vice Presidential address on *Genes and the Study of Behavior,* Caspari (1963) cautioned ". . . the analysis of $F_1$ species hybrids cannot give us, in principle, information about genetic units. This will only be possible through further crosses, $F_2$ and backcrosses." The step-by-step, often tedious analysis of the *D. paulistorum* situation, as presented here today, certainly indicates how correct he was. Understanding of the genetic basis of the sexual isolation among these races would have been impossible without a systematic study of the backcross individuals, in conjunction with a survey of the courtship and mating behavior in the $F_1$ hybrids and in the "pure" races. Surely "the genetic basis of behavioral characters is important, because it is the genes which are the basic units transmitted and reshuffled in the evolutionary process and which are arranged by selective forces into adaptive action patterns." (Caspari, 1963.)

## REFERENCES


Bateman, A. J. 1948. Intra-sexual selection in *Drosophila.* Heredity 2:349-368.

Carmody, G., A. Diaz Collazo, Th. Dobzhansky, L. Ehrman, J. S. Jaffrey, S. Kimball, S. Obrebski, S. Silagi, J. T. Tidwell, and R. Ullrich. 1962. Mating preferences and sexual isolation within and between the incipient species of *Drosophila paulistorum.* Am. Midl. Nat. 68:67-82.

Caspari, E. 1963. Genes and the study of behavior. Am. Zoologist 3:97-100.

Dobzhansky, Th. 1937. Genetics and the origin of species. First edition. Columbia Univer. Press, New York.

Dobzhansky, Th., L. Ehrman, O. Pavlovsky, and B. Spassky. 1964. The superspecies *Drosophila paulistorum.* Proc. Natl. Acad. Sci. 51:3-9.

Dobzhansky, Th., and O. Pavlovsky. 1962. A comparative study of the chromosomes in the incipient species of the *Drosophila paulistorum* complex. Chromosoma 13:196-218.

Dobzhansky, Th., and B. Spassky. 1959. *Drosophila paulistorum,* a cluster of species in *statu nascendi.* Proc. Natl. Acad. Sci. 45:419-428.

Ehrman, L. 1960a. The genetics of hybrid sterility in *Drosophila paulistorum.* Evolution 14:212-223.

———. 1960b. A genetic constitution frustrating the sexual drive in *Drosophila paulistorum.* Science 131:1381-1382.

———. 1961. The genetics of sexual isolation in *Drosophila paulistorum.* Genetics 46:1025-1038.

Knight, G. R., A. Robertson, and C. H. Waddington. 1956. Selection for sexual isolation within a species. Evolution 10:14-22.

Koopman, K. F. 1950. Natural selection for reproductive isolation between *Drosophila pseudoobscura* and *Drosophila persimilis.* Evolution 4: 135-148.

Mayr, E. 1963. Animal species and evolution. Harvard Univer. Press, Cambridge, Mass.

Muller, H. J. 1942. Isolating mechanisms, evolution, and temperature. Biol. Symp. 6:71-125.

Spieth, H. T. 1952. Mating behavior within the genus *Drosophila* (Diptera). Bull. Am. Mus. Nat. Hist. 99:395-474.

# Part VI: Mimicry

The adaptive resemblance of one organism (the **mimic**), usually in color, pattern, form, or behavior, to another organism (the **model**) is known as **mimicry**. The traits of the model which confer a degree of benefit to the mimic include the model's unpalatability to predators, the efficiency of its physical and chemical defenses or defensive postures, and its morphology (including color patterns). The resemblance of one organism to another can only be considered mimicry if both are found together. A circumstance in which similar organisms are not in the same area would most likely be an example of convergence rather than mimicry.

There are several types of mimicry. **Aggressive mimicry**, usually found in arthropod parasites and predators, *involves a resemblance which enables the mimic to approach its prey, the model.* The distinctive characteristic of this type of mimicry is the inherent conflict between the mimic and model. **Wasmannian mimicry** *involves resemblances which facilitate cohabitation with a mimic's host, its model.* This relationship may consist of some exploitation of the model by a mimic, but may also be a mutually advantageous relationship. Ant and termite mimics are common examples of this type of mimicry. Automimicry can be described as a type of Batesian mimicry (cf. below) within a species. Thus, some Monarch butterflies are unpalatable while others are not.

By far the most well known and easily illustrated types of mimicry are Batesian and Müllerian. **Batesian mimicry** results from *the resemblance of one organism (the mimic) to a protected model, thereby providing the mimic some protection from its enemies.* **Müllerian mimicry** results when *both model and mimic are unpalatable and their conspicuous coloration and pattern serve mutually as learning stimuli for predators.*

There are several underlying principles relevant to both Batesian and Müllerian mimicry. Many of these premises are important enough to be mentioned herein, although they are still in question and require further confirmation. In Batesian mimicry the model is unpalatable to predators, while the mimic is palatable but has evolved a protective resemblance to the model. The mimic and model habitats must be contiguous, but the mimic is generally found in smaller numbers. The importance of the abundance of the model compared to the mimic has long been an area of controversy. L. P. Brower and J. V. Z. Brower *(The Relative Abundance of Model and Mimic Butterflies in Natural Populations of the Battus philenor Mimicry Complex)* present field data illustrating the complex correlations between the range of abundance of mimic and model and the many factors affecting this relationship. In addition to the above requirements, predators must learn to associate the conspicuous color patterns of both model and mimic. In comparison, Müllerian mimicry assumes that both of the species involved are unpalatable. Individuals of each species would be expected to be eliminated by predators in similar proportions. Members of a Müllerian mimetic complex are found in the same area and their color patterns must be associated by predators with their unpalatability.

Batesian and Müllerian mimicry are distinct but not mutually exclusive types of mimicry. Elements of Batesian and Müllerian mimicry can occur in the same family of insects. E. G. Linsley *(Mimetic Form and Coloration in the Cerambycidae)* surveys the many examples of mimicry in the long-horned wood-boring beetles. Although most are Batesian mimics, Müllerian mimicry is also apparent. An even more complete example of the occurrence of both types of mimicry is found in the work by E. G. Linsley et al., *Mimetic Assemblages of Sibling Species of Lycid Beetles.* One of the lycid mimetic complexes discussed is composed of a lycid, a **sibling species** *(a closely related and nearly identical species)*, a long-horned (Cerambycidae) beetle, and moths from two different families. This Batesian complex contains elements of Müllerian mimicry because one of the mimetic moths is distasteful to predators.

# Mimetic Assemblages of Sibling Species of Lycid Beetles

E. G. Linsley, T. Eisner, and A. B. Klots[1]

Received April 4, 1960

Lycid beetles, throughout the tropical and subtropical regions of the world, serve as models for a great many Batesian mimics and participate in numerous Müllerian mimicry complexes. Their broad elytra, aposematic coloration, gregarious habits and sluggish behavior admirably suit them for this role. These characteristics are universal and unrelated species from various continents are often superficially identical in coloration and behavior. The family probably represents the most ancient living group of models for mimetic insects. Carpenter (in Carpenter and Ford, 1933) summarizes lycid mimicry as follows:

"They are resembled in colour and pattern by members of every order of Insects whose bodily form yields the foundation on which a Lycoid appearance can be built. Among *Lepidoptera* the Lycoid moths belong to families of sedentary habit which keep the wings flat over the back or even folded round the body. The *Hymenoptera* have not the shape of body which could produce a Lycoid appearance, but the necessary aposeme is shown by the wings, which in different families are orange with black tips and when lying over the back reproduce the Lycoid effect: there are Lycoid wasps of several families, Saw-flies, and Ichneumon-flies. Two-winged flies (*Diptera*) are known from several families, among which may be mentioned *Asilidae*.

"The sucking bugs (*Hemiptera*), in shape essentially similar to beetles, furnish abundant mimics from several families.

"Among the enormous assemblage of beetles themselves there are many families which have produced Lycoid species, and they all are sedentary in their habits: the writer cannot recall any really active beetle mimetic of the *Lycidae*. In some cases the modification has been considerable, so that a Lycoid Longicorn may require close examination to detect the characteristic antennae, modified in length to approximate to the shorter antennae of the model, although the numerous joints are there. The typical parallel-sided elytra are also modified by being widened posteriorly to be more like the expanded organs of the model.

"In all these mimics, which throughout the world must run into many hundreds of species, there is vast difference in the habits and food of the larvae, yet these agree in the one character of producing an adult with Lycoid colouring. They may be vegetarians or carnivores: among the former some feed on dead wood, taking years to mature and living in relatively constant temperature and humidity. Others live freely exposed and feed on leaves, etc., in the open. The carnivorous larvae may hunt their prey themselves in the open, or in dead wood, or may live underground, maturing in a few weeks, and have paralysed living insects supplied them to eat. Others live within the body of their host whose juices they absorb.

"The quality of the food, the manner in which it is obtained, and the environment in which the larvae live are so different that the superficial resemblance of the adults cannot be ascribed to these factors."

Our knowledge of such mimetic assem-

[1] University of California, Berkeley, Cornell University, Ithaca, N. Y., and the City College of New York, respectively. The writers are indebted to Dr. Mont Cazier, Director, Southwestern Research Station, American Museum of Natural History, Portal, Arizona, for making the facilities of the station available for these and related studies. Dr. Cazier, Miss Marjorie Statham, and Mrs. E. G. Linsley ably assisted in the taking of field samples and Mr. George M. Happ and Miss Frances A. McKittrick in tests with predators. Dr. W. L. Brown, Jr., kindly identified the ants, Dr. W. J. Gertsch the arachnids, Mr. J. W. Green the lycid beetles, Dr. Frederick Rindge the geometrid moth, and Dr. Jack McCormick the plants. The study was aided in part by a grant from the Bache Fund of the National Academy of Sciences, and by Grant E-2908 of the U. S. Public Health Service.

Reproduced with permission from *Evolution*, 15:15–29, 1961. Published by the Society for the Study of Evolution, Lawrence, Kansas.

blages as those described above, has been largely derived from the tropical regions of the Southern Hemisphere (cf. Marshall and Poulton, 1902; Shelford, 1901, 1917; Gahan, 1913; Carpenter, 1921; Arrow, 1928; Bates, 1892; etc.), and few quantitative data have been published relative to the numerical relationship between models and mimics.

In North America, the most conspicuous lycid beetles belong to the genus *Lycus*. Twelve species are known from within the boundaries of the United States (Green, 1949). Ten of these have been reported to occur in Arizona as follows: *Lycus* (*Neolycus*) *fernandezi* Dugés, *L.* (*N.*) *arizonensis* Green, *L.* (*N.*) *lecontei* Green, *L.* (*Lycostomus*) *sanguineus* (Gorham), *L.* (*L.*) *sanguinipennis* Say, *L.* (*L.*) *fulvellus* LeConte, *L.* (*L.*) *rubescens* (Schaeffer), *L.* (*L.*) *minutus* Green, *L.* (*L.*) *simulans* (Schaeffer), *L.* (*L.*) *loripes* (Chevrolat). Eight species have been recorded from the Huachuca Mts. (Schaeffer, 1908, 1911; Green, 1949) and at least six have now been found in the Chiricahua Mts.

During the last two weeks of July, 1959, the authors were privileged to study the composition of a mimetic assemblage associated with *Lycus fernandezi* Dugés and its sibling *L. arizonensis* Green and a parallel assemblage centered about *L. loripes* (Chevrolat) and its sibling *L. simulans* (Schaeffer). Both assemblages were found together on the grounds of the Southwestern Research Station of the American Museum of Natural History.

### The Locality

The Southwestern Research Station is located about 5 miles west of Portal, Arizona, above Cave Creek Canyon, Chiricahua Mountains, at an elevation of 5,400 feet. The area surrounding the station has been subjected to a great deal of ecologic disturbance and contains many plants not indigenous to the region. It is mostly comparatively xeric, but the strip adjacent to a small stream along one edge of the station grounds, which flows only during the rainy season, is markedly hydric. The area is, therefore, somewhat of an ecotone between xeric upland and the hydric canyon bottom. The former is characterized by Chihuahuan pine (*Pinus chihuahuana* Engelm.), alligator-barked juniper (*Juniperus deppeana* Staud.), Emory Oak (*Quercus emoryi* Torrey) and Arizona white oak (*Quercus arizonica* Sarge.). The latter is characterized by sycamore (*Platanus wrightii* Wats.), at least three willows (*Salix* spp.), cottonwood (*Populus fremontii* Wats.), box elder [*Acer negundo* var. *interius* (Britt.) Sarg.] and Arizona walnut [*Juglans major* (Torr.) Heller].

On the grounds of the Station, the principal laboratory building faces northward. Directly opposite the main entrance, across an access road and parking area, is a triangular-shaped meadow bounded on the south by about 160 feet of the access road, to the west by pathways, and to the northeast by a row of trees. In mid-July this meadow was dominated by masses of white sweet clover (*Melilotus alba*) at the peak of bloom. Some of the plants were six and one-half feet tall, or more, but most averaged about four and one-half feet. Four major congregating sites for lycids occurred in the meadow. The largest (Site A) was the more easterly and involved an area of about 1,800 square feet. The remaining sites (B, C, and D) occupied about 200, 150, and 400 square feet respectively. The last site was an experimental area enclosed by a fence to prevent plant injury by browsing deer. In each site aggregations of *L. fernandezi* and *L. loripes* occurred together on the same plants, along with their siblings and mimetic assemblages. However, the former was most numerous in site A, the latter in sites B, C, and D.

### Sampling and Estimating the Composition of Aggregations

In order to obtain some indication of the relative abundance of models and mimics in the two assemblages without seriously affecting the populations, it was necessary to take samples with a minimum of disturb-

TABLE 1. *Proportionate representation of various elements in samples of mimetic assemblages associated with* Lycus fernandezi *Dugés and* L. loripes (*Chevrolat*)

| *L. fernandezi* complex | Date of Sample | | | |
|---|---|---|---|---|
| | July 16 | | July 24–26 | |
| *Lycus fernandezi* | 43 ♂♂ | 19 ♀♀ | 48 ♂♂ | 24 ♀♀ |
| *Lycus arizonensis* | 5 ♂♂ | 2 ♀♀ | 13 ♂♂ | 4 ♀♀ |
| *Elytroleptus apicalis* | 3 ♂♂ | | 1 ♂ | 1 ♀ |
| *Seryda constans* | — | — | | 1 ♀ |
| *Ptychoglene coccinea* | | 2 ♀♀ | | 1 ♀ |
| *L. loripes* complex | | | | |
| *Lycus loripes* | 64 ♂♂ | 55 ♀♀ | 232 ♂♂ | 83 ♀♀ |
| *Lycus simulans* | 1 ♂ | 2 ♀♀ | 4 ♂♂ | 4 ♀♀ |
| *Elytroleptus ignitus* | 2 ♂♂ | | 1 ♂ | 1 ♀ |
| *Eubaphe unicolor* | 2 ♂♂ | 1 ♀ | — | — |

ance. On July 16, the field population of the *Lycus fernandezi* type was estimated by visual examination at about 250 individuals. By random sweeping, members of this complex were captured until it was estimated that about 75 had been taken (for the actual count, see below). The population of the *loripes* complex was larger, by visual estimate about 2,000, and an attempt was made to capture a sweep sample of about 150 individuals. During the next 10 days visual examinations of the assemblages were made, plant by plant, and the presence of mimics noted. On July 24 and 25, all of the mimics which could be found were removed from the aggregations for use in feeding experiments with predators. On July 26, all of the remaining lycids were collected. This procedure provided data on the relative abundance of the lycid siblings and on the relative numbers of the sexes. (table 1).

### THE *LYCUS FERNANDEZI* MIMETIC COMPLEX

*Lycus fernandezi* Dugés (figs. 1, 2) was originally described from Guanajuato, Mex-

FIG. 1. Members of the *Lycus fernandezi* mimetic complex photographed *in situ* while feeding at florets of *Melilotus alba*. Upper left, *L. fernandezi* Dugés; upper right, the moth *Seryda constans* (Henry Edwards); lower left, *Lycus arizonensis* Green; lower right, mating pair of *L. fernandezi*, with second male beneath right elytron of female, presumably feeding on secretions at base of abdomen.

FIG. 2. The *Lycus fernandezi* mimetic complex. Upper left, *L. fernandezi* Dugés (male left, female right); upper right, *L. arizonensis* Green (male left, female right); lower left, *Seryda constans* (Henry Edwards), lower center, *Ptychoglene coccinea* (Henry Edwards), lower right, *Elytroleptus apicalis* LeConte.

ico. It is a widespread species in northern Mexico, extending into southwestern United States at least as far as central Texas, southern New Mexico and central Arizona. As a model for mimicry it has the following superficial characteristics: (1) it is a large species, ranging in length from 10 to 18 mm, with the elytra very broadly dilated in the male, less so in the female, (2) the color is a conspicuous orange-yellow, with the apical one-fourth of the elytra, and the antennae, tibiae, and tarsi black (the apical black spot varies somewhat in different geographical regions, but in the Chiricahua Mountains the anterior margin is almost always distinctly notched on the humeral costa), (3) it gathers gregariously on flowering shrubs for feeding and mating, (4) it feeds on nectar, and thus is available as a model for a variety of potential mimics among other nectar (and pollen) feeding insects in both its aggregating and dispersed phases, (5) it is slow-moving as it crawls over the flowers, (6) the mating pairs remain together for long periods of time (the broader male usually completely covering the female), (7) it is slow and conspicuous in flight, the hind wings also being black-tipped, (8) it bleeds freely, the slightest injury bringing forth a large drop of blood, and (9) it appears to be avoided in varying degrees by a variety of vertebrate and invertebrate predators.

*Lycus arizonensis* Green (figs. 1, 2), presumed to be a species sibling of *L. fernandezi,* is slightly larger in average size (range 12–18.5 mm) but is almost identical in form and coloration. The only superficial difference between the two is the absence of a notch in the anterior margin of the elytral spot at the point where the humeral costa enters. In comparison with its sibling, it was present in much smaller numbers during the period in which we made our observations.

*Elytroleptus apicalis* LeConte (fig. 2), a longicorn beetle of the family Cerambycidae, is a little smaller in average size than *L. fernandezi* (our sample ranged from 10–13 mm in length), and the elytra are less broadly dilated. However, in coloration the two are practically identical, even to the notch in the anterior margin of the apical black spot on the elytra. The longicorns were seen crawling over the blossoms in the lycid aggregations, but were not abundant (see table). They are thought to be nectar and pollen feeders, although *Melilotus*, a rich nectar source, is generally regarded as a poor pollen plant.

*Seryda constans* (Henry Edwards) (fig. 2), a moth of the family Pyromorphidae, is a diurnal species which visits flowers, and apparently mimics *L. fernandezi* under dispersed as well as in congregated situations. When sipping nectar and crawling over the blossoms, it keeps the wings folded flat over the back, beetle-like. It approximates the lycid in size, our examples ranging from 12–15 mm in length with the wings folded. The ground color is very similar to that of the model. The appendages are also black, as are the distal margins of the forewings, although these last are usually narrower. Individuals, although rare in the aggregations (they were captured twice, observed three times), were found in small numbers in the general area and occasionally came to light.

*Ptychoglene coccinea* (Henry Edwards) (fig. 2) and *P. phrada* Druce, moths of the family Lithosiidae, are also diurnal flower visitors. Their ground color is bright red rather than orange-yellow; but, since we know little or nothing about the qualitative color vision of the predators involved, it may well be that this difference is of little or no significance. As in *Lycus fernandezi* the appendages and the distal margins of their forewings are black, but the black wing areas are usually more extensive than those of the beetles. The head and thorax of *P. coccinea* are red, of *P. phrada*, black. There appear to be constant, although slight, genitalic differences between them in both sexes. Both species range widely in Arizona, and both have been taken previously in some numbers at the Southwestern Research Station from late June to mid-August. Like the lycids described above, they constitute a pair of sibling species. However, we have no evidence as to whether or not there is a connection between the evolution of the sibling pairs of beetles and this sibling pair of mimetic moths. In 1959 *P. coccinea* was found several times crawling over the blossoms in the lycid aggregations, as well as in scattered situations; and its presence in the company of the beetles appears to be of mimetic significance. The subsequent discovery of still another Müllerian assemblage, which might well provide models for these moths suggests that in other areas they may be mimetically associated with a more appropriately colored group of lycids. These include *Lycus* (*Neolycus*) *lecontei* Green (fig. 4), the Arizona form of which is rufous or scarlet with the head, antennae, palpi, tibiae, tarsi and apical one-third of the elytra black. Although this species has not yet been reported from the Chiricahua Mountains, the superficially similar *Lycus* (*Lycostomus*) *sanguineus* (Gorham) (fig. 4) has been found there along with the gregarious, vinegar-smelling cantharid *Chauliognathus profundus* LeConte (fig. 4). This last, although narrower in form than the *Lycus*, is similar in size and coloration and gathers by the thousands on flowering shrubs such as *Baccharis*. However, in the vicinity of the Research Station, it did not appear in numbers until the first week of September, long after the lycid aggregations had dispersed. Whether these moths owe their coloration to the presence of any or all of these red and black beetles remains to be determined.

A number of other diurnal, flower visiting moths, presumably mimetic, are associated with lycids elsewhere in the Southwest and northern Mexico. These are discussed below.

FIG. 3. The *Lycus loripes* mimetic complex. Upper left, *L. loripes* (Chevrolat) (male left, female right); upper right, *L. simulans* (Schaeffer), (female left, male right); center left, *Elytroleptus ignitus* LeConte; lower left, *Eubaphe unicolor* (Robinson); center right, *Holomelina ostentata* (Henry Edwards) with wings folded; lower right, same with wings spread to show aposematic hind pair.

## THE *LYCUS LORIPES* MIMETIC COMPLEX

*Lycus loripes* (Chevrolat) (fig. 3) was described originally from Mexico. It is a widespread, common species, and is known from many localities in southern Arizona (especially in the Chiricahua, Huachuca, Santa Rita, Santa Catalina, Coyote, Dragoon, and Baboquivari mountains). The species is smaller on the average than *L. fernandezi*, ranging in length from 10–13 mm, and is concolorous yellow-orange with some black on the appendages. The elytra of the male are scarcely dilated. The beak is long and slender, being equal to about twice the distance between the eyes. Although less conspicuously marked than *L. fernandezi*, it was more numerous in our study sites and the individual aggregations were larger. Like *L. fernandezi*, these beetles are assumed to be nectar and pollen feeders, and apparently share with that spe-

FIG. 4. *Lycus sanguineus* (Gorham) (left), *L. lecontei* Green (center) and *Chauliognathus profundus* LeConte.

cies most of the characteristics which make it a suitable model for mimetic forms.

*L. simulans* (Schaeffer) (fig. 3) is a superficially identical sibling species of the same size and coloration as *L. loripes*. It differs by having a short beak, which is less than one and one-half times the distance between the eyes, and in proportions of the antennae and the shape of the palpi. In our observation plots it was represented in small numbers, like the sibling of *L. fernandezi*.

*Elytroleptus ignitus* LeConte (fig. 3), a longicorn beetle, is a little smaller in average size than *Elytroleptus apicalis* (our sample ranging in length from 8–11 mm), and the elytra are concolorous yellow-orange. However, it is the counterpart of *E. apicalis* in the *L. loripes* mimetic assemblage. It has similar habits and in our site the two species were present in about equal number.

*Eubaphe unicolor* (Robinson) (fig. 3), a yellow-orange geometrid moth, appears to be a lycid mimic in the *loripes* aggregations and under dispersed conditions. The wings are concolorous, the antennae black, and when at rest or crawling over the blossoms with wings folded it varies in length from about 10–14 mm. The moths were found seven times among the lycid aggregations, usually in the early morning or when the sky was overcast. Individuals also came to light in small numbers during the period when the lycids were congregating.

*Holomelina ostenta* (Henry Edwards) (fig. 3), a moth of the family Arctiidae, is slightly larger than the preceding (and considerably larger than the lycids) and differs by having bright red and black hindwings. However, when at rest the forewings, which are concolorous yellow-orange with darker fringes, are folded back flat, providing lycid coloration. This species was not found in the lycid aggregations and its abundance about the lights suggests that it may be primarily a nocturnal species. Possibly it benefits generally from the presence of the lycids in the area when it is at rest during the day. Various other species of *Holomelina* are widely distributed in North America; and many occur in regions lacking possible lycid models. Some have the forewings dark brown, others, as in *ostenta*, bright yellow-orange or orange-brown. Essentially nocturnal, many are active during the daytime, especially in grasslands and open habitats where they fly up in numbers when disturbed, and are then very conspicuous, and they sometimes visit flowers.

### OTHER *LYCUS* MIMETIC ASSEMBLAGES IN THE UNITED STATES

We have already alluded to the possible existence of a mimetic assemblage involving the red and black *Lycus lecontei* Green and *L. sanguineus* (Gorham). In the Huachuca Mountains another assemblage is suggested by the occurrence together of *Lycus sanguinipennis* Say and the mimetic longicorn *Elytroleptus rufipennis* (LeConte). Several lycid-like moths also occur in this region, including *Ptychoglene phrada* (Lithosiidae) and *Seryda constans*, *Tetraclonia dyari*, *Pyromorpha rata*, and *Pyromorpha fusca* (Pyromorphidae) which we have seen from Ramsey Canyon, Huachuca Mts., Arizona, 10–15 July 1941, leg. A. B. Klots, visiting *Melilotis alba* and Compositae, although no record was kept of the lycids associated with these moths. The first two species have already been discussed. *Tetraclonia dyari* and *Pyromorpha rata* have the forewings yellow-brown to orange-brown, with dark distal margins, and may fit into the *Lycus fernandezi* complex. *Pyromorpha fusca*, which has the wings solidly fuscous but the head, thorax and abdomen dull yellow, may mimic almost any of the small, dark lycids.

In eastern North America only one species of *Lycus* occurs, *L.* (*Lyconotus*) *lateralis* Melsheimer. It ranges from Maine south to Florida and west to Illinois, Missouri, Kansas, and Texas (Green, 1949). *L. lateralis* differs markedly in coloration from the red, orange or yellow and black patterns characteristic of the majority of the southwestern members of the genus, being black, with the elytral humeri and the sides and apex of the pronotum fulvous. However, it closely resembles lycids of other genera that occur in the same area. It has its complement of mimics, including the longicorns *Elytroleptus floridanus* (LeConte) and *Euryptera lateralis* (Olivier), which also differ from their brightly colored southwestern[2] congeners.

Also widespread in eastern North America is the extremely similar lycid *Caeniella dimidiata* (Fabricius), which has the basal fulvous area of the elytra somewhat more extensive than in *L. lateralis*. There can be little doubt that these lycids supplement each other aposematically. However, the most common eastern lycid is the somewhat larger *Calopteron reticulatum* (Fabricius), of very different but strikingly aposematic appearance, the elytra being cross-banded with black and orange.

A number of diurnal, flower-visiting pyromorphid moths occur with *L. lateralis* and *C. dimidiata*. Of these, *Pyromorpha dimidiata* (Herrich-Schäffer), (Holland, 1917, Pl. 47, fig. 33) which has smoky black forewings that are basally dull fulvous, matches *L. lateralis* very closely, *C. dimidiata* less well. *Acoloithus falsarious* Clemens, a smoky black moth with orange collars, has considerable resemblance to *L. lateralis*. Also, the ctenuchid moth *Lycomorpha pholus* (Drury) is an extremely good mimic of *C. dimidiata*, having about the same relative amount of bright orange basally on the forewings, the two apparently constituting an intimate Müllerian mimetic pair. Finally, a number of other pyromorphid moths in eastern and western North America, belonging to the genera *Pyromorpha* Herrich-Schäffer, *Triprocris* Grote, *Acoloithus* Clemens, and *Harrisina* Packard, are small, smoky-black winged, diurnal flower visitors that must be involved to a

---

[2] For example, *Euryptera ignita* Schaeffer and *E. cruenta* Martin, both from the Huachuca Mountains, are red or reddish-orange with black appendages. These species have retained their somewhat slender, lepturine ancestral form; but in tropical America a number of species have not only lycid-like markings but also flattened, expanded antennae and flared, costate elytra. Some of these, such as *E. latipennis* Serville and *E. lyciformis* Pascoe, occur sympatrically with more slender but identically marked species which appear to have the same lycid models. In some Brazilian localities, both *E. melanura* Redtenbacker and *E. dimidiata* Redtenbacker have counterparts with expanded elytra. In Mexico, *E. longipennis* Bates is slender; *E. patricia* Bates is broadened. Like many lycid mimics the species of *Euryptera* are mostly rare in collections and in great need of taxonomic study. Belon (1897) has provided a superficial color key for the separation of many of the described species, but it is incomplete and not definitive.

greater or lesser extent in the dark-winged lycid mimicry complexes.

### Relative Edibility of Lycids by Predators

A considerable amount of evidence has accumulated indicating that lycids are avoided by vertebrate predators, including lizards, birds, and monkeys. Marshall and Poulton (1902) offered six species of *Lycus* from Mashonaland to baboons and a kestrel and these were rejected, along with several other similarly marked beetles which made up a portion of an extensive Müllerian mimicry complex. Carpenter (1921) also offered African lycids to two young *Cercopithecus* monkeys. His experiments were conducted either by collecting insects at random from their natural surroundings and putting them down so that the monkey could study them, or by taking the monkey into open bush country on a long lead and making exact notes on what it did or did not do. Seven species of lycids were involved (of the genera *Lycus*, *Chlamydolycus*, and *Merolycus*). They all had a single scheme of coloration: orange or orange-brown with appendages black or black and orange and the tips of the elytra black. They lived freely exposed and large numbers were to be found congregated on a single flowering shrub. One of the monkeys put all lycids tested in the most distasteful class. Only twice in ten trials were they tasted, even under the influence of hunger, and no part was eaten (*Merolycus femoralis* Bourg. was tasted once and refused on the second occasion when the subject was less hungry). When an example of *Lycus constrictus* Fahr. was offered with the elytra pulled off, it gave off a strong odor and the subject would not accept it. Only two lycids were offered to the other monkey. One of them was eaten (*Chlamydolycus trabeatus* Guér.), but the other was tasted and then rejected.

In the West Indies, Darlington (1938) studied the mimetic assemblage associated with two blue-tipped yellow lycid beetles, *Thonalmus suavis* Jacquelin Duval, and *T. aulicus* Jacquelin Duval. These lycids were treated as inedible by the principal vertebrate predator of the region, the lizard *Anolis sagrei* (Duméril and Bibron). Not only were the lycids consistently refused, but so also were four species of *Thonalmus*-like Cerambycidae (although six other plainly marked species were eaten). Parsons (1940) tested one of the *Thonalmus*-like species (*Trichrous pilipennis* Chevrolat) by placing it in the mouth of an *Anolis* which had just rejected a *Thonalmus*. It was swallowed at once, suggesting that it is edible, although ignored in nature.

At Martha's Vincyard, Massachusetts, Jones (1932), in nine experiments offered 47 *Calopteron reticulatum* Fabricius to seven species of birds on feeding trays together with 519 other kinds of insects. After 141 visits by birds, 41 of 47 *Calopteron* remained, although 155 of 157 other beetles were eaten. Four of the six *Calopteron* were taken in one experiment when the only bird visitors were blue jays (*Cyanocitta cristata* Linnaeus). Jones also points out that this species does not appear among the 337,000 insects identified with the stomachs of 80,000 birds tabulated by McAtee (1932). Jones did not utilize moth mimics of *Calopteron* in his experiments, but his lists include two species of *Lycus* mimics, *Pyromorpha* (as *Malthaca*) *dimidiata* Herrich-Schäffer (Pyromorphidae) and *Lycomorpha pholus* Drury (Ctenuchidae). Although the presumed model for these moths, *Lycus lateralis* Melsheimer, was not offered on the feeding trays, the single example of *Lycomorpha pholus* offered was left on the tray, as were four of five *Pyromorpha dimidiata* offered in three experiments.

Virtually nothing seems to be known about how repellent lycids are to invertebrate predators. The single observation known to us is that by Darlington (1938), who saw "an Asilid fly catch a small black and red Lycid in the air, but immediately released it."

Our own experiments with predator-prey encounters took place within special observation cages, the confines of which were made to resemble when possible the natural surroundings of the particular predator being tested. None of the predators were starved for excessive periods of time prior to the experiments. In cases where a *Lycus* or one of its mimics was ignored or rejected, additional prey (usually dealate moths or scarab beetles) was offered immediately thereafter in order to determine whether the predator was actually discriminating or merely reluctant to take food.

Of the predators tested by us, it is likely that only the mantids, wasps, jays, and the lizard *Sceloporus jarrovi* are of significance as enemies of the particular lycid associations studied. The other test predators were ground foragers, and some, such as the horned toad, whiptail lizard, turtle, and grasshopper mice, were not found in the immediate vicinity of the station. Our reason for using such predators, whether or not they occurred in the special habitat under investigation, was to try to determine how general or predator-specific the repugnatorial effectiveness of lycids is.

### A. Invertebrate Predators

1. *Lycosid spiders* (*Lycosa carolinensis texana* Montgomery). Four female spiders were each offered a single *Lycus* (three of the concolorous *loripes* type, one of the black-tipped *fernandezi* type). During the following fifteen minutes only one (concolorous) was eaten. Although the other three were ignored, it is probable that, being of rather slow and even gait, they simply escaped detection. More vivacious prey offered as a choice thereafter was promptly eaten.

2. *Whip scorpions* [*Mastigoproctus giganteus* (Lucas)]. None of the six whip scorpions made attempts to seize the lycids offered, although all of them captured other insects of comparable size given them shortly thereafter. Perhaps again the *Lycus* simply failed to make themselves apparent. Whip scorpions resemble lycosids in that they respond best to jerkily moving prey.

3. *Solpugids* (*Eremobates* sp.). Three solpugids were each given a single *Lycus* (two concolorous, one black-tipped). They were all seized instantly, but were immediately released. Only one of the lycids showed noticeable injury as a result of the encounter. It received a marginal cut in an elytron from which it recovered fully, showing no noticeable abnormality as late as two days thereafter. It might be argued that the solpugids had not really been repelled, but had merely performed some kind of defensive maneuver. This is unlikely, however. The same solpugids were also offered an abundance of other arthropods, and of these, the only few that were rejected in a manner similar to the lycids possessed in all cases some kind of chemical defense mechanism (*Eleodes*, *Brachynus*, *Mastigoproctus*).

4. *Ant-lion larvae* (unidentified). Three concolorous and one black-tipped *Lycus* were tossed into pits and were instantly seized and eaten. The hollow, sucked-out carcasses were found just outside the pits several hours later. These ant-lions died before pupation, but this may have resulted from improper subsequent treatment.

5. *Preying mantids* (*Stagmomantis limbata* Hahn). In one series of tests, five encounters were witnessed, involving four different mantids striking at three concolorous and two black-tipped *Lycus*. In each case the beetles were released within seconds, and of the two that showed noticeable injury, only one died. Following these experiments, about one dozen *Lycus* were left in confinement with the four mantids for four days, at the end of which time none had been eaten. In another series of tests two half- and two full-grown mantids were offered three mimetic moths [two *Eubaphe unicolor* (Robinson), one *Seryda constans* (Henry Edwards), and one mimetic longicorn (*Elytroleptus apicalis* LeConte)]. All except the *Seryda* moth were devoured readily (this last was only partly eaten). Subsequently two were offered black-tipped *Lycus*, two concolorous. Three rejected the lycids immediately upon contact, one ate a black-tipped specimen and then refused contact with further lycid specimens.

The example that refused the black-tipped lycid after contact, subsequently became very agitated and refused and ran away from all prey subsequently offered, including flies and dull colored moths. After five days this specimen died. On another occasion, a nymph of this mantid was offered some lampyrids in Connecticut. It seized one, bit it, immediately dropped it, rubbed its mouthparts with a foreleg for a minute or two, and rejected regular food (tettigoniids) for half an hour. A second trial the following day produced the same result. No further tests were made, however, and the mantid survived and transformed successfully to the adult.

6. *Ants* [*Pogonomyrmex occidentalis* (Cresson)]. Six concolorous lycids were released one at a time a few inches away from the nest entrance of a natural colony. The ants attacked singly or in groups of two or three, but no ant remained in the vicinity of a beetle for more than a few seconds at a time. During this period they seized appendages and tugged at the wing margins, but as far as could be observed, they never persisted long enough to actually pierce the cuticle. The beetles walked away slowly, continually accosted, but never seriously hampered. Each eventually reached safety.

On the basis of comparable experiments by one of us (Eisner) with this ant and the related *P. badius*, involving a wide variety of insects, a rapid overpowering of the lycids would have been predicted, particularly near the nest entrance where the ants are highly aggressive. Ordinarily, when the ants grasp an insect with the mandibles, they do not relinquish their hold, but persist tenaciously, and as a rule also attempt to sting. They soon accumulate in swarms, and insects more vigorous than lycids (e.g., nymphs of *Periplaneta*) have been seen to be seriously hampered in their escape, and sometimes even completely stalled. Only in two exceptional instances were weak and sluggish insects observed to make an essentially unimpeded escape comparable to that of the lycids, and in both cases this involved species known to be protected by chemical means of defense (meloid and coccinellid beetles).

7. *Wasps* (*Polistes canadensis navajoe* Cresson). In a field observation, a female wasp twice seized and twice dropped a single concolorous lycid, after turning it about in its mandibles. The beetle was later examined for injuries but none were found. [Belt (1874) records a hunting wasp in Nicaragua that regularly preys upon "protected" heliconid butterflies, bites off the appendages, malaxates the body, and stores them for the young.]

B. Vertebrate Predators

1. *Lizards.* A mixed cluster of about 12 concolorous and black-tipped lycids was introduced to a cage housing a horned toad, [*Phrynosoma cornutum* (Harland)], two *Sceloporus jarrovi* Cope and one Whiptail lizard (*Cnemidophorus* sp.). Although the lizards watched the lycids crawling about, none attempted to eat them. Since all these lizards except the *Sceloporus* are ground foragers and were taken in dry habitats where the lycids did not occur, their indifference to the lycids must be attributable to reasons other than past encounters.

2. *Ornate Box turtle* [*Terrapene ornata* (Agassiz)]. About a dozen mixed concolorous and black-tipped lycids were offered and were all eaten in rapid succession. The turtle exhibited no behavioral abnormalities during the meal, and showed no detectable after-effects. This experiment was repeated twice on successive days with similar results.

3. *Jay* [*Cyanocitta stelleri* (Gmelin)]. Six *Lycus* (three of each type) were released one at a time into the cage. In each instance the bird descended from the perch, pecked the beetle to death and then pushed it into a crevice in a corner of the cage. This behavior was noted previously in captive *C. cristata* with both edible insects (*Tenebrio* larvae) and less palatable ones (*Brachynus*), particularly in cases where the bird had been well fed previously. However, *cristata* always returned and ate their hidden scraps at a later time; *stelleri* did not do so with the lycids.

4. *Grasshopper mice* (*Onychomys torridus* Coues). In the laboratory these mice regularly received a single daily meal of from 20 to 40 large-sized insects. These were usually eaten in half an hour. A typical meal consisted of about two dozen half inch long scarabs and three or four large sphinx moths.

(a) Mouse No. 1

On the first day, one concolorous and one black-tipped cerambycid mimic (*Elytroleptus ignitus* LeConte and *E. apicalis* LeConte), together with one black-tipped moth mimic [*Seryda constans* (Hy. Edwards)] were offered about midway during a normal meal of about 30 insects. They were all eaten, except for the moth, which was decapitated and rejected, suggesting that the cerambycids were edible, whereas the moth, a pyromorphid, was not.

(b) Mouse No. 2

(Given only concolorous *Lycus*). When offered *Lycus* for the first time, the mouse had not eaten since the previous meal 24 hours earlier. It promptly ate all 25 lycids given, showing no detectable abnormalities or aftereffects. On the following day, the mouse was given another five concolorous lycids, but only after it had already eaten about two dozen other insects. It repeatedly handled and smelled several of the *Lycus*, but released them all again, unharmed. When five more scarabs were then added, it selected them from among the lycids and ate them. No subsequent approach to the lycids was made.

This sequence of experiments was repeated on each of the next two days with the same results, except that on the second day one of five lycids offered was eaten.

(c) Mouse No. 3

On the first day the mouse was offered about ten concolorous *Lycus* and a concolorous *Elytroleptus* cerambycid mimic, after first having received about a quarter of its daily insect ration (1 sphinx, 7 scarabs). The first one seized was a *Lycus*. It was briefly inspected, but was released shortly, uninjured. After a second and finally a third *Lycus* was similarly inspected and released, the mouse ignored the remainder of the batch, including the *Elytroleptus*, which was not discriminated.

On the following day a concolorous lycid was introduced midway through the usual meal. It was ignored without previous inspection. The next day a concolorous *Elytroleptus* was offered in place of the model lycid. It was promptly seized and eaten.

(d) Mouse No. 4

Essentially the same series of experiments were intended as with mouse No. 3, but in this instance the black-tipped *L. fernandezi* complex was utilized rather than the concolorous *L. loripes* complex (a single black-tipped *Elytroleptus* cerambycid was available). Things did not proceed as far as with No. 3. When a group of seven *Lycus* and the cerambycid were offered midway through a meal, the mouse first proceeded to seize two lycids, rejecting them both. The third insect seized was the cerambycid and it was eaten without delay. The mouse proceeded to inspect and reject another three *Lycus*, before finally ignoring the entire group.

C. The "Free Hemorrhage" of Lycids

Lycids are known to exude fluid readily when handled, and it is generally assumed that this fluid contains the repugnatorial factors. Since the fluid can be emitted from virtually any part of the body, Darlington (1938) concluded that it is blood, rather than some special glandular secretion. In the particular Cuban lycids (*Thonalmus* spp.) studied by him, the fluid was found to be pink and produced "from between the abdominal segments ventrally, from lateral and ventral sutures of the thorax, from between the trochanter and femur and the leg, and, most frequently, from both the inner and outer margins of the elytra." The elytra, in fact, with their swollen, tube-like outer margins seemed especially adapted for bleeding.

Darlington further states that the emission of fluid is not autohemorrhage in a strict sense, since it never seemed to occur unless some membrane had first been ruptured. A surprisingly slight amout of pressure sufficed to bring about this rupture and initiate bleeding. The blood did not squirt out nor ooze, but merely flowed steadily to form large drops. He found the blood to be inoffensive when applied to his skin, lips, and tongue. Its taste and odor were mild, "reminiscent of crushed lettuce." Marshall and Poulton (1902) also make brief mention of the emission of the fluid by South African lycids, the fluid being "very strong-smelling" and white.

Our cursory observations confirm but add little to those of Darlington. Neither lycids of the *loripes* nor the *fernandezi* complexes produced detectable amounts of fluid unless injured, as judged by response to prodding and pinching with fine watchmaker's forceps under a microscope, although, a secretion might have been produced beneath the elytra on either the thorax or abdomen, since these regions were not scrutinized. Single viscous drops, light-grey in color, were collected at the site of injury. Three specimens that were chewed and swallowed by one of us (two of the *loripes* and one of the *fernandezi* type) had no apparent astringency or ill-flavor of any kind. A faint odor, vaguely resembling fresh hay, characterized the living beetles, and persisted for at least several days in pinned specimens.

A special effort was made to check on the survival of injured specimens. None of those with injured wings, with punctured appendages, or with single appendages amputated, showed conspicuous abnormalities as late as 48 hours after injury, except for anticipated difficulties in locomotion. Specimens with small single body punctures survived equally well. The only examples that died had been partly crushed with forceps, or had received lengthy incisions. Several specimens showing conspicuous imperfections of the elytra were collected in the field. Since these varied from a small marginal piece to almost an entire missing elytron, it is possible that the injuries were predator-inflicted.

Of special significance appears to be the finding that in many of the predator-prey encounters the lycids were rejected before fatal injury had been inflicted. In fact, often there was no noticeable injury whatever. Clearly, the repugnatorial principles are either not restricted to the blood alone, or if so restricted, they must pass somehow through the cuticle in effectively repellent concentrations. Also, there is likely to be considerable survival even under conditions where the predators involved are still naive, i.e., insufficiently experienced with lycids and hence aggressive toward them. Interestingly, similar observations have also been made frequently with highly protected butterflies (Danaidae, Ithomiidae, Heliconiidae, Acraeidae). Their bodies have a noticeably rubbery feel when pinched. They cease moving when pinched to an extent that would kill an ordinary butterfly, but recover fully in a few minutes. Collectors of these butterflies soon learn not to rely on pinching and immediately papering specimens of these groups. Under natural conditions involving predator attacks, such individuals are likely to survive as reproductive members of the species, while at the same time "teaching" the predator to shun that and similar mimetic species.

As a final comment on possible functions of lycid exudations, we add that during the summer of 1958, John A. Chemsak observing a species of *Calopteron* at Santo Domingo, 15 miles southeast of Simojovel, Chiapas, Mexico, noted the following:

"*Calopteron* (sp. near *reticulatum* Fabricius) were abundant in the Santo Domingo region and in most nearby areas. Adults of both sexes occurred on the upper surface of the leaves of various trees and shrubs. They were sluggish and quiescent, taking flight only when greatly disturbed. On several occasions two or more individuals were found together in what was at first assumed to be a mating posture. Closer examination revealed that some individuals were feeding on small globules of a light colored fluid secreted from the thorax immediately beneath the wings. The secreting individual elevated the elytra and wings to provide access to the site. Usually one individual fed at a time but on several occasions two were involved. In the sample taken a female was secreting for the benefit of two males. If this is generally the case the behavior is probably related to sexual activity."

In figure 1, a male of *Lycus fernandezi* may be seen with its head (and mouthparts) under the elytra of the female of a mating pair of the same species.

## Discussion and Summary

The feeding experiments reported above are not nearly as extensive nor as conclusive as we would have liked, largely because all categories of mimics were present in small numbers. They are nevertheless consistent with the generally accepted conclusion that lycids are distasteful to vertebrates, and they offer new evidence indicating that they are similarly distasteful or repellent to some common invertebrate predators. The widespread distribution of lycids, and their success in a wide range of predator-prey associations, would appear to be explained, in part at least, on this basis.

It would be interesting to know more about the inherent edibility or inedibility of the species associated with the lycids, since this data would enable the assemblages to be resolved into Batesian and Müllerian elements. Our own limited results are suggestive at best, and the discussion that follows is largely speculative, but might prove useful in setting the stage for further work of a more precise experimental nature.

The Pyromorphids seem to have repugnatorial properties of their own. In our experiments, the two specimens of *Seryda* used were rejected, one by a mantid, the other by a mouse. Jones (1932) and Carpenter (1921), in the experiments on birds and monkeys already cited above also found the pyromorphid, ctenuchid, and the zygaenids that they tested to be of low acceptability, and one zygaenid has been reported to have a strong odor (Marshall and Poulton, 1902). Whether odor and distastefulness are attributable to defensive secretions such as are known to occur in other moths (e.g., arctiids) remains to be determined. We made no attempt to look for secretory droplets in *Seryda*, nor did we check its odor.

None of our own trials included any of the arctiid or lithosiid mimics. There is circumstantial evidence, however, that many arctiids are distasteful and to be regarded as Müllerian elements. Arctiids are commonly aposematic, whether they are associated with mimetic complexes or not. Moreover, thoracic and tarsal glands, presumably defensive, are widespread throughout the family (Dethier, 1939; Rothschild, 1957). Jones (1932), with birds, tested two species congeneric with *Holomelina ostenta* (a presumed member of the *loripes* complex). In 14 experiments, 16 of 25 examples of *H. auriantica* Hübner were left on the trays; the other species, *H. opella* Grote, was only slightly more acceptable.

The cerambycids and the geometrid are probably Batesian rather than Müllerian mimics. *Elytroleptus* were eaten by both mantids and mice, the only predators to which they were offered. The single specimen that was left uneaten was ignored by a mouse (No. 3) that presumably discriminated against it on sight, having had previous experience with the model lycid. Many cerambycids possess noticeable odors and/or stridulate when disturbed, and this may conceivably serve in defense (Linsley, 1959). Neither species of *Elytroleptus*, so far as we noticed, had these properties. The cerambycids associated with the lycid complexes studied by Darlington (1938) and Parsons (1940) in Cuba, as already pointed out, also appear to be edible and hence are classed as Batesian mimics.

The geometrid *Eubaphe* was tested just twice, and then only with mantids, which accepted them. It appears, however, that edibility is the rule, rather than the exception, with geometrid moths. One of us (Eisner) in connection with other work has routinely fed an assortment of Eastern species to captive birds, lizards, mantids, spiders, and ants, without ever witnessing a single rejection or behavioral abnormality on the part of a predator. With few exceptions, most of the geometrids fed to birds by

Jones (1932) also had a relatively high acceptability ratio (at least those species that were tested in significant numbers). However, at least some geometrids (e.g., the Palaearctic *Abraxas*) are definitely protected and aposematic.

A number of moths other than those tested, belonging to "protected" families (Ctenuchidae, Pyromorphidae), occur widely in the same regions and environments as the lycids. They include species that may be functional members of the *L. fernandezi* or *L. loripes* mimetic complexes, or (because of an intermediate appearance) of both. Furthermore, the ctenuchid *Lycomorpha grotei*, because it has either red or yellow-brown forewings, may function as a Müllerian member of either the scarlet-winged or the yellow-winged lycid mimetic complexes, or of both, and thus may reinforce the protective effect of each.

In summary, it appears that each lycid assemblage is a composite of Müllerian and Batesian elements. The dominant form in each is a single species of lycid. This species and its sibling, together with an arctiid, and in one case (*fernandezi* group) a pyromorphid, make up the Müllerian framework of each complex, although other pyromorphid and ctenuchid moths are probably involved also. Edible longicorns, and in the *loripes* assemblage a geometrid, appear to participate as Batesian mimics. However, the division of these assemblages into Müllerian and Batesian elements according to traditional schemes should be approached with caution. In each assemblage the dominance of a single species is so strong, that the impression is given that it is the exclusive model, all other forms, Batesian and Müllerian, being subservient. The Müllerian associates (with the possible exception of *L. arizonensis*) would appear to be too scarce to contribute much on their own in the line of pattern reinforcement.

## Literature Cited


Arrow, G. J. 1928. Mimicry in beetles. Nat. Hist. Mag. (British Mus.), **1**: 166–172.

Bates, H. W. 1892. The Naturalist on the River Amazons. London, lxxxix + 395 pp.

Belon, R. P. 1897. Remarques sur le genre *Euryptera* Serv., du groupe des Lepturides et description d'une espècie nouvelle de Bolivie. Ann. Soc. Ent. Belgique, **41**: 339–343.

Belt, T. 1874. The Naturalist in Nicaragua. London.

Carpenter, G. D. H. 1921. Experiments on the relative edibility of insects, with special reference to their coloration. Trans. Ent. Soc. London, **1921**: 1–105.

—— and E. B. Ford. 1933. Mimicry. London, viii + 134 pp.

Darlington, P. J. 1938. Experiments on mimicry in Cuba, with suggestions for future study. Trans. Roy. Ent. Soc. London, **87**: 681–695.

Dethier, V. G. 1939. Prothoracic glands of adult Lepidoptera. Jour. New York Ent. Soc., **47**: 131 141.

Gahan, C. J. 1913. Mimicry in Coleoptera. Proc. So. London Ent. Nat. Hist. Soc., **1912–14**: 28–38.

Gorham, H. S. 1880. Fam. Lycidae. Biologia Centrali-Americana, Coleoptera, **3**: 1–28.

Green, J. W. 1949. The Lycidae of the United States and Canada. I. The tribe Lycini (Coleoptera). Trans. Amer. Ent. Soc., **75**: 53–70.

Holland, W. J. 1917. The Moth Book. Doubleday, Page & Co., Garden City. (Many variously dated printings.)

Jones, F. M. 1932. Insect coloration and the relative acceptability of insects to birds. Trans. Ent. Soc. London, **80**: 345–386.

Linsley, E. G. 1959. Ecology of Cerambycidae. Ann. Rev. Ent., **4**: 99–138.

Marshall, G. A. K., and E. B. Poulton. 1902. Five years' observation and experiments (1896–1901) on the bionomics of South African insects, chiefly directed to the investigation of mimicry and warning colours. Trans. Ent. Soc. London, **1902**: 287–584.

McAtee, W. L. 1932. Effectiveness in nature of the so-called protective adaptations in the animal kingdom, chiefly as illustrated by the food habits of Nearctic birds. Smithsonian Misc. Coll., **85**: 1–201.

Parsons, C. T. 1940. Observations in Cuba on mimicry and protective coloration. Psyche, **47**: 1–7.

Rothschild, M. 1957. Some observations on the garden tiger moth, *Arctic caia* L. (abstract). Proc. Royal Ent. Soc. London, Ser. C, **22**: 1.

Schaeffer, C. 1908. List of the Lampyridae from the Huachuca Mountains, Arizona, and descriptions of new species. Jour. New York Ent. Soc., **16**: 61–67.

——. 1911. New Coleoptera and miscellaneous notes. Jour. New York Ent. Soc., **19**: 113–126.

Shelford, R. 1901. Observations on some mimetic insects and spiders from Borneo and Singapore. Proc. Zool. Soc. London, **1901**: 230–284.

——. 1917. A Naturalist in Borneo. London, xxviii + 331.

# MIMETIC FORM AND COLORATION IN THE CERAMBYCIDAE (COLEOPTERA)[1]

E. GORTON LINSLEY
University of California, Berkeley

ABSTRACT

Batesian mimicry is widespread in the Cerambycidae, the commonest models being Aculeate Hymenoptera (ants, bees, wasps) and "protected" Coleoptera (especially lycids, cantharids, chrysomelids, and weevils). In some cases it is expressed as a general group mimicry, as in the wasp-mimicking Clytini, Lepturini, and Necydalini, the ant-mimicking Tillomorphini, and the lycid-mimicking Pteroplatini, although some members of these tribes have specific models. More commonly, mimicry is sporadic, involving marked departures from near relatives in form, coloration, and behavior, with striking resemblance to a specific model. Although evident in most subfamilies and a large number of tribes, Batesian mimicry is most fully developed in the Rhinotragini, where closely related forms mimic such diverse groups as bees, wasps, lycids, cantharids, and hispids. Müllerian mimicry has been less completely studied, although cerambycids are known to share Müllerian coloration with other insects in the West Indies and in Borneo.

Diurnally inactive Cerambycidae which secrete themselves during the day are usually dark colored, those which rest exposed on bark, twigs, or foliage, cryptic. Diurnally active forms generally have brilliant or contrasting coloration and may be conspicuous in flight, less so when at rest. Flower visitors are often colored to match prevalent pollen sources (Chevrolat, 1860; Linsley, 1957); or, if distasteful, they may be warningly colored. Mimicry of both Batesian and Müllerian types occurs irregularly throughout the family, especially among groups living in tropical regions. Unfortunately, however, few ecologists have carried out the necessary field studies to permit the proper interpretation of the individual cases occurring locally in a given region. The most significant contributions to the subject have been made by Bates (1870), Fowler (1901), Shelford (1902, 1916), and Gahan (1913). Shelford's studies are notable as an intensive analysis of the fauna of a small area, 20 miles in diameter, in which he encountered more than 70 species of mimetic Cerambycidae. No similar study has been made, to my knowledge, in any other comparable area, nor have Shelford's interpretations been confirmed by re-investigation. Nevertheless, his general conclusions appear to be valid although details may be subject to later revision. The present review is intended merely to consider mimicry in the Cerambycidae from a broad standpoint, with the hope of stimulating more field research on the subject.

Batesian mimicry, involving departures in form, coloration, and behavior, coinciding with those of such hymenopterous models as ants, bees, or wasps, or beetle models as lycids, cantharids, or chrysomelids, is widespread among the diurnal Cerambycidae, especially in the Cerambycinae, and exhibits varying degrees of complexity. In its simplest form it can be recognized as a "group mimicry," in which the species in a genus, in related genera, or in a tribe have a general but nonspecific resemblance to those of some other group of insects. Thus, the almost universally flower-visiting Lepturinae exhibit certain group characteristics commonly associated with this habit (e.g., an elongate and attenuated body form and bright coloration involving yellow and black, reds and black, etc.). These same characteristics are also common in flower-visiting wasps. Thus selection for wasplike flight and behavior has occurred repeatedly throughout much of the world, although the beetles, when not moving, may bear little superficial resemblance to wasps, as the North American *Strangalia bicolor* (Swederus) (Knab, 1896). The Clytini, also group mimics of wasps and also commonly but by no means universally flower visitors, mostly have short antennae, long legs, and very often yellow and black coloration. These have been selected not for wasplike flight, but for wasplike mannerisms in running about over logs and branches in the sunlight, as the well-known European wasp beetle *Clytus arietis* (Linnaeus) (Poulton, 1890) or the Australian *Aridaeus thoracicus* Donisthorpe (Tillyard, 1926).[2] Wasp mimics in this tribe occur in a large number of genera on all continents.

The closely related Tillomorphini, basically clytine in form, consists of species smaller in size. The genera in this tribe are group mimics of ants. The antlike form in this case is accomplished by a slight constriction of the elytra at the middle, usually accentuated by transverse ivory-like marks or oblique pubescent bands, a swollen, convex, pubescent elytral apex, and different coloration before and after the median constriction (fig. 1a-c). This type of mimicry, although basically similar in origin to that of the Lepturinae and Clytini, involves some morphological departure from the ancestral type, and is often perfected

[1]Accepted for publication June 4, 1958.

[2]The related *Aridaeus cleroides* McKeown (1942) is said to bear an "amazing resemblance" to the clerid *Trogodendron fasciculatum* Schr. which in turn simulates the movements of a wasp.

Reproduced with permission from the *Annals of the Entomological Society of America*, 52:125–131, 1959. Published by the Entomological Society of America, College Park, Maryland.

so as to coincide with that of models of diverse types, as in the Australian *Zoeidia* and *Ochrya* (Poll, 1892), and may include the development of antlike heads as in Macronini (*Brachyopes*) and Pseudocephalini (*Cyclocranium, Ametrocephala,* and *Formicomimus*) (Tillyard, 1926). Specific mimicry in the North American species of *Euderces* and *Eplophorus* may vary geographically and coincide with geographical color variation in the model, as *Eplophorus bicinctus* Linsley (1942) and its model *Leptalea gracilis* (Fabricius). The North African *Pseudomyrmecion* is said to live with ants of the genus *Crematogaster* (Grandi, 1951). Poulton (in Shelford, 1902) suggests that Clytini and Tillomorphini are protected groups and that their mimicry of wasps and ants is of the Müllerian type. However, more evidence is needed on this point (see below).

A few other cases of group mimicry may be mentioned. The ovoid, apterous Parmenini and their relatives bear a general resemblance to spiders (fig. 1d) in such widely separated areas as California (e.g., *Ipochus*), Chile (e.g., *Microcleptes*), and South Africa (e.g., *Spinosomatidia*). Likewise, the Old World Mesosini and Apomycini often bear a marked but nonspecific resemblance to Rhyncophora, the former to Anthribidae, the latter to Curculionidae (Shelford, 1902). Similarly, the Hippopsini and Agiini contain group mimics of brenthids, and the Australian species of *Microtragus* and *Athemistus*, dull colored, ground-frequenting forms, resemble ground weevils to a remarkable degree (McKeown, 1942). The broad, flattened, sedentary Pteroplatini are mimics of lycid beetles throughout tropical America, and in many cases this mimicry coincides in color and form with specific models in such genera as *Calopteron* and *Lycus*, to a high degree of perfection (figs. 2a-d). These mimics join aggregations of their models, with whom they regularly associate.

The Necydalini, on the other hand, are elongated, with short antennae and abbreviated elytra which expose the flying wings on the abdomen. These are group mimics of Hymenoptera. Thus, members of the Holarctic genus *Necydalis* look and fly like ichneumonids as they course over grassy areas. Both sexes of the Neotropical *Rhathymoscelis* are wasp-like, but the abdomen is more constricted and the mimicry more perfect in the male (fig. 3a) (Lane, 1936). The Australian *Hesthesis ferruginea* Boisduval closely resembles the wasp *Abispa ephippium* (Fabricius) both in form and coloration, visiting flowers and moving with the same jerkiness and flicking of the wings (Tillyard, 1926; McKeown, 1942); and the Indo-Malayan *Colorhombus* (*Nothopeus*), with the constricted abdomen appearing even narrower owing to a patch of silvery grey pubescence on each side, have colored wings like their models among various large wasps of the region (Shelford, 1902; Ritsema, 1911; Gahan, 1913; Kalshoven, 1935). Thus, *C. fasciatipennis* (Waterhouse) has a white band across dark wings, like its model *Mygnimia princeps* White; *C. hemipterus* (Olivier) is dark with glossy blue wings like its model *Macromerus splendida* Lepeletier; and *C. auricomus* (Ritsema) is golden pubescent with tawny wings, like *Salius ceylonicus* (Saussure). In New Zealand, the related spider wasp *S. wakefieldi* is mimicked by the extremely active, quick-flying, sunloving

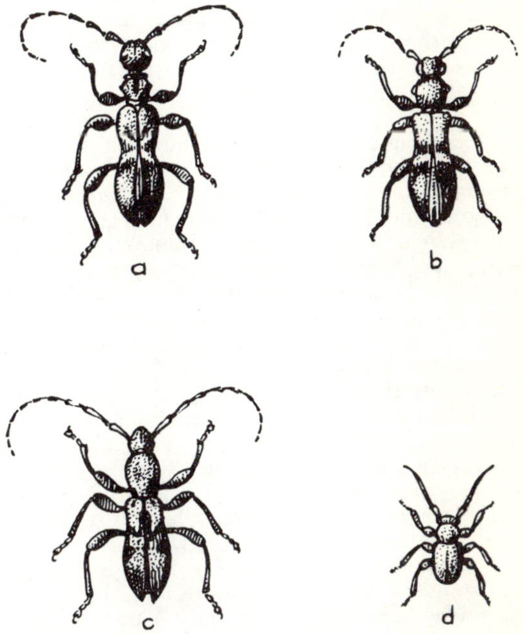

FIG. 1.—Australian ant mimics: (a) *Cyclocranium swierstrae* Poll, (b) *Ochrya nana* Poll, and (c) *Mesolita inermis* Poll. Chilean spider mimic: (d) *Microcleptes variolosus* Fairmaire and Germaine. Celeste Green del.

*Gastrosaurus nigricollis* Bates (Hudson, 1934). In North America, *Ulochaetes leoninus* LeConte is a robust species which exhibits the buzzing flight and habitus of a bumblebee (*Bombus*) and, when handled, the female makes a pretense of stinging with the ovipositor (Hardy and Preece, 1926).[3] The Chilean genus *Callisphyrus*, another member of this group, has enlarged legs with brushes of hair like the tibial scopa of anthophorine bees. However, cerambycids with an elongated body and abbreviated elytra are selected for wasplike behavior only if, as indicated above, they are diurnal in habit. Thus, the small, diurnal species of *Molorchus* exhibit wasplike mannerisms, but not the similarly formed nocturnal species of *Methia*. In the Methiini, only the diurnal *Coleomethia evaniiformis*

[3]By an odd coincidence, the larva of an asilid fly *Bombomimus astur* (Osten Sacken), which also mimics a bumblebee, preys upon the larva of *Ulochaetes* (Hardy and Preece, 1927).

Knull (1937) is mimetic, adopting the behavior of an evaniid wasp.

With regard to specific resemblance of the Batesian type, Shelford (1916) has appropriately commented, "A student of mimicry cannot fail to be struck with the fact that mimetic resemblances are due to a simulation of the model's characters and not to a direct imitation of them. A fly mimicking a wasp does not develop a second pair of wings to imitate the wasp's wings, but the wings of the fly may be so formed or so veined or grooved that they appear like the four wings of the wasp. Let us note one or two more instances of this from Borneo. A beetle of the distasteful family Endomychidae and belonging to the genus *Spathomeles* has a strong stout spine on each wing-case, and this character of spiny wing-cases or elytra is common to all the Amphisterni. This particular *Spathomeles* is closely mimicked by a Longicorn beetle *Zelota spathomelina;* the mimic is coloured like its model, and from each elytron springs what appears to be a spine, but on close examination is seen to be nothing but a finely pointed tuft of hairs. In other words, the spine of the model is simulated by the hair-tuft of the mimic. If natural selection is all-powerful, if it can guide variations into any and every direction, why has it not called into being a spine on the elytron of the Longicorn? Spinosity of the elytra is unknown amongst [Bornean] Longicorns, pubescence is common, and natural selection must use the material to hand and make the best of that. To take another instance: there occur in Borneo numbers of parasitic Hymenoptera of the family Braconidae. Some of these, mostly of the genus *Myosoma*, are black with reddish head and thorax, and they are mimicked by Longicorns of the genus *Oberea*. The Braconidae fly about very freely, they are wasp-waisted like so many Hymenoptera, and when they settle on a leaf they have the habit of walking about waving the antennae up and down in a very characteristic manner; the black wings are folded down over the back so that seen from above the wasp-waist is not visible. Now three species of *Oberea* resemble the Braconids in a very remarkable way; the colouring is the same, and a wasp-waisted effect is produced by a patch of silvery white pubescence on each side of the first and second abdominal segments (see fig. 3b); the part of the body covered with this pubescence tends to disappear from sight altogether, and the beetle looks as if the rest of the abdomen was slung on to the thorax by a mere pedicel as is really the case in the Braconidae. It may be added that all the Obereas are active fliers. all have the habit of agitating the antennae when walking about, and some of the species are so slender, and have the wing-cases so narrowed, that they look even more like Hymenoptera than the species with false wasp-waists.

"This kind of colouring whereby certain parts of the body, or even the whole body, may be made to disappear from sight is very common in the animal kingdom, but it actually occurs that an organ can be so constructed that part of it is practically invisible even when exposed. The little Longicorns of the genus *Xyaste* mimic beetles of the family Lycidae; now the Lycidae have short, thick, flattened antennae, but the Longicorns of the sub-family to which *Xyaste* belongs have the antennae long and slender. The effect of a short flattened antenna is produced in a very interesting way; the first few basal joints

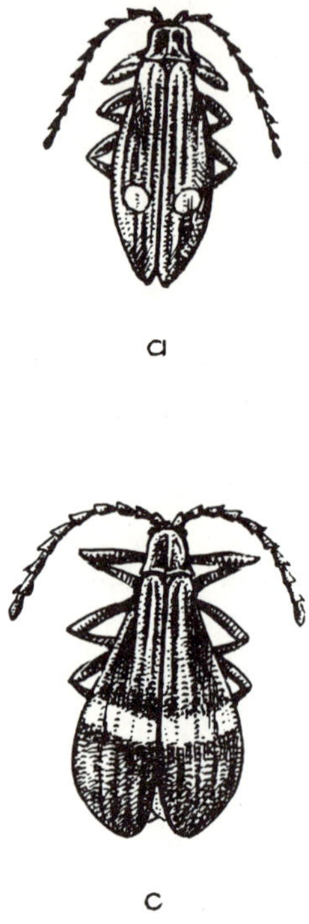

FIG. 2.—(a, c) Colombian lycids (*Calopteron* spp.) and (b, d) their pteroplatine mimics (*Pteroplatus* spp.). Celeste Green del.

are clothed thickly with long black hairs closely set together, whilst all the rest of the antenna is drawn out into a fine slender thread which is hardly visible; in fact at a short distance it is quite invisible, and the antennae appear to be exactly like those of the model."

Among the Lamiinae there are scattered ant mimics, such as *Falsovellida congolensis* (Hintz) which runs very rapidly and seldom takes flight (Duffy, 1957), and many are actually wingless forms, such as the North American *Michthysoma* and *Cyrtinus* and the Australian *Mesolita myrmecophila* Lea (1918). The genera *Glenea* and *Neonitocris* (Breuning, 1950) contain species resembling ichneumonids (Andrewes, 1904; Gahan, 1913) (fig. 3b) and other wasps (fig. 3c).

Lycid mimics appear in the Cermabycinae, Lepturinae, and Lamiinae (fig. 4a-d) (Bates, 1885; Shelford, 1902; Gahan, 1913; Chopard, 1949; Lepesme, 1950; Lane, 1954; Costa Lima, 1953). The Brazilian *Sternacanthus undatus* (Olivier) resembles *Erotylus sexfasciatus* Fabricius; the Burmese *Estigmenida variabilis* Gahan, the hispid *Estigmena chinensis* Hope; the Malayan *Ixais castelnaudi* (Thomson), the weevil *Episomus chlorostigma* Wied; and species of the Philippine genus *Doliops* mimic weevils of the genus *Pachyrrhynchus* (fig. 5c-d), and have a knoblike thickening corresponding to the terminal club of the weevil antennae. Similar mimicry occurs in South America, where various genera, including the remarkable *Zelliboria*, resemble hispids (Lane, 1951), *Ites*, cassidids (Lane, 1954), and *Probatius choliniformis* Lane the weevil *Naedius elegans* Lane and Moure (Lane, 1956). In Australia, the excessively slender species of *Lygesis* (Tillyard, 1926) resemble coreids of the genus *Leptocorixa*, but the most remarkable bug-mimic now known is the Brazilian *Melzerella lutzi* Costa Lima (1931), which is marked and colored like a pyrrhocorid or reduviid. On the other hand, *Compsosoma mutillarium* Klug, together with the similarly hairy and brightly colored weevil *Cyphus myrmosarius* Perty, are striking mimics of a mutillid wasp (F. Lane, *in litt.*).

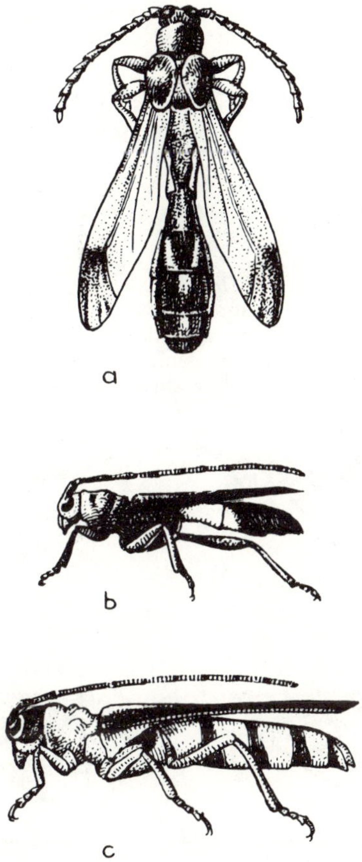

FIG. 3.—(a) *Rhathymoscelis peruibensis* Lane. (b, c) Wasp-mimicking African *Neonitocris*. Celeste Green del.

Shelford (1902, 1916) divided the mimetic Longicorns of Borneo into three groups: (i) those that mimic Hymenoptera; (ii) those that mimic other beetles; (iii) those that mimic other and presumably distasteful species of Longicorns. Among the numerous examples discussed by him, the following are deserving of special mention.

Stingless bees of the genus *Melipona*, which bite with vigor, swarm about the head, and have a disagreeable odor, are mimicked by many other insects, especially in the case of the Dammar bee, *M. vidua* Lepeletier, a black species with white wing-tips. Among these mimics is a similarly marked longicorn, *Epania singaporensis* Pascoe, whose near relative *E. sarawakensis* Pascoe resembles an ant. *Psebena brevipennis* Gahan mimics a braconid of the genus *Myosoma*, and *Glenea iresine* Pascoe a blue sawfly, *Hylotoma pruinosa*.

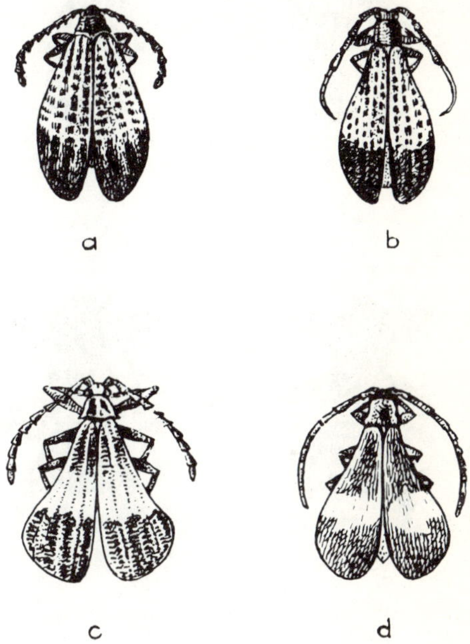

FIG. 4.—Neotropical lycids, (a) *Calopteron textile* Bourg., (c) *Calopteron* sp., and mimics, (b) *Lycidola retifera* Waterhouse, (d) *Lycomimus albocinctus* Melzer. Celeste Green del.

Among beetle mimics, the Saperdini and Astatheini contain many species resembling gallerucine Chrysomelidae. In Borneo these latter are all brilliantly or conspicuously colored, are excessively abundant in species as well as in individuals, and there is some evidence to show that they are unpalatable insects. Shelford regards it as not unlikely that their mimics are also unpalatable and that the mimicry is Müllerian rather than Batesian. He also states that in comparing collections of galerucids and longicorns of the two groups, if he picked out at random a galerucid of striking coloration he could almost invariably match it with a similarly colored longicorn. Among the striking patterns shared by the galerucids and longicorns were: (1) Bright blue with white antennae; (2) Yellow with blue band across basal half of elytra; (3) Dark, shining blue, with apical half of elytra yellow or red; (4) Elytra dark, shining blue, head and prothorax yellow or red; and (5) Fulvous yellow or brown. The parallelism extended to

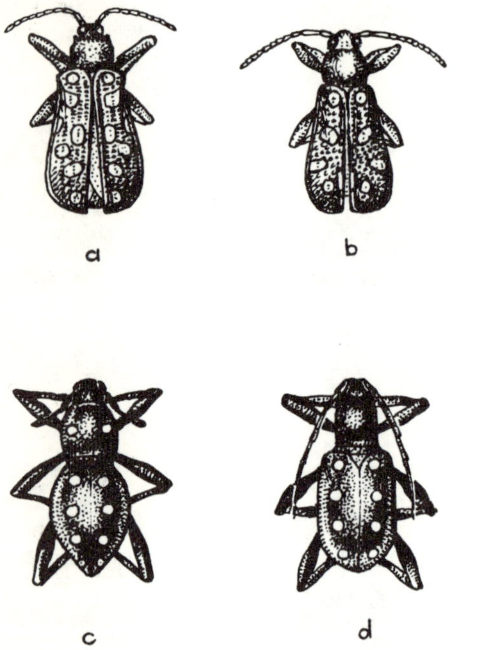

Fig. 5.—(a) A Neotropical hispid, *Cephalodonta maculata* (Olivier), and (b) a rhinotragine mimic, *Erythroplatys corallifer* White. (c) A Philippine weevil. *Pachyrrhynchus erischsoni* Waterhouse, and (d) its mimic, *Doliops curculionoides* Waterhouse. Celeste Green del.

quite minor details; e.g., the galerucids *Antipha nigra* and *A. abdominalis* were colored according to type 3, but the former has the abdomen beneath black, the latter yellow. Exactly the same difference distinguishes the two mimics *Astathes posticalis* and *A. flaviventris*. Other Bornean cerambycids are said to mimic coccinelids and cicindelids.

One of the most interesting and perhaps controversial results of Shelford's studies was the conclusion that Bornean longicorns mimic other longicorns. In his own words (Shelford, 1916) he states, "Just as amongst butterflies so amongst this group of beetles do we find that certain genera are protected and serve as models to many other genera. Amongst Longicorns the protected families are two only, both of the Cerambycidae, viz. the Callichromini and the Clytini. The species of the former which serve as models all belong to one genus, *Chloridolum;* they are all bright metallic green beetles, occurring in great abundance on felled timber; they emit a strong, but to human nostrils by no means an unpleasant, odor proceeding from glands in the thorax, which open to the exterior by two pores. They are mimicked by members of the Saperdini and Phytaeciini, and of the Oemiini, Disteniini, and Lepturini. The Clytini include the British *Clytus arietis*, a species which, as is well known, is marked like a wasp and when seized behaves as if it were about to sting. We cannot say for certain if this is a case of Müllerian or Batesian mimicry, but I am almost certain that the Oriental Clytini are for the most part protected. *Chlorophorus* (*Clytanthus*) *annularis*, a yellow beetle banded with black, swarmed in countless numbers on Mt. Penrisen, while around Kuching some species of *Demonax*, grey, black-banded beetles, were amongst the commonest of all Longicorns. Their abundance, combined with the fact that they are widely mimicked, forces us to the conclusion that these two genera, at any

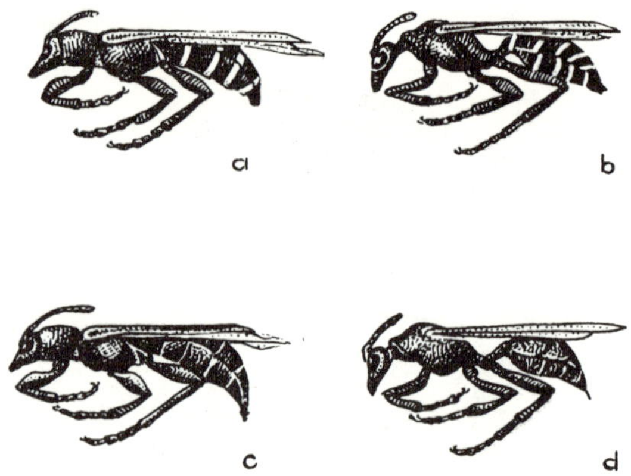

Fig. 6.—Brazilian wasps and rhinotragine mimics: (a) *Polybia nigra* Sauss., (b) *Sphecomorpha murina* Klug, (c) *Polybia liliacea* Sauss., and (d) *Acyphoderes odyneroides* White. Celeste Green del.

rate, are protected . . . These protected Clytini exhibit three types of coloration: (1) yellow with black bands or stripes; (2) grey with black bands; (3) as (2), but with red thorax. The Clytinae of type (1) are mimicked by a Lamiine, *Cylindrepomus comis*, by a Phytaeciine, *Daphisia* sp., and by a Lepturine, *Leptura* sp. A species of *Leptura* also mimics a *Demonax* coloured according to type (3). All the other mimics, including six species belonging to four subfamilies, resemble species of *Demonax* and *Xylotrechus*, coloured according to type (2)."

The most instructive case of Batesian mimicry from the ecological and evolutionary standpoint is that of the Neotropical Rhinotragini (figs, 5, 6, 7). Perhaps nowhere in the Animal

Kingdom is Batesian mimicry so highly developed. In this tribe closely related forms resemble, associate with, and behave like groups of insects which are very diverse in form, coloration, and habits. In fact, most of the genera in this tribe consist of species which are quite obviously mimetic. Some closely resemble protected groups of beetles; e.g., *Oxylymma*, *Rhinotragus*, *Ommata*, *Cantharoxylymma*, and *Oregostoma luridum* Klug look like species of *Chauliognathus* or other cantharids (fig. 6a, c, d); *Ornistomis* and certain other *Oregostoma* species mimic and congregate with lycids; and the mimetic analogue of *Erythroplatys corallifer* White is the hispine leaf beetle *Cephalodonta maculata* (Olivier) (fig. 5a-b).

Genera closely allied to these lycid-mimics contain flower-visiting species which are vespiform, as *Sphecomorpha*, *Phespia* (fig. 7b), and *Acyphoderes*. The resemblance of such groups to their models is very close. Thus Bates (1870) points out that *Acyphoderes odyneroides* White, which visits flowers of *Pitomba*, so closely suggests *Polybia nigra* Saussure, a wasp which visits the same flowers, that he was never sure whether he had captured the beetle or the wasp until he had closely examined the captured insect (fig. 6c-d). *Sphecomorpha murina* Klug, of the same region, likewise resembles *Polybia liliacea* Saussure (fig. 6a-b). Species of *Odontocera* are also vespiform and, according to Bates, the individuals fly nimbly from flower to flower, "deceiving the eye of the beholder" by their strong resemblance to wasps.

Other Rhinotragini resemble parasitic Hymenoptera. Among these may be mentioned the "striking resemblance" of the species of *Isthmidae* to braconid wasps, emphasized by the filiform abdomen of the female which resembles the ovipositor, and the yellow and black wings. Another species, *Physopoda albitarsis* (Klug), resembles a large mosquito of the genus *Culex*, and several groups mimic bees. Thus, the flower-frequenting species of *Epimelitta* and *Physopoda subvestita* (White) resemble species of the stingless meliponid bees, the latter species a common pale *Melipona* with black hairs on the hind tibiae. In *Tomopterus*, another group of bee-mimics, *T. larroides* White looks like a small bee of the genus *Megachile* (Bates, 1870).

Similar, but less diverse, groups of mimics occur in other sections of the family, as the lamiine tribes Tropidosomatini, in which the related genera *Tropidosoma* and *Pteracantha* are lyciform, *Ctenodes* and *Streptolabis* hispiform, and Gymnocerini, in which *Gymnocerus belti* Bates is weevil-like and *G. dulcissimus* Bates erotylid-like. As final examples of extreme mimetic development we mention the prionine genus *Anacolus*, in which the males are wasp-like and the females mimic *Chauliognathus* and other cantharids, and *Cycloprionus* Tippmann, in which the females are chrysomelid-like (Tippmann, 1953b).

In the West Indies there exists a widespread "Müllerian ring" of synaposomatic mimicry involving distasteful or inedible forms of various insect groups, particularly the Coleoptera and Hemiptera. Edible species in several orders, including beetles of various families, have taken advantage of the existence of this ring by developing "Batesian" mimics. Included in the Müllerian group are lycids of the genus *Thonalmus* which are bright orange or red, with the apex of the elytra blue. These are presumed to be models for numerous Batesian mimics, including cerambycids of the genera *Trichrous*, *Heterops*, and *Calocosmus* (this last, a relative of *Tetraopes*, may possibly be distasteful and function as part of the Müllerian ring).

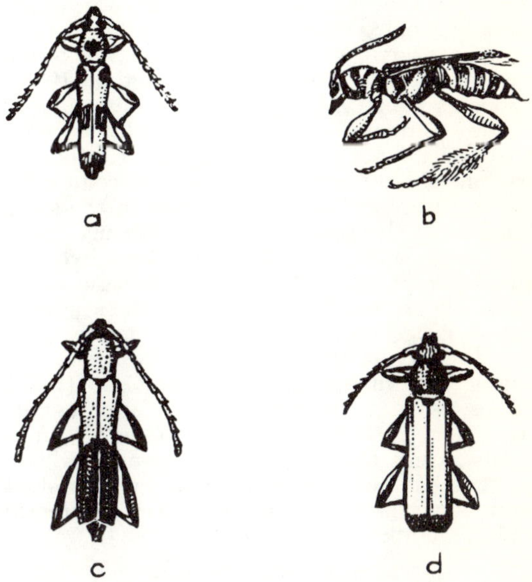


FIG. 7.—Mimetic Neotropical Rhinotragini. (a, c) *Ommata* spp., (b) *Phespia corinna* Pascoe, (d) *Rhinotragus opicalis* Guerin. Celeste Green del.

In a series of experiments, Darlington (1938) attempted to determine the relative edibility of some of these Cerambycidae. Seven specimens of four species of *Thonalmus*-like Cerambycidae were caged for periods of up to a week with the lizard *Anolis sagrei* (Duméril and Bibron), sometimes with and sometimes without *Thonalmus* in the cage. In no case did the lizards eat or, so far as Darlington observed, show any interest in the *Thonalmus*-like species. *Heterops dimidiata* (Chevrolat), with black-tipped elytra, survived as well as the blue-tipped species which resemble *Thonalmus* more closely. While the *Thonalmus*-like Cerambycidae were unmolested, the lizards did eat 11 plainly colored specimens of 6 rather diverse species (some Cerambycinae, some Lamiinae). They refused no plainly colored species. There was nothing that Darlington could find in the structure of these *Thonalmus*-like longicorns to make them less acceptable food than the plainly colored species—no difference in size or hardness or spinosity—nor did the *Thonalmus*-like forms exude a defensive fluid. Only one individual of *Calocosmus venustus* (Chevrolat) was available for introduction into the cages. It was found dead 2 days later, but apparently had not been killed by the lizards.

In conclusion, it may be mentioned that experimental evidence has been accumulated by many workers indicating that bees, wasps, and beetles of the families referred to above are generally avoided by vertebrate predators or have a certain degree of immunity from attack (see especially Jones, 1932, and Darlington, 1938). It appears significant that bee- and wasp-mimics of the Cermabycidae are generally flower-frequenting types, and that lycid-, cantharid-, chrysomelid-, and ant-mimics are most commonly found on trees and other vegetation in the same situations as their models. Also, although there is an abundance of species of Cerambycidae, individuals are often rare, and in mimetic forms are usually exceeded numerically by their models.

Most objections to the theory of mimicry have been directed at the mechanism by which the mimetic pattern is produced. Nicholson (1927) emphasizes that protective resemblance apparently offers a slight advantage to certain individuals rather than to the species as a whole, and thus has been brought about by a gradual accumulation of slight changes in the direction of an ultimate mimetic pattern.

REFERENCES CITED

**Andrewes, Leslie.** 1904. Mimicry of an ichneumonid by the longicorn, *Glenea pulchella* (Thomas.). Proc. Ent. Soc. London 1904: vi-vii.

**Bates, H. W.** 1870. Contributions to an insect fauna of the Amazon Valley (Coleoptera, Cerambycidae). Trans. Ent. Soc. London 1870: 243–55.

1879–1885. Prionidae, Cerambycidae, Lamiidae. Biologia Centrali-Americana, Coleoptera 5: 1–436, pls. 1–25.

**Breuning, St.** 1950. Nouvelles formes de Lamiaires du Congo Belge (Cerambycidae). Ann. Mus. Congo Belge, Tervuren (Sér. 8°), Sci. Zool. 4: 1–46.

**Chevrolat, L. A. A.** 1860. Description d'espèces de *Clytus* propres au Mexique. Ann. Soc. Ent. France (3) 8: 451–504.

**Chopard, L.** 1949. Le mimétisme. Paris: Payot. 335 pp.

**Costa Lima, A. da.** 1931. *Melzerella lutzi*, n. gen., n. sp. (Cerambycoidea-Lamiidae). Rev. Ent. [São Paulo] 1: 139–42.

1953. Insetos do Brasil. Vol. 8. Coleópteros, 2a. parte. Rio de Janeiro: Escola Nacional de Agronomia. 323 pp., 259 figs.

**Darlington, P. J., Jr.** 1938. Experiments on mimicry in Cuba, with suggestions for further study. Trans. Roy. Ent. Soc. London 87: 681–95, pl. 1.

**Duffy, E. A. J.** 1957. A monograph of the immature stages of African timber beetles. London: British Museum (Natural History). 338 pp.

**Fowler, W. W.** 1901. Protective resemblance and mimicry as affecting the Coleoptera. Proc. Ent. Soc. London 1901: xxxvi-lix.

**Gahan, C. J.** 1913. Mimicry in Coleoptera. Proc. South London Ent. Nat. Hist. Soc. 1912–13: 28–38.

**Grandi, G.** 1951. Introduzione allo studio dell'entomologia. Vol. II. Endopterigoti. Bologna, Italy. 1332 pp., 1198 figs.

**Hardy, G. A.,** and **W. H. A. Preece.** 1926. Notes on some species of Cerambycidae (Col.) from the southern portion of Vancouver Island, B. C. Pan-Pacific Ent. 3: 34–40.

1927. Additional notes on some Cerambycidae (Col.) from Vancouver Isalnd, B. C., with descriptions of some new varieties. Pan-Pacific Ent. 4: 61–67.

**Hudson, G. V.** 1934. New Zealand beetles and their larvae. Wellington, N. Z.: Ferguson & Osborn, Ltd. 236 pp., 17 pls.

**Jones, F. M.** 1932. Insect coloration and the relative acceptability of insects to birds. Trans. Ent. Soc. London 80: 345–86, 11 pls.

**Kalshoven, L. G. E.** 1935. Indomalayan *Nothopeus* species (Col., Cerambycidae), remarks on their food-plants, identity and mimicry. Ent. Mededel. Nederl.-Indie 1:50–54 (fig.).

**Knab, F.** 1896. [Coleoptera from Chicopee, Massachusetts.] Ent. News 7: 113.

**Knull, J. N.** 1937. New southwestern Buprestidae and Cerambycidae with notes. Ohio Jour. Sci. 37: 301–8, 1 pl.

**Kojima, T.** 1929. Immature stages of some Japanese cerambycid-beetles with notes on their habits. Jour. Coll. Agric. Univ. Tokyo 10: 101–28, pl. 6.

**Lane, Frederico.** 1936. Algumas notas sobre o genero *Rhathymoscelis* Thomas. e descripção de uma especie nova (Col. Cerambycidae). Rev. Mus. Paulista 20: 765–76, 2 pls.

1951. Cerambycoidea Neotropica nova II (Coleoptera). Dusenia 2(1): 1–20, 1 pl.

1954. Notas sinonímicas—I. *Lycodesmus* Melzer, 1927 —*Ites* Waterhouse, 1880 (Coleoptera, Lamiidae). Rev. Brasil. Ent. 1: 195–201, 2 figs.

1957. Cerambycoidea Neotropica nova IV (Coleoptera). Dusenia 7(1): 1–32, pl. 1.

**Lea, A. M.** 1918. On Australian Coleoptera. Rec. South Australian Mus. 1: 83–104, 1 pl.

**Lepesme, P.** 1950. Notes longicornesques. Longicornia 1: 587–90, 5 figs.

**Linsley, E. G.** 1942. Contributions toward a knowledge of the insect fauna of Lower California. No. 2. Coleoptera: Cerambycidae. Proc. California Acad. Sci. (4) 24: 21–96, pls. 4–5.

1957. Host relationships in the genus *Crossidius* (Coleoptera, Cerambycidae). Jour. Kansas Ent. Soc. 30: 81–89.

**McKeown, K. C.** 1942. Australian insects. Sydney: Royal Zool. Soc. New South Wales. 304 pp.

**Melzer, Julius.** 1928. Longicorneos do Brasil novos ou pouco conhecidos (Coleoptera, Cerambycidae). Arch. Inst. Biol. [São Paulo] 1: 143–58, pls. 24–26.

**Nicholson, A. J.** 1927. A new theory of mimicry in insects. Australian Zool. 5: 10–104, 14 pls.

**Poll, J. R. H. Neervoort van de.** 1892. On new or little known Australian Longicornia. I. Tijdschr. Ent. 34: 219–28, pl. 13.

**Paulton, E. B.** 1890. The colours of animals; their meaning and use especially considered in the case of insects. London. 360 pp.

**Ritsema, C.** 1911. Protective resemblance between the species of *Nothopeus* Pasc. and fossorial wasps. Notes Leiden Mus. 34: 22–24.

**Shelford, R. W. C.** 1902. Observations on some mimetic insects and spiders from Borneo and Singapore. Proc. Zool. Soc. London 1902(2): 230–84, pls. 19–23.

1916. A naturalist in Borneo. London: Unwin, Ltd. 331 pp.

**Tillyard, R. J.** 1926. The insects of Australia and New Zealand. Sydney: Angus & Robertson, Ltd. xi+560 pp., 44 pls., many text figs.

**Tippmann, F. F.** 1953a. Lamiinen mit Interferenz-Farbenspiel (Coleoptera: Cerambycidae, subfam. Lamiinae). Ent. Blätter 49: 148–53.

1953b. Studien über neotropische Longicornier—I. (Coleoptera: Cerambycidae). Dusenia 4: 181–228 pls. 13–17.

# The Relative Abundance of Model and Mimic Butterflies in Natural Populations of the *Battus Philenor* Mimicry Complex

Lincoln P. Brower and Jane Van Zandt Brower
*Biology Laboratory, Amherst College*

Generalizations concerning the relative numbers of model butterflies and their mimics have been made (Bates 1862, Edwards 1888, Poulton 1909, Remington 1958) but data from actual counts have seldom been recorded in the literature. Exceptions are the work of Carpenter (1920, 1949) on the frequency of the polymorphic mimic, *Pseudacraea eurytus* (Linné), and its models of the genus *Bematistes,* and the analysis by Ford (1936) of the numbers of polymorphic females of *Papilio dardanus* Brown and their respective models from Africa. These studies have recently been further analyzed by Sheppard (1959).

The purpose of this paper is to present new data from 4 localities in the southeastern United States on the relative numbers of butterflies of the *Battus philenor* (Linné) mimicry complex. The species concerned are: the model, *B. philenor,* and its mimics, *Papilio troilus* Linné, *P. polyxenes* (Fabricius), the black female form of *P. glaucus* (Linné), *Limenitis astyanax* (Fabricius), and the female of *Speyeria diana* (Cramer) (Poulton 1909). Several members of this complex have been investigated experimentally with regard to their palatability to the Florida Scrub Jay (*Cyanocitta c. coerulescens* (Bosc)) (J. Brower 1958b). It was found that *B. philenor* is distasteful and that the birds learned to reject it on sight alone. Mimicry was shown to be effective for *P. troilus* and *P. polyxenes,* and probably so for the black female form of *P. glaucus.*

## Areas of Study

### *Parker Islands, Highlands County, Florida*

Three miles east of the junction of Routes 27 and 70, at the base of the sand ridge of south central Florida, there is a bay swamp (*Magnolia virginiana* Linné) with adjacent cleared pastures containing thistles (*Cirsium* sp.) in spring, and swampy areas with abundant *Gyrotheca tinctoria* (Walt.) Salisb. in summer. Many species of Rhopalocera are attracted to the flowers of these plants and are easily netted while feeding. A combined total of 83 samples of butterflies in this mimicry complex was collected from 6 March to 13 June 1956 and from 13 June to 6 August 1959. All mimics were present except *L. astyanax* and *S. diana* which do not occur this far south.

The data for the 4 species in this locality are shown in Table I. It can be seen in the spring samples of 1956 and those of the summer of 1959 that the mimic, *P. troilus,* is prevalent in the area. The data excepting 6 March-3 April 1956, underestimate its abundance because many more were seen than recorded. This was because collecting was concentrated on *P. glaucus* which was needed to a greater extent in experimental studies of mimicry then being conducted. All *P. glaucus* males seen were caught and kept, but not all individuals of *P. troilus.* An accurate record was made of all *P. glaucus* females which were either collected or marked and released into the population so that they were not counted again. In 1959, *P. polyxenes* was sampled only from 13-30 June and was present but not collected in the rest of the samples. Therefore it is not included in the totals.

In all samples taken during the 2 periods, there is a striking rarity of the model, for only one male *B. philenor* was seen and collected. Finally, there is a high ratio of the yellow to black female form of *P. glaucus.* For 1956, the ratio was 31 black : 470 yellow (6% black), and for 1959, 40 black : 477 yellow (8% black). That the frequencies of these 2 forms should remain virtually the same in non-consecutive years and at different seasons has an important bearing on mimicry and will be discussed below.

### *The vicinity of Grahamsville, Marion County, Florida*

In order to learn the frequency of the model and mimics in a northern Florida locality, a day's collecting was done along an unpaved road in the Ocala National Forest, 15 miles east of Ocala on 12 August 1959. Although no comparison can be made between the absolute numbers of the single collection and the extensive Parker Islands studies, the relative numbers of each species within the respective samples can be considered. Unlike the selective techniques used in south Florida, the method employed in this sample, and in all others to be discussed, was to capture or record every individual seen during a collecting period. Table II

Reproduced with permission from *Ecology,* 43:154–158, 1962. Published by the Ecological Society of America and Duke University Press, Durham, North Carolina.

TABLE I. Numbers of individuals of the butterfly species in the *Battus philenor* mimicry complex at Parker Islands, Highlands Co., Fla., 6 Mar.-13 June 1956 and 13 June-6 Aug. 1959

| Dates | | 6 Mar.-3 Apr. '56 | 5-26 May '56 | 29 May-13 Jun. '56 | 13-30 Jun. '59 | 1-10 Jul. '59 | 17 Jul.-6 Aug. '59 | Totals 1956 | | Totals 1959 | |
|---|---|---|---|---|---|---|---|---|---|---|---|
| No. of samples | | 8 | 20 | 10 | 16 | 10 | 19 | | | | |
| *B. philenor* | ♂ | 0 | 0 | 0 | 0 | 1 | 0 | 0 | 0 | 1 | 1 |
| | ♀ | 0 | 0 | 0 | 0 | 0 | 0 | 0 | | 0 | |
| *P. troilus*† | ♂ | 118 | 218* | 177 | 321 | 101 | 239 | 513 | 758 | 661 | 1097 |
| | ♀ | 44 | 113* | 88 | 200 | 83 | 153 | 245 | | 436 | |
| *P. glaucus* | ♂ | 219 | 186 | 266 | 369 | 225 | 268 | 671 | 1172 | 862 | 1379 |
| black | ♀ | 11 | 4 | 16 | 16 | 14 | 10 | 31 | | 40 | |
| yellow | ♀ | 117 | 136 | 217 | 88 | 107 | 182 | 470 | | 477 | |
| *P. polyxenes* | ♂ | 1 | 15 | 13 | 1 | ** | ** | 29 | 41 | ** | ** |
| | ♀ | 1 | 1 | 10 | 5 | ** | ** | 12 | | ** | |
| | | | | | | | | | 1971 | | 2477 |

† Underestimated, except 6 Mar.-3 Apr.; see text.
* Based on 11 samples from 5-16 May.
** Not collected.

shows the data for this area. *B. philenor* was more abundant relative to *P. troilus* than in south Florida, although the numbers of the former are insufficient for statistical analysis. Although *P. polyxenes* and the black female form of *P. glaucus* were not observed, 3 males of *P. glaucus* were captured. The preponderance of males in all the species is due to the fact that most of the specimens were caught while they were drinking at damp muddy spots along the road. This phenomenon of "mudpuddling" is known to occur in butterflies almost solely among males.

TABLE II. Numbers of individuals of the butterfly species in the *B. philenor* mimicry complex in the vicinity of Grahamsville, Marion Co., Fla., 12 Aug. 1959

| | ♂ | ♀ | Total |
|---|---|---|---|
| *B. philenor* | 3 | 0 | 3 |
| *P. troilus* | 57 | 4 | 61 |
| *P. glaucus* | 3 | 0 black<br>0 yellow | 3 |

*The vicinity of Margret, Fannin and Union Counties, Georgia*

On 15 and 16 August 1959, collections of the *B. philenor* complex were made along an unpaved road from Margret to Youngcane in the mountains of northern Georgia. *P. polyxenes* and *L. astyanax* were not represented in the samples, but the female of the rare and local *S. diana* was present. It can be seen in Table III that there appears to be a slight increase in the proportion of the model, and a highly significant increase ($P < .001$) in the frequency of the mimetic female form of *P. glaucus* when compared with the south Florida area. *P. troilus* remains the most abundant mimic.

*The Great Smoky Mountains, Sevier County, Tennessee, and Swain and Haywood Counties, North Carolina*

From 6-21 August 1954, extensive samples were collected for use in tests of the palatability of the various butterfly species to vertebrate predators, especially mice (*Peromyscus*) (Brower and Brower 1961). These samples can be divided into 3 categories: (1) those collected at a single locality on or about a Mimosa tree (*Mimosa* sp.) or on cultivated Phlox (*Phlox paniculata* Linné) nearby; and (2) those sampled along roads in the mountainous parts of the 3 counties (a) at mudpuddle sites, and (b) at various flowers.

Samples of the first category were taken at Mount Sterling, Haywood County, N. C. For one hour on each of 4 afternoons, collections were made at this locality and the butterflies were marked with a dot of quick-drying cellulose paint and released. In this way no individual was counted twice. Only 2 recaptures were made on days other than the day of marking, one *B. philenor* female marked on 14 August and recaptured on 16 August, and one marked on 16 August and re-taken on 19 August. Table IV shows the numbers of the model and mimic. The counts made during a comparable period of continuous collecting on the 4 days were similar ranging from 55-72 for *B. philenor* and from 4-7 for *P. troilus*. In contrast with the Florida and north Georgia localities, here *B. philenor* relative to *P. troilus* comprised 92% of the combined populations of 268 butterflies. This difference is highly significant ($P < .001$).

TABLE III. Numbers of individuals of the butterfly species in the *B. philenor* mimicry complex in the vicinity of Margret, Fannin and Union Counties, Ga., 15 and 16 Aug. 1959

| | ♂ | ♀ | Total |
|---|---|---|---|
| *B. philenor* | 6 | 3 | 9 |
| *P. troilus* | 53 | 19 | 72 |
| *P. glaucus* | ** | 32 black<br>1 yellow | 33 |
| *S. diana* | ** | 19 | 19 |

** Not collected.

The records for males at mudpuddles are shown in

TABLE IV. *B. philenor* and *P. troilus* at *Mimosa*, Mount Sterling, Haywood Co., N. C., Aug. 1954

| | | 8 Aug. | 14 Aug. | 16 Aug. | 19 Aug. | Totals |
|---|---|---|---|---|---|---|
| *B. philenor* | ♂ | 29 | 46 | 47 | 38 | 160 |
| | ♀ | 26 | 19 | 25 | 17 | 87 |
| | | 55 (93%) | 65 (94%) | 72 (92%) | 55 (89%) | 247 (92%) |
| *P. troilus* | ♂ | 3 | 3 | 3 | 6 | 15 |
| | ♀ | 1 | 1 | 3 | 1 | 6 |
| | | 4 (7%) | 4 (6%) | 6 (8%) | 7 (11%) | 21 (8%) |
| Totals | | 59 | 69 | 78 | 62 | 268 |

Table V, and it can be seen that the proportion of *B. philenor* and *P. troilus* remains almost the same and that *L. astyanax* contributed to the total. These samples were made along State Routes 71 and 284 in the Great Smoky Mountains, and included groups ranging in number from 1 to 17 individuals. In all, *B. philenor* outnumbered *P. troilus* and *L. astyanax*. Of the 50 mudpuddle groups observed, 38 were composed solely of *B. philenor* and in the remaining 12, *P. troilus* and/or *L. astyanax* were in a minority. On 22 July 1959, similar collections were made along State Routes 284 and 73 between Cove Creek, Haywood County, N. C., and Greenbrier Cove, Sevier County, Tenn. At that time the proportion of the model was somewhat higher (Table V).

TABLE V. Male butterflies of the *B. philenor* mimicry complex at mudpuddles, in the Great Smoky Mountains, 1954 and 1959

| | 6-16 Aug. 1954 | 22 Jul. 1959 |
|---|---|---|
| *B. philenor* | 341 | 289 |
| *P. troilus* | 19 | 6 |
| *L. astyanax* | 11 | 2 |
| Totals | 371 | 297 |

The third category tabulated consisted of butterflies observed at flowers including Bergamot (*Monarda fistulosa* Linné), Thistle (*Cirsium* sp.), Purple Everlasting (*Antennaria* sp.), and cultivated Phlox. Five species of the complex including *S. diana* were found and their abundances are shown in Table VI. The yellow female

TABLE VI. Butterflies of the *B. philenor* mimicry complex at several species of flowers in the Great Smoky Mountains, 11-21 Aug. 1954

| | ♂ | ♀ | Sex undetermined | Total |
|---|---|---|---|---|
| *B. philenor* | 175 | 54 | 227 | 456 |
| *P. troilus* | 62 | 19 | 4 | 85 |
| *P. glaucus* | 75 | 14 black | — | 89 |
| | | 0 yellow | — | — |
| *L. astyanax* | 6 | 6 | 4 | 16 |
| *S. diana* | 30 | 7 | — | 37 |

form of *P. glaucus* was not observed in 1954, but on 22 July 1959, of 10 *P. glaucus* captured, 9 were males and the sole female was yellow.

DISCUSSION

The frequencies of *Battus philenor* and its mimics found in 4 areas of the southeastern United States are summarized in Table VII. These data demonstrate

TABLE VII. Frequencies of the model, *B. philenor*, and its mimics in 4 localities of the southeastern United States

| Species | S. Fla. | | N. Fla. | N. Ga. | Tenn. & N. C. | |
|---|---|---|---|---|---|---|
| | 1956 | 1959 | 1959 | 1959 | 1954 | 1959 |
| *B. philenor* | .00 | .0008 | .05 | .07 | .86 | .97 |
| *P. troilus* | .91 | .96 | .95 | .55 | .10 | .02 |
| *P. glaucus* (black ♀♀) | .04 | .04 | * | .24 | .01 | .00 |
| *P. polyxenes* | .05 | ** | .00 | .00 | .00 | .00 |
| *S. diana* (♀♀) | — | — | — | .14 | .006 | .00 |
| *L. astyanax* | — | — | .00 | .00 | .02 | .007 |
| Totals (N) | 830 | 1138 | 64 | 132 | 1217 | 297 |

* Not seen.
** Not collected.
— Does not occur in region.

quantitatively for the first time that *B. philenor* is extremely common in the Smoky Mountains and becomes increasingly more rare southward to south central Florida. There the situation is reversed and its mimic, *Papilio troilus*, is the most abundant species. The data also show that the frequency of the mimetic black female form of *P. glaucus* parallels the southward decline in numbers of the model so that at the southern extreme over 90% of its females are non-mimetic. (Table VIII).

TABLE VIII. Frequency of the dimorphic female forms of *P. glaucus* in 3 localities in the southeastern United States

| | S. Fla. | | N. Ga. | Tenn. & N. C. |
|---|---|---|---|---|
| | 1956 | 1959 | 1959 | 1954 & 1959 |
| Black | .06 | .08 | .97 | .93 |
| Yellow | .94 | .92 | .03 | .07 |
| Totals (N) | 501 | 517 | 33 | 15 |

Several of the authors cited above, (Edwards, Poulton, Remington), have emphasized the importance of the correlated geographic distribution of *B. philenor* and the black female form of *P. glaucus* from Georgia northward. Where *B. philenor* is abundant, the frequency of the black female form is usually high, but to the far north and northwest where *P. glaucus* extends beyond the range of its model, all the females are yellow. From the new data presented in this paper, it is clear that the frequency of the black female of *P. glaucus* declines outside the area of abundance of *B. philenor* not only to the north but also to the south. The fact that the yellow female form predominates in the extreme south greatly strengthens the idea that the black female is truly a mimic of *B. philenor*. Were this not so, it might be argued that the increasing frequency of the black form from north to south is an environmental adaptation to higher temperature and humidity. This latter notion is even more unlikely in view of our findings that there is almost no seasonal increase in the frequency of the black female form in southern Florida from spring to summer (Table I).

The low proportion of the black female form of *P. glaucus* where the model is rare or absent must be attributed to some disadvantage which does not affect the non-mimetic yellow female to as great an extent. There is reason to believe that this is sexual, the males being attracted less to the black than to the yellow females. In the western United States and southwestern Canada, almost entirely outside of the range of *B. philenor*, 3 species very closely related to *P. glaucus* replace the latter and in each there is no black female form, all being monomorphic (Brower 1959a). A field study of interspecific sexual behavior among these members of the *P. glaucus* group (Brower 1959b) showed that males of one of the western species are highly attracted to tethered yellow *P. glaucus* females but fail almost completely to respond to the black form. Thus, at least in an allopatric species very closely related to *P. glaucus*, there is a strong preference in courtship for a color pattern that is non-mimetic. Similar decrease in the intensity of male response to females of modified color-pattern has been shown by Stride (1956, 1957) to occur in the African mimetic butterfly, *Hypolimnas misippus* Linné, and by Crane (1955) in *Heliconius erato hydara* Hewitson of Trinidad. Moreover, Magnus (1958) has shown in *Argynnis paphia* Linné that the first sexual response of courting males depends on a visual stimulus of a very limited wave length which corresponds to the normal yellowish-brown coloration of their females.

In view of the almost complete absence of *B. philenor* in the south Florida area, the question arises as to why the polymorphism is still retained with 6%-8% of the females of *P. glaucus* being black. It seems doubtful that predators could encounter the model often enough to learn that a black butterfly is distasteful. That predation pressure is exerted was discovered in July 1960 when we observed and filmed adult and immature red-winged blackbirds (*Agelaius phoeniceus* Linné) catching and eating adult *P. glaucus* in the Parker Islands *Gyrotheca* swamp. An attempt was made to count the wings of the captured butterflies, but local flooding made it impossible to retrieve them.

The most probable explanation for the maintenance of the black female in this area is through immigration and gene flow from farther north. There is of course the possibility that it is being maintained by heterozygous advantage. However, Clarke and Sheppard (1959, 1962) and Sheppard (1961) have investigated the genetics of the 2 forms and conclude that the factor responsible is Y-linked or sex-limited and partially sex-linked. Cytoplasmic inheritance is unlikely but not excluded by the data. Their evidence from crosses of the two forms, and preliminary findings of our own that black females caught in the south Florida locality produce only black daughters, make it most unlikely that the black form is an autosomal sex-limited dominant as suggested by Remington (1954). This is because the fathers would have to have been homozygous for the black gene which is improbable where the frequency of the black female form is so low. This of course assumes random mating of males regardless of their genotypes. For these reasons, it seems doubtful that heterozygous advantage is responsible for maintaining the frequency of the black form. The decrease in proportion of the black form of *P. glaucus* in the absence or rarity of the model is analogous to the findings of Carpenter (1920) and Ford (1936) in their analyses of African mimetic butterflies. For *P. troilus*, it also could be expected that under similar circumstances non-mimetic characters might occur more commonly. Increased variability away from the mimetic color-pattern was in fact noted by us in the south Florida population of *P. troilus*, and a separate analysis is now being completed on specimens from most of the localities considered in this paper. This indicates that the degree of resemblance based on the size and coloration of the submarginal spots on the upper surface of the hind wings does vary according to the relative abundance of *B. philenor*. Where the latter is common, these spots of *P. troilus* are small and greenish in coloration, whereas in Florida where *B. philenor* is rare, they are larger and yellowish. This causes the mimic to look conspicuously different from the striking greenish upper surface of its model. The genetic mechanism underlying this situation has not yet been investigated.

The other 3 mimics of the complex are all geographically confined within the distribution of *B. philenor*. *Speyeria diana*, which is the only known mimic in its large genus, is strictly limited to the southeasern part of the United States where *B. philenor* is extremely common. *P. polyxenes* and *L. astyanax* are more generally distributed throughout the range of *B. philenor* in the east, although *L. astyanax*, on the basis of the results in this paper, does not occur as far south as does *P. polyxenes*. We have not analyzed possible mimetic breakdown of the latter on the fringes of its range where the model is rare. As Sheppard (1961) has pointed out, beyond the range of *B. philenor* 2 black species related to *P. polyxenes* occur, *P. brevicauda* (Saunders) and *P. nitra* (Edwards), in which mimicry is absent. This is analogous to the western species of the *P. glaucus* group discussed above. Similarly, 3 other species in the genus *Limenitis* occur almost entirely outside of the range of *B. philenor* (Hovanitz 1949), and all are non-mimetic. These are: *L. arthemis* (Drury) in the northeast, *L. weidemeyerii* (Edwards) in the central Rocky Mountains, and *L. lorquini* (Boisduval) in the northwest. The last species of the genus in North America is *L. archippus* (Cramer) which is sympatric with *B. philenor* over most of its range, but mimics orange to mahogany butterflies in a different family, the Danaidae (Brower 1958a, 1958c).

Thus, in this North American complex of butterflies there are several ways in which mimicry is strongly correlated with the geographic distribution of the model, *B. philenor*. Intraspecifically, this is shown by frequency changes in the mimetic and non-mimetic forms of one species, and quantitative variation in mimetic characteristics of another; interspecifically, by the containment of 3 mimetic species inside the range of abundance of their model with the occurrence of related but wholly non-mimetic species outside it. The fact that mimicry is manifested in these various ways makes it clear that the range of a mimetic species need not necessarily be limited by the range of its model.


## Summary

1. Four localities in the southeastern United States were sampled for 3,387 butterflies of the *Battus philenor* mimicry complex. These included the model, *B. philenor,* and its mimics, *Papilio troilus, P. polyxenes,* the black female form of *P. glaucus, Limenitis astyanax,* and the female of *Speyeria diana.*

2. It was found that as one progresses northward from south central Florida to the Great Smoky Mountains of Tennessee and North Carolina, the frequency of *B. philenor* to its mimics increases from nearly 0% to over 85%.

3. Correlated with this distribution are 3 ways in which the mimic species reflect the relative abundance of the model as a condition favorable to mimicry:

A. the high frequency of the mimetic black female form of *P. glaucus* coincides with areas of abundance of *B. philenor,* while the non-mimetic yellow female form prevails where the model is rare, irrespective of latitude. This excludes the possibility that the black form is an environmental adaptation to increasing temperature and humidity.

B. *P. troilus* is a closer mimic of *B. philenor* where the latter is relatively abundant and exhibits variability away from the mimetic color-pattern where the model is scarce.

C. *L. astyanax* and *S. diana* are limited in geographic distribution to those areas in which the model is common. To a lesser extent this is also true of *P. polyxenes.* This is not to say that mimicry necessarily regulates their ranges, but is coincident with them in these instances.



## Acknowledgments

We want to express our thanks to Drs. E. B. Ford, F.R.S., and P. M. Sheppard for reading and criticizing the manuscript, to Mr. Richard Archbold for providing the facilities of the Archbold Biological Station, and to Mr. F. T. Naumann for assistance in the field.

This work was supported in part by: a National Science Foundation pre-doctoral fellowship, a Fanny Bullock Workman Scholarship of Wellesley College, a Sigma Xi grant, the Higgins Fund of Yale University, and by N. S. F. Grant 8707.


## References


**Bates, H. W.** 1862. Contributions to an insect fauna of the Amazon Valley. Lepidoptera: Heliconidae. Trans. Linn. Soc. London **23**: 495-566.

**Brower, J. VZ.** 1958a. Experimental studies of mimicry in some North American butterflies. 1. *Danaus plexippus* and *Limenitis archippus archippus.* Evolution **12**: 32-47.

———. 1958b. Experimental studies of mimicry in some North American butterflies. 2. *Battus philenor* and *Papilio troilus, P. polyxenes,* and *P. glaucus.* Evolution **12**: 123-136.

———. 1958c. Experimental studies of mimicry in some North American butterflies. 3. *Danaus gilippus berenice* and *Limenitis archippus floridensis.* Evolution **12**: 273-285.

———, **and L. P. Brower.** 1961. Palatability of North American model and mimic butterflies to caged mice. Jour. Lepidopterists' Society **15**: 23-24.

**Brower, L. P.** 1959 a. Speciation in butterflies of the *Papilio glaucus* group. I. Morphological relationships and hybridization. Evolution **13**: 40-63.

———. 1959b. Speciation in butterflies of the *Papilio glaucus* group. II. Ecological relationships and interspecific sexual behavior. Evolution **13**: 212-228.

**Carpenter, G. D. H.** 1920. *Pseudacraea eurytus hobleyi,* its forms and its models on the islands of L. Victoria and the bearing of the facts on the explanation of mimicry by natural selection. Trans. Ent. Soc. London **1920**: 84-98.

———. 1949. *Pseudacraea eurytus* (L.) (Lep. Nymph.): A study of a polymorphic mimic in various stages of speciation. Trans. Royal Ent. Soc. London **100**: 71-133.

**Clarke, C. A. and P. M. Sheppard.** 1959. The genetics of some mimetic forms of *Papilio dardanus,* Brown, and *Papilio glaucus,* Linn. J. of Genetics **56**: 236-260.

———. 1962. The genetics of the mimetic butterfly *Papilio glaucus.* Ecology **43**: 159-161.

**Crane, Jocelyn.** 1955. Imaginal behavior of a Trinidad butterfly, *Heliconius erato hydara* Hewitson, with special reference to the social use of color. Zoologica **40**: 167-196.

**Edwards, W. H.** 1888. The butterflies of North America. First series. Boston: Houghton, Mifflin.

**Ford, E. B.** 1936. The genetics of *Papilio dardanus* Brown (Lep.) Trans. Royal Ent. Soc. London **85**: 435-466.

**Hovanitz, W.** 1949. Increased variability in populations following natural hybridization. In Genetics, paleontology, and evolution, ed. Jepsen, G. L., E. Mayr, and G. G. Simpson. Princeton: Princeton University Press.

**Magnus, D. B. E.** 1958. Experimental analysis of some "overoptimal" sign-stimuli in the mating-behaviour of the Fritillary Butterfly *Argynnis paphia* L. (Lepidoptera: Nymphalidae). Proc. Tenth Int. Congress of Entomology, 1956, **2**: 405-418.

**Poulton, E. B.** 1909. Mimicry in butterflies of North America. Ann. Ent. Soc. America **2**: 203-242.

**Remington, C. L.** 1954. The genetics of *Colias* (Lepidoptera). Advances in Genetics **6**: 403-450.

———. 1958. Genetics of populations of Lepidoptera. Proc. Tenth Int. Congress of Entomology, 1956, **2**: 787-805.

**Sheppard, P. M.** 1959. The evolution of mimicry; a problem in ecology and genetics. Cold Spring Harbor Symposia Quant. Biol. **24**: 131-140.

———. 1961. Some contributions to population genetics resulting from the study of the Lepidoptera. In press.

**Small, J. K.** 1933. Manual of the southeastern flora. Chapel Hill: University of North Carolina Press.

**Stride, G. O.** 1956. On the courtship behaviour of *Hypolimnas misippus* L., (Lepidoptera, Nymphalidae), with notes on the mimetic association with *Danaus chrysippus* L., (Lepidoptera, Danaidae). British J. Animal Behaviour **4**: 52-68.

———. 1957. Investigations into the courtship behaviour of the male of *Hypolimnas misippus* L. (Lepidoptera, Nymphalidae), with special reference to the role of visual stimuli. British J. Animal Behaviour **5**: 153-167.

# Part VII: Man and Insect

Man's survival is dependent on insects. The benefits derived from pollinators, predatory insects, and other insect "helpers" of man, such as honeybees and silkworms, are well known. The harmful effects of insects on man may be divided into two general aspects: health and resources. Medical aspects deal with *the effect of insects on man and his health through the direct action of parasites and the transmission of diseases to and between humans* (see the introduction to the section on HEALTH AND WELFARE). The study of insects that affect man in this manner and the methods of reducing such insect problems is termed **medical entomology.** The second harmful aspect comes about when insects and man compete for the same materials. *Man's attempts to reduce insect competition by various means* comprises the classical area of **economic entomology.**

The general public is aware of insects primarily as pests: aphids on roses, larvae in apples, flour beetles in cereal or crackers. As a result, the beginning student often thinks of entomology strictly in terms of **insect control,** *the suppression of pest insect populations.* Invariably he thinks of insect control by means of **insecticides,** *substances that kill insects by their chemical action.* This association of insect control and insecticides has developed because of man's heavy reliance upon these chemicals in recent decades. The insecticides developed by industry have provided such spectacular results that they have been relied upon almost exclusively for pest control by the general public. Entomologists also have relied heavily on insecticides, with many adopting the attitude that insecticidal control procedures were the ultimate solution for all pest problems. However, insecticides have proven to be only another of the pest controls which are used individually to counter specific problems. Each of the many types of controls, such as **cultural control** *(variations in farming operations),* **biological control** *(introduction and establishment of insect enemies), and* **legislative control** *(regulation of commerce and farming to reduce prevalence and distribution of pests),* has proved to give some relief when applied to specific pest problems; but no single type of control has proved satisfactory in all cases.

Because entomologists relying on one type of control have encountered difficulty in controlling pests, we have come to recognize the need for a greater understanding of an insect pest's ecology. Research toward this understanding has initiated a trend away from attempts at complete pest eradication through repeated non-selective techniques and toward tolerance of pests of low population levels. Such tolerance tends to support an ecological continuum within the **food web** *(the interlocking pattern of energy flow from plants through a series of organisms eating and in turn being eaten)* and allows for maximum regulation of insect populations through natural enemies and changes in physical conditions, with a minimum of interference by man.

*The Integrated Control Concept* by V. M. Stern, et al., sets the stage for the entire section on MAN AND INSECT by providing a frame of reference within which all of the other papers can be discussed. The economics and dynamics of pest population management as expressed in *The Integrated Control Concept* provide the basic goals and theories of economic entomology. Since the "text book approach" primarily involves the mechanics of control procedures and their application, the discussions of goals, theories, and principles of control in this article will provide a greater understanding of economic entomology than is provided by most texts.

Insects become pests not only because they attack materials of interest to man, but because they typically occur in such large numbers that they make a significant economic impact. Because entomologists cannot examine an entire population at any one time, samples of a population must be taken and inferences about the population made from those samples. The inferences are only as meaningful as the accuracy of the samples. *A Comparison of Five Methods of Sampling Alfalfa Fields for Arthropod Populations* by F. A. Fenton and D. E. Howell provides an examination of the effectiveness of sweeping, plant clippings, pit traps, a type of sticky trap, and Berlese funnels (which separate live insects from soil litter by a combination of heat and light). This study demonstrates that the efficiency of these common sampling methods varies with the type and habits of the insect. The study also illustrates the kinds of inadequacies that may be encountered in sampling technology.

AUTOCIDE – The potential expansion of population is a function of the species' **innate capacity for increase,** *the maximal rate of increase under essentially ideal conditions,* restricted by the resistance of environmental factors (food availability, abundance of natural enemies, unsuitable weather conditions, etc.). Until recently, all of man's attempts to reduce or eliminate populations of pest insects were designed to increase environmental resistance, either by changing the quality and availability of food, encouraging natural enemies, or changing the physical or chemical environment of the pest. The other aspect contributing to population size, that of the pest's reproductive capacities, remained unassailable until man learned to produce sterility in insects without interfering with their reproductive activities. E. F. Knipling *(Use of Insects for Their Own Destruction)* provides an overview of **autocidal control,** *death or reduced reproductive capacity of insects as caused by individuals of their own species.* The individuals neutralizing the population have been altered in some way by man (for example, by gamma-ray sterilization). In following the ramifications of the basic autocidal technique, Knipling provides an outstanding example of how entomologists can exploit potentially valuable areas of control.

*Screw-worm Control through Release of Sterilized Flies* by A. H. Baumhover, et al., the pioneering program with this technique, provides a spectacular example of developmental research and coordinated execution of a complex research plan. In *Radiosterilization vs. Chemosterilization in House Flies and Mosquitoes,* C. H. Schmidt, et al., explore the potential for autocidal control with other Diptera. Not only do the two test species react differently to sterilization, but the choice of the method of sterilization, whether chemical or by ionizing radiation, also yields differences. The work amply demonstrates the kinds of difficulties encountered by entomologists in the seemingly simple process of applying techniques of control, developed for one pest, to other insects.

H. Laven's work on *Eradication of Culex pipiens fatigans through Cytoplasmic Incompatibility* provides experimental proof that lethal genetic combinations can result in reduction of pest populations. This approach to autocidal control, as earlier suggested by Knipling, also points up another and even more exciting suggestion: that we have the capability to build innocuous strains that can competitively displace undesirable strains.

BIOLOGICAL CONTROL – The introduction, establishment, and manipulation of the natural enemies of insects has attracted much enthusiasm as a desirable alternative to the use of pesticides. In fact, we have now reached the stage where the professional entomologist is frequently told by the public to "use the natural enemies" to solve pest problems. *Suppression of Dung-Inhabiting Fly Populations by Pupal Parasites* by E. F. Legner and H. W. Brydon briefly introduces the range of organisms that attack house flies. From this potential arsenal of natural enemies, several species and strains of native and imported hymenopterous parasites are evaluated for the degree of control provided throughout the year. In their discussion of results and conclusions it is easy to see the difficulty biological control researchers have in unraveling the interrelated population dynamics of a pest species and its natural enemies.

Although the development of resistant varieties is not usually included under the category of biological control, *Developing Resistant Plants – The Ideal Method of Controlling Insects,* by P. Luginbill, Jr., is included in this section because here again we are manipulating one biotic factor in the ecosystem to alter the pest levels of infestation. The only difference in this situation is that the biotic factor is the food supply (host plant or animal), rather than the complex of natural enemies.

Plant resistance usually is based on one of three different mechanisms: **antibiosis** *(an adverse effect of the plant on the biology of the insect),* **non-preference** *lack of or masked plant characters required for host recognition by a pest),* and **tolerance** *(the plant's ability to withstand an infestation and still produce the crop).* Through incorporation of these mechanisms with other desirable crop characteristics, entomologists and agronomic scientists have developed crop plants with remarkably increased yields.

CULTURAL CONTROL – Control through modification of farming operations is seldom a completely satisfactory method of controlling insects, although cultural practices can directly and often significantly affect population levels of pests. *Effect of Certain Cultural Practices on the Abundance of Tobacco Hornworms, Tobacco Budworms, and Corn Earworms on Tobacco after Harvest* by R. L. Rabb, et al., adequately demonstrates how pest population levels may be altered by modifying the growth pattern of the host plant. The use of crop rotation as a highly effective cultural control technique is amply justified by O. L. Barnes in *Effects of Cultural Practices on Grasshopper Populations in Alfalfa and Cotton.* The possible influence of weed populations on levels of grasshopper abundance provides an unusual example of higher plant diversity resulting in increased pest population levels. This is opposed to the usual influence of **monoculture** *(extensive use of land for a single crop)* where high pest populations are encouraged by extensive availability of a single kind of food.

CHEMICAL CONTROL – The four papers in this section cover very distinct though interrelated topics that must be considered in our use of chemicals to suppress populations of pest insects. *Laboratory Tests of 55 Insecticides on Douglas Fir Tussock Moth Larvae* by R. L. Lyon, et al., presents the kinds of testing and information required to screen chemicals of known insecticidal activity for possible use against a specific pest. In addition to analyses of toxicity, the effect of carriers (solvent or diluent) and residual actions of chemicals on foliage illustrate some of the aspects that must be researched before a known pesticide can be recommended for a given pest situation.

The frustration of an economic entomology that has relied very heavily on pesticides is evident in A.W.A. Brown's paper, *Insecticide Resistance Comes of Age.* The pesticide as a new selective force in the environment provides a striking example of the evolutionary process in operation as pest populations evolve **insecticide resistance,** *in which populations of a formerly susceptible species can no longer be controlled by a given insecticide at the rates normally recommended.* The magnitude and types of resistance reported by Brown, as well as the ease with which resistance develops, reinforces the need for careful, considered, conservative attitudes in pesticide use.

The introduction of pesticides into the environment has caused changes in the target organism, as seen in Brown's paper. However, pesticidal use has created the additional problem of environmental contamination, especially when persistent chemicals such as DDT are used. These long-lasting substances are often concentrated through food chains so that large carnivores (e.g., penguins, lake trout, and eagles) suffer reproductive failures. As these problems have developed, we have found it necessary to monitor segments of the environment to determine levels of suspect chemicals present in water and in a variety of plant and animal tissues. Monitoring requires procedures that identify specific molecules. *Biological Interaction between Plasticizers and Insecticides* (E. P. Lichtenstein, et al.) points up a number of problems inherent in identifying compounds that may have significant biological activity and determining the extent to which they are present in our surroundings. The ubiquitous presence of polychlorinated biphenyl compounds and the confusion in distinguishing between them and various pesticides emphasizes the need for objective re-examination and re-evaluation of the entire pesticidal contamination problem.

The newest form of chemical control involves the use of chemicals extracted from insects themselves or synthetically produced molecules that mimic the biological activity of those naturally produced. The natural products include the hormones that insects secrete internally for control of molting (ecdysone) and growth, development, and egg maturation (juvenile hormone). They also include the **pheromones,** *chemical signals that are exchanged between members of the same species,* and other externally emitted substances responsible for repellency and defense, sex attraction, alarm, and suppression of development. Many of the natural materials have not been synthesized by man, but their potential uses when available are envisioned easily after reading *Chemical Insect Attractants* by M. Jacobson and M. Beroza.

INTEGRATED CONTROL—At this point the diversity of control procedures available for suppression of pest populations by man should be evident, as should the complex nature of research on control methodology. However, an even more important consideration is that none of these procedures by themselves have provided entirely satisfactory results. It is easy to see where E. F. Knipling's *(Some Basic Principles in Insect Population Suppression)* suggested combination of control techniques based on sound ecological study of the pests can lead to much more efficient use of the resources required for insect control than traditional, uncoordinated control programs operated simultaneously by many agencies.

A basic difference exists between the goals Knipling envisions for integrated control and the type of integrated control described in the earlier article by V. M. Stern, et al. *(The Integrated Control Concept).* The latter authors recommend combining control techniques on a closely monitored pest complex, using any given procedure only when it is necessary to suppress the population back to a subeconomic level. Knipling advocates massive attack on the few most important pest species. This attack is to be a coordinated effort using a selected combination of control techniques applied simultaneously or in sequence with the goal of completely eliminating the population from major geographic regions.

Although their goals differ, either of the two approaches to integrated control has the potential to yield more satisfactory results than any of the single-factor methods previously described. Because of the increased complexity of applying such combined control procedures and because of the need to monitor pest populations more closely, many more entomologists will need to be trained for such work.

HEALTH AND WELFARE – Whether insects are more important as competitors of man or as man's parasites would depend on your outlook. A resident of urban areas in North America or Europe might view the importance of insects in terms of food and clothing costs as affected by insect contamination or injury. A resident of more tropical areas, locked in the throes of a fatal malarial fever, might think differently, since his disease can be traced directly to the bite of a mosquito. Not only is man exposed to the direct attack of insect parasites such as fleas, lice, mosquitoes, and bed bugs, but the problem is greatly compounded when the insects also transmit microbial and other diseases to man as a result of the attack.

Control of the insects that attack man follows the same concepts of population management previously discussed for control of insect competitors. These control procedures are now practiced on a world-wide scale and will undoubtedly be intensified. As the human population continues to grow, increasing percentages of the population are exposed to insect **vectors** *(insect carriers of disease-producing organisms)*. This factor, in combination with insecticide resistance in vectors and mobility of the human population, greatly increases the possibility of epidemics.

One of the most important problem areas in medical entomology is the mechanism or mechanisms used by the insect parasite in locating its host. A. W. A. Brown's paper, *The Attraction of Mosquitoes to Hosts,* demonstrates the complex and tortuous track of investigations required in solving this problem. The variety and interaction of factors involved in host attraction have prevented us from completely understanding this relatively "simple" behavior. *L-Lactic Acid: A Mosquito Attractant Isolated from Humans* (F. Acree, Jr., et al.) adds pieces to the puzzle but in so doing indicates the rather rudimentary state of our knowledge. Each step in the advancement of our knowledge about host attraction, whether the host be plant, man, or some other animal, provides us with new vistas of opportunity in insect control. Once we understand what the physical and chemical conditions are that attract, for example, a mosquito, it will be far easier to mask or eliminate those conditions.

*The Pig-Mosquito Cycle of Japanese Encephalitis Virus in Taiwan* by H. S. Hurlbut ranks as a classic in its comprehensive approach to disease transmission by insects. Many of the basic host-vector-pathogen relationships are demonstrated, including population abundance of suspected vectors, sources of blood meals taken by field populations, ability of mosquitoes to become infective after feeding on infected hosts, and ability of infective mosquitoes to transmit viruses. Hurlbut also includes results from many of the other techniques used in virus and transmission studies. Among them are virus identification by virus dilution neutralization tests, determination and development of virus antibody in pigs, loss of virus neutralizing antibody, viremia (presence of virus in the blood), and the results of vaccination as a method of protecting host animals.

# ENVIRONMENTAL RESOURCES OF MAN

## A. Introduction

### The Integrated Control Concept[1]

**VERNON M. STERN, RAY F. SMITH, ROBERT van den BOSCH, and KENNETH S. HAGEN**[2]

All organisms are subjected to the physical and biotic pressures of the environments in which they live, and these factors, together with the genetic make-up of the species, determine their abundance and existence in any given area. Without natural control, a species which reproduces more than the parent stock could increase to infinite numbers. Man is subjected to environmental pressures just as other forms of life are, and he competes with other organisms for food and space.

Utilizing the traits that sharply differentiate him from other species, man has developed a technology permitting him to modify environments to meet his needs. Over the past several centuries, the competition has been almost completely in favor of man, as is attested by decimation of vast vertebrate populations, as well as populations of other forms of life (Thomas, 1956). But while eliminating many species, as he changed the environment of various regions to fit his needs for food and space, a number of species, particularly among the Arthropoda, became his direct competitors. Thus, when he subsisted as a huntsman or foraged for food from uncultivated sources, early man was largely content to share his subsistence and habitat with the lower organisms. Today, by contrast, as his population continues to increase (Hertzler, 1956) and his civilization to advance, he numbers his arthropod enemies in the thousands of species (Sabrosky, 1952).

The increase to pest status of a particular species may be the result of a single factor or a combination of factors. In the last century, the most significant factors have been the following.

First, by changing or manipulating the environment, man has created conditions that permit certain species to increase their population densities

[1] Received for publication October 24, 1958. This investigation (including those parts reported in the other two papers of this issue) was supported in part by research grants nos. E 885 and E 2192 from the National Institute of Allergy and Infectious Diseases of the National Institutes of Health, Public Health Service.

[2] Mr. Stern is Assistant Entomologist in Entomology, Citrus Experiment Station, Riverside; Mr. Smith is Associate Professor of Entomology and Associate Entomologist in the Experiment Station, Berkeley; Mr. van den Bosch is Associate Entomologist in Biological Control, Citrus Experiment Station, Riverside; and Mr. Hagen is Associate Entomologist in Biological Control in the Experiment Station, Berkeley.

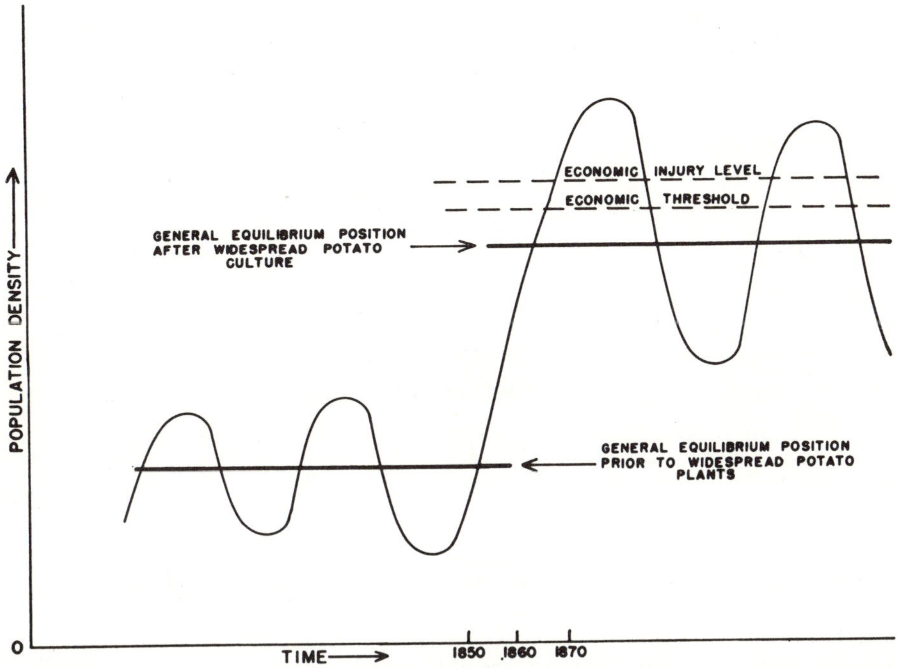


Fig. 1. Schematic graph of the change in general equilibrium position of the Colorado potato beetle, *Leptinotarsa decemlineata* following the development of widespread potato culture in the United States. For a discussion of the significance of economic-injury levels and economic thresholds in relation to the general equilibrium position, see p. 89.

(Ullyett, 1951). The rise of the Colorado potato beetle, *Leptinotarsa decemlineata* (Say), to pest status occurred in this manner (see fig. 1). When the potato, as well as other solanaceous plants, was brought under widespread cultivation in the United States, a change favorable to the beetle occurred in the environment, which enabled it to become very quickly an important pest (Trouvelot, 1936). Similarly, when alfalfa, *Medicago sativa* L., was introduced into California about 1850, the alfalfa butterfly, *Colias philodice eurytheme* Boisduval, which had previously occurred in low numbers on native legumes, found a widespread and favorable new host plant in its environment, and it subsequently became an economic pest (Smith and Allen, 1954).

A second way in which arthropods have risen to pest status has been through their transportation across geographical barriers while leaving their specific predators, parasites, and diseases behind (Smith, 1959). The increase in importance through such transportation is illustrated by the cottony cushion scale, *Icerya purchasi* Maskell (see fig. 2). This scale insect was introduced into California from Australia on acacia in 1868. Within the following two decades, it increased in abundance to the point where it threatened economic disaster to the entire citrus industry in California. Fortunately, the timely importation and establishment of two of its natural ene-

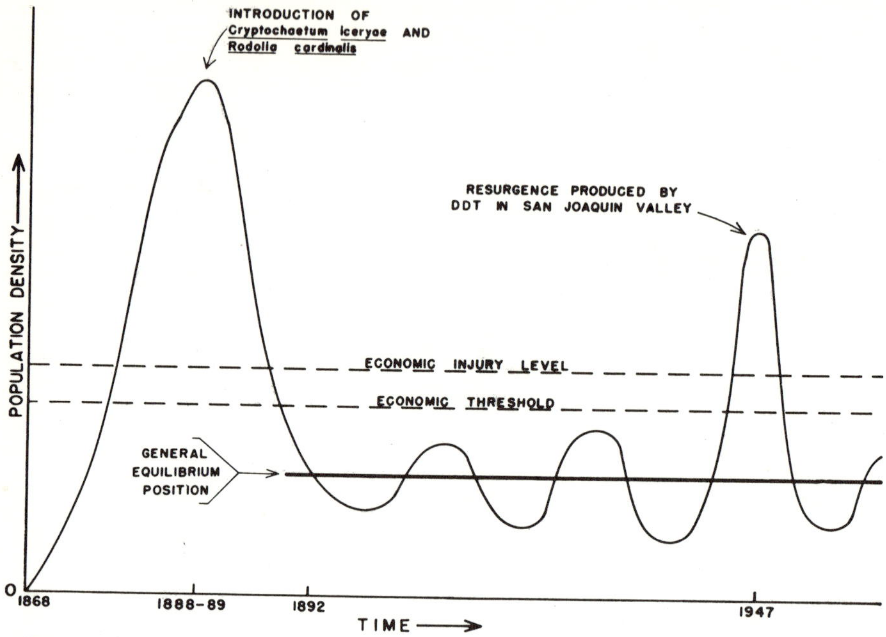


Fig. 2. Schematic graph of the fluctuations in population density of the cottony cushion scale, *Icerya purchasi,* on citrus from the time of its introduction into California in 1868. Following the successful introduction of two of its natural enemies in 1888 this scale was reduced to noneconomic status except for a local resurgence produced by DDT treatments.

mies, *Rodolia cardinalis* (Mulsant) and *Cryptochaetum iceryae* (Williston), resulted in the complete suppression of *I. purchasi* as a citrus pest (Doutt, 1958).

The cottony cushion scale again achieved the status of a major pest when the widespread use of DDT on citrus in the San Joaquin Valley eliminated the vedalia (Ewart and DeBach, 1947).

A third cause for the increasing number of pest arthropods has been the establishment of progressively lower economic thresholds (see p. 89 for definition and discussion). This can be illustrated by lygus bugs (*Lygus* spp.) on lima beans. Not too many years ago the blotches caused by lygus bugs feeding on an occasional lima bean were of little concern, and lygus bugs were considered a minor pest on this crop. However, with the emphasis on product appearance in the frozen-food industry, a demand was created for a near-perfect bean. For this reason, economic-injury thresholds were established and lygus bugs are now considered serious pests of lima beans.

In the face of this increased number of arthropod pests man has made remarkable advances in their control, and economic entomology has become a complex technical field. Of major importance have been new developments in pesticide chemistry and application.

The discovery of the insecticidal properties of DDT, and its spectacularly successful application to arthropod-borne disease and agricultural pest problems, spurred research in chlorinated hydrocarbon chemistry and stimulated the development of other organic pesticides. On a national scale, the experiment stations, state and governmental agencies, and commercial companies, all searching for new or better answers to old insect-pest problems, eagerly accepted the new chemicals. Within a short period many became an integral part of public health and agricultural pest-control programs. Without question, the rapid and widespread adoption of organic insecticides brought incalculable benefits to mankind, but it has now become apparent that this was not an unmixed blessing. Through the widespread and sometimes indiscriminate use of pesticides, the components and intricate relations of crop environments have been drastically altered, and as a result a number of serious problems have arisen (Wigglesworth, 1945; Michelbacher, 1954; Pickett, 1949; Pickett and Patterson, 1953; Solomon, 1953; and others). Among these new problems and old ones which have been aggravated are:

1. Arthropod resistance to insecticides. This phenomenon relating to the genetic plasticity of the arthropods has been reviewed by Metcalf (1955), Hoskins and Gordon (1956), Crow (1957), Brown (1959), and others. In many cases, resistance is already drastic enough to have eliminated certain insecticides from important pest-control programs. There are today in excess of 70 demonstrated cases of arthropod resistance. Actually, a much larger number of pest species exist which are developing resistance or have already done so, but there has not been time to evaluate these cases.

2. Secondary outbreaks of arthropods other than those against which control was originally directed (Massee, 1954; DeBach and Bartlett, 1951; Ripper, 1956; and others). These outbreaks usually result from the interference of the insecticide with biological control (Lord, 1947; Bartlett and Ortega, 1952; Michelbacher, 1954; Michelbacher and Hitchcock, 1958; and others). This may also occur through the effect of the insecticide on the plant, which, in turn, affects the development of the secondary pest (Fleschner and Scriven, 1957). An example is the increase in mites on plants growing in soil receiving certain chemical treatments (Klostemeyer and Rasmussen, 1953).

3. The rapid resurgence of treated species necessitating repetitious insecticide applications (Holloway and Young, 1943; Bovey, 1955; Schneider, 1955; Stern and van den Bosch, 1959; and others). These flarebacks occur from individuals surviving treatment or from individuals migrating into the treated area, where they can reproduce unhindered because their natural enemies have been eliminated.

4. The toxic insecticide residues on food and forage crops (Brown, 1951; Linsley, 1956). This problem may result from two sources. First, untimely applications or accidental increases in dosage may result in residues above the tolerance limits. Second, the first three problems mentioned above are interrelated and by aggravating one another may lead to excessive treatment and a residue problem. For example, where the level of resistance is increasing, it requires either more frequent applications or higher insecticide dosages to control the pest, or both. This increased insecticide program may in turn have a drastic effect on the ecosystem, which frequently results in outbreaks of secondary pests or rapid resurgence of the resistant pest

for which control was originally intended. Often, under such conditions, where insects threaten the crop or marketability of a crop close to harvest, the grower is faced with the problem of suffering a severe monetary loss or of making an insecticide application closer to harvest than is ordinarily permissible. In many instances, the end result is a residue far above the accepted tolerance limit at harvest time.

5. Hazards to insecticide handlers and to persons, livestock, and wildlife subjected to contamination by drift (Hayes, 1954; Petty, 1957; Upholt, 1955).

6. Legal complications from suits and other actions pertaining to the above problem.

Unquestionably, some of these problems have arisen from our limited knowledge of biological science; others are the result of a narrow approach to insect control. Few studies have included basic investigations on the effects the chemicals might have on other components of the ecosystems to which the pests belong. The entomologist may recognize the desirability of a thorough investigation of these aspects, but because of the need for immediate answers to pressing problems and because of other pressures, he does not have the necessary time. In other instances because fundamental knowledge is lacking, the investigator may be unaware of the intricate nature of the biotic complex with which he is dealing, and of the destructive potential that many chemicals in use today have on the environment of the pests. Finally, and most unfortunately, there are workers who are highly skeptical that biotic factors are of any consequence in the control of pest population densities and thus choose to ignore any approach to pest control other than the use of chemicals.

Whatever the reasons for our increased pest problems, it is becoming more and more evident that an integrated approach, utilizing both biological and chemical control, must be developed in many of our pest problems if we are to rectify the mistakes of the past and avoid similar ones in the future (DeBach, 1951, 1958*a*; Pickett, Putman, and Roux, 1958; Ripper, 1944; Huffaker and Kennett, 1956; Wille, 1951; Michelbacher and Middlekauff, 1950; and others).

## TERMINOLOGY

To clarify the discussion in other parts of this paper some definitions and explanations of terms are here given:

**Biological control.** *The action of parasites, predators, or pathogens on a host or prey population which produces a lower general equilibrium position than would prevail in the absence of these agents.* Biological control is a part of natural control (*q.v.*) and in many cases it may be the key mechanism governing the population levels within the framework set by the environment. If the host or prey population is a pest species, biological control may or may not result in economic control. Biological control may apply to any species whether it is a pest or not, and regardless of whether or not man deliberately introduces, manipulates, or modifies the biological-control agents.

**Biotic insecticide.** *A biotic mortality agent applied to suppress a local insect pest population temporarily.* The effects of the agent usually do not persist and they are similar to those resulting from the use of a chemical insecticide in that they do not produce a permanent change in the general equilibrium position. A polyhedrosis virus applied as a spray to control the alfalfa caterpillar is a typical example of a biotic insecticide. Preparations of microörganisms used in this manner are sometimes referred to as *microbial insecticides.* Predators, such as lady beetles, or parasites, when they are released in large numbers, can also act, in some instances, as biotic insecticides.

**Biotic reduction.** *Deaths or other losses to the population* (e.g., *dispersal, reduced fecundity*) *caused or induced by biotic elements of the environment in a given period of time.*

**Economic control.** *The reduction or maintenance of a pest density below the economic-injury level* (q.v.).

**Economic-injury level.** *The lowest population density that will cause economic damage.* Economic damage is the amount of injury which will justify the cost of artificial control measures; consequently, the economic-injury level may vary from area to area, season to season, or with man's changing scale of economic values.

**Economic threshold.** *The density at which control measures should be determined to prevent an increasing pest population from reaching the economic-injury level.* The economic threshold is lower than the economic-injury level to permit sufficient time for the initiation of control measures and for these measures to take effect before the population reaches the economic-injury level.

**Ecosystem.** *The interacting system comprised of all the living organisms of an area and their nonliving environment.* The size of area must be extensive enough to permit the paths and rates of exchange of matter and energy which are characteristic of any ecosystem.

**General equilibrium position.** *The average density of a population over a period of time* (*usually lengthy*) *in the absence of permanent environmental change.* The size of the area involved and the length of the period of time will vary with the species under consideration. Temporary artificial modifications of the environment may produce a temporary alteration of the general equilibrium position (*i.e.,* a temporary equilibrium).

**Governing mechanism.** *The actions of environmental factors, collectively or singly, which so intensify as the population density increases and relax as this density falls that population increase beyond a characteristic high level is prevented and decrease to extinction is made unlikely.* The governing mechanisms operate within the framework or potential set by the other environmental elements.

**Integrated control.** *Applied pest control which combines and integrates biological and chemical control.* Chemical control is used as necessary and in a manner which is least disruptive to biological control. Integrated control may make use of naturally occurring biological control as well as biological control effected by manipulated or introduced biotic agents.

**Microbial control.** *Biological control that is effected by microörganisms (including viruses).*

**Natural control.** *The maintenance of a more or less fluctuating population density within certain definable upper and lower limits over a period of time by the combined actions of abiotic and biotic elements of the environment.* Natural control involves all aspects of the environment, not just those immediate or direct factors producing premature mortality, retarded development, or reduced fecundity; but remote or indirect factors as well. For most situations, governing mechanisms (*q.v.*) are present and determine the population levels within the framework or potential set by the other environmental elements. In the case of a pest population, natural control may or may not be sufficient to provide economic control.

**Natural reduction.** *Deaths or other losses to the population caused by naturally existing abiotic and biotic elements of the environment in a given period of time.*

**Population.** *A group of individuals of the same species that occupies a given area.* A population must have at least a minimum size and occupy an area containing all its ecological requisites to display fully such characteristics as growth, dispersion, fluctuation, turnover, dispersal, genetic variability, and continuity in time. The minimum population and the requisites in occupied area will vary from species to species.

**Population dispersion.** *The pattern of spacing shown by members of a population within its occupied habitat and the total area over which the given population may be spread.*

**Selective insecticide.** *An insecticide which while killing the pest individuals spares much or most of the other fauna, including beneficial species, either through differential toxic action or through the manner in which the insecticide is utilized (formulation, dosage, timing, etc.).*

**Supervised insect control.** *Control of insects and related organisms supervised by qualified entomologists and based on conclusions reached from periodically measured population densities of pests and beneficial species.* Ideally, supervised control is based on a sound knowledge of the ecology of the organisms involved and projected future population trends of pests and natural enemies.

**Temporary equilibrium position.** *The average density of a population over a large area temporarily modified by a procedure such as continued use of insecticides.* The modified average density of the population will revert to the previous or normal density level when the modifying agent is removed or expended (*cf.* "general equilibrium position").

## THE NATURE AND WORKING PRINCIPLES OF BIOLOGICAL AND CHEMICAL CONTROL

**Biological Control.** Biological controls are part of natural control which governs the population density of pest species. On the other hand, with certain exceptions, chemical controls involve only immediate and temporary decimation of localized populations and do not contribute to permanent density regulation. This distinction is not always clearly made, and biological control is often thought of as being similar in its action to chemical control. Perhaps one reason for the misunderstanding is that in spectacular instances a biotic agent may act in the manner of a chemical in eliminating a local pest population. For example, this may occur when weather conditions are favorable and disease pathogens eliminate a localized pest population. Parasites and predators may sometimes act in a similar manner. However, the important prevailing characteristic of biological control is one of permanent population-density regulation. Usually these governing mechanisms occur over such a large area and are so subtle or intricate in their action that they are not easily observed and recorded; thus they tend to be overlooked.

A principal phase of applied biological control is the importation and establishment of natural enemies of pests that accidentally gain entry into new geographical regions. These new pests frequently escape the natural enemies that help to regulate their densities in the areas to which they are indigenous (Elton, 1958). Under satisfactory conditions in the new environment, the pest may flourish and reach damaging abundance. As a counter measure, the natural enemies are obtained from the native home of the pest and transplanted into the new environment to increase the biotic resistance of the environment to the pest.

Biological control is thus utilized to permanently increase environmental resistance to an introduced pest. The hope is that the introduced enemies will lower the general equilibrium position of the pest sufficiently to maintain it permanently below the economic threshold. Most often the introduction of a biotic agent is not so spectacular, and it is an exception when the general equilibrium position of the introduced pest is lowered sufficiently to prevent its occasionally or even commonly reaching economic abundance at certain times or places (Clausen, 1956; Simmonds, 1956). This, of course, is precisely the status of a native pest which, though attacked by a complex of parasites and predators, still has a general equilibrium position high enough to permit it occasionally to cause damage of greater or lesser severity. Thus, in any geographic area the governing mechanisms in the environment are constantly at work to counteract the inherent natality of plant and animal pest species. In terms of crop protection, these regulating factors actually keep thousands of potentially harmful arthropod species permanently below economic thresholds. Moreover, these environmental pressures tend to localize the outbreaks of those forms which on occasion are capable of rising above economic thresholds. A biological control agent is self-perpetuating and capable of response to fluctuations in the population density of the pest it attacks. Biological controls, whether imported or native, are permanent characteristics of a given environment.

**Chemical Control.** Chemical control of an arthropod pest is employed to reduce populations of pest species which rise to dangerous levels when the environmental pressures are inadequate. When chemicals are used, the damage from the pest species must be sufficiently great to cover not only the cost of the insecticidal treatment but also the possible deleterious effects, such as the harmful influence of the chemical on the ecosystem. On some occasions, the pest outbreak may cover a wide area; in other instances, damaging numbers occur in very restricted locations. These outbreaks occur during the season favorable to the pest, with the relaxed environmental pressures occurring some time before the outbreak. Chemical control is only needed at those times and places where natural control is inadequate. Chemical control should act as a complement to the biological control.

An insecticide must always be manipulated by man, who adds it to a restricted segment of the pest's environment to decimate a localized pest population. Because chemical insecticides are nonreproductive, have no searching capacity, and are nonpersistent, they constitute short-term, retricted pressures. They cannot permanently change the general equilibrium position of the pest population nor can they restrain an increase in abundance of the pest without repeated applications. Therefore, they must be added to the environment at varying intervals of time.

In certain pest-control programs, the insecticide is applied over extensive geographical areas. In some areas, after application, the pest population density may be far below the economic threshold and below its general equilibrium position; but since the insecticide is not a permanent part of the environment, the pest may return to a high level when the effects of the insecticide are gone.

The effectiveness of a chemical insecticide or a biotic insecticide is measured in per cent of kill or in per cent of clean fruits, uninjured cotton bolls, and so forth, in the area of application. Such applications have little influence on the pest in adjoining areas except as localized population depressants. In general, this contrasts sharply with the role of the permanent biotic mortality agent, whose effectiveness is best measured by its influence on the general equilibrium position of the pest species over an entire geographical region or a long period of time.

## ECONOMIC THRESHOLDS AND THE GENERAL EQUILIBRIUM POSITION

Chemical control should be used only when the economic threshold is reached and when the natural mortality factors present in the environment are not capable of preventing the pest population from reaching the economic-injury level. The economic-injury level is a slightly greater density than the economic threshold. This difference in densities provides a margin of safety for the time that elapses between the detection of the threatening infestation and the actual application of an insecticide. The economic threshold and the economic-injury level of a pest species can vary depending upon the crop, season, area, and desire of man; the general equilibrium position, on the other hand, barring "permanent" changes in the environment, is a fixed population level (Griffiths, 1951; Strickland, 1954).

A species population is plastic and is undergoing constant change within the limits imposed upon it by its genetic constitution and the characteristics of its environment. Typical fluctuations in population and dispersion are shown in figure 3. The population dispersions shown at the three points in time A, B, and C are not static but rather are instantaneous phases of a continuously changing dispersion.

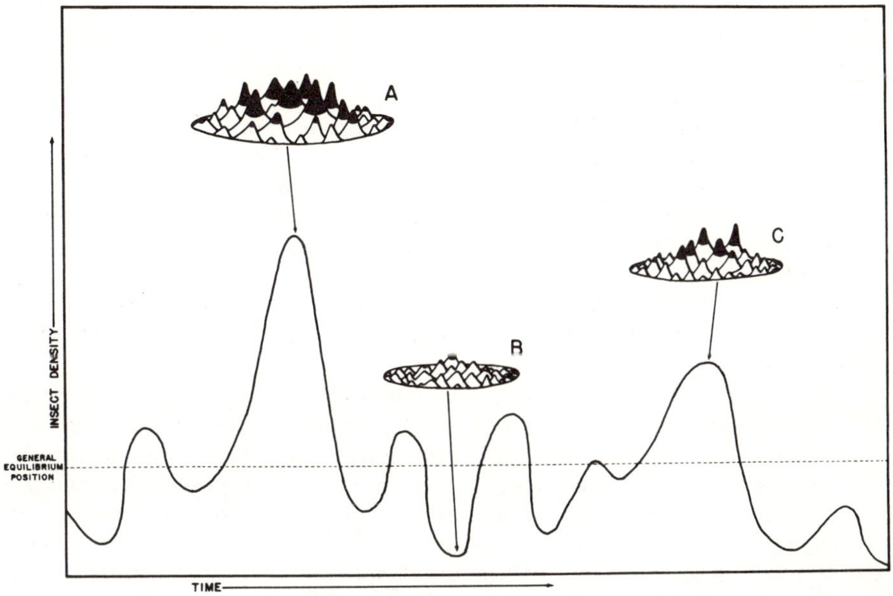


Fig. 3. Schematic graph of the population trend and population dispersion of a pest species over a long period of time. The solid line depicts the fluctuations in the population density with time. The broken line depicts the general equilibrium position. The population dispersion is indicated at the specific times *A*, *B*, and *C*. The basal area of these models reflects the distributional range, the height indicates population density. Population densities above the economic threshold are black.

Thus at point A, when the population is of greatest numerical abundance, it also has its widest distributional range (as depicted by the maximum diameter of the base of the model), and is of maximum economic status (as depicted by the number and magnitude of the blackened pinnacles representing penetrations of the economic threshold). At point B, on the other hand, when the species population is at its lowest numerical abundance, it is also most restricted in geographical range and is of only minor economic status. Point C represents an intermediate condition between points A and B.

In order to determine the relative economic importance of pest species, both the economic threshold and general equilibrium position of the pests must be considered. It is the general equilibrium position and its relation to the economic threshold, in conjunction with the frequency and amplitude of fluctuations about the general equilibrium position, that determine the severity of a particular pest problem.

In the absence of permanent modifications in the composition of the environment, the density of a species tends to fluctuate about the general equilibrium position as changes occur in the biotic and physical components of the environment. As the population density increases, the density-governing factors respond with greater and greater intensity to check the increase; as the population density decreases, these factors relax in their effects. The general equilibrium position is thus determined by the interaction of the species population, these density-governing factors, and the other natural factors of the environment. A permanent alteration of any factor of the environment—either physical or biotic—or the introduction of new factors may alter the general equilibrium position.

The economic threshold of a pest species can be at any level above or below the general equilibrium position or it can be at the same level. Some phytophagous species may utilize our crops as a food source but even at their highest attainable density are of little or no significance to man (see fig. 4, *A*). Such species can be found associated with nearly every crop of commercial concern.

Another group of arthropods rarely exceeds the economic thresholds and these consequently are occasional pests. Only at their highest population density will chemical control be necessary (see fig. 4, *B*).

When the general equilibrium position is close to the economic threshold, the population density will reach the threshold frequently (see fig. 4, *C*). In some cases, the general equilibrium position and the economic threshold are at essentially the same level. Thus, each time the population fluctuates up to the level of the general equilibrium position insecticidal treatment is necessary. In such species the frequency of chemical treatments is determined by the fluctuation rate about the general equilibrium position, which in some cases necessitates almost continuous treatment.

Finally, there are pest species in which the economic threshold lies below the general equilibrium position; these constitute the most severe pest problems in entomology (see fig. 4, *D*). The economic threshold may be lower than the level of the lowest population depression caused by the physical and biotic factors of the environment, *e.g.*, many insect vectors of viruses. In such cases, particularly where human health is concerned, there is a widespread and almost constant need for chemical control. This produces conditions favorable for development of insecticide resistance and other problems associated with heavy treatments.

One solution to pest problems and particularly those in this last category is to change the environment permanently so that the general equilibrium position will be lowered. For example, this might be accomplished through the introduction of a new biological control agent or through the permanent modification of a large portion of a required habitat. This has been done in certain areas with malaria-vector mosquitoes and similar pests by the draining of swamps and the destruction of other favorable habitats. Such methods may completely eliminate the species from some areas.

Environmental changes unfavorable to the pest may also be made through the use of plants and animals resistant to the pest species. This control method may involve three different aspects—tolerance, preference, and antibiosis (Painter, 1951). If tolerance alone is involved, the general equilibrium position may not be changed but the economic threshold is raised. Where preference or antibiosis is involved, the ability of the pest to reproduce upon the host is reduced, so that the general equilibrium position is lowered.

The lack of a sound measure of economic thresholds, in many cases, has been a major stumbling block to the development of integrated pest-control programs. Our changing economy, variations in natural governing mecha-

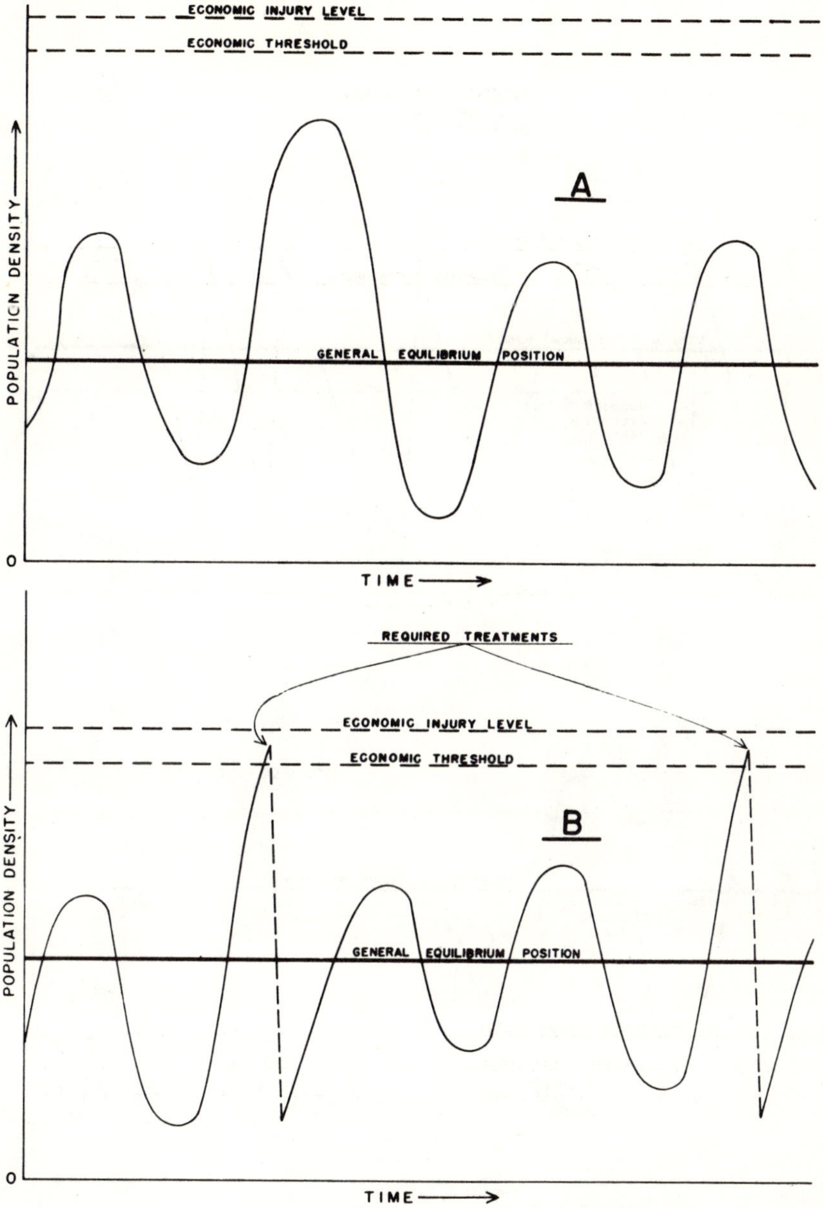


Fig. 4. Schematic graphs of the fluctuations of theoretical arthropod populations in relation to their general equilibrium position, economic thresholds, and economic-injury levels. *A,* Noneconomic population whose general equilibrium position and highest fluctuations are below the economic threshold, *e.g., Aphis medicaginis* Koch on alfalfa in California. *B,* Occasional pest whose general equilibrium position is below the economic threshold but whose highest population fluctuations exceed the economic threshold, *e.g.,*

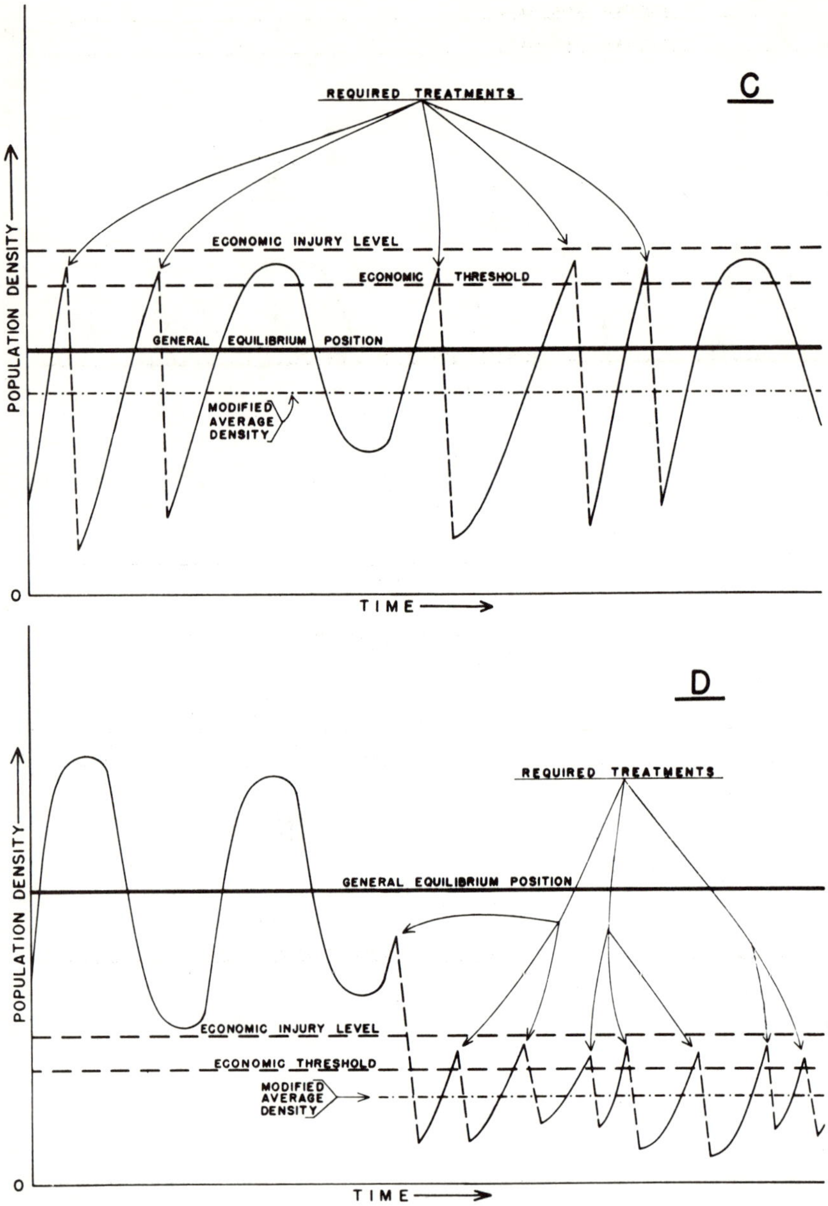


*Grapholitha molesta* Busck on peaches in California. *C,* Perennial pest whose general equilibrium position is below the economic threshold but whose population fluctuations frequently exceed the economic threshold, *e.g., Lygus* spp. on alfalfa seed in the western United States. *D,* Severe pest whose general equilibrium position is above the economic threshold and for which frequent and often widespread use of insecticides is required to prevent economic damage, *e.g., Musca domestica* in Grade A milking sheds.

nisms from one geographical area to another, differences in consumer demands, and the complexity of measuring the total effect of insects on yield and quality often make the assessment of economic damage extremely difficult. Yet the economic threshold and the economic-injury level must be determined reasonably and realistically before integrated pest control can develop to its fullest. Success in any well-balanced pest-control program is dependent on the aim of holding insect populations below experimentally established economic levels rather than attempting to eliminate all the insects.

## THE INTEGRATION OF BIOLOGICAL AND CHEMICAL CONTROL

Biological control and chemical control are not necessarily alternative methods; in many cases they may be complementary, and, with adequate understanding, can be made to augment one another. One reason for the apparent incompatibility of biological and chemical control is our failure to recognize that the control of arthropod populations is a complex ecological problem. This leads to the error of imposing insecticides on the ecosystem, rather than fitting them into it. It is short-sighted to develop a chemical control program for the elimination of one insect pest and ignore the impact of that program on the other arthropods, both beneficial and harmful, in the ecosystem. On the other hand, this approach is no worse than the other extreme which would eliminate chemicals to preserve the biological control even in the face of serious economic damage. For we must recognize that modern agriculture could not exist without the use of insecticides. The evidence that biological and chemical control can be integrated is mounting. It has come from many sources involving many kinds of pests in various situations: see Ullyett (1947), Pickett and Patterson (1953), Ripper (1956), Huffaker and Kennett (1956), DeBach (1958*a*), Stern and van den Bosch (1959), and many others.

In approaching an integrated control program, we must realize that man has developed huge monocultures, he has eliminated forests and grasslands, selected special strains of plants and animals, moved them about, and in other ways altered the natural control that had developed over thousands upon thousands of years. We could not return to those original conditions if it were desirable. We may, however, utilize some of the mechanisms that existed before man's modifications, to establish new balances in our favor.

**Recognition of the Ecosystem.** To establish new, favorable balances, it is first necessary to recognize the "oneness" of any environment, natural or man-made. The populations of plants and animals (including man) and the nonliving environment together make up an integrated unit, the ecosystem. If an attempt is made to reduce the population level of one kind of animal (for example, a pest insect) by chemical treatment, modification of cultural practices, or by other means, other parts of the ecosystem will be affected as well. For this reason, the production of a given food or fiber must be considered in its entirety. This includes simultaneous consideration of insects, diseases, plant nutrition, plant physiology, and plant resistance, as well as the economics of the crops (Forbes, 1880; Ullyett, 1947; Pickett, 1949; DeBach, 1951; Solomon, 1953; Pickett and Patterson, 1953;

Glen, 1954; Michelbacher, 1954; Huffaker and Kennett, 1956; Simmonds, 1956; Balch, 1958; Decker, 1958; and others).

In most agricultural ecosystems, some potentially harmful organisms are continually held at subeconomic levels by natural controlling forces. In others, the pests are held below economic levels only part of the time. A pest species may be under satisfactory biological control over a large area or a long period of time, but not in all individual fields or during all periods. In a single field or orchard, or during a portion of a year, the pest population may rise to economic levels, while elsewhere or at other times the pest may be subeconomic. It is in such situations that integrated control programs are especially important. These intermittently destructive populations must be reduced in a manner that permits the biological control which prevailed before or prevails elsewhere to take over again. If a chemical treatment destroys the biotic agents without eradicating the pest, then repeated treatments may become necessary.

**Population Sampling and Prediction.** The sampling methods utilized by most research investigators for experimental plots are usually too time-consuming and tedious to be of practical value in establishing pest population levels in commercial crops. Special index methods are needed that are rapid and simple to use. Ideally, these should be of such nature that they can be easily utilized by the person examining the crop. But in many cases the grower is not able to evaluate all situations because of the difficulties and complexities involved in determining the status of some pest populations at the times of the year when they must be controlled. Then qualified entomologists will be required to evaluate the populations (Ripper, 1958).

One answer to this problem has been the development of supervised control in California, Arkansas, Arizona, and elsewhere. In a supervised control program the farmer, or a group of farmers, contracts with a professional entomologist who determines the status of the insect populations. On the basis of his population counts, other conditions peculiar to each situation, and his knowledge of the ecology of the pests and their biological controls, the entomologist makes predictions as to the course of the population trends and advises as to when controls should be applied and what kind. For instance, in the case of the alfalfa caterpillar, *Colias philodice eurytheme*, when economic thresholds are reached, the recommended procedure may involve immediate cutting of the hay crop without treatment, application of the polyhedrosis virus (Steinhaus and Thompson, 1949; Thompson and Steinhaus, 1950), or treatment with an insecticide. The course to be taken depends on the characteristics of the particular infestation (Smith and Allen, 1954).

Wherever possible, knowledge must be developed so that we can predict the times when occasional pests will be present in outbreak numbers. This will eliminate unnecessary and environment-disturbing "insurance" treatments. When this is not possible, the treatments can be timed according to the actual pest population levels, as is now done with many field-crop pests.

With those crops that do not yet have fixed chemical control schedules, every effort should be made to plan programs dependent upon pest population levels and to avoid dependence upon insurance and prophylactic treatments. If this is not done there is real danger that on these crops, too, pest-control problems will become increasingly complex.

**Augmentation of Natural Enemies.** In some situations, the development of integrated control requires the augmentation of the natural-enemy complex (DeBach, 1958*a*). The introduction of additional natural enemies is usually the simplest and best solution. This may not be possible or effective with some pest species, and methods of overcoming the inefficiency of the natural-enemy complex must be sought. This can be done by periodic colonization of parasite or predators (Doutt and Hagen, 1950; DeBach, Landi, and White, 1955), artificial inoculation of the host at times of low density (Smith and DeBach, 1953; Huffaker and Kennett, 1956), modification of the environment, or selective breeding of parasites and predators.

The modification of the environment may involve changes in irrigation, introduction of a covercrop, or development of greater plant heterogeneity. Refuges for beneficial forms can be produced by strip treatments with chemicals (DeBach, 1958*a*) or by the development of uncultivated and untreated areas (Grison and Biliotti, 1953). Modification of the environment may also involve the control of ants or other organisms which curtail parasite and predator activity (Flanders, 1945; DeBach, Fleschner, and Dietrick, 1951). The selective breeding of parasites and predators may be directed toward increased or modified tolerance ranges to physical conditions (DeBach, 1958*b*) or insecticides (Robertson, 1957).

Where prophylactic treatments are proved to be necessary for a perennial pest, selective materials must be developed and utilized to foster biological control both of other pests and of the pest of direct concern at other times.

**Selective Insecticides.** Chemical control programs are limited by the nature of the available materials. In the past, nonselective insecticides applied for one insect in a pest complex often have eliminated the biotic factors holding other pests in check. More recently, we have had available a greater variety of materials, some of them selective in their action (Ripper, 1956).

The selective use of insecticides may be accomplished in at least four ways First, the insecticide itself may be selective in its toxicological action. Narrow-range toxicants may be utilized to reduce a pest of concern and at the same time spare the beneficial forms (Ripper, 1944; Ripper, Greenslade, and Hartley, 1951). A particular material may be selective in one situation and not in another; or it may be selective at low dosages but not at high dosages. Furthermore, the manner of application (Ripper, 1956) and especially the type of carrier and residue deposit may produce differential effects on the insect complex (Flanders, 1941; Holloway and Young, 1943; DeBach and Bartlett, 1951).

Second, we can produce a selective action on a pest-parasite complex by treating only those areas where the pest-parasite ratio is unfavorable. This method is one of the bases of supervised control of the alfalfa caterpillar in California (Smith and Allen, 1954). Population levels of both the host caterpillar and its parasite, *Apanteles medicaginis* Muesebeck, are determined at appropriate intervals in all fields. A prediction of possible damage is made on the basis of these population levels, and only those infestations which are potentially damaging are treated. In this way, on an area-wide basis, the balance is shifted in favor of the parasites, even though parasite adults and parasite larvae within the host caterpillars are often destroyed in the treated fields. The success of such programs will depend on the exact nature of the local problem and the quality of supervision. The rates of dispersal of parasites, predators, and pests are complicating factors.

Third, proper timing of insecticides can produce a selective action on the pest and natural-enemy complex (Ewart and DeBach, 1947; Michelbacher and Middlekauff, 1950; Bartlett and Ewart, 1951; Jeppson, Jesser, and Complin, 1953; Massee, 1954; DeBach, 1955). In such situations, an intimate knowledge of the behavior patterns of the pests and their natural enemies is required.

Fourth, nonselective materials with short residual action may be used if the beneficial forms can survive in a resistant stage or in an untreated reservoir area. Stern and van den Bosch (1959) have demonstrated that parasites of the spotted alfalfa aphid can survive nonselective treatments if they are in the more resistant pupal stage. DeBach (1958*a*) reports successful integration of biological and chemical control of purple scale on citrus where alternate pairs of tree rows were sprayed at 6-month intervals with a nonselective oil treatment.

For some pests a disease pathogen may be used as a selective insecticide (Steinhaus, 1954; 1956). For example, under supervised control in the Dos Palos area of California, the polyhedrosis virus affecting the alfalfa caterpillar has been used successfully either alone or in combination with a selective insecticide to avoid the use of a nonselective treatment. More recently, interest has developed in the commercial use of virulent strains of *Bacillus thuringiensis* Berliner, for the control of certain truck- and field-crop pests in California. The use of disease pathogens as selective insecticides is in its infancy, but can be expected to increase in importance with additional research (Steinhaus, 1951, 1957).

The ideal selective material is not one that eliminates all individuals of the pest species while leaving all of the natural enemies. Use of such a material would force the predators and parasites to leave the treated area or starve (Clausen, 1936; Flanders, 1940). The ideal material is one that shifts the balance back in favor of the natural enemies (Boyce, 1936; Ripper, 1944; Wigglesworth, 1945).

**Future of Integrated Control.** If our knowledge were adequate today to outline an ideal integrated control program for a crop now utilizing an intensive fixed spray program, it would not be possible to switch to such a program immediately. The effects of the previous treatments may last several years. In some instances, effective biological control no longer exists and would have to be reëstablished. This may be a slow process (DeBach, 1951; DeBach and Bartlett, 1951; Pickett and Patterson, 1953; and others).

It should be emphasized also that the development of integrated control is not a panacea that can be applied blindly to all situations, for it will not work if biotic mortality agents are inadequate or if low economic thresholds preclude utilizing biological control (Barnes, 1959). However, it has worked so well in some appropriate situations that there can be no doubt as to its enormous advantages and its promise for the future.

## LITERATURE CITED

BALCH, R. E.

1958. Control of forest insects. Ann. Rev. Ent. **3**:449–68.

BARNES, M. M.

1959. Deciduous fruit insects and their control. Ann. Rev. Ent. **4**:343–62.

BARTLETT, B., and W. H. EWART

1951. Effect of parathion on parasites of *Coccus hesperidum*. Jour. Econ. Ent. **44**: 344–347.

BARTLETT, B. R., and J. C. ORTEGA

1952. Relation between natural enemies and DDT-induced increases in frosted scale and other pests of walnuts. Jour. Econ. Ent. **45**(5):783–85.

BOVEY, PAUL

1955. Les actions secondaires des traitements antiparasitaires sur les populations d'insectes et d'acariens nuisibles. Schweiz. Landw. Monatsh. **33** (9/10):369–79.

BOYCE, A. M.

1936. The citrus red mite, *Paratetranychus citri* McG. in California, and its control. Jour Econ. Ent. **29** (1):125–30.

BROWN, A. W. A.

1951. Insect control by chemicals. John Wiley & Sons, Inc., New York, N.Y. 817 pp.

1959. Spread of insecticide resistance. Adv. Pest Control Res. **2**:351–414. Interscience Publishers, Inc., New York, N.Y.

CLAUSEN, C. P.

1936. Insect parasitism and biological control. Ent. Soc. Amer. Ann. **29**:201–23.

1956. Biological control of insect pests in the continental United States. U.S. Dept. Agr. Tech. Bul. 1139. 151 pp.

CROW, J. F.

1957. Genetics of insect resistance to chemicals. Ann. Rev. Ent. **2**:227–46.

DEBACH, P.

1951. The necessity for an ecological approach to pest control on citrus in California. Jour. Econ. Ent. **44**(4):443–47.

1955. Validity of the insecticidal check method as a measure of the effectiveness of natural enemies of diaspine scale insects. Jour. Econ. Ent. **48**(5):584–88.

1958*a*. Application of ecological information to control of citrus pests in California. Tenth Internatl. Congr. Ent. Proc. **3**:185–97.

1958*b*. Selective breeding to improve adaptations of parasitic insects. Tenth Internatl. Congr. Ent. Proc. **4**:759–67.

DEBACH, P., and B. BARTLETT

1951. Effects of insecticides on biological control of insect pests of citrus. Jour. Econ. Ent. **44**(3):372–83.

DEBACH, P., C. A. FLESCHNER, and E. J. DIETRICK

1951. A biological check method for evaluating the effectiveness of entomophagous insects. Jour. Econ. Ent. **44**(5):763–66.

DEBACH, P., J. H. LANDI, and E. B. WHITE

1955. Biological control of red scale. California Citrog. **40**:254, 271–72, 274–75.

DECKER, G. C.

1958. Don't let the insects rule. Agr. Food Chem. **6**(2):98–103.

DOUTT, R. L.

1958. Vice, virtue and the vedalia. Ent. Soc. Amer. Bul. **4**(4):119–23.

DOUTT, R. L., and K. S. HAGEN

1950. Biological control measures applied against *Pseudococcus maritimus* on pears. Jour. Econ. Ent. **43** (1):94–96.

ELTON, CHARLES S.

1958. The ecology of invasions by animals and plants. 181 pp. Methuen & Co. Ltd., London; John Wiley & Sons Inc., New York, N.Y.

EWART, W. H., and P. DEBACH

1947. DDT for control of citrus thrips and citricola scale. California Citrog. **32**:242–45.

FLANDERS, S. E.

1940. Environmental resistance to the establishment of parasitic Hymenoptera. Ent. Soc. Amer. Ann., **33**:245–53.

1941. Dust as an inhibiting factor in the reproduction of insects. Jour. Econ. Ent. **34**(3):470–1.

1945. Coincident infestations of *Aonidiella atrana* and *Coccus hesperidum*, a result of ant activity. Jour. Econ. Ent. **38**:711–12.

FLESCHNER, C. A., and G. T. SCRIVEN

1957. Effects of soil-type and DDT on ovipositional response of *Chrysopa californica* (Coq.) on lemon trees. Jour. Econ. Ent. **50**(2):221–22.

FORBES, S. A.

1880. On some interactions of organisms. Illinois State Lab. Nat. Hist. Bul. **3**:3–17.

GLEN, R.

1954. Factors that affect insect abundance. Jour. Econ. Ent. **47**:398–405.

GRIFFITHS, J. T.

1951. Possibilities for better citrus insect control through the study of the ecological effects of spray programs. Jour. Econ. Ent. **44**:464–68.

GRISON, P., and E. BILIOTTI

1953. La signification agricole des "stations refuges" pour la faune entomologique. Acad. d'Agr. de France Compt. Rend. **39**(2):106–9.

HAYES, W. J.

1954. Agricultural chemicals and public health. U. S. Public Health Reports **69**(10): 893–98.

HERTZLER, J. O.

1956. The crisis in world population. 279 pp. Univ. Nebraska Press, Lincoln.

HOLLOWAY, J. K., and T. ROY YOUNG, JR.

1943. The influence of fungicidal sprays on entomogenous fungi and on the purple scale in Florida. Jour. Econ. Ent. **36**(3):453–57.

HOSKINS, W. M., and H. T. GORDON

1956. Arthropod resistance to chemicals. Ann. Rev. Ent. **1**:89–148.

HUFFAKER, C. B., and C. E. KENNETT

1956. Experimental studies on predation: Predation and cyclamen-mite populations on strawberries in California. Hilgardia **26**(4):191–222.

JEPPSON, L. R., M. J. JESSER, and J. O. COMPLIN

1953. Timing of treatments for control of citrus red mite on orange trees in coastal districts of California. Jour. Econ. Ent. **46**:10–14.

KLOSTERMEYER, E. C., and W. B. RASMUSSEN

1953. The effect of soil insecticide treatments on mite populations and damage. Jour. Econ. Ent. **46**:910–12.

LINSLEY, E. G., editor

1956. Evaluation of certain acaricides and insecticides for effectiveness, residues, and influence on crop flavor. Hilgardia **26**(1):1–106.

LORD, F. T.

1947 The influence of spray programs on the fauna of apple orchards in Nova Scotia: II. Oystershell scale, *Lepidosaphis ulmi* (L.) Canad. Ent. **79**:196–209.

MASSEE, A. M.

1954. Problems arising from the use of insecticides: effect on the balance of animal populations. 6th Commonwealth Entomol. Conf. Rept. Pp. 53–57. London, England.

METCALF, R. L.

1955. Physiological basis for insect resistance. Physiol. Rev. **35**:197–232.

MICHELBACHER, A. E.

1954. Natural control of insect pests. Jour. Econ. Ent. **47**(1):192–94.

MICHELBACHER, A. E., and S. HITCHCOCK
1958. Induced increase of soft scales on walnut. Jour. Econ. Ent. **51**(4):427–31.
MICHELBACHER, A. E., and W. W. MIDDLEKAUFF
1950. Control of the melon aphid in northern California. Jour. Econ. Ent. **43**(4): 444–47.
PAINTER, R. H.
1951. Insect resistance in crop plants. xi + 520 pp. Illus. Macmillan Company, New York, N.Y.
PETTY, C. S.
1957. Organic phosphate insecticide poisoning; an agricultural occupational hazard. Louisiana St. Med. Soc. Jour. **109**(5):158–64.
PICKETT, A. D.
1949. A critique on insect chemical control methods. Canad. Ent. **81**(3):67–76.
PICKETT, A. D., and N. A. PATTERSON
1953. The influence of spray programs on the fauna of apple orchards in Nova Scotia. IV. A Review. Canad. Ent. **85**(12):472–78.
PICKETT, A. D., W. L. PUTMAN, and E. J. LE ROUX
1958. Progress in harmonizing biological and chemical control of orchard pests in eastern Canada. Tenth Internatl. Congr. Ent. Proc. **3**:169–74.
RIPPER, W. E.
1944. Biological control as a supplement to chemical control of insect pests. Nature **153**:448–52.
1956. Effect of pesticides on balance of arthropod populations. Ann. Rev. Ent. **1**:403–38.
1958. The place of contracting organizations and professional supervision in the application of pest control methods. Tenth Internatl. Congr. Ent. Proc. **3**:93–97.
RIPPER, W. E., R. M. GREENSLADE, and G. S. HARTLEY
1951. Selective insecticides and biological control. Jour. Econ. Ent. **44**(4):448–59.
ROBERTSON, J. G.
1957. Changes in resistance to DDT in *Macrocentrus ancylivorus* Rohw. Canad. Jour. Zool. **35**(5):629–633.
SABROSKY, C. W.
1952. How many insects are there? Pp. 1–7 *in:* Insects, 1952 Yearbook of Agriculture. 780 pp. U. S. Dept. Agr., Washington, D.C.
SCHNEIDER, F.
1955. Beziehungen zwischen nützlingen und chemischer Schädlingsbekämpfung. Deutsche Gesell. f. Angw. Ent. Verhandl. (13[te] Mitglied versamm.). **1955**:18–29.
SIMMONDS, F. J.
1956. The present status of biological control. Canad. Ent. **88**(9):553–63.
SMITH, H. S., and P. DEBACH
1953. Artificial infestation of plants with pest insects as an aid in biological control. Seventh Pacific Sci. Congr. Zoology (1949) Proc. **4**:255–59.
SMITH, RAY F.
1959. The spread of the spotted alfalfa aphid, *Therioaphis maculata* (Buckton), in California. Hilgardia **28**(21):647–91.
SMITH, R. F., and W. W. ALLEN
1954. Insect control and the balance of nature. Sci. Amer. **190**(6):38–42.
SOLOMON, M. E.
1953. Insect population balance and chemical control of pests. Chem. and Indus. **1953**: 1143–47.
STEINHAUS, E. A.
1951. Possible use of *Bacillus thuringiensis* Berliner as an aid in the biological control of the alfalfa caterpillar. Hilgardia **20**(18):359–81.
1954. The effects of disease on insect populations. Hilgardia **23**(9):197–261.
1956. Potentialities for microbial control of insects. Agr. Food Chem. **4**(8):676–80.
1957. Concerning the harmlessness of insect pathogens and the standardization of microbial control products. Jour. Econ. Ent. **50**(6): 715–20.

STEINHAUS, E. A., and C. G. THOMPSON

1949. Preliminary field tests using a polyhedrosis virus in the control of the alfalfa caterpillar. Jour. Econ. Ent. **42**(2):301–5.

STERN, V. M., and R. VAN DEN BOSCH

1959. Field experiments on the effects of insecticides. Hilgardia **29**(2):103–30.

STRICKLAND, A. H.

1954. The assessment of insect pest density in relation to crop losses. 6th Commonwealth Ent. Conf. Rept., pp. 78–83. London, England.

THOMAS, W. L., JR., EDITOR

1956. Man's role in changing the face of the earth. 1193 pp. Univ. Chicago Press, Chicago, Ill.

THOMPSON, C. G., and E. A. STEINHAUS

1950. Further tests using a polyhedrosis virus to control the alfalfa caterpillar. Hilgardia **19**(14):411–45.

TROUVELOT, B.

1936. Le doryphore de la pomme de terre (*Leptinotarsa decemlineata* Say) en Amerique de Nord. Ann. Epiphyt. (n.s.) **1**:277–336.

ULLYETT, G. C.

1947. Mortality factors in populations of *Plutella maculipennis* (Tineidae: Lep.) and their relation to the problem of control. South African Dept. Agr. and Forestry Ent. Mem. **2**(6):77–202.

1951. Insects, man and the environment. Jour. Econ. Ent. **44**(4):459–64.

UPHOLT, W. M.

1955. Evaluating hazards in pesticides use. Agr. Food Chem. **3**(12):1000–6.

WILLE, J. E.

1951. Biological control of certain cotton insects and the application of new organic insecticides in Peru. Jour. Econ. Ent. **44**:13–18.

WIGGLESWORTH, V. B.

1945. DDT and the balance of nature. Atlantic Monthly **176**(6):107–13.

# A COMPARISON OF FIVE METHODS OF SAMPLING ALFALFA FIELDS FOR ARTHROPOD POPULATIONS[1]

F. A. FENTON AND D. E. HOWELL
Oklahoma State University, Stillwater

In 1954 a study was undertaken to determine the effects of several insecticides on the insect, spider, and mite populations in an alfalfa field. One of the first problems was to develop a standard method of sampling which would have a minimum amount of experimental error. A review of literature showed that several methods had been used by other workers. Five methods were tested and some were modified. This paper is a report on these tests, most of which were made in 1955.

### METHODS USED IN THIS STUDY

*The sweep net.*—Delong (1932), Gray and Treloar (1933), Beall (1935), Romney (1945), and Whittaker (1952), experimented with and evaluated the sweep net method of sampling for insect populations in various types of habitats from low-growing crops such as alfalfa to trees and shrubs. Their conclusions varied from unfavorable (DeLong, Gray and Treloar) to favorable provided the catch was correlated with temperature and wind velocity taken at the time sweeps were made (Romney).

In this study the sweep net had a diameter of 13 inches and a handle length of 19 inches. Only one pass was made with the net while the next and succeeding strokes were always made in parts of the field not previously disturbed by the net. In short sparse growths most of the plants were included in a single stroke but as the plants grew taller, less and less of them were hit with the net. Five samples were taken per field, each one in a different area and at each point from 10 to 50 sweeps constituted a sample. The insects and other arthropods collected were immobilized at once. This was accomplished by pouring a few drops of carbon tetrachloride on a piece of absorbent towelling paper at the bottom of a cardboard container. The net was then partially everted so that the insects and other animals in the net were inside the container. Finally a few more drops of the fumigant were added to the net. At least one minute was allowed to elapse before emptying the contents of the net into the container. Each collection was either examined within 24 hours or preserved in 80 percent alcohol for later examination.

The proportion of insects resting on the plants which were caught by the net was determined by the following tests: First, the number of tillers included in one stroke of the net was found to range from 124 to 177 with an average of 155 in those parts of the field where the stand was perfect. This meant in an average stand of alfalfa, having few or no skips in the rows, that 19,375 tillers were sampled for each 125 strokes of the net. Since, however, the stand was not perfect in areas of this field which were sampled, probably 15,000 tillers is a conservative estimate and this number has been used in our compilations.

Secondly, one-row plots were set up having the approximate number of tillers that would be swept with one stroke of the net. Such plots were located in various parts of the field and were prepared for sampling by the simple method of cutting down all of the other plants on both sides of each row as well as at both ends. This was done the day before the samples were taken. The procedure followed in making the tests was to make one pass with the net so as to include all of the plants in the row. The insects and spiders collected in this one net stroke were then removed from the net and preserved according to the procedure previously described, following which the row was repeatedly swept until all insects remaining on the plants were taken. These tests were made in an alfalfa field gone to seed and in a plot where the growth was rank and continuing. Except for thrips and alfalfa aphids, from 44.2 to 71.4 percent of all the insects and spiders resting on the plants were taken by the first net stroke.

*Plant clippings.*—Shortly after the sweeps were made, five tillers were clipped at random from each of five points, also selected at random, in the field. Each clipping was immediately placed in a wide-mouth one-quart fruit canning jar to which a little carbon tetrachloride had been added. Thus 25 clippings constituted a sample for each date of collection. In tall alfalfa plants only the upper portion of a tiller was taken. The insects were separated from the clipping by placing the plants in a flask of 60% alcohol, and thoroughly shaking the mixture. Two or three rinses were necessary to remove all insects and spiders.

*Berlese funnels.*—A modification of the Berlese funnel as first built by Berlese (1905) was used to process surface trash and soil to separate the

[1]Accepted for publication April 26, 1957.

Reproduced with permission from the *Annals of the Entomological Society of America*, 50:606–611, 1957. Published by the Entomological Society of America, College Park, Maryland.

TABLE I

NUMBER OF SPOTTED ALFALFA APHIDS AND THRIPS COLLECTED IN 125 SWEEPS WITH ESTIMATED PERCENTAGES OF POSSIBLE TOTALS PRESENT ON THE PLANTS AS DETERMINED BY PLANT CLIPPING COUNTS, STILLWATER, OKLAHOMA, 1955

| Dates | Theoretical Number of Aphids and Thrips per 15,000 Tillers Based Upon Plant Clipping Counts | | Actual Number of Aphids and Thrips Collected From 15,000 Tillers by 125 Sweeps | | Percent Taken By Sweeps | | Temp. °F. | Wind mph. |
|---|---|---|---|---|---|---|---|---|
| | Aphids | Thrips | Aphids | Thrips | Aphids | Thrips | | |
| June 15–17 | 27,000 | 104,000 | 551 | 8 | 2 | 0.007 | 90 | 7½ |
| June 25–27 | 38,400 | 105,600 | 1,078 | 64 | 2.8 | 0.06 | 83 | 5 |
| July 11–13 | 40,200 | 94,200 | 2,980 | 73 | 7.4 | 0.08 | 91 | 5 |
| July 25–27 | 0 | 40,800 | 755 | 14 | ? | 0.03 | 86 | 0 |
| Aug. 22–24 | 7,200 | 49,200 | 474 | 64 | 6.5 | 0.13 | 84 | 5 |
| Sept. 6– 8 | 1,200 | 105,000 | 269 | 101 | 22.4 | 0.96 | 77 | 5 |
| Sept. 21–23 | 7,200 | 28,800 | 591 | 23 | 8.2 | 0.08 | 95 | 4½ |
| Oct. 10–12 | 1,800 | 20,400 | 118 | 20 | 6.5 | 0.098 | 78 | 5 |
| Oct. 26–28 | 8,400 | 54,000 | 845 | 10 | 10.0 | 0.02 | 81 | 12 |

TABLE II

COMPARATIVE NUMBERS OF COLLEMBOLA[a] AND MITES TAKEN IN CLIPPINGS, SWEEPS, AND PIT TRAPS, STILLWATER, OKLAHOMA, 1955

| Dates | Number per 25 Clippings | | Number per 125 Sweeps | | Number per 4 Pit Traps | |
|---|---|---|---|---|---|---|
| | Collembola | Mites | Collembola | Mites | Collembola | Mites |
| June 15–17 | 0 | 2 | 0 | 0 | 38 | 52 |
| June 25–27 | 0 | 0 | 0 | 0 | 140 | 92 |
| July 11–13 | 0 | 0 | 2 | 0 | 45 | 81 |
| July 25–27 | 0 | 1 | 2 | 0 | 169 | 93 |
| Aug. 22–24 | 0 | 0 | 0 | 0 | 117 | 80 |
| Sept. 6– 8 | 0 | 3 | 0 | 34 | 68 | 61 |
| Sept. 21–23 | 0 | 8 | 1 | 1 | 75 | 47 |
| Oct. 10–13 | 0 | 0 | 2 | 0 | 145 | 23 |
| Oct 26–28 | 0 | 0 | 0 | 0 | 19 | 30 |
| Totals.... | 0 | 14 | 7 | 35 | 816 | 559 |

[a]*Pseudosinella violenta* Fols., *Achorutes humi* Fols., *Entomobrya purpurascens* Pack. *Entomobrya multifasciata* Tull., *Lepidocyrtus cyaneus* var. *anescens* Guthrie, *Lepidocyrtus cyaneus* var. *cinereus* Fols., *Lepidocyrtus cyaneus* Tull., *Lepidocyrtus unifasciatus* James, *Isotoma viridis* var. *delta* MacGil., *Depranura sabulicola* Mills, *Folsomia elongata* MacGill., *Sminthurides pumilis* Kraus.

contained insects. The one used in this experiment was a metal funnel 11 inches deep attached to a cylinder with a diameter of 11 inches and a depth of 14 inches. The lower opening was attached to the screw top of a glass fruit preserving jar. The latter, screwed to the funnel, was used to trap the arthropods which were forced out of the sample by the heat generated by a 200-watt electric light bulb. The socket of this bulb was soldered to the center of a metal cover which fitted tightly over the top end of the cylinder. The sample, which was composed of surface soil, trash, and plant roots from one-half square foot area, was supported on a 14-mesh wire screen at the bottom of the cylinder.

*Pit traps.*—Barnes (1953) and Fichter (1954) used pit traps to measure surface inhabiting arthropod populations. In our study one-half pint wide-mouth glass fruit preserving jars were used in groups of four at approximately three-foot intervals. Each jar was charged with about one inch of water to which a teaspoonful of 40 percent formalin and a pinch of detergent had been added. The period of exposure was 48 hours.

*Henderson sampler.*—A fork-type sampling device, designed by Henderson (1956), was used concurrently with sweeping as a means of comparing the two methods of sampling for the spotted alfalfa aphid, *Therioaphis maculata* (Buck.). This device consists of a fork-like hand tool which holds four narrow strips of window glass covered with an adhesive material to hold the trapped aphids in place for later examination. The fork is placed in position on the ground beneath the plants and the plants are shaken to dislodge the aphids.

## RESULTS

*Effectiveness of the sweep net in collecting aphids and thrips.*—Assuming that most of the aphids and thrips remained on the plants until the clipping was transferred to the jar gives a base for calculating the percentage taken by the sweep net, since the average number of tillers taken in by 125 sweeps is given as 15,000. In this calculation, however, two things must be taken into consideration: first, in many samples only the upper part of the tiller was clipped; and secondly, it is well known that some of the spotted alfalfa aphids drop from the plant upon the slightest disturbance. Therefore, the figures given in Table I must be taken as the ***minimum*** number of thrips and spotted alfalfa aphids calculated to be present on 15,000 tillers. It is seen that the sweep net collected from 2 to 22.4 percent of the minimum number of aphids present on the tillers, except for the period of July 25–27 when none were taken. It is evident that usually less than 10 percent of the aphids are taken from the plants in the net, and considerably less than one percent of the thrips in most cases. The discrepancy shown in the July 25–27 counts can be accounted for by the fact that the sweep net covered much more of the field being sampled than the spot clipping method.

*Mite and Collembola populations best sampled by pit traps.*—Table II shows the numbers of Collembola and mites collected by sweeps, clippings and pit traps. It appears that most of these small arthropods are inhabitants of the soil surface in an alfalfa field and that comparatively few occur on plants. In the case of mites, the few collected in sweeping were mostly plant feeders. Collembola were occasionally collected by sweeping, and one species, *Deuterosmynthurus* sp.[2] was swept from alfalfa plants in large numbers on certain dates. However, sampling by pit traps was necessary to record the mite and Collembola populations found in alfalfa fields.

TABLE III

SUMMARY OF NINE 48-HOUR PIT TRAP COLLECTIONS, JUNE 15–OCTOBER 28, 1956, STILLWATER, OKLAHOMA

| Item | Total Collected | Maximum Collected During 48 Hour |
|---|---|---|
| Coleoptera (misc.) | 52 | 23 |
| Carabidae (adults) | 36 | 13 |
| Carabidae (larvae) | 32 | 13 |
| Cicindellidae | 15 | 6 |
| Monotomidae | 52 | 19 |
| Meloidae | 315 | 255 |
| Hymenoptera (misc.) | 91 | 32 |
| Formicidae | 228 | 68 |
| Diptera (misc.) | 22 | 11 |
| Chloropidae | 84 | 18 |
| Phoridae | 39 | 20 |
| Lycoridae | 86 | 35 |
| Lepidoptera (misc.) | 5 | 2 |
| Hemiptera (misc.) | 41 | 35 |
| Homoptera (misc.) | 16 | 5 |
| *T. maculata* | 196 | 121 |
| Orthoptera (misc.) | 19 | 10 |
| Collembola (misc.) | 816 | 169 |
| Thysanoptera | 10 | 5 |
| Acarina (mites) | 559 | 93 |
| Araneida | 165 | 7 |

*Other species taken in pit traps, rarely in sweeping.*—As might be expected, beetles of the family Carabidae were commonly collected in the pit traps, but most of them rarely, if ever, by sweeping. In two years of collecting we obtained only one species regularly by sweeping, namely *Lebia viridis* Say. In sixteen 48-hour periods of collection Carabidae were taken on 13 dates totaling 189 adult specimens (Table III). Of these, the following three have been identified: *Poecilus chalcites* Say, *Scarites substriatus* Hald., and *S. subterraneus* Fab. Rarely was a *Lycosa* spider taken in sweeping, but quite a number were found in the pit traps. The same was true for two species of Cicindellidae, namely, *Cicindella punctulata* Oliv., and *Taracha virginica* L.

*Other records from pit traps.*—Ants of several

[2]Found to be a new species by D. L. Wray.

TABLE IV

List of Identified Species Collected by Sweeping Alfalfa Fields During 1955 Based on 100 Sweeps per Date

| Species | 6/15 | 6/27 | 7/13 | 7/27 | 8/24 | 9/7 | 9/21 | 10/10 | 10/26 | Totals |
|---|---|---|---|---|---|---|---|---|---|---|
| COLEOPTERA | | | | | | | | | | |
| *Collops 4-maculatus* | 1 | 1 | 6 | 18 | 8 | 9 | 3 | 0 | 0 | 46 |
| *Scymnus natchezianus* | 1 | 0 | 0 | 1 | 1 | 0 | 1 | 0 | 2 | 6 |
| *Diabrotica 12-punctata howardi* | 2 | 1 | 13 | 8 | 3 | 164 | 32 | 8 | 16 | 247 |
| *Lebia viridis* | 1 | 0 | 1 | 1 | 0 | 2 | 2 | 4 | 2 | 13 |
| *Haltica foliacea* | 1 | 0 | 5 | 1 | 3 | 1 | 0 | 2 | 0 | 13 |
| *Collops balteatus* | 0 | 0 | 1 | 1 | 1 | 7 | 1 | 0 | 0 | 11 |
| *Epicauta lemniscata* | 0 | 0 | 0 | 74 | 0 | 48 | 1 | 0 | 0 | 123 |
| *Epicauta unicolor* | 0 | 0 | 0 | 4 | 7 | 0 | 0 | 0 | 0 | 11 |
| *Hippodamia convergens* | 19 | 29 | 209 | 101 | 33 | 12 | 2 | 4 | 0 | 409 |
| ORTHOPTERA | | | | | | | | | | |
| *Melanoplus mexicanus* | 1 | 0 | 7 | 3 | 0 | 1 | 0 | 1 | 0 | 13 |
| *Oecanthus nigricornis* | 2 | 1 | 29 | 32 | 2 | 5 | 4 | 2 | 0 | 77 |
| HYMENOPTERA | | | | | | | | | | |
| *Chelonus texanus* | 0 | 1 | 9 | 0 | 3 | 0 | 2 | 0 | 0 | 15 |
| *Aphidius testaceipes* | 0 | 17 | 0 | 0 | 0 | 0 | 0 | 0 | 0 | 17 |
| *Cremnops vulgaris* | 0 | 5 | 0 | 1 | 0 | 2 | 1 | 0 | 0 | 9 |
| *Diplazon laetatorius* | 0 | 3 | 9 | 0 | 0 | 0 | 0 | 0 | 0 | 12 |
| *Apis mellifera* | 0 | 0 | 9 | 0 | 11 | 1 | 0 | 0 | 0 | 21 |
| *Agathis gibbosus* | 0 | 0 | 77 | 20 | 3 | 14 | 482 | 64 | 26 | 686 |
| *Liodontomerus perplexus* | 0 | 0 | 4 | 5 | 1 | 0 | 102 | 248 | 330 | 690 |
| DIPTERA | | | | | | | | | | |
| *Thaumatomyia glabra* | 1 | 3 | 196 | 40 | 56 | 155 | 61 | 0 | 14 | 526 |
| *Hylemya cilicrura* | 5 | 9 | 0 | 0 | 0 | 0 | 0 | 14 | 14 | 42 |
| HEMIPTERA | | | | | | | | | | |
| *Psallus seriatus* | 1 | 1 | 20 | 7 | 22 | 8 | 6 | 0 | 0 | 65 |
| *Lygus lineolaris* | 29 | 222 | 500 | 158 | 13 | 26 | 18 | 42 | 72 | 1080 |
| *Orius insidiosus; O. tristicolor* | 17 | 122 | 474 | 189 | 114 | 197 | 11 | 6 | 12 | 1142 |
| *Thyanta pallido-virens accerra* | 2 | 0 | 0 | 13 | 5 | 18 | 10 | 16 | 18 | 82 |
| *Nabis* species | 2 | 39 | 39 | 24 | 22 | 25 | 4 | 4 | 12 | 171 |
| *Nysius raphanus* | 1 | 4 | 16 | 173 | 27 | 0 | 1 | 6 | 0 | 228 |
| *Nysius californicus* | 0 | 0 | 14 | 6 | 0 | 2 | 1 | 0 | 0 | 23 |
| *Jalysus wickhami* | 1 | 1 | 5 | 4 | 6 | 0 | 0 | 2 | 0 | 19 |
| *Geocoris pallens; G. punctipes* | 2 | 0 | 5 | 2 | 21 | 1 | 0 | 0 | 0 | 31 |
| *Liorhyssus hyalinus* | 0 | 0 | 10 | 20 | 9 | 1 | 0 | 0 | 0 | 40 |
| *Sinea diadema* | 0 | 0 | 5 | 4 | 0 | 8 | 6 | 8 | 0 | 31 |
| *Aceratagallia uhleri* | 7 | 14 | 14 | 21 | 5 | 23 | 8 | 0 | 14 | 106 |
| *Empoasca fabae* | 52 | 156 | 88 | 27 | 62 | 135 | 10 | 34 | 4 | 568 |
| *Macrosteles fascifrons* | 1 | 4 | 2 | 1 | 8 | 30 | 1 | 42 | 16 | 105 |
| *Vanduzea triguttata* | 0 | 0 | 0 | 0 | 7 | 5 | 5 | 2 | 2 | 21 |
| *Spissistilus festinus* | 1 | 4 | 28 | 25 | 8 | 24 | 30 | 26 | 26 | 172 |
| *Nesosteles species* | 2 | 3 | 5 | 11 | 3 | 13 | 1 | 102 | 2 | 142 |
| *Macrosiphum pisi* | 106 | 523 | 0 | 0 | 0 | 0 | 0 | 0 | 0 | 629 |
| *Therioaphis maculata* | 441 | 862 | 2384 | 604 | 379 | 215 | 473 | 94 | 676 | 6128 |
| LEPIDOPTERA | | | | | | | | | | |
| *Caenurgina erechtea* | 1 | 2 | 0 | 1 | 6 | 4 | 1 | 4 | 2 | 21 |
| *Euerythema* species | 2 | 0 | 1 | 0 | 12 | 11 | 1 | 0 | 2 | 29 |
| *Loxostege similalis* | 1 | 0 | 1 | 1 | 11 | 2 | 2 | 0 | 0 | 18 |
| THYSANOPTERA | | | | | | | | | | |
| *Frankliniella occidentalis* | 6 | 51 | 58 | 11 | 51 | 81 | 18 | 16 | 8 | 300 |

species were usually present in the collections and, at certain times of the year, they were also taken in sweeps. Adults and larvae of the convergent lady beetle, *Hippodamia convergens* Guer., were also collected and in one year an elaterid beetle, *Monocrepidius vespertinus* (Fab.) was taken in large numbers. Both of these species were also taken in large numbers by sweeping. The three-striped blister beetle, *Epicauta lemniscata* Fab. is another example of a species commonly taken by both methods. The spotted alfalfa aphid was frequently found, and in one 48–hour period more of these aphids were collected in traps than in sweeping.

One of the surprising developments of the pit trap collections was that Diptera and Hymenoptera were frequently represented by several species. These were usually very small species such as the Chloropidae and Phoridae in the Diptera and representitives of several families of parasitic Hymenoptera not as yet identified. Among the small beetles were the families Monotomidae and Pselaphidae. In the case of Diptera the specimens were preserved in much better condition than when swept from the plants with a net.

*Sweep net best for most species.*—The sweep net was by far the best equipment for collecting the most species of insects found in alfalfa, and for determining their comparative abundance during the season. It was very difficult to determine the effects of temperature and wind

TABLE V

NUMBER OF SPOTTED ALFALFA APHIDS TAKEN BY SWEEPS COMPARED TO NUMBERS CAUGHT BY FORK TOOL, PAYNE COUNTY, OKLAHOMA, MAY 29, 1955

| Average Number Per | | | Equivalent | | | |
|---|---|---|---|---|---|---|
| Field Number | 25 Sweeps | Fork Tool[a] | Sq. Inch Fork Tool | Sq. Inches[b] In 25 Sweeps | Temp.[c] °F. | Wind m.p.h. |
| 1 | 967 | 226 | 53.2 | 18.2 | 74 | 12 |
| 2 | 222 | 72 | 16.9 | 13.1 | 68 | 12 |
| 3 | 591 | 110 | 25.9 | 22.8 | 70 | 5 |
| 4 | 52 | 3 | 0.7 | 74.3 | 64 | 8 |

[a]One for equals 1.25 square inches.
[b]Based on aphids per square inch in each field.
[c]On ground shaded by alfalfa plants.

conditions on the numbers collected because of natural weekly variations in the abundance of a species occurred at different times during a season. Table IV shows the records for 45 species very few of which, with the exceptions noted, were collected by other methods. This list is by no means complete, but includes all of those collected in sufficient numbers to be of interest. Certain of the Diptera and Hymenoptera were very abundant at times, but have not been identified so far. On the basis of 100 sweeps, which is the number taken for comparison in Table IV, each date represents approximately 12,500 tillers sampled. At no time during this period was any of these species, except aphids and thrips, sufficiently abundant to have occurred at the rate of one per tiller.

*Comparison of sweeping and Henderson sampler.*—A comparison of the sweep method with the fork sampling device of Henderson is shown in Table V. Four fields were sampled simultaneously with the net and the Henderson sampler. It was found that the latter was much faster. The spotted alfalfa aphid population varied from light to heavy. The numbers caught by the two methods varied in three fields to very nearly the same degree. However, the numbers taken in the field showing the lightest infestation were quite different as determined by the two methods. The fork device showed that the infestation in field number one was 73 times as heavy as in field number four, whereas the net showed it to be only 18 times as heavy. Much more of this field was covered by the 125 sweeps than by the ten-spot samples taken by Henderson's sampler. Thus, the chances of missing a more heavily infested area by the latter device were much greater.

*Comparison of sweeping with Berlese funnel samples.*—The sweep net method of sampling was also compared with the one where surface soil, trash, and plant roots were processed with Berlese funnels. Comparisons were made (Table VI) of the numbers of spotted alfalfa aphids, thrips, *Frankliniella occidentalis* (Perg.), Collembola and mites, to show certain differences. More aphids and thrips were collected in the Berlese funnel samples than in the sweeps. However, one species of thrips, *Aelothrips bicolor* Hinds, was not taken in the soil and trash samples. The differences in numbers of mites and Collembola were striking, by far the most being collected by means of the Berlese funnel.

TABLE VI

COMPARISON OF SWEEP NET WITH BERLESE FUNNELS, NORTH ENTOMOLOGY FIELD, PAYNE COUNTY, OKLAHOMA, 1955

| Dates and Meteorological Conditions | Insects and Mites | Number per 125 Sweeps | Number per 17–20[a] Samples |
|---|---|---|---|
| Mar. 29–31 Temp. 68°F. Wind–12 m.p.h. | aphids | 2956 | 801 |
| | thrips | 8 | 59 |
| | Collembola | 0 | 51 |
| | mites | 2 | 67 |
| Apr. 14 Temp. 66°F. Wind–5 m.p.h. | aphids | 6624 | 1601 |
| | thrips | 93 | 171 |
| | Collembola | 0 | 47 |
| | mites | 1 | 97 |
| Totals | aphids | 9580 | 2402 |
| | thrips | 101 | 230 |
| | Collembola | 0 | 98 |
| | mites | 3 | 164 |

[a]Each sample consisted of one-half square foot.

## SUMMARY AND CONCLUSIONS

One stroke with a sweep net on a handle 19 inches long sampled an average of 155 tillers in an alfalfa field where the stand was good. With the exception of the spotted alfalfa aphid and a species of thrips, from 44 to 77 percent of the insects resting on the plants were collected by this method. On the other hand not over 10 percent of these aphids resting on the plants were picked up by the net. Comparative figures also show that a maximum average of not over one percent of the thrips were collected by the net. This was the best method of the five tested to collect the greatest number of species, particularly those which are winged and very active. The sweep net was an excellent tool for locating extremely light infestations. In somewhat heavier ones where the population distribution was very uneven this method proved to be very valuable.

Clipping plant tillers and immediately placing them in a jar for later examination was far more accurate than the net for determining the number of aphids and thrips present on the alfalfa plants. There was a large population of various species of insects and mites living on the surface of an alfalfa field which was best sampled with the pit trap or Berlese funnel methods. Especially abundant in this microhabitat were the ground beetles, Collembola and mites.

The Berlese funnel method was needed to determine larvae and other stages inhabiting the soil, and it also measured arthropod populations quantitatively which the pit trap did not. On the other hand, the pit trap sampled the soil surface of an alfalfa field for a much longer period than was possible by means of the Berlese funnel, and there is evidence that it will disclose the presence of certain very delicate forms which would not be found by the Berlese funnel.

The Henderson sampler was very well suited for taking samples of the spotted alfalfa aphid and usually caught them in the same proportionate numbers as the sweep net. It is concluded that all methods have their special uses and that a combination of several is needed to make a thorough and complete study of the arthropod ecology in an alfalfa field.

## REFERENCES CITED

**Barnes, V. S.** 1953. The ecological distribution of spiders in non-forest maritime communities at Beaufort, North Carolina. Ecol. Monogr. 24: 315–37.

**Berlese, A.** 1905.- Apparecchio per raccogliere presto ed in gran numero piccoli artropodi. Redia 2: 85–9.

**Beall, G.** 1935. Study of arthropods by the method of sweeping. Ecology, 16: 216–25.

**DeLong, D. M.** 1932. Some problems encountered in the estimation of insect populations by the sweeping method. Ann. Ent. Soc. America 25: 13–17.

**Gray, H. E.,** and **A. E. Treloar.** 1933. On the enumeration of insect populations by the method of net collection. Ecology, 14: 356–67.

**Fichter, E.** 1954. An ecological study of invertebrates of grassland and deciduous shrub savanna in eastern Nebraska. Amer. Midland Nat. 51.

**Henderson, C. F.** 1956. A sampling fork for estimating populations of small arthropods. U. S. Dept. Agric. ARS–33–18.

**Romney, V. X.** 1945. The effect of physical factors upon catch of the beet leafhopper by a cylinder and two sweep-net methods. Ecology, 26: 135–47.

**Whittaker, R. H.** 1952. A study of summer foliage insect communities in the Great Smokey Mountains. Ecol. Monogr. 1: 1–44.

# B. Autocide

## Use of Insects for their Own Destruction[1]

E. F. Knipling, *Entomology Research Division, Agric. Res. Serv., U.S.D.A.*

Abstract

In considering ways of controlling or eradicating insect populations, the importance of a fuller understanding of the population density in an area is stressed. Knowledge of the rate of increase of population from one generation to the next is basic to an understanding of the degree of control that is required to hold insects to noneconomic levels. Theoretical calculations are presented to show that a low-level mortality that is constant and superimposed on mortality produced by normal environmental resistance can in time lead to a greatly reduced population. The importance of applying control measures against the total population, rather than against segments of the population is pointed out.

Four ways of using an insect species to destroy its own kind are: (1) The release of insects made sexually sterile by gamma radiation, the only one that has already been employed successfully; (2) the use of chemicals that produce sexual sterility in the natural population of insects. The advantage of the desired type of chemical sterilants over chemicals that kill the insects is shown by theoretical trends in the population of a species subjected to the two types of control; (3) the release of adults infected with pathogens that would destroy larval progeny by transmitting the disease to other adults and contaminating the environment where the insect reproduces and develops; (4) the development and release of insect strains that carry deficient genetic characteristics. Such genetic deficiencies should be of a type that would not prevent rearing of the insect under controlled conditions or seriously interfere with the ability of the released males to compete with normal males in mating. Such genetic lethal factors would have to be expressed in the immature progeny or become lethal in the adult during hibernation.

The success of the sterile-male method of eradicating the screw-worm (*Callitroga hominivorax* (Cqrl.)) from Curacao and the southeastern United States has suggested the possibilities of utilizing sexually sterile insects for controlling or eradicating other species. Also, I have given considerable thought to other ways of employing an insect to destroy its own kind. Four ways of utilizing the autocidal principle of insect control are proposed for exploration in further research:

(1) Release of males that have been made sexually sterile by gamma irradiation or other means.

(2) Development of chemicals or other means of inducing sexual sterility in the natural population of an insect.

(3) Development and release of strains of an insect that possesses inferior or lethal genetic characteristics.

(4) Release of insects to spread pathogenic organisms.

Considerable research wil be necessary to determine if these approaches will be effective and practical for controlling various insects. From a theoretical standpoint, all the methods could be highly effective, and I think they are worthy of exploration. I might add that I have enough confidence in research by competent determined scientists to believe that seemingly impossible obstacles can be overcome if the basic principles are sound.

The basic principle of releasing sexually sterile males to control insects and other organisms has been proved sound. Several publications relate to this subject (Bushland & Hopkins 1951, 1953, Baumhover *et al.* 1955, Knipling 1955, Lindquist 1955, Knipling 1959). It therefore seems reasonable to assume that other means of using insects to control their own kind could also be developed. One point in favor of these methods is the almost certain premise that release of males carrying some lethal factor will be the most efficient procedure that can be devised for locating and destroying females. It therefore seems worthwhile to consider conducting research on any potentially feasible way of utilizing an insect to destroy its own kind.

To develop this concept it will be necessary to give much more attention to nutrition and culture methods for insects, dispersal and mating habits, population dynamics, and factors influencing the rate of development. A basic consideration of the population dynamics of an insect species is so important in applying the autocidal approach to insect control that I should like to present some of my views on this matter.

[1] Presented at the meeting of the Eastern Branch, Entomological Society of America on October 30, 1959, at Atlantic City, New Jersey, under the title "Novel Approaches to Insect Control" Accepted for publication November 30, 1959.

**Table 1.—Estimated boll weevil population and square infestation trends starting with winter-surviving population of 20 per acre under low, moderate, and high biotic increase rates per generation.**

| Fruiting Period[a] | Number of Squares/Acre | Low Increase (2.5-Fold) | | Moderate Increase (5-Fold) | | High Increase (7.5-Fold) | |
|---|---|---|---|---|---|---|---|
| | | Number of Weevils/Acre | % Squares Infested | Number of Weevils/Acre | % Squares Infested | Number of Weevils/Acre | % Squares Infested |
| First | 100,000 | 20 | 2.5 | 20 | 2.5 | 20 | 2.5 |
| Second | 200,000 | 50 | 3.7 | 100 | 6.3 | 150 | 9.5 |
| Third | 100,000 | 125 | 15.7 | 500 | 63.0 | 1,125[b] | 100.0 |

[a] A period of 3 weeks is considered a fruiting period, which is about the lengthof time required for one generation of the insect.
[b] This number could furnish enough migrants to infest all the squares on about another half acre.

Population Dynamics in Relation to Insect Control.—Before we can utilize insects to destroy themselves, we must have information on the total number in the environment at the different levels in their abundance cycle. We need to know the norm and extremes of the rate of increase from one generation to the next and also the effect on the total population when various levels of control are applied over a period of several generations.

It is my opinion that the overall population of species can be drastically reduced in time if a new hazard to its survival is applied uniformly and constantly even at a relatively low level. The level of control that will cause a significant decline can of course be expected to vary with the species and with ecological conditions.

I have done considerable theorizing on the probable rates of increase of different insect species from one generation to the next under normal conditions, and also on the rate of increase when new hazards to survival are introduced over and above usual environmental hazards. I have given considerable attention to the boll weevil (*Anthonomus grandis* Boh.) in such studies, the population dynamics of which I believe would be fairly representative of those of a considerable number of our major pests. Table 1 gives the estimated rate of increase of this insect starting from a low level during seasons that would be unfavorable, average, and favorable for the boll wevil. On the basis of available information on the feeding and oviposition habits of this insect and the fruiting behavior of the cotton plant producing moderate yields of cotton, the percentage of cotton squares that would be punctured at the different population levels has also been estimated. In these calculations I have assumed that each boll weevil will puncture six squares per day. Cotton yielding about $1\frac{1}{4}$ bales per acre could be expected to produce about 400,000 squares per acre in one season.

If we accept the estimate of a fivefold increase rate per generation as representative for many insect species, we can speculate on the effect on the total population over a period of several generations if we introduce new hazards that destroy various percentages of the insect above those resulting from the normal hazards. Assuming that the hazards during the winter are reasonably constant we can project the effects of such hazards into the subsequent season. As examples of new hazards we might assume that the rate of increase is reduced each generation by introducing a new parasite, by developing and planting a resistant variety, by releasing sterile males, or by any other procedure that reduces survival at a uniform and constant level.

In order to appraise the long-range effect that is possible by introducing such new hazards, I have made some theoretical calculations that are presented in table 2. These calculations are based on my estimates of the population dynamics of the boll weevil but are presented in efforts to establish principles applicable to insects in general.

I wish to emphasize that the figures are theoretical. It is recognized that the number of insects in a given area will vary from year to year and the rate of increase will vary from generation to generation. My purpose in presenting the figures is to suggest what might be achieved under average conditions if a new low-level control measure were applied and superimposed on other control factors for an extended period of time. Although theoretical, these figures may be fairly realistic. We know that an upset in the balance between destructive and beneficial insects through the use of insecticides can cause outbreaks of the host under circumstances when rates of parasitism are normally not above 50%. Conversely, newly introduced insects often develop high populations for several years but subside to a much lower population level as new parasites or predators become established, even though the rate of parasitism may not exceed 25% to 50%.

In the example given in table 2, several assumptions are made: (1) that the insect has attained stability in its environment and is subjected to hazards in development that are reasonably constant from year to year; (2) that the normal increase rate averages fivefold from one generation to the next; and (3) that 8% of the hibernating generation survives the winter and enters the field; and (4) that the growers in the area use insecticides that destroy 90% of the weevils that survive hibernation.

**Table 2.—Theoretical effect on the total population of boll weevils when subjected to hazards that raise the mortality by 50% above normal on the assumption of 8% survival of hibernating insects and 90% destruction of hibernating survivors by insecticides. Number of insects in 50,000 acres of cotton.**

| Generation | Under Normal Conditions | Following Introduction of New Hazards That Produce 50% Kill Above Normal | | |
|---|---|---|---|---|
| | | First Year | Second Year | Third Year |
| Survivors | 1,000,000 | 1,000,000 | 125,000 | 15,000 |
| First | 5,000,000 | 2,500,000 | 312,000 | 38,000 |
| Second | 25,000,000 | 6,250,000 | 780,000 | 95,000 |
| Third | 125,000,000 | 15,625,000 | 1,950,000 | 237,000 |

Although hypothetical, I believe that these assumptions and the insect population trends for an average community are not too unrealistic for the boll weevils on the basis of studies that I have made utilizing information from published reports of various investigators and information obtained through inquiries from a number of authorities. I have concluded that 200 overwintered boll weevils per acre of cotton is a fairly normal level in the spring, also that 90% mortality of overwintered boll weevils can be expected in a community practicing early-season control. Thus, 20 insects per acre is suggested as the average population level in the spring in a cotton-growing community practicing early-season boll weevil control. An assumed population level of about 20 per acre would also be fairly realistic in northern fringe survival areas, where mortality due to cold weather is generally high.

The theoretical population trends under conditions of assumed normal hazards and cotton-growing practices, as compared with the trends when the rate of increase is further reduced by 50%, are very interesting and may be highly significant in their implications.

It is recognized that the population trend will not continue downward indefinitely, nor is it likely even to approach the level indicated for the second or third seasons. Various population density factors will no doubt come into play and permit a higher percentage survival as the result of the overall decline. However, I do believe that under the circumstances cited, the average population would become stabilized at a level considerably below the former level because of the introduction of a new hazard that is constant in its effect on the total population. A new constant hazard that destroys about 50% of the insects in excess of those destroyed by the normal environmental resistance factors could in a few years have such a drastic effect on the population that the insect would become of minor economic significance. I would like to urge research on various insect species to determine what effect would be produced on total populations in isolated areas when subjected to such low levels of control. It would be desirable, of course, to achieve higher levels of control if possible. In this connection it is interesting to note the relative population trends of an insect subjected to high levels of control on the entire population as compared with the population trend when the level of control is near perfect on all but a small percentage of the host plants in a community.

Table 3 shows the assumed normal trend of an insect population starting from a low level in the absence of control; 100% control on 90% of the acreage each generation; and 90% control each generation on 100% of the acreage in a community. A casual consideration of the relative effect on population trends produced by the two types of control programs would suggest that the results would be essentially the same. However, on the assumption that population increases are largely confined to the original infested fields until the insects are forced to disperse because of lack of food or to seek hibernation quarters, the total population in the community obtaining moderately high control on all acreage could be reduced a hundred times as much as in the community where 10% of the acreage received no control but 100% control is achieved on the remaining 90% of the acreage.

**Table 3.—Theoretical population trend of an insect species in a community subjected to: (1) no control, (2) treatments that destroy all insects on 90% of acreage each generation with no control on remaining 10%, (3) community-wide treatments that destroy 90% of the insects each generation on all the acreage.**

| | Number of Insects in 50,000 Acres of Crop | | |
|---|---|---|---|
| Generation | Condition 1 (no control) | Condition 2 | Condition 3 |
| Winter survivors | 1,000,000 | 1,000,000 | 1,000,000 |
| $F_1$ | 5,000,000 | 500,000 | 500,000 |
| $F_2$ | 25,000,000 | 2,500,000 | 250,000 |
| $F_3$ | 125,000,000 | 12,500,000 | 125,000 |

Although several methods of insect control such as chemicals, resistant varieties, or cultural control procedures might be used to obtain the desired results, I believe that insects themselves might provide the most practical and efficient way to obtain maximum results if ways to achieve autocidal control can be developed.

The four ways of utilizing the autocidal principle of insect control suggested herein follow:

(1) Release of Sterile Males for Insect Control.—The production and release of insects made sterile by gamma radiation might be developed to control numerous other important species in addition to the screw-worm. The method has shown sufficient promise, based on laboratory tests, to warrant pilot eradication experiments against certain tropical fruit flies. The fruit fly laboratory at Honolulu, Hawaii, is making plans for eradication experiments against the oriental fruit fly (*Dacus dorsalis* Hendel) and melon fruit flies on the island of Rota in the Pacific in cooperation with the U. S. Navy and Trust Territory. The use of sterile males of the Mexican fruit fly (*Anastrepha ludens* (Loew)) to prevent establishment of invading insects from Mexico also offers great promise. In this connection it should be pointed out that fruit flies have multiple mating habits.

Excessive radiation damage which reduces the ability of the males to compete with normal males may prove to be the most significant obstacle to the practical application of this method of insect control. If radiation or other suitable ways of inducing sexual sterility can be developed, and if it is feasible to rear large numbers of a particular insect at a reasonable cost, I believe there will be many situations where the production and release of sterile males could provide a practical method of control. Workers have a tendency to be discouraged from considering this approach because of the large number of insects assumed to be present in the natural population. It is my opinion that the rearing of adequate numbers of insects for sterilization and release will not be the most formidable obstacle. If we had effective methods of producing sterility without serious damage to the males, we could for many of our major pests first reduce the natural population to a very low level through the use of chemicals or cultural practices and then release enough sexually vigorous sterile males to reduce the surviving population to the vanishing point. Once the natural population is eliminated, it might then be possible to keep the insect from becoming re-established through continuous releases of a relatively few sterile insects, even in areas subjected to repeated reinfestation by natural dispersion or by accidental reintroductions. The cost of maintaining control by organized and continuous release programs throughout the favorable season might be only a fraction of the initial eradication cost and could be well below the annual cost of control by current methods.

In line with the earlier discussion on population trends of an insect subjected to low-level control procedures, the use of sterile insects might be the most effective way to achieve such control. A chemical control procedure that destroys 50% of the insect in each generation above that achieved by usual insect control practices, or the planting of a resistant variety that would result in a similar reduction might have cumulative effects that would be highly significant. This supposition is indicated in table 2. However, the maximum effects would be the same each generation and each year. The release of sterile males in sufficient numbers to reduce the biotic increase rate by 50% each generation during the first year, if repeated the second year using the same number of insects, would be much more effective. As the natural population declines, the release of sterile males at a constant level would provide such a high ratio of sterile to fertile insects that eradication might be expected. The release of just enough sterile insects to effect 50% reduction in the normal increase rate each generation during the first season and then the continuance of such rate of release each season might be a more practical and economical way to achieve insect eradication than to attempt to release enough sterile insects to dominate the population by 90% or more during the first generation.

Research is urgently needed to develop means of causing sterility in insects without affecting their sexual vigor. Exposure to gamma rays is one way, but much research needs to be done on various stages of insects and under different conditions to determine the best procedure to follow. Other methods of inducing sterility need to be investigated; perhaps chemical sterilants for use under controlled conditions might be developed. Mitlin *et al.* (1957) and Mitlin & Baroody (1958) have shown that a number of chemicals will prevent ovarian development in the house fly (*Musca domestica* L). The production of sterile hybrids as a result of crossing two related species or races might be possible for some insects without changing their mating behavior or ability to find the females of the natural population. The possibility of this approach for tsetse fly (*Glossina* spp.) control was suggested by Vanderplank (1947) and for the gypsy moth (*Porthetria dispar* (L.)) by Downes (1959).

(2) Potential Value of a Chemical Causing Sexual Sterility in Natural Populations of Insects.—Preventive programs with chemicals now depend on immediate kill of enough of the insects to prevent increase of the population to damaging levels. Let us consider the theoretical possibilities of a chemical that would cause sexual sterility in all exposed individuals without otherwise affecting their mating vigor. I do not know whether it is possible to find such a chemical, but if such could be developed, we would have available a new approach to insect control that could be truly revolutionary.

The theoretical possibilities of this approach to insect control are shown in table 4. The same general assumptions discussed on normal population trends have been made. If the normal rate of increase from the winter survivors is fivefold per generation, the population reduction by a chemical that destroys 90% of the insects each generation by killing is much more gradual than that caused by a chemical that would produce sexual sterility. With such a chemical it would probably be easier to eradicate certain insects than to obtain economic control with current insecticides that depend on immediate kill. The reason for the greater effect of the sterilizing chemical is readily apparent. Insects that are made sexually sterile cannot contribute to further increases in the population. In turn, however, if 90% are sterile, they can theoretically reduce the biotic potential of the 10% survivors by 90%.

**Table 4.—Theoretical population trends of an insect species subjected to no treatment and to a treatment that affects 90% of the population by killing or sexual sterilization.**

| Generation | No Treatment | Treatment that Kills | Treatment that Induces Sterility |
|---|---|---|---|
| Parent | 1,000,000 | 1,000,000 | 1,000,000 |
| $F_1$ | 5,000,000 | 500,000 | 50,000 |
| $F_2$ | 25,000,000 | 250,000 | 2,500 |
| $F_3$ | 125,000,000 | 125,000 | 125 |

In my opinion the potential advantages of sexual sterilants over killing agents for use in insect control are so great that intensive research to find such chemicals is warranted. Scientists are making progress in developing chemicals that cause sexual sterility in higher animals. I feel confident that chemicals having such action could also be found for insect control and eradication, if intensive basic and applied programs of research were initiated on a variety of major insect pests.

(3) Development and Release of Strains of Insects that Possess Genetic Lethal Factors.—Through variations in the genetic makeup of insects nature permits them to become better adapted to survival in a changing environment. Environmental resistance factors through the ages have improved the adaptability of species by eliminating the least adaptable individuals in a population. Man has taken advantage of genetic variations in organisms and through selection has produced greatly improved plant and animal species with characteristics that increase their chances of surviving natural hazards.

To my knowledge man has not set out to produce organisms that are inferior. What are the opportunities for deliberately producing strains of destructive insects that are inferior in one or a number of ways and thus are more susceptible to normal environmental resistant factors? Personally, I believe this approach is possible. It might be far easier, through selection, to produce inferior organisms than it is to produce superior organisms.

If the development of inferior insect strains is feasible, how can we utilize them to control insects? I again propose their production and sustained release in sufficient numbers during favorable periods to dominate the natural population and in this way transfer the undesirable traits into the natural population in sufficient frequency to lead to its destruction owing to environmental resistance factors which eventually take effect.

It is the general concept of biologists that nature will quickly reduce the frequency or eliminate undesirable genes from an insect population and naturally occurring or even introduced undesirable mutant genes will have little effect on the population as a whole. However, I believe the concept of dominating the natural population through initial and sustained releases of large numbers of inferior organisms has not been considered. It seems entirely feasible through selection of undesirable genetic traits to develop strains of an insect that can be reared under controlled conditions and in the adult stage will survive long enough when released in the natural environment to compete effectively with the native population in mating and thus transmit the undesirable traits. If so, then the problem is to produce and release enough insects to dominate the natural population to the degree required. Undoubtedly there are many types of genetic defects in an insect species that could be considered.

The genetic deficiencies should not interfere to any great degree with the production and survival of the insect under controlled laboratory conditions or reduce survival or ability of the released males to compete with the males in the natural population. However, the genetically deficient genes must be of such nature that they will prevent survival of the progeny in nature during some period in the seasonal cycle. The following types of genetic deficiencies might meet these requirements:

*Inability to Diapause.*—If a strain of a normally diapausing insect could be developed that could not diapause, such a trait would be of no disadvantage in rearing under controlled conditions; in fact, it could be a distinct advantage. Moreover, such a trait would not limit survival of the insect when released during the regular period of reproduction in the growing season. However, all progeny carrying such deficiency would fail to survive the winter because they could not hibernate. Biologists have already shown that the diapause characteristic in certain insects is genetically controlled, in part at least. Thus, this deficiency would seem to be a desirable lethal character.

*Inability to Fly.*—Certain insects after reaching their host plants are capable of feeding and reproducing during the growing season without flying. However, ability to fly would be vital to survival both before and after hibernation. The boll weevil is such an insect. It might be possible to develop strains of the boll weevil that would be incapable of flying. Perhaps several genes, independent in action, could be found and bred into strains that could not fly. Inability to fly would not interfere with production of the insect under controlled conditions. When adequately dispersed, such deficiency would probably not reduce the ability of males released in cotton fields to compete with normal males in mating with females. However, such defect would lead to death of most or all individuals so affected during the fall, winter, or spring because they could not fly to hibernation areas or to cotton fields in the spring.

*Deficiencies in Immature Stages.*—It would seem possible to develop strains of insects with deficiencies in immature stages which would not be serious handicaps to survival in artificial media in the laboratory, but might be lethal to individuals attempting to develop on their normal host. The inability of an insect to glue its eggs on a plant or the absence of vital spines on larvae and deformed mouth parts in larvae are examples of deficiencies that might interfere with development of the insect in nature but would not seriously affect their survival in the laboratory. Adults of such strains might survive in their natural environment and be fully competitive in mating with the natural population.

If a dominant or several independent recessive characteristics of the type suggested could be found and fixed in an insect strain, such traits might be more useful for control or eradication than the induced damage to genetic material caused by exposure to gamma irradiation. As previously stated, gamma irradiation has a serious effect on the longevity and mating vigor of certain insects. Among multiple mating species, sperms from irradiated males apparently do not compete fully with sperms from normal males. On the basis of current knowledge, radiation damage might be the most important obstacle to the use of this sterility approach. One dominant or three or more recessive lethal characteristics independent in action might be equally as effective as induced sterility among monogamous insects and be much more effective among polygamous species.

(4) Release of Insects to Spread Pathogenic Organisms.—Insect pathology is a growing phase of entomology, and the use of pathogens to control destructive insects is receiving more and more study. In view of their importance as natural control factors for numerous major pests, there is every reason to believe that through research and experience diseases of insects will prove useful for the control of many species. How to apply insect pathogens to produce the desired effect is one of the major problems. I wish to propose for consideration that research be undertaken to determine if insects can be used to disseminate diseases among their own kind.

Several factors seem to favor this approach:

(1) Certain pathogens are nonvirulent when present in or on the adults but are highly virulent to the larvae.

(2) For some disease organisms the amount or number necessary to cause infection in the early larval stages is extremely small.

(3) The adult of any given species in seeking a mate or oviposition site will naturally find the environment where the disease is most likely to have the best opportunity to spread.

(4) If spread of an organism to adults in the natural population can be achieved through mating by releasing infected adults, it seems safe to assume that the insects released will be capable of finding individuals of their own kind more effectively than can be done by any other method.

At my suggestion, C. G. Thompson, of the Insect Pathology Laboratory of the Entomology Research Division, has undertaken investigations that have already shown results that justify further study.

There seem to be two ways in which adult insects might be used to spread an organism. Certain viruses are much more pathogenic in the early larval instars than in the late instars. According to Dr. Thompson, there are indications that larvae infected late in the larval stage will transmit the disease in lethal amounts through the adult to newly hatched larvae of the next generation. The other possibility is to contaminate the adults directly with virus material so that the eggs or plants on which eggs are deposited will become sufficiently contaminated to infect newly hatched larvae.

Both methods seem to warrant study. If it could be shown that adults infected systemically or by external contamination would cause 100% infection early in the larval stage, the released insects themselves would not damage the crop or cause an increase in the overall population. If in turn, however, the contaminated insects that are released caused a high incidence of infection among the progeny of the natural population, the overall population would be reduced. There are several ways this might occur: (1) by spreading the disease to adult females through mating with infected and released males, thus exposing the progeny to infection; (2) by contaminating plants where the released insects rest, oviposit, or feed which in turn could lead to infection of larvae from normal uninfected adults in the natural population; and (3) by increasing the amount of virus in the environment if larval progeny of released insects become infected.

As with other ways to utilize insects to destroy their own kind, the method may be practical only when the natural population is at a low level so that the number needed to dominate it will not be too great. For most species this might be a requirement in order for it to be economically feasible to rear and release the required number of insects. In some instances, however, the release of a relatively few infected adults might be the most effective and practical way to introduce a new disease into a high natural population of an insect or cause an epizootic of a common disease before it would normally occur.

The spread of insect pathogens by deliberately releasing infected adults seems to offer unusual possibilities, especially if it can be shown that the released insects could spread the disease to adults in the natural population through mating. If mating of an infected male to an uninfected female would assure fatal disease among all the larval progeny, the final effect on the total population could be greater than the release of gamma-irradiated sterile males even among insects that have single mating habits. The matings would not only destroy the reproductive potential of females actually mated to released infected males, but the diseased larval progeny and the infected adults might cause further spread of the disease by contaminating the plants.

The greatest advantage over the sterile-male method could be among multiple-mating species. For example, among these species, such as tropical fruit flies and mosquitoes, it is apparently necessary to release up to three sterile males to compete with one fertile unirradiated male. On the same basis, a disease-contaminated male could theoretically destroy the reproductive capacity of several females if the organism is transmitted to all females with which it mates. In applying the male-sterility method among polygamous species, it might be necessary to release 900 gamma-irradiated sterile insects to reduce the biotic potential of a natural population of 100 insects per acre by 75%. To achieve the same reduction with disease-contaminated insects it would be necessary to release only about 55 insects if mating and infection of mates take place an average of three times.

Extensive investigations with different insects and pathogenic organisms under various ecological situations might be required to demonstrate practical control by this method. However, on theoretical grounds, this approach offers unusual opportunities. As with other techniques involving the release of insects, ways to rear insects in large numbers and effective dispersal procedures are prerequisites to success of the method. It is also likely that success would depend on application of the method to the entire natural populations in large areas.

## References Cited


**Baumhover, A. H.,** A. J. Graham, D. E. Hopkins, F. H. Dudley, W. D. New, and R. C. Bushland. 1955. Control of screw-worms through release of sterilized flies. Jour. Econ. Ent. **48**(4): 462–6.

**Bushland, R. C.,** and D. E. Hopkins. 1951. Experiments with screw-worm flies sterilized by X-rays. Jour. Econ. Ent. **44**(5): 725–31.

**Bushland, R. C.,** and D. E. Hopkins. 1953. Sterilization of screw-worm flies with X-rays and gamma rays. Jour. Econ. Ent. **46**(4): 467–9.

**Downes, J. A.** 1959. The gypsy moth and some possibilities of the control of insects by genetic means. Canadian Ent. **91**(10): 661–4.

**Knipling, E. F.** 1955. Possibilities of insect control or eradication through the use of sexually sterile males. Jour. Econ. Ent. **48**(4): 459–62.

**Knipling, E. F.** 1959. Sterile-male method of population control. Science **130**(3380): 902–4.

**Lindquist, A. W.** 1955. Possibilities in the use of gamma radiation for control or eradication of the screw-worm. Jour. Econ. Ent. **48**(4): 467–9.

**Mitlin, Norman,** B. A. Butt, and Thomas J. Shortino. 1957. Effect of mitotic poisons on house fly oviposition. Physiol. Zool. **30**(2): 133–6.

**Mitlin, Norman,** and Anne M. Baroody. 1958. The effect of some biologically active compounds on growth of house fly ovaries. Jour. Econ. Ent. **51**(3): 384–5.

**Vanderplank, F. L.** 1947. Experiments in the hybridization of tsetse flies (*Glossina*, Diptera) and the possibility of a new method of control. Trans. Roy. Ent. Soc. London **98**: 1–18.

# Screw-Worm Control Through Release of Sterilized Flies[1]

A. H. Baumhover, A. J. Graham, B. A. Bitter[2] D. E. Hopkins, W. D. New, F. H. Dudley, and R. C. Bushland
*Entomology Research Branch, Agr. Res. Serv., U.S.D.A.*

Screw-worms, *Callitroga hominivorax* (Cqrl.), did not exist in the southeastern United States until about 20 years ago, and it is probable that, if the present infestation could be eradicated, the area might be kept free of infestation through inspection of livestock shipments originating in infested areas.

E. F. Knipling has suggested the possibility that screw-worms might be eradicated from the Southeastern States through the release of sterilized males. In average winters screw-worms survive only in peninsular Florida, and during the summer months infest areas to the north. Flies do not normally migrate from Texas to the Southeastern States.

Laboratory experiments (Bushland & Hopkins 1951, 1953) at Kerrville, Tex., showed that screw-worm flies could be sterilized by irradiation in the pupal stage with x-rays or gamma rays. Under cage conditions it was found that male screw-worm flies mated repeatedly but that females mated only once. If a female mated with a sterilized male she laid only infertile eggs. When mixed populations of sterilized and normal insects were caged together, the ratio of fertile to infertile egg masses deposited by the normal females was about the same as the ratio of normal to sterile males in the cage.

These laboratory experiments were followed by field trials on Sanibel Island off the west coast of Florida. The results supported the laboratory findings. When sterilized males were released at the rate of 100 per square mile per week, the fertility of egg masses laid by the normal flies declined from the 100 per cent recorded prior to release to less than 25 per cent. After the eighth week of release the original fly population on the island appeared to have been eradicated. However, in the twelfth week a single fertile egg mass was collected.

Since Sanibel Island is only about 2 miles off the coast and is part of a chain of islands, it is within the flight range of insects from untreated areas. Therefore, efforts were made to find a properly isolated screw-worm population for a test to determine whether such a population could be eradicated through release of sterilized flies.

Officials of the Curacao Veterinary Service had sent specimens for identification by the Entomology Research Branch and requested advice concerning the screw-worm problem on the island. Since correspondence indicated that screw-worms were abundant there, the senior author made a trip to Curacao to conduct a survey. As a result of that visit the island administration and the Entomology Research Branch agreed to cooperate in this field experiment.

Test Area.—Curacao is an island of approximately 170 square miles about 40 miles off the coast of Venezuela. A constant trade wind blows out of the east. It was believed that the distance over water and the resistance of the wind were sufficient to prevent screw-worms from flying to Curacao from the mainland, and that if proper precautions were taken to prevent accidental importation of flies on cattle boats a valid experiment could be conducted.

Screw-worms were found to be extremely abundant on the island. They probably breed chiefly in wounds on neglected domestic animals. The island is estimated to have about 25,000 goats and 5,000 sheep, which roam through the brush and cactus largely unattended. Other potential hosts are a few hundred roving pigs, some neglected burros, about 300 deer, and numerous rabbits similar to our cottontails. There are also dairy cattle and horses on the island, but those animals, being fairly well cared for, are not considered important in breeding screw-worms.

The island is inhabited by about 115,000 people, most of whom work in town or for the oil refinery. Many of these workers live in the country and own small flocks of goats, which roam about in unfenced pastures. As there are no permanent streams on the island, during dry weather the goats must come to the houses for water except on those plantations having shallow wells in the pastures. Probably half the goats are seen daily when they come near the houses for water, but they do not receive much attention.

As the island is well served by roads, it was convenient

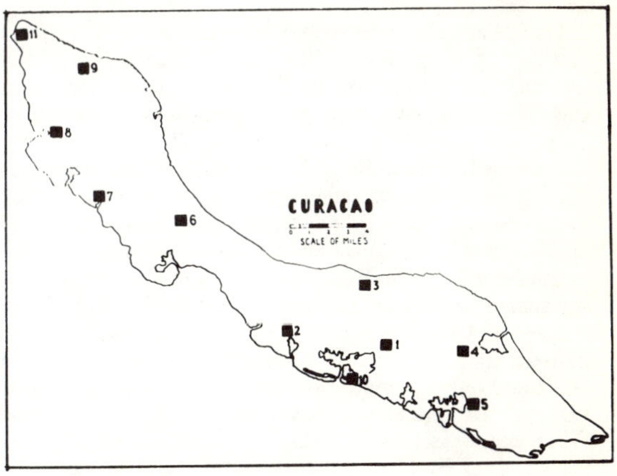


Fig. 1.—Map of Curacao showing location of goat pens.

to travel by automobile in conducting the field tests. There is an excellent airport. An airplane belonging to a flying club could be rented, and skilled pilots were available for distributing sterilized flies by air.

[1] This experiment was planned in conference with E. F. Knipling and A. W. Lindquist. The authors are further indebted to them for suggestions and guidance during the conduct of the work. Accepted for publication April 5, 1955.
[2] Veterinary Service, Government of Netherlands Antilles.

**Table 1.—Numbers of fertile (F) and sterile (S) screw-worm egg masses collected in goat pens on Curacao from March 17 through May 18, 1954.**

| Date | No. 1 | | No. 2 | | No. 3 | | No. 4 | | No. 5 | | No. 6 | | No. 7 | | No. 8 | | No. 9 | | No. 10 | | Totals | | Per Cent Sterile |
|---|---|---|---|---|---|---|---|---|---|---|---|---|---|---|---|---|---|---|---|---|---|---|---|
| | F | S | F | S | F | S | F | S | F | S | F | S | F | S | F | S | F | S | F | S | F | S | |
| Oviposition before release of sterilized males | | | | | | | | | | | | | | | | | | | | | | | |
| March 17–23 | 12 | 0 | 37 | 0 | 8 | 0 | 6 | 0 | 13 | 0 | 11 | 0 | 8 | 0 | 14 | 0 | 15 | 0 | 9 | 0 | 133 | 0 | 0 |
| 24–30 | 9 | 0 | 32 | 0 | 8 | 0 | 8 | 0 | 18 | 0 | 20 | 0 | 15 | 0 | 21 | 0 | 16 | 0 | 8 | 0 | 155 | 0 | 0 |
| Transition period | | | | | | | | | | | | | | | | | | | | | | | |
| March 31–April 6 | 11 | 0 | 48 | 1 | 7 | 1 | 2 | 0 | 13 | 2 | 20 | 0 | 7 | 0 | 19 | 0 | 6 | 0 | 1 | 0 | 134 | 4 | 3.0 |
| Oviposition following release of sterilized males | | | | | | | | | | | | | | | | | | | | | | | |
| April 7–13 | 5 | 2 | 26 | 2 | 9 | 4 | 12 | 4 | 12 | 1 | 14 | 3 | 4 | 0 | 10 | 0 | 4 | 0 | 8 | 1 | 104 | 17 | 14.0 |
| 14–20 | 18 | 6 | 24 | 8 | 30 | 8 | 11 | 2 | 9 | 4 | 25 | 4 | 14 | 4 | 16 | 5 | 22 | 4 | 14 | 1 | 183 | 46 | 20.1 |
| 21–27 | 22 | 2 | 8 | 1 | 35 | 5 | 6 | 0 | 6 | 1 | 24 | 1 | 21 | 4 | 30 | 7 | 28 | 4 | 5 | 3 | 185 | 28 | 13.1 |
| April 28–May 4 | 27 | 1 | 11 | 2 | 23 | 3 | 24 | 3 | 13 | 1 | 15 | 0 | 4 | 1 | 23 | 6 | 14 | 2 | 8 | 4 | 162 | 23 | 12.4 |
| May 5–11 | 26 | 6 | 20 | 4 | 39 | 7 | 6 | 3 | 11 | 2 | 24 | 0 | 8 | 4 | 26 | 1 | 33 | 3 | 7 | 9 | 200 | 39 | 16.3 |
| 12–18 | 41 | 7 | 29 | 5 | 23 | 8 | 9 | 1 | 11 | 3 | 19 | 1 | 34 | 2 | 40 | 7 | 30 | 3 | 4 | 0 | 240 | 37 | 14.4 |

Estimation of Fly Activity.—In this experiment procedures developed in the Florida work were followed. Experience had shown that fairly reliable estimates of fly activity could be made by exposing wounded goats to oviposition by flies in the natural population.

Goat pens were established in 11 locations distributed over the island, as shown in figure 1. Each pen consisted of a fenced enclosure about 20 feet square, in which eight goats were maintained. Each week two goats in each pen were wounded and artificially infested with approximately 100 newly hatched screw-worm larvae. When the larvae were 3 days old, the wounds were fumigated with benzol to destroy the larvae. The wounds were attractive to ovipositing flies from the time of infestation until several days after the benzol treatment.

At the beginning of the experiment in each goat pen there were always at least two fairly fresh wounds and two healing wounds to attract ovipositing flies. Later in the work infested goats were not kept in the pens to avoid the risk of escape. From that time goats were wounded and infested in a locked enclosure and only brought to the field after larvae had been removed from the wounds on the third day.

The goat pens were visited late in the afternoon each day for the purpose of collecting egg masses. Individual egg masses were removed and placed on moist filter paper and held in a petri dish for observations on hatching. Sometimes eggs were laid at a pen after the observer had made his visit. Early in the work, when fertile egg masses were common, such egg masses hatched and the shells usually fell off before the next inspection. The observer could recognize such missed egg masses by the presence of newly hatched larvae in the wound. Infertile egg masses, which did not hatch, remained intact.

Early in the experiment there were undoubtedly errors in estimating egg masses. If eggs hatched prior to a visit, one could only estimate the number of egg masses on the basis of the number of larvae he found in the wound. When egg masses were thickly deposited around a wound, the observer could not always count them accurately if he did not know whether a given clump of eggs represented one large or two small masses. However, it is believed that those sources of error were minor.

Rearing and Sterilization of Flies.—Insects for release were reared and sterilized at the Orlando, Florida, laboratory. The flies used were of a strain originally started at Kerrville, Texas.

The larvae were reared on a mixture of ground horse meat, blood, and water by a modification of the procedure described by Melvin & Bushland (1941). The pupae were held at a constant temperature of 80° F., and when the pupae were 5 days old they were sterilized with gamma rays, using the cobalt-60 irradiation unit described by Darden *et al.* (1954). During the experiment three different doses of sterilizing radiation were employed—5,000 6,000 and 7,500 r. Although flies sterilized with 5,000 r could not produce normal egg masses, they sometimes laid a very few eggs, which caused confusion in record taking when the normal population was near destruction. When the sterilizing dose was increased to 7,500 r, females incapable of ovipositing more than one egg were produced.

Immediately after irradiation the pupae were packed with a little excelsior in No. 1 kraft paper sacks, 130 insects in each sack, and shipped by air freight to Curacao, where they emerged in the sacks and expanded their wings on the excelsior provided.

Within a few hours after emergence was completed, the flies were distributed by airplane. The pilot flew at a height of about 400 feet in flight lanes 1 mile apart. An entomologist riding in the airplane tore open the sacks and dropped them through a pipe which extended below the fuselage. The bottoms of the sacks were smeared with a paste of sand and flour to weight them so that they would drop without drifting unduly. This method of dispersal seemed quite satisfactory, as flies dropped in intact sacks appeared unharmed by the fall.

In the first phases of the experiment the flies were not fed prior to release, but later it was thought desirable to provide the insects with food and water. This was done by packing the insects in sacks with wire-screen bottoms through which they could be fed.

Flies were distributed twice each week. On alternate flights the flight lanes were shifted so that the flies were distributed in lanes ½ mile apart each week.

The packaging rate of 130 pupae per sack was selected on the expectation that out of that number about 100 flies would emerge. Thus, when 50 males and 50 females were expected in each sack, sacks dropped at 1-mile intervals represented a distribution of 50 males per square mile. The actual percentage emergence was determined on sacks retained in the laboratory, and the exact rate of release was calculated subsequently. Thus the rate of fly distribution was regulated by the number of sacks dropped per mile in flight lanes 1 mile apart.

Effects of Releasing 100 Sterilized Males Per Square Mile per Week.—By March 17 ten goat pens were established over the island, and during the following week 133 fertile egg masses were recorded. Just what adult fly density such a figure represents is not known, but that rate of egg-mass deposition was seldom equaled in any experiment in Florida. Beginning on March 26,

**Table 2.—A comparison of releasing sterilized male screw-worm flies at rates of about 100 and 400 per square mile per week.**

| Date | 100 Males per Week | | 400 Males per Week | |
|---|---|---|---|---|
| | Average Number of Egg Masses per Pen | Per Cent Sterile | Average Number of Egg Masses per Pen | Per Cent Sterile |
| July 19–25 | 10.5 | 30.2 | 16.0 | 37.5 |
| July 26–Aug. 1 | 9.5 | 33.3 | 11.8 | 55.9 |
| Aug. 2–8 | 10.0 | 28.3 | 6.0 | 53.3 |
| Average | 10.0 | 30.6 | 11.3 | 48.9 |

flies sterilized with 7,500 r of gamma rays were distributed. During the second week 155 egg masses, all of them fertile, were collected. The week of March 31 to April 6 represented a transition period, since some of the ovipositing flies might be young females that had mated since release of sterilized flies began, whereas others would be older insects that had already mated. During that third week 138 egg masses were collected, and of these 4 were sterile. Semiweekly releases of sterilized flies at the rate of 100 per square mile per week were continued until May 7.

The records on egg masses are summarized in table 1. The data are remarkably uniform. They definitely show that some of the released males mated with females of the native population, but since egg-mass collections increased rather than declined it appeared that the proportion of sterile matings was not sufficient to cause a decline in the population.

Releasing sterilized males at the rate of approximately 100 (actually 99) per square mile per week along with an equal number of sterile females caused about 15 per cent of the native flies to lay egg masses that failed to hatch. It should not be concluded that those sterile males re presented only 15 per cent of the male population. Undoubtedly many males died soon after distribution before they could find water and food. In observations in the United States normal males lived longer in the laboratory than did sterilized males, and this was probably true in Curacao. Therefore, it should not be assumed that, because only about one-sixth of the Curacao females laid infertile eggs during the period of release at the rate of 100 males per square mile per week, there were about 500 normal males per square mile in the normal population. Probably the normal population was much smaller but merely lived longer and competed more effectively for mates than the released flies.

Effect of Weather on Fly Activity.—From the middle of May until July 12 there were no regular releases of sterilized flies on the island. During the first month some special tests were performed which indicated that males sterilized with 7,500 r mated as efficiently as those treated with 5,000 r. During the next 4 weeks no flies were released. Also during this time the eleventh goat pen was established so that the northwest end of the island could be adequately sampled for egg-mass data. From mid-May to mid-July the egg-mass collections indicated a natural decline in the fly population, as collections dropped from 188 to 70 per week.

The decline in the fly population is attributed to the dry weather. After the first week in May there was almost no rainfall until the middle of June. After June 14 rains were frequent, and for the remainder of the year the weather was very favorable for screw-worms, so that any further decline in the fly population could be attributed to sterile matings. In previous years screw-worms had been observed to increase from July to January.

**Table 3.—Egg mass records during release of sterilized males over island of Curacao.**

| Date | Males Released per Square Mile | Total Egg Masses in 11 Goat Pens | | |
|---|---|---|---|---|
| | | Fertile | Sterile | % Sterile |
| Aug. 9–15 | 491 | 15 | 34 | 69 |
| 16–22 | 224 | 17 | 38 | 69 |
| 23–29 | 175 | 17 | 36 | 68 |
| Aug. 30–Sept. 5 | 381 | 10 | 37 | 79 |
| Sept. 6–12 | 451 | 7 | 42 | 86 |
| 13–19 | 701 | 3 | 23 | 88 |
| 20–26 | 450 | 0 | 10 | 100 |
| Sept. 27–Oct. 3 | 607 | 0 | 12 | 100 |
| Oct. 4–10 | 484 | 0 | 0 | — |
| 11–17 | 727 | 0 | 0 | — |
| 18–24 | 577 | 0 | 0 | — |
| Oct. 25–31 | 400 | 0 | 0 | |
| Nov. 1–7 | 400 | 0 | 1 | 100 |
| 8–14 | 485 | 0 | 1 | 100 |
| 15–21 | 393 | 0 | 0 | — |
| 22–28 | 400 | 0 | 0 | — |
| Nov. 29–Dec. 5 | 421 | 0 | 0 | — |
| Dec. 6–12 | 498 | 0 | 0 | — |
| 13–19 | 313 | 0 | 0 | — |
| 20–26 | 265 | 0 | 0 | — |
| Dec. 27–Jan. 2 | 400 | 0 | 0 | — |
| Jan. 3–6 | 188 | 0 | 0 | — |

Comparison of 100 and 400 Rates of Release.—For 4 weeks beginning July 12 a comparison was made between two rates of releasing males. In the southern half of the island the flies were released twice weekly at the rate of 100 per square mile per week and in the northern half at 400 per square mile per week. When emergence figures were calculated from final laboratory records, the actual releases were found to be 107 and 450 per square mile per week.

From the results of this experiment, shown in table 2, it will be noted that the lower rate was more effective than it was in the spring when the natural population was higher. In the first experiment (Table 1) releasing at the rate of approximately 100 per square mile caused 22 per cent sterility among egg masses collected on the southern half of the island. In the second experiment the sterility from a similar release rate was 31 per cent, and from the 400 release rate it was 49 per cent.

The record of 49 per cent sterility of egg masses collected on the northern half of the island indicated that such a reduction in breeding potential should cause a decline in the fly population.

Eradication Experiment.—Beginning on August 9 sterile flies were released over the entire island at the approximate rate of 400 males plus 400 females per square mile per week. Attaining this release rate required the actual release of 68,000 of each sex over the entire island, or 136,000 flies per week. It was necessary to ship twice that many flies to compensate for losses due to conditions beyond our control. Occasionally the insects emerged en route, either because the shipments were delayed or because the packages were exposed to unusually high temperatures. Rearing, irradiating, and packing the insects at the rate of 130 pupae per release sack strained the capacity of the Orlando laboratory. Some weeks it was impossible to meet the release quota because of rearing or shipping mishaps. In order to attain the desired average it was necessary to release practically all the insects that reached Curacao in good condition. For that reason weekly releases varied greatly. During the first 8 weeks, in which screw-worms were practically eliminated from Curacao, the weekly releases ranged from 175 to 701 males per square mile and the average was 435. The figures for each week's release are shown in table 3.

From October 3 until November 4 no egg masses were collected, but a single infertile egg mass was taken at goat pen No. 6 on November 4 and another at goat pen No. 4 on November 11. After November 11 there was no oviposition on the island detectable by the sampling procedure.

Under weather conditions prevailing on Curacao, about the shortest time for a complete life cycle from egg to egg is 3 weeks, and 4 weeks is about the average. The egg-mass data show that fertile flies were active until mid-September. As there are about 25,000 goats and 5,000 sheep roaming on the island, it would be expected that, if fertile females were ovipositing on observed wounds at the goat pens, many more should be ovipositing on unattended wounds in nature. Flies ovipositing on neglected sheep or goats or on wild mammals should have produced progeny to oviposit a month later in mid-October. The fact that no egg masses were collected after October 3 is a good indication that the native fly population was greatly reduced, but it also suggests that the 11 goat pens under observation did not provide an adequate sample of a declining fly population when the number of ovipositing females was very small.

The females that oviposited on November 4 and 11 probably were the progeny of flies ovipositing early in October. It is possible, but not likely, that the two egg masses were laid by flies 3 or 4 weeks old and hence from parents that oviposited in early September. Since the records were taken at widely separated pens, they probably represented different centers of survival.

Maintenance of the goat pens and systematic releases of sterilized flies were continued through January 6, 1955. Thus, observations were made for 8 weeks after the last record of oviposition. This was ample time for two complete life cycles.

It is quite probable that the two flies that oviposited on November 4 and 11 had brothers and sisters active at the same time. It seems, though, that there were no fertile eggs laid on animals in nature in November, as 6 months have elapsed (May 1955) without further evidence of normal flies.

In addition to taking the egg mass records at goat pens, livestock owners were questioned regarding the occurrence of screw-worm cases in their animals. Very few cases of myiasis were reported after October 1, and every one proved to be due to an infestation of *Callitroga macellaria* (F.) rather than *hominivorax.* This checking with livestock owners is being continued by one of us (B. A. Bitter), but to this writing (May 1955) there has been no evidence of renewed screw-worm activity.

During the course of this experiment there was an obvious increase in the number of surviving kid goats. At the beginning of the study less than half the nannies had kids and twins were seldom seen, but at the conclusion most of the nannies had kids and twins were common. Livestock owners remarked that it had been their practice to seek out new-born kids for treatment with larvicides to prevent almost certain death through navel infestation. As the screw-worm incidence declined the owners abandoned that practice and all remarked at the great increase in the kid crop.

Conclusion.—It appears that screw-worms were eradicated from the island of Curacao through the release of sterilized flies. The success of this experiment strongly supports the theory that it may be practical to eradicate screw-worms from the Southeastern States by this method.

Summary.—To test whether an isolated population of screw-worms, *Callitroga hominivorax* (Cqrl.) might be eradicated through the release of sterilized flies, experiments were made on the island of Curacao.

Screw-worms were reared in the laboratory at Orlando, Florida. When the pupae were 5 days old they were sterilized with gamma rays and shipped by air freight to Curacao. The adults emerged 2 days after irradiation. Within 24 hours of emergence the insects were distributed by airplane over the island.

When sterilized flies were released at the rate of approximately 100 of each sex per square mile per week, some of the males mated with females of the normal population, as indicated by the deposition of infertile egg masses on wounded goats.

Egg masses collected from wounded goats prior to release of sterilized flies hatched 100 per cent.

During April and May, when weather favored fly activity, releasing sterilized males at the rate of about 100 per square mile per week caused 15 per cent of the females to lay infertile eggs on wounded goats in pens. This rate of sterility was not sufficient to cause a reduction in the fly population.

At the end of the dry season in July and August releasing at 107 per square mile per week over half the island caused 30.6 per cent sterility, and 450 per square mile per week on the other half caused 48.9 per cent sterility.

From August 9 to October 3 the entire island was treated at a rate that averaged 435 males per square mile per week. Although frequent rains made conditions suitable for fly increase, the releases were highly effective. In 8 weeks the percentage of sterile egg masses increased from 69 to 100 and the population declined more than 99 per cent. The releases were continued. Except for two sterile egg masses collected after 12 and 13 weeks, there was no evidence of normal fly activity after the eighth week of releases. The releases and observations were discontinued 8 weeks after the last egg mass was collected.

The apparent eradication of screw-worms from Curacao supports the theory that screw-worms may be eradicated from the Southeastern States by releasing sterilized flies.

### References Cited

**Bushland, R. C.,** and D. E. Hopkins. 1951. Experiments with screw-worm flies sterilized by x-rays. Jour. Econ. Ent. **44**: 725–31.

**Bushland, R. C.,** and D. E. Hopkins. 1953. Sterilization of screw-worm flies with x-rays and gamma-rays. Jour. Econ. Ent. **46**: 648–56.

**Darden, E. B.,** E. Maeyens, and R. C. Bushland. 1954. A gamma-ray source for sterilizing insects. Nucleonics **12**(10): 60–2.

**Melvin, Roy,** and Raymond C. Bushland. 1941. The nutritional requirements of screw-worm larvae. Jour. Econ. Ent. **33**: 850–2.

# Radiosterilization vs. Chemosterilization in House Flies and Mosquitoes [1]

C. H. Schmidt, D. A. Dame,[2] and D. E. Weidhaas[3]

Entomology Research Division, Agr. Res. Serv., USDA, Gainesville, Florida

ABSTRACT

Laboratory tests were conducted during 1961 and 1962 to compare the effectiveness of males of the house fly, *Musca domestica* L., and the common malaria mosquito, *Anopheles quadrimaculatus* Say, sterilized by apholate or tepa and by gamma radiation. For radiosterilization, house fly pupae were treated 31–54 hr prior to eclosion with 2850 r and mosquito pupae at least 24 hr old with 10000 or 12000 r from a cobalt-60 source supplying 736 ± 33 r/min. For chemosterilization, house fly male adults 1 day old were fed 1% apholate in the diet for 3 days and mosquito male adults for 3–4 days after emergence. Mosquito male adults 0–24 hr old were also exposed to contact residues containing 7 mg/ft$^2$ of tepa. The most important criterion of effectiveness was the reduction in number of viable egg batches from females confined with both treated and untreated males. Results from chemosterilization equaled or surpassed those from radiosterilization. With mosquitoes, highly competitive, yet permanently sterile, males were more attainable with chemo- than with radiosterilization. With house flies either method seemed adequate, but the slightly greater degree of recovery after irradiation than after chemosterilization must be considered.

Entomologists have at their disposal both chemicals and ionizing radiation for the sexual sterilization of insects. The sterile-male technique of insect control envisioned by Knipling was successfully demonstrated (Knipling 1960) with gamma-irradiated screw-worm flies, *Cochliomyia hominivorax* (Coquerel). Further investigations showed that this species is also susceptible to chemosterilants (Chamberlain 1962, Crystal 1963). The common malaria mosquito, *Anopheles quadrimaculatus* Say, was sterilized by gamma radiation, with a resulting reduction in mating vigor (Davis et al. 1959), and by chemosterilants (Weidhaas et al. 1961). The inability to control this species through the release of males sterilized by gamma radiation (Weidhaas et al. 1962) was shown to be due to behavioral deficiencies in the males of the colonized strain released (Dame et al. 1964). Morlan et al. (1962) tried unsuccessfully to control *Aedes aegypti* (L.) by releasing irradiated males; Weidhaas and Schmidt (1963) reported that although the mating vigor of irradiated males of this species was severely reduced, the mating vigor of chemosterilized males appeared unaffected. The house fly, *Musca domestica* L., extensively used as a test organism for screening insect chemosterilants, was reported to have an apparent increase in mating vigor following chemosterilant treatment (LaBrecque et al. 1962). Although the use of chemosterilants has unique advantages over irradiation for the production of sterile males, it is important to compare the effectiveness of males sterilized by both techniques. The research reported here was conducted to compare the effect of chemosterilants and ionizing radiation on the house fly and the common malaria mosquito in terms of the effectiveness of the resulting sterile males.

Methods.—In these investigations the most important criterion for effectiveness of sterile house fly and mosquito males was the reduction in the numbers of viable egg batches from females caged with both treated and untreated males. Mixed populations of treated and untreated males caged with 50 or 60 virgin females at ratios of 4:1:1, 2:1:1, or 1:1:1, respectively, formed the bases of the competition; e.g., 240 treated ♂ : 60 untreated ♂ : 60 untreated ♀. Treated males were tested for sterility at a 1:1 ratio by caging 60 with 60 untreated ♀. Normal fertility was determined by caging 60 untreated ♂ with 60 untreated ♀. In the mosquito tests a further criterion was used; effectiveness of the treated males was sought by ascertaining the percentages of females inseminated in noncompetitive crosses (60 treated ♂ : 60 ♀). The numbers inseminated were determined by adding the numbers of females ovipositing (since this species seldom oviposits when uninseminated) to the numbers of nonovipositing females found by phase microscopy to contain sperm in their spermathecae.

Sterilization was accomplished with laboratory-reared house flies and mosquitoes either by irradiating the pupae or by treating the adults with chemosterilants. Pupae were exposed to gamma radiation from a cobalt-60 source at 736 ± 33 r/min between 9 AM and 12 noon. Virgin adult males were fed apholate-treated food or exposed to residues of tepa on glass. Tests were conducted at 80° ± 4°F and 70–80% RH, but for handling purposes insects were immobilized in a cold room at 34° ± 2°F.

Dry house fly pupae were exposed at 9 AM to 2850 r, 31–54 hr prior to eclosion. Emerging males were fed untreated food composed of 6 parts sugar, 6 parts powdered milk, and 1 part powdered egg. One-day-old virgin males, not irradiated, were given similar food containing 1% apholate. This diet was prepared by thoroughly mixing 12.5 ml of acetone containing 0.25 g apholate into 25 g of fly food; after drying, the food was pulverized and given to the flies for 3 days and then replaced with untreated food. Untreated males and females were fed untreated food. Water was supplied to all flies by a wick bottle.

All flies, both treated and untreated, were of uniform age and sexed within 8 hr after emergence to insure virginity. Males and females of all groups were kept separately in organdy-covered 10×10×7-in. holding cages for 4 days following emergence. On the 5th day, treated and untreated males were transferred to cylindrical wire mating cages, about 18 in. long and 12 in. in diameter, closed at one end with plywood and at the other with a stockinette sleeve. On the 6th

[1] Received February 10, 1964; accepted for publication May 26, 1964.

[2] Now at Salisbury, Southern Rhodesia, South Africa.

[3] Now at Beltsville, Md.

day, after males had become oriented in the cages, untreated virgin females were introduced. On the 11th day females were removed and placed individually into 3-dr plastic oviposition vials, covered at each end with black cloth. One end of each vial was inserted into 1-day-old uninfested larval rearing medium. The black cloth covering the embedded end had been soaked in fresh milk and contained an overlapping fold in which the flies usually oviposited. On the 13th day the vials were opened and the total number of eggs and the number of hatched eggs recorded for each female.

Irradiated house flies were also observed for longevity and recovery of fertility. Mortality records were maintained on treated and untreated virgin males for 30 days. Recovery of fertility was investigated by caging groups of males with virgin females for 3 days and repeating the process 10 days later with other virgin females. Females were allowed to oviposit on larval rearing medium en masse and the subsequent production of pupae was observed.

Unsexed mosquito pupae at least 24 hr old were exposed from 11 AM to 12 noon to 10000 or 12000 r in about 75 cc of tapwater, replaced immediately after irradiation. Resulting males, as well as untreated males and females, were maintained on 20% honey solution supplied underneath a wire resting screen in a small soufflé cup. Other males were sterilized by feeding for the first 3 to 4 days after emergence on 1% apholate in 20% honey solution, and thereafter fed untreated honey solution. Contact sterilization was effected by exposure to residues of 7 mg/ft$^2$ of tepa on glass. The insides of quart jars and their glass covers were coated with a measured quantity of a 0.1% solution of tepa in methanol; jars were rolled until apparent dryness, and then allowed to dry overnight; 50 to 60 0-to-24-hr-old males were then held in the jar for 2 hr starting at 9 AM. These males were fed untreated 20% honey solution. Tapwater was supplied all mosquitoes in a saturated cotton pad.

All mosquitoes, both treated and untreated, were of uniform age and sexed within 24 hr after emergence to insure virginity. Males and females were kept separately in the cylindrical wire cages. When the treated and untreated males were at least 1 to 2 days old, they were placed in an empty cage about 1 hour prior to the introduction of virgin females. Since in this mosquito mating is a crepuscular occurrence (Stahler and Terzian 1956, Killough and Weidhaas 1963), males and females were caged together in the morning so that sufficient orientation time could elapse before the evening mating period to prevent either the sterile or the fertile male from having a mating advantage. After 3 to 5 days a guinea pig was placed in the cage for 3–4 hr and 2 days later the blood-fed females were individually bottled for egg deposition in 10-dr glass vials containing water. At 3–5 days after oviposition the number of eggs and the number of hatched eggs were recorded for each female.

To determine the effects of chemosterilized sperm on embryogenesis, we observed the state of embryonic development in mosquito eggs of known age from females mated with treated or untreated males. The eggs were cleared with 30% hydrogen peroxide for 10 min and the floats removed with 1% sodium hypochlorite; immediately after rinsing, eggs were placed in Ringer's solution and observed by phase microscopy; the eggs were photographed, discarded, and other eggs utilized for subsequent observations.

RESULTS AND DISCUSSION.—The record of sterility of house fly and mosquito females caged with treated and/or untreated males is given in Table 1. The record of actual and expected degree of sterility induced under competitive conditions is given in Table 2. Only small percentages, 9% and 5%, of the house fly males treated with apholate or gamma radiation, respectively, were not completely sterilized. Maximum hatch among individual egg batches from females mated to irradiated males was 4%, and to chemosterilant treated males 14%. In the competitive tests treated flies performed as well as untreated flies. With chemosterilized flies more sterile egg batches than expected were found, but the overall percentage of sterile eggs did not appreciably exceed the expected value. Recovery of fertility in the irradiated male resulted in an increase from 1 pupa/25 ♀ in the first mating to 2 pupae/♀ in the subsequent matings. Longevity of males did not appear to be materially affected by irradiation (Table 3). LaBrecque (1963) reported that apholate caused permanent male sterility without decreasing longevity.

All treatments, except radiation at 10000 r, produced complete sterility in most male mosquitoes. However, except for the tepa treatment which produced full sterility in all males, there were occasional instances of normal hatch in the noncompetitive crosses. Under competitive conditions only the tepa-treated males and males incompletely sterilized at 10000 r performed as well as untreated males. These findings coincided with those from field studies in which tepa-treated males appeared to be more aggressive than irradiated or apholate-fed males (Dame et al. 1964). The performance of the apholate-treated males and males irradiated at 12000 r was similar—each suffered about 26% reduction in mating vigor. The effectiveness of males irradiated at 12000 r, a marked improvement over that noted in 1959 by Davis et al., was due to delaying irradiation of pupae until they were at least 24 hr old. Little difference was noted between results of treatments causing slight or no reduction in male competitiveness and those that did cause some reduction. Insemination rates of male mosquitoes exposed to tepa residues or irradiated at 10000 r varied little from those of males fed apholate or irradiated at 12000 r. Although comparative data are not available on recovery of fertility, the tepa treatment did produce permanent sterility (Dame and Ford 1964).

Embryogenesis in eggs from female mosquitoes mated to chemosterilized males followed 3 different patterns. Most eggs that failed to hatch were similar to those from untreated parents in that embryonic development was not evident. Many of the remaining eggs were highly disorganized in development; in some, a few well-formed setae were seen amidst an apparently otherwise undifferentiated mass of cellular material. Finally, some of the embryos developed normally at first, although often at a reduced rate, and then became disorganized at various stages of development. The lack of embryonic development and death of the developing embryo appeared to fit the criterion of Type I and Type III dominant lethality, respectively, as defined by von Borstel and Rikemeyer (1959).

The utilization of dominant lethality produced by chemical or physical mutagens for insect control merits continued attention. It is pertinent to note that in each of the 3 comparative studies conducted in our laboratory chemosterilization methods produced results equaling or surpassing those obtained by radiosterilization. For mosquito sterilization it would appear that the desired goal of highly competitive, yet permanently sterile, males is more readily attainable with chemosterilization than radiosterilization. With house flies either method seems adequate, but it is necessary to consider the slightly greater degree of re-

**Table 1.—Fertility of untreated females caged with untreated males, males treated with apholate, tepa, or gamma rays, or mixed populations of treated and untreated males.**

| Treatment | Ratio[a] | Number of tests | Females inseminated (%) | Average number of females ovipositing per test | Egg batches (%) from individual females giving indicated percent hatch: 60–100 | 1–59 | 0 | Egg hatch (%) |
|---|---|---|---|---|---|---|---|---|
| | | | *Anopheles quadrimaculatus* | | | | | |
| Apholate[b] | 1:0:1 | 5 | 64 | 25 | 1 | 3 | 96 | — |
| | 1:1:1 | 3 | — | 35 | 75 | 3 | 22 | — |
| | 2:1:1 | 3 | — | 29 | 70 | 4 | 26 | — |
| | 4:1:1 | 5 | — | 30 | 42 | 1 | 57 | — |
| Tepa[c] | 1:0:1 | 2 | 71 | 15 | 0 | 0 | 100 | — |
| | 1:1:1 | 2 | — | 25 | 51 | 4 | 45 | — |
| 12000 r | 1:0:1 | 2 | 71 | 30 | 4 | 3 | 93 | — |
| | 4:1:1 | 2 | — | 27 | 36 | 9 | 55 | — |
| 10000 r | 1:0:1 | 3 | 76 | 28 | 13 | 20 | 67 | — |
| | 4:1:1 | 3 | — | 24 | 37 | 9 | 54 | — |
| Control | 0:1:1 | 6 | 85 | 28 | 98 | 2 | 0 | — |
| | | | *Musca domestica* | | | | | |
| Apholate[b] | 1:0:1 | 2 | — | 38 | 0 | 9 | 91 | 0.7 |
| | 4:1:1 | 2 | — | 35 | 13 | 5 | 82 | 15.7 |
| 2850 r | 1:0:1 | 2 | — | 39 | 0 | 5 | 95 | .1 |
| | 4:1:1 | 2 | — | 38 | 20 | 6 | 74 | 18.6 |
| Control | 0:1:1 | 2 | — | 38 | 88 | 11 | 1 | 84.3 |

[a] Treated males : untreated males : untreated females.
[b] Offered at 1% in diet.
[c] Contact residue of 7 mg/ft².

**Table 2.—Theoretical, expected,[a] and actual sterility in competitive mating tests.**

| Treatment | Ratio[b] | Sterile egg batches (%): Theoretical | Expected | Actual | Sterile eggs (%): Theoretical | Expected | Actual |
|---|---|---|---|---|---|---|---|
| | | | *Anopheles quadrimaculatus* | | | | |
| Apholate[c] | 1:1:1 | 50 | 48 | 22 | — | — | — |
| | 2:1:1 | 67 | 64 | 26 | — | — | — |
| | 4:1:1 | 80 | 77 | 57 | — | — | — |
| Tepa[d] | 1:1:1 | 50 | 50 | 45 | — | — | — |
| 12000 r | 4:1:1 | 80 | 74 | 55 | — | — | — |
| 10000 r | 4:1:1 | 80 | 54 | 54 | — | — | — |
| | | | *Musca domestica* | | | | |
| Apholate[c] | 4:1:1 | 80 | 73 | 82 | 80 | 83 | 84 |
| 2850 r | 4:1:1 | 80 | 76 | 74 | 80 | 83 | 81 |

[a] Based on corrections for normal sterility in untreated flies and incomplete sterility in treated flies.
[b] Treated males : untreated males : untreated females.
[c] Offered at 1% in diet.
[d] Contact residue of 7 mg/ft².

**Table 3.—Effect of 2850 r of gamma radiation on longevity of male house flies irradiated 24 to 48 hr before eclosion (2 replicates).**

| Treatment | Mean number of flies per test | Mortality (%) within indicated number of days after emergence | | | | |
|---|---|---|---|---|---|---|
| | | 12 | 16 | 23 | 26 | 30 |
| 2850 r | 97 | 16 | 30 | 66 | 87 | 100 |
| Control | 51 | 0 | 15 | 72 | 94 | 100 |

covery following irradiation than that following chemosterilization. Thus, it is important to recognize that each species to be considered must be evaluated in the light of factors applicable, and perhaps unique, to it alone. One also must be aware that the chemosterilants currently available are not selective in their mutagenic activity. Therefore, to reduce the potential hazards involved in their use, field investigations with chemosterilants should be directed toward selective application to insure that the pest species is the only organism which contacts the sterilant.

ACKNOWLEDGMENT.—The authors acknowledge the assistance of H. R. Ford and M. E. Williams of this Division in conducting these experiments.

## REFERENCES CITED

**von Borstel, R. C.,** and **M. L. Rikemeyer.** 1959. Radiation-induced and genetically contrived dominant lethality in *Habrobracon* and *Drosphila.* Genetics 44 (6) : 1053–74.

**Chamberlain, W. F.** 1962. Chemical sterilization of the screw-worm. J. Econ. Entomol. 55 (2) :240–8.

**Crystal, M. M.** 1963. The induction of sexual sterility in the screw-worm fly by antimetabolites and alkylating agents. J. Econ. Entomol. 56: 468–73.

**Dame, D. A.,** and **H. R. Ford.** 1964. Chemosterilization and its permanency in mosquitoes. Nature 201(4920): 733–4.

**Dame, D. A., D. B. Woodard, H. R. Ford,** and **D. E. Weidhaas.** 1964. Field behavior of sexually sterile *Anopheles quadrimaculatus* males. Mosquito News 24 (1) : 6–14.

**Davis, A. N., J. B. Gahan, D. E. Weidhaas,** and **C. N. Smith.** 1959. Exploratory studies on gamma radiation for the sterilization and control of *Anopheles quadrimaculatus.* J. Econ. Entomol. 52: 868–70.

**Killough, R. A.,** and **D. E. Weidhaas.** 1963. The use of light in rearing *Anopheles quadrimaculatus* Say for behavioral studies. Mosquito News 23 (3) : 222–5.

**Knipling, E. F.** 1960. The eradication of the screw-worm fly. Sci. Am. 203 (4) :54–61.

**LaBrecque, G. C.** 1963. Chemosterilants for the control of house flies. Advan. in Chem. Ser. 41: 42–46.

**LaBrecque, G. C., D. W. Meifert,** and **C. N. Smith.** 1962. Mating competitiveness of chemosterilized and normal male house flies. Science 136 (3514) : 388–9.

**Morlan, H. B., E. M. McCray, Jr.,** and **J. W. Kilpatrick.** 1962. Field tests with sexually sterile males for control of *Aedes aegypti.* Mosquito News 22(3): 295–300.

**Stahler, N.,** and **L. A. Terzian.** 1956. Some studies on the influence of light on the mating activity of *Anopheles quadrimaculatus* Say. Ann. Entomol. Soc. Am. 49 (5) : 429–35.

**Weidhaas, D. E.,** and **C. H. Schmidt.** 1963. Mating ability of male mosquitoes, *Aedes aegypti* (L.), sterilized chemically or by gamma radiation. Mosquito News 23 (1) : 32–34.

**Weidhaas, D. E., H. R. Ford, J. B. Gahan,** and **C. N. Smith.** 1961. Preliminary observations on chemosterilization of mosquitoes. Proc. 48th Ann. Mtg. New Jersey Mosquito Exterm. Ass., p. 106–9.

**Weidhaas, D. E., C. H. Schmidt,** and **E. L. Seabrook.** 1962. Field studies on the release of sterile males for the control of *Anopheles quadrimaculatus.* Mosquito News 22 (3) : 283–91.

# Eradication of *Culex pipiens fatigans* through Cytoplasmic Incompatibility

*Culex pipiens fatigans* is the chief vector of filariasis in south-east Asia. Urbanization has often caused the numbers of this mosquito—and with it the danger of filariasis infection—to increase alarmingly[1]. The natural vigour, tolerance and fast development of resistance to insecticides of this mosquito necessitate the development of other control methods, and cytoplasmic incompatibility[2] seems to be an ideal means.

Crossing between members of allopatric populations of the *Culex pipiens* complex can produce four different results. Most populations will produce normal offspring in reciprocal crosses, while some give offspring in one direction and embryos which will not hatch in the opposite direction. Other crosses are infertile in both directions. This lack of offspring is due to cytoplasmic incompatibility[3], which is inherited cytoplasmically. It remains constant for indefinite numbers of generations in the female line. In an incompatible cross the sperm is blocked before it can fuse with the haploid egg nucleus, and if the embryos develop they do so from the haploid egg nucleus and die before hatching (unpublished work of E. Jost).

Twenty different crossing types are known in the *Culex pipiens* complex[4], and it is possible to specify at least one, sometimes several, strains of the complex which are incompatible with a certain population of *Culex pipiens* anywhere in the world. Desirable genic traits can be introduced into an incompatible strain without changing the cytoplasmic incompatibility, and so strains from temperate regions can be adapted to tropical environments.

Release of incompatible males into a natural population should be suitable for the control of *Culex pipiens*. Cage experiments show that females cannot discriminate between normal and incompatible males, and the competitiveness of males depends largely on nutritional conditions during larval development. Cage populations with an initial ratio of incompatible and normal males of 1 : 1 have been eradicated after three or four generations.

The World Health Organization sponsored a pilot experiment at the Filariasis Research Unit at Rangoon.

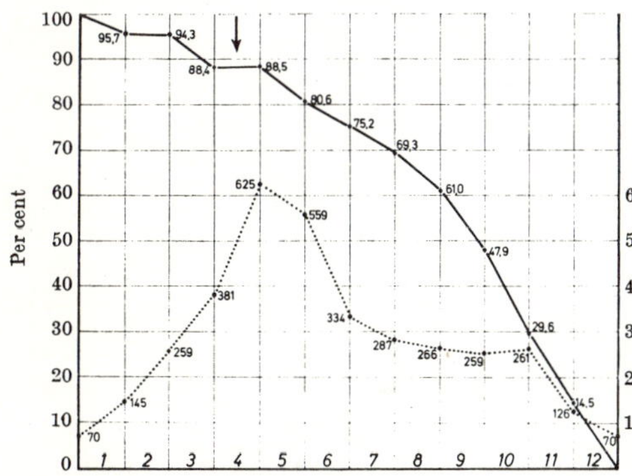


Fig. 1. Eradication of *C. fatigans* in the village of Okpo, Burma, February–May 1967. ——, Daily average percentage of hatching rafts per week 1–12; . . . . , daily average number of egg rafts per week 1–12.

A strain of *Culex pipiens fatigans* with the cytoplasm from a strain from Paris and the genome from a strain from Fresno, California—considered to be better adapted to ecological conditions in Burma than the pure Paris strain—was used. During August to October 1966, males were tested for incompatibility with females from twenty-five natural populations in Rangoon and its surroundings.

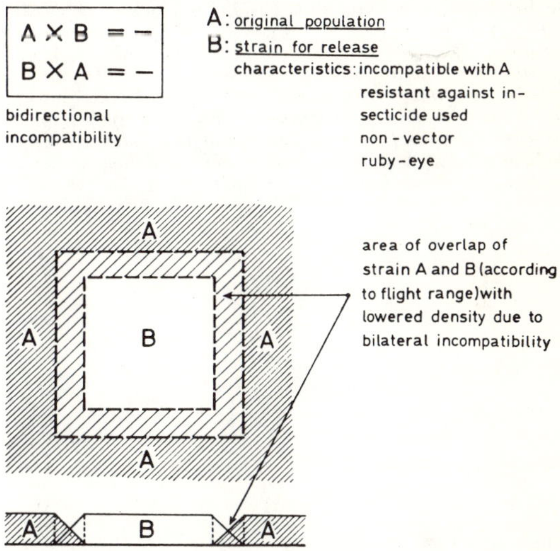


Fig. 2. Self-operating system. Replacement of strain *A* by strain *B* with subsequent equilibrium between *A* and *B*.

These females laid a total of 1,472 rafts with 130,445 eggs. Only 180 larvae hatched (0·14 per cent).

The field work was carried out at Okpo, a village about 10 miles north of Rangoon, surrounded by completely dry rice fields during winter. The size of the mosquito population was estimated to vary from a minimum of 2,000 males and 2,000 females to as many as 10,000 males and 10,000 females on a given day, depending mostly on fluctuations in the number and extent of breeding places.

Incompatible males were released between February and May 1967 (Fig. 1). From March 16 until May 6, 5,000 incompatible males were released daily. Because of the initially small number released the percentage of non-hatching rafts from incompatible crosses remained low (week 1–4, 4·3–11·6 per cent). Soon after the release of the optimum number of 5,000 incompatible males/day (arrow in Fig. 1) the percentage of inviable rafts increased to an average of 19·4 per cent (fifth week, March 21–27) in spite of an almost ten-fold increase in the size of the population during the first 4 weeks (seventy rafts/day during the first week, 625 rafts/day during the fourth week). During the next seven days (March 28 to April 3) a slight increase to 24·8 per cent occurred; during the following week (April 4–10) 30·7 per cent was observed. From then on the percentage of inviable rafts increased more rapidly. It reached 39·0 per cent during April 11–17 (week 8) and 70·4 per cent during April 25 to May 1 (week 10). A new wave of higher production in the breeding areas, beginning in week 10, was obviously cut off by the high percentage of non-hatching rafts. During May 2–8 (week 11) the percentage increased to an average of 85·5 per cent and finally, on May 9 and 10, 100 per cent inviable rafts were obtained. The production of new larvae was therefore stopped and no more adults were expected to emerge after about 10 days. Unfortunately, the monsoons started on May 11 and the experiment had to stop.

The results of this experiment show that the incompatibility principle can operate in nature and that, after the release of an adequate number of incompatible males, a *C. p. fatigans* population was eradicated in about 3 months or five or six generations. If in the future a population in an extended area is completely suppressed, the problem of filling the vacuum created could be serious. With the removal of *C. p. fatigans* with its high density in most breeding places an opening will be available for aquatic insects with similar ecological requirements. Another species of mosquito could come to occupy the niche. It might be harmless or a vector of the same importance. A strain of *C. p. fatigans* unable to transmit filariasis, or even a strain with no desire to bite man, could be developed and liberated to fill the empty niche. This strain should be incompatible with the surrounding population. Fig. 2 shows a model for the eradication and replacement of a natural population (*A*) based on the possibility of creating an equilibrium between a new population (*B*) and the surrounding original population (*A*). The model requires the replacement strain *B* to be incompatible with the original population *A* in two reciprocal crosses. Strain *B* must be a non-vector, and it would be useful if it were also resistant to an insecticide if such is used initially to depress strain *A*. A visible, non-deleterious genetic marker, for example, ruby-eye, included in strain *B* could be very helpful in further observations of whether equilibrium between *A* and *B* is maintained.

This work was supported by the World Health Organization and was carried out while I was a consultant to their Vector Biology and Control Unit.

H. Laven

Institut für Genetik,
Johannes Gutenberg-Universität,
Mainz.

Received June 22; revised September 15, 1967.

[1] Mattingly, P. F., *Bull. WHO*, **27**, 579 (1962).
[2] Laven, H., *Z.f. indukt. Abst. Vererbl.*, **88**, 443, 478 (1957).
[3] Laven, H., *Nature*, **177**, 141 (1956).
[4] Laven, H., in *Genetics of Insect Vectors of Diseases* (edit. by Wright, J. W., and Pal, R.) (Elsevier Publ. Co., Amsterdam, in the press, 1967).

# C. Biological Control

## Suppression of Dung-Inhabiting Fly Populations by Pupal Parasites[1]

E. F. LEGNER[2] AND H. W. BRYDON[3]

ABSTRACT

Populations of hymenopterous pupal parasites of house flies were charted on *Fannia femoralis* Stein and *Ophyra leucostoma* (Wiedemann) in 2 coastal southern California poultry ranches with weekly random samples made over a period of 18 months. Although 6 parasite species were active in the area, *Muscidifurax raptor* Girault and Sanders and *Spalangia endius* Walker accounted for more than 95% of the observed parasitism on both hosts. Maximum parasite activity (more than 60% parasitism) was correlated with weather, *M. raptor* being prominent in the cooler, more humid, summer months (June–August), and *S. endius* in the hotter and drier months (September–November). Activity was sharply reduced with the advent of winter and remained low until the following spring, although host populations were highest during this colder period. The mass release of 4 imported West Indian and Midwestern parasite strains produced a higher host parasitism 2 months earlier in the year than occurred under natural conditions, being effected again by *M. raptor* and *S. endius,* with their periods of maximum activity coinciding with that of the native strains. Results are compared with other areas in southern California. Possible ways to improve biological control of house flies in southern California are discussed.

A whole array of synantrophic flies (those associated with man) breeding in filth habitats have been called house flies, but the term usually refers to *Musca domestica* L. and to *Fannia* species. Considerable information is now available regarding the origin of the genus *Musca* in the Eastern Hemisphere and its "recent" invasion of the Western Hemisphere (Legner and McCoy 1966; Sacca 1964; Sacca and Chillcott, personal communications). The large number of species of *Fannia* in the Holartic Region, however, attest to the endemicity of some species there, although a major portion of species appears to be endemic forms in the Palearctic (Chillcott 1961). The present worldwide distribution of species in both genera most likely came about largely through the activities of man (shipping, migration, etc.), since life cycles could continue uninterrupted in his presence. It is also feasible that shipping especially was, and is, responsible for the dispersal of many of the natural enemies attacking house flies, especially parasitic species of the genus *Spalangia,* which are known to have a worldwide distribution (Boucek 1963).

Considering the great reproductive potential of house flies (in the case of *M. domestica* more than 900 eggs/female), it is little wonder that many natural regulations have been recognized which attack and destroy them. It is astonishing, however, that arguments still continue as to the relative merits of natural control in suppressing fly populations. Hodge (1911) provided a key to the tremendous breeding potential of house flies when he stated that, "A pair of flies beginning operations in April may be progenitors, if all were to live, of $191{,}010\times10^{15}$ flies by August," a number which would just cover the earth, although not to a depth of 47 ft as reported by Hodge.

That the natural regulation involved is of a high and rather constant magnitude under most situations is readily evident when Richards' (1961) statement is considered regarding an insect of considerably smaller reproductive potential:

"If the sexes are produced in equal numbers, it is clear that a species that lays 100 eggs, not a high number for an insect, must have a 98 per cent mortality if the population is to be stable. A deviation of only 0.5 per cent in the mortality will result in a 25 per cent increase or decrease in the population in the next generation. How is this to be reconciled with the observed fact that large variations in such mortality factors as parasitism do not necessarily lead to the expected large alteration in the level of the population? Most of this excess stability in the population can be accounted for by the homoiostatic effect

[1] Paper no. 1667, University of California Citrus Research Center and Agricultural Experiment Station, Riverside, California. Accepted for publication October 4, 1965.
This study was carried out in part during the tenure of Grant no. GM 12699-01 of the National Institutes of Health, United States Public Health Service, and supported in part by California Department of Public Health Research Grant no. 0135, Fly Control Research Project.
[2] Assistant Entomologist, Department of Biological Control, University of California, Riverside.
[3] Director, Fly Control Research, Orange County Health Department, Santa Ana.

of overlapping causes of death, so that individuals that in any particular year escape one source of mortality merely survive to encounter one of the others. No other arrangement could account for the relative stability of many populations in most years; adjustments by density-dependent factors are usually too slow and in some cases are actually observed to be absent. In general, only a part, sometimes a small part, of the mortality attributable to any one cause is 'irreplaceable mortality,' in the sense of Thompson" (Thompson 1929).

Slobodkin (1962) went further in claiming an upper limit to the population density of *all* animal species in each habitat by the combined action of many regulatory mechanisms when he stated that, "Despite [the] enormous variation in reproductive patterns [among animals], each female animal alive now—in every species, in almost every location—will be replaced by precisely one female alive a generation from now." However, even if the author refers to univoltine species only in this example, it would be difficult to interpret his reasoning.

Physical factors (temperature, suitability of breeding media, etc.) will obviously exert a strong influence on the development of a fly population. However, in addition to physical restraint, of all the reported biotic regulatory agents operating against house flies, predators appear to be the greatest reducers of total numbers, having been observed to lower populations during every stage of the flies' life history. Attempts at biological control, which differs from natural regulation through man's intervention in *altering* the breeding situation in some way to directly favor natural regulation, have been varied and to a degree successful under certain situations. A review of house fly natural enemies and their manipulation in biological control is, therefore, appropriate to place the present study in proper perspective.

### REVIEW OF NATURAL BIOTIC REGULATORY AGENTS AND BIOLOGICAL CONTROL OF HOUSE FLIES

Beginning with the predators, which are without question some of the most numerous biotic agents that reduce house fly numbers, Axtell (1962) and Rodriguez and Wade (1961) have studied the predaceous habits of certain dung-infesting machrochelid mites known to feed on the eggs of house flies. Simmonds (1940) gave an account of certain dung beetles of the families Copridae and Histeridae that have been effective in reducing house fly breeding in cattle dung in Java, Sumatra, and South Africa. Simmonds also referred to an ant, *Pheidole megacephala* (F.), which, when accidentally introduced into Fiji about 1910, brought about noticeable fly control for several years. This same ant was reported very effective in the control of house flies in the Hawaiian Islands (Phillips 1934). *Pheidole* is no longer completely effective against the house fly in Hawaii, however, and it has since been reported destructive to certain other beneficial insects (C. J. Davis, personal communication). Other ants exert a high degree of natural regulation through their activities in feeding on house fly eggs as revealed by Pimentel's work in Puerto Rico (Pimentel 1955). This was the first truly quantitative evaluation of any of these biotic agents and it emphasized the importance of nonphysical regulatory factors in fly suppression. Portchinsky (1913) showed that the larvae of other related Diptera (e.g., *Muscina*) are effective in killing house fly larvae; and recently Anderson and Poorbaugh (1964) discussed the beneficial aspects of *Ophyra leucostoma* (Wiedemann), also predaceous in the larval stage on *Fannia* and *Musca*.

Little is known of the effectiveness of a group of small staphylinid beetles of the genus *Aleochara* that attack house flies both as parasites and predators. The immature stages of these beetles develop inside the puparium of the fly, and either pupate there or in the surrounding medium. Adult beetles feed voraciously on eggs and young larvae (Casey 1916, Colhoun 1953, Kemner 1926, Lesne and Mercier 1922, Lindquist 1936, White and Legner 1966).

West (1951) lists several pathogens fatal to *M. domestica*, some of which, i.e. *Bacterium delendae-muscae* R. & D., are specific for house flies and closely related Diptera. Recently Briggs (1959), Burns et al. (1961), Dunn (1960), and Hall and Arakawa (1959), have studied the use of *Bacillus thuringiensis* Berliner in the control of house flies, and variable success has been achieved where it has been used as a poultry-feed additive.

Some organisms have been known to alter the breeding habitat to render it unsuitable for house fly development. Furman et al. (1959) discussed the soldier fly, *Hermetia illucens* (L.), in the natural regulation of *M. domestica* both in the laboratory and in droppings beneath poultry cages. In some circumstances house fly emergence was completely prevented by the larval activities of *H. illucens*. Kilpatrick and Schoof (1959) give evidence for the effectiveness of *H. illucens* in fly control. To favor this fly in biological control, however, may not be desirable in view of its undesirability as a pest species (Meleney and Harwood 1935, Herms and James 1961). A dung beetle, *Canthon pilularius* (L.), is listed as having been introduced to Puerto Rico from Texas to aid control of horn fly, *Haematobia irritans* (L.), of cattle (Bartlett 1939). This beetle breaks the dung clumps into small units, which it buries in the ground as food for its young.

West (1951) summarizes the effects of other vertebrate and arthropod predators which take varying numbers of the immature and adult stages of flies for food. The effectiveness of predators attacking the adult stage in natural regulation is limited, however, primarily because an adult fly may deposit a substantial number of eggs before serving as food for the predator. Increasing the effectiveness of predators on this stage would, therefore, be difficult and probably useless as a means of producing any significant reduction in a fly population. In this respect, poison baits

Table 1.—Parasites released at 3 sites on Ranch I from April 10 to July 7, 1964.

| Parasite species | Approximate number released[a] | | | | |
|---|---|---|---|---|---|
| | April | May | June | July | Total |
| *Muscidifurax raptor* | 2,000 | 13,000 | 2,100 | 3,000 | 20,100 |
| *Spalangia endius* | 1,300 | 10,000 | 25,000 | 15,000 | 51,300 |
| *Spalangia cameroni* | 500 | 1,500 | 3,000 | 1,500 | 6,500 |
| *Spalangia nigroaenea* | 700 | 700 | 4,000 | 6,000 | 11,400 |

[a] Estimate of number of parasites that entered the release environment based on the total number that emerged in a check of 50 parasitized pupae placed next to the release container.

and autosterilization techniques could be potentially of greater significance.

House fly pupae are attacked by numerous predatory and parasitic agents that evoke a significant reduction of the fly population by attacking the final host developmental stage; consequently, they rank as some of the most effective regulatory agents. One can expect that the pupal stage often represents a residue of less than 1% of the total eggs deposited by the previous generation of flies. If this were not the case, as Richards (1961) and Herms and James (1961) theorize, then Hodge's (1911) predictions would be realized. Therefore, any biotic agent destroying the pupal stage can be thought of as exerting a vastly greater natural control effect *per individual* than one destroying any previous stage. In the case of house flies, this hypothesis may be significant since it is the adult stage that one is trying to eliminate. In a biological control sense, one could effectively employ a pupal parasite and at the same time take advantage of all the natural regulation that occurred on preceding stages. In nature, it has been found that the pupal stage, because of its inaccessibility as well as its hard protective covering, is not readily attacked by a large number of organisms. It is rarely bothered by predaceous ants (Pimentel 1955), pathogens do not penetrate it, and predaceous beetles and mites have a much greater preference for eggs and young larvae.

Hymenopterous parasites appear to be of real value in the natural control of house flies, particularly species of the genus *Spalangia,* which have been used in field tests in Pacific Islands and Puerto Rico with some success (Jenkins 1960). In West's (1951) literature review of hymenopterous parasites of house flies, references were made to parasite introductions into several areas, including Hawaii, for house fly control; but no true evaluation of effectiveness of these introductions is known.

Since West's review in 1951, additional reports referring to house fly parasites are those by Steve (1959), McCoy (1963), Sychevskaya (1963), Wylie (1963), Sacca (1964), and Legner et al. (1965). In many cases, the percent parasitization varied considerably and data on true effectiveness is lacking. The problems of sampling and evaluating effectiveness are manifold, and some of these difficulties are discussed by Simmonds (1934).

Numerous reports exist on the parasites of closely related species such as the blow fly, face fly, and stable fly, which are known to have parasites in common with the house fly (Altson 1920, Bishopp 1913, Crandall 1939, Froggatt and McCarthy 1914, Fukuda 1939, Lindquist 1936, Pemberton 1948, Pinkus 1913, and Vandenburg 1931).

Since 1962 we have observed in southern California that numerous hymenopterous parasites attacked house flies to varying degrees, although quantitative percent values were not obtained. It became obvious, however, that 2 species, *Muscidifurax raptor* Girault and Sanders and *Spalangia endius* Walker, seemed most widespread in the southern California subtropical belt. However, the latter species was rarely abundant before July, while the former was scarce after that month on a wide array of synantrophic flies. Parasitization was generally low in all areas in winter. It was felt that this was a typical response of these parasites to certain environmental conditions, but uniform data were lacking to establish their suspected behavioral characteristics.

The present study was designed to evaluate the effectiveness of a native population of hymenopterous pupal parasites attacking house flies breeding in chicken dung, and to study the effects of mass releasing introduced West Indian and Midwestern parasitic strains. The comparative uniformity of fly breeding in chicken dung, which is dynamically replenished by

FIG. 1.—Placing parasite release container in position over chicken droppings at Ranch I. Parasitized pupae are contained within the screened polyethylene quart equipped with honey and water dispensed through a cotton wick and glass vial at the top. Adult parasites emerge and many mate inside the container before emerging into the surrounding environment.

FIG. 2.—Aerial photo of Ranch I, giving locations of parasite release sites (outlined in circles on the photo).

the bird droppings, was chosen as the most suitable habitat for charting a fly-parasite population, in order to characterize the natural behavioral patterns of the species involved.

METHODS AND MATERIALS

Natural enemies attacking the pupae of house flies were secured in the West Indies (Legner et al. 1965) and the Midwestern United States during 1963. Several parasitic Hymenoptera were found to be very active on 3 islands in the West Indies and in central Wisconsin; and living cultures of these were established in the Department of Biological Control, University of California, Riverside. Beginning April 10 and extending to July 7, 1964, periodic releases of 4 species of the introduced pupal parasites were begun on 1 poultry ranch near the coast in southern California. Species released were biparental strains of *S. endius* from Puerto Rico; *S. cameroni* Perkins from Trinidad; and *S. nigroaenea* Curtis from Wisconsin; and 1 uniparental strain of *M. raptor* from Puerto Rico also was released. The dates on which these species were released and their total numbers are shown in Table 1. Parasites were liberated as newly emerged adults in specially designed containers equipped with honey and water (Fig. 1) and placed at 3 release sites (shown as circles in Fig. 2). The duration of the life cycle for all 4 species released was similar, about 19–24 days at 78°F and 55% RH.

The ranch had 65 houses which contained 250,000 birds (Fig. 2). Each house was 516 ft long with 1 row of 2.7×6-ft cages (20–25 birds/cage), 1 each side of a 5½-ft aisle. Both the aisle and cage floors were concrete. A corrugated metal roof with a 2-in. north-to-south slope covered each house and joined the slatted sides. Parasite release sites were spaced 300 ft apart in the center of the ranch (Fig. 2).

Thirty 1-quart samples of poultry manure were taken weekly from February through April 1964 on a random selection basis. Following release of parasites in April, biased sampling procedures were initiated, whereby 30 random samples were taken from within 100 ft of each of the 3 parasite release sites, as opposed to 30 samples taken in areas more than 100 ft from a release site. Such sampling was continued until July 1965.

The depth of manure from which samples were collected ranged from approximately 1 in. to 10 or 12 in. when cleaning occurred. Samples were taken from both sides of the aisle and in the center of the deposit where the character of the habitat was most homogenous throughout the year, being dynamically replenished by droppings from the caged birds above. The margins, where considerable pupation occurred, were not included. This sampling method consequently overcame some of the difficulties referred to by Simmonds (1934) in charting parasite frequency.

As manure accumulated, it varied from a flat-top

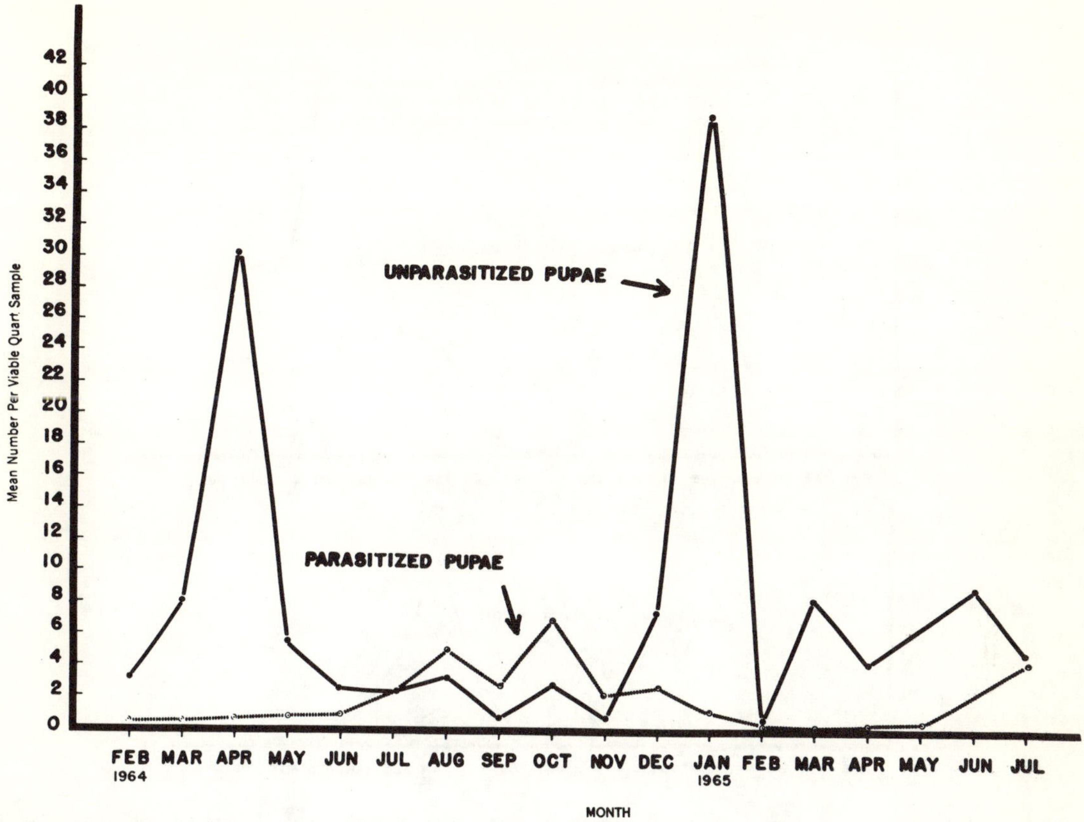


FIG. 3.—Population of parasitized and unparasitized pupae of *Fannia femoralis* in the check environment of Ranch I in survey periods of 1964–65.

surface as wide as the cages to a gently rounded mound, the crest of which ran the length of the house, and upon which a continuous breeding site was available to synantrophic flies. Hand cleaning operations were conducted once in February, August, and November of 1964. One-half of a house was cleaned out, and more than 1 month was required to complete all houses, thus minimizing the disrupting influence of such an operation on the fly population.

Until a core-sampling device was developed in April, samples were collected by plunging 1-quart ice cream cartons down from the manure surface to the underlying concrete. Whenever the manure was less in depth than the 6¾-in. height of the sampling carton, a hand trowel was used to scrape the balance of the 1-quart sample from the side of the sampling hole and add it to the carton. In the event samples were taken deeper than the height of the carton, the balance of the depth was not removed for this study.

The sampling device that replaced the crude ice-cream-carton method consisted of a thin-walled, stainless-steel cylinder 3½ inches in diameter. A plunger plate and a 24-in. plunger rod were supported and guided by a ½-in. diameter U-shaped metal tube attached to opposite sides of the cylinder. The metal tube, which was approximately 16 in. high, also functioned as a handle while the cylinder was pushed into the poultry manure. As the core sample was collected in the device, it compacted sufficiently to remain intact while the cylinder was withdrawn and inserted into a marked, quart-size, polyethylene sack. The cylinder was removed by pushing on the plunger rod handle, which also freed the sample inside the sack. Although fly breeding of different magnitude admittedly occurred in other portions of the manure deposit (e.g., the edges of the mound), samples were necessarily restricted to the middle to eliminate depth variation among the samples and between the checks and parasite release sites.

Immediately upon completion of sampling, the samples were placed in Berlese funnels in the laboratory where they were exposed to approximately 20 hr of white light from 100-w frosted bulbs. This process caused fly larvae, which were saved for other investigations, to disperse from the manure and be trapped in receptacles containing 70% alcohol. The larval separation process did not injure pupae, which remained buried in the manure, but allowed for some drying which facilitated pupal extraction. Pupae of primarily 6 species of flies, *Fannia canicularis* (L.), *F. femoralis* Stein, *Musca domestica, Ophyra leucostoma, Phormia regina* (Meigen), and *Muscina stabu-*

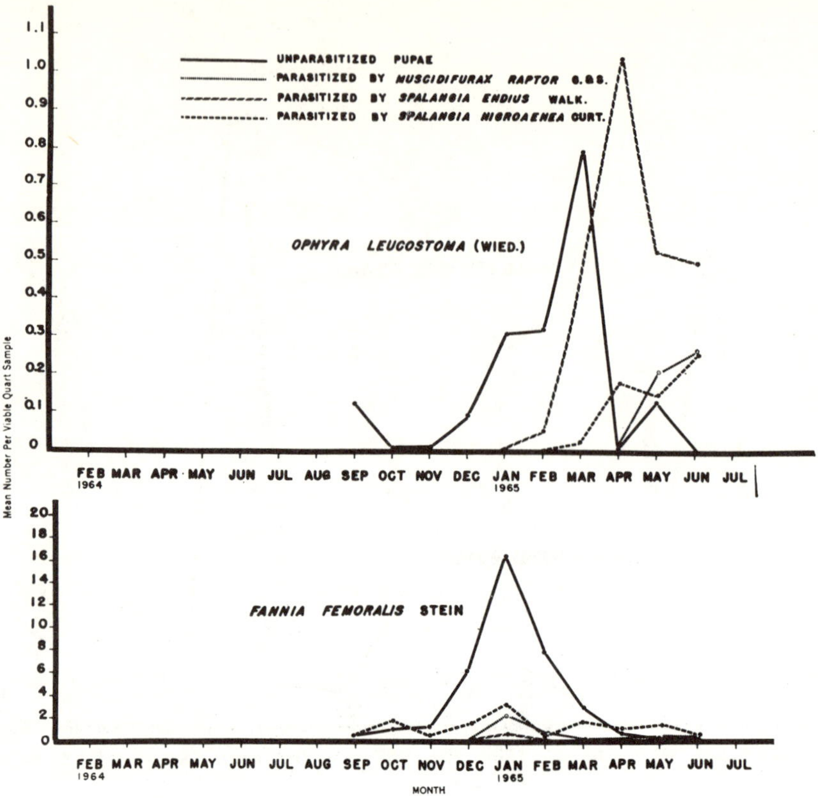


FIG. 4.—Population of parasitized and unparasitized pupae of *Fannia femoralis* and *Ophyra leucostoma* on Ranch II in survey periods of 1964–65.

*lans* (Fallén), were removed by placing the partially dried samples in large white-enamel trays and sorting them out under a magnifying glass. Pupae were transferred to a glass petri dish, where viable individuals were separated to species, using a dissecting microscope with back lighting. They were then placed in 15-dram plastic vials with 100-mesh brass-screen lids, and incubated in a protected area near the ranch. Emerged parasites were determined to species; after 1 month unemerged hosts were dissected for diapausing parasites. Records were kept of the maximum and minimum daily temperature and rainfall at Ranch I.

A smaller ranch situated about 10 mi. more inland from the coast in Orange County, which was also the site of a parasite study, had 26 houses with 1 row of 3-bird cages on either side of a 3-ft-wide concrete aisle. In this case, however, the bird droppings fell on a dirt floor and the sides of the houses were open. Each month, 15 1-quart samples were taken at random, and processed in the same manner as that of the first ranch. Meteorological data collected at Ranch I was essentially similar for Ranch II, which was situated in the same coastal belt.

## RESULTS AND DISCUSSION

The predominant synantrophic fly species found breeding in sample sites in the chicken droppings on both ranches during the experimental period was *F. femoralis,* comprising over 95% of all such flies sampled at any given time of year. The residual 5% included *F. canicularis,* collected from early spring until early summer, and *Musca domestica,* present only in the late summer. Isolated collections of *Phormia regina, Stomoxys calcitrans* (L.), and *F. scalaris* (F.) also were made, but these species were so infrequent that no significance could be attached to their seasonal presence.

Parasitic species already present on both ranches prior to release of the introduced strains in April 1964 were *Spalangia endius, S. cameroni, S. nigroaenea, Nasonia vitripennis* (Walker), *Muscidifurax raptor,* and *Stilpnus anthomyidiperda* (Viereck). All these species attacked pupae of *F. femoralis* externally, except *S. anthomyidiperda,* which parasitized larvae of *F. canicularis,* no definite relationship with *F. femoralis* having been established. *O. leucostoma* served as host for 3 species, *S. endius, M. raptor,* and *S. nigroaenea.*

In Fig. 3 and the lower half of Fig. 4, the viable *F. femoralis* pupal population (pupae which gave rise to adult flies) is compared with the population of total parasitized pupae in the check environment of Ranch I and Ranch II, respectively. Frequency of hosts and parasites is expressed as the mean number of indi-

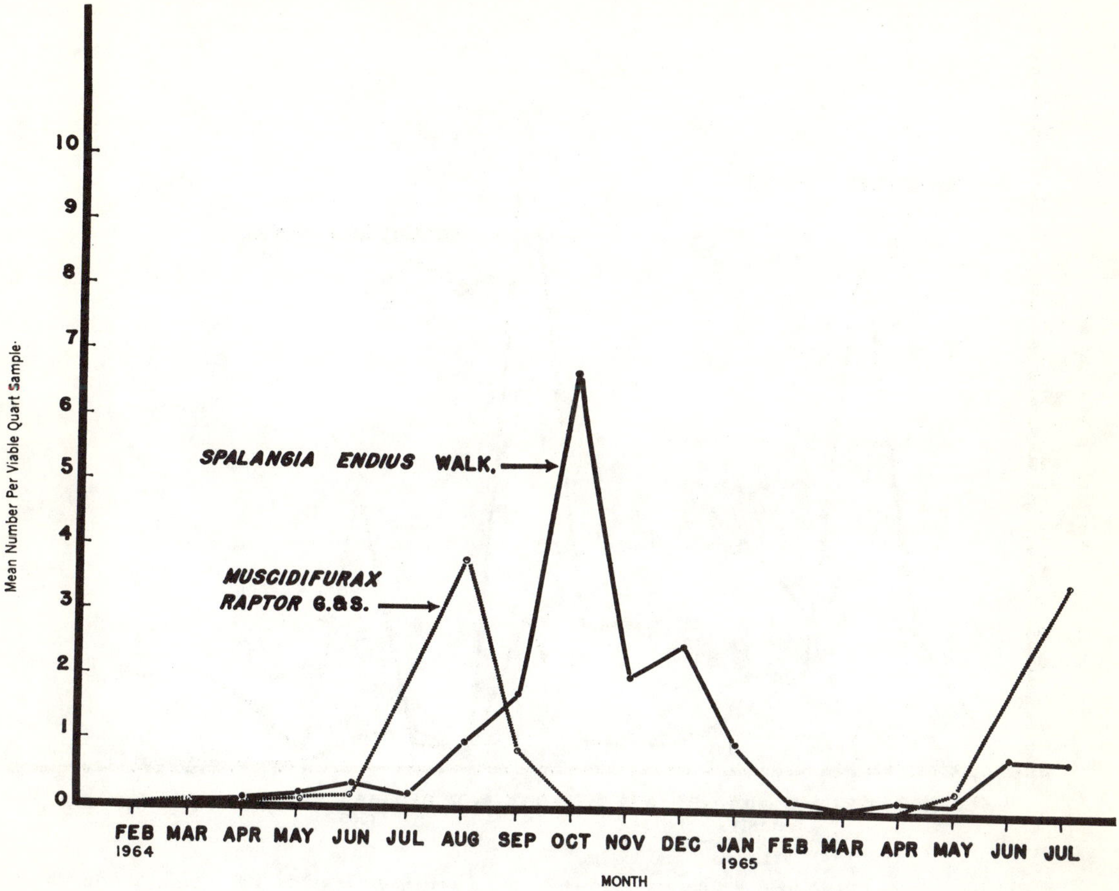


FIG. 5.—Population of *Fannia femoralis* pupae parasitized by *Spalangia endius* and *Muscidifurax raptor* in the check environment of Ranch I in survey periods of 1964–65 (represents more than 95% of total observed parasitism in Fig. 3).

viduals per viable quart sample; that is, samples in which either viable parasitized pupae, viable unparasitized pupae, or both were present. These figures then relate to sites in the chicken droppings at which fly breeding actively occurred; and, therefore, loci suitable for parasite activity within a given sampling interval. A consideration of the host and parasite populations in this way, then, further eliminated some of the problems expressed in detail by Simmonds (1934) concerning evaluation of parasite effectiveness.

Fig. 5 is a breakdown of the total parasitism shown in Fig. 3 (check environment Ranch I) for the principal parasitic species involved, *M. raptor* and *S. endius,* these 2 species accounting for more than 95% of all the observed parasitization of *F. femoralis.*

In Fig. 6, the *F. femoralis* population of viable pupae within 100 ft of 3 parasite release sites is compared with the frequency of its total pupal parasitism. Fig. 7 is a breakdown of this parasitism into its principal components, which also involved *M. raptor* and *S. endius* in more than 95% of the recoveries.

Parasitism frequency of *O. leucostoma,* together with its unparasitized pupal density, is shown in Fig. 8 for Ranch I and in the top half of Fig. 4 for Ranch II. Fig. 9 gives the mean high and low temperatures and rainfall at Ranch I during the experimental period.

*Parasitism of Fannia femoralis.*—In the check environment of Ranch I, *F. femoralis* had 1 large population peak early in the year, followed in the same year by 2 smaller peaks approximately ⅙ as great (Fig. 3). The highest populations occurred characteristically in the coldest months of each respective year. A close parallel was observed on Ranch II in 1964 (Fig. 4), where the highest population of *F. femoralis* occurred at the same time of year.

Parasitization by hymenopterous parasites, on the other hand, was lowest during these coldest months when the host population was at its peak (Fig. 3), suggesting that the activity of such parasites was curtailed by low temperatures. During the warmer months (July–November 1964), parasitization ex-

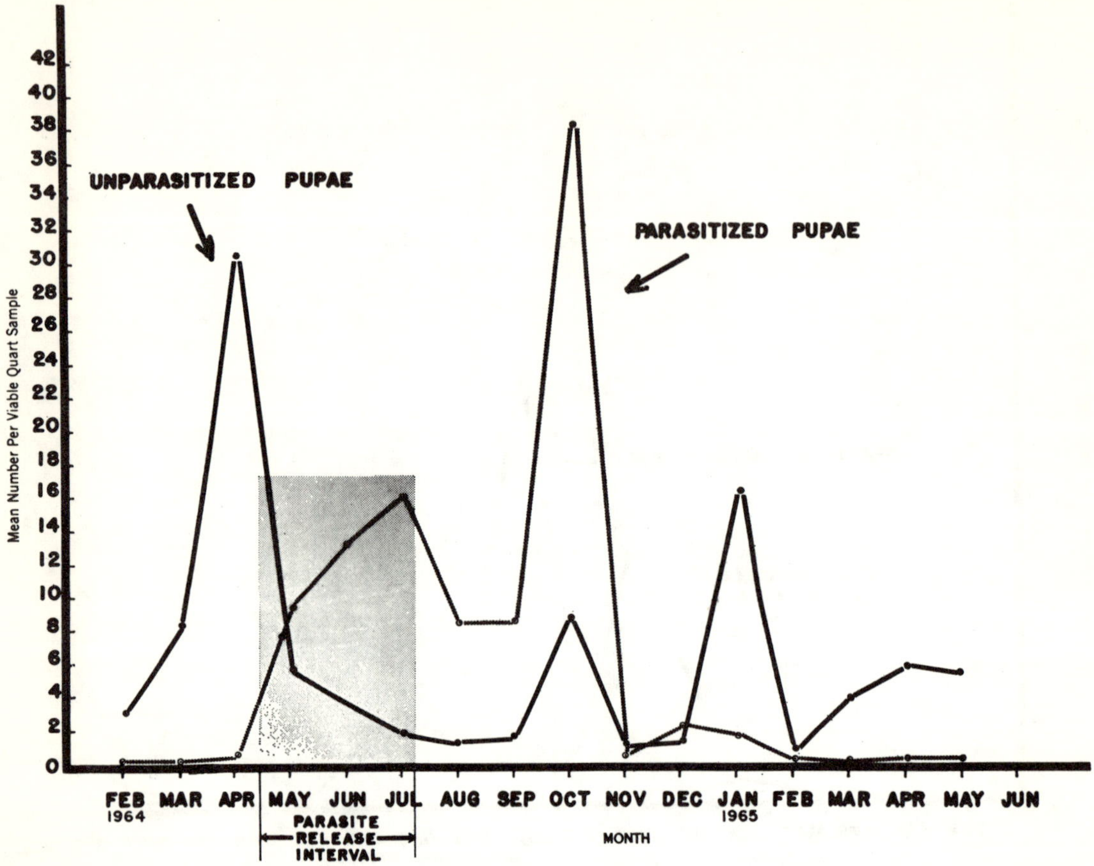


FIG. 6.—Population of parasitized and unparasitized pupae of *Fannia femoralis* within 100 ft of 3 parasite release sites on Ranch I in survey periods of 1964–65.

ceeded the level of the viable host pupal population (Fig. 3 and 4).

Fig. 5 shows that *M. raptor* was the predominant parasite active during the cooler and damper portions of the summer (June–August), while *S. endius* exceeded and finally prevailed in the warmer and drier months (August through October) (refer to Fig. 9 for temperature data).

All parasite and host activity was sharply reduced within a period of 1 month with the advent of a period of very cold weather in mid-November 1964. The simultaneous decline of the host population in November cannot be correlated with parasite activity, but was probably also the result of the very sudden drop in temperature to below freezing in the middle of the month. Although cool weather is optimum for *F. femoralis* development, it is probable that the lower temperature limit was reached at this time. Subsequent recovery and tremendous buildup of this host was shown in December 1964 and January 1965, when mean temperatures rose (Fig. 3, 6, 9). It can be postulated that this resultant soaring of the population was at least in part due to the general absence of parasite activity during these colder months.

An examination of the host and parasitized pupae populations of *F. femoralis* within 100 ft of the 3 release sites can be made by referring to Fig. 6. During the warm part of the year, the total parasitized and unparasitized pupal population was generally higher in the central areas of Ranch I where the parasite releases were made than on the edges (compare Fig. 3 and 6). This fact was correlated significantly (5% level) with higher larval counts made during this period for the central areas of the ranch. The higher density may have been due to more humid conditions of manure at the central points than existed over other portions of the ranch during the hotter and drier months of September and October. This condition favored *Fannia* development. Nonetheless, the greater host pupal production did not give rise to a proportionally high number of adults, but was held down in the pupal stage by the actions of parasites (Fig. 6).

Relative to this high parasite activity in release sites when compared with the checks are the results of laboratory studies with *M. raptor* and *S. endius,* wherein these species were observed to parasitize successfully host pupae that would have been unable to complete their development to the adult stage, this inability pre-

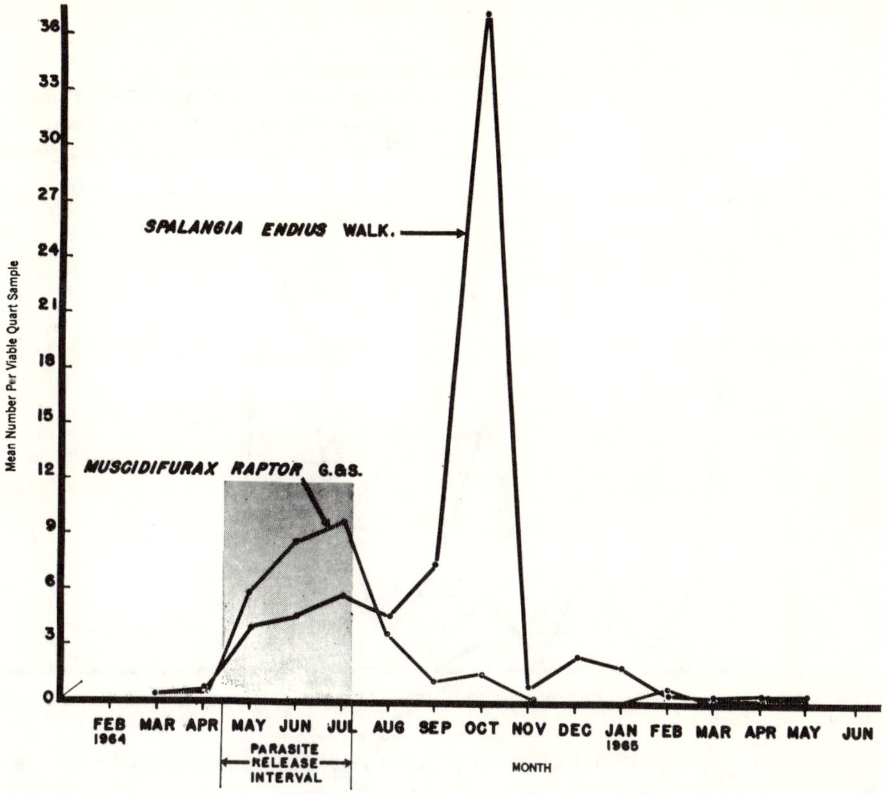


FIG. 7.—Population of *Fannia femoralis* pupae parasitized by *Spalangia endius* and *Muscidifurax raptor* within 100 ft of 3 parasite-release sites on Ranch I in survey periods of 1964–65 (represents more than 95% of total observed parasitism in Fig. 5).

sumably being due to weakness of individuals. The highest frequency of this kind of parasitism in the laboratory has never exceeded 6% with either species; therefore, its occurrence in the experimental area can be assumed to have been correspondingly low.

Parasitized pupae in the release areas were more prevalent earlier in the season and occurred at a higher density during the rest of the season than in the checks. This higher density occurred with reference to the time that releases were being made of the introduced strains (Fig. 6). For example, the buildup of *M. raptor* in the check sections of Ranch I (Fig. 5) did not begin until after June 1964, whereas a significant increase was noticeable in the release sites after April of that year (Fig. 7). Also, the incidence of *S. endius* in the check sections (Fig. 5) was low before July, whereas increased activity of this species was evident 2 months earlier in the release sites (Fig. 7). Although the native strain of *S. endius* could not be morphologically separated from the introduced strain, there is some indication from data presented in Fig. 7 that the high activity of this species in the release sites in September and October 1964 might have been due to the releases and that it probably resulted from the combined action of P-1 and F-1 generations.

A comparison of the slopes of the parasite curves in the check and release sites (compare Fig. 3 with 6 and 5 with 7) shows a distinct escalation of parasite activity during the period of parasite releases in April to July 1964 in the release sites. This phenomenon is strongly indicative of an increase in parasite activity caused by parasite releases.

The lag of several months after releases were begun before *S. endius* showed maximum activity in the release sites was probably largely due to its preference for a warmer and drier environment. It is presumed that a high percentage of released adults simply remained in the immediate release sites until the environment became more favorable. Random net collections and visual observations around the release and check sites of Ranch I during the release period confirmed the existence of a larger number of *S. endius* adults in the vicinity of the release sites, even a month after releases had ceased. It was erroneously concluded at that time that *M. raptor* had dispersed from the release sites with greater rapidity than *S. endius.* The true case, as it now appears, was that *M. raptor* was active below the surface of the deposits while *S. endius* was awaiting more favorable conditions to become fully active. Judging from the net

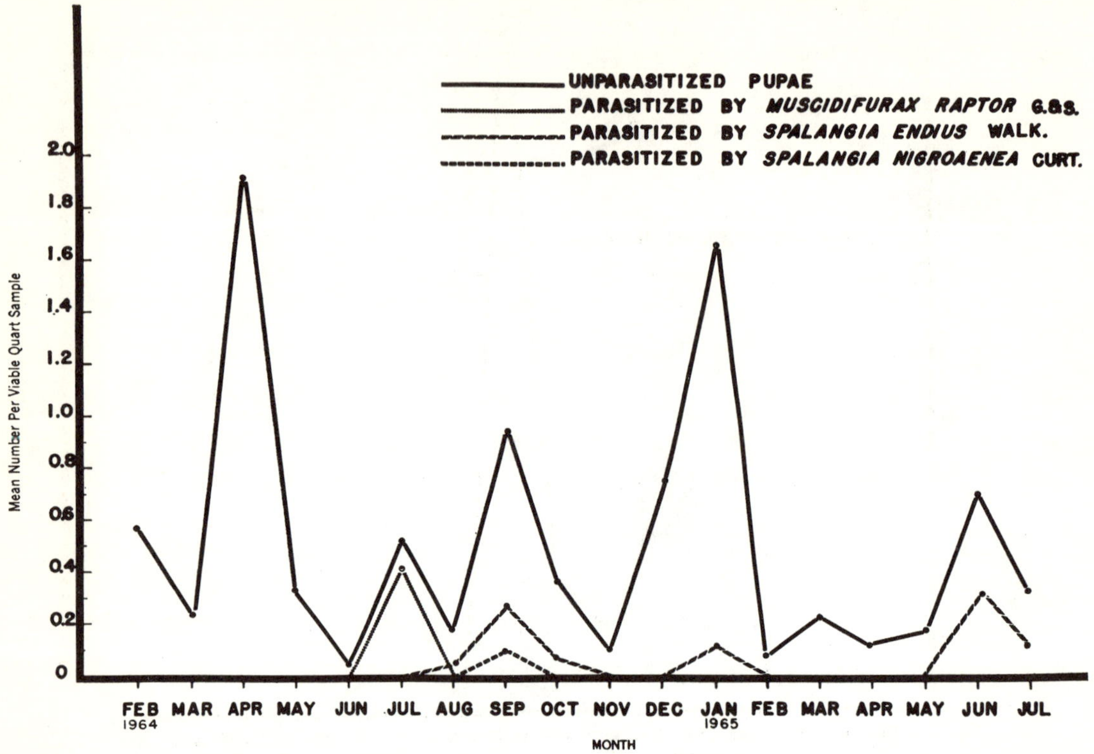


FIG. 8.—Population of parasitized and unparasitized pupae of *Ophyra leucostoma* in the check environment on Ranch I in survey periods of 1964–65.

collections, this waiting period was spent largely on the surface and at the edges of the manure deposits.

Assuming that some mortality of adults occurred during this waiting period, it might have been advantageous to wait until August for release of *S. endius,* at which time conditions were more suitable for parasitizing its host. On the other hand, presence in the host environment for a considerable length of time before maximum activity could be realized, might have, in a conditioning sense, been favorable for this species. Table 2 shows a significantly higher (5% level) proportion of females of *S. endius* recovered from the release sites than from the checks in the present study. This could have been influenced by the prolonged exposure of P-1 individuals to the environment, especially since no noticeable difference in sex ratio occurred in laboratory culture of both domestic and imported strains of this species.

Judged by data presented in Fig. 5 and 7, it would most likely have been useless to expect that *M. raptor* released in August would perform with any high degree of efficiency until cooler weather prevailed.

Activity of the released uniparental strain of *M. raptor* could be compared with that of the native biparental strain of the same species already present on the ranch. Although uniparental in reproduction, Puerto Rican *M. raptor* is not unisexual since male eggs oviposited by females give rise to nonfunctional males. The frequency of this male production in the laboratory has never exceeded 10% when rearing was at temperatures of 70°–85°F. Therefore, field recoveries of this strain probably would not have included more than this percentage of males. The female:male sex ratio within 100 ft of the release sites favored females almost 2:1 over the checks, after releases were begun in April 1964, for the rest of that year (Table 3). The average female:male sex ratio for check sites from June through September was 0.95:1 and for release sites 1.91:1. This was a significant difference (1% level), and it can, therefore, be considered that the Puerto Rican strain of *M. raptor* was responsible for at least half of the observed parasitism in the release sites by this species during the cited period.

The peaks of the *F. femoralis* viable pupal population in August and October 1964, following the high winter peak in the check sections (Fig. 3), was reduced to 1 peak in October in the release sites (Fig. 6). The absent peak was still represented in the total viable individuals collected; however, parasitism compensated for the difference. This phenomenon appears to be correlated with parasite releases earlier in the season.

The behavior of parasite species in the release sites (Fig. 7) was similar to that in the check sections (Fig. 5) in that *M. raptor* and *S. endius* were domi-

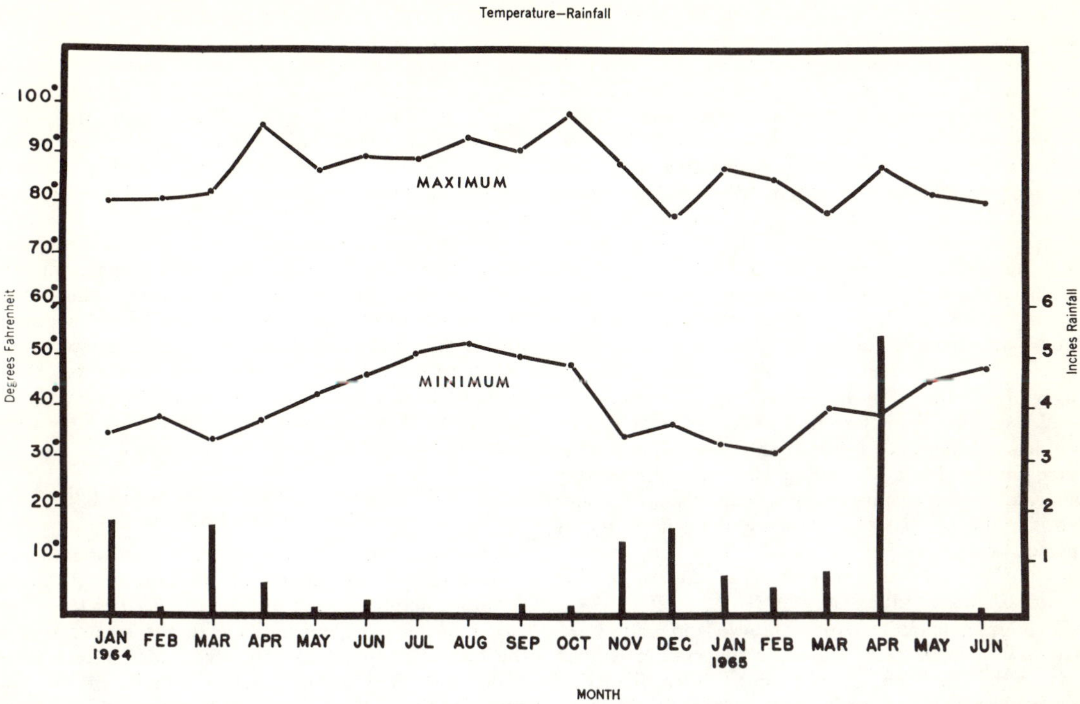


FIG. 9.—Mean high and low temperatures and total inches of rainfall at Ranch I during study periods of 1964–65.

nant, representing more than 95% of the parasitism. No increase in effectiveness in the release sites over the checks was noticeable for *S. cameroni* or *S. nigroaenea,* although thousands of individuals were liberated during the release period (Table 1).

These species probably do not prefer *F. femoralis* as a host; they may be unsuited for high activity under the existing climatic situation, or they are unable to compete with the other species in the same habitat. Collections by the senior author in various portions of the Western Hemisphere during the past 3 years tend to bear out the conclusion that climate might be largely responsible for determining which of the 3 species predominates. Although all 3 *Spalangia* have been found to co-exist on *M. domestica* throughout the hemisphere, *S. endius* is usually most abundant in a xerophilic environment, as opposed to the other 2 species which dominate in more humid regions. *S. nigroaenea* is found principally in the temperate humid regions while *S. cameroni* is most abundant in the tropical humid regions.

Table 2.—Sex ratio of *Spalangia endius* from *Fannia femoralis* host pupae at check and release sites on Ranch I.

| Months | Ratio of females :males | |
|---|---|---|
| | Check sites | Release sites |
| June–August 1964 | 1.37 :1 | 1.51 :1[a] |
| September–December 1964 | 1.35 :1 | 1.59 :1[a] |

[a] Significantly higher than the check at 5% level.

The 3 species of *Spalangia* presumably fit into the "ecological homologue" category as outlined by DeBach and Sundby (1963). They attack the same hosts in the same habitat, yet they coexist. As far as is known, they are similar in their ability to search through the habitat for hosts, in their adult food requirements, length of immature development, etc. Unless some subtle difference in their behavior is yet to be discovered, it appears that seasonal changes in temperature and humidity (even in the tropics) might influence them differentially and account for their coexistence. A similar phenomenon has been recorded by Park (1954) for 2 species of *Tribolium.* Hutchinson (1957) also has stated that habitat changes (e.g., humid and dry manure) can account for the coexistence of homologous species.

Another similarity between the release and check sites was the time at which each parasite species as-

Table 3.—Sex ratio of *Muscidifurax raptor* from *Fannia femoralis* host pupae at check and release sites on Ranch I.

| Month (1964) | Ratio of females :males | |
|---|---|---|
| | Release sites | Check sites |
| June | 2.40 :1 | 2.00 :1 |
| July | 1.90 :1 | 1.65 :1 |
| August | 1.34 :1 | 0.64 :1 |
| September | 1.83 :1 | 0.82 :1 |
| October | 1.75 :1 | — |

sumed prominence, *Muscidifurax raptor* being found most active again during the cooler summer months, while *S. endius* predominated during the hottest period (Fig. 7 and 9). *M. raptor* did not cease its activity entirely during the hot months in the release sites, but was recovered in low numbers until the cold weather in November. It is suggestive that this lingering was due at least in part to the Puerto Rican strain of *M. raptor* released, since females predominated in the collections 1.75:1 (Table 3). It is recognized, however, that a good part of the activity might have been due to the native strain and that the central position in the ranch was influential in sustaining a higher parasite population.

The introduced strain of *S. endius* apparently no longer predominated in the release sites after the onset of cold weather. This fact was evidenced by a lack of significant differences between the abundance of this species in the check and release sites after November. Laboratory temperature studies (unpublished) substantiate this in part by showing that mortality of Puerto Rican *S. endius* adults is high at temperatures below 55°F, whereas temperate-zone strains of this species are more cold resistant.

Diapausing parasites were observed on only 3 occasions in puparia of *F. femoralis* collected in August 1964 in the check and release sites. Since dissection was necessary to determine diapause, it is not definitely known what parasite species was involved; however, from a comparison with larvae of cultured specimens it appears more than likely that this was *M. raptor*. This, then, gives evidence for at least 1 mechanism for carrying parasites over an unfavorable season. Even though the frequency of diapause observed in the current study was low, the possibility should not be overlooked that diapausing individuals may have been more common in other portions of the manure deposit than were sampled here. Similarly, the disappearance of parasites during some portions of the winter observed from these samples does not necessarily imply their complete absence from the environment; they might, in fact, have occurred at other loci in the manure which were not included in the sampling. It was observed, for example, that *F. femoralis* pupae formed in large numbers on the margins of the deposit during the damp late winter and early spring months. Random collections and incubation of pupae from these sites did not turn up a large number of parasitized individuals, and it was concluded that the parasite abundance picture would not have been changed appreciably if these sites had been included in the general survey.

*Parasitism of Ophyra leucostoma.*—Populations of *O. leucostoma* followed closely those of *F. femoralis,* and the latter can be considered the principal synantrophic fly host of this predator on Ranch I, since other sizable Diptera were generally absent. Laboratory investigations by Anderson and Poorbaugh (1964) on *F. canicularis* and by the authors on *F. femoralis* showed that these were suitable hosts for *O. leucostoma,* although when given a choice, *O. leucostoma* greatly preferred *Musca domestica* in our experiments.

Percentage parasitization of *O. leucostoma* pupae was lower than that of *F. femoralis* at any given time of year on Ranch I (Fig. 8). It was higher than on *F. femoralis* on Ranch II, however, where a lower density of *Ophyra* was present (Fig. 4). The peaks of parasite activity on *Ophyra* for the 2 predominant parasite species occurred at the same time of year as it did on *F. femoralis* at Ranch I, with *Muscidifurax raptor* active most during the cooler summer months and *S. endius* during the hotter months. The overall low density of *O. leucostoma* at Ranch I did not allow for a significant comparison to be made between parasite release sites and the check environments. However, there did not appear to be any appreciable increase of *Ophyra* parasitism in the release sites when compared with the checks.

### CONCLUSIONS AND SUGGESTIONS FOR BIOLOGICAL CONTROL

The occurrence of a significant parasitization of *F. femoralis* by hymenopterous parasites on 2 ranches during a major portion of the year verifies the active role that such organisms have in the suppression of the total fly population. However, their effectiveness being restricted to the warmer months of the year suggests that existing parasitism might be effectively supplemented through the importation and release of a cold-resistant strain (or strains) just prior to the characteristic buildup of the host population in the late fall (such a buildup is characteristic in the coastal area as judged from observations made from 1962 to 1965). Presumably such a cold-resistant parasite might either be produced artificially in the laboratory and released in large numbers at the appropriate time, or it might be procured from another area. The latter approach presumably would be more economically practical. The release of artificially produced cold-hardy strains might have to be made periodically since this resistance could not be expected to endure with the wild types of the species still present in the environment.

Modifications of housing structure might improve the environment to favor parasite activity, especially in winter where protection might be afforded from the cold through screening or other devices.

In the present study, manure clean-out was performed in such a way as to favor parasite buildup over the entire ranch, in that manure from only ½ of a poultry house was removed at any 1 time. Serious consideration should be given to duplicating this operation as a means of augmenting parasitism. This type of alternate clean-out has been found in other studies greatly to favor predator populations of the beetles, ants, and mites that contribute significantly to the 90%+ reduction of flies prior to their reaching the pupal stage. Such a reduction is essential, especially for a species with a high reproductive potential, as was pointed out in the introductory paragraphs. We have also observed that leaving a pad of dry manure 2 in. thick at clean-out time also greatly favors other factors contributing to natural control, such as the physical drying-out of manure and provision of shelter for predaceous fauna.

Observations in the present study and those made at other fly-breeding sites in southern California during the past 3 years on all principal species of synanthrophic flies show that the 2 principal parasite species involved in natural control in this area are *M. raptor* and *S. endius*. Other species, as previously noted, occur at lower population densities throughout most of this subtropical southern California region.

These 2 species in other areas show similar activity to that in this study, although the time of activity may be related to phenological events in each separate locality. For example, in the inland areas where mean temperatures are higher earlier in the year than on the coast, *S. endius* overtakes *M. raptor* correspondingly earlier, which is apparently a climatic influence. The principal host species in the inland areas is *M. domestica* and percent parasitization never reaches the same high level as demonstrated with *Fannia* in this study.

The effectiveness of these 2 principal species apparently is proportional to the availability of host pupae in the breeding site; therefore, *M. domestica* pupae, which are generally found deeper in manure deposits than *Fannia* pupae, are not attacked to as high a degree. Modifications of poultry-house construction and bird caging could help to enhance the effectiveness of these parasites against *Musca* by eliminating pupation in the soil with a concrete base and by dispersing the droppings over a wider area, respectively.

The introduction of exotic species of larval and pupal parasites would appear to offer the most likely chance for more successful biological control in southern California, especially against *Musca*. There is considerable evidence that *Musca* was introduced into the Western Hemisphere (Legner and McCoy 1966) and that more effective natural enemies might be sought in the East.

*Fannia,* which are represented by native species in the West, but which includes numerous non-native forms that are now considered pest species in the West that presumably invaded from the East and South America poses a bigger problem in that existing natural enemies have already had a better chance to invade the breeding sites of these very closely related species. Using the premise that this invasion did, in fact, not occur to its fullest extent, it might be advantageous to seek out additional natural enemies of invaded *Fannia* species in their endemic areas in other portions of the Holarctic and Neotropical Regions.

Another way to favor the activity of hymenopterous parasites in fly-breeding sites is through avoiding the use of nonselective toxicants. There is ample evidence from work by Bartlett (1963) on related parasitic Hymenoptera that toxicants used in fly control are extremely lethal to this group of parasites. The long-range value of such materials in fly control should be reviewed.

## REFERENCES CITED


**Altson, A. M.** 1920. The life history and habits of two parasites of blow flies. Proc. Zool. Soc. London, Pt. 3: 195–243.

**Anderson, J. R., and J. K. Poorbaugh.** 1964. Biological control possibility for house flies. California Agric. 18(9): 2–4.

**Axtell, R. C.** 1962. Macrochelidae (Acarina: Mesostigmata) inhabiting manure and their effect on housefly production. Ph.D. Thesis, Cornell Univ., Ithaca, N. Y.

**Bartlett, B. R.** 1963. The contact toxicity of some pesticide residues to hymenopterous parasites and coccinellid predators. J. Econ. Entomol. 56(5): 694–8.

**Bartlett, K. A.** 1939. Introduction into Puerto Rico of beneficial insects for control of the horn fly. Puerto Rico Agr. Exp. Sta., Agr. Notes no. 88. 6 p.

**Bishopp, F. G.** 1913. The stable fly (*Stomoxys calcitrans*) an important livestock pest. J. Econ. Entomol. 6: 112–26.

**Boucek, Z.** 1963. A taxonomic study in *Spalangia* Latr. (Hymenoptera, Chalcidoidea). Acta Entomol. Mus. Nat. Pragae 35: 429–512.

**Briggs, J. D.** 1959. Feeding layers *Bacillus thuringiensis* Berliner to control *Musca domestica* L. developing in chicken feces. Bull. Entomol. Soc. Amer. 5(3): 115.

**Burns, E. C., B. H. Wilson, and B. A. Tower.** 1961. Effect of feeding *Bacillus thuringiensis* to caged layers for control. J. Econ. Entomol. 54(5): 913–5.

**Casey, T. L.** 1916. A new species of *Baryodma*. Can. Entomol. 48: 70–71.

**Chillcott, J. G.** 1961. A revision of the Nearctic species of Fanniinae (Diptera: Muscidae). Can. Entomol. 92, Suppl. 14. 295 p.

**Colhoun, E. H.** 1953. Notes on the stages and the biology of *Baryodma ontarionis* Casey (Coleoptera: Staphylinidae), a parasite of the cabbage maggot, *Hylemya brassicae* Bouche (Diptera: Anthomyiidae). Can. Entomol. 85: 1–8.

**Crandall, H. A.** 1939. The biology of *Pachycrepoideus dubius* Ashmead (Hymenoptera), a pteromalid parasite of *Piophila casei* Linne. Ann. Entomol. Soc. Amer. 32: 632–52.

**DeBach, P., and R. A. Sundby.** 1963. Competitive displacement between ecological homologues. Hilgardia 34(5): 105–66.

**Dunn, P. H.** 1960. Control of houseflies in bovine feces by feed additive containing *Bacillus thuringiensis* Berliner. J. Insect Pathol. 2(1): 13–16.

**Froggatt, W. W., and T. McCarthy.** 1914. The parasites of the sheep maggot fly (*Nasonia brevicornis*). Notes and observations in the field and laboratory. Agr. Gaz. New South Wales 25(9): 759–64.

**Fukuda, S.** 1939. Biology of *Mormoniella vitripennis* (Wlk.), a parasite of *Sturmia sericariae* Corn. [in Japanese]. Zool. Mag. 51: 185–94.

**Furman, D. P., R. D. Young, and E. P. Catts.** 1959. *Hermetia illucens* (Linn.) in natural control of *Musca domestica*. J. Econ. Entomol. 52(5): 917–21.

**Hall, I. M., and K. Y. Arakawa.** 1959. The susceptibility of the house fly, *Musca domestica* Linn., to *Bacillus thuringiensis* Berliner. J. Insect Pathol. 1(4): 351–5.

**Herms, W. B., and M. T. James.** 1961. Medical Entomology. The Macmillan Co., N. Y. p. 364.

**Hodge, C. F.** 1911. How can you make your home town or city flyless. Nature and Culture, July 1911, p. 9–23.

**Hutchinson, G. E.** 1957. Concluding remarks. Cold Spring Harbor Symp. Quant. Biol. 22: 415–27.

**Jenkins, D. W.** 1960. Pathogens, parasites and predators of medically important insects. Biological Control of Insects of Medical Importance. Amer. Inst. Biol. Sci. Tech. Rep. Nov. 1960: 6–20.

**Kemner, N. A.** 1926. Die Lebensweise und die Parasitische Entwicklung der echten Aleochariden. Entomol. Tidskr. 47: 133–70.

**Kilpatrick, J. W., and H. F. Schoof.** 1959. Interrelationship of water and *Hermetia illucens* breeding to *Musca domestica* production in human excrement. Amer. J. Trop. Med. Hyg. 8(5): 597–602.

**Legner, E. F., and C. W. McCoy.** 1966. The house fly, *Musca domestica* Linneaus, as an exotic species in the Western Hemisphere incites biological control studies. Can. Entomol. 98(3): 243–8.

**Legner, E. F., E. C. Bay, and C. W. McCoy.** 1965. Parasitic natural regulatory agents attacking *Musca domestica* L. in Puerto Rico. J. Agr. Puerto Rico. 49(3): 368–76.

**Lesne, P., and L. Mercier.** 1922. Un Staphylinidae des muscides fucicoles *Aleochara* (polystoma) *algarum* Fauvel. Caractères adaptatifs de la larve à la vie parasitaire. Ann. Soc. Entomol. France 91: 351–8.

**Lindquist, A. W.** 1936. Parasites of horn fly and other flies breeding in dung. J. Econ. Entomol. 29: 1154–8.

**McCoy, C. W.** 1963. Mass liberation of laboratory reared parasites, *Spalangia muscidarum* (Richardson), for control of *Stomoxys calcitrans* (L.) in Lancaster County, Nebraska. M.S. Thesis, Univ. Nebraska, Lincoln.

**Meleney, H. E., and P. D. Harwood.** 1935. Human intestinal myiasis due to the larvae of the soldier fly, *Hermetia illucens* Linné (Diptera: Stratiomyidae). Amer. J. Trop. Med. 15: 45–49.

**Park, T.** 1954. Experimental studies of interspecies competition. II. Temperature, humidity and competition in two species of *Tribolium*. Physiol. Zool. 27: 177–238.

**Pemberton, C. E.** 1948. History of the Entomology Department Experiment Station, Hawaiian Sugar Planters Assoc. 1904–1945. The Hawaiian Planters' Record 52(1): 53–90.

**Phillips, J. S.** 1934. The biology and distribution of ants in Hawaiian pineapple fields. Hawaii Pineapple Prod. Sta. Bull. 15: 1–57.

**Pimentel, D.** 1955. Relationship of ants to fly control in Puerto Rico. J. Econ. Entomol. 48(1): 28–30.

**Pinkus, H.** 1913. The life history and habits of *Spalangia muscidarum* Richardson, a parasite of the stable fly. Psyche 20: 148–58.

**Portchinsky, I. A.** 1913. *Muscina stabulans* Fall, mouche nuisible à l'homme et à son ménage, en état larvaire destructeuse des larves de *Musca domestica*. [In Russian, French title, English translation by J. Mullman] Bur. Entomol. Sci. Com. Cent. Bd. Land Adm. Agric. Mem. 10(1): 39 p. (Abstract in Rev. Appl. Entomol. (B) 1: 108–13. 1913.)

**Richards, O. W.** 1961. The theoretical and practical study of natural insect populations. Annu. Rev. Entomol. 6: 147–62.

**Rodriguez, J. G., and C. F. Wade.** 1961. The nutrition of *Macrocheles muscaedomesticae* (Acarina: Macrochelidae) in relation to its predatory action on the house fly egg. Ann. Entomol. Soc. Amer. 54(6): 782–8.

**Sacca, G.** 1964. Comparative bionomics in the genus *Musca*. Annu. Rev. Entomol. 9: 341–58.

**Simmonds, H. W.** 1934. Some difficulties in determining by means of field samples the true value of parasitic control. Bull. Entomol. Res. 39: 435–40.

1940. Investigations with a view to the biological control of houseflies in Fiji. Trop. Agric. 17(10): 197–9.

**Slobodkin, L. B.** 1962. Growth and Regulation of Animal Populations. Holt, Rinehart & Winston, N. Y. 184 p.

**Steve, P. C.** 1959. Parasites and predators of *Fannia canicularis* (L.) and *Fannia scalaris* (F.). J. Econ. Entomol. 52(3): 530–1.

**Sychevskaya, V. I.** 1963. Nekotorye Chalcidoidea, vyvedennye iz pupariev sinanthropnykh mukh v uzbekistane. [Some Chalcidoidea reared from puparia of synantrophic flies in Uzbekistan.] Uzbek Inst. Med. Parasitol. and Helminthol. Samarkand, Uzbek 42(6): 858–64.

**Thompson, W. R.** 1929. On natural control. Parasitology 21: 269–81.

**Vandenburg, S. R.** 1931. Report of the Entomologist. Rep. Guam Agr. Exp. Sta. (1930), p. 23–25.

**West, L. S.** 1951. The Housefly, its Natural History, Medical Importance, and Control. Comstock Pub. Co., Cornell Univ. Press, Ithaca, N. Y. 584 p.

**White, E. B., and E. F. Legner.** 1966. Notes on the life history of *Aleochara taeniata*, a staphylinid parasite of the house fly, *Musca domestica*. Ann. Entomol. Soc. Amer. 59(3): 573–7.

**Wylie, H. G.** 1963. Some effects of host age on parasitism by *Nasonia vitripennis* (Walk.) (Hymenoptera: Pteromalidae). Can. Entomol. 95(8): 881–6.

# Developing Resistant Plants - The Ideal Method of Controlling Insects

By PHILIP LUGINBILL, JR., *Entomology Research Division, Agricultural Research Service*

The most effective and ideal method of combating insects that attack plants is to grow insect-resistant varieties.

E. F. Knipling, U.S. Department of Agriculture, and others have placed much emphasis on the need for nonchemical control of insect pests. The late Rachel Carson in her controversial book, "Silent Spring," emphasized the hazards to man and animals of using chemical pesticides and encouraged the development of other methods for controlling insects. Some of the alternatives she mentioned, which have been investigated for many years by entomologists, include (1) using biological control agents, that is, releasing insects and diseases that kill destructive insects of economic importance; (2) using attractants to lure insects to their death; (3) treatment with radiation or chemosterilants to sterilize either one of a mating pair so the resulting eggs are infertile; and (4) applying electromagnetic energy, that is, destroying insects by using light or sound in some manner.

Miss Carson did not mention the development of plants with natural resistance as a method, perhaps because it is a hidden control. Natural resistance is one of the safest, most practical, and at the same time one of the most economical approaches to biological control yet conceived Indeed, the potential of this method of controlling insect pests staggers the imagination.

What is an insect-resistant plant? A plant has a degree of resistance to insect attack if it sustains less damage at a given level of infestation than the average of other plants grown in the same environment. The reasons a plant is resistant are usually complex, and they vary depending on the insect and the crop. The late R. H. Painter of Kansas State University was an internationally known authority on host plant resistance to insect attack. In his book "Insect Resistance in Crop Plants," he suggested that three interrelated factors affect resistance: (1) Plants may not be acceptable to the insect as food, oviposition site, or shelter. (2) Plants may adversely affect development of the insect, and feeding on them may kill the insect (this is called antibiosis). (3) Plants may be tolerant to insect attack, that is, they may show less damage than susceptible plants or may survive infestations that would severely injure or kill more susceptible plants.

How is a resistant variety obtained? Scientists use three general methods: (1) They introduce a variety that through natural selection already possesses a higher-than-usual level of resistance to a given pest. (2) They expose plants of a given variety to pests and select the resistant individual plants for propagation. (3) They hybridize plants to transfer the demonstrated resistant characteristics or germ plasm to desirable varieties.

Scientists have known for many years that plants and animals differ in their response to the stresses and strains of their environment. Some individuals within the same species or strain are resistant to insects, diseases, and an adverse environment; others are susceptible. Also, when conditions are adverse, according to Darwin's law of survival of the fittest, many individual plants and animals are eliminated from the environment; others survive and perpetuate their kind. Those that survive as a group are then more resistant to the specific environmental hazard than were their parents. This process of natural selection is constantly at work. If nature had been left undisturbed by man from the beginning so that plants had been allowed to develop in a completely wild environment, many species would undoubtedly have evolved with some resistance to insects and diseases.

W. H. Skrdla of the U.S. Department of Agriculture noted that such a resistant wild species actually evolved in the tomato family. Some years ago when scientists were studying plants in Peru, where the tomato originated, they found a wild tomato plant growing in a sugarcane field. This plant, now called the Trujillo tomato, was nearly immune to the diseases caused by the organisms *Fusarium* and *Alternaria* spp.

The environmental factors that caused such a plant to evolve are not known. Perhaps it was the lone survivor of many attacked by *Fusarium*, *Alternaria*, or similar organism. Perhaps it was the only one that survived stress factors other than disease that were present in the environment. Regardless of the nature of the resistance and how it was formed, the resistant germ plasm in this one tomato plant has been a blessing to the American farmer. From the one source of resistance, more than 30 commercial varieties of tomatoes have been developed. How many wild plants of other culti-

Reproduced with permission from *United States Department of Agriculture, Agric. Res. Serv. Prod. Res. Rept.* No. 111:1-14, 1969. Published by the United States Department of Agriculture.

vated crops contain germ plasm that is resistant to disease or insects or both and are growing isolated and unrecognized in their country of origin?

In agriculture today, natural selection is seldom allowed to take place. Man persists in moving plants from place to place so they are exposed to new pest species. Man also is responsible for spreading insects and diseases into new areas before the native plants can develop resistance. Moreover, natural selection and development of resistance to pests, according to Darwin's law of survival of the fittest, do not necessarily assure the evolution of plant varieties that provide the food, fiber, or esthetic features that man may require or desire in a plant variety. Generally the plants that survive pressures, though they possess some resistance to stress factors such as insects, diseases, and drought, usually have few desirable agronomic characteristics. For example, the monococcum group of wheats possesses some resistance to the wheat stem sawfly (*Cephus cinctus* Norton) and to the Hessian fly (*Mayetiola destructor* (Say)) but has few, if any, desirable characteristics from the standpoint of agronomy or food.

Agricultural scientists, therefore, have had to assume the part once taken by nature in breeding and developing plant varieties. In a few years they try to breed selections that might take nature hundreds or even thousands of years to develop. Working in their experimental nurseries, they use sophisticated techniques of breeding and cell analysis. However, they place primary emphasis on plant varieties with desirable agronomic characteristics. Previously they had given scant attention to qualities that render a plant resistant to insects, primarily because most breeders do not have available the necessary insect populations or techniques to test the plants for resistance or susceptibility. As a result, many varieties grown today are acceptable agronomically but have little or no built-in resistance to insects.

The urgent need to find methods other than chemical to control insect pests has now prompted scientists to place increased emphasis on the breeding and development of varieties with insect resistance. The primary goal of this research is to find, identify, and transfer natural insect resistance into plants being developed for commercial production. This research has been very successful. Scientists have found sources of insect-resistant germ plasm, have bred this resistance into commercially acceptable varieties, and have released these resistant plants to farmers.

The impact of this remarkable method of insect control on American agriculture can best be understood by noting a few of the many outstanding accomplishments obtained by using varieties with built-in resistance to insect attack.

## Hessian Fly

The Hessian fly, a black gnatlike insect one-fourth inch long, lays eggs on leaves of wheat plants. Maggots hatching from these eggs kill the wheat by sucking the plant juices. At one time the Hessian fly was considered the most serious pest of wheat in the United States and caused annual losses of millions of dollars. Today such losses are comparatively minor. Rarely, if ever, do Hessian flies destroy an entire field.

The dramatic shift in the economic importance of the Hessian fly was due largely to the development of insect-resistant wheat varieties (fig. 1).

BN–34181

FIGURE 1.—Hessian fly damage in (fall) winter wheat at Lafayette, Ind. Standing rows are W–38 resistant; other rows are susceptible varieties.

Also delaying the seeding of wheat in the fall until most of the fall brood of flies disappeared helped reduce populations. One of the first varieties resistant to the Hessian fly was Pawnee, a hard red winter wheat released in 1942. Twenty-two other varieties with resistance to the Hessian fly are now available: Gage, Omaha, Ottawa, Ponca, Warrior (hard red winter wheats); Ace, Arthur, Benhur, Dual, Georgia 1123, Knox 62, Monon, Redcoat, Reed, Riley, Riley 67, and Todd (soft red winter wheats); Big Club 43 and Poso 42 (soft white spring wheats); and Lathrop, Parker, and Russell (hard red spring wheats).

Many State experiment stations are so convinced of the value of built-in insect resistance that they no longer release a new wheat variety unless it is resistant to the Hessian fly in its genetic makeup.

As an example of the economic importance of wheat varieties resistant to the Hessian fly, let us consider the results of two studies. R. L. Gallun, entomologist, U.S. Department of Agriculture, in cooperation with plant scientists from the various

States conducted a survey, which indicated that $8\frac{1}{2}$ million acres of wheat resistant to the Hessian fly were grown during 1964. R. H. Painter in 1942 estimated that the Pawnee variety yielded about 14 bushels per acre more wheat when the Hessian fly infestation was heavy than did the susceptible Tenmarq. (Heavy infestations of this fly have been known to destroy an entire crop.) If we assume that farmers are growing wheat resistant to the Hessian fly because the infestation is heavy and that the fly can damage and reduce yields by as much as 14 bushels per acre, the growing of resistant rather than susceptible varieties in 1964 prevented a loss of 119 million bushels of grain valued at about $238 million. This loss is probably excessive, because heavy infestations of the Hessian fly would probably not prevail throughout the entire $8\frac{1}{2}$ million acres of resistant wheat planted. However, even if infestations were light and spotty, the savings to the grower would certainly have to be estimated at half this amount or $119 million.

Growing wheat varieties resistant to the Hessian fly is so effective in some areas that reduced infestations approach eradication. After Big Club 43 and Poso 42 were released in California, W. B. Noble, formerly of the U.S. Department of Agriculture, reported in 1950 that as the acreage of these two wheats increased, the fly population decrease was such that the insect forms were difficult to find. Less than 1 percent of the wheat grown in California's Butte and Solano Counties was infested with the Hessian fly, where previously it had been heavily infested. Thus the Hessian fly is no longer a problem in this area of California, even though susceptible wheats are again grown widely.

According to R. H. Painter, a similar situation prevailed in central Kansas just after Pawnee was released.

The successes indicate that populations of the Hessian fly could probably be reduced to minimum levels throughout all infested areas of the United States if the entire wheat acreage were seeded exclusively to resistant varieties.

The reader may then conclude that the Hessian fly is no longer a problem. Unfortunately this is not so. Although planting varieties resistant to insect attack effectively controls the biological forms of this fly in most fields today, these same varieties will not necessarily eliminate the biological forms that may appear tomorrow. Like all living creatures, the Hessian fly in most areas of the country is constantly changing biologically, and it is therefore adjusting to the types of plants now grown within its environment. Its adjustments require a compensating adjustment in resistance of the wheat plant.

W. B. Cartwright, W. B. Noble, and R. L. Gallun of the U.S. Department of Agriculture and other scientists have demonstrated that biological strains of the Hessian fly can successfully attack some resistant wheat varieties now in commercial production (fig. 2). Dr. Gallun reported four such

BN–34182

FIGURE 2.—Hessian fly biological race A developed on susceptible wheat variety Michigan Amber but not on resistant wheats Dual, Purdue 39153, and Purdue 4217.

biological races in Indiana, designated for experiment purposes as races A, B, C, and D. Each was selected from samples of the native field population of the State, and each race was identified on the basis of the insect's ability to infest and reproduce successfully on varieties of wheat with different types of resistance to the Hessian fly.

In the past, biological race A has predominated in wheatfields in Indiana. However, during 1955–65 as the acreage planted to varieties resistant to race A (Dual, Monon, Redcoat, and Reed) increased, populations of race A declined or disappeared in some areas. Race B, which can survive on these varieties, has now become dominant. Fortunately scientists anticipated this change and developed Benhur and Knox 62, which are resistant to this race. These varieties are now being introduced into areas infested with race B.

Growing Benhur, Knox 62, or any other wheat with resistance to one or possibly two biological races will not permanently solve the problem of the Hessian fly. To be really effective, a variety should be resistant to all known races. Experimental multirace-resistant varieties have already been developed and will soon be ready for the farmer. It is only through such a continuing research program that scientists will be able to control the ever-shifting complex of Hessian fly races found throughout the wheat-growing areas of the United States.

## Wheat Stem Sawfly

The wheat stem sawfly is a pest of primary economic importance to wheat growers throughout the Great Plains area of western Canada and the United States. In 1941 the late C. W. Farstad of the Canadian Department of Agriculture estimated that wheat losses caused by the sawfly in Canada amounted to several million bushels annually. In some fields losses ranged from 75 to 85 percent. The North Dakota Department of Agriculture estimated that losses for the State in 1944 approached 3 million bushels of grain valued at about $6 million.

The adult sawfly is a black wasplike insect about one-fourth to one-half inch long. However, it is the larva, which is cream colored, black headed, and about one-half inch long, that causes damage by boring up and down inside the wheat stem and cutting it at the base as the plant matures.

Control of the sawfly is complicated because the insect spends most of its life cycle—from egg to adult—within the wheat stem. Most insecticides do not control the sawfly, but certain cultural practices reduce damage. These practices include rotating the wheat crop with oats and barley, planting nonpreferred host plants in or near infested fields, shallow plowing to expose infested overwintering stubble to the weather, and early harvesting or swathing to remove the grain from the field before it is damaged. However, the most effective control is to grow wheat varieties resistant to sawfly attack.

The story behind the development of Rescue, the first sawfly-resistant wheat, and the manner in which this variety reduced sawfly populations in Montana from epidemic to minimum numbers in only a few years reads like a melodrama. It might well be entitled, "Rescue to the Rescue."

As early as 1920, the wheat stem sawfly was a serious pest of wheat in Canada. In 1924 researchers began to look for a satisfactory control method. They tried biological control, insecticides, and cultural practices. None of these methods, which are usually effective in controlling other insect pests, were entirely satisfactory. Just when it appeared that no satisfactory control could be found, C. W. Farstad, entomologist, and A. W. Platt, agronomist, Canadian Department of Agriculture, working as a research team in Alberta, Canada, noticed that solid-stemmed wheat suffered less damage than hollow-stemmed wheat because the sawfly larva could not complete its development in the solid-stemmed plants. The pith in this wheat provided a mechanical barrier to the feeding and movement of the insect within the stem. This observation proved to be the key to developing varieties resistant to sawflies.

One group of solid-stemmed wheats from Portugal, originally developed to resist lodging along the windy coastal area of this country, was especially resistant to the wheat stem sawfly. Farstad and Platt crossed one of these varieties (S–615) with the variety Apex, a commercial Canadian wheat of high quality and yield. From the progeny of this cross they developed the first wheat variety resistant to the sawfly and appropriately named it Rescue. Rescue proved to be effective in reducing grain losses caused by the sawfly. When it was planted in experimental nurseries at several locations in Montana and Canada, the cutting damage by the sawfly was less than 10 percent in the solid-stemmed variety but as high as 90 percent in the hollow-stemmed variety Thatcher (fig. 3). Therefore in the fall of 1944 when control became unusually urgent, an immediate increase was planned in the small amount of experimental Rescue seed available.

One bushel of Rescue seed was given to the United States in October 1944. The Montana Agricultural Experiment Station and the extension service reported that this 1 bushel was increased to 60,000 bushels in only 24 months. This 1 bushel of seed was shipped to the Arizona Agricultural Experiment Station for planting during the winter of 1944–45. By the spring of 1945, 35 bushels of Rescue seed were sent to and planted

BN–34183

FIGURE 3.—Wheat stem sawfly cutting damage in heavily infested spring wheat: Left, solid-stemmed, resistant Rescue; right, hollow-stemmed, susceptible Thatcher. (Courtesy of Raymond Kelly, Choteau, Mont.)

by the Montana Agricultural Experiment Station. In the fall of 1945, 877 bushels of seed were harvested. These 877 bushels were then shipped to Arizona. In the spring of 1946, 3,870 bushels of seed were harvested and sent to Montana. The train carrying the seed was delayed by a railroad strike, but special authorization was obtained from railroad and labor officials to permit the shipment to proceed. At each designated dispersal point in the infested area, Rescue seed was rationed to farmers as if it were gold. Less than 24 hours after delivery, all the seed had been planted, but planting was at least a month later than usual for this area. As if by a miracle, rain fell soon afterward, and an excellent crop of about 60,000 bushels of grain was obtained in the fall of 1946. By the end of the 1947 season enough Rescue seed was available to plant all the sawfly-infested acreage in Montana. In 1948 the losses due to sawfly damage declined an estimated $4 million. In the following years sawfly populations and wheat losses were reduced to a minimum in all areas where Rescue was grown.

It is difficult to estimate the value of that original bushel of Rescue seed to the Montana farmer. Its development cost only a few thousand dollars for the salaries of a part-time plant breeder and entomologist. Any estimate of the value would probably be too low. However, scientists in this country would have taken at least 10 years to develop a resistant wheat variety similar to Rescue. Also, losses from sawfly damage dropped an estimated $4 million the first year Rescue was widely grown. Thus a conservative estimate for a 10-year period would be that the 1 bushel of wheat saved Montana farmers at least $40 million, and savings to the Canadian farmers were probably many times this amount.

The value of Rescue cannot be measured entirely in terms of dollars saved by preventing damage. Like Hessian fly-resistant wheats, Rescue is also efficient in suppressing sawfly populations. Scientists have observed many times that when Rescue has been grown for 5 or 6 successive years in a field or area, sawfly populations are reduced to such a low level that susceptible varieties can again be grown for several years without damage. However, also like the Hessian fly-resistant wheats, Rescue did not completely solve the sawfly problem. For all its hardiness, Rescue has some undesirable agronomic characteristics. Too, it is not rust resistant and cannot be used in North Dakota and eastern Montana, where both the wheat stem sawfly and rust are serious problems. After the release of Rescue, American and Canadian scientists coordinated their research to develop resistant wheats for all areas infested with sawflies and rust.

Some spring wheat varieties developed by this cooperative research program were Chinook and Cypress from Canada and Sawtana and Fortuna from the United States. The development and release of Fortuna by K. Lebsock and W. B. Noble of the U.S. Department of Agriculture and by the North Dakota Agricultural Experiment Station in 1966 is a milestone, since this is the first variety to have both sawfly and rust resistance. Researchers also developed resistant winter wheats. Rego, a hard red winter wheat resistant to the sawfly, was the first of its kind. The credit for its development goes to E. R. Hehn and C. R. Haun of the Montana Agricultural Experiment Station and F. H. McNeal and P. Luginbill, Jr., of the U.S. Department of Agriculture. Other experimental spring and winter wheat selections with improved resistance to sawflies and rust are in the final stages of testing.

The success of past research efforts in developing agronomically acceptable wheats with good sawfly and rust resistance indicates that perhaps in 5 or at most 10 years resistant varieties will be available for all areas where the sawfly is a problem. The farmer can then control this insect by simply planting the resistant variety recommended for his area.

## Spotted Alfalfa Aphid

The spotted alfalfa aphid (*Therioaphis maculata* (Buckton)) is soft bodied, about one-sixteenth inch long, and straw colored except for six or more rows of dark spots along its back. Both the nymphs and adults suck the juice from leaves or stems of alfalfa and thus reduce stands, lower yields, and increase harvesting costs. This pest was first discovered in southeastern New Mexico early in 1954. It spread rapidly the next 2 years into

30 States across the southern two-thirds of the United States. Damage during 1954–56 was estimated at $81 million.

Soon after its discovery, researchers began to look for a satisfactory method of control. They tried biological control, insecticides, and alfalfa varieties resistant to the aphid. The most successful method was using resistant varieties.

In 1955 State and Federal entomologists and plant breeders, working in teams, began to locate sources of germ plasm that were resistant to the spotted alfalfa aphid. They used this germ plasm to develop resistant varieties that could be adapted to various alfalfa-growing areas. Almost immediately they found that the variety Lahontan was resistant to the aphid. Lahontan was originally developed for resistance to the stem nematode (*Ditylenchus dipsaci* (Kühn)). This resistance to two plant pests was fortunate. Not only did the farmer now have a readily available and efficient method of combating the aphid, but also the entomologist and plant breeder had a valuable germ plasm that could be used to develop other resistant varieties. Unfortunately Lahontan, though highly resistant to the aphid, is only moderately hardy in cold weather and is not adapted to some areas in the extreme Southwest where the aphid is most abundant.

In 1955 a team of Federal and State scientists began a crash program to develop a resistant variety that would meet all the requirements. They searched for resistant plants of African alfalfa, the variety most commonly grown in the Southwest. After much testing and selecting, they located seven clones from Nevada and two from California. From these they developed a new synthetic variety, Moapa, and released it to the growers in 1957. Moapa, in addition to having all the desirable agronomic qualities of the African varieties, was also aphid resistant. Its development and release in 3 years (1955–57) was a remarkable accomplishment, since a new variety usually takes about 10 years to develop.

Subsequently scientists have developed many more resistant varieties. Zia, adapted to New Mexico and parts of Texas, was released in 1958. Cody was released in 1959. It was developed for Kansas and adjacent areas where more than 3 million acres of Buffalo, an aphid-susceptible variety, were formerly grown. Sonora, adapted to Arizona, Nevada, and California, was released in 1963. In 1963 the total land planted to these five varieties was estimated at 1,700,000 acres, and scientists conducting tests in Arizona found that damage in fields planted to susceptible varieties was more than 20 times that in fields planted to resistant varieties. Also at Bakersfield, Calif., large-scale plots were seeded with Moapa and Lahontan and with several susceptible varieties (fig. 4). The results indicated that the resistant varieties outyielded the susceptible by as much as 50 percent.

Thus the annual savings to the grower in dollars must run into millions. For example, nearly 2 million acres of resistant varieties were planted in 1963. The yield on irrigated, suitable land should be about 2 tons per acre. Then since heavy aphid populations throughout this area could reduce the yield 50 percent and since top-quality hay sells for about $35 a ton, the savings for 1963, based on yield alone, could have been $70 million. These loss estimates are probably excessive, because heavy aphid infestations would most likely not occur throughout the entire area planted to resistant varieties. However, even if infestations were light and spotty, the annual savings to the growers would certainly have to be estimated at half this amount or a conservative $35 million.

Perhaps the most amazing aspect of developing varieties resistant to the spotted alfalfa aphid was the low cost. C. H. Hanson of the U.S. Department of Agriculture wrote in the May 1961 issue of "Crops and Soils" that he estimated the cost of de-

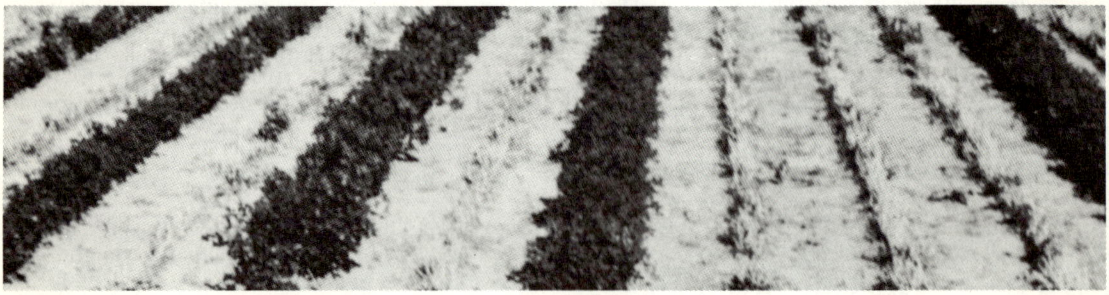

BN–34184

FIGURE 4.—Spotted alfalfa aphid damage in alfalfa varieties at Bakersfield, Calif.: Susceptible Calverde and Arhims (rows killed) and resistant Moapa and Lahontan (uninjured rows).

veloping Moapa to be less than $30,000, the amount paid in salaries to part-time plant breeders and entomologists.

Researchers must continue to develop new alfalfa varieties that are resistant to aphid attack and adaptable to all areas of the country where the spotted alfalfa aphid is a problem. They must also constantly search for insect biotypes, since it is always possible that a biologically different strain will develop in nature that might successfully attack and reproduce on resistant varieties. In California several years ago aphids that appeared to be a biotype developed on one of the parent clones of Moapa growing in a nursery. Although this biotype has not become established in the field and on Moapa, researchers are already trying to find resistant clones that can be substituted for susceptible ones if necessary.

What about other insects that attack alfalfa? Scientists have begun to develop synthetics (varieties) with some degree of resistance to such other major insect pests as lygus bugs (*Lygus* spp.), the clover seed chalcid (*Bruchophagus roddi* Gussakovskii), the alfalfa weevil (*Hypera postica* (Gyllenhal)), the meadow spittlebug (*Philaenus spumarius* (L.)), the potato leafhopper (*Empoasca fabae* (Harris)), and the pea aphid (*Acyrthosiphon pisum* (Harris)). Team, a new high-yielding alfalfa variety with resistance to the alfalfa weevil, pea aphid, and several plant diseases, is a product of this research. Its name honors the cooperative efforts of the Agricultural Research Service and the agricultural experiment stations of Maryland, North Carolina, and other States. The ultimate goal is to provide synthetics with resistance to all major insect pests of alfalfa.

## European Corn Borer

The European corn borer (*Ostrinia nubilalis* (Hübner)) is the most destructive insect pest of corn in the United States. In 1949 alone it caused damage estimated at $350 million.

The borers, believed to have been introduced in broomcorn from Hungary or Italy, were first observed in this country in 1917. Since then the insect has spread rapidly and is now found throughout the corn-growing areas of the United States.

Like many insects, the European corn borer has four stages of development—egg, larva (caterpillar), pupa, and adult (moth). It overwinters as a larva, usually in the cornstalk, changes to a pupa in May, and emerges as a moth from late May to July. The adult moth deposits eggs in irregular white clusters on the underside of corn leaves. They hatch in 5 to 7 days, and the young larvae emerge. They work inside the cornstalk, where they cannot be reached by chemical or other control agents. Here they destroy the food channel of the plant. The resulting plant starvation causes incomplete development of ears and weakens the stalk or ear shank. Eggs of the second brood of borers are laid between July and September and the larvae appear from August through late October.

We have four ways to control this pest. The recommended cultural practices include feeding infested plants to livestock and plowing under the corn so that the adult moth cannot emerge. Insecticides can be used. Insect parasites and diseases can be released to destroy the borer. Corn can be grown that has built-in immunity or resistance to the borer.

The most promising control method is using resistant hybrids. The search for such varieties really began when hybrid corn was introduced, because with open-pollinated corn, borer resistance was difficult to obtain without losing some of the many good agronomic qualities. Also, it was difficult to incorporate insect resistance in open-pollinated corn. However, with hybrid corn, resistance could be rapidly and easily incorporated from one or several resistant inbreds.

The term "resistant" applied to plants attacked by the European corn borer means possessing qualities that inhibit the borer's establishment and survival in or on the plant. Against very young larvae, this type of resistance has its chief effect during the whorl stage of plant growth, when the larvae become established and feed primarily in the whorl. Resistance to feeding on leaves in this area of the plant is usually associated with first-generation corn borers. Inbred lines with high resistance to such first-brood infestations have been released from several State experiment stations. Some inbred lines contributing resistance to hybrids are A392, B2, L317, Oh7, Oh40B, Oh41, Oh43, Oh45, Oh51A, R4, R61, W10, W22, W23, and WR3. (The number of resistant lines in a hybrid determines its level of resistance. For effective resistance, a hybrid should contain at least three.)

The term "tolerance" is usually applied to resistance associated with infestation by the second-generation brood. A line or hybrid with borer tolerance is one that stands up well to infestation, has no shank breakage, and the corn can be harvested efficiently with mechanical equipment. Corn inbreds B14, B15, and B30, all developed in Iowa, contribute this type of resistance.

Ideally the resistant hybrid should be impervious to leaf feeding by the first generation of borers and tolerant to damage by the second generation. Many corn hybrids contain some degree of each or perhaps both types, but they are commercial hybrids with closed or secret pedigrees. Because we do not know the pedigrees, we have difficulty in determining how widely these inbreds or others developed by private industry are used. We also have trouble measuring their effectiveness in reducing borer damage and populations in areas infested with this insect. However, both can be estimated to some degree by referring to survey and research data compiled by scientists who are developing inbreds resistant to the corn borer.

In 1949 when all known control practices, both natural (diseases and parasites) and artificial (cultural and insecticides), were in effect to reduce borer populations, the European corn borer caused the heaviest losses, $350 million. Resistant hybrids were not generally used during 1949, and two were grown for the first time on a limited acreage in Ohio. In recent years annual losses have averaged about $100 million (in 1965, $59 million), and the acres planted to resistant corn hybrids have increased to 30 million (1962 estimate compiled with assistance of G. F. Sprague, U.S. Dept. of Agr.). Meanwhile, of course, DDT and other hydrocarbon insecticides developed after World War II for corn borer control have helped reduce losses. However, the amount of resistant corn now being grown and the scientific data available on its effectiveness in reducing damage indicate that using resistant hybrids has been a major factor in this reduction.

In 1959 F. F. Dicke, formerly with the U.S. Department of Agriculture, an international authority on corn borer resistance, determined at Ankeny, Iowa, that hybrids resistant to corn borer attack reduced losses by at least 20 bushels per acre when borer infestations were heavy (fig. 5). An estimated savings of a fourth this amount, or 5 bushels per acre (allowing for less efficient hybrids and spotted or light borer infestations), on the 30 million acres of resistant corn grown in 1962 would amount to a savings of $150 million. Also, the savings for each year since 1962 would be even greater, since the acreage planted to resistant hybrids increases each year.

Growing resistant varieties not only reduces losses in the current year's crop by preventing feeding damage but also indirectly reduces losses in subsequent crops by reducing borer populations. Tests conducted by F. F. Dicke and L. H. Penny of the U.S. Department of Agriculture indicated that growing resistant corn effectively reduced or suppressed borer populations by as much as 50 to 60 percent. The cumulative effect of an annual reduction in population of this magnitude would soon almost eradicate this insect. Apparently this kind of reduction has already taken place in some areas.

In 1956, G. H. Stringfield, an agronomist in Ohio who pioneered in developing resistance to corn borers, stated that the European corn borer had been reduced to minor importance on farms where hybrids with resistance to borer attack were growing. The 1967 corn borer control recommendation listed in the "Suggested Guide for the Use of Insecticides to Control Insects Affecting Field

BN–34185, BN–34186

FIGURE 5.—European corn borer damage in dent corn at Ankeny, Iowa: Above, susceptible WF9; below, resistant C.I. 31A (WF9 × 458–1).

Corn, Forage, Small Grains, and Soybeans," published by Ohio State University, reads in part as follows: "Prevent damage by selecting resistant hybrids (refer to "Agronomy Guide for Ohio"). Insecticidal treatment in Ohio usually not necessary for level of infestation in field corn. . . ."

T. A. Brindley, U.S. Department of Agriculture, an international authority on the corn borer, stated in an interview by a staff member of the Des Moines Register in 1964 that "we don't have the corn borer losses once common in Iowa . . . because today the farmer is planting improved varieties which are more resistant to the borer. If we had the same kind of corn farmers planted in Iowa back in the early 1950's, this would be a serious infestation . . . now there isn't cause for too much alarm."

Resistance to the corn borer is of primary importance in marketing and selling hybrid seed corn, because seed companies have been forced to develop hybrids that are resistant to borer damage in order to survive in the highly competitive seed corn business. Undoubtedly the three major elements affecting American agriculture—continuing scientific research, competition among growers and seed producers, and desire to increase yields and income—eventually will reduce this most serious pest of corn to one of minor importance throughout the entire corn-growing area of the United States.

## Other Insect Pests

Scientists are also making progress in reducing losses and suppressing populations of other insect pests by developing resistant hybrids and varieties.

W. A. Douglas, U.S. Department of Agriculture, estimated that in Mississippi alone the development of resistant dent corn hybrids—Dixie 18 (yellow), Coker 811 (white), and others—has reduced losses from the corn earworm (*Heliothis zea* (Boddie)) and the rice weevil (*Sitophilus oryzae* (L.)) by $10 million during 1958–68. Damage to dent corn for the entire United States has been estimated at $170 million a year. Much of this loss could probably be prevented by developing and growing resistant hybrids. Many sweet corn inbreds with resistance to attack by the corn earworm have been developed by E. V. Walter, formerly of the Department. These inbreds and others developed by scientists at State experiment stations and in private industry have already been incorporated into some commercial sweet corn hybrids (fig. 6).

In the United States the corn earworm causes an estimated annual loss to sweet corn growers of more than $12 million, even when insecticides are applied for its control. If we add to this amount the estimated cost of chemical sprays and the value of corn forage lost because of chemical residues, the total estimated savings to be realized by using resistant hybrids are more than $17 million annually.

Resistance to insect damage has been built into several crops. Will barley, a variety recently released by Oklahoma State University and the U.S. Department of Agriculture, is resistant to attack by the greenbug (*Schizaphis graminum* (Rondani)) (fig. 7). This variety yields over 50 bushels per acre more than susceptible varieties heavily infested by this insect. The potential savings in grain to the farmer from one such resistant barley alone will therefore be tremendous in years of heavy greenbug populations. Also, wheat varieties resistant to this insect are in advanced stages of development in the breeding program (fig. 8).

For some crops, breeding programs to develop resistant varieties are just starting. The U.S. Department of Agriculture has begun an all-out effort to develop agronomically acceptable cotton varieties with built-in resistance to insects based on scientists' observations that plants without nectar are less attractive than those with nectar to the following insects: Cotton leafworm (*Alabama argillacea* (Hübner)), bollworm (*Heliothis zea* (Boddie)), cabbage looper (*Trichoplusia ni* (Hübner)), pink bollworm (*Pectinophora gossypiella* (Saunders)), and boll weevil (*Anthonomus grandis* (Boheman)) (fig. 9).

Some corn plants are resistant to damage by the corn rootworm (*Diabrotica* spp.) (fig. 10). These plants are being used to develop resistant inbreds and hybrids that can be grown commercially in the infested area. It is hoped that commercial hybrids resistant to rootworms will be available soon, because these beetles are building up resistance to chemical control at an alarming rate.

Researchers are discovering many varieties of commercial sugarcane that are more tolerant than others to attack by the sugarcane borer (*Diatraea saccharalis* (F.)). Growing these varieties will help reduce losses. Scientists are also developing sweetclover varieties resistant to the sweetclover weevil (*Sitona cylindricollis* Fåhraeus) and small grain varieties resistant to the cereal leaf beetle (*Oulema melanopus* (L.)), a pest recently introduced into this country.

Scientists of the U.S. Forest Service have conducted preliminary studies to determine whether some trees are more resistant to insect attack than others. In the Midwest the American elm is commonly infested by the woolly elm aphid (*Eriosoma americanum* (Riley)), and infestations have been observed on the same tree from year to year. Adjoining trees with interlocking limbs are sometimes not infested. The same phenomenon has been observed on Norway spruce infested with the spruce aphid (*Elatobium abietinum* (Walker)). The aphid prefers to infest certain trees rather than others.

BN–34187, BN–34188

FIGURE 6.—Corn earworm damage in sweet corn at Lafayette, Ind. Above, susceptible Golden Cross; below, resistant La2W × 14S.

BN–34189

FIGURE 7.—Greenbug damage in barley at Stillwater, Okla., 1964: Will, resistant; Rogers, susceptible.

BN–34190

FIGURE 8.—Greenbug damage in wheat at Stillwater, Okla.: Left, resistant selection DS28A × Ponca; center, susceptible Ponca; right, resistant DS28A.

BN–34191, BN–34192

FIGURE 9.—Cotton variety Empire (above) completely defoliated by feeding of cotton leafworm under caged conditions at Brownsville, Tex.; nectarless cotton strain (below) undamaged.

Some species of elm are apparently more susceptible to the elm leaf beetle (*Galerucella xanthomelaena* (Schrank)) than others (fig. 11). During the summer of 1965 A. F. Kenyon of Oklahoma State University observed that the Christine Buisman elm tree was nearly defoliated by the elm leaf beetle, but winged, Chinese, and cedar elms suffered little or no damage. The cedar elm also showed resistance to the elm bark beetle (*Hylurgopinus rufipes* (Eichh.)), one of the insect vectors of Dutch elm disease.

In the Midwest the locust borer (*Megacyllene robiniae* (Forster)) is destructive on black locust. However, certain stands of this tree, even in a generally infested area, suffer little damage. The reason some trees remain undamaged could very well be because of natural immunity to the locust borer.

BN–34193

FIGURE 10.—Corn rootworm damage in sweet corn at Lafayette, Ind.: Texas inbred 5014–5–1–1 (leaning) and Texas inbred 5014–2–1–1 (standing), apparently resistant.

BN–34194, BN–34195

FIGURE 11.—Christine Buisman elm tree (left) nearly defoliated by elm leaf beetle at Stillwater, Okla.; winged elm tree (right) in same field shows little or no damage.

C. H. Brett of North Carolina State University found that some vegetable varieties are more resistant to insect feeding than others. The snap bean variety Wade, for example, is more resistant to damage by the Mexican bean beetle (*Epilachna varivestis* Mulsant) than the variety Bountiful (fig. 12). The summer crookneck squash is more resistant to the pickleworm (*Diaphania nitidalis* (Stoll)) than the squash variety Benning Green Tint Scallop (fig. 13).

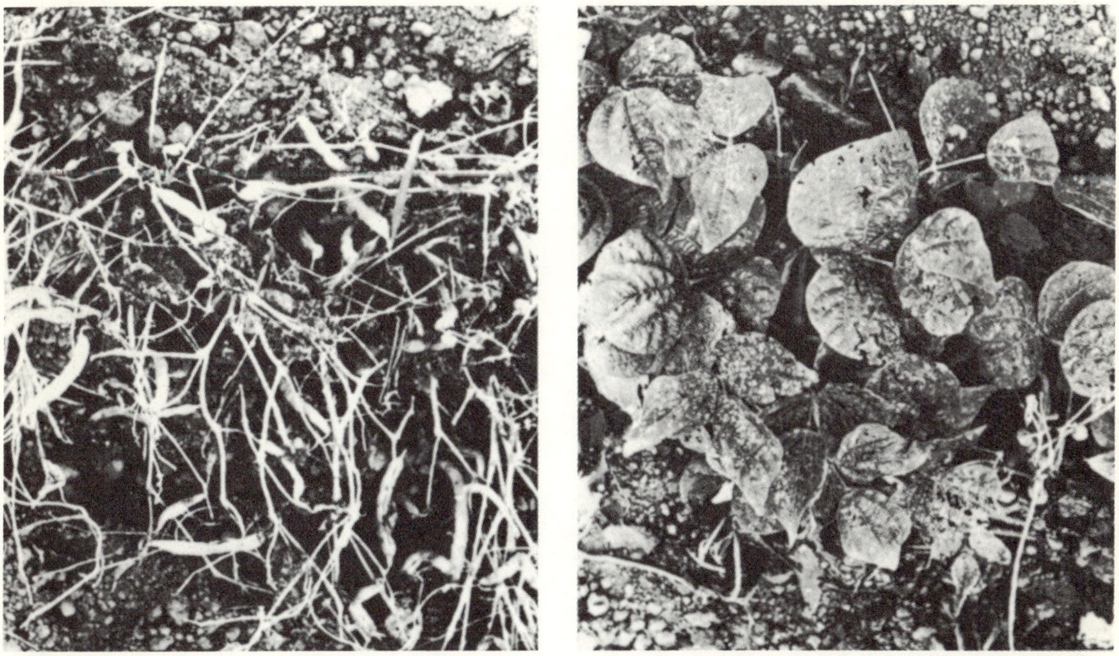

BN–34196, BN–34197

FIGURE 12.—Bountiful snap bean (left), susceptible to Mexican bean beetle; Wade (right), resistant. (Courtesy of C. H. Brett, N.C. State University.)

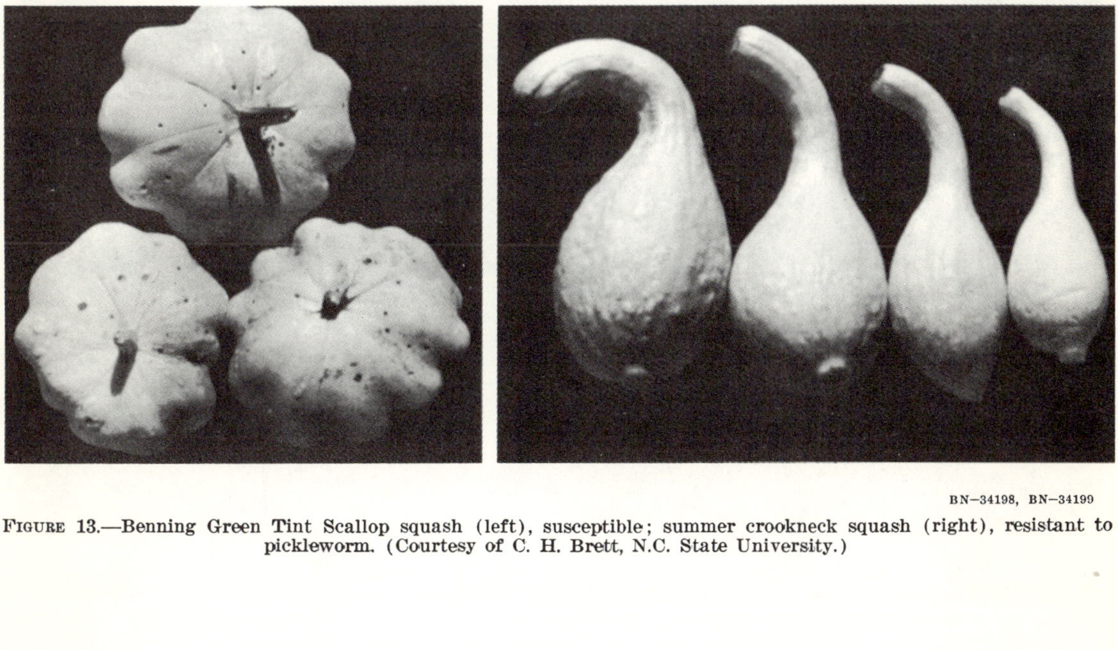

BN–34198, BN–34199

FIGURE 13.—Benning Green Tint Scallop squash (left), susceptible; summer crookneck squash (right), resistant to pickleworm. (Courtesy of C. H. Brett, N.C. State University.)

## Discussion

Let us consider the total estimated value of the research on resistant plants compared to its cost for four of the insects discussed. About 115 professional man-years for the Hessian fly, 92 for the sawfly, 119 for the alfalfa aphid, and 136 for the corn borer have cost about $20,000 a man-year for salary and other expenses, totaling $9.3 million. Federal, State, and private agencies have invested this amount to develop resistant varieties for these four insect pests. The savings to the farmer are about $308 million a year. After a variety or inbred is developed, it is usually grown successfully for about 10 years before it is discarded for biological, agronomic, or other reasons. Thus the cost of research ($9.3 million), the annual savings ($308 million), and the 10 years' use of a variety give a total net monetary value of about $3 billion, a 300:1 return for each research dollar invested. This cost, of course, is based only on preventing damage and yield loss. The estimated value of the bonus effects—eradication or suppression of insect populations and elimination of chemical sprays and residues—is not included.

Research dealing with the resistance of host plants to attack by insects is a challenging new frontier of science. With proper research, many major insect pests causing economic damage to crops will probably be controlled by using multiresistant varieties, perhaps by the end of the century.

To accomplish this, much basic and applied research must be conducted (1) to find new sources of insect-resistant germ plasm in plants and build this resistance into agronomically acceptable plants; (2) to combine the insect-resistant germ plasms for the different insect species into the same crop plant; (3) to study insect biotypes; and (4) to learn more about the nature of plant resistance.

Developing plants with resistance to insect attack is a slow process. Ten years or more are usually required to develop a variety with resistance to a single insect species and probably twice as long for varieties resistant to two or more species. Building multiresistance to pests into our many cultivated crops is not simple. To succeed, entomologists, agronomists, pathologists, geneticists, and others must work together as a research team.

Resistance to pests exists throughout nature. It is all around us in animals and plants. It needs only to be discovered, identified, and put to work to solve many of our most serious pest problems.

# D. Cultural Control

## Effect of Certain Cultural Practices on the Abundance of Tobacco Hornworms, Tobacco Budworms and Corn Earworms on Tobacco After Harvest[1]

R. L. Rabb, H. H. Neunzig, and H. V. Marshall, Jr.
North Carolina Aricultural Experiment Station, Raleigh

Populations of the tobacco hornworm, *Protoparce sexta* (Johannson), the tobacco budworm, *Heliothis virescens* (F.), and the corn earworm, *H. zea* (Boddie), often reach high levels in tobacco fields on secondary or sucker growth after harvest. Stalk destruction as a means of decreasing populations of tobacco insects has been recommended for many years (Metcalf 1909); however, no experimental data have been published. Other cultural practices affecting the amount of sucker growth might also affect the level of insect populations. Also, the possibility exists that the degree of parasitism of tobacco insects may vary on plants having large amounts of foliage as compared with plants having scant foliage affording little protective cover for the host insect.

Shortly after flowering, tobacco plants are usually topped: i.e. the terminal portion of the plants including floral structures is removed. As a consequence, sucker growth in leaf axils is stimulated, and growers remove or control these suckers to permit nutrients to be more available to the marketable leaves. Suckers are removed by hand or controlled with an oil emulsion or with the growth inhibitor MH-30. Hand suckering or control with oil emulsions is only partially effective, and many secondary suckers are produced on the plants. MH-30 is more effective, and usually only small, narrow, malformed leaves develop. Sucker growth is usually not profuse when plants are untopped, and consequently only small amounts of foliage or seed capsules are produced late in the season. Many growers cut tobacco stalks at or near

[1] Contribution from the Entomology Department, North Carolina Agricultural Experiment Station, Raleigh, North Carolina. Published with the approval of the Director of Research as Paper No. 1673 of the Journal Series. Received March 30, 1964; accepted for publication June 8, 1964.

Table 1.—Influence of certain cultural practices on the abundance of late-instar tobacco hornworms, tobacco budworms, and corn earworms on tobacco after harvest. Clayton, N. C.

| Treatment | Number of late-instar larvae per 1/20 acre | | | | | |
|---|---|---|---|---|---|---|
| | 1962 | | | 1963 | | |
| | With parasites[a] | Without parasites | Total | With parasites[a] | Without parasites | Total |
| | *Hornworms* | | | | | |
| 1. Topped—manual suckering | 128 | 28 | 156 | 100 | 15 | 115 |
| 2. Topped—MH-30 | 92 | 3 | 95 | 28 | 3 | 31 |
| 3. Not topped | 77 | 49 | 126 | 63 | 4 | 67 |
| 4. Same as (1)—stalks cut | 44 | 3 | 47 | 15 | 18 | 33 |
| | *Budworms* | | | | | |
| 1. Topped—manual suckering | — | — | 27 | 29 | 122 | 151 |
| 2. Topped—MH-30 | — | — | 3 | 0 | 1 | 1 |
| 3. Not topped | — | — | 20 | 19 | 42 | 61 |
| 4. Same as (1)—stalks cut | — | — | 1 | 0 | 10 | 10 |
| | *Earworms* | | | | | |
| 1. Topped—manual suckering | — | — | 5 | 16 | 120 | 136 |
| 2. Topped—MH-30 | — | — | 1 | 0 | 11 | 11 |
| 3. Not topped | — | — | 7 | 13 | 287 | 300 |
| 4. Same as (1)—stalks cut | — | — | 1 | 0 | 11 | 11 |

[a] Parasites of hornworms included *Apanteles congregatus* (Say) and tachinids; *Cardiochiles nigriceps* (Viereck) and tachinids parasitized budworms, and *Netelia* sp. and tachinids parasitized earworms.

Reproduced with permission from the *Journal of Economic Entomology*, 57:791–792, 1964.
Published by the Entomological Society of America, College Park Maryland.

ground level after harvest; however, these plants frequently produce "ground suckers" on the portion of the plant remaining in the soil.

METHODS.—In 1962 and 1963, experiments were conducted to study the effects of topping, manual and chemical sucker control, and stalk cutting on the production of late-instar hornworms, budworms, and earworms. Each year, 4 cultural treatments (Table 1) were replicated 4 times in a randomized block, each plot consisting of four 22-plant rows. Topping and manual sucker control were accomplished as recommended by the N. C. Agricultural Extension Service. MH-30 was applied as directed by the manufacturer. Immediately after harvest (August 20, 1962, and August 28, 1963) stalks were cut in treatment 4; therefore, all larvae in these plots were produced on ground suckers. Fifth-instar hornworms, except those newly molted, were collected each Monday, Wednesday, and Friday from date of harvest until October 24 in 1962 and October 14 in 1963. In 1962 last-instar budworms and earworms were collected approximately every 3 days from September 13 to October 23. In 1963 all late-instar budworms and earworms were removed from the plants every 3 days from August 30 to November 4.

The incidence of parasitism among hornworms was determined by observation in the field. In 1963, budworms and earworms were confined in containers and allowed to continue their development to establish whether larvae were parasitized or free of parasites.

RESULTS.—Where MH-30 was used or stalks were cut, populations of the 3 species of insects were consistently lower (Table 1). The highest populations of hornworms and budworms were found in both years on plants topped and manually suckered. Earworm populations, however, were heaviest in both years on untopped plants.

The degree of parasitism varied, but no correlation between the level of parasitism and type of cultural practice was detected.

## REFERENCE CITED

**Metcalf, Z. I.** 1909. Insect enemies to tobacco. North Carolina Dep. Agr. Spec. Bull. 1–72.

# Effect of Cultural Practices on Grasshopper Populations in Alfalfa and Cotton[1]

O. L. Barnes, *Entomology Research Division, Agric. Res. Serv., U.S.D.A.*[2]

Abstract

Surveys made in the Salt River Valley of Arizona in 1954–57 indicate that cultural practices, such as shortening the period alfalfa is grown on the same land and rotation with cotton or other cultivated crops, will reduce grasshopper populations in alfalfa. They also suggest that weed control would aid in controlling populations in alfalfa.

In the course of ecological studies on grasshoppers in central and southern Arizona during the period 1936–45, it was observed that populations increased with the age of the alfalfa stand. Plowing up the alfalfa and following it with a cultivated crop such as cotton usually reduced the population drastically. These trends were most noticeable in the Salt River Valley, where cotton and alfalfa are major crops and the migratory grasshopper (*Melanoplus bilituratus* (Walk.)) and the differential grasshopper (*M. differentialis* (Thos.)) are the dominant economic species. It was also noticed that the weediness of alfalfa fields often increased with the age and thinning of the stand.

Alfalfa fields are the leading habitat of economic grasshoppers in the warm valleys of southern Arizona, but an exclusive diet of alfalfa is inadequate for normal development of either the migratory grasshopper, as observed by Brett (1947), Pfadt (1949), and Barnes (1955), or the differential grasshopper, as found in rearing tests at Tempe. Other plants are needed to supplement alfalfa in the diet. Feeding tests at Tempe in 1953 and 1954 showed that diets consisting mostly of alfalfa but including one or more of the weeds commonly found growing in alfalfa fields were adequate for the migratory grasshopper.

Cultural Control.—Grasshopper control by cultural practices has often been mentioned in the literature. Shotwell (1930) found that heavy egg infestations of the migratory grasshopper were virtually destroyed by late-fall plowing and that populations were greatly reduced by summer fallow of infested land. Parker (1957) recommended cultural practices to destroy grasshopper eggs, restrain nymphs, and reduce damage. He also suggested grasshopper-immune crops and alteration of plant habitats as means of reducing infestations. Davis (1949) reported reduction of grasshopper damage by regrassing weedy field margins. On the other hand, Isely (1942) observed that infestations of the differential grasshopper increased in Arkansas areas where soybeans largely replaced cotton in the farming program.

Further investigation of the effects of cultural practices on grasshopper populations in the Salt River Valley of Arizona was started in the spring of 1954 and continued, as time permitted, through the summer of 1957. The studies were limited to fields of alfalfa and cotton, and the main objectives were to determine relative grasshopper populations in alfalfa and cotton, in alfalfa stands of different ages, in clean and weedy alfalfa, and to observe the effect of the age of alfalfa stands on the percentage of weedy fields.

Unfortunately for these studies grasshopper populations were unusually low in the Salt River Valley during this period. Already low in most fields in 1954 from application of insecticides to control grasshoppers or other insects during the preceding 10-year period, the numbers were reduced still further in the next 3 years by greatly increased use of malathion and other chemicals for control of the spotted alfalfa aphid. The summer grasshopper populations in alfalfa declined from 0.31 per square yard in 1954 to 0.068 per square yard in 1957, whereas during the period 1936–44 they averaged 0.6 to 2.65 per square yard. Notwithstanding the low populations, the type of crop, age of alfalfa stand, and whether fields were clean or weedy influenced grasshopper abundance.

Surveys.—Five surveys were made—a spring survey in alfalfa fields in 1954, summer surveys in alfalfa and cotton in 1954 and 1955, and summer surveys in alfalfa in 1956 and 1957. The alfalfa fields were grouped according to the age of the stand, as 1, 2, and 3 years or older. The grasshopper population in each field was estimated by visual inspection while walking across the field or counts by the pointer method (Barnes *et al.* 1953) on 20 one-square-yard areas usually about 10 to 15 steps apart and extending from near the margin into the central part of the field. The method of estimate and the spacing of samples depended on the height and density of the vegetation; the size, activity, and abundance of the grasshoppers; and the size of the fields. Alfalfa fields were designated as clean or weedy by visual inspection. Insofar as practicable, the fields of each crop or age group were representative of different local areas, soil types, and sizes.

Results.—The results of these surveys are summarized in tables 1–3. All averages in these tables are weighted averages.

The two surveys that included both alfalfa and cotton fields were made from August 23 to September 1, 1954, and from July 7 to August 10, 1955. As shown in table 1, the average population in alfalfa was about 15 times that in cotton. Evidently, a rotation from alfalfa to cotton, a crop that requires deep plowing in preparation for plant-

[1] Accepted for publication October 27, 1958.

[2] C. W. McLellan assisted in making grasshopper surveys and tabulating field data.

**Table 1.—Comparison of grasshopper populations in alfalfa and cotton.**

| Year | Number of Fields | Number of Grasshoppers per Square Yard |
|---|---|---|
| | *Alfalfa* | |
| 1954 | 48 | 0.310 |
| 1955 | 50 | .235 |
| Average | — | .272 |
| | *Cotton* | |
| 1954 | 24 | 0.016 |
| 1955 | 17 | .020 |
| Average | — | .018 |

**Table 2.—Relation between the age of alfalfa stands and grasshopper populations.**

| | 1 Year | | 2 Years | | 3–5+ Years | |
|---|---|---|---|---|---|---|
| | Number of | | Number of | | Number of | |
| Date of Survey | Fields | Grasshoppers per Square Yard | Fields | Grasshoppers per Square Yard | Fields | Grasshoppers per Square Yard |
| May 3 to June 14, 1954 | 13 | 0.077 | 9 | 0.611 | 14 | 1.528 |
| Aug. 23 to Sept. 1, 1954[a] | 23 | .293 | 11 | .550 | 14 | .150 |
| July 7 to Aug. 10, 1955 | 17 | .116 | 19 | .276 | 14 | .326 |
| July 3 to July 26, 1956 | 19 | .084 | 17 | .083 | 21 | .177 |
| July 8 to Aug. 13, 1957 | 9 | .037 | 17 | .049 | 28 | .090 |
| Total or average | 81 | .143 | 73 | .261 | 91 | .377 |

[a] In this survey one 1-year stand had four grasshoppers and one 2-year stand five per square yard, caused mainly by migration from an adjacent third-year stand that had recently been plowed. The respective populations in the remaining fields were 0.125 and 0.105 per square yard.

ing and clean cultivation during the growth period, is effective in reducing grasshopper populations.

Table 2 shows that the average population per square yard increased with the age of the stand. Table 3 gives grasshopper populations of alfalfa stands of different ages in both clean and weedy fields, based on data obtained in five surveys between May 1954 and August 1957. Regardless of the age of the stand, populations were decidedly higher in weedy fields than in clean fields. In all fields the average population increased as the age of the stand increased. The percentage of weedy fields also increased with the age of the stand.

In 1954, prior to heavy use of insecticides for control of the spotted alfalfa aphid, *Melanoplus bilituratus* and *M. differentialis* were clearly dominant in alfalfa fields in the Salt River Valley. In 1955 *Trimerotropis pallidipennis* (Burm.) was the leading species, followed by *Encoptolophus pallidus subgracilis* Caudell and *M. differentialis.* In 1956 the leading species were *M. bilituratus, T. pallidipennis,* and *E. p. subgracilis,* in the order given. The same species were dominant in 1957, but *pallidipennis* was more abundant than *bilituratus.* The decline in relative abundance of *bilituratus* and *differentialis* was probably caused by mortality from insecticides applied for aphid control in alfalfa fields and their margins. The favored habitats of *T. pallidipennis* and *E. p. subgracilis* are less restricted, and many nymphs or adults of these species could move into alfalfa fields from other crops or uncultivated land.

**Table 3.—Relation between age of alfalfa stand and alfalfa populations in clean and weedy fields and percentage of weedy fields.**

| | Clean Fields | | Weedy Fields | | |
|---|---|---|---|---|---|
| Age of Stand (Years) | Number | Number of Grasshoppers per Square Yard | Number | Per Cent of Total | Number of Grasshoppers per Square Yard |
| 1 | 50 | 0.039 | 31 | 38 | 0.261 |
| 2 | 38 | .153 | 35 | 48 | .377 |
| 3–5+ | 27 | .207 | 64 | 70 | .453 |
| Average | — | .128 | — | — | .387 |

These surveys indicate that grasshopper populations can be reduced, and the need for chemical control decreased, by cultural practices. Such practices would include shortening the period alfalfa is grown continuously on the same land and rotation of alfalfa with cotton or other cultivated crops. It is likely that weed control in alfalfa would aid in restraining grasshopper infestations.

## References Cited


**Barnes, O. L.** 1955. Effect of food plants on the lesser migratory grasshopper. Jour. Econ. Ent. **48**(2): 119–24.

**Barnes, O. L.,** F. T. Cowan, E. J. Hinman, N. J. Nerney, and Lee Seaton. 1953. Tests of insecticides for grasshopper control—1950, 1951, and 1952. U. S. Bur. Ent. and Plant Quar. **E-860,** 18 pp.

**Brett, C. H.** 1947. Interrelated effects of food, temperature, and humidity on the development of the lesser migratory grasshopper, *Melanoplus mexicanus mexicanus* (Saussure). Oklahoma Agric. Expt. Sta. Tech. Bull. **T-26.**

**Davis, E. G.** 1949. Reducing grasshopper damage by regrassing weedy roadsides and fence rows. U. S. Dept. Agric. Circ. **813.**

**Isely, Dwight.** 1942. Insect problems resulting from changes in agriculture in Arkansas. Jour. Econ. Ent. **35**(4): 473–7.

**Parker, J. R.** 1957. Grasshoppers. A new look at an ancient enemy. U. S. Dept. Agric. Farmers' Bull. **2064.**

**Pfadt, R. E.** 1949. Food plants as factors in the ecology of the lesser migratory grasshopper (*Melanoplus mexicanus* (Sauss.)). Wyoming Agric. Expt. Sta. Bull. **290.**

**Shotwell, R. L.** 1930. A study of the lesser migratory grasshopper. U. S. Dept. Agric. Tech. Bull. **190.**

# E. Chemical Control

## Laboratory Tests of 55 Insecticides on Douglas-Fir Tussock Moth[1] Larvae[2]

ROBERT L. LYON, HAROLD W. FLAKE, JR.[3] and LINDA BALL[4]

Pacific Southwest Forest and Range Experiment Station, Forest Service,
U.S. Department of Agriculture, Berkeley, Calif. 94701

ABSTRACT

Fifty-five insecticides were tested for their toxicity to larvae of the Douglas-fir tussock moth, *Hemerocampa pseudotsugata* McDunnough. The 10 most active insecticides in descending order of average toxicity at $LD_{90}$ were: pyrethrins, Dursban® (*O,O*-diethyl *O*-3,5,6-trichloro-2-pyridyl phosphorothioate), Zectran® (4-dimethylamino-3,5-xylyl methylcarbamate), dichlorvos, GC-6506 (dimethyl *p*-(methythio) phenyl phosphate), aminocarb, tetramethrin, allethrin, Bayer 37298 (*O*-ethyl *O*-2,4,5-trichlorophenyl ethylphosphonothioate), and DDT. Four of the 10 insecticides—pyrethrins, Dursban, Zectran, and DDT—have been field-tested recently with varying results. Several insecticides were examined in the laboratory in more detail in different formulations for such characteristics as rate of toxic action and residual activity.

The U.S. Forest Service's insecticide evaluation project headquartered in Berkeley, Calif., is charged with evaluating and developing safe, effective chemicals for controlling destructive forest insects. Present emphasis is on defoliators previously controlled mainly by DDT. This paper reports laboratory studies made to date of insecticides to control the Douglas-fir tussock moth, *Hemerocampa pseudotsugata* McDunnough, including some preliminary work on the effect of carriers on toxicity. Results of some recent field tests of selected insecticides are summarized.

The Douglas-fir tussock moth defoliates true firs and Douglas-fir. Its outbreaks are generally short-lived but often grow explosively and can reduce tree growth severely (Wickman 1963). Such outbreaks can result in as much as 75% mortality in forest stands (Dodge and Trostle 1964).

METHODS AND MATERIALS.—*Insect*.—Test insects were obtained from eggs collected from northern California. The larvae were reared on an artificial (meridic) diet developed by Lyon and Flake (1966). Most testing was done with 4th-instars, but occasionally the 5th instar also was used.

Treated insects were held for posttreatment observations at room temperature in 8-oz clear plastic containers.[5] An ×-shaped incision through the container lid made it possible to probe the insect to assesss its condition without lifting the lid. In some tests, the insects were held after treatment on cut foliage or potted trees.

*Insecticide Formulation*.—The 55 insecticides tested (Table 1) were formulated as needed and kept under refrigeration (1–3°C) for no longer than 3 days before being replaced by fresh samples. All references to concentration (wt/v) denote the active ingredient except where the purity of the technical product was unknown. The solvents used are identified with the data.

*Treatment Procedure*.—Tests by topical application were performed with an ISCO Model M micro-applicator and a ¼ ml tuberculin syringe fitted with a 27-gauge hypodermic needle. The insects were anesthetized with $CO_2$ and were treated in batches of 10. They were subsequently held for 1-, 3-, 5-, or 7-day mortality counts. Moribund and dead insects were combined in estimates of mortality which were corrected where necessary by Abbott's formula.

The insecticides were applied, usually in acetone, to the thoracic tergites just anterior to the dorsal tussocks. The accuracy of placement of the measured droplet was not critical as determined by a test of pyrethrins and Zectran applied to the dorsum of 3 body regions: (1) The thorax, (2) the anterior abdomen, and (3) the posterior abdomen. Two-sample Wilcoxon tests of the null hypothesis that treatments were the same showed that any differences were not significant at the 5% level.

Most testing by topical application was done by dosing with 1 $\mu$liter of insecticide solution per insect, and carefully assigning insects of comparable weight to each insecticide and concentration. To help reduce variability in our data, we switched later to dosing with 1 $\mu$liter/100 mg body weight. This change would reduce variability if toxicity was unaffected by or was dependent on insect weight; specifically, if toxicity varied inversely with insect weight: toxicity decreased as insect weight increased. Pyrethrins are an example. They tended to be less toxic as insect weight increased. The toxicity of Zectran, on the other hand,

[1] Lepidoptera: Lymantriidae.

[2] Received for publication July 22, 1969.

[3] Present address: Southwestern Region, U. S. Forest Service, 517 Gold Ave., S.W., Albuquerque, N. Mex. 87101.

[4] Current mailing address: 1234 Savoy Street, San Diego, Calif. 92107.

[5] Mutual Plastic Mold Corp., LaMirada, Calif. Cat. no. CP 508.

varied directly with insect weight. Where this relationship held, variability would actually increase if one switched from a constant dosage (1 μliter/insect) to a dosage based on weight (1 μliter/100 mg body weight). However, we assumed the inverse relationship between toxicity and insect weight is the more common. By selecting a relatively narrow weight range in our test insects, we believe either method of dosing will not differ in its effect.

Aerosol tests were conducted in a chamber, 12 in. square and 30 in. high, consisting of a wooden frame covered by a disposable, transparent plastic bag. On each day of testing, the bag was discarded each time a different insecticide or formulation was tested. The aerosol (deodorized kerosene) was produced by a DeVilbiss no. 40 nebulizer using an air pressure of 5 psi. The aerosol was introduced from 1 side 4 in. from the top of the chamber. The aerosol deposited at the base of the chamber was measured with $MgO_2$-coated slides (May 1950). It had an MMD of 49 μ. Dosage was expressed as the amount of spray introduced as determined by pre- and postspray weighing of the nebulizer with its insecticide charge.

**Table 1.—Insecticides, chemical names of proprietary compounds, purity (%), and sources of 55 insecticides tested on Douglas-fir tussock moth.**

Abate®; *O,O*-dimethyl phosphorothioate *O,O*-diester with 4,4′-thiodiphenol; 93.7; American Cyanamid
ACD-7029; 3-isopropyl-4-(methylthio) phenyl methylcarbamate[a]; Allied
Allethrin; 90; City Chemical
Aminocarb (Matacil); 96; Chemagro
Azinphosmethyl; 89; Chemagro
Bayer 25141; *O,O*-diethyl *O*-[*p*-(methylsulfinyl) phenyl] phosphorothioate; 89; Chemagro
Bayer 37289; *O*-ethyl *O*-2,4,5-trichlorophenyl ethylphosphonothioate; 96; Chemagro
Bromophos; 92; Stauffer
Carbanolate (Banol®); 100; Upjohn
Carbaryl; 99; Union Carbide
Carbofuran (NIA 10,242)[a]; Niagara
Carbophenothion; 95; Stauffer
Diazinon; 99; City Chemical
Dichlorvos; 93; Shell
Dicrotophos (Bidrin®); 90; Shell
Dimethoate; 95; American Cyanamid
Dimethrin; 100; MGK
Dimetilan; 90.8; Geigy
Dursban®; *O,O*-diethyl *O*-3,5,6-trichloro-2-pyridyl phosphorothioate; 99; Dow
Gardona®; 2-chloro-1-(2,4,5-trichlorophenyl) vinyl dimethyl phosphate[a]; Shell
GC-3707; dimethyl 3-hydroxyglutaconate dimethyl phosphate[a]; Allied
GC-6506; dimethyl *p*-(methythio) phenyl phosphate[a]; Allied
GC-9879; 4-hydroxy-2-mercaptobutyric acid *gamma*-lactone *S*-ester with *O,O*-diethyl phosphorodithioate[a]; Allied
GC-10101; 4-hydroxy-2-mercaptobutyric acid *gamma*-lactone *S*-ester with *O,O*,-diethyl phosphorothioate[a]; Allied
GC-10284; *O,O*-diethyl phosphorodithioate *S*-ester with *N*-(2-cyanoethyl)-2-mercaptoacetanilide[a]; Allied
GS-10133; 10-chloro-7-(*p*-chlorophenoxy)-6*H*-dibenzo [*c,e*] [1,2] thiazine 5,5-dioxide[b]; Geigy
GS-13005; *O,O*-dimethyl phosphorodithioate *S*-ester with 4-(mercaptomethyl)-2-methoxy-$\Delta^2$-1,3,4-thiadiazolin-5-one; 97; Geigy
GS-13798; 2-methyl-8-quinolyl methylcarbamate; 99.4; Geigy

**Table 1.—(Continued)**

Imidan®; *O,O*-dimethyl *S*-phthalimidomethyl phosphorodithioate; 97; Stauffer
Malaoxon; diethyl mercaptosuccinate *S*-ester with *O,O*-dimethyl phosphorothioate; 99; American Cyanamid
Malathion; 95; American Cyanamid
Mesurol®; 4-(methylthio) 3,5-xylyl methylcarbamate; 99; Chemagro
Mobam®; benzeo[*b*]thien-4-yl methylcarbamate; 98; Mobil
Monocrotophos (Azodrin®); 90; Shell
Naled[a]; Chevron
Phosphamidon; 89.5; Chevron
*p,p′*-DDT; 1,1,1-trichloro-2,2-bis(*p*-chlorophenyl) ethane; 77.2; City Chemical
Propoxur (Baygon®); 100; Chemagro
Pyrethrins; esters of chrysanthemum mono- and dicarboxylic acids; 20; Fairfield
Pyrethrins (OTI); Official test insecticide of Chemical Specialties Manufacturers Association[c]; CSMA
R-3422-S; ethyl-(2-mercaptoethyl) carbamate *S*-ester with *O,O*-dimethyl phosphorodithioate; 98; Stauffer
Ronnel; 95; Dow
S-4084; *O*-*p*-cyanophenyl *O,O*-dimethyl phosphorothioate[a]; Sumitomo
S-4087; *O*-*p*-cyanophenyl *O*-ethyl phenylphosphonothioate[a]; Sumitomo
SD-8530; 3,4,5-trimethylphenyl methylcarbamate[d]; Shell
SD-8786; 2,3,5-trimethylphenyl methylcarbamate[d]; Shell
SD-9077; 4-methoxy-3,5-xylyl methylcarbamate[a]; Shell
Sumithion®; *O,O*-dimethyl *O*-4-nitro-*m*-tolyl phosphorothioate[a]; Chemagro
Tetramethrin (Neopynamin); 90.3; Sumitomo
Thanite®; a mixture of isobornyl thiocyanatoacetate (82%) and related compounds; 82; City Chemical
Trichlorfon; 99; Chemagro
Zectran®; 4-dimethylamino-3,5-xylyl methylcarbamate; 90; Dow
Zectran (MF); 4-methylformamido-3,5-xylyl methylcarbamate[a]; Dow
Zectran (F); 4-formamido-3,5-xylyl methylcarbamate[a]; Dow
Zectran (MA); 4-methylamino-3,5-xylyl methylcarbamate[a]; Dow
Zectran (A); 4-amino-3,5-xylyl methylcarbamate[a]; Dow

[a] Technical; purity unknown.
[b] 0.8 lb/gal concentrate.
[c] 1.148 mg/ml.
[d] 1/4 lb/gal concentrate.

The short-term residual life of naled and Zectran was examined by applying aerosols to freshly cut Douglas-fir foliage (in 2 cases 2-year-old potted trees were used) and caging insects on the treated foliage (1) immediately after spraying, and (2) after aging outdoors in shade. Insects were caged on the watered foliage in quart-size cylindrical paper containers with 40-mesh cloth tops.

The treated foliage was aged in the shade rather than in sunlight because uniform sunlight conditions in the San Francisco Bay Area could not be relied upon. The toxic life of insecticide deposits under shade conditions would be longer than that in sunlight.

RESULTS.—*Screening.*—Complete dosage-mortality regressions were computed for 34 compounds by a modified version of the BMD 03S probit-analysis program (Dixon 1967). The $LD_{50}$ and $LD_{90}$ values along with 95% confidence intervals (Fig. 1)[6] show that only 4 of the 34 insecticides were significantly more toxic

[6] Tables summarizing data are available on request to the Director of the Pacific Southwest Forest and Range Experiment Station, P. O. Box 245, Berkeley, Calif. 94701

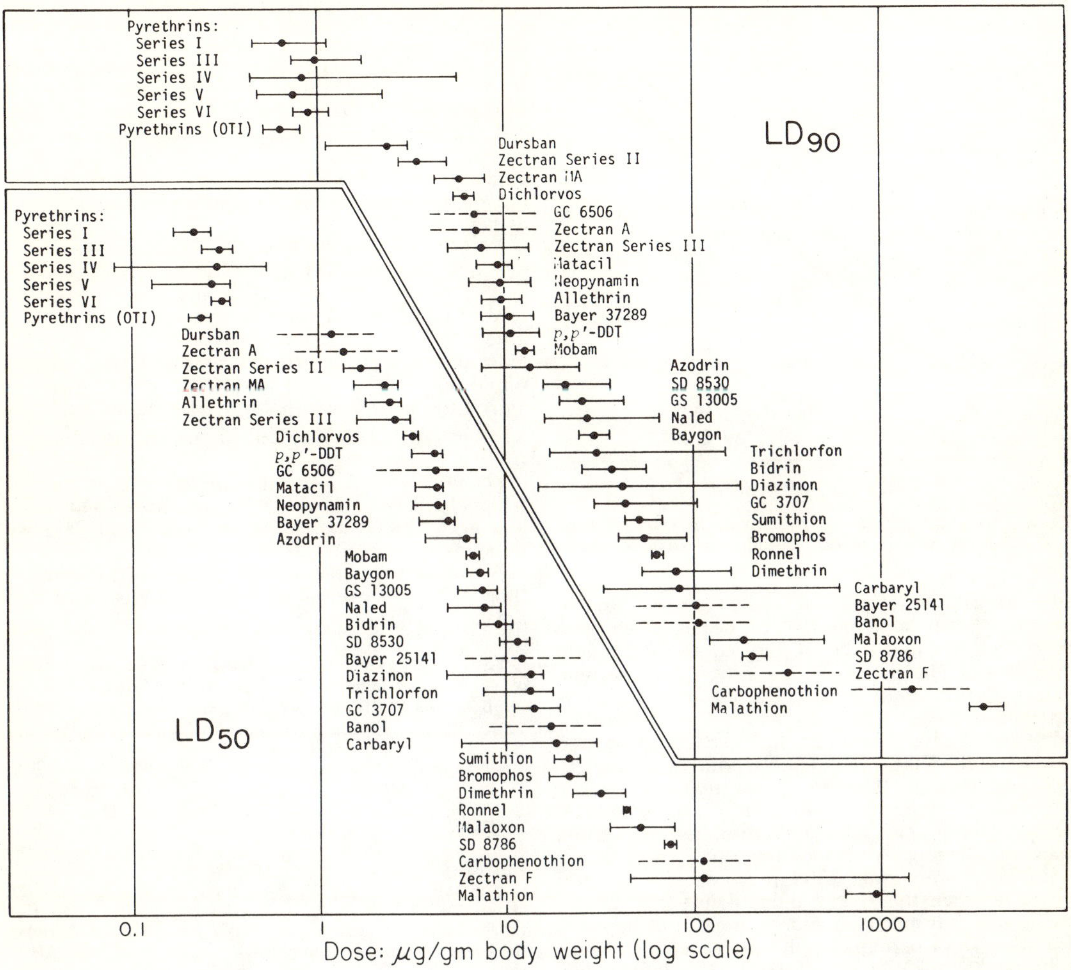


FIG. 1.—Toxicity of candidate insecticides applied by topical application to 4th-instar Douglas-fir tussock moth larvae—$LD_{50}$ and $LD_{90}$ values with 95% confidence limits. Dashed lines indicate data were too skewed or heterogeneous to give useful confidence limits.

than DDT at $LD_{90}$. The compounds along with their toxicity index ($LD_{90}$ DDT/$LD_{90}$ candidate) were: pyrethrins (Test I), 16.4; Dursban, 4.5; Zectran, 3.2 in Test II, and 1.4 in Test III; and dichlorvos, 1.8. At $LD_{50}$, allethrin replaces dichlorvos as more toxic, with an index of 1.7.

In 5 of the 6 tests, conducted over a several months' period, pyrethrins were used as a standard. Each time they gave a consistent estimate of toxicity (Fig. 2). In 2 tests Zectran showed a marked difference in toxicity (significant at $LD_{90}$ but not at $LD_{50}$) which we are unable to explain. In no case were the breakdown products of Zectran identified by Abdel-Wahab et al. (1966) (designated as Zectran A, MA, F, and MF, see Table 1) significantly more toxic than Zectran itself.

Seven additional compounds were given a preliminary test. Three concentrations at 20 insects/concentration were tested with a dose of 1 μliter/100 mg body weight. None of the compounds proved to be more toxic than DDT. These were, in descending order of toxicity: carbofuran, SD-9077, S-4087, azinphosmethyl, S-4084, ACD-9029, and Mesurol.

Fourteen compounds showed little or no activity to the tussock moth at the highest dose tested (100 μg/g body weight): Abate, dimethoate, dimetilan, GC-9879, GC-10101, GC-10284, GS-10133, GS-13798, Imidan, phosphamidon, R-3422-S, Gardona, Thanite, and Zectran MF. Malathion would also fall in this group because its $LD_{10}$ was 372 μg/g body weight.

*Rate of Toxic Action.*—Posttreatment mortality counts were made at 1, 3, and 5 (or 7) days to assess the rate of toxic action of selected insecticides. These counts allowed us to estimate the approximate end point of acute toxic action. They also provide a useful guide for judging an adequate posttreatment holding period for laboratory tests and for helping to select a suitable posttreatment waiting period for mortality checks in field trials. In field tests, residual life and other characteristics of the spray and environment also help influence endpoint mortality. For ex-

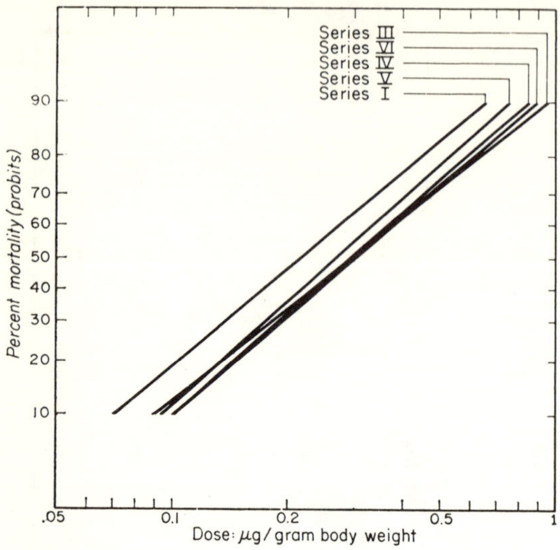


**Fig. 2.—Standard curves for pyrethrins in tests by topical application to 4th-instar Douglas-fir tussock moth larvae.**

ample, laboratory studies at room temperature of compounds with large positive or negative temperature coefficients would be of limited value in field predictions of end point of toxic action because temperature can fluctuate widely during insecticide application.

Many of the insecticides at the 1-day count showed mortality considerably below end point. All insecticides came close to their end point mortality at the 3-day count except the following 10: allethrin, Mobam, carbanolate, carbaryl, dicrotophos, malathion, malaoxon, Zectran A, carbophenothion, and Bayer 25141. There were some shifts in the relative order of toxicity between the 1- and 3-day counts. These results suggest that mortality counts should not be made before the 5th or perhaps 7th day after treatment.

The rate of toxic action of DDT, Dursban, naled, pyrethrins, and Zectran was examined further by using more frequent observations (1 hr to 7 days). They all reached end point mortality, or very nearly so, at the 3-day posttreatment count. Their rate of toxic action increased as dose increased. Dursban, Zectran, and DDT usually acted very slowly during the first few hours.

*Effect of Carrier.*—The choice of carrier can be important in bringing out the full potential of an insecticide. Preliminary work was done on the effect of carrier on toxicity of selected insecticides (Table 2). Carbaryl was 20 times more toxic in an oil carrier than in acetone. The toxicity of allethrin and pyrethrins was much less affected by the type of carrier than was carbaryl. When pyrethrins and Zectran were tested in acetone with 5% corn oil matched against acetone alone, the toxicity of pyrethrins was greatly reduced (probably by degradation). But Zectran toxicity was only moderately reduced. This unequal effect convinced us to abandon a 19:1, acetone: corn oil carrier system for general screening work. Other workers sometimes add a small amount of oil to acetone for screening. We had felt that this would be a worthy idea since field formulations for forest insect control are generally in oil.

Naled by topical application in the carriers, Ortho 38B and Ortho CS-5086,[7] was almost without activity compared to formulation in acetone, cycle oil or diesel oil:xylene (8:2). However, all 4 formulations were equally toxic when applied as aerosols. The Ortho solvents showed little affinity for the cuticle. Topically applied droplets of Ortho solvents remained discreet whereas droplets of acetone and oils spread easily. The danger of overlooking a solvent's potential on the basis of topical application data alone is evident.

*Residual Action on Foliage.*—The toxicity of naled declined quickly (Table 3). Ortho 38B appeared to extend residual life more than any other carrier. The solvent later showed promise as a carrier for naled in a field test on spruce budworm in New Mexico.[8] Foxlene[9] did not appear to aid residual life. An additional test with naled in Ortho 38B and in diesel oil: xylene (8:2) showed that the Douglas-fir condition, whether cut twigs or potted plants, did not influence results.

Zectran showed a longer though still quite short residual life when applied to cut Douglas-fir foliage and aged. Deposit toxicity dropped from 78% with no aging to 52% after 24 hr.

A similar experiment using potted Douglas-fir seedlings showed a complete loss of activity after aging 3 days.

**Table 3.—Toxicity to Douglas-fir tussock moth larvae[a] of aged and unaged naled deposits [b] in different carriers applied to cut Douglas-fir foliage.[c]**

| | 5-day mortality (%) after aging in shade: | | |
|---|---|---|---|
| Carrier | 0 hr | ½ hr | 1 hr |
| Untreated controls | 1 (270) [d] | | |
| Ortho 38B | 100 | | 67 (180) |
| Ortho CS-5086 | 97 | | 36 (120) |
| Ortho 25A | 67 | 57 | 23 ( 90) |
| Diesel oil/xylene/Foxlene (7/2/1) | 97 | 13 | 22 ( 90) |
| Cycle oil | | | 21 (150) |
| Ortho 38D | | | 20 |
| Ortho 38C | | | 10 |
| Diesel oil/xylene/Foxlene (2/1/1) | | | 10 |
| Diesel oil/xylene (8/2) | 97 | 37 | 4 (210) |
| Ortho 25B | 60 | 7 | 0 |
| Ortho 25C | 40 | 0 | 0 |
| Ortho 25D | 33 | 0 | 0 |
| Ortho 38E | | | 0 |

[a] Mixture of 4th and 5th-instars; average weight 134 mg.
[b] Dosage: 150–200 µliter introduced into aerosol chamber; concentration 40 µg/µliter.
[c] Arranged in descending order of effectiveness at 1-hr aging period.
[d] 30 insects/entry or as indicated in figure in parentheses.

[7] The Ortho solvents used in our tests were supplied by Chevron Chemical Co., Richmond, Calif.

[8] P. E. Buffam, G. L. Downing, A. D. Moore, J. E. Larson, H. E. Meyer, and C. F. Tiernan. 1966. Results of aerial spray trials with 5 insecticides against the spruce budworm in New Mexico, June 1965. Unpublished report on file at Pacific SW Forest & Range Exp. Sta., Berkeley, Calif.

[9] Foxlene no. M-466, a terpene polymer supplied by Foxlyn Laboratories, Foxyl Farms Co., Mountain Lakes, N. J.

**Table 2.—Effect of carrier on toxicity of selected insecticides applied topically to Douglas-fir tussock moth larvae[a].**

| Carrier | No. insects each concn | 24-hr mortality (%) at concentration of (μg/μliter) | | | | | | | | | | | |
|---|---|---|---|---|---|---|---|---|---|---|---|---|---|
| | | 0 | 0.01 | 0.02 | 0.05 | 0.1 | 0.2 | 0.5 | 1 | 5 | 10 | 50 | 100 |
| *Carbaryl (5th-instar)* | | | | | | | | | | | | | |
| Acetone | 40 | 0 | | | | 0 | | | 0 | | 5 | 65 | 60 |
| Diesel oil:acetone (1:1) | 40 | 4 | | | | | | | 25 | 100 | 92 | | |
| Dimethylformamide | 40 | 2 | | | | 5 | | | 20 | 80 | 88 | | |
| *Allethrin (5th-instar)* | | | | | | | | | | | | | |
| Acetone | 10 | 0 | | | | 0 | | | 40 | | 100 | | |
| Acetone:DMSO (1:1) | 20 | 0 | | | | 10 | | | 45 | | 90 | | |
| Diesel oil | 20 | 15 | | | | 15 | | | 60 | | 80 | | |
| Cycle oil | 10 | 20 | | | | 40 | | | 100 | | 100 | | |
| *Pyrethrins (5th-instar)* | | | | | | | | | | | | | |
| Acetone | 80 | 0 | | | 10 | 50 | | 100 | | | | | |
| Acetone:DMSO (8:2) | 20 | 0 | | | 25 | 45 | | 95 | | | | | |
| Diesel oil | 20 | 0 | | | 5 | 10 | | 85 | | | | | |
| Cycle oil | 20 | 40 | | | 60 | 60 | | 90 | | | | | |
| *Pyrethrins (4th-instar)* | | | | | | | | | | | | | |
| Acetone | 50 | 0 | 30 | 52 | 86 | | | | | | | | |
| Acetone:corn oil (19:1) | 20 | 0 | 0 | 0 | 5 | | | | | | | | |
| *Zectran (4th-instar)* | | | | | | | | | | | | | |
| Acetone | 70 | 3 | | | | 44 | 80 | 100 | | | | | |
| Acetone:corn oil (19:1) | 70 | 3 | | | | 26 | 48 | 100 | | | | | |

[a] Dose: 1 μliter insect.

*Field Tests*.—Eight of the insecticides reported here were recently field-tested on Douglas-fir tussock moth. They include 4 of the 10 insecticides found most active in laboratory tests. Malathion was tested by helicopter on half-grown larvae at 1 lb/gal no. 2 stove oil per acre, but had little effect on the population.[10] This outcome would have been anticipated from our laboratory data, which show malathion to be barely insecticidal to this insect.

A field test was conducted in 1965[11] on 20-acre plots by helicopter spraying Dursban and Zectran in cycle oil at the rate of 0.25 lb/gal per acre. The insects treated were mainly 3rd instar. The interpretation of the results was complicated by the onset of a polyhedral virus disease which resulted in complete absence of new-generation egg masses in the study area. Nevertheless, observations of surviving larvae 1 day after treatment, before the virus disease began to show up, suggested that the insecticides caused less than 70% mortalities.

SD-8530, Zectran, azinphosmethyl, DDT, and trichlorfon were field tested in 1966[12] by hand application with a Paache Model 62-6 atomizer to simulate 0.15 lb/gal per acre to 1st- and 2nd-instar larvae. Zectran and DDT were applied in oil formulation, the others in water. There was an 84–86% reduction by all insecticides tested except trichlorfon at the 2-day posttreatment observation. SD-8530, Zectran, and azinphosmethyl proved as effective as DDT. Our laboratory tests would not have predicted such good performance by SD-8530 or azinphosmethyl.

[10] C. L. Smola. 1964. Stowe Reservoir Campground aerial spray project for the Douglas-fir tussock moth—Modoc National Forest, 1964. Unpublished report on file at U. S. Forest Serv., Div. of Timber Management, San Francisco, Calif.

[11] D. McComb. 1966. Dursban and Zectran pilot control study for Douglas-fir tussock moth. Unpublished report on file at U. S. Forest Serv., Div. of Timber Management, Insect & Disease Control Branch, Portland, Oreg.

[12] N. E. Johnson. 1967. Douglas-fir tussock moth studies—1966. Unpublished report on file at Weyerhaeuser Co., Forestry Res. Center, Centralia, Wash.

In a small field-test[13] Zectran and pyrethrins were applied by helicopter to 4th-instar Douglas-fir tussock moth larvae set out on an airport runway in petri dishes covered by coarse nylon mesh. A kerosene formulation of Zectran appiled at 0.15 lb/gal per acre gave 86% kill of test larvae. Pyrethrins were not effective at the 0.04 lb/gal per acre rate.

ACKNOWLEDGMENTS.—We express our gratitude to Keturah Pennington for assisting in rearing insects, to Gerald S. Walton for providing the statistical tests and computer programs, and to George L. Downing for providing field samples of Douglas-fir tussock moth eggs.

[13] P. E. Buffam, G. L. Downing, H. E. Meyer, H. W. Flake, Jr., P. E. Letchworth, and T. L. Andrews. 1966. Results of the April 1965 pyrethrins and Zectran spray tests against Douglas-fir tussock moth larvae at the Stockton Airport. Unpublished report on file at Pacific SW Forest & Range Exp. Sta., Berkeley, Calif.

## REFERENCES CITED

**Abdel-Wahab, A. M., R. J. Kuhr, and J. E. Casida 1966.** Fate of $C^{14}$-Carbonyl-labeled aryl methylcarbamate insecticide chemicals in and on bean plants. J. Agr. Food Chem. 14 (3) : 290–8.

**Dixon, W. J. [ed.] 1967.** BMD O3S biological assay: probit analysis, p. 357–368. *In* BMD biomedical computer programs. University of California School of Medicine, Los Angeles.

**Dodge, H. R., and G. C. Trostle. 1964.** Douglas-fir tussock moth. USDA Forest Pest Leafl. 86, 7 p.

**Lyon, R. L., and H. W. Flake, Jr. 1966.** Rearing Douglas-fir tussock moth larvae on synthetic media. J. Econ. Entomol. 59 (3) : 696–8.

**May, K. R. 1950.** The measurement of airborne droplets by the magnesium oxide method. J. Sci. Instr. 27: 128–30.

**Wickman, B. E. 1963.** Mortality and growth reduction of white fir following defoliation by the Douglas-fir tussock moth. Pacific SW Forest & Range Expt. Sta., Berkeley, Calif. U.S. Forest Serv. Res. Paper PSW-7, 15 p.

# Insecticide Resistance Comes of Age

*By* A. W. A. Brown

***Department of Zoology, University of Western Ontario, London, Canada***

It is now 21 years since DDT-resistant house flies were first discovered in northern Sweden, and since the report of Wilson and Gahan that a DDT-resistant strain could be produced by laboratory selection at Orlando. Although the resistance of the San Jose scale to lime-sulfur was detected by Melander in Washington state nearly 60 years ago, the resistance problem as we know it came on the scene with the organic residual insecticides in 1946. Now that this year it has come of age, we might well examine how the economic entomologist has fared with this cuckoo in the nest; is it a clear case of defeat, or is it peaceful coexistence.

A. W. A. Brown

At least we know the nature of the enemy. It is produced by selection, not habituation. You can expose flies to non-killing doses of DDT for 150 generations, as Lüers has done with *Drosophila,* and you don't get resistance. But as soon as the dose begins to kill a segment of the population, as Gaines Eddy and coworkers have shown with body lice, the resistance to DDT develops. As the susceptible individuals are selected out, the more resistant genotypes are left to produce the succeeding generations. The insecticide cannot produce postadaptations; it can only favour the proportion of the preadaptations for resistance already just barely present in the population.

Moreover resistance is specific. DDT selection produces resistance to DDT and its relatives such as methoxychlor and TDE, and not to the cyclodiene derivatives. On the other hand, selection with aldrin or dieldrin induces resistance to the cyclodiene derivatives but not DDT-resistance. Thus there are 2 types of resistance within the organochlorine insecticides, gamma-BHC belonging with the cyclodienes, as Busvine pointed out in 1953. Neither of them carry any cross-resistance to the organophosphorus compounds, at least in insects, although selection with OP compounds usually induces or preserves the organochlorine resistances. Our troubles of today, therefore, involve DDT-resistance, cyclodiene-resistance and OP-resistance, with carbamate-resistance a possible fourth.

Before 1946, our experience with the problem was limited to 10 species (Table 1), with resistance to hydrogen cyanide in 3 scale insects, to arsenicals in 2 cattle ticks, the codling moth and peach twig borer, and to tartar emetic in 2 species of thrips.

Now in 1967 we know resistance to have developed in 224 species of insects and acarines (Table 2). Of these 97 are of public-health or veterinary importance, and 127 attack field or forest crops or stored products. The most widespread type is cyclodiene-resistance, often called dieldrin-resistance, involving 135 species, as compared to 91 with DDT-resistance. The total of OP-resistant species now reaches 54 insects, mites and ticks. Additional to this tabulation is carbamate-resistance, -now developed in the light-brown apple moth in Tasmania, the tobacco budworm in Texas, and the cotton leafworm in Egypt-, and 2 other resistance types. Pyrethrin-resistance has developed in the tobacco moth in Georgia, the house fly in Sweden, the mushroom phorid in England, and certain populations of the bed bug and body louse. Rotenone-resistance developed in the Mexican bean beetle in the northeast USA as long ago as 1949, but nicotine-resistance has never made an appearance.

That makes a total of 9 resistance types developed by insects; the tetranychids or red mites have developed 9 more, to 9 different groups of acaricides (Table 3). The two-spotted mite and the European red mite have become resistant in succession to 5 groups of chlorinated acaricides. *Tetranychus urticae* (formerly *T. bimaculatus* or *telarius*) has gone on to develop resistance to the sulfur-containing Aramite® as well as the nitrogenous azobenzene, having developed resistance to compounds of selenium long ago; the McDaniel mite last year developed resistance to the dinitro compound binapacryl in Washington state. At least 13 species of mites, to be listed later, have developed resistance to OP compounds; moreover in mites, in contrast to insects, selection with chlorinated acaricides produces OP-resistance, and not vice versa.

Among insects of agricultural importance, DDT-resistance made its appearance in 1951 (Table 4), developing in the codling moth and oriental peach moth, the cabbage worm and cabbage looper, the potato beetle and potato tuber-worm, the cotton bollworm and pink bollworm, and the tobacco budworm and northern tobacco hornworm. On grape and apple, 4 species of leafhoppers have become DDT-resistant. On apple the red-banded leaf roller has developed resistance to TDE. The spruce budworm and others not listed here raise the total of DDT-resistant plant-feeding insects to 29.

No less than 53 species of phytophagous insects have developed resistance to the cyclodiene insecticides and gamma-BHC (Table 5). Treatment of cotton with toxaphene and endrin led to resistance in the boll weevil, leafworm, fleahopper and aphid in the southeastern states, and the leaf-perforator and salt-marsh caterpillar in the southwest, besides the leafworm *Spodoptera* in Egypt, the spiny bollworm *Earias* in Israel, and the cotton-stainer *Dysdercus* in Peru. On tobacco, 5 lepidopterous species have developed cyclodiene-resistance, namely the budworm and northern hornworm in the southeast, 2 kinds of *Euxoa* cutworms in Ontario and the splitworm *Phthorimaea* in Queensland. On rice, resistance to BHC has been developed by the stem-borer *Chilo* and the leaf beetle *Lema* in Japan, and by 3 species of leafhoppers in East Asia, while in the Mississippi Valley the rice water weevil has developed resistance to aldrin. On sugar-cane, BHC-resistance in the froghopper *Aeneolamia* in Trinidad was followed by endrin-resistance in the borer *Diatraea* in

[1] Presidential address delivered at the opening session of the annual meeting of the Entomological Society of America in New York City, November 27, 1967.

Table 1.—Resistance to inorganic and botanical insecticides, 1945.

| | | |
|---|---|---|
| Lime-sulphur | 1 | *Aspidiotus* |
| Hydrogen cyanide | 3 | *Saissetia, Aonidiella, Coccus* |
| Lead arsenate | 2 | *Carpocapsa, Anarsia* |
| Sodium arsenite | 2 | *B. microplus, B. decoloratus* |
| Tartar emetic | 2 | *Scirtothrips, Taeniothrips* |
| | 10 | |

Louisiana. Other cyclodiene-resistant insects not listed here include 3 species of thrips, and endosulfan-resistant strawberry aphids in Washington state.

It is with soil insects that this type of resistance has become so striking in the past decade (Table 6), involving 2 species of wireworms in the Carolinas and 1 in Washington state, the white-fringed beetle *Graphognathus* in Alabama, and 4 species of *Diabrotica* rootworms, mainly in corn fields, in the states east of the Rockies. The alfalfa weevil now shows this resistance in areas stretching from California to New York. Among the root maggots, cyclodiene-resistance has developed in 3 species on tobacco, and in the cabbage maggot, turnip maggot, onion maggot and carrot rust fly, and overseas in the barley fly and large bulb fly.

Resistance to organophosphorus compounds is now present in 32 species of plant-feeding arthropods (Table 7). Of these, 10 are tetranychids and 3 others are also mites. Of the 12 hemipterous species, 9 are aphids and 2 are leafhoppers. Caterpillars now OP-resistant include the rice borer in Japan, the apple moth in Tasmania, and the cotton leafworm in Egypt, as well as 2 stored-products insects to be discussed later. Two fruit maggots complete the list.

Too often now has insecticide-resistance endangered the principal crop of an entire country (Table 8). In Ghana the staple export crop is cocoa and its principal pest is the cocoa capsid, combatted almost exclusively with lindane until BHC-resistance spelled its failure in 1961. The same thing happened to toxaphene applied against the leafworm on the cotton crop of Egypt, to parathion and BHC applied against the stem borer of the rice crop in Japan, and to BHC and dioxathion which were already the last resort in cattle dips against ticks in Queensland. The boll weevil, codling moth, sugarcane froghopper and spruce budworm are other examples of similar importance.

Stored-products insects have not yet challenged fumigants with control failures, but significant resistance to BHC has developed in local populations of the rice weevil, granary weevil, maize weevil and red flour beetle (Table 9). Resistance to malathion has developed in the red flour beetle, the tobacco moth and the Indian meal moth. Laboratory selection of the granary weevil has induced resistance to pyrethrins and very recently to the carbamate Baygon or arprocarb.

Table 2.—Numbers of species with resistances of various types, 1967.

| | DDT | Dld | OP | Other | Total |
|---|---|---|---|---|---|
| Diptera | 44 | 68 | 14 | 3 | 88 |
| Lepidoptera | 14 | 14 | 6 | 5 | 34 |
| Hemiptera | 10 | 15 | 14 | 4 | 36 |
| Acarina | 3 | 7 | 16 | 4 | 25 |
| Coleoptera | 5 | 19 | 1 | 1 | 22 |
| Other orders | 15 | 12 | 3 | 2 | 19 |
| Total | 91 | 135 | 54 | 20 | 224 |

Table 3.—Resistance of greenhouse and orchard mites to acaricides.

| | *T. urticae* | *P. ulmi* |
|---|---|---|
| Ovex (Ovotran®) | E. USA 53 | Calif. 54 |
| Genite® and fenson | Calif. 54 | Calif. 54 |
| Chlorobenzilate & chlorobenside | Calif. 55 | Calif. 54 |
| Dicofol (Kelthane®) | Connecticut 55 | W. Va., Penn. 61 |
| Tetradifon (Tedion®) | USA & Europe 64 | Penn. 61 |
| Aramite® | Ohio 55 | None |
| Azobenzene | England 65 | — |
| Selenium | E. USA 43 | — |
| Binapacryl (*T. mcdanieli*) | — | Washington 66 |

Several beneficial species have been found to have developed resistance to insecticides (Table 10), and there may well be more. The predacious mite *Typhlodromus* became as parathion-resistant as its spider mite prey in B.C. orchards. Two species of mayflies have developed more DDT-resistance than the budworm in northeastern spruce forests. The braconid *Macrocentrus* parasitic on the oriental peach moth can develop DDT-resistance, and the boll weevil parasite *Bracon mellitor* has recently been proved to be equally adaptable. Colonies of honey bees at Riverside and Baton Rouge have been found to be more DDT-tolerant than they used to be. Among the vertebrates (Table 11) inhabiting the cotton-growing areas of Mississippi, strong resistance to cyclodiene insecticides has now been developed by 3 species of fish and 2 cricket frogs. Even the pine mouse in Virginia has developed resistance to the endrin used to control it, and considerable DDT-tolerance has been induced by laboratory selection of white mice.

Resistance in insects of medical and veterinary importance has so often been described that we'll simply say that one or both types of organochlorine-resistance is known in 5 species of fleas, 5 ticks, 2 cockroaches, 2 bed bugs, the body louse and 4 lice on livestock (Table 12). The German cockroach, bed bug and Australian cattle tick have gone on to develop OP-resistance. Among the flies (Table 13), some type of organochlorine-resistance has been developed by 20 species, indeed in all genera except *Glossina* tsetse flies and *Phlebotomus* sand flies, and OP-resistance by the house fly, horn fly, Australian sheep blow fly, African faeces fly and a pest midge in Florida.

Among the culicine mosquitoes, 11 species of *Aedes* have developed resistance, 4 of them to OP compounds (Table 14). DDT-resistance has been a problem in the eradication of *A. aegypti* in the Caribbean area. The 2 salt-marsh mosquitoes of Florida were developing DDT-resistance by 1946, and now OP-resistance is becoming significant in *A. taeniorhynchus*. The 2 irrigation-water mosquitoes of California were strongly organochlorine-resistant by 1951, and *A. nigromaculis* has gone on to develop resistance in succession to parathion, methyl parathion and fenthion. Among the 9 species of *Culex* and 2 of *Psorophora* that have developed resistance (Table 15), *C. quinquefasciatus* is especially important since it transmits filariasis; but although it rapidly develops resistance

Table 4.—Plant-feeding insects resistant to DDT and TDE.

| | APPLE AND PEACH | | CABBAGE | | TOBACCO |
|---|---|---|---|---|---|
| LEP. | *Carpocapsa pomonella* | LEP. | *Pieris rapae* | LEP. | *Heliothis virescens*[3] |
| | *Grapholitha molesta* | | *Trichoplusia ni*[1] | | *Protoparce quinquemaculata* |
| | *Argyrotaenia velutinana* | | POTATO | | GRAPE |
| | *Epiphyas postvittana* | LEP. | *Phthorimaea operculella* | HEM. | *Erythroneura variabilis* |
| HEM. | *Erythroneura lawsoniana*[2] | COL. | *Leptinotarsa decemlineata*[1] | | *E. elegantula* |
| | *Typhlocyba pomaria* | | *Epitrix cucumeris* | | |
| | *Myzus persicae* | | COTTON | | |
| | | LEP. | *Heliothis zea* | | |
| | | | *Pectinophora gossypiella* | | |

[1] Also Dld-resistant.
[2] Also OP-resistant.
[3] Also carbamate-resistant.

Table 5.—Plant-feeding insects resistant to cyclodienes and BHC.

| | COTTON | | SUGAR CANE | | RICE |
|---|---|---|---|---|---|
| LEP. | *Alabama argillacea* | LEP. | *Diatraea saccharalis* | LEP. | *Chilo suppressalis* |
| | *Spodoptera littoralis* | | *Aeneolamia varia* | COL. | *Lema oryzae* |
| | *Estigmene acraea* | | TOBACCO | | *Lissorhoptrus oryzophilus* |
| | *Earias insulana* | LEP. | *Protoparce sexta* | HEM. | *Delphacodes striatella* |
| | *Bucculatrix thurberiella* | | *Heliothis virescens* | | *Leptocoris varicornis* |
| COL. | *Anthonomus grandis* | | *Euxoa messoria* | | *Scotinophora lurida* |
| HEM. | *Dysdercus peruvianus* | | *Euxoa detersa* | | |
| | *Psallus serialus* | | *Phthorimaea operculella* | | |
| | *Aphis gossypii* | | | | |

to DDT or to BHC-dieldrin, it is slow to develop OP-resistance and this readily reverts.

During the global campaign to eradicate the transmission of malaria, no less than 35 species of anopheline mosquitoes have developed resistance, 34 of them to dieldrin and 12 to DDT (Table 16). The situation is worst where the same population has developed both resistances, as with *A. sacharovi* in Greece, *A. pharoensis* in Egypt, *A. stephensi* around the Persian Gulf, *A. sundaicus* in Java and *A. albimanus* in Central America; this last species has now shown increased tolerance to malathion. Two species in Mexico have shown behavioristic resistance deriving from an increase in irritability to DDT. That such changes in response can be brought about by selection and may be based on a decrease in detoxication has been proved in the laboratory.

The mechanism of physiological resistance, -what makes a fly now resist DDT-, has been found to be mainly the detoxication of DDT to DDE. Although oxidation to dicofol is possible, the principal process is dehydrochlorination. The DDT-dehydrochlorinase enzyme requires glutathione for its activity, and has been found in DDT-resistant strains of house flies and mosquitoes, and in normally-tolerant species like the Mexican bean beetle and southern tobacco hornworm. DDT-resistant strains of the pink bollworm also detoxify it to DDE, while some resistant strains of the cotton bollworm have increased detoxication, others decreased absorption. Attempts to counteract DDT-resistance by adding DMC or WARF-antiresistant as DDT synergists to inhibit the detoxifying enzyme, or by substituting Dilan® which lacks the aliphatic chlorine to be dehydrochlorinated, were ultimately rendered unsuccessful by the house flies developing resistance to them too. And so DDT-resistance has been met simply by substituting a cyclodiene insecticide, or more recently some OP compound.

Table 6.—Soil insects resistant to dieldrin and heptachlor.

| COLEOPTERA | DIPTERA |
|---|---|
| *Conoderus fallii* | *Hylemya liturata* |
| *C. vespertinus* | *H. platura* |
| *Limonius californicus* | *Euxesta notata* |
| *Diabrotica virgifera* | *Hylemya brassicae* |
| *D. balteata* | *H. floralis* |
| *D. longicornis* | *H. antiqua* |
| *D. undecimpunctata* | *H. arambourgi* |
| *Graphognathus leucoloma* | *Merodon equestris* |
| *Hypera postica* | *Psila rosae* |

The mechanism of cyclodiene-resistance remains unknown, detoxication being negligible or lacking. Some resistant strains and individuals have more body fat than normal, but this is not consistent and cannot account for the characteristic intensity of cyclodiene-resistance. House flies resistant to gamma-BHC absorb less and detoxify more of this compound, to PCCH and thiophenols, but this strong resistance seems to lie in the ganglion itself. Some dieldrin-resistant populations have been successfully combatted by switching back to DDT, as in *Anopheles sundaicus* on the south coast of Java; but usually the next step is to OP compounds with the consequent avoidance of the persistent cyclodiene residues.

The mechanism of OP-resistance is mainly detoxication, by esterase enzymes hydrolysing the insecticide (e.g. parathion) and the real toxicant produced by oxidation in the insect (e.g. paraoxon). Most OP-resistant strains of house flies have a lower aliesterase activity to match the increase in the phosphatase-like detoxication enzyme. Malathion-resistant strains of insects and mites are however characterized by increased carboxyesterase, to attack the unique weak-point of this OP molecule. A mechanism deriving from a less-sensitive cholinesterase target enzyme has been found in cattle ticks and one strain of the two-spotted mite, but in other strains and in all insects studied the mechanism is detoxication. Only malathion seems to be set apart from the other OP compounds in the cross-resistance picture. Although inhibitors for the enzymic detoxication are known, notably EPN for carboxyesterase, they have not seen use as OP synergists. Instead an empirical search is continued for OP compounds to which minimal cross-resistance is shown and minimal resistance is developed, such as dimethoate in house flies, and bidrin and certain phosphoroamidothioates in red mites.

Carbamate-resistance also derives from esterase detoxication, and varying cross-resistance is shown between OP compounds and carbamates. Oxidative enzymes in the microsome cellular fraction are also involved in the detoxication of carbamates and also OP compounds.

Genetical studies of the different resistance types have shown that most of them are mainly caused by a single gene (Table 17); this was found for DDT-resistance in 13 species, for dieldrin-resistance in 19 species and for OP-resistance in 6 species. Multiple-factor inheritance was found only in arsenic-resistance and low-order tolerances to carbamates, malathion and DDT in 4 species.

Table 7.—Resistance of plant-feeding pests to OP compounds.

| ACARINA | HEMIPTERA | LEPIDOPTERA |
|---|---|---|
| *Tetranychus urticae* | *Myzus persicae* | *Chilo suppressalis* |
| *T. cinnabarinus* | *M. cerasi* | *Epiphyas postvittana* |
| *T. tumidus* | *Sapaphis pyri* | *Spodoptera littoralis* |
| *T. mcdanieli* | *S. plantaginis* | *Ephestia cautella* |
| *T. schoenei* | *Therioaphis maculata* | *Plodia interpunctella* |
| *T. canadensis* | *Chromaphis juglandicola* | DIPTERA |
| *T. atlanticus* | *Aphis pomi* | |
| *T. pacificus* | *Phorodon humuli* | *Dasyneura pyri* |
| *Panonychus ulmi* | *Toxoptera graminum* | *Tryporyza incertula* |
| *P. citri* | *Erythroneura elegantula* | |
| *Vasates schlechtendali* | *Nephotettix cincticeps* | |
| *Aculus cornutus* | *Blissus leucopterus* | |
| *Bryobia praetiosa* | | |

Table 8.—Resistant insects of principal crops of countries.

| | | | |
|---|---|---|---|
| Ghana | Cocoa | *Distantiella theobroma* | $\gamma$-BHC |
| U.A.R. | Cotton | *Spodoptera littoralis* | Toxaphene |
| S-E. U.S.A. | Cotton | *Anthonomus grandis* | Endrin |
| Trinidad | Sugar-Cane | *Aeneolamia varia* | $\gamma$-BHC |
| Japan | Rice | *Chilo suppressalis* | BHC, OP |
| Wash. & B.C. | Apple | *Carpocapsa pomonella* | DDT |
| New Brunswick | Spruce | *Choristoneura fumiferana* | DDT |
| Queensland | Cattle | *Boophilus microplus* | Dld, OP |

Table 9.—Resistance developed by stored-products insects.

| | | | |
|---|---|---|---|
| Field | Gamma-BHC | *Sitophilus oryzae* | Queensland & England |
| | | *S. granarius* | South Africa |
| | | *S. zea-mays* | Kenya |
| | | *Tribolium castaneum* | Kenya |
| | DDT | *Sitophilus oryzae* | Queensland |
| | Malathion | *Tribolium castaneum* | Nigeria, England, U.S.A. |
| | | *Ephestia cautella* | Georgia, Florida |
| | | *Plodia interpunctella* | Georgia |
| Laboratory | DDT | *Tribolium castaneum* | PIL, Slough |
| | Pyrethrins | *Sitophilus granarius* | Parkin & Lloyd |
| | Arprocarb | *Sitophilus granarius* | Kumar & Morrison |

Table 10.—Beneficial invertebrates resistant to insecticides.

| | | | |
|---|---|---|---|
| Honey bee | *Apis mellifera* | DDT | California |
| Parasitic wasp | *Macrocentrus ancylivorus* | DDT | Ontario |
| Mayfly nymphs | *Heptagenia hebe* & *Stenonema fuscum* | DDT | New Brunswick |
| Predacious mite | *Typhlodromus occidentalis* | Parathion | Brit. Columbia |

Table 11.—Vertebrates resistant to chlorohydrocarbon insecticides.

| | |
|---|---|
| Fish | Mosquito-fish, *Gambusia affinis*<br>Golden Shiner, *Notemigonus crysoleucas*<br>Green Sunfish, *Lepomis cyanellus*<br>Bluegill Sunfish, *Lepomis macrochirus* |
| Amphibians | Northern Cricket-Frog, *Acris crepitans*<br>Southern Cricket-Frog, *Acris gryllus* |
| Mammals | Laboratory Mouse, *Mus musculus*<br>Pine Mouse, *Pitymys pinetorum* |

The OP-resistance gene has always proved to be dominant, and so has that for carbamate-resistance. DDT has usually been found to be recessive, except in the house fly. Dieldrin-resistance is nearly always intermediate, the dosage-mortality lines of the heterozygotes lying midway between those of either parental strain and sharply distinct from both.

Many resistance genes have now been precisely located on certain chromosomes by linkage studies with marker strains. In *Drosophila melanogaster,* the principal resistance gene is located near the middle of chromosome 2, with a second gene in the middle of chromosome 3. In the house fly, the dominant gene that causes strong DDT-resistance, called *Deh* because it determines the dehydrochlorinase enzyme, is located on chromosome 5, while a second gene ***kdr*** causing knockdown-resistance and determining nerve sensitivity to DDT is on chromosome 2. The main OP-resistance gene, called *a* from its association with reduced aliesterase activity, is located on chromosome 5, as also the gene for carbamate-resistance. The genes for resistance to dieldrin and BHC have been found in different linkage groups in different strains. In culicine mosquitoes, the genes for DDT-resistance and for dieldrin-resistance in *Aedes aegypti* lie very close together on chromosome 2, whereas in *Culex quinquefasciatus* the dieldrin-resistance gene is on chromosome 3. In the malaria mosquito *Anopheles quadrimaculatus* the dieldrin-resistance gene is on its chromosome 3, and in *A. pharoensis* the DDT-resistance is linked with dieldrin-resistance on the same chromosome.

These genetical studies shed light on the observation that *Aedes aegypti* and *A. taeniorhynchus* develop both organochlorine resistances when treated either with DDT or with dieldrin, whereas *C. quinquefasciatus* develops them separately; and on the common experience that OP selection induces or maintains DDT-resistance in the house fly. They suggest that the reason the European corn borer in North America has not developed DDT-resistance is a lack of the gene for it. The stable fly has also failed to develop DDT-resistance in North America, although it has done so in Europe. Dieldrin-resistance in *Anopheles gambiae,* due to a single powerful gene, has developed in West Africa but not in continental East Africa. One species or subspecies may readily develop a resistance, while another close relative may not; examples with respect to dieldrin-resistance are *Musca domestica* versus *nebulo* in India, *Hylemya liturata* versus *platura* in North America, and the cocoa capsid versus the brown capsid in Ghana; as a result the first to develop resistance became the dominant species.

DDT-resistance usually develops after an initial latent period of several generations before it steeply increases. Evidently the genome as a whole must be remodelled so that insects with the DDT-resistance gene are no longer handicapped. They must have been at first, for otherwise they would have been DDT-resistant from the first. Cyclodiene-resistance on the other hand develops without delay, not needing new supporting alleles. Indeed, dieldrin-resistant strains of the cabbage maggot in eastern Canada live twice as long and produce twice as many eggs, due to an associated multifactorial system. OP-resistance in

Table 12.—Resistance to arthropods of medical and veterinary importance.

| | | | |
|---|---|---|---|
| Lice | | Fleas | |
| *Pediculus corporis* | DDT, DL | *Zenopsylla cheopis* | DDT, DL |
| *Linognathus vituli* | DDT | *Z. astia* | DDT, DL |
| *Haematopinus eurysternus* | DDT | *Pulex irritans* | DDT, DL |
| *Bovicola limbata* | DL | *Ctenocephalides canis* | DDT, DL |
| *B. caprae* | DL | *C. felis* | DDT, DL |
| Cockroaches | | Ticks | |
| *Blattella germanica* | DDT, DL, OP | *Boophilus microplus* | DDT, DL, OP |
| *Blatta orientalis* | DL | *B. decoloratus* | DDT, DL |
| Bed bugs | | *Amblyomma americana* | DL |
| *Cimex lectularius* | DDT, DL, OP | *Rhipicephalus sanguineus* | DL |
| *C. hemipterus* | DDT, DL | *R. evertsi* | DL |
| | | *R. appendiculatus* | DL |
| | | *Dermacentor variabilis* | DDT, DL |

Table 13.—Resistance of the 3 types developed in flies.

| | | | |
|---|---|---|---|
| *Musca domestica* | DDT, DL, OP | *Culicoides furens* | DL |
| *Chrysomyia putoria* | DDT, DL, OP | *Hippelates collusor* | DL |
| *Glyptotendipes paripes* | DL, OP | *Fannia canicularis* | DDT |
| *Chironomus zealandicus* | DL | *Phormia terraenovae* | DDT |
| *Haematobia irritans* | DL, OP | *Simulium aokii* | DDT |
| *Phaenicia cuprina* | DL, OP | *S. venustum* | DDT |
| *P. sericata* | DL | *Drosophila melanogaster* | DDT |
| *Stomoxys calcitrans* | DDT, DL | *D. virilis* | DDT |
| *Psychoda alternata* | DDT, DL | *Leptoconops kerteszii* | DDT |
| *Leptocera hirtula* | DDT, DL | *Chaoborus astictopus* | DDT |

Table 14.—Resistance in aedes mosquitoes.

| DDT | | Dieldrin | Org. Phos. |
|---|---|---|---|
| *aegypti* | *cantator* | *aegypti* | *taeniorhynchus* |
| *sollicitans* | *detritus* | *sollicitans* | *nigromaculis* |
| *taeniorhynchus* | *cantans* | *taeniorhynchus* | *melanimon* |
| *nigromaculis* | *albopictus* | *nigromaculis* | *dorsalis* |
| *melanimon* | *vittatus* | *melanimon* | |
| *atropalpus* | *vexans* | *cantator* | |

insects develops very slowly, even the house fly requiring more than 100 generations of laboratory selection. It usually first develops as a weak polyfactorial system before the strong monofactorial resistance eventually declares itself. Even then it is liable to revert, since the OP-resistant individuals are usually handicapped by low fertility. Carbamate-resistance also develops very slowly. This slowness cuts both ways, for when the apple maggot fails to drift to parathion, the body louse to carbaryl, and the boll weevil to methyl parathion, we cannot know whether they won't get there eventually. By contrast, DDT-resistance may reach an upper plateau of intensity, as it has in some anopheline mosquitoes, where an increase in dosage or frequency of applications is sufficient to regain practical effectiveness.

Resistance may arise simultaneously at many points which then coalesce, as in house flies, mosquitoes and body lice; but in some plant-feeding insects it appears in certain fields or orchards and then spreads from them. Aldrin-resistance in the western corn rootworm diffused from a single spot in Nebraska into 7 states in 5 years. Resistance develops and spreads faster not only the higher the selection pressure but also the wider the area covered by the insecticide, since fewer susceptible individuals are left to dilute the resistant survivors of the applications. Residual insecticides are more liable to induce resistance, because they continue to select long after they have been applied. Resistance also develops faster in subtropical climates than temperate, because there are more generations per year. The boll weevil in the southern USA, the cotton leafworm in the Nile Delta, and mosquitoes in interior California and coastal Florida are perfect examples of such consequences.

The substitution of measures which achieve 100% control over circumscribed areas and leave no penumbra of partial kill may be expected to avoid or delay resistance; such would be the use of pyrethrins or nicotine instead of organochlorine insecticides, or root maggot control by side dressings instead of broadcast treatments, or mosquito control at isolated bases in Canada surrounded by large untouched populations. Mixtures of 2 different insecticides have been advocated, but laboratory model experiments on the house fly, bed bug and the German cockroach have

Table 15.—Resistance in culex and psorophora mosquitoes.

| DDT | Dieldrin | Org. Phos. |
|---|---|---|
| *quinquefasciatus* | *quinquefasciatus* | *quinquefasciatus* |
| *pipiens* | *pipiens* | *tarsalis* |
| *tarsalis* | *tarsalis* | |
| *coronator* | *tritaeniorhynchus* | |
| *tritaeniorhynchus* | *restuans* | |
| *peus* | *Ps. confinnis* | |
| *salinarius* | *Ps. discolor* | |
| *erythrothorax* | | |

Table 16.—Resistant anophelines, in order of first appearance.

| DDT | Dieldrin | |
|---|---|---|
| *sacharovi* | *sacharovi* | *m. flavirostris* |
| *sundaicus* | *quadrimaculatus* | *pharoensis* |
| *stephensi* | *gambiae* | *albitarsis* |
| *subpictus* | *subpictus* | *labranchiae* |
| *albimanus* | *coustani* | *strodei* |
| *pharoensis* | *pulcherrimus* | *triannulatus* |
| *quadrimaculatus* | *albimanus* | *sundaicus* |
| *annularis* | *pseudopunctipennis* | *aconitus* |
| *culicifacies* | *aquasalis* | *neomaculipalpus* |
| *albitarsis* | *culicifacies* | *crucians* |
| *nunez-tovari* | *vagus* | *filipinae* |
| *aconitus* | *barbirostris* | *maculipennis* |
| | *annularis* | *l. atroparvus* |
| | *sergenti* | *m. messeae* |
| | *fluviatilis* | *rangeli* |
| | *splendidus* | *philippinensis* |
| | *stephensi* | *funestus* |

found that resistance to both is developed as fast as to one or the other used separately. Alternations or rotations of sulfur-containing, chlorinated and OP acaricides have delayed resistance to any one of them by the European red mite in eastern orchards. Selection of the boll weevil with toxaphene mixed with DDT has not led to resistance to the mixture, but has induced a high tolerance to OP compounds and carbamates. Pairs of negatively-correlated insecticides, where resistance to the one is accompanied by an increase in susceptibility to the other, have been discovered in *Drosophila* between DDT and phenylthiourea, the DDT-resistance gene allele giving the increased PTU-susceptibility. Examples of negative correlation have turned up in the house fly, body louse and boll weevil, but they have not held good for all populations, being fortuitous rather than tied to the same gene allele.

As insecticide-resistance comes of age, we find that there are now certain species that have run the whole gamut of resistance to DDT, cyclodienes, OP compounds and sometimes carbamates; examples are the house fly, the cattle tick and the sheep blow fly in Australia, the cotton leafworm in Egypt, and the two-spotted mite. Although there are few populations even of these species that cannot be controlled by some compound or other, there is a real need for insecticides of entirely new chemical configuration and toxic action. Such are the new acaricides Morestan (a dithiol quinoxaline) and Phenoflurazole (a fluorinated benzimidazole), and the substituted ureas and salicylanilides appearing as candidate insecticides. They may not be resistance-proof, for even *Bacillus thuringiensis* toxin has induced resistance in house flies (Table 18). The new chemosterilants in laboratory selection experiments have had successively diminishing sterilizing effect, although the situation often reverses due to the inheritance of deleterious mutations. We would therefore not assume that any chemical would never induce resistance, even if it were an insect hormone.

At present our real need is a means of knowing where each pest population stands with respect to resistance, since the practical danger is not the lack of some insecticide to do the job, but of incurring a control failure due to unsuspected resistance to the compound being used. Standard test methods for resistance, judiciously applied, will tell us the answer by asking the insects themselves, as they have done over the past decade for malaria mosquitoes and other insects of concern to WHO. It is a

Table 17. Instances of monofactorial inheritance of insecticide-resistances.

| Dld-Res. | DDT-Res. | OP-Res. | Carbamate-R. | HCN-Res. |
|---|---|---|---|---|
| I *Musca* | D *Musca*[3] | D *Musca* | D *Musca* | I *Aonidiella* |
| I *Anopheles,* 7 spp.[1] | R *Anopheles,* 5 spp. | D *Chrysomya* | D *Drosophila* | |
| I *C. fatigans* | D *C. fatigans* | D *C. tarsalis* | | |
| I *A. aegypti* | I *A. aegypti* | D *Tetranychus,* 2 spp. | | |
| I *Phaenicia* | I *Euxesta* | D *Boophilus* | | |
| I *Chrysomyia* | D *Drosophila* | | | |
| I *Hylemya,* 2 spp. | R *Blattella* | | | |
| I *Euxesta* | R *Boophilus* | | | |
| I *Blattella* | I *Pediculus* | | | |
| I *Cimex* | | | R recessive | |
| I *Pediculus* | | | I intermediate | |
| D *Boophilus*[2] | | | D dominant | |

[1] D in Ivory-coast *A. gambiae.*
[2] Some overdominance.
[3] 2ndary *kdr* gene is R.

pleasure to report that the first test methods for insects of agricultural importance were drawn up by an E.S.A. special conference last month; consisting of a standard method for the boll weevil and a tentative method for red mites, they constitute a start which will be pursued by FAO itself as well as E.S.A. These 21 years have taught us that there is no single doctrinaire answer to the resistance problem; we can only deal with it in detail, by having a good intelligence system, and by understanding the biochemical and genetical principles underlying it. Insecticide-resistance has been a challenge that has involved us in every facet of modern biology, and thereby has enriched our professional life immeasurably. With the chemical synthetic activities of industry, the strategic recommendations of government, and the basic investigations of universities, we shall continue to meet that challenge.

NOTE

Full citations for the instances of resistance described above may be found in: Brown, A. W. A. (1968) *In* Pesticides, ed. R. H. White-Stevens, Vol II, Chapt. 12. Marcel Dekker, New York.

Table 18.—Laboratory induction of resistance to "non-insecticides."

| | | |
|---|---|---|
| Chemosterilant | | |
| Apholate | *Aedes aegypti* | Hazard et al. 1964 |
| Metepa | *Aedes aegypti* | Klassen & Matsumura 1966 |
| Metepa | *Musca domestica* | Sacca & Scirocchi 1966 |
| Hempa | *Aedes aegypti* | George & Brown 1967 |
| Bacterial Toxin | | |
| Thuricide | *Musca domestica* | Harvey & Howell 1965 |

## ESA MEETING ATTENDANCE

The total registration at each meeting of the Society is always a source of interest. Figures, which include invited guests and non-members, as well as the members and their wives for the last 13 meetings are assembled here. We do not have accurate figures for the 1953 or the 1954 meeting.

| Date | Place | Total Attendance |
|---|---|---|
| Nov. 28–Dec. 1, 1955 | Cincinnati, Ohio | 625 |
| Dec. 27–30, 1956 | New York, New York | 650 |
| Dec. 2–5, 1957 | Memphis, Tennessee | 717 |
| Dec. 1–4, 1958 | Salt Lake City, Utah | 679 |
| Nov. 30–Dec. 3, 1959 | Detroit, Michigan | 985 |
| Nov. 28–Dec. 1, 1960 | Atlantic City, New Jersey | 858 |
| Nov. 27–30, 1961 | Miami, Florida | 1,130 |
| Dec. 3–6, 1962 | Phoenix, Arizona | 1,024 |
| Dec. 2–5, 1963 | St. Louis, Missouri | 1,084 |
| Nov. 30–Dec. 3, 1964 | Philadelphia, Pennsylvania | 1,211 |
| Nov. 29–Dec. 2, 1965 | New Orleans, Louisiana | 1,620 |
| Nov. 28–Dec. 1, 1966 | Portland, Oregon | 1,210 |
| Nov. 27–30, 1967 | New York, New York | 1,446 |

## MICHIGAN ENTOMOLOGICAL SOCIETY

Recently a copy of the December 1967 *Newsletter, Michigan Entomological Society* was received at the ESA office. This quarterly publication is offset printed and is well edited. An illustrated article on drawing figures directly from a stereomicroscope was of particular interest. Entomologists in Michigan, and elsewhere as well, may wish to write:

Julian P. Donohue
*Executive Secretary*
Department of Entomology
Michigan State University
East Lansing, Michigan 48823

## AMERICAN MOSQUITO CONTROL ASSOCIATION

The annual meeting of the American Mosquito Control Association will be held at the Jung Hotel, New Orleans, Louisiana on March 31–April 3, 1968. For additional information write:

T. G. Raley
*Executive Secretary*
Consolidated M. A. D.
P. O. Box 278
Selma, California 93662

# Biological Interaction Between Plasticizers and Insecticides[1]

E. P. LICHTENSTEIN, K. R. SCHULZ, T. W. FUHREMANN, and T. T. LIANG
Department of Entomology, University of Wisconsin, Madison 53506

ABSTRACT

Because of the structural similarity to DDT-related compounds, 11 plasticizers were tested for their toxicological interaction with dieldrin and DDT. Many of these polychlorinated biphenyl compounds were toxic to *Drosophila melanogaster* Meigen and house flies, *Musca domestica* L., but to a lesser extent than dieldrin or DDT. Their toxicity increased with a decrease in their chlorine content. Moreover, sublethal dosages of several of the plasticizers increased the toxicity of dieldrin and DDT. Since plasticizers are in the environment, their potential effects on biological systems, especially in combination with other synthetic chemicals, should not be disregarded.

Plasticizers such as polychlorinated biphenyls are widely distributed in the environment (Risebrough et al. 1968). Widmark (1967a) reported that extracts of sea eagles, pike, and salmon contained polychlorinated biphenyl plasticizers (PCB). Holmes et al. (1967) reported that polychlorinated biphenyl compounds had been detected in British wildlife. They also stated that "in birds' livers and eggs they are often in greater quantities than organochlorine pesticide residues." Koeman et al. (1969) found chlorinated biphenyls in fish, mussels, and birds from the River Rhine and the Netherlands coastal areas and stated that the PCB compounds, "besides being a hazard in their own right" also "interfere with the detection of organochlorine insecticides such as DDT."

Polychlorinated biphenyls are produced by various manufacturers in different countries. Some of them are marketed under the name of Aroclor® by the Monsanto Chemical Co., St. Louis. They represent "a series of inert, chemically-resistant, fire-retarding plasticizers compatible with a wide variety of resins. They vary from mobile, oily liquids to white crystals and hard transparent resins" (Monsanto 1968). They are used in synthetic resins, synthetic and natural rubbers, cellulose resins, paint, varnish, wax, asphalt, and in allyl starch. They also find applications for dust prevention, moisture proofing, sealing, impregnation, and vapor suppression to prolong the residual life of insecticides. "Chlorinated polyphenyls" were recommended in 1953 for "improving lindane residues" (Hornstein and Sullivan 1953). The authors stated that "lindane residues can be made to last longer and look better" if applied with chlorinated polyphenyls.

The toxicity of polychlorinated biphenyl compounds was briefly reviewed by Von Oettingen (1955) who reported that "feeding rats about 0.3 gm daily was highly toxic when given over a period of six days and that even 0.05 gm per rat was definitely toxic, causing injury to the liver." According to the same source, the toxicity to rats of chlorodiphenyl by inhalation was already studied in 1937 and some hepatoxic action was found. The threshold limits (maximum allowable concentration for an 8-hr working day) was set by the American Conference of Government Hygienists (1953) as 1.0 to 0.5 mg of polychlorinated biphenyl compounds/m$^3$ of air. The present study was conducted to investigate the effects of some plasticizers on insects, both singly and in combination with dieldrin or DDT.

MATERIALS AND METHODS.—The insecticides and insecticide metabolites used in this study were analytical grade lindane, heptachlor, heptachlor epoxide, aldrin, dieldrin, *o,p'*-DDT, *p,p'*-DDT, *p,p'*-DDE, and *p,p'*-TDE. Eleven polychlorinated bi- or triphenyl standards were obtained through the courtesy of Dr. R. F. Thomas, USDA, Beltsville, Md. They are referred to in this study as plasticizers A through K and are identified in Table 3 with various Aroclors according to their Monsanto identification number. The plasticizers A through H contain 2 phenyl rings, and their chlorine content increases from A to H. Plasticizer I is a mixture of chlorinated biphenyls and chlorinated triphenyls (60:40), and J and K are chlorinated triphenyls.

*Bioassay Procedures.*—The effects of the 11 plasticizers, dieldrin, and DDT on insects were tested with *Drosophila melanogaster* Meigen and with house flies, *Musca domestica* L., DDT-susceptible, CSMA-1948 strain.

(a) In 1 series of experiments, the effects of each chemical on insect mortalities were investigated. When *D. melanogaster* were used, 8 ml of hexane containing either 0.1 μg of dieldrin, 3.0 μg of DDT, 200 or 800 μg of plasticizers A, B, C, and D, and 2000 μg of plasticizers E through K were pipetted into 4-oz

[1] Approved for publication by the Director of the Wisconsin Agricultural Experiment Station. This study was supported in part by Public Health Service Research Grant CC-00297 from the National Communicable Disease Center, Atlanta, Ga. Contribution from the Wisconsin Agricultural Experiment Station as a collaborator under North Central Regional Cooperative Research Project 85, entitled, Reduction of Hazards Associated with the Presence of Residues of Insecticidal Chemicals in the Environment. The cost of printing this paper was paid by the Department of Entomology, University of Wisconsin. Accepted for publication Apr. 24, 1969.

bioassay jars (Edwards et al. 1957). Solvents were then evaporated and fifty 3-day-old *D. melanogaster* were introduced into each of duplicated test jars that now contained the dry residue of the various chemicals. Mortality counts were performed after 24- and 48- hr exposure times.

In addition to these tests, 2-day-old adult female house flies were treated topically on the ventral portion of the abdomen with 2 μliter of acetone containing either 0.02 μg of dieldrin, 0.5 μg of DDT, or 10 and 20 μg of each of the plasticizers. Controls were treated with 2 μliter of acetone only. After 30 house flies had been treated with each of the chemicals as described, they were held in 2 groups of 15 within 2 bioassay jars for 24 hr, then mortality counts were made.

(b) The 2nd series of bioassays was designed to test the combined, and possibly synergistic, effects of the plasticizers with either dieldrin or DDT. For this purpose, *D. melanogaster* flies were exposed to either 200 μg each of the plasticizers, 0.1 μg of dieldrin, or 3.0 μg of DDT. When insecticides only were used (controls), 0.1 ml of hexane containing 250 μg of corn oil was added to the insecticide solution in the bioassay jars to eliminate loss of dieldrin or DDT from the glass surfaces.

In addition, flies were exposed to the combined dry residue of 0.1 μg of dieldrin and 200 μg of either 1 of the plasticizers, or to the combined residue of 3.0 μg of DDT and 200 μg of 1 of the 11 plasticizers. These experiments were replicated 4 times, each replicate consisting of 50 *D. melanogaster* flies.

With house flies, 2 μliter of acetone containing either 0.015 μg of dieldrin, 0.3 μg of DDT or 10 μg of the plasticizers A, B, C, D, and E were applied topically to the ventral portion of 45 insects which were kept in groups of 15 within 3 bioassay jars. In addition, 2 μliter of acetone containing both dieldrin (0.015 μg) and 1 of the plasticizers (10 μg), or DDT (0.3 μg) and 1 of the plasticizers (10 μg) were applied as described. The treated house flies were then held for 24 hr when mortality counts were made. Statistical analyses of the data were performed with the t-test.

*Chromatographic Behavior of Plasticizers.*—To obtain some knowledge about the chromatographic behavior of the 11 plasticizers, they were dissolved in redistilled hexane and analyzed by gas-liquid chromatography (GLC) and thin layer chromatography (TLC). For comparison purposes the above-mentioned insecticides and metabolites were handled the same way.

For tests by GLC, 2 instruments with 2 different columns were employed in these investigations. One instrument was a Packard gas chromatograph, Model 7834, equipped with a 150-mCi tritium electron affinity ionization detector operated at 50 v. A 1.83-m glass column (4.0 mm ID) containing 5% SE 30 on 60/80 Chromosorb W AW-DMCS was conditioned for 7 days at 250°C before use. A column pressure of 25 psi of nitrogen resulted in a flow rate of 125 ml/min. The injector temperature was maintained at 240°C, the detector cell at 215°C, and the oven temperature at 190°C.

The 2nd instrument was a Jarrell-Ash gas chromatograph, Model 28-700 equipped with a 100-mCi tritium electron affinity ionization detector and operated at 20 v. A 1.22-m glass column (4.0 mm ID) containing a 1:1 mixture of 5% QF 1 and 5% DC 200 coated on Anakrom AS, 80- to 90-mesh, was conditioned for 7 days at 250°C before use. A column pressure of 16 psi of nitrogen gave a flow rate of 100 ml/min. The injector temperature was maintained at 250°C, the detector cell at 210°C, and the oven temperature at 190°C.

One μliter of hexane containing a mixture of insecticides ($3\times10^{-5}$ μg of lindane, $5\times10^{-5}$ μg of heptachlor and aldrin, $1\times10^{-4}$ μg of heptachlor epoxide, DDE, and dieldrin, $2\times10^{-4}$ μg of DDT and TDE) was injected onto the column in each of the 2 gas chromatographs. After all of the insecticide peaks had been recorded, 2 μliter of hexane containing 1 of the 11 plasticizers ($2\times10^{-2}$ μg of A, B, C, D, E, or F, $5\times10^{-3}$ μg of G or H, $5\times10^{-2}$ μg of I, J, or K) were injected and tested in the same way. To compare the resulting chromatograms, retention times of the various peaks were calculated in reference to the retention time observed with aldrin.

For tests by TLC, 10 μg of each of the 11 plasticizers were spotted on aluminum oxide G-coated glass plates (20×20 cm), 2.5 cm above the lower edge. In addition, 10 μg of each of the insecticides and some of their metabolites were also spotted for reference. The chromatograms were then developed with 1% acetone in *n*-heptane, sprayed with reagent as described by Mitchell (1957) and exposed to UV light for 30 min.

RESULTS AND DISCUSSION.—*The Effect of Plasticizers on Insects.*—Since the polychlorinated biphenyls are of toxicological interest, both their toxicity to insects and their interaction with insecticides were studied in 2 series of experiments, as previously described.

(a) Dieldrin, DDT, or any one of the less chlorinated and more volatile plasticizers (A, B, C, D, E) were toxic to both test insects (Table 1). While these polychlorinated biphenyls contain increasing amounts of chlorine, their toxicity towards the insects decreased (A>B>C>D>E). Dieldrin was the most toxic of

**Table 1.—Effect of plasticizers or insecticides on insect mortalities.**

| | *D. melanogaster* | | | House flies | |
|---|---|---|---|---|---|
| | Applied[a] | % mortality | | Applied[b] | % mortality |
| Material | (μg) | 24 hr | 48 hr | (μg/g fly) | 24 hr |
| None[c] | | 0 | 0 | | 0 |
| Dieldrin | 0.1 | 31 | 84 | 1.0 | 73 |
| *p,p'*-DDT | 3.0 | 24 | 58 | 25.0 | 57 |
| Plasticizers[d] | | | | | |
| A | 200 | 0 | 0 | 500 | 17 |
| | 800 | 57 | 92 | 1000 | 43 |
| B | 200 | 0 | 0 | 500 | 10 |
| | 800 | 35 | 64 | 1000 | 30 |
| C | 200 | 0 | 4 | 500 | 17 |
| | 800 | 40 | 73 | 1000 | 20 |
| D | 200 | 0 | 0 | 500 | 13 |
| | 800 | 15 | 45 | 1000 | 13 |
| E | 2000 | 0 | 0 | 1000 | 10 |
| F–K | 2000 | 0 | 0 | 1000 | 0 |

[a] Applied in 8 ml hexane to 4-oz glass jars. *D. melanogaster* exposed to dry residue.
[b] Applied topically to house flies in 2 μliter of acetone. Weight of 1 fly ≅ 20 mg.
[c] Controls were treated with solvents only.
[d] Letters A to K refer to Aroclors as listed in Table 3.

the compounds, as measured by the percent mortality of the insects within a 24- or 48-hr exposure period. Approximately 25 and 30 times more DDT than dieldrin, and 8000 and 1000 times more plasticizers than dieldrin had to be used to obtain similar mortalities of *D. melanogaster* or of house flies, respectively, within a given time of exposure to these chemicals. The more highly chlorinated and less volatile plasticizers were nontoxic even when used at dosages of 2000 μg with *D. melanogaster* and 20 μg/fly (1000 μg/g fly) with house flies.

(b) Table 2 summarizes the results obtained after the insects had been exposed to sublethal dosages (200 μg for *D. melanogaster,* 10 μg/house fly) of either 1 of the plasticizers A, B, C, D, or E and in combination with dieldrin or DDT. Though no appreciable insect mortalities were observed with plasticizers alone, they significantly increased the toxic effects of both dieldrin and DDT. This effect was most noticeable with DDT on house flies and differences observed were highly significant. After a 24-hr exposure period, 38% of the house flies had died after only DDT had been applied. However, all the insects were killed when DDT was applied in combination with any one of these plasticizers. The polychlorinated biphenyls also increased the toxic effect of DDT with *D. melanogaster,* but to a lesser extent. The most striking effect was observed with compound A. Plasticizers F, G, H, I, J, and K applied with dieldrin or DDT to glass surfaces to which *D. melanogaster* were exposed reduced the toxicity of the insecticide, resulting in mortalities of 30% to none within 48 hr.

In the absence of additional data, no conclusions should be drawn concerning the combined effects of plasticizers and insecticides in tissues of higher animals. However, plasticizers are in the environment (Holmes et al. 1967, Risebrough et al. 1968, Koeman et al. 1969) and their potential hazards, especially in combination with other synthetic chemicals, should not be disregarded.

*Chromatographic Behavior of Plasticizers.*—Results obtained after solutions of the 11 plasticizers had been analyzed with 2 different columns in separate gas chromatographs are summarized in Table 3. The observed retention times were compared as described with those obtained after the injection of lindane, heptachlor, heptachlor epoxide, aldrin, dieldrin, DDT, DDE, and TDE. The injection of nearly all of the plasticizers resulted in chromatograms with a multitude of peaks, sometimes as many as 15. Photographs of thin layer plates, prepared as previously described, are presented in Fig. 1 and 2. Since analyses of most plasticizers by GLC resulted in chromatograms with many peaks, it is conceivable that each of these plasticizers represents a multitude of compounds that were not separated by TLC as described, thus resulting in 1 or 2 spots only.

**Table 2.—Combined effects of plasticizers and insecticides on insects.**

| Plasticizers[a] | % mortalities of insects after exposure to: No insecticide | Dieldrin | *p,p'*-DDT |
|---|---|---|---|
| | *D. melanogaster after 48 hr*[b] | | |
| None[c] | 0 | 66± 6 | 59±6 |
| A | 0 | 87± 4[d] | 93±5[e] |
| B | 0 | 81± 3[d] | 84±6[d] |
| C | 5±2 | 77± 9 | 83±6[d] |
| D | 0 | 72± 4 | 77±3[d] |
| E | 0 | 10± 3[d] | 47±3[f] |
| | *House flies*[g] *after 24 hr* | | |
| None | 0 | 27± 5 | 38±8 |
| A | 4 | 49± 3[d] | 100[e] |
| B | 4 | 27± 5 | 100[e] |
| C | 0 | 47± 5[f] | 98±3[e] |
| D | 6 | 58±11[f] | 100[e] |
| E | 0 | 24± 6 | 100[e] |

[a] Refers to Aroclors (A to E) as listed in Table 3.
[b] *D. melanogaster* were exposed in 4-oz glass jars to the dry surface deposit of the chemicals: plasticizers, 200 μg; dieldrin, 0.1 μg; DDT, 3.0 μg.
[c] Bioassay jars without plasticizers contained 250 μg of corn oil each.
[d, e, f] Differences observed between treatments with the insecticide only and those with insecticide plus plasticizer are significant at the d = 1%, e = 0.1%, or f = 5% level.
[g] Chemicals were applied topically in 2 μliter of acetone to house flies: plasticizers, 500 μg/g fly; dieldrin, 0.75 μg/g fly; DDT, 15 μg/g fly. Control: only acetone was applied. Weight of 1 fly ≅ 20 mg.

**Table 3.—Comparison of plasticizers and insecticides by GLC.**

| Plasticizer Aroclor | | % weight chlorine | GLC I[a]: No. of peaks | GLC I[a]: Identical with[c] | GLC II[b]: No. of peaks | GLC II[b]: Identical with[c] |
|---|---|---|---|---|---|---|
| 1221–2[d] | A | 21 | 1 | None | 5 | LI |
| 1232–2 | B | 32 | 15 | LI, HE, AL, DE, DI, TD, DT | 15 | LI, HE, HO, DE |
| 1242–2 | C | 42 | 14 | LI, HE, AL, DE, DI, TD, | 14 | LI, HE, HO, DE |
| 1248–2 | D | 48 | 15 | LI, HE, AL, DE, DI, TD, DT | 15 | LI, HE, HO, DE, DT |
| 1254–2 | E | 54 | 11 | AL, DE, DI, TD, DT | 10 | HO, DE, TD, DT |
| 1260–2 | F | 60 | 15 | TD, DT | 13 | DI, TD, DT |
| 1262–2 | G | 62 | 12 | TD | 13 | DI, TD, DT |
| 1268–2 | H | 68 | 7 | None | 8 | None |
| 4465–2, 3 | I | 65 | 11 | TD | 11 | DI |
| 5442–3 | J | 42 | 3 | None | 7 | LI, HE, HO |
| 5460–3 | K | 60 | 3 | None | 0 | None |

[a] Packard gas chromatograph, column 5% SE 30 on 60/80 Chromosorb W. Oven temp = 190°C.
[b] Jarrell-Ash gas chromatograph, column 5% QF-1, 5% DC 200 (1:1) on 80/90 Anakrom AS. Oven temp = 190°C.
[c] LI = Lindane, HE = Heptachlor, HO = Heptachlor epoxide, AL = Aldrin, DI = Dieldrin, DT = *p,p'*-DDT, DE = DDE, TD = TDE.
[d] 2 = chlorinated biphenyls, 3 = chlorinated triphenyls.

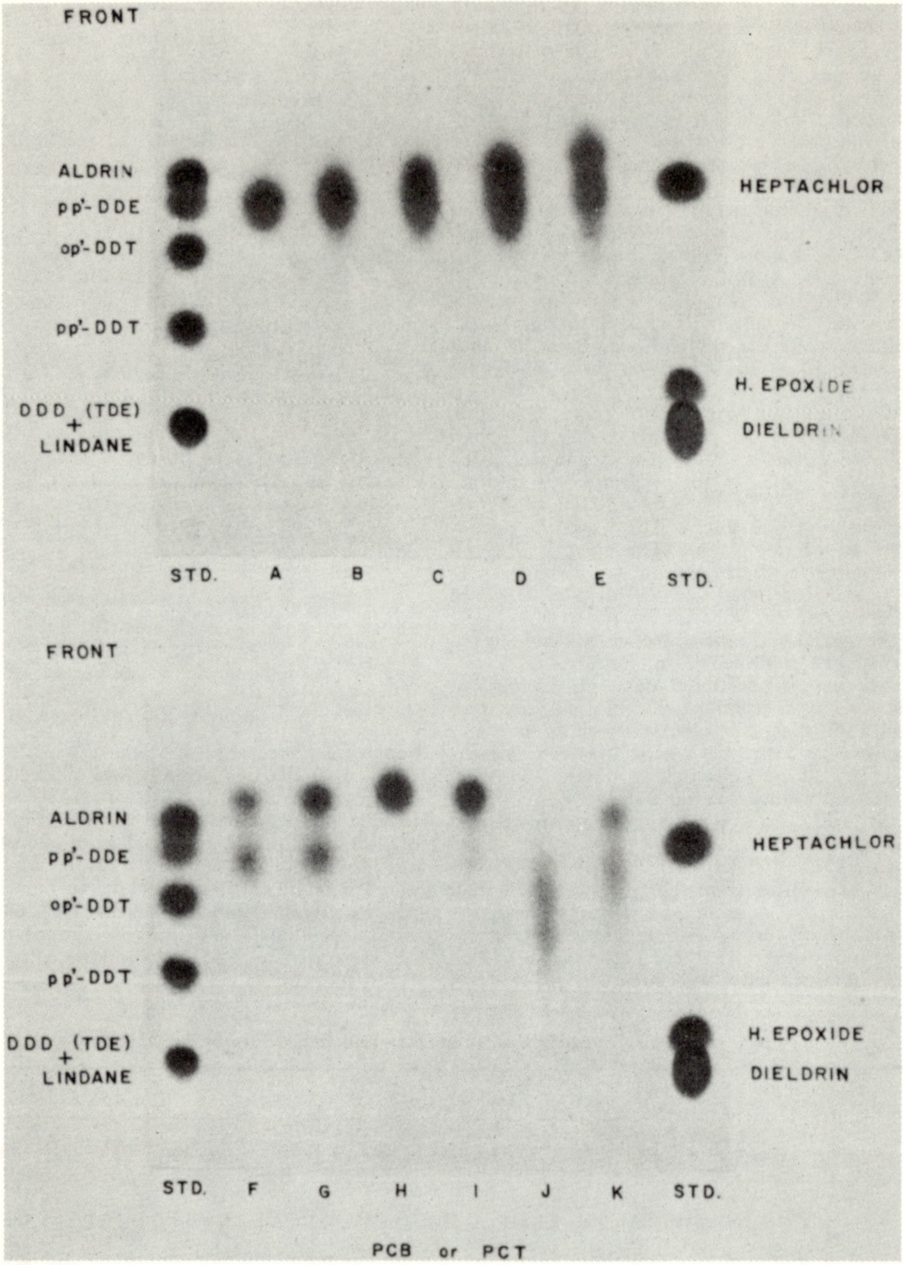


FIG. 1 (above).—Thin layer chromatogram of plasticizers A, B, C, D, and E on aluminum oxide G. Solvent: 1% acetone in *n*-heptane. STD = Insecticide standards for comparison.

FIG. 2 (below).—Thin layer chromatogram of plasticizers F, G, H, I, J, and K on aluminum oxide G. Solvent: 1% acetone in *n-heptane*. STD = Insecticide standards for comparison.

REFERENCES CITED

**Amer. Conf. Gov. Ind. Hygienists, Los Angeles. 1953.** Arch. Ind. Hyg. Occup. Med. 8: 296.

**Edwards, C. A., S. D. Beck, and E. P. Lichtenstein. 1957.** Bioassay of aldrin and lindane in soil. J. Econ. Entomol. 50: 622–6.

**Holmes, D. C., J. H. Simmons, and J. O. G. Tatton. 1967.** Chlorinated hydrocarbons in British wildlife. Nature 216: 227.

**Hornstein, I., and W. N. Sullivan. 1953.** The role of chlorinated polyphenyls in improving lindane residues. J. Econ. Entomol. 46: 937–40.

**Koeman, J. H., M. C. Ten Noever de Brauw, and R. H. de Vos. 1969.** Chlorinated biphenyls in fish, mussels and birds from the river Rhine and the Netherlands coastal area. Nature 221: 1126.

**Mitchell, L. C. 1957.** Ascending paper chromatography: A way to do it. J. Ass. Agr. Chem. 40: 999.

**Monsanto. 1968.** Aroclor® plasticizers. Tech. Bull.. O PL–306. Monsanto Organic Chemicals Division, St Louis.

**Risebrough, R. W., P. Rieche, D. P. Peakall, S. G. Herman, and M. N. Kirven. 1968.** Polychlorinated biphenyls in the global ecosystem. Nature 220: 1098.

**Von Oettingen, W. F. 1955.** The halogenated hydrocarbons—toxicity and potential dangers. U. S. Public Health Serv. Publ. no. 414. p. 306.

**Widmark, G. 1967.** Residue analysis of pesticides with the aid of gas chromatography. Sixth Int. Congr. Plant Protect. (Abstr.)

# Chemical Insect Attractants

Insects promote their own destruction in responding to traps baited with specific lures.

Martin Jacobson and Morton Beroza

Insects have managed to persist in hostile surroundings because they have developed extraordinary adaptations or abilities, one of which is a highly specialized sense of smell. Some insects can follow unerringly an odor trail to a source of food, to host plants and animals, to the opposite sex, or to the right place to lay eggs. Because many insects depend for their survival on these odors, frequently they can be attracted by means of a chemical to a trap for detection purposes, or to a toxicant that destroys the insect, or to a substance which makes them incapable of fertile mating.

The naturally occurring sex attractants, emitted by the insect in fantastically minute amounts, are among the most potent, physiologically active substances known today. For example, only $10^{-7}$ microgram of sex attractant released by a virgin female gypsy moth will lure to her numerous males flying upwind from a distance of ¼ mile or more. Traps baited with infinitesimal amounts of such attractants are being used to delineate areas of infestation, making it unnecessary to spray with pesticides large areas that are uninfested. These lures are usually highly specific, attracting only one or a few closely related species, and then only males. To our knowledge there are no powerful general attractants.

Attractants may be classified as sex, food, or oviposition lures. The type of lure is inferred or deduced from insect behavior, and the assignment is frequently uncertain. If exposure to a chemical causes a male insect to assume a mating position, the chemical is probably a sex attractant.

The authors are chemists in the Pesticide Chemicals Research Branch, Entomology Research Division, of the U.S. Department of Agriculture, Agricultural Research Service, Beltsville, Md.

### Sex Attractants

The key to the survival of insects is their high reproductive potential, which depends on the ability of opposite sexes to find each other and mate. An important, if not essential link of this process is the sex attractant, usually released by the female to lure the male. These chemical messengers have been called "female assembling scents" (*1*) and "sex pheromones," from the Greek "pherein" (to carry) and "horman" (to excite, stimulate) (*2*).

The extent to which sex attractants occur in the insect world is unknown. The more obvious examples of insects attracting mates, as the gypsy moth does, were discovered when males responded to caged or immobilized females. Although such observations are comparatively few, the interesting possibility remains that sex lures may be prevalent among insect species. Uncertainty here arises from our lack of knowledge of insect mating habits. In the case of the European corn borer, *Pyrausta nubilalis* (Hübner), one of our most destructive insect pests, not until observations were made with infrared light was the existence of a sex attractant disclosed (*3*). The honey bee (*Apis mellifera L.*) has a sex attractant which may be extracted with ethyl ether from the mandibular glands of queens (*4*). The drone (males) did not respond to these extracts until materials impregnated with the extracts were suspended from a pole at least 15 feet above ground level. In this manner, the olfactory nature of this lure was demonstrated.

Chemical fractionation of the lipids from the honey bee gland yields "queen substance" (*trans*-9-oxo-2-decenoic acid) (I), which is attractive to drones at 0.1 mg per assay tube, and at least two other substances with some attractiveness. Since sex lures are usually much more potent, the queen substance may not be the true honey-bee sex lure. Indeed, extirpation of the mandibular glands does not necessarily render a virgin queen incapable of mating (*5*), and reconstitution of the lipid complex results in considerably more attractiveness than is shown by individual fractions.

$$\underset{\displaystyle \overset{\|}{O}}{CH_3C}(CH_2)_5CH{=}\underset{\displaystyle \overset{\|}{O}}{CHC}{-}OH$$

(I)

The sex attractants of some species of bumble bee are released by the female abdomen (*6*). Although the odorous substance was not isolated, dummies, rubbed mechanically with freshly caught females, attracted males that attempted copulation with these dummies.

The virgin female of the introduced pine sawfly, *Diprion similis* (Hartig), is capable of attracting exceptionally large numbers of males; one caged female placed in a field attracted well over 11,000 males. Column chromatography of extracts of females yielded a substance which, though impure, in amounts as small as 0.004 microgram, attracted 500 to 1000 males over a distance of 100 to 200 feet within 5 minutes. Activity of an attractant in some types of extracts is often masked, but a potent extract may be recovered by chromatography (*7*).

Reproduced with permission from *Science,* 140:1367–1373, 1963. Published by the American Association for the Advancement of Science, Washington, D.C.

The elucidation of the chemical structure of sex attractants has been hampered by the infinitesimal amounts of lure emitted by the insect. Thus far the true sex lures of only three insect species have been identified—those from the gypsy moth, *Porthetria dispar* (L.); the silkworm moth, *Bombyx mori* (L.); and the American cockroach, *Periplaneta americana* (L.). These insects have been studied because of their tremendous economic importance. In *P. dispar* and *B. mori* the attractant is formed in the lateral glands of the abdomen of the virgin female. The female is able to protrude and retract these glands, and thus regulate the release of the attractant.

The moth *B. mori*, a native of Europe, is not considered to be a pest, since it is extremely valuable to the silk industry. But *P. dispar* is very destructive to forest and shade trees in New England, causing millions of dollars worth of damage. The female gypsy moth is unable to fly, but the male is a strong flier and is lured upwind by the female scent from a distance of approximately ¼ mile. Although it had long been known that a benzene extract of the virgin female abdominal tips would lure males in the field, it took a great deal of painstaking chemical study over a period of about 30 years to isolate and identify the pure sex attractant (*8, 9*). It was possible to isolate, from an extract of 500,000 virgin female gypsy moths, 20 mg of the pure attractant. A colorless oily liquid showing blue fluorescence in ultraviolet light, it has been identified as *d*-10-acetoxy-*cis*-7-hexadecen-1-ol (IIa) and synthesized (*8, 10*).

$$CH_3(CH_2)_5CHCH_2CH=CH(CH_2)_nCH_2OH$$
$$\quad | \quad OCCH_3 \quad (\text{C}=O)$$

(II, a, b, and c)

(In IIa, IIb, and IIc, *n* is equal to 5, 7, and 9, respectively.) Fractions were bio-assayed in field traps (*8, 9*) or in the laboratory (*11*). In the latter method, male gypsy moths, clipped by their wings to a rack and exposed to the scent on a rod or filter paper, exhibited curving of the abdomen and copulatory attempts. The *dl*-form (synthetic) of IIa was successfully resolved by treating the acid succinate with *l*-brucine, separating the brucine salts by fractional crystallization from acetone, decomposing the salts, and saponifying the acid succinates with ethanolic alkali (*12*). Characterization of the natural attractant of the gypsy moth led to the synthesis of a homolog, *d*-12-acetoxy-*cis*-9-octadecen-1-ol (IIb), which has been designated "gyplure" and is almost as attractive to males as the natural lure (*13*); a higher homolog (*d*-14-acetoxy-*cis*-11-eicosen-1-ol) (IIc) was much less attractive to males (*14*). Conversion at the double bond of gyplure to the *trans* form causes a tremendous decrease in activity. Comparisons of the activities of these gypsy-moth attractants are shown in Table 1. The preparation and use of gyplure as a gypsy-moth attractant have recently been patented (*15*).

Table 1. Comparative attractancy of gyplure and its homologs to male gypsy moths.

| Compound | Attractancy ($\mu$g) | |
|---|---|---|
| | Laboratory | Field |
| *d*-IIa | $10^{-12}$ | $10^{-7}$ |
| *l*-IIa | $10^{-12}$ | $10^{-6}$ |
| *d*-IIb (*cis*-gyplure) | $10^{-12}$ | $10^{-5}$ |
| *d*-IIb (*trans*-gyplure) | $10^{4}$ | $2.5\times10^{5}$ |
| *d*-IIc | $10^{-2}$ | 10 |

"Bombykol," the sex attractant of the female silkworm moth, has been isolated from an ethanol-ethyl ether extract (3:1) of the abdominal tips; 12 mg of pure attractant was obtained from 500,000 virgin females (*16*), and identified as 10,12-hexadecadien-1-ol (III) (*17, 18*). Although the configuration

$$CH_3(CH_2)_2CH=CHCH=$$
$$=CH(CH_2)_8CH_2OH$$

(III)

in bombykol was at first thought to be *cis*-10, *trans*-12 (*17*), subsequent synthesis of the four possible geometric isomers of III showed that *trans*-10, *cis*-12 was the correct structure (*19*). The four geometric isomers showed the following attractiveness to male silkworm moths (in micrograms per milliliter) in laboratory tests (*20*): *cis*-10, *cis*-12, 1; *cis*-10, *trans*-12, $10^{-3}$; *trans*-10, *cis*-12 (bombykol), $10^{-12}$; *trans*-10, *trans*-12, 10. The laboratory bioassay was based on the behavior of the male moth when confronted close to the antennae with the attractant placed on a glass rod. A whirring vibration of the wings and typical circling dance ("schwirrtanz") are elicited by exposure to the odor of an active compound (*16*). The similarity in structure of the sex attractants of the gypsy moth and silkworm moth is remarkable. The synthesis of straight-chain, primary alcohols containing 14 to 18 carbon atoms with a pair of conjugated double bonds and their use as insect sex attractants have been patented in Germany (*21*).

The female American cockroach [*Periplaneta americana* (L.)] produces a specific sex attractant which has not been obtained by extraction from the insect but which has been obtained in extremely poor yield by extracts of filter papers over which virgin females had crawled (*22*). The attractant was collected in much larger amounts by drawing a stream of air through large milk cans containing thousands of virgin female roaches and collecting the volatile substances in a trap cooled with dry-ice (*23*). Fractionation by chromatography and steam distillation of the combined volatile fraction, that was collected continuously from approximately 10,000 female roaches for 9 months, yielded 12.2 mg of the pure, odorous attractant as a yellow liquid. Identified as 2,2-dimethyl-3-isopropylidenecyclopropyl propionate (IV) (*24*), it elicits typical intense excitement, wing-raising, and copulatory attempts in males at less than $10^{-14}$ microgram; this amount is equivalent to approximately 30 molecules! The attractant has not yet been synthesized, and the glands that produce this attractant have not been located.

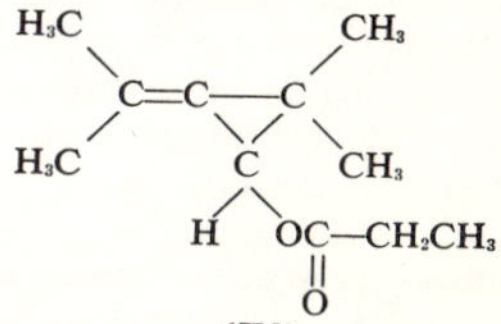

(IV)

From the abdomen of the female pink bollworm moth, *Pectinophora gossypiella* (Saunders)—a pest of cotton—a sex attractant for males can be extracted with ether or methylene chloride (*25*), and the same solvents will remove an attractant from the abdomens of female tobacco hornworm moths, *Protoparce sexta* (Johannson) (*26*).

Although most of the insect sex attractants are species specific, there are some exceptions. For example, an extract of female gypsy moths (*Porthetria dispar*) attracts male gypsy moths and male European nun moths, *P. monacha* (L.), but an extract of female nun moths attracts only males of this species (*27*).

The release of sex scents is not constant throughout the day. A maximum is reported in the early morning hours for *Polychrosis botrana* (Schiffermüller) (*28*) and *Protoparce sexta* (*27*), but with the gypsy moth mating takes place only during daylight hours (*9*).

Males of the tropical water bug,

*Lethocerus indicus* (Lepetier and Serville), produce a scent that may possibly belong to the group of sex attractants. The extract, a clear fluid with a cinnamon-like odor, is used by the natives of Southeast Asia as a spice for greasy foods. The active compound, identified as *trans*-2-hexenyl acetate (V), is believed to make the female more receptive to the male (*29*).

$$CH_3(CH_2)_2CH{=}CHCH_2O{-}\underset{\underset{\displaystyle O}{\|}}{C}CH_3$$

(V)

**Food and Oviposition Lures**

Since insects live in almost every conceivable environment, their food requirements are extraordinarily diverse. With the exception of microorganisms, almost no form of life (including insects themselves) or organic matter of natural origin is immune to attack by this voracious group of animals. Because of the sheer numbers of species (well over 900,000), our knowledge of their food habits is limited to some of the disease-bearing, economically important, or pestiferous species. Very little is known about the stimuli that draw insects to their food supply. This area of research promises to be a fertile one in the future because the tools necessary for analysis of the infinitesimal quantities of chemicals associated with odors have already been devised. Basic knowledge gained here is bound to lead to more selective and efficient insect control.

There is still little information on the exact nature of insect attractants. However, entomologists have faithfully observed and reported the attraction of insects to certain foods, extracts of plants, animals, and chemicals. Although the chemist can derive little from references to attraction of foods, such as sliced cucumbers, crushed bananas, and fish meal, because of their complex composition, certain observations of attraction to chemicals fall into a pattern. For example, ammonia, amines, sulfides, and fatty acids have been reported to attract many species (*30*). These compounds are products of decomposing organic matter and apparently represent a source of food to the insect. Many insects, especially in the order Diptera, instinctively oviposit in the vicinity of these chemicals (*31, 32*), probably to provide food for their young on emergence; for example 91 percent of the house flies, *Musca domestica* L., attracted to ammonium carbonate were females, and nearby food traps captured equal numbers of both sexes (*32*).

That fruit flies respond to, and consume protein hydrolyzates (*33*), has led to the wide use of the latter in insecticide preparations for the control of pests. This combination obviates the need for complete coverage because the insect comes to the lure. Tests on the attraction of constituent individual amino acids have usually been disappointing, although lysine attracts the mosquito (*Aedes* and *Culex* spp.) (*34*); Mexican fruit fly, *Anastrepha ludens* (Loew); oriental fruit fly, *Dacus dorsalis* Hendel; Mediterranean fruit fly, *Ceratitis capitata* (Wiedemann), commonly known as "medfly"; and melon fly, *Dacus cucurbitae* Coquillett (*35*). Of the other amino acids tested, only glutamine attracted the medfly. One report states that certain mixtures of amino acids attract mosquitoes but the individual amino acids do not (*36*). One would not normally expect amino acids to be effective attractants because of their low vapor pressure, and, in general, they are not *strong* attractants.

Essential oils and various plant extracts are promising as sources of attractants, possibly because they may contain compounds which draw an insect to a favored food or ovipositional site. Methyleugenol, a constituent of citronella and of Huon pine oil, is a powerful lure for the oriental fruit fly (*37*). Some entomologists believe it to be a sex attractant because it attracts only the male. Only relatively small amounts are required for its action. This compound can also be considered a food lure because the male flies devour the chemical and some avidly gorge themselves to the bursting point (*38*). A mixture of eugenol and geraniol (9:1) is an effective lure for the Japanese beetle; it attracts equal numbers of each sex (*39*).

Finding a lure in nature and isolating the active substance is one approach to finding insect attractants. Another way is to screen a large number of chemicals to obtain a "lead," that is, a weak attractant; related compounds are then examined in an attempt to find a better lure. Initial screening may be hit or miss; but when a large number of chemicals of many different types are tested, some leads do turn up. To pursue a lead, a program of systematic chemical synthesis is needed since the desired compounds are seldom available. Scientists of Entomology Research Division of the U.S. Department of Agriculture were the first group to pursue a synthesis program in the search for insect lures. The synthesis of "siglure" (*sec*-butyl ester in Table 2), which in 1957 helped eradicate the medfly from Florida (*40*), was a result of their systematic search.

Table 2. Olfactometer and field ratings of siglure and related compounds. The esters of 6-methyl-3-cyclohexene-1-carboxylic acid are designated by *R*. The isopropyl ester is arbitrarily rated as 100.

| *R* | Olfactometer | Field |
|---|---|---|
| $CH_3$ | 48 | |
| $CH_2CH_3$ | 122 | 38 |
| $CH_2CH_2CH_3$ | 96 | 58 |
| $CH(CH_3)_2$ | 100 | 100 |
| $CH_2CH_2CH_2CH_3$ | 71 | 98 |
| $CH_2CH(CH_3)_2$ | 99 | 48 |
| $CH(CH_3)CH_2CH_3$* | 87 | 279 |
| $CH(C_2H_5)CH_2CH_3$ | 83 | 231 |
| $CH_2CH{=}CH_2$ | 107 | 53 |
| $CH_2C{\equiv}CH$ | 83 | 40 |
| $CH_2CH_2Cl$ | 86 | 29 |
| $CH_2CH_2OCH_3$ | 0 | |
| $CH_2C_6H_5$ | 0 | |

*Siglure.

In seeking medfly lures the ethyl ester of 6-methyl-3-cyclohexene-1-carboxylic acid was one of the first synthetics to prove attractive in laboratory (olfactometer) tests. Thirty additional esters with the same acid moiety were synthesized and tested (*40*). The olfactometer ratings of 13 of these esters are given in Table 2. A striking feature of the data is that a small difference in *R* may give a large difference in attractiveness. Table 2 also gives the results of testing the most attractive compounds in field trials. Although agreement between olfactometer and field ratings is poor, the olfactometer permits rapid screening of chemicals and thus eliminates those compounds unlikely to be attractants so that they are not used for time-consuming field tests. Final evaluation of a lure must be made in field tests, preferably under actual conditions of use.

The commercial production of siglure brought new problems since field tests indicated that the commercial product was less appealing to the medfly than the original laboratory preparations. Further investigation disclosed that the laboratory chemical had a *trans* configuration (VI) and that the commercial product contained appreciable amounts of the *cis* isomer (*41*), which proved to be much less attractive. The commercial method of synthesis had to be changed, and an infrared method of analysis was written into the specification to insure the procurement of the more attractive *trans* isomer.

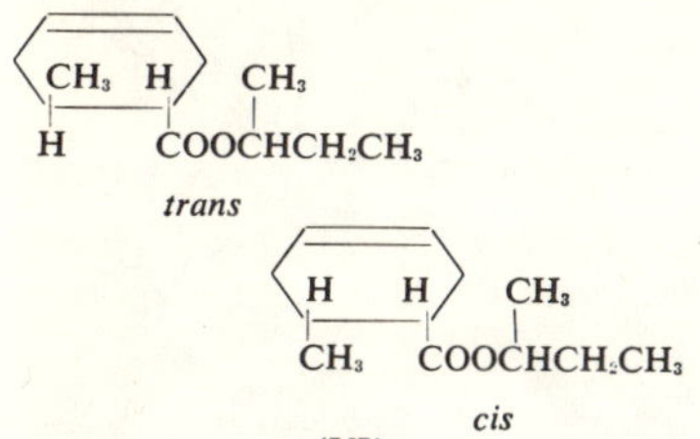

(VI)

It is thus clear that subtle stereochemical or minute structural alterations can cause great differences in the insects' response. In order to avoid the epimerization (change of *trans* to *cis*) of *trans*-siglure, conditions that induce this change were studied. In one of these experiments the hydrogen chloride adduct of siglure was produced. When the compound was found to exceed the attractiveness of siglure, 45 other esters of the hydrohalogenated acid were synthesized (*42*). Medlure and trimedlure, the secondary and tertiary butyl esters, were most effective. Trimedlure (VII), now in commercial production, is the lure that rapidly helped eradicate, in the summer of 1962, a reinfestation of the medfly in Florida.

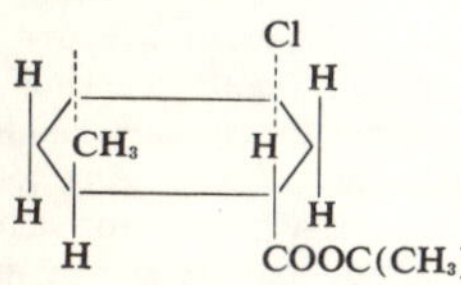

(VII)

Preparations of trimedlure are mixtures of isomers—theoretically there are eight—depending on the position of the chlorine atom (one of the four dashed lines of VII), and whether the methyl and ester groups are *cis* or *trans*. Two solid substances of correct elemental composition and a liquid portion have been separated from the mixture. The liquid and one of the solids are attractive, but the other solid is not (*43*).

The best medfly attractants in both the siglure and trimedlure series contain alcohol moieties having three to five carbon atoms with methyl groups (branching) on the oxygen-linked carbon atom. Attempts to vary the structure of the acid portion have yielded only inferior products.

We do not know why the male medfly finds siglure, medlure, and trimedlure so attractive, and we know of no such chemicals in nature. Before the discovery of the synthetics, male medflies were found to be lured to angelica seed oil, which they consume. Presumably the insect may also associate the odor of the synthetic medfly lures with a source of food. However, most of the synthetics do not cause feeding, nor should it be supposed that food attractants necessarily induce feeding. Some substances, termed arrestants (*44*), trigger a feeding response, but do not attract; sucrose, of negligible vapor pressure, does not actually attract the house fly, but upon finding sucrose the fly remains to feed. The protein hydrolyzate is an example of a material that attracts the medfly and other tropical fruit flies (although weakly) and also induces a feeding response.

The discovery of the specific medfly attractants illustrates the success of the volume-screening approach, which depends heavily on the development of a rapid, reliable screening procedure. A potent lure for the melon fly, cue-lure, was found by this method (*45, 46*). Compounds containing the 4-phenyl-2-butanone structure were attractive to the melon fly; the best of these were para-substituted through an oxygen (*47*). Table 3 gives the results of testing cue-lure and related compounds in the field (*45*).

We tested about 2000 synthetics for attracting the gypsy moth before the structure of the natural sex lure of the female gypsy moth was determined. Even though virtually nothing was known about the chemical structure of this natural lure, several attractive compounds were found. Although they were far less potent than the insect product, they were, like the natural attractant, 16-carbon, straight-chain compounds with a primary alcohol group, or an epoxy group which could be hydrolyzed to an alcohol (*48*).

## Biological Testing and Survey

The testing of materials for attraction to insects requires considerably more finesse than the testing of insecticides, and no general means suitable for testing against all insects has been devised and probably never will be. A suitable bioassay must be built around the natural habits and idiosyncrasies of each species. Few gypsy moths, for example, seek a mate on a cloudy day, and even very attractive preparations are ineffective on such days. The southern armyworm, *Prodenia eridania* (Cramer), which responds to its sex attractant in the early morning hours (between 3:00 and 5:00 A.M.), is indifferent to the same stimulus at other times. An olfactometer for fruit flies is an 8-foot cubical screened enclosure containing about a dozen traps suspended from a slowly rotating wheel (*49*); as in nature, the flies respond to the lures while in free flight. Crawling insects have been offered a choice of two air streams of identical composition except that one contains a chemical vapor (*50*).

Table 3. Melon flies attracted by cue-lure and related compounds in field tests. *R*, the para-substituent in phenyl butanone 2.

| *R* | Flies caught |
|---|---|
| $CH_3C(=O)O$ —* | 30,752 |
| $CH_3CH_2C(=O)O$ — | 22,985 |
| HO — | 14,574 |
| $CH_3CH_2CH_2C(=O)O$ — | 12,508 |
| $(CH_3)_2CHCH_2C(=O)O$ — | 6,894 |
| $CH_3O$ — | 2,408 |

*Cue-lure.

Testing is further complicated by the effect of vapor concentration on attraction. Too low a concentration will not be detected, and too high a concentration may repel. Thus several concentrations of a chemical should be tested in order to avoid missing a good attractant. Of course, the most desirable lures attract at both high and low concentrations.

Laboratory testing of lures, which is possible with certain species, has some advantages. Conditions, such as light, temperature, and humidity, may be standardized; insects may be reared in a controlled environment; tests may be carried out throughout the year; tests may be rapid and well replicated. However, an important prerequisite for efficient laboratory testing is a method of rearing large numbers of insects. Such rearing is often stymied by the vulnerability of some species to disease or parasites. Daily and seasonal variations are other obstacles. Some species respond to an attractant only at a definite stage of maturity.

The method of exposing the chemical may be important. It may be spotted on paper, dispersed in solution or emulsion, or impregnated on a wick. The latter (a dental roll) has been used to measure a lure's persistence by determining, under standard conditions, the number of days a given amount of chemical will attract.

A new and unique means of testing attractants in the laboratory may soon be practicable. The sex attractants of the silkworm moth and gypsy moth and probably those of most Lepidoptera (moths and butterflies), as well as insects of other orders, are perceived by olfactory receptors located in the male antennae (*51*). On this basis, an electrophysiological method for the bioassay of the sex lures has been developed which is as much as 1000 times more sensitive than any available behavioral method. By means of tiny silver electrodes inserted into the sensory hairs of the male antenna, it was found that a local electrical potential is set up by exposure to the female sex scent; the released nerve impulses are amplified and recorded from an oscilloscope to give characteristic patterns, called "electroantennograms" (*52*). The shape of the patterns, which depend on the stimulus, is reproducible.

The organs of olfaction are not always on the antennae; in the blow fly their location could not be found (*53*). Presumably, in the foregoing electrophysiological method, the electrodes would have to be inserted into the appropriate sensory sites. Location of the sites may be determined by causing loss of response to an odor on extirpating or covering the olfactory receptor.

Ultimately the best attractants must be tested in the field since laboratory and field data all too often do not agree. In the field, attractants must prove themselves in the presence of a multitude of natural odors, colors, light conditions, and weather. For insects that cannot be reared in the laboratory, direct field testing of chemicals, although entailing considerable labor, is a valid recourse. Such tests, limited to the season when the insects abound, are subject to such variables as light, temperature, humidity, weather conditions, insect density, age of population, predators, and natural competing attractants. It is desirable to include in these tests, as well as in laboratory trials, a standard lure, if one is available. Comparison with the standard helps minimize the many variables.

In traps the attractants assure early detection of an infestation before it can enlarge or spread. By delineating an infestation, the attractants also make control and eradication measures very efficient, help locate those last few hard-to-find insects, and show when insecticide applications may be terminated in a given area. Since the high volume of fast international trade and traffic has made the introduction of foreign injurious species much more likely, if not inevitable, chemically baited traps have become an important means of preventing an accidentally introduced species from becoming established and assaulting the products of our agricultural community.

The number of insects caught in the field depends to a large extent on the design of traps and their placement. In an attempt to improve the efficiency of the gypsy-moth trapping program, trap design, trap color, bait dispensers, duration of bait effectiveness, bait strength requirements, trap height, trap placement, and related parameters were studied (*9*). The latest gypsy moth trap, shown in Fig. 1, depends on a sticky adhesive to ensnare moths responding to the sex lure. Another type of trap, extensively used for detecting fruit fly infestations, is shown in Fig. 2 (*54*). The chemically baited wick attracts insects inside the trap, where they are overcome by a fast-acting volatile insecticide, such as DDVP (2,2-dichloro-

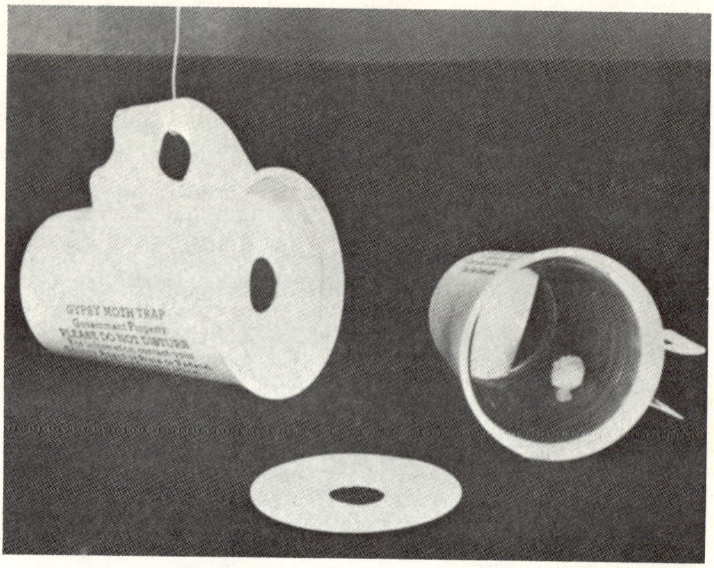

Fig. 1. Recent design of gypsy moth trap.

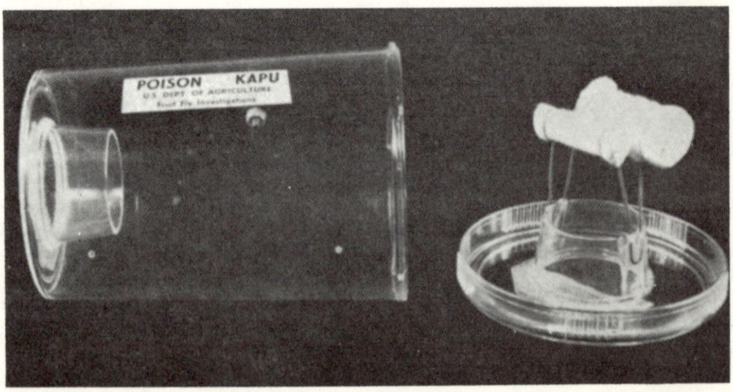


Fig. 2. Plastic trap for fruit flies.

vinyl dimethyl phosphate), lindane, or naled (Dibrom). Birds and ants sometimes consume trapped insects, and means of discouraging these raids have been devised; for example, chlordane keeps ants out of traps. The McPhail trap (Fig. 3) traps the Mexican fruit fly when baited with a fermenting mixture or a protein hydrolyzate. Oddly enough, the same lures in the trap of Fig. 2 catch no Mexican fruit flies.

Air movement is probably the most important parameter to consider in trap placement (*9*). Traps situated upwind from an infestation are more likely to catch insects than those down or across wind. Traps in hollows or surrounded by dense growth do not catch well because of restricted air movement. Odors of chemicals tend to sink to the ground because they are almost always much heavier than air. The male gypsy moth apparently has adapted its behavior to this property since it flies long distances close to the ground. Catches of this moth in traps placed between ground level and 6 feet above showed little difference but fell off at the 12-foot level.

The gypsy moth is an excellent example of an insect whose sex attractant may be used for survey and possibly for control. Males fly into the wind and pick up the scent as they approach the nonflying females. Early experiments with marked males showed that they may be attracted by the female scent from distances of up to ¼ mile. For many years entomologists of the USDA made use of hydrogenated benzene extract of the abdominal tips of virgin females (*55*) in metal field traps to locate infested areas and to determine the size of the infestation by the numbers of males caught in the traps. Synthetic gyplure has now replaced the natural sex lure; and the disposable, inexpensive cup-type paper trap shown in Fig. 1 has completely supplanted the costly, unwieldy metal traps previously used. The gyplure is impregnated on small wicks of dental roll at a concentration of 25 micrograms per trap, and an adhesive substance ("Tanglefoot") is used to line the inside walls of the cup. Baited traps are hung on the limbs of trees and attracted males enter through an opening in each end of the cup. Each year, immediately before the male flight season (July and August), approximately 50,000 traps are placed in the infested New England area. Gyplure has such potency that a single pound of this attractant, which may be made quite inexpensively, is sufficient to last for more than 300 years if used for survey alone.

Fig. 3. McPhail trap.

The results of field trapping completed very recently in cotton-growing areas of the United States and Mexico have shown that a crude extract of the abdomens of the virgin female pink bollworm moth may be used satisfactorily as a survey tool and possibly for control purposes. The sex attractants of other insect species, especially those that fly, have the potential for equally valuable use.

One of the largest insect-detection programs is aimed at keeping out three destructive species of fruit fly not present on the U.S. mainland—the oriental fruit fly, Mediterranean fruit fly, and melon fly. A unique feature of the program is a trap (Fig. 2) baited with three different attractants to lure the three species.

The utility of gyplure-adhesive combinations coated on fiber boards is being explored for control. In limited field trials, boards coated with Tanglefoot containing 5 percent gyplure resulted in large catches of moths and appeared to be effective. Consideration is also being given to limited and large-scale field tests aimed at annihilating male moths through the use of the lure combined with toxicants. The lure may suffice to confuse males seeking mates, making a toxicant unnecessary. Since the males are multiple maters, chemosterilants are worth trying (*56*).

## Summary

Many insects depend on odors to lead them to sources that supply their basic needs. Knowledge of these behavioral patterns helps to combat insects of economic importance. For example, attractants in traps offer a remarkably simple means of detecting insects and may even be useful in direct control. They are increasing the efficiency of insecticide applications because treatments need be applied only where insects are caught and only as long as they are caught.

The sex attractants are the most sophisticated of the lures and the most desirable for survey or control because of their great potency and specificity. The sex lures of the gypsy moth and silkworm moth have been identified and synthesized; the lure of the American cockroach has just been identified. Sex lures have recently been demonstrated in the European corn borer, honey bee, pink bollworm, southern armyworm, tobacco hornworm, and introduced pine sawfly. Additional study of insect mating habits is needed to ascertain the prevalence of sex lures in the insect world.

Food and oviposition lures have been found among foodstuffs and essential oils, and by screening chemicals. Chemical screening, accompanied by programs of synthesis, has produced excellent lures: trimedlure for the Mediterranean fruit fly and cue-lure for the melon fly. In this effort a means of rapidly evaluating the attractiveness of many chemicals is most important. The bioassay must be tailored to the idiosyncrasies of each species and should preferably be a laboratory procedure. However, the final judgment on lures must be based on performance under field conditions.

### References and Notes

1. H. B. D. Kettlewell, *Entomologist* **79**, 8 (1946).
2. P. Karlson and A. Butenandt, *Ann. Rev. Entomol.* **4**, 39 (1959).
3. A. Sparks, "The use of infrared photography to study mating habits of the European corn borer," presented at the 18th Ann. Conf. North Central Branch, Entomol. Soc. Am., 22 March 1963.
4. N. E. Gary, *Science* **135**, 773 (1962).
5. R. A. Morse, N. E. Gary, T. S. K. Johannson, *Nature* **194**, 605 (1962).
6. B. Kullenberg, *Zool. Bidrag. fran Uppsala* **31**, 253 (1956).
7. H. C. Coppel, J. E. Casida, W. C. Dauterman, *Ann. Entomol. Soc. Am.* **53**, 510 (1960); J. E. Casida, H. C. Coppel, T. Watanabe, *J. Econ. Entomol.* **56**, 18 (1963).
8. For reviews of the pioneering chemical and entomological studies see M. Jacobson, M. Beroza, W. A. Jones, *J. Am. Chem. Soc.* **83**, 4819 (1961); R. F. Holbrook *et al.* (*9*).
9. R. F. Holbrook, M. Beroza, E. D. Burgess, *J. Econ. Entomol.* **53**, 751 (1960).
10. M. Jacobson, M. Beroza, W. A. Jones, *Science* **132**, 1011 (1960).
11. B. C. Block, *J. Econ. Entomol.* **53**, 172 (1960).
12. M. Jacobson, *J. Org. Chem.* **27**, 2670 (1962).
13. ———, *ibid.* **25**, 2074 (1960).
14. ——— and W. A. Jones, *ibid.* **27**, 2523 (1962).
15. M. Jacobson (to Secretary of Agriculture), U.S. patent 3,050,551; U.S. patent 3,018,219.
16. E. Hecker, *Umschau* 499 (1959). The chemistry of the sex attractant reported before 1959

has been reviewed by Karlson and Butenandt (*2*).
17. A. Butenandt, R. Beckmann, D. Stamm, E. Hecker, *Z. Naturforsch.* **14b**, 283 (1959).
18. A. Butenandt, R. Beckmann, E. Hecker, *Z. Physiol. Chem.* **324**, 71 (1961).
19. A. Butenandt, R. Beckmann, D. Stamm, *ibid.*, p. 84; E. Hecker, *Verhandl. 11th Intern. Kongr. Entomol. (Vienna) 1960* **3**, 69 (1961); A. Butenandt, E. Hecker, M. Hopp, W. Koch, *Ann. Chem.* **658**, 39 (1962); E. Truscheit and K. Eiter, *ibid.*, p. 65.
20. A. Butenandt and E. Hecker, *Angew. Chem.* **73**, 349 (1961).
21. A. Butenandt, W. Guex, E. Hecker, R. Rüegg, U. Schwieter, German patent 1,108,976; W. Guex, R. Rüegg, U. Schwieter, German patent 1,111,615.
22. L. M. Roth and E. R. Willis, *Am. Midland Naturalist* **47**, 66 (1952); D. R. A. Wharton, E. D. Black, C. Merritt, Jr., M. L. Wharton, M. Bazinet, T. T. Walsh, *Science* **137**, 1062 (1962).
23. R. Yamamoto, *J. Econ. Entomol.* **56**, 119 (1963).
24. M. Jacobson, M. Beroza, R. T. Yamamoto, *Science* **139**, 48 (1963).
25. M. T. Ouye and B. A. Butt, *J. Econ. Entomol.* **55**, 419 (1962).
26. N. Allen, W. R. Kinard, M. Jacobson, *ibid.*, p. 347.
27. I. Schwinck, *Z. Angew. Entomol.* **37**, 349 (1955).
28. B. Götz, *Z. Angew. Zool.* **26**, 243 (1939).
29. A. Butenandt and N.-D. Tam, *Z. Physiol. Chem.* **308**, 277 (1957).
30. N. Green, M. Beroza, S. A. Hall, in *Advances in Pest Control* (Interscience, New York, 1960), vol. 3, pp. 129–179.
31. J. B. Cragg and G. R. Ramage, *Parasitology* **36**, 168 (1945); R. W. Dean, *J. Econ. Entomol.* **34**, 123 (1941); R. A. Vanskaya, *Rev. Appl. Entomol.* **31B**, 225 (1943).
32. C. H. Richardson, *New Jersey Agr. Expt. Sta. Bull. No. 292* (1916); *Ann. Entomol. Soc. Am.* **9**, 408 (1916).
33. L. F. Steiner, *Agr. Chem.* **10**, No. 11, 32 (1955).
34. A. W. A. Brown and A. G. Carmichael, *J. Econ. Entomol.* **54**, 317 (1961); *Nature* **189**, 508 (1961).
35. L. D. Christenson, USDA Entomol. Research Div., Beltsville, Md., unpublished data.
36. B. Schaerffenberg and E. Kupka, *Naturwissenschaften* **46**, 457 (1959).
37. F. M. Howlett, *Entomol. Soc. London, Trans., Pt. II* (1912), p. 412; *Bull. Entomol. Research* **6**, 297 (1915).
38. L. F. Steiner, *J. Econ. Entomol.* **45**, 241 (1952).
39. E. A. Richmond, *Proc. Entomol. Soc. Wash.* **29**, 36 (1927); W. E. Fleming, E. D. Burgess, W. W. Maines, *U.S. Dept. Agr. Circ. No. 594* (1940).
40. S. I. Gertler, L. F. Steiner, W. C. Mitchell, W. F. Barthel, *J. Agr. Food Chem.* **6**, 592 (1958).
41. N. Green and M. Beroza, *J. Org. Chem.* **24**, 761 (1959); L. F. Steiner, W. C. Mitchell, N. Green, M. Beroza, *J. Econ. Entomol.* **51**, 921 (1958).
42. M. Beroza, N. Green, S. I. Gertler, L. F. Steiner, D. H. Miyashita, *J. Agr. Food Chem.* **9**, 361 (1961); M. Beroza, N. Green, S. I. Gertler (to Secretary of Agriculture), U.S. patent 3,016,329.
43. M. Beroza and N. Green, *Advan. Chem. Ser.* No. 41, 11 (1963).
44. V. C. Dethier, L. Browne, C. N. Smith, *J. Econ. Entomol.* **53**, 134 (1960).
45. M. Beroza, B. H. Alexander, L. F. Steiner, W. C. Mitchell, D. H. Miyashita, *Science* **131**, 1044 (1960).
46. M. Beroza and B. H. Alexander (to Secretary of Agriculture), U.S. patent 2,974,086.
47. B. H. Alexander, M. Beroza, T. A. Oda, L. F. Steiner, D. H. Miyashita, *J. Agr. Food Chem.* **10**, 270 (1962).
48. M. Jacobson (to Secretary of Agriculture), U.S. patent 2,900,756.
49. P. Gow, *J. Econ. Entomol.* **47**, 153 (1954).
50. N. E. McIndoo, *ibid.* **19**, 545 (1926).
51. I. Schwinck, *Z. Vergleich. Physiol.* **37**, 439 (1955).
52. D. Schneider, *J. Insect Physiol.* **8**, 15 (1962).
53. N. E. McIndoo, *J. Agr. Research* **46**, 607 (1933).
54. L. F. Steiner, *J. Econ. Entomol.* **50**, 508 (1957).
55. H. L. Haller, F. Acree, Jr., S. F. Potts, *J. Am. Chem. Soc.* **66**, 1659 (1944).
56. E. F. Knipling, *Science* **130**, 902 (1959).

# F. Integrated Control

## Some Basic Principles in Insect Population Suppression[1]

*By* E. F. KNIPLING

*Entomology Research Division, Agricultural Research Service, U. S. Department of Agriculture*

I deeply appreciate being selected as the recipient of the Founder's Memorial Award this year. It is an honor to deliver a lecture in recognition of James Fletcher, one of the great pioneers in entomology in the Americas. The topic I have chosen involves some complex calculations. My problem has been to condense the material to stay within the time limitation and to present it so that it will be meaningful.

*Left:* R. L. USINGER; *right:* E. F. KNIPLING

Many of you have heard me speak on certain aspects of the subject before. Some of my views are theoretical, and based on hypothetical situations that have not been substantiated experimentally. Others are based on well-established principles that are well supported by scientific observations and practical experiences.

The methods of insect control now generally practiced make it possible to virtually eliminate the spread of insect borne diseases of man. Losses to crops, livestocks, forests, and stored items, although still substantial in terms of reduced yields, reduced values of affected commodities, and costs of control, can be held to levels that help make possible the production and protection of an abundance of food never before enjoyed by man.

In spite of the great progress, the methods involving the use of insecticides in particular leave much to be desired. Constant vigilance is required to maintain the advantage gained in the fight against the many hundreds of destructive insects. Adverse side effects result from the use of certain insecticides. There is little permanency to the control methods employed. If we relax in our control efforts for one or two insect generations, the insect populations return to normal or actually exceed normal numbers causing continuing losses to crops and livestock. Moreover, due to the ability of insects to develop resistance to insecticides, we cannot count on our leading insecticides to serve their purpose against certain major pests for an indefinite period of time.

Entomologists recognize the limitations of many of the current insect control methods. The character of research on insects has changed drastically during the last decade in efforts to improve insect control procedures. Virtually every reasonable method of insect control is now being explored. Substantial progress is being made on insect control chemicals that are highly selective against the target species. Ways of utilizing natural biological control agents more effectively are being investigated. Exciting and challenging research is underway to fully understand and utilize the factors that govern insect behavior, especially in connection with insect reproduction. A renewed effort, with the cooperation of plant breeders, is being made to take full advantage of the germ plasm in plants that permit the selection of varieties tolerant or resistant to insect attack. Novel ways are being found to employ insects themselves for their own destruction.

We must know the merits and limitations of these methods against various insects and under various circumstances. In order to understand the full capabilities of some of the newer insect control techniques, or even for some of the currently employed insect control methods, we must have a better understanding of some of the basic principles of insect population suppression. Some of the newer techniques now under investigation will have little practical value when applied to small segments of large insect populations. However, when applied uniformly to insect populations in large segments of non-isolated areas or against the total population in small isolated areas, such techniques may be more effective and more economical than current methods of control.

The basic principles of insect population suppression and how these principles can be applied alone or properly integrated is the topic I wish to discuss with you on this occasion.

I am confident that with new techniques of control and with new concepts in their use we can develop methods of regulating many of the major insect populations to the extent that all damage can be prevented, in a way that will avoid or greatly minimize hazards to other forms of life in the environment, and at a continuing cost that will be much lower than the procedures we now follow. An all out approach to insect population control will not be practical or justified for all the destructive insects. But in advocating research on this approach, I have in mind some of our key insect species—those that are responsible for our greatest losses; those that now require expenditures in terms of millions for control year after year; and those that are now responsible for the most extensive use of insecticides. I have in mind such insects as the cotton boll weevil (*Anthonomus grandis* Boheman), screw-worm (*Cochliomyia hominivorax* (Coquerel), corn earworm (*Heliothis zea* (Boddie)), to-

[1] More complete and a more detailed version of the 1965 Founder's Memorial Award Lecture. Entomological Society of America meeting, New Orleans, Louisiana, November 29, 1965.

Reproduced with permission from the *Bulletin of the Entomological Society of America,* 12:7–15, 1966. Published by the Entomological Society of America, College Park, Maryland.

bacco hornworm (*Manduca sexta* (Johannson)), tobacco budworm (*Heliothis virescens* (F.)), cabbage looper (*Trichoplusia ni* (Hübner)), codling moth (*Carpocapsa pomonella* L.), sugarcane borer (*Diatraea saccharalis* (F.)), European corn borer (*Ostrinia nubilalis* (Hübner)), pink bollworm (*Pectinophora gossypiella* (Saunders)), tropical fruit flies (Tephretidae), cattle grubs (*Hypoderma* spp.), face fly *Musca autumnalis* De Geer), and perhaps other important agricultural pests that in the aggregate take a major toll or represent major threats to our agricultural resources. If we can eventually take care of some of these major pest problems by the total population control concept without adversely affecting the beneficial insect complex in the environment, we would at the same time make it possible for natural biological agents to do a better job controlling other important insect problems.

I will attempt to discuss some basic principles of insect population suppression and suggest ways that we might approach this objective. Many of my projections are based on theoretical calculations. It will require difficult and costly research to confirm or reject some of the hypotheses that will be advanced. However, it is my view that the projected effects and practicability of some of the approaches to be outlined, when fully perfected, are as likely to be too conservative as they are to be too optimistic.

### Information Gained from the Screw-worm Population Control Programs

The screw-worm problem in the United States and Mexico represents the first effort to apply the total population control concept by the use of the sterile insect release system. The method has since been applied successfully against four important species of tropical fruit flies. The screw-worm eradication and control programs were and are being conducted through the cooperation and financial support of the livestock industry, and State and Federal agencies. The current program now also involves international cooperation between two countries, Mexico and the United States.

The screw-worm program has proven remarkably successful in spite of the long-range movement of previously mated fertile flies. The use of sterile insects not only succeeds in eliminating established populations but their presence in a screw-worm free area prevents the reestablishment of populations from long-range migrants. A good understanding of the ecological situations favorable and unfavorable for screw-worm survival is vital to the success of such undertaking. The secret to maximum efficiency is to take full advantage of normal fluctuations in abundance and distribution of the natural population during different times of the year; to conduct good surveys; and to promptly report screw-worm cases, followed by the immediate release of sterile flies. Basic information obtained by A. H. Baumhover, B. G. Hightower, and others on the research team, on the long-range movement of screw-worm flies, on the scope of the area where infestations occur, on the ecological requirements of the insect, and on the fluctuations in abundance of the insects during different times of the year is not only of vital importance in the conduct of the screw-worm control efforts but it should be useful in the application of the total control concept for other insect species.

If the factor of long-range migration of the insect had not been uppermost in the minds of the research workers during the initial phases of the development of the technique, the concept of complete screw-worm population control would never have been demonstrated. Experiments on total populations of the insect on small islands was the only feasible way to demonstrate the concepts with available research resources. In view of the long-range migration of the insect an eradication experiment against a nonisolated population would have failed even in an area involving 40,000 square miles. The importance of conducting research on isolated insect populations cannot be overemphasized when one wishes to fully appraise the potentialities of various methods of insect population suppression.

### Characteristic Trends of Insect Populations Subjected to Different Systems of Control

In order to apply successfully the insect population control concept, a good understanding of the inherent capabilities of different systems of control is essential. In order to appraise the potential value of different methods of control, I have followed the practice of establishing hypothetical insect population models that are judged to be reasonably representative of the type of insect, the population density, and the population trends that must be dealt with. As a strating point, it is desirable to estimate the trends of a normal uncontrolled population. It is recognized that the trends of insect populations vary depending on the species and the circumstances. Nevertheless, the establishment of hypothetical models in which all basic assumptions remain constant, makes it possible to compare the relative merits of different systems of control. Table 1 represents the assumed trend of an uncontrolled population starting from a low level in the population cycle. Many of our major insects, after winter hazards or following other unfavorable periods, experience a period of scarcity. I would regard the hypothetical model shown to be reasonably representative of population trends of such pests as the boll weevil, pink bollworm, tobacco hornworm, tobacco budworm, codling moth, tropical fruit flies, screw-worm, house fly (*Musca domestica* L.), and even for scale insects, aphids, and mites. The trend shown may also be representative of insects that have only one generation a year, but the increase would take place over a period of several years, rather than during one season.

Table 1.—Trend of a normal uncontrolled insect population. Fivefold increase rate.

| Generation | Number of Insects |
|---|---|
| Parent | 1,000,000 |
| $F_1$ | 5,000,000 |
| $F_2$ | 25,000,000 |
| $F_3$ | 125,000,000 |

It is well known that most insects have the capacity to increase at a much higher rate than the level indicated. But many hazards, including parasitism and predation, greatly limit the actual rate of increase. The rate of increase can be expected to exceed 5-fold under favorable conditions or be lower than 5-fold under unfavorable conditions. However, the basic model will serve to explain certain principles of insect population control.

### Trends of Insect Populations Subjected to Control with Insecticides

If the normal increase of an insect population is known, we can readily determine what effect different levels of control will have, *if applied uniformly to the total population.* Any significant degree of control will have a great impact on the abundance of the insect if the control procedure operates independently and is superimposed on all other natural hazards. For example, if the normal increase rate is 5-fold, only 50% control above normal hazards each generation would result in a population of 15,625,000 by the $F_3$ generation, instead of 125,000,000. This level of control applied continuously and uniformly to a total population might solve many insect problems. A resistant variety of a crop that reduces the reproductive capacity of an insect population by 50% could be the solution to an insect problem if all growers used such variety. An 80% control level by any method would keep an insect population stable. Control at the 90% level with insecticides would cause a downward trend and have the effect shown in Table 2.

Table 2.—Trend of an insect population subjected to 90% kill with insecticides each generation, when the normal increase rate is 5-fold per generation.

| Generation | Number of Insects |
|---|---|
| Parent | 1,000,000 |
| $F_1$ | 500,000 |
| $F_2$ | 250,000 |
| $F_3$ | 125,000 |
| $F_4$ | 62,500 |

If 90% of the original population of 1,000,000 is destroyed, the 100,000 survivors could be expected to produce 500,000 progeny. Thus, the initial population would be reduced by half. Each generation, treated in the same way, would be further reduced by one-half. If we continued the treatments, we would find that theoretical zero would not be reached until the 18th generation. From a practical standpoint, employing usual methods of insect control, the elimination of a well-established insect population is difficult. From the standpoint of numbers of insects remaining in a population, there often may be little difference between a high degree of control and the complete elimination of the total population. Yet, final elimination is extremely difficult when conventional methods of control are employed.

In actual control operations, the population of an insect might respond quite differently from the theoretical, especially if broad spectrum insecticides are employed. The presence of a complex of insect parasites and predators may be the chief reason why a normal uncontrolled population will not normally increase at a rate higher than 5-fold. In the absence of such natural hazards the insect population might increase 10-fold or even 20-fold per generation. A 90% level of control with insecticides could under such circumstances result in a sudden drop in the population with a subsequent gradual rise to a higher level in spite of continued control efforts. For example, we will assume that the capacity for increase after most of the parasites and predators are destroyed with insecticides becomes 20-fold per generation. In such case the population trend would be as shown in Table 3.

I believe that such trends are actually experienced in our efforts to control certain insects. This may be the situation in cotton fields when the use of insecticides destroys virtually all parasites and predators of the bollworms, *Heliothis* spp. Insecticides as usually employed may have the capacity to control about 90% of the larvae. In the absence of the normal insect parasite and predator

Table 3.—Trend of an insect population subjected to 90% kill each generation when the capacity for increase changes from 5-fold to 20-fold due to the destruction of natural parasites and predators.

| Generation | Numbers of Insects | |
|---|---|---|
| Parent | 1,000,000 → | 100,000 |
| | ↙ | |
| $F_1$ | 2,000,000 → | 200,000 |
| | ↙ | |
| $F_2$ | 4,000,000 → | 400,000 |
| | ↙ | |
| $F_3$ | 8,000,000 | |

complex, an increase of 20-fold for the bollworm population does not seem unreasonable. Consequently, we could expect a gradual increase in bollworms, in spite of control efforts, not unlike that shown for the hypothetical population model. In such event the grower will tend to apply higher dosages of insecticides and may make more frequent applications in order to keep the insects under satisfactory control.

In spite of the limitations of the conventional insecticide method this system of insect control has outstanding merits for insect population control, as we well know from experience. When insect populations are high, insecticide treatments are highly efficient in terms of numbers of insects killed. Their use under such circumstances provides the only practical way to prevent damage by many of our insect species. A kill of 90% of an insect population numbering 1,000,000 would mean the destruction of 900,000. It should be noted, however, that in each subsequent generation the insecticide treatments become less efficient in terms of number of insects destroyed (Table 2). In the $F_4$ generation only 56,000 insects would be killed. If the population reached a level of 100 which would occur on about the 14th generation, only 90 insects would be killed employing the same dosage of insecticide.

Thus, we can draw the first highly significant conclusion: *The system of insect control by the use of conventional insecticides is highly efficient in terms of numbers of insects destroyed when the insect population is high, but highly inefficient when the population is low.*

We know from experience that the elimination of insect populations by the use of insecticides is often a slow, difficult and costly process. Even if 98% control were achieved each generation, it would be necessary to apply treatments for 5 generations to expect the elimination of an initial population of 1,000,000 insects. Generally, as many insecticide treatments are required to eliminate the last 1% or even the last 0.1% of an insect population as is required to destroy the first 99% or 99.9%. The hypothetical model (Table 2) clearly supports practical experience.

### Trends of an Insect Population Subjected to Control by the Release of Sterile Insects

The use of sterile insect releases has an effect on insect population trends that is entirely different from that exhibited by insecticides. Table 4 shows the theoretical effect of sterile insect releases when the initial release rate causes a downward trend in the natural population and releases of sterile insects are continued and maintained at a constant level. The model assumes

Table 4.—Trend of an insect population subjected to sterile insect releases.

| Generation | Number of Insects Natural Population | Number of Sterile Insects | Ratio Sterile to Fertile | Number of Progeny |
|---|---|---|---|---|
| Parent | 1,000,000 | 9,000,000 | 9:1 | 500,000 |
| $F_1$ | 500,000 | 9,000,000 | 18:1 | 131,580 |
| $F_2$ | 131,580 | 9,000,000 | 68:1 | 9,535 |
| $F_3$ | 9,535 | 9,000,000 | 942:1 | 50 |
| $F_4$ | 50 | 9,000,000 | 180,000:1 | 0 |

full competitiveness of the sterile insects and a 5-fold increase for all insects successful in fertile matings.

The significant feature of this method of insect control is that the sterile insect releases become progressively more effective as the natural population declines. With a 9:1 ratio, the sterile insects provide 90% control, or the same as the initial insecticide treatment. However, in the $F_1$ generation when the same number of sterile insects is released, the ratio of sterile to fertile insects increases to 18:1. Thus, the rate of population decline accelerates. Such acceleration continues until the theoretical ratio of 180,000 sterile to 1 fertile insect is obtained and no chance of further reproduction would be expected. A total of 45,000,000 sterile insects would be required as employed in this model to achieve theoretical elimination of the population.

If 1,000,000 insects represented a high population level for the infested area, the sterile insect release rate required would in all probability be impractical. However, if the initial population represented a low level in a large segment of the infested area, and low cost rearing methods for the insect were available, we might expect the sterile insect release method alone to be entirely practical for the elimination of the natural population. However, for most well-established insect populations, it will be impractical to adequately overflood the natural population even at the lowest level of seasonal abundance. In such event prior reduction of the natural population by a method that is more efficient when populations are high, such as the use of insecticides or cultural methods would be indicated. Thus, we might draw a second conclusion: *The sterile insect release system is inefficient as a means of insect population control when the natural population density of the insect is high, but it becomes highly efficient when the natural population density is low.*

Thus, we have two systems that should be highly complementary in achieving insect population control. We will establish a model to test this hypothesis.

Table 5 projects the trend of the same type of insect population when it is subjected to control by the combined use of insecticides and sterile insect releases. It is assumed that each system operates independently. The insecticide might be used against the immature stages and the sterile insects released to compete with adults that emerged or insecticides might be employed to control the adults in the fall and sterile insects used for survivors the next spring.

Calculations for the integrated program show complete elimination of a population of 1,000,000 insects by the use of insecticides for one generation, plus the liberation of 3,600,000 sterile insects. These requirements are extremely favorable in comparison with those for insecticides alone, which would require treatments for 18 generations to achieve complete population control. The requirements are equally favorable in comparison with the need for 45,000,000 sterile insects to achieve theoretical elimination of the population by this system alone. Thus, we can draw another highly significant conclusion: *The integration of the conventional insecticide control system and the release of sterile insects provides a system of insect population control that is much more efficient than either system employed alone.*

We might emphasize the significance of the sterile insects in such an integrated program by making the point that *the sterile insect release system, when used in conjunction with the insecticide system of insect control, provides a means of reversing the law of diminishing returns in dealing with the elimination of insect populations.*

### Application of the Basic Principles of Integrating Chemical Treatments and Sterile Insect Releases for the Control of a Specific Insect

The theoretical models establish basic principles of insect population suppression that can be put to practical use in dealing with specific insect problems. The boll weevil is among our most destructive insects that should yield to the integrated concept of insect population control.

A number of years ago I advanced the opinion that the sterile insect release method could prove useful in boll weevil eradication when integrated with the use of insecticides or other control methods such as cultural measures. It was obvious from the beginning that the use of sterile male boll weevils alone would not be practical. Elimination of the insect by the use of insecticides alone would prove costly and might cause objectionable side effects. However, the integrated approach along

Table 5.—Trend of an insect population subject to treatment by an integrated program of insecticide applications and sterile insect releases.

| Generation | Natural Population | | Sterile Insects Released | Ratio Sterile to Fertile | Expected Progeny |
|---|---|---|---|---|---|
| Parent | 1,000,000 insecticides kill 90% | 100,000: | 900,000 | 9:1 | 50,000 |
| $F_1$ | | 50,000: | 900,000 | 18:1 | 13,158 |
| $F_2$ | | 13,158: | 900,000 | 68:1 | 954 |
| $F_3$ | | 954: | 900,000 | 942:1 | 0 |

lines projected seemed to offer a method of eradication that would be well within the realm of technical feasibility. We needed answers to three vitally important questions, however, before we could make an appraisal of the potentialities of the integrated approach. These were: (1) can the boll weevil be sexually sterilized without serious adverse effects on the competitiveness and behavior of the males, (2) can the natural population be reduced to a low level at a cost that would then make it practical and advantageous to overflood the natural population to the required degree with sterile males, and (3) will it be feasible to mass produce the required number of the insects.

Investigations were first started on methods of inducing sterility in the boll weevil. Initial research was undertaken by T. B. Davich and later by D. A. Lindquist at the College Station Laboratory in cooperation with the Texas Agricultural Experiment Station. Research has continued by Davich and associates at the Boll Weevil Research Laboratory in cooperation with the Mississippi State Agricultural Experiment Station. Only partial success has been achieved thus far in producing competitive sterile males.

Research on mass rearing methods for the boll weevil have been underway by a number of investigators. The initial cooperative work was also undertaken at College Station by E. S. Vanderzant and T. B. Davich. Subsequently, important contributions have been made by a number of investigators, particularly by R. T. Gast of the Boll Weevil Research Laboratory. Excellent progress has been made in mass rearing the insect in unlimited numbers at a cost that would be entirely practical.

The third line of investigation, a key factor equally as important as the two already mentioned, required information on the natural population density of the boll weevil and the extent to which the population could be reduced by the use of insecticides or cultural measures. A number of entomologists, both Federal and State, have obtained important information on this vital point during recent years. The use of conventional insecticides can be employed at various seasons of the year to achieve a high degree of population suppression. However, the most desirable time appears to be in the fall. The basic research by L. D. Newsome and J. R. Brazzel of Louisiana State University on the diapause phenomenon in the boll weevil, and subsequent research by J. R. Brazzel, T. B. Davich and L. D. Harris in Texas demonstrated that the use of insecticides to interrupt the development of diapausing boll weevils provides a promising means of boll weevil control. A number of investigators in other areas have confirmed the findings of Brazzel and associates. About 4 insecticide treatments applied in the fall, as advocated by Brazzel, can be expected to achieve about 90% population control. A 90% level of control would go a long way toward reducing natural populations to manageable levels with sterile males. However, it is my opinion that a considerably higher level of control will be necessary, especially in view of the difficulties encountered in achieving sterility in male boll weevils without seriously affecting mating competitiveness.

Accordingly, I proposed another type of fall program based on a schedule of treatments designed to limit both reproducing and diapausing boll weevils at the most critical period in the fall. Calculations based on hypothetical boll weevil population models, taking into consideration the biology of the insect during the fall, showed that if a 4 treatment diapause schedule would reduce populations by 90%, a 7 treatment schedule designed to limit both reproduction and diapause would provide 99% control.

The reproduction-diapause control schedule makes possible two opportunities to prevent or destroy fall progeny that are designed to overwinter. Potentially, it should equal the combined effect of a fall and spring program. It is my belief that the estimated 99% control resulting from properly timed treatments in relation to the life history of boll weevils in the fall, would be conservative under circumstances where 4 diapause treatments would achieve 90% population control. The validity of the theoretical calculations have been confirmed in Mississippi by field studies of E. P. Lloyd and associates of the Boll Weevil Research Laboratory, and by Perry L. Adkisson of the Texas Agricultural Experiment Station. A high level of control within the range of 99% was obtained by M. E. Merkl and T. B. Davich in a well-isolated cotton growing area in the vicinity of Mobile, Alabama, when the fall treatment schedule was modified.

In view of the progress already made, there is every reason to believe that the integrated insecticide-sterile insect method of boll weevil population represents a promising approach to the problem.

In order to project more clearly the principles involved in relation to the basic requirements discussed, I will establish a hypothetical boll weevil population and determine the theoretical effect of an integrated program. The results of this projection are shown in Table 6, which involves an isolated cotton growing area of 1,000 acres. A 98% reduction of an estimated normal uncon-

Table 6.—Hypothetical boll weevil population control program on 1,000 acres by integrating the use of a fall insecticide reproduction-diapause control program and the release of sterile males the following season. Five-fold increase for each fertile mated pair.

| Generation | Uncontrolled Population | Integrated Control Program | | | |
|---|---|---|---|---|---|
| Parent | 200,000 | Insecticides Kill 98% ——→ | 4,000 ——→ | 200,000 sterile ♂ / Ratio 100:1 ↘ | 200 progeny |
| $F_1$ | 1,000,000 | | 200 ——→ | 200,000 sterile ♂ / Ratio 2,000:1 ↘ | No progeny |
| $F_2$ | 5,000,000 | | | | |
| $F_3$ | 25,000,000 | | | | |

trolled spring population averaging 200 boll weevils per acre will be assumed for the fall program, instead of the theoretical 99% already discussed. The surviving spring population assumed to average 4 boll weevils per acre (2 males and 2 females) will then be subjected to the release of 200 sterile males per acre or a total of 200,000 for the hypothetical 1,000 acre cotton growing area. This would provide an overflooding ratio of 100 sterile to 1 fertile male competing for mates with the low level female population. In my opinion no insect species in existence can maintain its original population level in the face of a starting ratio of 100 reasonably competitive sterile to one fertile male. However, the reduced $F_1$ progeny would again be subjected to sterile males at the same rate of 200 per acre. The ratio of sterile to fertile males would then theoretically increase to 2,000:1, which should eliminate reproduction.

The type of control program projected would theoretically eliminate the population by making 7 insecticide treatments and by releasing 400 sterile males per acre. The model assumes complete competitiveness of the sterile male boll weevils. Unfortunately, the production of even reasonably competitive sterile male boll weevils has not yet been achieved. However, T. B. Davich and associates of the Boll Weevil Research Laboratory have shown that a high ratio of sterile males having greatly reduced competitiveness can nevertheless exert a marked effect on the reproductive potential of low natural populations. Elimination of natural populations has been achieved in small plot tests. Thus, the principles discussed have already been demonstrated. Any significant improvements in techniques for producing sterility in male boll weevils should make the proposed integrated approach entirely feasible as a means of eliminating boll weevil populations. Attainment of the two other basic requirements, practical means for mass production of the boll weevil and practical means for reduction of natural populations to manageable levels, seems assured.

What would be the cost of a program of the nature proposed? A series of 7 insecticide treatments using the low volume technique would cost about $7.00 per acre. The production and release of 400 sterile males per acre would cost about $2.00 per acre. However, to be conservative we might assume the need for 1,000 sterile males per acre at a cost of $5.00. Thus, the total cost might be of the order of $12.00 per acre. It should be pointed out that from the standpoint of population dynamics, the sterile males would contribute more to complete elimination of the population than would the insecticide treatments. Theoretically, it would require insecticide treatments for three additional generations at the 98% control level, involving about 16 treatments to expect complete population control following the fall program. This would cost of the order of $16.00 per acre, as compared with the estimated $5.00 per acre for the elimination of the low level population by the use of sterile males. Based on research by Knox Walker of the Texas Agricultural Experiment Station, the rate of increase of boll weevils when the population is low may be greater than the assumed 5-fold rate of increase. However, even with a 10-fold increase theoretical elimination would be achieved if the sterile males are reasonably competitive.

If the boll weevil occurs on 10,000,000 acres of cotton in the United States, the total cost for an eradication program of the nature projected might be within the range of $100–$200,000,000. We are not accustomed to thinking in such terms to provide permanent solutions to insect problems. But let us analyze the costs of control methods farmers now use. The uncoordinated use of insecticides directed against segments of the insect population costs cotton growers an estimated $70,000,000 every year. Yet, each spring we still have an abundance of the insect that threatens the next seasons crop. Many farmers spend more than $12.00 per acre each year to protect the cotton crop. In spite of this expenditure of funds, annual damage due to the boll weevil is estimated to be about $300,000,000 based on normal cotton prices.

The proposed eradication procedure has not been perfected. However, the basic approach is sound. Substantial progress has already been made in meeting the requirements for achieving complete population control. Research on any method that could conceivably be used to achieve complete elimination of an insect pest as damaging as the boll weevil at a cost less than the annual losses should be vigorously pursued. Aside from the economic benefits, the elimination of the boll weevil would make the greatest single contribution that could be made toward a reduction in environmental pollution due to insecticides.

### The Integration of Predator Releases and Sterile Insect Releases

The inundation of an insect population by the mass production and sustained release of parasites or predators may offer opportunities for insect population control similar to that afforded by the sterile insect release method. However, few, if any, efforts have been made to apply this technique against the total population of an insect when large and sustained releases were made over a period of several pest generations. I recognize that fundamental information is lacking on the efficiency of various parasites and predators in destroying specific insect hosts at different density levels for both the parasites or predators and the host. However, it would seem logical to assume that from the wide variety of parasites and predators available for major insects, certain species should be capable of exerting a drastic impact on the reproductive potential of an insect population if adequate numbers are released and such releases are sustained over a period of several generations.

In my studies on the fundamentals of insect population suppression I have attempted to appraise the value of mass and sustained releases of insect predators, especially when integrated with the sterile insect release method. The term predators in this sense will also include insect parasites.

As with other systems of insect population control, I have made use of hypothetical insect population models to estimate the potential value of predator releases.

It is probable that the efficiency of released predators will be governed both by the density of the predator and the density of the hosts. I know that many complex factors are involved in predator-host relationships. On the basis of current knowledge, it is difficult to predict the efficiency of various predator-prey density levels. Nevertheless, starting with certain basic assumptions, it should be possible to make a reasonably valid appraisal of the inherent potential of the predator inundation system of insect population control and then determine the advantages of an integration of this system with other methods of control, such as by means of conventional insecticides, attractants, sterile insect releases, or by other methods.

In this appraisal, I wish to consider the integration of predator releases and sterile insect releases. The requirements for achieving various levels of suppression in the reproductive potential of insect populations by the release of sterile insects can be estimated with good confidence. The theoretical effects have been confirmed experimentally. Table 7 shows the number of fully competitive insects that must be released to achieve different levels of reduction in the reproduction potential of a given insect population.

Table 7.—Calculated effects of sterile insect releases on the reproductive potential of a hypothetical insect population.

| Sterile Population | Natural Population | Percent Control |
|---|---|---|
| 1,000 | 1,000 | 50 |
| 3,000 | 1,000 | 75 |
| 9,000 | 1,000 | 90 |
| 99,000 | 1,000 | 99 |

It may be noted that in order to achieve 90% control, it it necessary to release 9 times as many sterile insects than is necessary to achieve 50% control. It will require 99 times as many to achieve 99% control than 50% control, and 11 times as many to achieve 99% than is required to achieve 90% control.

It is my opinion that a similar relationship might prevail when we attempt to control insect populations by inundation with parasites or predators.

We will make the arbitrary assumption that the release of 2,000 adult predators will destroy 50% of the progeny of a population of 1,000 adult host insects above the normal hazards encountered by the host insect. With this basic assumption, we might postulate the effects of higher predator release rates as shown for Table 8.

Table 8.—The assumed effect, above natural hazards, of different levels of predator releases on the reproductive potential of an insect prey population.

| Predator Density | Prey Density | Percent Control |
|---|---|---|
| 2,000 | 1,000 | 50 |
| 4,000 | 1,000 | 75 |
| 8,000 | 1,000 | 87.5 |
| 16,000 | 1,000 | 93.75 |
| 32,000 | 1,000 | 96.88 |
| 64,000 | 1,000 | 98.44 |
| 128,000 | 1,000 | 99.22 |

I have attempted to obtain data, from literature, on the degree of efficiency of various ratios of predators to prey. Information supplied by S. E. Flanders, of the University of California; J. M. Franz, of Dermstad, Germany; and F. W. Lawson, of the Entomology Research Division, indicates that the assumed effects of different ratios of predator to prey may be valid in principle. At least we can safely say that the findings of specialists in the field of biological control show that the ratio of a predator population to the prey population must be very high to achieve virtual extinction of a prey population.

Since we can conclude with confidence that the projected effects of sterile insect releases are valid, and if we can assume that the inundation of a prey population with a select predator population would be essentially as efficient, insect for insect, as inundation by sterile insects we can then calculate the merits of integrating the two systems of insect population control. The results of this appraisal are shown in Table 9.

Table 9.—Calculated effects of an integrated program of sterile insect and predator releases on the reproductive potential of a hypothetical insect population consisting of 1,000 insects, in comparison with the effects of each system employed alone.

| Sterile Insects Alone | | Predators Alone | | Both Systems Combined |
|---|---|---|---|---|
| No. Sterile | % Control | No. Predators | % Control | % Control |
| 1,000 | 50 | 2,000 | 50 | 75 |
| 3,000 | 75 | 4,000 | 75 | 93.75 |
| 9,000 | 90 | 10,000 | 90 | 99 |
| 99,000 | 99 | 100,000 | 99 | 99.99 |
| 9,999,000 | 99.99 | 10,000,000 | 99.99 | 99.999999 |

This hypothetical model assumes that the sterile insects and the predator releases produce their effects independently and without interaction against different stages of the host insect. This should be a safe assumption to make if the released predators attack the immature stages of the host insect and the sterile insects affect the natural adult population. It should be noted that in this hypothetical model the release of 1,000 sterile insects and 2,000 predators—a total of 3,000 released insects—would theoretically achieve 75% control. Referring to Tables 8 and 9, we will note that in order to achieve this level of control by each system alone, it would be necessary to release about 3,000 sterile insects or 4,000 predators. There would be little advantage in the integration of the two systems for this level of control. At the 93.75% control level, however, the integrated program becomes about two times as efficient as either method alone. The theoretical efficiency of the integrated program steadily increases at high levels of control. At the 99% control level for the integrated program, 9,000 sterile insects and 10,000 predators would be required, each system alone providing 90% theoretical control. The total requirements would be 19,000 insects. The sterile insect method alone would require the release of 99,000 insects to achieve 99% control, and the predator release method alone, according to the assumptions, would require 100,000 insects. Thus, the integrated program would be better than 5 times as efficient as either system employed alone. This would mean, for example, that if complete control of an insect population would cost $50.00 per acre by either system used alone, the same result could be achieved for $10.00 per acre, if the two systems are combined.

It may only be of academic interest, but further projections in the calculations show that at the 99.99% control level, the integrated program becomes about 50 times as efficient as either system employed alone.

On the basis of the above analysis, we may advance the hypothesis that: *The integration of predator (or parasite) releases and sterile insect releases, which independently affect different stages of the host insect, provides a more efficient system of insect population control than either system employed alone.*

Although this proposition needs to be confirmed, there is every reason to believe that the integration of parasite or predator releases and sterile insect releases offers a great opportunity to develop practical ways to achieve and maintain complete population control on a regional basis for some of our major insects. Insects such as the tobacco budworm, tobacco hornworm, corn earworm, and cabbage looper might be cited as good candidates among crop pests for such integrated control procedure on a regional basis without the use of insecticides.

It should be emphasized that the proposed system of control should become progressively more effective as the natural population declines. For example, if an integrated program reduced the reproductive rate of a parent population by 99% and the expected decline in the natural population occurred, the same level of sterile insect and predator releases would, theoretically, have a much greater impact on the population during the next generation, because the ratio of sterile to fertile insects would increase and so would the ratio of the predator to the prey. If increasingly higher ratios of predator to prey will function against a diminishing prey population in the way we envision the effects of sterile to fertile insects, we can project the results of the integrated program into subsequent generations.

It may be worthwhile to project the theoretical effect of such integrated program through two generations. If the initial parent population consisted of 1,000,000 insects and 9,000,000 sterile insects were released for a 9:1 ratio of sterile to fertile insects a 90% reduction in reproduction would result. According to earlier assumptions the release of 10,000,000 predators would also cause a 90% reduction in the reproductive potential of the population. Thus, the combined effect would be 99%. If the insect population was stable, this would mean that only 10,000 $F_1$ adult progeny would be produced. The original release rate of 9,000,000 sterile insects would then result in a sterile to fertile ratio of 900:1 or a theoretical 99.89% suppression of reproduction. If the same relationship holds for the predator releases, a similar suppression of reproduction would be achieved. The combined effect would therefore result in suppression of reproduction by 99.9879% in the $F_1$ generation. This would mean that less than 2 of the 10,000 $F_1$ progeny could reproduce, which in turn would mean theoretical extinction in the $F_1$ generation. In actual practice allowance would have to be made for the expected number of progeny produced by the insects that survive and are successful in fertile matings. However, this would not alter the principle.

Once complete dominance of an insect population is achieved by systems of control such as those discussed, it should be possible to maintain complete control in large nonisolated areas by the continuous release of relatively few sterile insects alone or by continuing an integrated program at a much lower release rate than was required to achieve complete control of the initially higher established population.

### The Integration of Cultural Control and Release of Sterile Insects for Tobacco Hornworm Control

Both the tobacco hornworm and the tomato hornworm, (*Manduca quinquemaculata* (Haworth)), are major pests of tobacco and tomatoes. For the purpose of this report, only the tobacco hornworm, the more important and the more abundant of the two species, will be considered, although both species would have to be dealt with in any effort to protect tobacco from hornworms. Losses caused by tobacco hornworms on tobacco based on usual prices for tobacco are estimated to be of the order of $35,000,000 annually. The amount spent for their control by tobacco growers is difficult to estimate, but would probably amount to $10,000,000 annually.

The total population control concept for these insects is being given attention in current research efforts. The use of light traps on a community-wide basis is showing considerable promise. The sex pheromone method may prove effective using virgin females in conjunction with light traps, or hopefully by employing a synthetic sex attractant if such attractant can be produced. However, on the basis of theoretical calculations, it should be entirely feasible to achieve complete control of the tobacco hornworm on a regional basis by instituting an integrated program. This program would consist of a rigid cultural control program in the fall to limit overwintered hornworms plus the release of sterile insects the following season. Then the continued release of sterile males alone in low numbers should provide a way to maintain complete population control of this insect at a cost of only a fraction of the cost of achieving control of the initial normal population. The possibilities of this approach to tobacco hornworm control have been discussed in previous papers that I have presented. However, I am again including this method in this discussion because I feel that it provides one of the best examples of the feasibility of applying the total population control concept to one of our major insects by employing the newer methods of insect control.

The total acreage of tobacco and tomatoes east of the Mississippi River amounts to about 1,250,000 acres. Based on actual field data obtained by F. R. Lawson and his associates of the Oxford, North Carolina, laboratory it is reasonable and probably conservative to assume a normal spring population density of 40 moths per acre. Lawson and associates estimate that through a fully coordinated fall cultural program, it should be feasible to reduce the normal population by 80%. According to the basic assumption, this would mean an average of about 8 moths per acre for the spring brood, following the previous fall cultural program. If 8 moths per acre of host crops is a valid estimate, the total spring brood of moths would consist of about 10,000,000 moths, east of the Mississippi River.

If a population of hornworms under normal tobacco-growing practices will increase at a 5-fold rate per generation, it would be necessary to overflood the natural population by more than 4:1 to start a downward population trend. A 10:1 ratio will be established as the initial release rate. This would require 100,000,000 moths for the first brood.

If the cost of rearing moths is calculated at $5.00 per million, the total cost of moths for brood one would be $500,000. It is assumed that the same number of moths would be required for brood two.

If the expected effect is achieved during the first year, the number of moths required for the second year could be reduced by ½ and still achieve complete domination of the population. A mortality of 75% is assumed for the population entering hibernation during the first winter after the program is initiated.

The calculated trend of the hornworm population sub-

Table 10.—Estimated trend of the hornworm population east of the Mississippi River when subjected to an integrated program involving cultural control and sterile insect releases.

| Broods | Natural Moth Population | Sterile Moth Population | Ratio of Sterile to Fertile Moths | Number of Moths Reproducing |
|---|---|---|---|---|
| *First Year* | | | | |
| 1 | 10,000,000 | 100,000,000 | 10:1 | 909,090 |
| 2 | 4,545,450 | 100,000,000 | 22:1 | 197,454 |
| *Second Year* | | | | |
| 1 | 246,818[1] | 50,000,000 | 202:1 | 1,215 |
| 2 | 6,075 | 50,000,000 | 8,230:1 | 0 |

[1] A mortality of 75% owing to hazards during hibernation is assumed. Theoretically, a moth population of 987,270 would have resulted from the 197,454 reproducing moths in the second brood in the absence of winter hazards.

jected to the 2-year program as outlined, is shown in Table 10. Theoretical elimination of the population would be achieved by the end of the second season.

It is not possible to make an accurate estimate of the cost for a program of the type projected. However, the cost for producing a total of 300,000,000 moths would probably amount to about $1,500,000. If all other costs for such a program were equivalent to the cost for rearing moths, the overall costs would be $3,000,000 to achieve complete domination and elimination of the original established population during a 2-year program.

In all probability some moths would have to be released each year in a continuing program to prevent reestablishment of long-range migrants from other areas. Such maintenance program should, however, cost much less than the yearly operations during the first 2 years, which are designed to achieve complete control over the original population. If 50,000,000 sterile moths were adequate each year to prevent the reestablishment of damaging tobacco hornworm populations east of the Mississippi River such program probably would not cost more than a half million per year. This should be compared with the estimated annual costs of $10,000,000 for control by current methods. Of greater economic significance, the estimated damage caused by the insect would be eliminated. Moreover, the need for insecticides to control this pest would be avoided, thus contributing to a reduction of environmental pollution due to insecticides and the elimination of residues on tobacco resulting from hornworm control by current methods.

Again, these projections for controlling one of our important insects are mere estimates based on a hypothetical situation. There is reason to believe, however, that they reasonably reflect what could be achieved in the development and application of the proposed technique.

New developments in the possible use of sex attractants in combination with light traps may eventually provide an even more promising way to achieve and maintain complete control of tobacco hornworm. However, the projected integrated cultural and sterile insect release method, based on theoretical estimates, offer promise as an effective, desirable, and economical way to solve this major insect problem.

## General Comments

I have attempted to show how different systems of insect population control can effect population trends, with special emphasis on the theoretical advantages of integrating two or more systems. Time does not permit a discussion of basic principles that might be involved in population control by the use of insect attractants and chemosterilants when applied to the natural population. However, these methods of insect control also have characteristic effects on insect population trends that need to be considered in establishing their merits and limitations when employed alone or when integrated with other systems of insect control.

The development of procedures for achieving and maintaining complete control of specific insect populations will not be easy. A satisfactory solution to each major insect problem will require imagination and the best scientific talent that we can muster. Research costs will be high just to develop the basic information needed. In such research, special facilities and costly experiments such as those conducted on isolated islands during recent years will be necessary in many cases. The application of the methods in practice will not be simple. The costs involved in applying the control measures will not be low. However, we are dealing with many specific insect problems that cost our agricultural industries and our Nation's economy 10, 25, 100, or even several hundred million dollars year after year, with no immediate prospect in sight of substantially reducing such losses by following the conventional approach to insect control. In fact, due to objections and possible restrictions in the use of broad, spectrum insecticides we might expect increasing costs in dealing with major insect problems. The high cost of control, the high losses in spite of control efforts, and the undesirable side effects of current methods of control obligate us to take an entirely new look at some of the most costly and most troublesome of our insect problems. There is ample justification for taking bold and positive steps in our research efforts in order to fully explore the possibilities of any method of control that may be more desirable, more effective, or more economical in our long-range effort to find more permanent and acceptable solutions to major insect problems. The current losses to our agricultural economy resulting from the key insect species together with the public concern over the side effects resulting from current methods of control provide wide latitude for the development of practical means for their control by other means. These are the reasons for my interest, my confidence, and my enthusiasm for research on methods that eventually can be employed to meet many of our most important insect problems by applying the basic principles of insect population suppression discussed with you on this occasion.

# HEALTH AND WELFARE

## The Attraction of Mosquitoes to Hosts

*Anthony W. A. Brown, PhD*

When a man is standing in mosquito-infested territory in a light breeze, the mosquitoes will pass him by, but those which enter the downwind cone of his emanations will turn and work upwind to him from as much as 30 feet away, as closely observed by Hocking[1] of *Aedes dorsalis* on the Alberta prairie. When a man is standing in a closed room, the mosquitoes in flight alongside ignore him, but let him lie down and he will activate all those which experience the ascending convection currents above him, as found by Mer et al[2] for *Anopheles sacharovi* in Israel. The photographs obtained by Kellogg and Wright[3] for *Ae aegypti* exposed to a smoke-marked moving cone of warm, moist air show that these avid females proceed upwind to its source by a behavioral response known as anemotaxis, supplemented by klinokinesis on the boundary of the cone that makes them turn when finding themselves in a decreasing concentration gradient and go straight in an increasing one, and thus become trapped in this one-way route.[4]

It was Roth[5] who demonstrated by extirpation experiments that the antennae and not the eyes were indispensable for *Ae aegypti* and *An quadrimaculatus* to find their human host. The 13 segments of *Ae aegypti's* antenna carry, in addition to spines and tactile hairs, an abundance of thin-walled sensilla trichoidea of two types, the first very long and pointed and most frequent in the distal segments of the antenna; the second shorter and blunt and more common at the base of this organ. Steward and Atwood[6] also found thin-walled sensilla basiconica, "peg organs," particularly towards the end of the antenna. In *An atroparvus* and *An quadrimaculatus* there are also a few sensilla coeloconica or sunken peg organs, as discovered by Ismail.[7] Among the three types of thin-walled sensilla in *Aedes aegypti* are the hygrometers for water vapor[8] and the detectors for carbon dioxide.[9] Steward and Atwood obtained evidence from progressive extirpation of antennal segments that it was the type-1 trichoid sensilla that were mainly responsible for detecting the emanation from a human hand.[6]

From the Department of Zoology, University of Western Ontario, London, Canada.

Read as part of the Symposium on Insects and Disease during a joint meeting of the Section on Dermatology with the AMA committee on Cutaneous Health and Cosmetics and the Entomology Research Division, US Department of Agriculture, at the 114th annual convention of the American Medical Association, New York, June 23, 1965.

Reprint requests to 275 Victoria St, London, Ontario, Canada, (Dr. Brown).

Over the years a great many workers have made empirical experiments to determine what factors are attractive to mosquitoes by counting the numbers coming to a treated source as compared to a control source, and in my department we have been doing this kind of thing since 1948. Many experiments were done in the laboratory with airstreams of the cascade type which are removed through an annulus around the port of issuance[10] Others were performed in the Canadian forest with robots clothed and heated to maintain 150 lb of water at 98 F with the expenditure of 50 watts which characterizes the output of a resting man.[11]

Of the airborne factors investigated (Table 1), moisture proved to be the most powerful single factor, both in airstreams and on clothed bodies, being especially strong in a hot, dry environment, but becoming unattractive when the ambient air was cooler than 60 F. All workers agree that moist air is attractive to mosquitoes, so long as 100% relative humidity is not reached, for this has been frequently found to be unattractive.

In our experiments, warmth was the next factor in importance, while the combination of warmth and moisture was highly attractive, as also found by van Thiel[12] for *An atroparvus*, by Parker,[13] and later by Wright[14] for *Ae aegypti*. Parker and Wright considered that warmth was not attractive in itself. This contrasted with the earlier conclusions of Howlett,[15] Reuter,[16] DeLong et al,[17] and Christophers.[18] We have found perfectly dry objects such as billiard balls to be attractive if raised to temperatures not exceeding 110 F. Moreover, the attractiveness is due to convective and not radiant heat, for it is abolished by the interposition of an airtight window of thallium bromoiodide, fully transparent to infrared radiation, between the warm body and the mosquitoes.[19]

Carbon dioxide, the third airborne factor, was considered by Rudolfs,[20] DeLong et al,[17] and Willis[21] to be an activator rather than an attractant. More recently, Wright[14] accepts this view and considers that carbon dioxide only activates the mosquito to the attractant effect of moisture. But we found that carbon dioxide was attractive to *Ae aegypti* when simply added to a dry airstream,[10] while Reeves[22] discovered that dry ice provides an effective trap for field populations of *Culex tarsalis* and other culicine vectors. Nevertheless, it was concluded by Laarman[23] from extensive experi-

Table 1.—Multiples by Which the Attractiveness of Bodies or Airstreams can be Increased

| | | Multiple |
|---|---|---|
| Airborne factors | Moisture | 4 |
| | Convective heat | 3 |
| | $CO_2$ | 2 |
| Visual factors | Nonreflectivity | 5 |
| | Movement | 2 |
| | Contour | 1.6 |

ments on *An atroparvus* that the essential role of carbon dioxide in orientation is that of an activating stimulus.

Of the visual factors governing the attractiveness of bodies, dark colors were found to be more attractive than bright ones for *Ae aegypti* by Brett,[24] for *C fatigans* by Howlett[15] and Ko,[25] and for the western American species of *Aedes* by Gjullin.[26] But, for anopheline mosquitoes, the investigations of Ko in Formosa and the US Department of Agriculture laboratory (when it was located in Orlando, Fla), found the situation to be the reverse. When we compared different colored cloths for attractiveness to the Canadian species of *Aedes*, mostly *Ae punctor* and *Ae communis*, we found the brighter clothes to be less attractive, the inverse relationship holding good for reflected light between 625 and 475 m$\mu$ wavelength, ie, between the orange and the blue-green.[27] It appears that the other half of the spectrum visible to the human eye, red on the one hand and violet on the other, is black to the eyes of *Aedes* mosquitoes and therefore attractive. Movement of a black body, whether a cylinder or a mouse, roughly doubled its attractiveness.[28] An increase in black-white interface in a checkerboard pattern, or indeed any flicker effect, made an object yet more attractive to *Ae aegypti*. For approaches to a mouse exposed in a 10-foot-square cage of mosquitoes, the ensemble of visual factors was no less powerful than the airborne ones.

Three of these six attractive factors were then investigated in the hands of men exposed to *Ae aegypti* (Table 2). In pairs equated for skin temperature, dark hands proved to be more attractive than pale[29] in the order Negroes, Orientals, dark Caucasians, light Caucasians. In general however Freyvogel[30] has reported that Africans are no more attractive than Europeans to *Ae aegypti*, the comparisons having been performed without regard to skin temperature. Among men equated for skin hue, we found warm hands to be more attractive than cool ones, and recent tests at Gainesville, Fla, have shown this to hold good among women as well as men. The Gainesville group also found that men were more attractive than women in eliciting approaches by this species of mosquito, as Rahm[31] had reported for Europeans and Freyvogel[30] for Africans. As for the factor of moisture output, we were surprised to discover that hands with greater water-vapor output were less attractive than drier ones, a finding subsequently reported by Gouck and Bowman[32] and recently confirmed in a large number of subjects at Gainesville. The suggestion that this was due to a greater evaporation producing a cooler hand[14] is not the explanation since we had already selected each pair of hands for equality in surface temperature.[29]

The attractiveness of sweat to mosquitoes is marginal. It was reported to exert no attraction by early investigators such as Howlett,[15] Crumb,[33] Rudolfs,[20] and Reuter,[16] but was later reported to be attractive by DeLong et al[17] and Parker.[13] We found that sebaceous armpit sweat was slightly attractive to *Ae aegypti*,[34] while sudoriferous forehead sweat was without effect. Compared to the emanations from the human body at rest, sweat is of minor importance for this mosquito, as the experiments of Rahm[35] have established.

It became evident to us, in contrast to the view advanced by Wright,[14] that there must be principles present in body emanations which are attractive to mosquitoes over and above moisture and carbon dioxide. As Reuter[16] had previously reported, we also found mammalian blood to be attractive[36]; the activity was completely removed by submitting the blood to a mild vacuum for several hours, but was not completely recoverable by restoring the carbon dioxide content to its initial level, thus suggesting that other volatile attractive principles had been in the blood. Laarman[23] found that an airstream carrying the body emanations of a rabbit from which the carbon dioxide had been removed by passage through soda lime was still highly attractive to *An atroparvus*.

In our initial search for chemicals attractive to *Ae aegypti*, it became evident that, as DeLong et al[17] had noted previously, this mosquito was completely unresponsive to the esters, alcohols, acids, aldehydes, and amines to which the housefly is so sensitive. In the field we had found vapors of gasoline and diethyl ether to be attractive to forest culicines,[11] just as recently the Gainesville group has discovered that chloroform and certain other narcotics are attractive to *Ae aegypti*. These vapors could scarcely be considered as physiological in terms of a mammalian host. However, we took a lead from the fact that Rudolfs[20] found that among hundreds of compounds tested, peptone and the amino acids alanine and cystine were attractive to *Ae sollicitans* and *Ae cantator*. We therefore tested a number of protein hydrolysates, including those used in fruit-fly baits, and found most of them to be attractive. Meanwhile, during the progress of the work Schaerffenberg and Kupka[37] had reported alanine and cystine to be attractive to *C pipiens* when mixed together and with nine other compounds in unstated propor-

Table 2.—Multiples by Which Attractiveness can be Raised by Biological Factors

| | Multiple |
|---|---|
| Dark vs light skins | 1.6 |
| Warm vs cool skins | 1.3 |
| Dry vs moist skins | 1.5 |
| Armpit sweat | 1.25 |
| Beef blood | 5 |

tions, and Roessler[38] was finding tyrosine and threonine to be individually attractive to *Ae aegypti*. By making up hydrolysates lacking in certain single amino acids, by testing the amino acids individually in a free-flight cage type of olfactometer, and by submitting our protein hydrolysates to column chromatography and testing eluate fractions for attractiveness, we found the most attractive one to be lysine. Alanine and arginine were also active.[39] Certain commercial samples of lysine that were initially unattractive could be activated when they were relieved of repellent contaminants by exposure to a mild vacuum. Lysine prepared pure through its picrate salt and then its dihydrochloride was found to be attractive in direct proportion to its content of carbon dioxide, either adsorbed onto it or combined in what is probably a carbaminoyl compound, or both.[40] The full range of 26 amino acids was tested in a glass-enclosed free-flight cage, and it was found that five of them which resembled lysine in being alkaline and combining with carbon dioxide were almost as attractive, five were like alanine and threonine in adsorbing carbon dioxide and being moderately attractive, and four were like tyrosine in being fairly attractive without carrying any carbon dioxide[41] (Figure). The most attractive amino acid is L-lysine, the unnatural D-isomer being equally attractive; one of our samples of L-lysine, when tested on *Ae aegypti* in the olfactometer at Gainesville, outdrew in turn chloroform and the vapor from dry ice.

Meanwhile certain estrogens, such as estriol, estradiol, and estrone, had been discovered by Roessler[42] to be attractive to *Ae aegypti* when minute amounts were volatilized in an olfactometer. In experiments performed by Dr. Roessler when he was a guest scientist in our laboratory, estriol also declared its attractiveness in the free-flight cage provided it was enclosed in glass to isolate it from extraneous vapors.[41] Lysine was also attractive in the olfactometer, but estriol remained attractive at vapor dilutions beyond the threshold of activity of lysine. Roessler[43] had already found that the attractiveness of women to *Ae aegypti* over the menstrual cycle showed peaks and troughs corresponding to the urinary excretion, and presumably blood concentration, of estrogens. The Gainesville group could confirm the trough in the three days immediately following menstruation, but the peaks they obtained were not congruent with those found by Roessler. The comparative attractiveness of two Panamanian men to *An albimanus* and *Mansonia nigricans* was found by Woke[44] to agree with their comparative blood and urinary estrogen levels.

It is difficult to accept the idea that substances such as amino-acids and estrogens are attractants for mosquitoes, particularly because they have such low vapor pressures, and in the case of a compound as difficult to prepare pure as lysine, it is hard to consider the attractiveness not to be

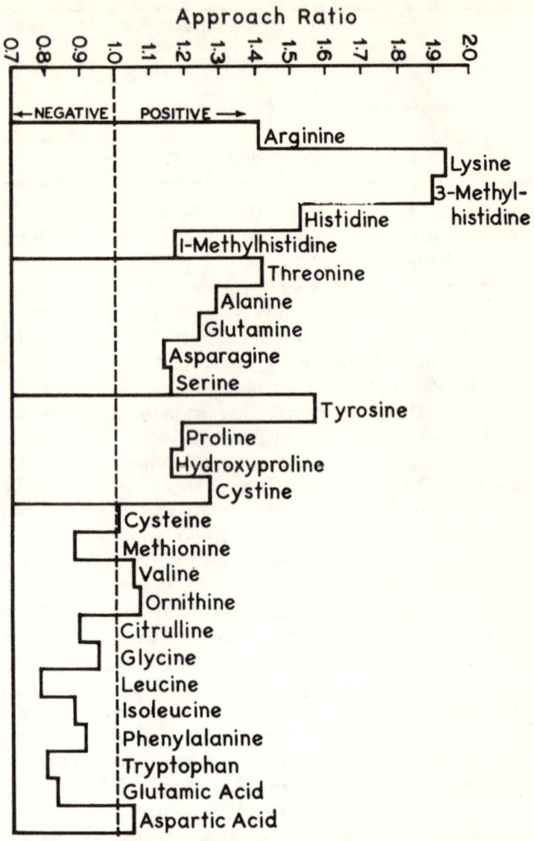


**Response of *Aedes aegypti* to 26 amino acids. (Data from Roessler and Brown[41])**

due to contaminant impurities. A unitary theory where only moisture and warmth are effective attractants, with a common mechanism involving infrared absorption, is much more appealing, since such a theory can be extended to cover the action of repellents. But the experiments of Johnston[45] with *Phormia* flies and honeybees proved that they did not detect sugary attractants through thallium bromoiodide windows, just as Peterson and Brown[19] had previously found regarding the perception of a warm body by *Ae aegypti*.

From the responses of mosquitoes with successive antennal segments extirpated, Steward and Atwood[6] concluded that the blunt type-2 trichoid sensilla are particularly affected by repellents, while the pointed type-1 sensilla are sensitive to the emanation of the human hand. Daykin et al[4] have concluded that the type-1 sensilla are receptors for carbon dioxide, and agree with Steward and Atwood that the type-3 peg organs are probably thermoreceptors; they suggest that the type-2 sensilla are hygroreceptors which are stimulated by dry air. It is further suggested by Wright et al[46] that repellents also have this effect since they have the property (indicated by their infrared absorption) of readily attaching to the surface of these receptors. Therefore, repellents should become less effective when the air is dry and these type-2 sensilla are already stimulated. Wright et al[46] tested this point and found that the vapor of ethohexadiol (Rutgers 612) became less repellent in air of less than 5% relative humidity. On the other hand, Hocking and Khan[47] found that indalone gradually increased in repellency as humidities decreased down to 3%.

From the empirical and pragmatic standpoint, we take it as highly significant that in traps for *An atroparvus* in the Netherlands, van Thiel and Weurman[48] found that the effects of carbon dioxide, moisture, and warmth were additive to each other but did not yet add up to the attractiveness of pig. Moreover for *Aedes* and *Culex* mosquitoes in southern Germany, Roessler[43] found that traps carrying these three factors could be made more effective by lysine or tyrosine, and by estriol, again in additive steps. This suggests that amino acids and estrogens each constitute a group of compounds capable of enhancing the base attractiveness of moisture, warmth, and carbon dioxide, and that the spectrum of amino acids and estrogens put out by different host types might largely determine the host selection exhibited by various species and races of mosquitoes, and the difference in attractiveness between the sexes and individual humans.

## Summary

There are numerous possible factors which affect the attraction of mosquitoes to man and animals, and corollary influences which repel them. The effects of moisture, warmth, carbon dioxide, and estrogens seem to be of importance, and it is significant that the mosquito does not respond to agents to which the housefly is sensitive. The amino acids which attract mosquitoes probably are of more significance than active organic substances not present physiologically in effective quantities in the hosts.

## References

1. Hocking, B.: The Use of Attractants and Repellents in Vector Control, *Bull WHO*, suppl 29, pp 121-126, 1963
2. Mer, G.; Birnbaum, D.; and Aioub, A.: The Attraction of Mosquitoes by Human Beings, *Parasitology* **38**:1-2 (May) 1947.
3. Kellogg, F.E., and Wright, R.H.: The Guidance of Flying Insects: V. Mosquito Attraction, *Canad Ent* **94**:1009-1016 (Oct) 1962.
4. Daykin, P.N.; Kellogg, F.E.; and Wright, R.H.: Host-Finding and Repulsion of *Aedes aegypti*, *Canad Ent* **97**:239-262 (March) 1965.
5. Roth, L.M.: Loci of Sensory End-Organs Used by Mosquitoes (*Aedes aegypti* L. and *Anopheles quadrimaculatus* Say) in Receiving Host Stimuli, *Ann Ent Soc Amer* **44**:59-74 (March) 1951.
6. Steward, C.C., and Atwood, C.E.: The Sensory Organs of the Mosquito Antenna, *Canad J Zool* **41**:577-594 (May) 1963.
7. Ismail, I.A.H.: Sense Organs in the Antennae of *Anopheles maculipennis atroparvus* (v. Thiel), and Their Possible Function in Relation to the Attraction of Female Mosquito to Man, *Acta Trop* **19**:1-58, 1962.
8. Roth, L.M., and Willis, E.R.: Possible Hygroreceptors in *Aedes aegypti* (L.) and *Blattella germanica* (L.), *J Morph* **91**:1-14 (July) 1952.
9. Willis, E.R., and Roth, L.M.: Reactions of *Aedes aegypti* (L.) to Carbon Dioxide, *J Exp Zool* **121**:149-179 (Oct) 1952.
10. Brown, A.W.A.; Sarkaria, D.S.; and Thompson, R.P.: Studies on the Responses of the Female *Aedes* Mosquito: I. The Search for Attractant Vapours, *Bull Entomol Res* **42**:105-114 (July) 1951.
11. Brown, A.W.A.: Studies on the Responses of the Female *Aedes* Mosquito: IV. Field Experiments on Canadian Species, *Bull Entomol Res* **42**: 575-582 (Nov) 1951.
12. Van Thiel, P.H.: Quelles sont les Excitations incitant l'*Anopheles maculipennis atroparvus* à Visiter et Piquer l'Homme ou le Bétail?, *Bull Soc Path Exot* **30**:193-203, 1937.
13. Parker, A.H.: Stimuli Involved in the Attraction of *Aedes aegypti* L. to Man, *Bull Entomol Res* **39**:387-397 (Dec) 1948.
14. Wright, R.H.: The Attraction and Repulsion of Mosquitoes, *World Rev Pest Control* **1**:20-30 (winter) 1962.
15. Howlett, F.M.: The Influence of Temperature Upon the Biting of Mosquitoes, *Parasitology* **3**:479-484 (Nov) 1910.
16. Reuter, J.: *Orienterend Onderzoek naar de Oorzaak van het Gedrag van* Anopheles maculipennis *Meig. by de Voedselkeuze* (thesis), Acad. Proefschr. Rijksuniv., Leiden (The Netherlands), 1936.
17. DeLong, D.M., et al: *Final Summary Report on Project* 272, Office of the Quartermaster-General, Washington, DC (contract W44-1099m-1185), March 31, 1949.
18. Christophers, S.R.: Mosquito Repellents: Being a Report of the Work of the Mosquito Repellent Inquiry, Cambridge, 1943-1945, *J Hyg* **45**:176-231 (May) 1947.
19. Peterson, D.G., and Brown, A W.A.: Studies on the Responses of the Female *Aedes* Mosquito: IV. The Response of *Aedes aegypti* (L.) to a Warm Body and Its Radiation, *Bull Entomol Res* **42**:535-541 (Nov) 1951.
20. Rudolfs, W.: *Chemotropism of Mosquitoes*, bulletin 367, New Jersey Agricultural Experimental Station, Rutgers University, New Brunswick, NJ, 1922.
21. Willis, E.R.: The Olfactory Response of Female Mosquitoes, *J Econ Entomol* **40**:769-778 (Dec) 1947.
22. Reeves, W.C.: Quantitative Field Studies on a Carbon Dioxide Chemotropism of Mosquitoes, *Amer J Trop Med Hyg* **2**:325-331 (March) 1953.
23. Laarman, J.J.: The Host-Seeking Behaviour of the Malaria Mosquito *Anopheles maculipennis atroparvus*, *Acta Leidensia* **25**:1-144, 1955.
24. Brett, G.A.: On the Relative Attractiveness to *Aedes aegypti* of Certain Coloured Cloths, *Trans Roy Soc Trop Med Hyg* **32**:113-124 (Sept) 1938.
25. Ko, R.: On the Colour-Preferences of Mosquitoes, *J Formosa Med Soc* (Taihoku) No. 244, 1925.
26. Gjullin, C.M.: Effect of Clothing Color on the Rate of Attack of *Aedes* Mosquitoes, *J Econ Entomol* **40**:326-327 (April) 1947.
27. Brown, A.W.A.: Studies on the Responses of the Female *Aedes* Mosquito: VI. The Attractiveness of Coloured Cloths to Canadian Species, *Bull Entomol Res* **45**:67-78 (March) 1954.
28. Sippell, W.L., and Brown, A.W.A.: Studies on the Responses of the Female *Aedes* Mosquito: V. The Role of Visual Factors, *Bull Entomol Res* **43**:567-574 (Jan) 1953.
29. Smart, M.R., and Brown, A.W.A.: Studies on the Responses of the Female *Aedes* Mosquito: VII. The Effect of Skin Temperature, Hue, and Moisture on the Attractiveness of the Human Hand, *Bull Entomol Res* **47**:89-100 (March) 1956.
30. Freyvogel, T.A.: Ein Beitrag zu den Problemen um die Blutmahlzeit von Stechmücken, *Acta Trop* **18**:201-251, 1961.
31. Rahm, U.: Zum Problem der Attraktion von Stechmücken durch den Menschen, *Acta Trop* **13**:319-344, 1956.
32. Gouck, H.K., and Bowman, M.C.: Effect of Repellents on the Evolution of Carbon Dioxide and Moisture From Human Arms, *J Econ Entomol* **52**:1157-1159 (Dec) 1959.
33. Crumb, S.E.: A Mosquito Attractant, *Science* **55**:446-447 (April 28) 1922.
34. Thompson, R.P., and Brown, A.W.A.: The Attractiveness of Human Sweat to Mosquitoes and the Role of Carbon Dioxide, *Mosquito News* **15**:80-84 (June) 1955.
35. Rahm, U.: Zur Bedeutung des Duftes und des Schweisses bei der Attraktion von *Aedes aegypti* durch den Menschen, *Acta Trop* **14**:208-217, 1957.
36. Burgess, L., and Brown, A.W.A.: Studies on the Responses of the Female *Aedes* Mosquito: VIII. The Attractiveness of Beef Blood to *Aedes aegypti*, *Bull Entomol Res* **48**:783-793 (Dec) 1957.
37. Schaerffenberg, B., and Kupka, E.: Der attractiv Factor des Blutes für blutsaugenden Insekten, *Naturwissenschaften* **46**:457-458, 1959.
38. Roessler, P.: The Attractions of Female Mosquitoes (*Aedes aegypti* L.) to Olfactory Stimuli, *Ver XI Int Kongr Entomol* (Wien) **3**:74-78, 1962.
39. Brown, A.W.A., and Carmichael, A.G.: Lysine and Alanine as Mosquito Attractants, *J Econ Entomol* **54**:317-324 (April) 1961.
40. Lipsitz, E.Y., and Brown, A.W.A.: Studies on the Responses of the Female *Aedes* Mosquito: IX. The Mode of Attractiveness of Lysine and Other Amino-Acids, *Bull Entomol Res* **54**:675-687 (Jan) 1964.
41. Roessler, P., and Brown, A.W.A.: Studies on the Responses of the Female *Aedes* Mosquito: X. Comparison of Oestrogens and Amino-Acids as Attractants, *Bull Entomol Res* **55**:395-403 (Dec) 1964.
42. Roessler, P.: Anlockung weiblicher Stechmücken (*Aedes aegypti* L., *Culicidae*) mit Duftstoffen, *Naturwissenschaften* **47**:549-550, 1960.
43. Roessler, P.: Zur Stechmückenanlockung durch den Menschen, *Med Welt*, Aug 3, 1963, pp 1579-1581.
44. Woke, P.A.: Observations on Differential Attraction Among Humans for *Mansonia nigricans* and *Anopheles albimanus* (*Diptera: Culicidae*), *Proc NJ Mosq Ext Assoc* **49**:173-181, 1962.
45. Johnston, J.W.: Infrared Loss Theory of Olfaction Untenable, *Physiol Zool* **26**:266-273 (July) 1953.
46. Wright, R.H.; Kellogg, F.E.; and Daykin, P.N.: *Report on Research Contract*, B.C. Research Council, Vancouver, Canada, Feb 15, 1965.
47. Hocking, B., and Khan, A.A.: Host-Finding by Blood-Sucking Flies and the Mode of Action of Repellent Chemicals, *Canad Ent* to be published.
48. Van Thiel, P.H., and Weurman, C.: L'Attraction exercée sur *Anopheles maculipennis atroparvus* par l'Acide Carbonique dans L' Appareil de Choix, *Acta Trop* **4**:1-9, 1947.

# L-Lactic Acid: A Mosquito Attractant Isolated from Humans

Abstract. L-*Lactic acid was the major component in material isolated from humans that was active as an attractant for female yellow fever mosquitoes,* Aedes aegypti *(L.). The* L*(+)-isomer was several times as attractive as the* D*-isomer. Good correlation was found between the attractiveness of an individual to mosquitoes and the quantity of lactic acid present in an acetone washing of his hand.*

The nature of the mosquito attractant in emanations from warm-blooded animals has been the subject of investigation and conjecture for many years. We have now identified L-lactic acid as the major component in material isolated from humans that attracts female yellow fever mosquitoes, *Aedes aegypti* (L.).

Attractive material was first isolated from condensate obtained from a flowing nitrogen stream passing over a human arm and through a cold trap (about −15°C); the quantity of attractant obtained was too small for identification. Larger amounts were then obtained by washing the forearms of subjects with acetone. When these solutions were treated with ammonia and evaporated to dryness under reduced pressure, they yielded a residue that contained the attractive material.

The attractive substances in the acidified residue from arm washings were separated by thin-layer chromatography on silica gel with a developing solvent containing butanol, acetic acid, and water (75 : 18 : 75) and were located by spraying with a solution of bromocresol green indicator adjusted to *p*H 9.8. The major attractive component was found in the zone having the same $R_F$ value (0.60) as L-lactic acid. The acid was recovered by eluting this zone with acidified acetone, adding ammonia to alkalinity, and evaporating to dryness. A minor amount of an unidentified substance showing slight attractant activity was also found at $R_F$ 0.1 to 0.2. Material eluted from other sections of the chromatogram showed no appreciable activity in the mosquito bioassay.

The identity of the attractant eluted from the zone at $R_F$ 0.60 was confirmed as lactic acid by thin-layer chromatography on silica gel developed with an ethanol, ammonia, water (80 : 5 : 15) system and with an acetone, methanol, water (4 : 4 : 2) system, and by chromatography on paper developed with an ethanol, ammonia, water (80 : 5 : 15) system and with an ethyl acetate, acetic acid, water (3 : 1 : 1) system; spots were eluted from the paper with ethanol. In all tests, the attractant gave a single spot with the same $R_F$ value as authentic L-lactic acid. Further confirmation was obtained from the infrared spectrum of the ammonium salt (KBr disc), which agreed with that of ammonium L-lactate. The gas chromatographic retention time (*1*) of the methyl ester obtained by treating the acidified extractive from the zone at $R_F$ 0.60 with diazomethane agreed with that of L-lactic acid similarly methylated. The presence of lactic acid in the attractant material trapped from a gas stream that had passed over a human arm was then tested for and found by paper chromatography.

An enzymatic assay (*2*) specific for L-lactic acid demonstrated the presence of 14 mg of L-lactic acid in the acidified residue from acetone washings of 800 arms. A portion of this material equivalent to 10 μg of L-lactic acid had essentially the same attraction for mosquitoes as 10 μg of authentic L-lactic acid. In three tests, L-lactic acid was five times as attractive to mosquitoes as an equal quantity of D-lactic acid.

The olfactometer used in the bioassays was a Plexiglas (*1*) enclosure (173 by 65 by 100 cm), designed to bring purified outside air (free of human emanations) into a chamber containing 200 to 300 female mosquitoes via two traps (*3*). A subject put his arm through one of two rubber sleeves located just ahead of the traps, and the outside air was allowed to sweep over the forearm and hand and attract mosquitoes into the traps. Degree of attraction was gauged by the percentage of mosquitoes attracted by the arm as compared with the percentage attracted by no arm (check trap). For the bioassays of isolated fractions, the olfactometer was modified; the rubber sleeves were replaced with two Plexiglas disks, each having a centrally located, polyethylene tube (1.25 cm, outside diameter) through which an auxiliary flow of 2.5 liters of dry, tank nitrogen and 2.5 ml of carbon dioxide per minute was maintained to introduce the sample (see Fig. 1). The sample tube was an assembled male and female 24/40 ground glass joint with the outside

Table 1. Percentage of female *A. aegypti* attracted by tube containing 10 μg of L-lactic acid as a stream of nitrogen (2.5 liter/min) passed through, with and without the addition of 2.5 ml of $CO_2$ per minute.

| Cage | Mosquitoes attracted (%) | | | |
|---|---|---|---|---|
| | No $CO_2$ added | | $CO_2$ added | |
| | Lactic acid | No lactic acid | Lactic acid | No lactic acid |
| A | 0.1 | 0.0 | 13 | 0.1 |
| B | 0.0 | 0.0 | 9 | 0.0 |
| E | 0.0 | 0.0 | 20 | 0.2 |
| F | 0.0 | 0.0 | 32 | 0.3 |
| G | 0.0 | 0.0 | 15 | 0.0 |
| H | 0.1 | 0.1 | 20 | 5.5 |

Table 2. Amount of L-lactic acid found in acetone rinses of skin of individuals compared with the percentage of female *A. aegypti* attracted by the rinses.

| Subject | L-Lactic acid found (μg/ml) | Attracted by 0.5 ml of rinse (%) |
|---|---|---|
| | *Test No. 1* | |
| HG | 30 | 33 |
| NS | 54 | 52 |
| TN | 15 | 26 |
| | *Test No. 2* | |
| HG | 44 | 53 |
| NS | 25 | 41 |
| TN | 12 | 41 |

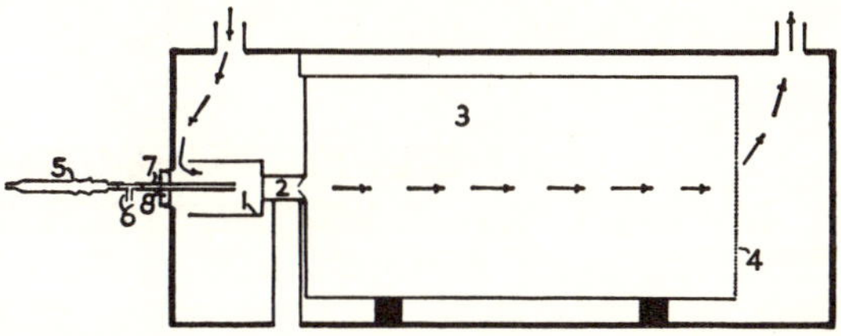

Fig. 1. Schematic side view of olfactometer with arrows showing flow of purified outside air through (1) plastic sleeve, (2) mosquito trap, (3) mosquito chamber, and (4) aluminum screening. An auxiliary flow of 2.5 liters of $N_2$ per minute passes through (5) sample tube, (6) T-tube (through which 2.5 ml of $CO_2$ per minute is added), (7) polyethylene tube in (8) Plexiglas disk and then into (3) via (1) and (2). A duplicate sample inlet system (5, 6, 7, 8, 1, 2), behind the one shown, also leads into the mosquito chamber (3). Mosquitoes attracted by vapors from sample tube enter trap (2) from chamber (3) through opening in screen funnel and cannot return.

diameter of the unground ends reduced to 0.9 cm to fit easily into the line. Before the test, the sample, usually in 0.5 ml of acidified acetone, was plated out onto the inner walls of the sample tube by allowing the solvent to evaporate as the tube was rotated and tilted back and forth. Attractiveness of samples was based on the percentage of insects trapped in 3 minutes after the sample tube was connected into one of the two lines (the other was the check). All parts were handled with disposable gloves to avoid contamination of the system. Before each test, mosquitoes were exposed to the nitrogen–carbon dioxide mixture; if more than 2.5 percent of the mosquitoes responded, the sleeves and traps were replaced.

With mosquitoes of average avidity, 10 $\mu g$ of L-lactic acid attracted 29 to 75 percent of the mosquitoes in 3 minutes, and three to four consecutive tests could be made (subjected to air flow for 9 to 12 minutes) before the sample tube lost its attractiveness. The attractiveness of lactic acid did not appear to be enhanced by the addition of any other component isolated from the acid fraction, including the unidentified minor attractant. However, the presence of carbon dioxide is essential (Table 1). Carbon dioxide has long been recognized as a mosquito activator (*4*), but alone in nitrogen or in purified air it does not attract mosquitoes.

Other carboxylic acids containing as many as five carbon atoms were tested for attractiveness, but the response was never more than feeble compared with the responses to L-lactic acid.

The quantity of lactic acid in the acetone rinses of the hands of individuals who differed in their attractiveness to mosquitoes was determined by enzymatic assay. Overall there was good correlation between the amount of lactic acid found in the rinses of these subjects and the percentage of mosquitoes attracted by the rinses from each (Table 2). The rinses showed the same order of attractiveness between subjects that had been shown in many previous tests of the hands of the same individuals.

Lactic acid occurs in sweat and on skin (*5*); as a product of animal muscle metabolism it is invariably in the L(+)-configuration (*6*) (sometimes referred to as "sarcolactic acid"). An old report (*7*) indicated that the compound is slightly attractive to *A. aegypti*. The compound appears to be more volatile than is generally recognized, and conversion to the less volatile ammonium salt was required to obtain satisfactory recoveries in the evaporations of biological material and eluates from chromatograms.

A good attractant in traps is an invaluable aid to the detection of insect infestations. Such an attractant is needed for the worldwide eradication of the yellow-fever mosquito; lactic acid may be useful for this purpose.

*Note added in proof*: A recent report (*8*) stated that formic, acetic, propionic, and lactic acids attracted *A. aegypti* and that lactic acid (L- or D- not specified) was the most attractive. The acids tested were not isolated from humans, and the concurrent need of carbon dioxide for attraction was not noted.

FRED ACREE, JR., R. B. TURNER
H. K. GOUCK, MORTON BEROZA*
NELSON SMITH
*Entomology Research Division, Agricultural Research Service, Gainesville, Florida 32601*

**References and Notes**

1. The gas chromatography was carried out on an F & M Model 810 instrument equipped with a flame-ionization detector and stainless steel columns, 180 by 0.3 cm (OD), packed with 5 percent Carbowax 20M on 60/80 mesh Diatoport S (F & M Scientific Corp.) and maintained at 100°C. The nitrogen flow rate was 45 ml/minute.
2. J. C. Rosenberg and B. F. Rush, *J. Clinical Chem.* **12**, 299 (1966).
3. C. E. Schreck, H. K. Gouck, N. Smith, *J. Econ. Entomol.* **60**, 1188 (1967).
4. H. K. Gouck and M. C. Bowman, *ibid.* **52**, 1157 (1959); C. N. Smith, I. H. Gilbert, H. K. Gouck, *Proc. Int. Congr. Entomol. 11th, Sec. 3, Symp. 5* (1962), pp. 82–86; H. K. Gouck and I. H. Gilbert, *J. Econ. Entomol.* **55**, 386 (1962).
5. M. J. Marples, *The Ecology of the Human Skin* (Thomas, Springfield, Ill., 1965), pp. 115–116.
6. H. R. Mahler and E. H. Cordes, *Biological Chemistry* (Harper and Row, New York, 1966), p. 430.
7. D. M. DeLong *et al.*, *Final Summary Report, Project 272, Office of the Quartermaster-General*, Washington, 31 March, 1949. This report was cited by R. P. Thompson and A. W. A. Brown, *Mosquito News* **15**, 80 (1955).
8. W. Müller, *Z. Vergl. Physiol.* **58**, 241 (1968).
9. We thank A. F. Hofbauer, D. R. Godwin, K. Posey, C. E. Schreck, and J. Jackson for technical assistance.
* Agricultural Research Center, Beltsville, Md. 20705.

# THE PIG-MOSQUITO CYCLE OF JAPANESE ENCEPHALITIS VIRUS IN TAIWAN[1]

By Herbert S. Hurlbut[2]

*Abstract:* Vector populations were measured and virus isolations were made in order to follow the annual cycle of infection. The cycle was correlated with the appearance of infection in pigs. Laboratory transmission tests with pigs and mosquitoes measured the level and duration of viremia in relation to pig age and antibody level, and related these factors to the proportion of mosquitoes infected.

Hale *et al* (1957) presented concrete evidence of a pig-mosquito cycle for Japanese encephalitis virus on Singapore I., and Scherer *et al* (1959 a) and Gresser *et al* (1958) in a more extensive study in Japan, presented strong evidence of such a cycle. Virus was isolated from pigs exposed in their natural environment; mosquitoes were infected under laboratory conditions by allowing them to feed upon viremic pigs and the virus was transmitted to normal pigs when infected mosquitoes fed upon them. The level and duration of viremia in pigs was also determined experimentally by these workers in a small series of animals. Antibody surveys in these studies indicated that a high percentage of pigs become infected. Wang *et al* (1962 b) reported isolation of the virus from mosquitoes in Taiwan, and Hurlbut & Nibley (1964), a similar finding in Okinawa. Wang *et al* (1962 a) reported antibody in pigs in Taiwan and Okinawa.

All of these reports name *Culex tritaeniorhynchus* as the vector. The oriental form of this mosquito, *Culex tritaeniorhynchus summorosus* Dyar, to which these reports presumably refer, was given subspecific status by Colless (1957). Colless recognized this subspecies on Singapore I. together with four closely related species comprising a *Culex vishnui* group. Lien (1962 & pers. comm.) reports that Taiwan's *C. vishnui* group includes *Culex annulus* Theobald, a species not readily distinguishable, in the female, from *C. tritaeniorhynchus*. The importance of this mosquito in the transmission of Japanese encephalitis virus has not yet been determined.

The present report deals with a larger series of pig and mosquito infections than has been hitherto reported, making it possible to obtain more information on the level and duration of viremia in relation to pig age and antibody level, and to relate these factors directly to the proportion of mosquitoes infected. The development of vector populations was measured in relation to virus isolations from mosquitoes and infections acquired by sentinel pigs. Mosquitoes were infected by feeding upon sentinel pigs during these naturally acquired infections. These studies have been made in order to evaluate the importance of the pig-mosquito cycle in generating the summer epidemics of Japanese encephalitis in Taiwan.

## MATERIALS AND METHODS

The pigs were first generation hybrids from Taoyuan sows and Berkshire boars, purchased at two months of age from farms in the vicinity of Taipei, between November and April. The laboratory colony (Newson *et al* 1956) of *C. tritaeniorhynchus* was obtained from the 406 Medical General Laboratory, United States Army Medical Command, Japan, through the courtesy of Lt. Colonel Gordon Field. The colony of *Aedes albopictus* (Skuse) came from wild specimens captured in the vicinity of Taipei in 1960. Wild specimens of *C. tritaeniorhynchus* were collected near Taipei and reared to adult mosquitoes in the laboratory after specific identification in the larval stage.

The virus used in the laboratory experiments was isolated from *C. tritaeniorhynchus* collected near Taipei

Table 1. Identification of virus T-143, isolated from *Culex tritaeniorhynchus*, Taipei, Taiwan, 1960

| Virus | Immune serum: Log neutralization index | |
|---|---|---|
| | JE-Nak | T-143 |
| JE-Nak | 2.4 | 4.3 |
| WN-Ug | 1.1 | 1.3 |
| SLE | 0.8 | 0.9 |
| T-143 | 2.4 | 3.8 |

JE-Nak = Nakayama strain of Japanese encephalitis virus. WN-Ug = Uganda strain of West Nile virus. SLE = Hubbard strain of St. Louis encephalitis virus.

1. This study was supported in part by funding under Public Law 480, Section 104 (c).

2. From the United States Naval Medical Research Unit. No. 2 (NAMRU-2), Taipei, Taiwan. The opinions and assertions contained herein are those of the author and are not to be construed as official or reflecting the views of the Navy Department or the Naval Service at large.

Reproduced with permission from the *Journal of Medical Entomology*, 1:301–307, 1964. Published by the Entomology Department, Bishop Museum, Hawaii.

in 1960. It was used in the second mouse brain passage after identification by virus dilution neutralization tests in mice (Table 1).

The vector populations were measured by means of New Jersey light traps operated throughout the year, four nights per week from sunset to sunrise, in pig pens on farms in the suburbs of Taipei. On the day following the night of capture the mosquitoes were taken to the laboratory, identified, triturated in pools not exceeding 200 specimens, and the centrifuged fluid suspension inoculated into 3-day old mice. Details of the procedure have been presented elsewhere (Hurlbut & Nibley 1964). Living and dead specimens having no visible blood in the gut were selected for testing. Blood samples for identification were taken from specimens in which the blood had not receded anterior to the posterior margin of the fifth tergite. Bloods were identified by precipitin tests according to the method described by Weitz (1956). It is probable that a small proportion of *C. annulus* was inadvertently included with the *C. tritaeniorhynchus* due to the difficulty of distinguishing the females of these two species in mass sorting. The importance of *C. annulus* as a vector must await further study.

In the laboratory tests, pigs were infected by the bite of infected mosquitoes that had been incubated for at least 7 days at 30°C after virus was injected into the hemocoele. This procedure assured virtually 100% infection of the mosquitoes, and pig infections could be established routinely by the bite of a single mosquito allowed to engorge in the normal manner. Virus introduced in the salivary secretion is probably significantly more concentrated, due to the small volume of fluid, than when introduced by the artificial method of injection with a needle. Cell infection at the site of injection would probably be favored by the natural method. Just prior to infection, a blood sample was taken from the pig for a neutralization test performed without heat inactivation. Serum was separated from the clot after about 18 hours at 4°C and stored immediately at −20°C. It was usually tested within 7 days for neutralizing antibody by mixing the undiluted serum with equal volumes of the serially diluted virus and incubating for one hour at 37°C. Six mice 21–28 days old were inoculated intracerebrally with 0.03 ml of each of 5 serial decimal dilutions. Control titrations were done in a similar manner by substituting 0.5% bovine serum albumen for the test serum. The heparinized whole blood of viremic animals was titrated daily in 3-day old mice by injecting 0.01 ml intracerebrally and 0.02 ml subcutaneously, using serial fivefold dilutions. Virus titers are expressed as the logarithms of the $LD_{50}$ dilution calculated according to the method of Reed & Muench.

Normal mosquitoes were fed upon the pig for 30 minutes immediately after the blood sample was taken. Prior to feeding, mosquitoes were kept on a fixed schedule of 13 hours of artificial light alternating with 11 hours of darkness. Mosquitoes about 3–5 days old were exposed to the pig during the dark period of the cycle after being deprived of food, but provided with water, for the preceding 24 hours. Lots of 40 mosquitoes in glass jars of one quart capacity, with the mouth covered by good quality netting, were applied to the abdomen of the pig while the jar was wrapped in black cloth. After feeding, each lot was anesthetized with $CO_2$ and the engorged specimens recovered; unengorged mosquitoes were discarded. The mosquitoes, fed a diet of 10% dextrose, were incubated at 30°C for 14 days. On the 14th day, the mosquito's ability to excrete virus was tested by placing it, anesthetized with $CO_2$, in a small cotton-stoppered glass tube containing 1 ml of fluid suitable for sustaining the virus. The mosquito was held in the fluid for 15 minutes with care being taken to immerse the proboscis. It is assumed that the mosquito expelled salivary secretion into the fluid during the agonistic stage of recovery from $CO_2$ anesthesia. The mosquito was then removed from the fluid and triturated separately. Mice were inoculated intracerebrally with the respective fluids.

## RESULTS

*Field Observations.* Fig. 1 gives the weekly mosquito counts from May to November 1961, based on the operation of 9 traps in pig pens in the Taipei suburban area. The graph is typical of the yearly population curve, showing an explosive increase to a yearly peak in midsummer. *C. tritaeniorhynchus* is a warm weather mosquito with the population remaining at a very low level between November and April; however, breeding is continuous throughout the year in the latitude of Taipei. The mosquito population data are presented in more detail in Sun (1964). Summer epi-

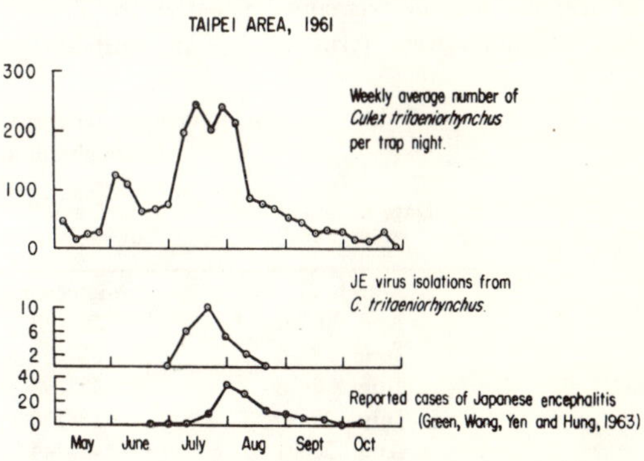


Fig. 1. Vector population curves and reported cases of Japanese encephalitis. Taipei area, 1961.

demics come at the population peak and the virus can be isolated from the mosquito at this time, fig. 1 (Green *et al* 1963). The essential features of this pattern are repeated yearly. Evidence that the mosquitoes feed upon the pig is presented in Table 2.

Table 2. Stomach bloods from *Culex tritaeniorhynchus* taken in light traps operated at pig shelters, Taipei, Taiwan, 1961. Identified by precipitin tests.

| | Percent of total | | | | | Total No. |
|---|---|---|---|---|---|---|
| | Pig | Man | Bovid | Other mammal | Avian | |
| June | 95.2 | 3.6 | 0.0 | 1.2 | 0.0 | 167 |
| July | 93.5 | 0.0 | 4.9 | 1.6 | 0.0 | 183 |
| August | 93.6 | 0.5 | 4.3 | 1.6 | 0.0 | 184 |
| Total | 94.0 | 1.3 | 3.2 | 1.5 | 0.0 | 534 |

In April, 1963, a sentinel pig was placed in each of 4 of the pig pens where mosquito traps were being operated. The pigs were 4.5 months old and antibody-free when they were placed in these typical farm environments on April 1st. Two were housed with other pigs and two were solitary. Blood samples were taken monthly for antibody determination, and more frequently for virus isolation. Laboratory mosquitoes were fed upon the pigs when blood was taken for the latter purpose. The blood remained negative for both antibody and virus during April, May, June and July. However, on 1 and 2 August, respectively, virus was isolated from two of the pigs, and from a third on 13 August. The virus titer of the blood was determined by inoculating 3-day old mice in the field at the time the samples were taken, and mosquitoes fed upon the pigs at the same time permitted subsequent determination of the percentage infected. The data are presented in Tables 3 and 4. A second virus isolation, in low concentration, was obtained from pig #22 four days after the first. Blood titration, twice a week, after the original infection remained negative for 2 weeks, when titrations were terminated. An antibody rise

Table 4. Japanese encephalitis virus isolations from pigs in their natural habitat: Taipei, Taiwan, 1963.

| | Pig number: 20 | 21 | 22 | 23 |
|---|---|---|---|---|
| * Virus titer, blood, $Log_{10}$, $LD_{50}$ | 0.9 | 2.8 | **3.0 | Neg. |
| Number and percent of *C. tritaeniorhynchus* infected | 0/7<br>0% | 17/24<br>71% | 19/22<br>86% | Neg. |

* titration in 3-day-old mice, ic-sc 0.03 ml.

** estimated from average survival time in terminal dilution.

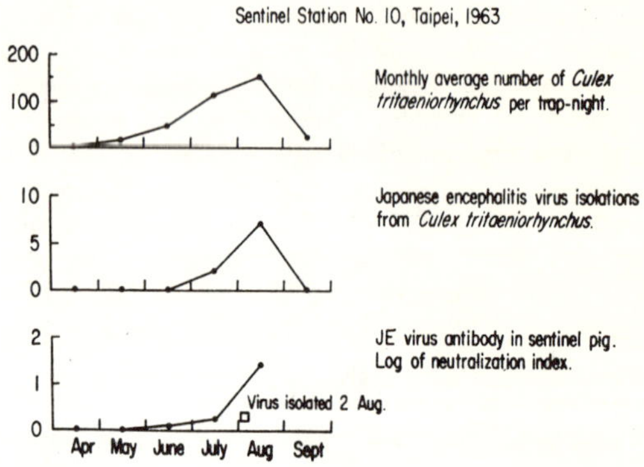


Fig. 2. Sentinel station record of pig-mosquito cycle. Taipei, 1963.

in pig #20 was detected by neutralization test 8 days after the virus isolation. The other pigs were tested 19 days after virus isolation and found positive. Virus was not obtained from pig #23, but an antibody rise indicated that it had been infected. The percentages in Table 4 are for excreted virus, presumably from salivary gland infections, after 14 days of incubation at 30°C. This was confirmed in 2 cases through transmission by bite to suckling mice. The scheduling of these observations was made possible by sampling the vector population as indicated in fig. 2.

*Laboratory Transmission Tests.* Table 5 summarizes 22 tests performed to determine the susceptibility of

Table 3. Development of Japanese encephalitis virus antibody in pigs in their natural habitat, Taipei, Taiwan, 1963.

| Date of blood collection | Days of exposure | Pig number: 20 | 21 | 22 | 23 |
|---|---|---|---|---|---|
| | | Log neutralization index | | | |
| March 27 | Pre-exposure | 0.0 | 0.0 | 0.0 | 0.0 |
| Apr. 30-May 1 | 28—30 | 0.4 | 0.0 | 0.0 | 0.1 |
| June 3—4 | 63—65 | 0.3 | 0.1 | 0.5 | 0.0 |
| July 2—8 | 92—98 | 0.3 | 0.3 | 0.0 | 0.0 |
| July 29-30 | 118-120 | 0.0 | 0.2 | 0.0 | 0.1 |
| Aug. 20-21 | 142-144 | 1.2 | 1.4 | 1.3 | 1.3 |
| Virus isolated: | | Aug. 13 | Aug. 2 | Aug. 1 & 5 | none |

mosquitoes to infection when fed upon viremic pigs. Seven pigs were used, and the mosquito groups ranged from 13–52 individuals with 50% having 30 or more. The 50% infective dose for *C. tritaeniorhynchus* evidently lies between virus titers 2.1 and 3.0. However, the infection rate varied within wide limits for the same virus titer in different tests. For example, the log titer 1.2 yielded rates of 3.8, 10 and 17, and log 2.5 rates of 48 and 96%. The laboratory colony and wild *C. tritaeniorhynchus* are compared in Table 6 in 5 paired simultaneous feedings on viremic pigs. The differences do not appear to be of practical significance.

Table 5. Susceptibility of colonized *Culex tritaeniorhyhchus* to different concentrations of Japanese encephalitis virus, when fed upon viremic pigs.

| Virus titer* | No. mosq. tested | No. mosq. infected | Per cent infected | Range per cent |
|---|---|---|---|---|
| 0.1-1.0 | 293 | 17 | 5.8 | 0-14 |
| 1.1-2.0 | 237 | 34 | 14.3 | 3.3-37 |
| 2.1-3.0 | 107 | 48 | 44.9 | 37-96 |

* Log $LD_{50}$ dil. in 3-day old mice.
22 tests, 7 pigs; Virus T-143, 2nd mouse brain passage.

Table 6. Susceptibility to Japanese encephalitis virus of colonized and wild *Culex tritaeniorhynchus.* Simultaneous feedings on infected pigs—Taipei, 1963

| Paired groups | Colony | | Wild | | Log virus titer of blood |
|---|---|---|---|---|---|
| | Number tested | Per cent infected | Number tested | Per cent infected | |
| 1 | 43 | 14 | 20 | 20 | 1.0 |
| 2 | 22 | 4.5 | 11 | 36 | 1.2 |
| 3 | 24 | 96 | 28 | 82 | 2.5 |
| 4 | 25 | 48 | 40 | 65 | 2.5 |
| 5 | 13 | 0 | 19 | 5.3 | no test |

Table 7. Susceptibility to Japanese encephalitis virus of two species of colonized mosquitoes. Simultaneous feedings on infected pigs—Taipei, 1963

| Paired groups | *C. tritaeniorhynchus* | | *Aedes albopictus* | | Log virus titer of blood |
|---|---|---|---|---|---|
| | Number tested | Per cent infected | Number tested | Per cent infected | |
| 1 | 26 | 15 | 10 | 0 | 1.1 |
| 2 | 19 | 5 | 12 | 0 | 1.3 |
| 3 | 25 | 48 | 13 | 0 | 2.5 |
| 4 | 24 | 96 | 12 | 0 | 2.5 |
| 5 | 22 | 4 | 10 | 0 | 1.2 |
| 6 | 27 | 15 | 27 | 0 | 1.6 |
| 7 | 30 | 37 | 14 | 0 | 2.0 |
| 8 | 28 | 7 | 16 | 0 | 1.0 |

In Table 7 *C. tritaeniorhynchus* and *A. albopictus* are compared in a similar manner. It is obvious that *A. albopictus* is not infected at the virus titers to be expected in viremic pigs. Similar tests with *Culex pipiens quinquefasciatus* Say and *C. annulus* were not successful because these mosquitoes could not be induced to feed upon pigs in sufficient numbers under laboratory conditions.

The duration of effective viremia in these tests (arbitrarily defined as yielding over 30% infection of mosquitoes) was usually not longer than 2 days. Virus could often be detected in low concentration for 4 days. Effective viremia was not observed earlier than 2 days after infection, nor later than 4 days. Apparently age did not influence the level of viremia in a group of 5 pigs ranging from age 193–331 days. Two animals of 200 and 331 days old had essentially the same level of viremia and each had a mosquito infection rate of over 90%.

*Neutralizing Antibody.* Neutralizing antibody specific for Japanese encephalitis virus was found in diminishing concentration from weaning at 2 months until 6 months of age when many pigs had become completely antibody-free (Table 8). This antibody is

Table 8. Post-weaning loss of neutralizing antibody in pigs. Japanese encephalitis virus, Taipei, Taiwan.

| Age of pigs in months | Log neutralization index | | |
|---|---|---|---|
| | 0.0-0.9 | 1.0-1.9 | 2.0-3.9 |
| 2 - 4 | *20 | 16 | 64 |
| 4 - 6 | 60 | 32 | 8 |

* Per cent of pigs tested. Based on paired sera from 25 pigs.

presumably of maternal origin. One litter of 8 from a mother having a high level of antibody showed high titers at weaning time, except one completely negative individual. Another instance revealed 2 litter mates with one positive at a high level and the other with less than one log unit when weaned. However, the negative individuals developed antibody in a normal manner when infected.

Pigs having various antibody levels were infected by mosquito bite and normal mosquitoes were fed upon them during their viremic period. The results are presented in Table 9. Virus was recovered from the blood of all pigs except one in the highest antibody titer group; a rise in antibody titer was detected in all individuals. Pigs having a pre-infection neutralizing index of log 1.0 or higher, infected not more than 2.6% of the mosquitoes. Those having an index greater than log 2.0 did not infect the mosquitoes. In the absence of pre-infection antibody the maximum level of viremia ranged from log 1.1–2.5. The range in

Table 9. Mosquito-Pig infection experiments: Japanese encephalitis virus

| | | | |
|---|---|---|---|
| Pre-infection neutralizing antibody. Log neutralization index | 0.0 | 1.0–1.9 | 2.0–2.9 |
| Per cent of mosquitoes infected (highest per pig) | 10–96 | 0.0–2.6 | 0.0 |
| Post-infection antibody | 3.2–4.1 | 2.4–3.7 | 3.1–3.7 |
| Pig age in days at time of infection | 193–331 | 193–197 | 193–197 |
| Number of pigs tested | 4 | 4 | 3 |

Pig: Hybrid, Berkshire boar/Taoyuan sow
Mosquito: *Culex tritaeniorhynchus*
Virus: Strain T-143, 2nd mouse brain passage, isolated from *Culex tritaeniorhynchus*, Taipei, Taiwan, 1960.

highest % of mosquitoes infected per pig was 9–96.

*Vaccination.* Two pigs, approximately one year old and free of neutralizing antibody, were given single, subcutaneous 1 ml injections of formalinized mouse brain vaccine to test the effect upon the level of viremia, and the mosquito infection rate. The vaccine was prepared and standardized by the Taiwan Serum and Vaccine Lab. according to the minimal requirement for Japanese encephalitis vaccine issued by the Nat. Instit. of Health of Japan. It was made available through the courtesy of Dr S. T. Hsu. The results are presented in Table 10. Tests for neutralizing antibody, 2 and 3 months post-vaccination, were negative or within the equivocal range. After challenge by infected mosquito bite, one pig developed a maximum virus titer of log 1.2 for which the mosquito infection rate was 16.7%. The other pig had low-level viremia at a log titer of 0.8 and the mosquitoes that fed upon it did not become infected. Four unvaccinated pigs without pre-infection neutralizing antibody developed maximum virus titers of 1.1–2.5 and infected 9–96% of the test mosquitoes. These extremes in the unvaccinated animals, together with the apparent absence of neutralizing antibody in those that were vaccinated, cast doubt upon the effectiveness of vaccination in these tests. However, both virus titers and mosquito infection rates were low in the vaccinated pigs and further investigation is justified to determine whether these could be maintained in a larger series.

Table 10. Viremia and mosquito infectivity of vaccinated and unvaccinated pigs.

| | Pre-infection Log neut. index | Log virus titer* | Per cent mosq. infected |
|---|---|---|---|
| Vaccinated pig #33 | 0.6** | 1.2 | 5/30=16.7 |
| Vaccinated pig #35 | 0.0*** | 0.8 | 0/28= 0.0 |
| 4 Unvaccinated pigs | 0.0 | 1.1–2.5 | 9–96 |

* Maximum per pig. ** 3 months post-vaccination.
*** 2 months post-vaccination.

DISCUSSION

The findings on the duration and level of viremia and the % of mosquitoes infected are in essential agreement with those of Hale *et al* (1957) and Gresser *et al* (1958) where the data permit comparison. Both the mosquito and the porcine host exhibited wide variability in their resistance to the virus. This may be due in part to natural physiologic changes influenced by environment, as well as genetic differences. However, no evidence was seen to support this speculation.

The statement by Hale *et al* (1957) that 8-month old pigs do not become viremic is in disagreement with the findings from numerous specimens in the present study. A pig 11 months old may infect 96% of the mosquitoes that feed upon it. This is a fact of primary importance in temperate and subtropical climates where, as Scherer *et al* (1959 a) have pointed out, the progressive loss of maternal antibody results in the accumulation of large numbers of susceptible animals before the encephalitis season. More precisely, this accumulation of susceptibles is possible because the populations of the vector mosquito remain at low level until midsummer and then increase explosively bringing about rapid dissemination of virus throughout the pig population. The young weanling pigs that are antibody free are not all from sows that are antibody free as Hale *et al* (1957) supposed. Many are aberrant individuals that do not acquire or do not retain maternal antibody. Individual differences in decay rates of maternal antibody were observed by Scherer *et al* (1959 a).

Pigs having a serum neutralizing index of log 1.0–2.8 can be infected by mosquito bite and will circulate virus in low concentration. This was seen in 6 of 7 animals as shown in Table 9. This may be related to the mode of introduction of the virus, i.e., by mosquito bite. It suggests that the primary site of infection is the site of injection where there is an initial high concentration of virus. Coexisting viremia and neutralizing antibody were observed in a naturally infected pig also by Scherer *et al* (1959 b).

The evidence supporting the hypothesis of a natural pig-mosquito cycle in Taiwan may be summarized as follows: 1) virus isolations from *C. tritaeniorhynchus*; 2) specific antibody in a high percentage of pigs; 3) *C. tritaeniorhynchus* feeding upon pigs in nature; 4) the experimental demonstration of a pig-mosquito cycle; 5) virus isolations from sentinel pigs; and 6) infection of mosquitoes fed upon sentinel pigs during naturally acquired viremia.

There can be no doubt that such a cycle takes place. The question remaining to be answered is whether this cycle plays an essential role in generating the summer epidemics. A final answer is probably not possible without carefully controlled field studies in which the pig is made ineffective as a source of virus to the mosquito vector. However, certain factors make an affirmative answer reasonable. The yearly turnover in the pig population introduces susceptible animals into almost every rural community. The vector mosquito, breeding in and around cultivated fields, exhibits regular summer cycles of abundance and feeds upon the pig in large numbers. Thus, the elements are present for explosive dissemination of the virus. It appears that the agricultural environment, as it exists in Taiwan, is responsible for the epidemics since both the pig and the mosquito vector are products of agriculture, causing what might otherwise be an obscure sylvan virus to spread rapidly through the farm population, both human and animal.

Vaccination of pigs is a possible method of reducing significantly the incidence of the human disease. A log neutralization index between 1.0 and 1.9 would keep the viremia at an extremely low mosquito infecting level, perhaps lower than necessary for the practical purpose of breaking the transmission cycle. Such a procedure, in contrast to vaccination of the human host, may largely eliminate the virus from the environment. Human vaccination alone, on the other hand, may not be effective under conditions of undiminished exposure to the natural virus.

Another approach to the problem is suggested by the wide variation in maximum level of viremia in pigs seen in these experiments. A breed of pig might be selected that is not an effective source of virus to the vector mosquito.

Selective measures for the control of the mosquito vector may be devised that are economically feasible. For example, the explosive development of large populations of the vector in midsummer may be due to some agricultural practice that could be modified.

## SUMMARY

Sentinel pigs kept in typical farm environments remained antibody free from April until August, when coincident with an increase in the vector population and the appearance of infected mosquitoes, the pigs became infected and the mosquitoes that fed upon them developed infections and transmitted the virus.

Laboratory transmission tests showed that the 50% infective dose for *C. tritaeniorhynchus*, when fed upon viremic pigs, lies between log titers 2.1–3.0. Wide variations were observed in both the maximum level of viremia and in the per cent of mosquitoes infected. Apparently pig age did not influence the result, since animals 6 and 11 months old had maximum mosquito infection rates over 90%.

Maternal antibody was present at a high level in over 60% of the pigs at 2 months of age but had disappeared or was at a very low level in over 60% at 6 months of age. Pigs having a pre-infection neutralization index greater than log 1.0 infected not more than 2.6% of the mosquitoes that fed upon them. Pigs having a serum neutralizing index of log 1.0–2.8 were infected by mosquito bite and circulated virus in low concentration. In all cases there was a post-infection rise in antibody.

Vaccination with one injection of formalin inactivated mouse brain vaccine did not yield neutralizing antibody in 2 pigs tested.

*Acknowledgement:* The author is pleased to acknowledge the important contributions of senior technicians, HMC I.E. Fishburne and Mr W. C. Huang, and the valued assistance of Messrs M. J. Lo, H. H. Chu, S. C. Lien, C. K. Hsieh, T. E. Liu, C. I. Cheng and W. L. Cheng.

## REFERENCES

**Colless, D. H.** 1957. Notes on the culicine mosquitoes of Singapore. II: The *Culex vishnui* group (Diptera, Culicidae), with descriptions of two new species. Ann. Trop. Med. & Parasit. **51**: 87–101.

**Green, I. J., S. P. Wang, C. H. Yen, & S. C. Hung.** 1963. The epidemiology of Japanese encephalitis virus on Taiwan in 1961. Amer. J. Trop. Med. & Hyg. **12**: 668–74.

**Gresser, I., J. L. Hardy, S. M. K. Hu, & W.F. Scherer.** 1958. Factors influencing transmission of Japanese B encephalitis virus by a colonized strain of *Culex tritaeniorhynchus* Giles, from infected pigs and chicks to susceptible pigs and birds. Amer. J. Trop. Med. & Hyg. **7**: 365–73.

**Hale, J. H., K. A. Lim, & D. H. Colless.** 1957. Investigation of domestic pigs as a potential reservoir of Japanese B encephalitis virus on Singapore Is. Ann. Trop. Med. & Parasit. **51**: 374–79.

**Hurlbut, H. S. & C. Nibley.** 1964. Virus isolations from mosquitoes in Okinawa. J. Med. Ent. **1**: 78–82.

**Lien, J. C.** 1962. Non-anopheline mosquitoes of Taiwan. Annotated catalog and bibliography. Pacific Insects **4**: 615–49.

**Lien, J. C.** 1964. Personal communication.

**Newson, H. D., T. E. Blakeslee, S. Toshioka. M. Sakai, C. M. Wheeler, T. Shimada, & J. Akiyama.** 1956. A preliminary report on the laboratory colonization of the mosquito *Culex tritaeniorhynchus* Giles. Mosq. News **16**: 282.

**Scherer, W. F., J. T. Moyer, T. Izumi, I. Gresser, & J. McCown.** 1959a. Ecologic studies of Japanese encephalitis virus in Japan; VI. Swine infection. Amer. J. Trop. Med. & Hyg. **8**:698–706.

**Scherer, W. F., J. T. Moyer, & T. Izumi.** 1959b. Immunologic studies of Japanese encephalitis virus in Japan; V. Maternal antibodies, antibody responses and viremia following infection of swine. J. Immunol. **83**: 620–26.

**Sun, W. K. C.** 1964. The seasonal succession of mosquitoes in Taiwan. J. Med. Ent. **1**: 277–84.

**Wang, S. P., J. T. Grayston, & I. H. Chu.** 1962a. Encephalitis on Taiwan; V. Animal & bird serology. Amer. J. Trop. Med. & Hyg. **11**: 155–58.

**Wang, S. P., J. T. Grayston, & S. M. K. Hu.** 1962b. Encephalitis on Taiwan; III. Virus isolations from mosquitoes. Amer. J. Trop. Med. & Hyg. **11**: 141–48.

**Weitz, B.** 1956. Identification of blood meals of blood-sucking arthropods. Bull. Wld. Hlth. Org. **15**: 473–90.